MODERN
DIGITAL ELECTRONICS

Series Foreword

The McGraw-Hill Companies has been a leader in providing trusted information and analysis for well over a century. From the Industrial Revolution to the Internet Revolution, The McGraw-Hill Companies have filled a critical need for information and insight by helping individuals and businesses in the field of Engineering.

As early as 1910, the McGraw-Hill book company was making a difference on college campuses with the publication of its first series, Electrical Engineering Texts, outlined and edited by Professor Harry E. Clifford of Harvard University. McGraw-Hill's Electrical Engineering textbooks have shaped engineering curricula worldwide. I am pleased that I have been invited to be the Global Series Editor in Electrical Engineering, helping to shape how the next generation of electrical engineering students around the globe will learn.

As advances in networking and communications bring the global academic community even closer together, it is essential that textbooks recognize and respond to this shift. It is in this spirit that we will publish textbooks in the McGraw-Hill Core Concepts in Electrical Engineering Series. The series will offer textbooks for the global electrical engineering curriculum that are reasonably priced, innovative, dynamic, and will cover fundamental subject areas studied by Electrical and Computer Engineering students. Written with a global perspective and presenting the latest in technological advances, these books will give students of all backgrounds a solid foundation in key engineering subjects.

Modern Digital Electronics by R.P. Jain of B.M. Institute of Engineering and Technology, India, is a textbook for a course on digital electronics at the undergraduate level in a college of engineering. The book deals with basic digital circuits using integrated gate circuits. It uses the hardware description language VHDL and introduces computer-aided design. Semiconductor memories, programmable logic devices and microprocessors are fully described. Each chapter includes review questions as well as problems, and answers are provided for 112 problems.

This book has been reviewed and assessed for use in engineering classrooms at all levels. Like *Modern Digital Electronics*, each book in the Core Concepts series presents a comprehensive, straightforward, and accurate treatment of an important subject in Electrical & Computer Engineering. With their clear approach, contemporary technology, and international perspective, Core Concepts books are an unmistakable choice for professors wanting understandable, concise engineering textbooks that adhere to the standards of The McGraw-Hill Companies.

—Richard C. Dorf,
University of California, Davis
Series Editor, The Core Concepts Series in Electrical and Computer Engineering

MODERN DIGITAL ELECTRONICS

R P JAIN
Director
B M Institute of Engineering and Technology
Sonepat

McGraw Hill Higher Education

Boston Burr Ridge, IL Dubuque, IA New York
San Francisco St. Louis Bangkok Bogotá Caracas Kuala Lumpur
Lisbon London Madrid Mexico City Milan Montreal New Delhi
Santiago Seoul Singapore Sydney Taipei Toronto

The McGraw·Hill Companies

Mc Graw Hill Higher Education

MODERN DIGITAL ELECTRONICS

Published by McGraw-Hill, a business unit of The McGraw-Hill Companies, Inc., 1221 Avenue of the Americas, New York, NY 10020.

This book is printed on acid-free paper.

1 2 3 4 5 6 7 8 9 0 BKM/BKM 0 9 8 7 6

ISBN 978–0–07–340457–8
MHID 0–07–340457–8

Publisher: *Suzanne Jeans*
Senior Sponsoring Editor: *Michael S. Hackett*
Developmental Editor: *Rebecca Olson*
Executive Marketing Manager: *Michael Weitz*
Project Coordinator: *Melissa M. Leick*
Senior Production Supervisor: *Kara Kudronowicz*
Associate Media Producer: *Christina Nelson*
Senior Designer: *David W. Hash*
Cover Designer: *Rokusek Design*
(USE) Cover Image: *©PhotoDisc*
Compositor: *Techbooks*
Printer: *Book-mart Press*

This book was previously published by Tata McGraw-Hill Publishing Company Limited, New Delhi, India, copyright © 2003, 1997, 1984.

Library of Congress Cataloging-in-Publication Data

Jain, R. P. (Rajendra Prasad), 1942–
Modern digital electronics / R. P. Jain — 1st ed.
p. cm.
Includes index.
ISBN 978–0–07–340457–8 — ISBN 0–07–340457–8 (alk. paper)
1. Digital electronics. I. Title.

TK7868.D5J34 2008
621.381—dc22

2006041929
CIP

www.mhhe.com

PREFACE

The tremendous power and usefulness of digital techniques and systems can be seen from the wide variety of industrial machinery, computers, microprocessors, household appliances (washing machines, refrigerators, digital TVs, etc.), medical equipment, internet, e-banking, e-business, e-governance, etc. which are based on the principles of digital electronics. The areas of applications of digital electronics have been increasing day by day, resulting in an unprecedented interest in the subject. In fact, digital systems have invaded all walks of life and have created a digital revolution.

One of the most important reasons for the unprecedented growth of digital electronics is the advent of integrated circuits (ICs). Developments in the IC technology have made it possible to fabricate complex digital circuits, such as microprocessors, memories, complex programmable logic devices (CPLDs) and field programmable gate arrays (FPGAs), etc.

The wonderchip **microprocessor** has been the most fantastic development of the recent years. No other single development has affected our lives as much as the microprocessor in such a short time. Its ever-increasing applications have resulted in developments which were simply unheard of until a few years ago. The emergence of various programmable logic devices have resulted in significant changes in the design methodologies of digital systems. The designers of modern digital systems in industry rarely use conventional manual techniques. Instead, computer aided design (CAD) tools are used. But this has not made the basic concepts and the manual techiques of digital theory and practice obsolete. Rather, the manual techniques are the foundation of CAD tools and they provide a clear insight into the CAD tools. Therefore, it is very essential to have a strong foundation of the basic digital techniques in order to make effective use of automation in design.

This has made it imperative for all those who aspire to design, develop, test, and maintain various electronic systems to learn the principles of modern digital devices and systems.

This book deals with the subject of digital techniques and systems from the basic circuits (gates) to small scale integrated circuits (SSI), medium scale integrated circuits (MSI), large scale integrated circuits (LSI), and very large scale integrated circuits (VLSI). Computer aided design concepts, CAD tools and the hardware description language VHDL have been introduced to familiarize the readers with the CAD techniques.

This book is self contained and is suitable for a course in digital electronics and logic design for electrical, electronics, computer and other engineering disciplines and computer science programmes. Students of physics specializing in electronics will also find the book useful.

The book has been systematically organized and the presentation has been kept at a level suitable for a student with the basic knowledge of circuit theory and electronics.

Chapter 1 introduces the fundamental concepts of digital electronics, advantages of digital systems, and the basic digital circuits. Various number systems and commonly used codes in digitial systems and microprocessors have been discussed in Chapter 2. Error-detecting and error-correcting codes have also been included. Chapter 3 reviews semiconductor devices from the point of view of their applications in digital circuits. Based on these devices, various digital circuits, referred to as logic families, have been discussed in Chapter 4. Availability of various digital functions in ICs have changed the teaching of digital electronics from the good old style using discrete devices to a new style using modern digital ICs. For example, now it is no longer important to minimize the number of gates for the design of a digital circuit, since a number of similar gates are available in a single IC chip; rather it has become necessary to minimize the number of IC packages. Thus the designers of digital systems have to be thoroughly familiar with the principles of operation and flexibilities available in various ICs in order to optimize the design of digital systems from the point of view of cost, space, power requirement, speed of operation, etc. This chapter also deals with the interfacing problems between ICs of the same logic families and between those of different logic families to obtain maximum benefits.

Chapter 5 deals with the conventional methods of combinational circuit design. The Quine-McCluskey method has also been included.

Combinational logic design using MSI circuits is covered in Chapter 6, which is important for the design of digital systems considering the simplicity in design, cost, space, power requirement, speed and other factors.

Chapter 7 introduces the basic building block of a sequential circuit—the FLIP-FLOP. All types of FLIP-FLOPs with their excitation tables and triggering methods have been discussed in detail. Sequential logic design has been discussed in Chapter 8. Here again, both the approaches, namely conventional design using FLIP-FLOPs and the modern approach using available MSI circuits, have been discussed. Design of synchronous sequential as well as asynchronous sequential circuits have been discussed.

Chapter 9 deals with timing circuits and their applications which are essential to a digital system.

The analog-to-digital (A/D) and digital-to-analog (D/A) converters form an important part of many digital systems and the commonly used techniques for such conversions have been discussed in Chapter 10.

Chapter 11 deals with semiconductor memories which have assumed an important role in present-day digital systems. Various semiconductor memories, such as static and dynamic shift registers, static and dynamic RAMs, ROM, PROM, EPROM, EAROM, CAM, and CCD, have been discussed in detail. Programming techniques used for programmable ROMs and erasing techniques used for erasable programmable ROMs have also been discussed.

Chapter 12 presents various programmable logic devices (PLDs), such as programmable logic arrays (PLA), programmable array logic (PAL), Complex Programmable logic devices

(CPLDs) and field programmable gate array (FPGA) devices. These devices are extremely useful for the design of complex digital circuits.

Microprocessors have been introduced in Chapter 13. The fundamentals of microprocessors have been presented in a manner so that even a novice would understand this highly sophisticated device. The most widely used Intel 8085A 8-bit microprocessor has been chosen for discussion. Its organization, operation and programming have been discussed in detail, which will help the students learn the use of microprocessors. The Intel's 16-bit microprocessor 8086 has also been introduced briefly.

Chapter 14 introduces the computer aided design (CAD) approach to digital system design. CAD tools needed for this purpose have been discussed. The VHDL, a hardware description language, has been introduced, which is the basic requirement of designing using CAD tools.

A Glossary of the important terms used in the book and Review Questions with answers for each chapter have been included to enhance the understanding of the users.

The solution manual is also available for the teachers who adopt this book.

R P JAIN

About the Author

R P Jain is presently the Director of the B M Institute of Engineering and Technology, Sonepat. He was the founder and Director-Principal of C R State College of Engineering, Murthal (Sonepat) Haryana, a Haryana Government Engineering College. Before taking up this assignment in 1987, he was the Head of the Electrical and Electronics Engineering Department at the Birla Institute of Technology and Science, Pilani. He obtained his Ph. D. from BITS, Pilani and has been teaching courses in Network Analysis and Synthesis, Analog and Digital Electronic Circuits and Systems, Electronic Measurements, Control Systems, Microprocessors, etc. for the last 40 years. His areas of interest include digital systems, microprocessors, optoelectronics and reliability engineering, planning and development of technical institutions. He is the co-author of *Digital Electronics Practice Using ICs* (TMH, 1983), author of *Digital Electronics and Microprocessors—Problems and Solutions* (TMH, 1987), Switching Theory and Logic Design (TMH, 2003) and Editor-in-Chief of the Encyclopaedia of Science and Technology (CPH, 1994).

ACKNOWLEDGEMENTS

No words are sufficient to express my gratitude to my wife Sushila Jain for her exemplary patience, understanding, and co-operation during the preparation of the book.

Any constructive comments, suggestions, criticism from students and colleagues will be highly appreciated and gratefully acknowledged.

R P JAIN

CONTENTS

3. SEMICONDUCTOR DEVICES—SWITCHING MODE OPERATION 58

4. DIGITAL LOGIC FAMILIES 89

7. FLIP-FLOPs 237

8. SEQUENTIAL LOGIC DESIGN 267

1

FUNDAMENTAL CONCEPTS

1.1 INTRODUCTION

All of us are familiar with the impact of modern digital computers, communication systems, digital display systems, internet, email etc. on society. One of the main causes of this revolution is the advent of *integrated circuits* (*ICs*), which became possible because of the tremendous progress in semiconductor technology in recent years. Most of us may not be familiar with the principles of working of computers, communication systems, internet, email, etc. even though these have become an important part of our daily life. The operation of these systems, and many other systems, is based on the principles of digital techniques and these systems are referred to as *digital systems.*

Some of us are familiar with electronic amplifiers. These are used to amplify electrical signals. This type of signals are continuous signals and can have any value in a limited range and are known as *analog signals.* The electronic circuits used to process (amplify) these signals are known as *analog circuits* and the systems built around this kind of operation are known as *analog systems.*

On the other hand, in an electronic calculator, the input is given with the help of switches. This is converted into electrical signals which have two discrete values or levels. One of these may be called as LOW level and the other one as HIGH level. The signal will always be of one of the two levels. Here, the actual value of the signal is immaterial as long as it is within the specified range of LOW or HIGH level. This type of signal is known as a *digital signal* and the circuits inside the calculator used to process these signals are known as *digital circuits.* A calculator is an example of a *digital system.*

There has been an unprecedented growth in digital techniques since Claude Shannon systematized and adapted George Boole's theoretical work in 1938. Developments in semiconductor technology, together with the progress in digital techniques, brought about a revolution in digital electronics when a single chip device known as *microprocessor* was introduced by Intel Corporation of America in 1971.

Since the introduction of microprocessors, the digital systems have gained tremendous power and importance. There is no field of knowledge which has affected our lives as much as the digital theory and applications, in such a short span of time. It has, infact, created a culture which nobody could have imagined till a few years ago. The rate of growth in this field has been unprecedented and the digital technology has become the most powerful technology for all future innovations in every walk of human endeavour whether it is computers, communication systems, information systems, entertainment and consumer products, business, banking and finance, office machines, homes, cars, education, industrial control systems, scientific and medical instruments, and defence equipment, etc.

Some of the principal reasons for the widespread use of digital techniques and systems are:

- The devices used in digital circuits generally operate in one of the two *states*, known as ON and OFF resulting in a very simple operation.
- There are only a few basic operations in digital circuits which are very easy to understand.
- Digital techniques require Boolean algebra which is very simple and can easily be learnt even in schools.
- Digital circuits require basic concepts of electric network analysis which can easily be learnt at the junior level in colleges. The principal electrical characteristics required are *switching speed* and *loading* considerations. On the other hand, analog circuits and systems involve frequency and time domain concepts, complicated circuit analysis techniques, etc. which make the understanding of these circuits much more difficult than the digital circuits.
- A large number of ICs are available for performing various operations. These are highly reliable, accurate, small in size and the speed of operation is very high. A number of programmable ICs are also available.
- Various ICs are available in a *logic family* with similar electrical characteristics which make the design and development of digital systems very simple and also reduces interfacing problems. Also, a number of logic families based on different technologies are available which help in optimizing the system design from the point of view of power requirement and speed of operation.
- The effect of fluctuations in the characteristics of the components, ageing of components, temperature, and noise, etc. is very small in digital circuits.
- Digital circuits have capability of memory which makes these circuits highly suitable for computers, calculators, watches, telephones, etc.
- The display of data and other information is very convenient, accurate and elegant using digital techniques.
- Many students have an opportunity to learn programming of digital computers, hence they have a strong motivation to study the way the digital hardware works.
- It is a very fascinating and challenging field of study because most of the latest electronic systems are digital in nature.

1.2 DIGITAL SIGNALS

As mentioned above, a digital signal has two discrete levels or values. Two different representations of digital signals are shown in Fig. 1.1. In each case there are two discrete levels. These levels can be represented using the terms LOW and HIGH. In Fig. 1.1*a*, lower of the two levels has been designated as LOW level and the higher as HIGH level. In contrast to this, in Fig. 1.1*b*, higher of the two levels has been designated as LOW level and the lower as HIGH level. Digital systems using the representation of signal shown in Fig. 1.1*a* are said to employ ***positive logic system*** and those using the other representation of the signal shown in Fig. 1.1*b* are said to employ ***negative logic system***. The genesis of the term *logic* is given later. In each of the two signals we observe that the voltage corresponding to a given level is not fixed, rather voltages in a limited range are designated as a level. As long as the voltage belongs to a level it will be taken as that level and the exact value of the voltage is immaterial. For example, any voltage in the range of 3.5 to 5 V will be considered as HIGH level in the positive logic system and LOW level in the negative logic system. Similarly, any voltage in the range of 0 to 1 V will be considered as LOW level in the positive logic system and HIGH level in the negative logic system. The actual voltage ranges corresponding to LOW and HIGH level are not same for all types of circuits and are different for different logic families (see Chapter 4). Unless otherwise specified, we shall be dealing with positive logic system.

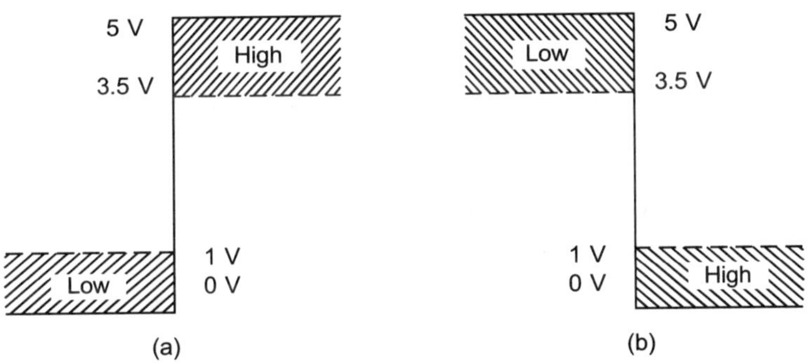

Fig. 1.1
Digital signal representation (a) Positive logic (b) Negative logic.

The above discussion brings out one of the main advantages of digital systems, viz. they are less susceptible to noise, fluctuations in the characteristics of components, etc.

The two discrete signal levels HIGH and LOW can also be represented by the *binary digits* 1 and 0 respectively. A *b*inary digi*t* (0 or 1) is referred to as a *bit*. Since a digital signal can have only one of the two possible levels 1 or 0, the *binary number system* (see Chapter 2) can be used for the analysis and design of digital systems. The two levels (or states) can also be designated as ON and OFF or TRUE and FALSE. George Boole introduced the concept of binary number system in the studies of the mathematical theory of LOGIC in the work entitled *An Investigation of the Laws of Thought* in 1854 and developed its algebra known as *Boolean algebra*. These logic concepts have been adapted for the design of digital hardware since 1938 when Claude Shannon-organised and systematized Boole's work in *Symbolic Analysis of Relay and Switching Circuits*.

1.3 BASIC DIGITAL CIRCUITS

In a digital system there are only a few basic operations performed, irrespective of the complexities of the system. These operations may be required to be performed a number of times in a large digital system like digital computer or a digital control system, etc. The basic operations are AND, OR, NOT, and FLIP-FLOP. The AND, OR, and NOT operations are discussed here and the FLIP-FLOP, which is a basic memory element used to store binary information (one bit is stored in one FLIP-FLOP), will be introduced in Chapter 7.

1.3.1 The AND Operation

A circuit which performs an AND operation is shown in Fig. 1.2. It has N inputs ($N \geq 2$) and one output. Digital signals are applied at the input terminals marked $A, B, ..., N$, the other terminal being ground, which is not shown in the diagram. The output is obtained at the output terminal marked Y (the other terminal being ground) and it is also a digital signal. The AND operation is defined as : the output of an AND gate is 1 if and only if all the inputs are 1. Mathematically, it is written as

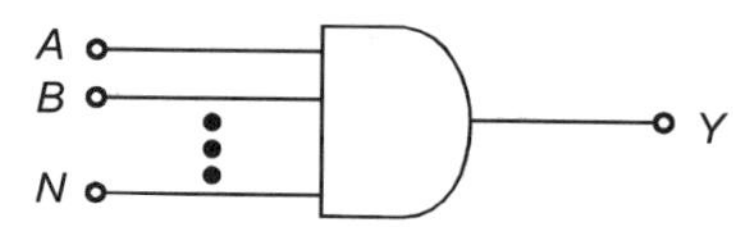

Fig. 1.2
The standard symbol for an AND gate.

$$\begin{aligned} Y &= A \text{ AND } B \text{ AND } C \ldots \text{ AND } N \\ &= A \cdot B \cdot C \cdot \ldots \cdot N \\ &= ABC \ldots N \end{aligned} \tag{1.1}$$

where $A, B, C, \ldots N$ are the input variables and Y is the output variable. The variables are binary, i.e. each variable can assume only one of the two possible values, 0 or 1. The *binary variables* are also referred to as *logical variables.*

Equation (1.1) is known as the *Boolean equation* or the *logical equation* of the AND *gate.* The term gate is used because of the similarity between the operation of a digital circuit and a gate. For example, for an AND operation the gate opens ($Y = 1$) only when all the inputs are present, i.e. at logic 1 level.

Truth Table

Since a logical variable can assume only two possible values (0 and 1), therefore, any logical operation can also be defined in the form of a table containing all possible input combinations (2^N combinations for N inputs) and their corresponding outputs. This is known as a *truth table* and it contains one row for each one of the input combinations.

For an AND gate with two inputs A, B and the output Y, the truth table is given in Table 1.1. Its logical equation is $Y = AB$ and is read as "Y equals A AND B".

Table 1.1 Truth table of a 2-input AND Gate

Inputs		*Output*
A	*B*	*Y*
0	0	0
0	1	0
1	0	0
1	1	1

Since, there are only two inputs, A and B, therefore, the possible number of input combinations is four. It may be observed from the truth table that the input–output relationship for a digital circuit is completely specified by this table in contrast to the input–output relationship for an analog circuit. The pattern in which the inputs are entered in the truth table may also be observed carefully, which is in the ascending order of binary numbers formed by the input variables (See Chapter 2).

1.3.2 The OR Operation

Figure 1.3 shows an OR gate with N inputs ($N \geq 2$) and one output. The OR operation is defined as: the output of an OR gate is 1 if and only if one or more inputs are 1. Its logical equation is given by

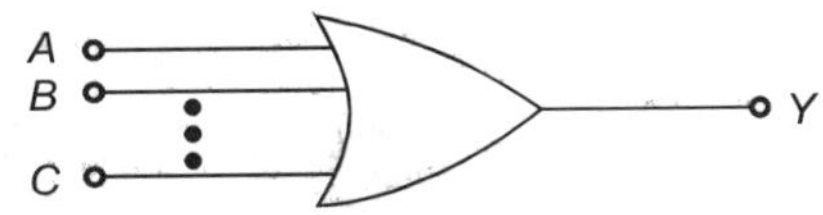

Fig. 1.3
The standard symbol for an OR gate.

$$Y = A \text{ OR } B \text{ OR } C \ldots \text{ OR } N$$
$$= A + B + C + \ldots + N \qquad (1.2)$$

The truth table of a 2-input OR gate is given in Table 1.2. Its logic equation is $Y = A + B$ and is read as "Y equals A OR B".

Table 1.2 Truth table of a 2-input OR Gate

Inputs		*Output*
A	*B*	*Y*
0	0	0
0	1	1
1	0	1
1	1	1

1.3.3 The NOT Operation

Figure 1.4 shows a NOT gate, which is also known as an *inverter*. It has one input (A) and one output (Y). Its logic equation is written as

$$Y = \text{NOT } A$$
$$= \bar{A} \qquad (1.3)$$

Fig. 1.4
The standard symbols for a NOT gate.

and is read as "Y equals NOT A" or "Y equals complement of A". The truth table of a NOT gate is given in Table 1.3.

The NOT operation is also referred to as an *inversion* or *complementation.* The presence of a small circle, known as the *bubble*, always denotes inversion in digital circuits.

Table 1.3 Truth table of a NOT gate

Input A	*Output* Y
0	1
1	0

1.4 NAND AND NOR OPERATIONS

Any Boolean (or logic) expression can be realized by using the AND, OR and NOT gates discussed above. From these three operations, two more operations have been derived: the NAND operation and NOR operation. These operations have become very popular and are widely used, the reason being the only one type of gates, either NAND or NOR are sufficient for the realization of any logical expression. Because of this reason, NAND and NOR gates are known as *universal gates.*

1.4.1 The NAND Operation

The NOT–AND operation is known as the NAND operation. Figure 1.5*a* shows an N input ($N \geq 2$) AND gate followed by a NOT gate. The operation of this circuit can be described in the following way:

The output of the AND gate (Y') can be written using Eq. (1.1)

$$Y' = AB \ldots N \tag{1.4}$$

Now, the output of the NOT gate (Y) can be written using Eq. (1.3)

$$Y = \overline{Y}' = \left(\overline{AB \ldots N}\right) \tag{1.5}$$

The logical operation represented by Eq. (1.5) is known as the NAND operation. The standard symbol of the NAND gate is shown in Fig. 1.5*b*. Here, a bubble on the output side of the NAND gate represents NOT operation, inversion or complementation.

The truth table of a 2-input NAND gate is given in Table 1.4. Its logic equation is $Y = \overline{A \cdot B}$ and, is read as "Y equals NOT (A AND B)".

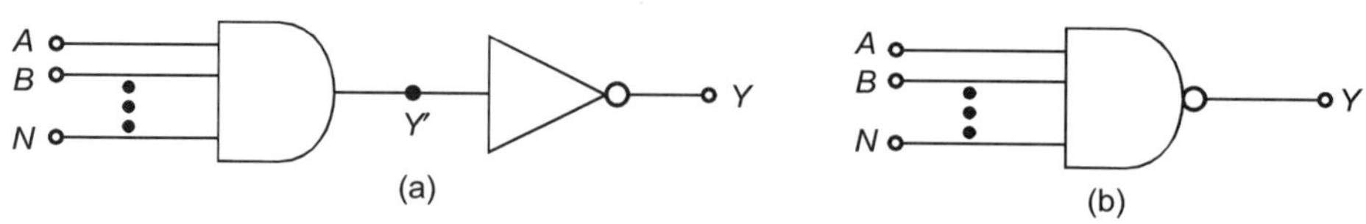

Fig. 1.5
(a) NAND operation as NOT–AND operation,
(b) Standard symbol for the NAND gate.

Table 1.4 Truth table of a 2-input NAND gate

Inputs		*Output*
A	*B*	*Y*
0	0	1
0	1	1
1	0	1
1	1	0

The three basic logic operations, AND, OR and NOT can be performed by using only NAND gates. These are given in Fig. 1.6.

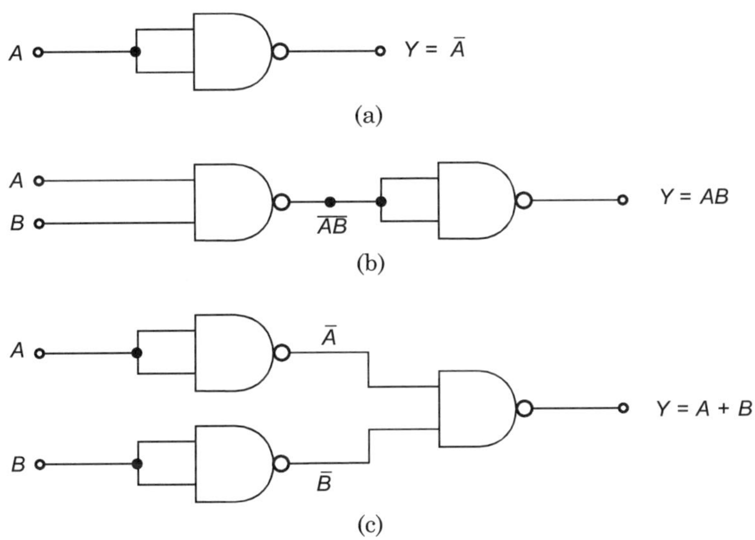

Fig. 1.6
Realization of basic logic operations using NAND gates (*a*) NOT (*b*) AND (*c*) OR.

1.4.2 The NOR Operation

The NOT–OR operation is known as the NOR operation. Figure 1.7*a* shows an *N* input ($N \geq 2$) OR gate followed by a NOT gate. The operation of this circuit can be described in the following way:
The output of the OR gate Y' can be written using Eq. (1.2) as

$$Y' = A + B + \ldots + N \tag{1.6}$$

and the output of the NOT gate (Y) can be written using Eq. (1.3)

$$Y = \bar{Y}' = \overline{A + B + \ldots + N} \tag{1.7}$$

The logic operation represented by Eq. (1.7) is known as the NOR operation.

The standard symbol of the NOR gate is shown in Fig. 1.7*b*. Similar to the NAND gate, a bubble on the output side of the NOR gate represents the NOT operation.

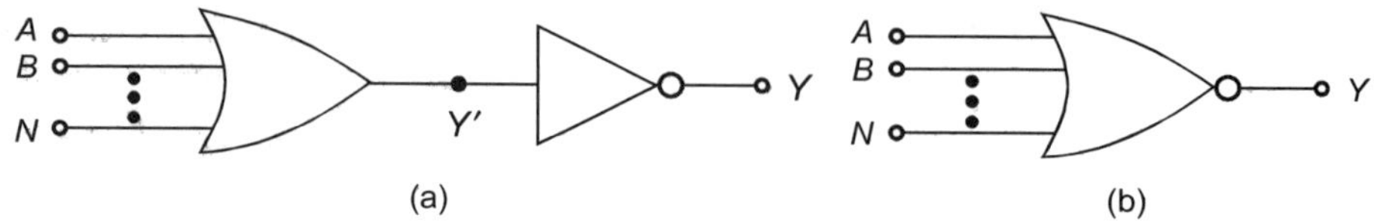

Fig. 1.7

(*a*) NOR operation as NOT–OR operation
(*b*) Standard symbol for the NOR gate

Table 1.5 gives the truth table of a 2-input NOR gate. Its logic equation is $Y = \overline{A+B}$ and is read as "*Y* equals NOT (*A* OR *B*)".

Table 1.5 Truth table of a 2-input NOR gate

Inputs		*Output*
A	*B*	*Y*
0	0	1
0	1	0
1	0	0
1	1	0

The three basic logic operations, AND, OR, and NOT can be performed by using only the NOR gates. These are given in Fig. 1.8.

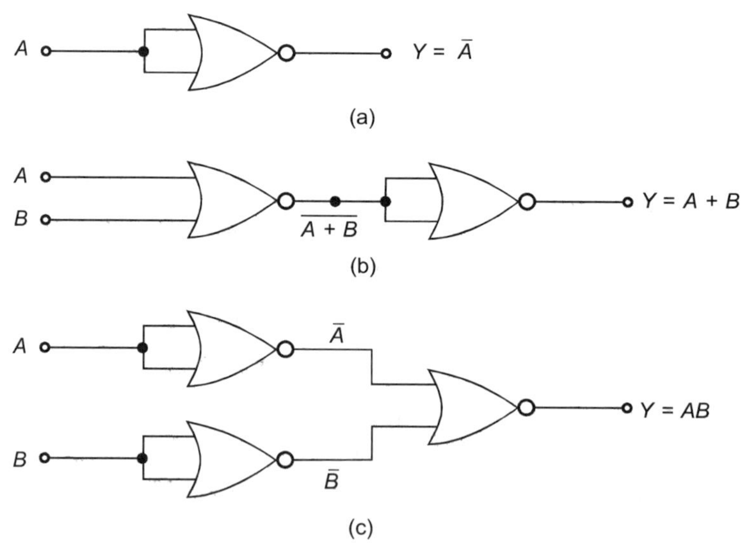

Fig. 1.8

Realization of basic logic operations using NOR gates
(a) NOT (b) OR (c) AND.

1.5 EXCLUSIVE-OR OPERATION

The EXCLUSIVE-OR (EX-OR) operation is widely used in digital circuits. It is not a basic operation and can be performed using the basic gates–AND, OR and NOT or universal gates NAND or NOR. Because of its importance, the standard symbol shown in Fig. 1.9 is used for this operation.

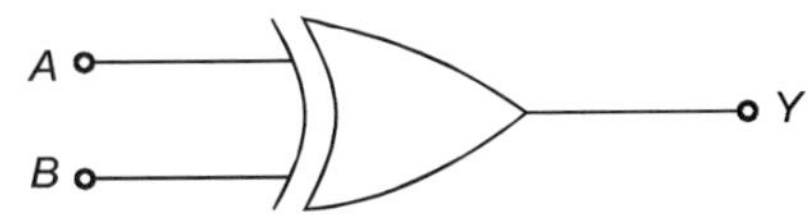

Fig. 1.9
Standard symbol for EX-OR gate.

The truth table of an EX-OR gate is given in Table 1.6 and its logic equation is written as

$$Y = A \text{ EX-OR } B = A \oplus B \tag{1.8}$$

Table 1.6 Truth table of a 2-input EX-OR gate

Inputs		*Output*
A	*B*	*Y*
0	0	0
0	1	1
1	0	1
1	1	0

If we compare the truth table of an EX-OR gate with that of an OR gate given in Table 1.2, we find that the first three rows are same in both. Only the fourth row is different. This circuit finds application where two digital signals are to be compared. From the truth table we observe that when both the inputs are same (0 or 1) the output is 0, whereas when the inputs are not same (one of them is 0 and the other one is 1) the output is 1.

1.6 BOOLEAN ALGEBRA

As discussed above, the digital signals are discrete in nature and can only assume one of the two values 0 or 1. A number system based on these two digits is known as *binary number system.* In the middle of 19th century, an English mathematician George Boole developed rules for manipulations of binary variables, known as *Boolean algebra.* This is the basis of all digital systems like computers, calculators, etc.

Binary variables can be represented by a letter symbol such as *A, B, X, Y, ...* The variable can have only one of the two possible values at any time, viz. 0 or 1. The Boolean algebraic theorems are given in Table 1.7.

From these theorems, we observe that the even numbered theorems can be obtained from their preceding odd numbered theorems by (*i*) interchanging + and · signs, and (*ii*) interchanging 0 and 1.

Theorems which are related in this way are called *duals.*

Table 1.7 Boolean algebraic theorems

Theorem No.	*Theorem*
1.1	$A + 0 = A$
1.2	$A \cdot 1 = A$
1.3	$A + 1 = 1$
1.4	$A \cdot 0 = 0$
1.5	$A + A = A$
1.6	$A \cdot A = A$
1.7	$A + \bar{A} = 1$
1.8	$A \cdot \bar{A} = 0$
1.9	$A \cdot (B + C) = AB + AC$
1.10	$A + BC = (A + B)(A + C)$
1.11	$A + AB = A$
1.12	$A(A + B) = A$
1.13	$A + \bar{A}B = (A + B)$
1.14	$A(\bar{A} + B) = AB$
1.15	$AB + A\bar{B} = A$
1.16	$(A + B) \cdot (A + \bar{B}) = A$
1.17	$AB + \bar{A}C = (A + C)(\bar{A} + B)$
1.18	$(A + B)(\bar{A} + C) = AC + \bar{A}B$
1.19	$AB + \bar{A}C + BC = AB + \bar{A}C$
1.20	$(A + B)(\bar{A} + C)(B + C) = (A + B)(\bar{A} + C)$
1.21	$\overline{A \cdot B \cdot C \cdot \ldots} = \bar{A} + \bar{B} + \bar{C} + \ldots$
1.22	$\overline{A + B + C + \ldots} = \bar{A} \cdot \bar{B} \cdot \bar{C} \ldots$

Theorems 1.1 to 1.8 involve a single variable only. Each of these theorems can be proved by considering every possible value of the variable. For example, in Theorem 1.1,
if $A = 0$ then $0 + 0 = 0 = A$
and if $A = 1$ then $1 + 0 = 1 = A$
and hence the theorem is proved.

Theorems 1.9 to 1.20 involve more than one variable and can be proved by making a truth table. For example, Theorem 1.10 can be proved by making the truth table given in Table 1.8.

Table 1.8 Truth table to prove Theorem 1.10

A	B	C	BC	$A + BC$	$A + B$	$A + C$	$(A + B)(A + C)$
0	0	0	0	0	0	0	0
0	0	1	0	0	0	1	0
0	1	0	0	0	1	0	0
0	1	1	1	1	1	1	1
1	0	0	0	1	1	1	1
1	0	1	0	1	1	1	1
1	1	0	0	1	1	1	1
1	1	1	1	1	1	1	1

From the table we observe that there are 8 (= 2^3) possible combinations of the three variables *A*, *B*, and *C*. For each combination, the value of $A + BC$ is the same as that of $(A + B)(A + C)$, which proves the theorem. Theorems 21 and 22 are known as De Morgan's theorems. These theorems can be proved by first considering the two variable case and then extending this result. From the truth table given in Table 1.9, we get the relations

$$\overline{A \cdot B} = \bar{A} + \bar{B} \tag{1.9}$$

and

$$\overline{A + B} = \bar{A} \cdot \bar{B} \tag{1.10}$$

Table 1.9 Truth table to prove De Morgan's theorems

A	B	$\bar{A}$	$\bar{B}$	$\overline{AB}$	$\bar{A} + \bar{B}$	$\overline{A+B}$	$\bar{A} \cdot \bar{B}$
0	0	1	1	1	1	1	1
0	1	1	0	1	1	0	0
1	0	0	1	1	1	0	0
1	1	0	0	0	0	0	0

Now consider the NAND operation of three variables,

$$\begin{aligned}\overline{ABC} &= \overline{(AB) \cdot C} \\ &= \overline{(A \cdot B)} + \bar{C} \\ &= \bar{A} + \bar{B} + \bar{C} \text{ using Eq. (1.9)}\end{aligned} \tag{1.11}$$

In a similar way, the NOR operation of three variables gives

$$\begin{aligned}\overline{A + B + C} &= \overline{(A + B) + C} \\ &= \overline{(A + B)} \cdot \bar{C} \\ &= \bar{A} \cdot \bar{B} \cdot \bar{C} \text{ using Eq. (1.10)}\end{aligned} \tag{1.12}$$

The above results can be easily extended to any number of variables.

A logic problem can be specified in terms of a set of statements. This set of statements can be represented in terms of an equation called the logic equation or in terms of a truth table. A digital circuit using the gates discussed above can be designed to realize a logic equation. In general, it is possible to simplify (minimize) a logic equation. The minimized logic equation will probably need less number of gates and/or less number of inputs for the gates. The techniques used to minimize logic equations will be discussed in Chapter 5. An example of the realization of a circuit for a given logic equation is given below:

Example 1.1 Realize (design) a digital circuit for the logic equation

$$Y = \bar{A} \cdot B + A \cdot \bar{B} \tag{1.13}$$

Solution

This equation consists of two input variables *A* and *B* and one output variable *Y*. The variable *A* appears as $\bar{A}$ in the term $\bar{A} \cdot B$ and as *A* in the term $A \cdot \bar{B}$. Similarly the variable *B* appears as *B* in the term $\bar{A} \cdot B$ and as $\bar{B}$ in the term $A \cdot \bar{B}$. The two terms $\bar{A} \cdot B$ and $A \cdot \bar{B}$ are obtained by AND operations and the output *Y* is the result of OR operation performed on $\bar{A} \cdot B$ and $A \cdot \bar{B}$ terms. The realization of this equation is obtained using NOT, AND and OR gates as shown in Fig. 1.10.

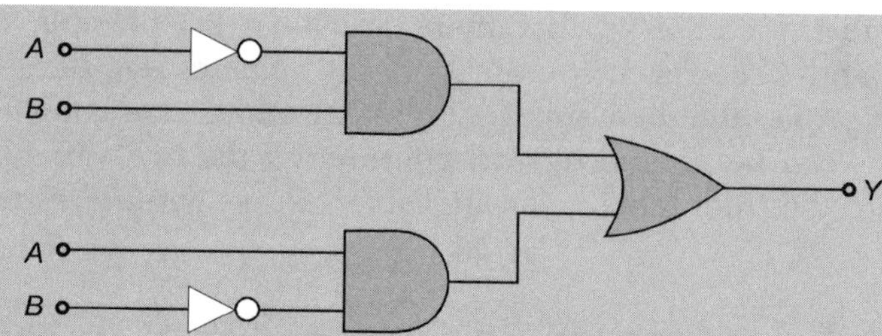

Fig. 1.10
Realization of Eq. (1.13).

1.7 EXAMPLES OF IC GATES

All the logic functions introduced in this chapter are commercially available in integrated circuit (IC) form. For example, 7400 IC chip is a quadruple 2-input NAND gate available in 14-pin DIP. It has four identical, independent 2-input NAND gates arranged as shown in Fig. 1.11. It requires a + 5 V d.c. supply (to be connected between V_{CC} and GND pins) for the operation of the gates. Table 1.10 gives some of the available gate ICs. The details of their pin connections, electrical characteristics, etc. can be obtained from the manufacturers' data catalogues. The necessary information along with the practical applications of these chips are available in Jain and Anand (see Bibliography).

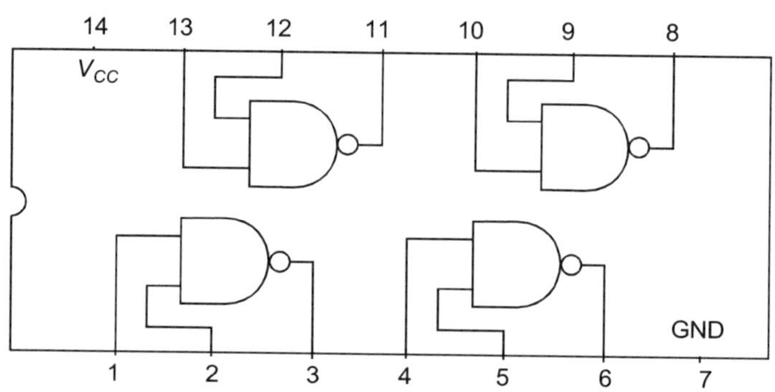

Fig. 1.11
Block diagram of 7400 IC.

Table 1.10 Some of the available IC gates

IC No.	*Description*
7400	Quad 2-input NAND gates
7402	Quad 2-input NOR gates
7404	Hex inverters
7408	Quad 2-input AND gates
7410	Triple 3-input NAND gates

(*Contd.*)

Table 1.10 (*Contd.*)

IC No.	*Description*
7411	Triple 3-input AND gates
7420	Dual 4-input NAND gates
7421	Dual 4-input AND gates
7427	Triple 3-input NOR gates
7430	8-input NAND gate
7432	Quad 2-input OR gates
7486, 74386	Quad EX-OR gates
74133	13-input NAND gate
74135	Quad EX-OR/NOR gates
74260	Dual 5-input NOR gates

1.8 SUMMARY

In this chapter, the basic concepts of the digital systems have been discussed. The basic features and advantages of these systems have been given briefly. The level of the treatment has been kept low to avoid any confusion. Table 1.11 summarizes the operation of all the gates introduced in this chapter. For convenience, two input gates have been taken and the different symbols used for various operations are also given. A brief exposure to Boolean algebra has also been given. The techniques for the simplification of logic equations will be discussed in Chapter 5.

Table 1.11 Summary of logic gates

Gate	*Logic diagram*	*Function*	*Truth table*
AND	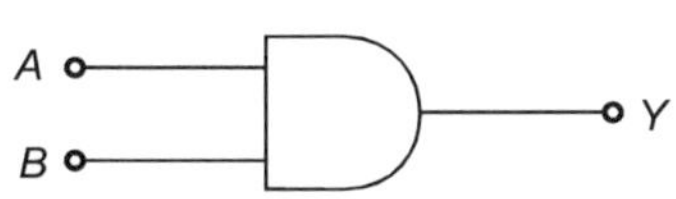	$Y = A$ AND B $= A \cdot B$ $= A \cap B$ $= A \wedge B$ $= AB$	*Inputs* A B, *Output* Y 0 0 0 0 1 0 1 0 0 1 1 1
OR	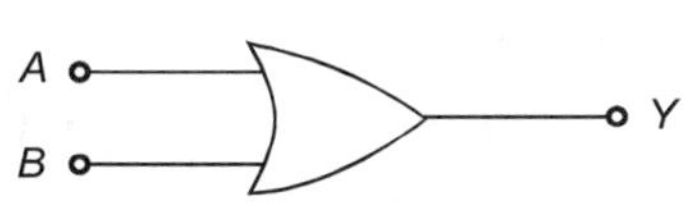	$Y = A$ OR B $= A + B$ $= A \cup B$ $= A \vee B$	*Inputs* A B, *Output* Y 0 0 0 0 1 1 1 0 1 1 1 1
NOT (inverter)	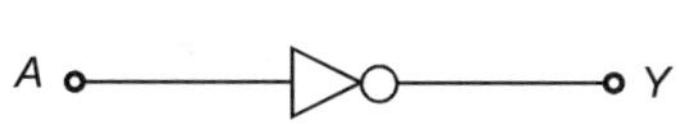	$Y =$ NOT A $= \bar{A}$	*Input* A, *Output* Y 0 1 1 0

(*Contd.*)

Table 1.11 *(Contd.)*

Gate	Logic diagram	Function	Truth table		
			Inputs		*Output*
			A	*B*	*Y*
NAND	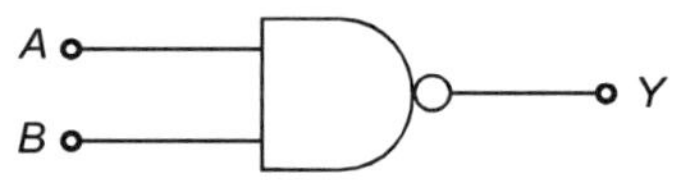	$Y = A$ NOT AND B	0	0	1
		$= A$ NAND B	0	1	1
		$= \overline{A \cdot B}$	1	0	1
		$= \overline{A \cap B}$	1	1	0
		$= \overline{A \wedge B}$			
		$= A \uparrow B$			
		$= \overline{AB}$			
NOR	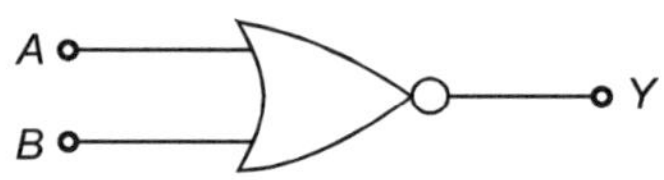	$Y = A$ NOT OR B	0	0	1
		$= A$ NOR B	0	1	0
		$= \overline{A + B}$	1	0	0
		$= \overline{A \cup B}$	1	1	0
		$= \overline{A \vee B}$			
		$= A \downarrow B$			
EX-OR	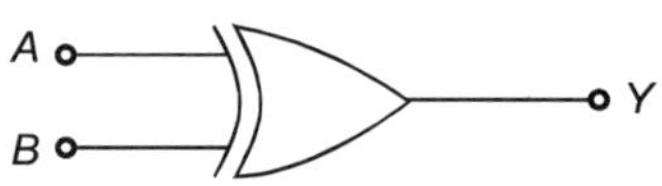	$Y = A$ EX-OR B	0	0	0
		$= A \oplus B$	0	1	1
		$= A\bar{B} + \bar{A}B$	1	0	1
			1	1	0
EX-NOR	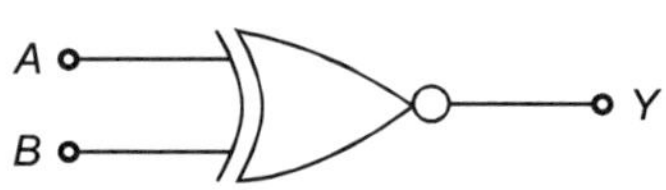	$Y = \overline{A \text{ EX-OR } B}$	0	0	1
		$= A$ EX-NOR B	0	1	0
		$= A \odot B$	1	0	0
		$= \overline{A\bar{B} + \bar{A}B}$	1	1	1
		$= \bar{A}\,\bar{B} + AB$			

Glossary

Active-high input The input terminal is active (or enabled) when held at HIGH logic level.

Active-high output The output terminal is at HIGH logic level when active (or enabled).

Active-low input The input terminal is active (or enabled) when held at LOW logic level.

Active-low output The output terminal is at low logic level when active (or enabled).

Analog circuit An electronic circuit that processes analog signals.

Analog signal A continuous signal that can have any value in a given range. It is also known as continuous signal.

Analog system An electronic system that consists of analog circuits and/or devices.

AND gate A logic circuit whose output is 1 if and only if all its inputs are 1.

Binary Having two states.

Binary number system A number system with base (or radix) 2, i.e only two symbols 0 and 1 are used to represent any number.

Binary variable A variable that can have only two values 0 and 1.

Bit A binary digit 0 or 1.

Boolean algebra The algebra of binary variables.

Boolean function A well defined relationship between binary variables specified by either a Boolean equation or a truth table.

Boolean variable Same as the binary variable.

Bubble A small circle indicating inversion operation or active-low condition.

Chip A piece of silicon (or any semiconductor material) on which an integrated circuit is fabricated.

Computer An electronic digital system that processes data. See also digital computer.

Digital circuit An electronic circuit that processes digital signals.

Digital computer A programmable digital system capable of performing various arithmetic and logical operations.

Digital electronics Branch of electronics which deals with the digital devices, circuits and systems.

Digital signal A signal with only two discrete values. These are usually represented by LOW and HIGH or 0 and 1.

Digital system An electronic system consisting of digital circuits and/or devices.

DIP (Dual-in-line package) A semiconductor chip package having two rows of pins perpendicular to the edges of the package.

Disable Prohibits an activity from proceeding or output(s) to be produced in a digital circuit. Also known as INHIBIT.

Dual Two boolean functions are said to be dual of each other if any one of these is obtained from the other one by (i) interchanging + and · signs, and (ii) interchanging 0 and 1.

Enable Allows activity to proceed or output(s) to be produced in a digital circuit.

EX-NOR (Exclusive-NOR) gate A two input gate that produces a high output only when both the inputs are same.

EX-OR (Exclusive-OR) gate A two input gate whose output is logic 0 when both the inputs are equal and logic 1 when they are unequal.

False One of the states of a logic circuit, the other state is true.

FLIP-FLOP It is a basic memory element in digital systems. It is same as bistable multivibrator.

Gate A logic circuit in which the output depends upon the inputs according to some logic rules.

High level The voltage corresponding to logic 1.

IC package An integrated circuit chip packaged as a single multilead component. For example DIP.

Inhibit Same as disable.

Integrated circuit (IC) A small semiconductor chip containing several electronic circuits.

Inversion The process of complementing a logic signal.

Inversion circle Same as bubble.

Inverter A logic gate whose output is the complement of its input.

Logic circuit An electronic circuit that operates on digital signals in accordance with a logic function.

Logic family A group of logic circuits built around a standardized integrated circuit technology. For example, transistor-transistor logic (TTL), complementary metal-oxide-semiconductor logic (CMOS) etc.

Logical variable Same as binary variable.

Logic gate Same as gate.

LOW-level The voltage corresponding to logic 0.

Memory A device that stores binary information.

Microprocessor A semiconductor IC chip with the capabilities of CPU of a computer.

NAND gate A logic gate whose output is logic 0 if and only if all of its inputs are logic 1.

Negative logic system Logic system in which the lower of the two levels is represented by 1 and the higher level is represented by 0.

Noise Unwanted electrical signals which are random in nature.

NOR gate A logic gate whose output is logic 1 if and only if all of its inputs are logic 0.

NOT gate Same as inverter.

OFF One of the states of a logic circuit, the other state is ON.

ON One of the states of a logic circuit, the other state is OFF.

OR gate A logic gate whose output is logic 0 if and only if all its inputs are logic 0.

Positive logic system Logic system in which the higher of the two levels is represented by 1 and the lower level is represented by 0.

Programmable ICs ICs which can be programmed for desired purpose by a user.

Switching speed Speed with which an electronic switch can change from ON to OFF and vice-versa. It is usually expressed in terms of the propagation delay time.

True One of the states of a logic circuit, the other state is *False.*

Truth table A table that gives outputs for all possible combinations of inputs to a logic circuit.

Universal gate A gate that can perform all the basic logical operations, such as NAND, and NOR.

Review Questions

1.1 Ordinary electrical switch is __________ device. (analog/digital)

1.2 A train of pulses is __________ signal. (analog/digital)

1.3 The output of an AND gate is *high* if and only if all its inputs are __________ . (high/low)

1.4 If one of the inputs to an OR gate is *high* its output will be __________ . (high/low)

1.5 If one of the inputs of an EX-OR gate is *high*, its output will be __________. (same as other input/inverse of other input)

1.6 To INHIBIT (or DISABLE) an AND gate one of its inputs is connected to logic level _____. (0/1)

1.7 To ENABLE an AND gate one of its inputs is connected to logic level __________ . (0/1)

1.8 A AND gate is ENABLED by connecting one of its inputs to logic level __________ . (0/1)

1.9 To DISABLE a NOR gate one of its inputs needs to be connected to logic level __________ . (0/1)

1.10 One of the inputs of an AND gate is labelled as ENABLE. This control input is __________ . (active-low/active-high)

1.11 One of the inputs of a NOR gate is labelled as ENABLE. This control input is __________ . (active-low/active-high)

1.12 The number of rows in a truth table of 4 variables is __________ .

1.13 A 3-input NOR gate is required to detect the simultaneous occurrence of all the inputs in the LOW state. Its output is __________ . (active-low/active-high)

1.14 The number of 3-input NAND gates in a 14-pin IC is __________ .

1.15 The minimum member of bits required to distinguish 108 distinct objects is __________ .

Problems

1.1 Which of the following systems are analog and which are digital? Why?
- (a) Pressure gauge
- (b) An electronic counter used to count persons entering an exhibition.
- (c) Clinical thermometer
- (d) Electronic calculator
- (e) Transistor radio receiver
- (f) Ordinary electric switch.
- (g) Electronic Voting Machine (EVM)

1.2 In the circuits of Fig. 1.12, the switches may be ON (1) or OFF (0) and will cause the bulb to be ON (1) or OFF (0).
- (a) Determine all possible conditions of the switches for the bulb to be ON (1)/OFF (0) in each of the circuits.
- (b) Represent the information obtained in part (*a*) in the form of truth table.
- (c) Name the operation performed by each circuit (refer to Table 1.11).

1.3 The voltage waveforms shown in Fig. 1.13 are applied at the inputs of 2-input AND, OR, NAND, NOR, and EX-OR gates. Determine the output waveform in each case.

1.4 Find the relationship between the inputs and output for each of the gates shown in Fig. 1.14. Name the operation performed in each case (refer to Table 1.11).

1.5 Make the truth tables for each of the circuits of Figs 1.6 and 1.8 and verify the operation performed.

1.6 For each of the following statements indicate the logic gate(s) AND, OR, NAND, NOR for which it is true.
- (a) All LOW inputs produce a HIGH output.
- (b) Output is HIGH if and only if all inputs are HIGH.

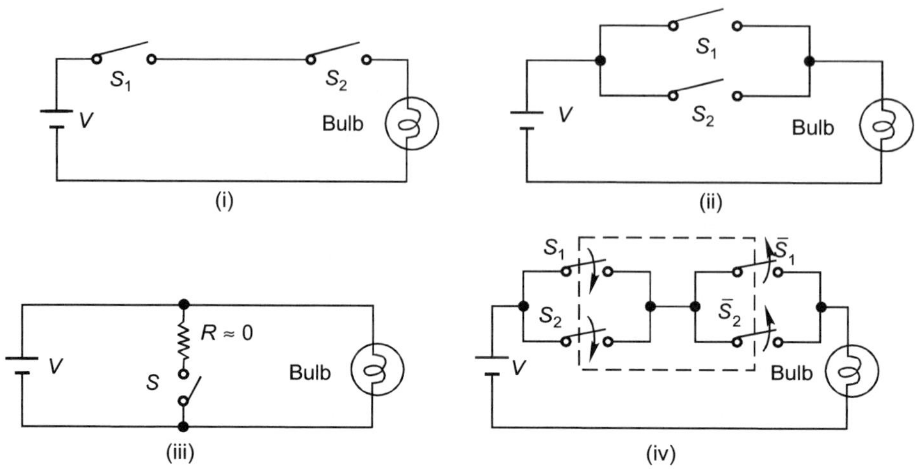

Fig. 1.12
Circuits for Problem 1.2

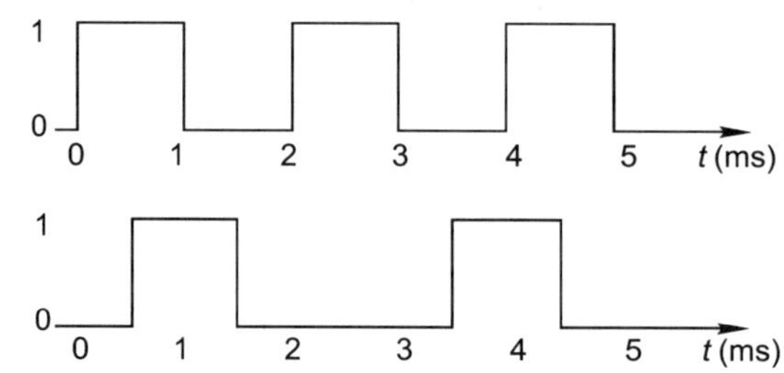

Fig. 1.13
Waveforms for Problem 1.3

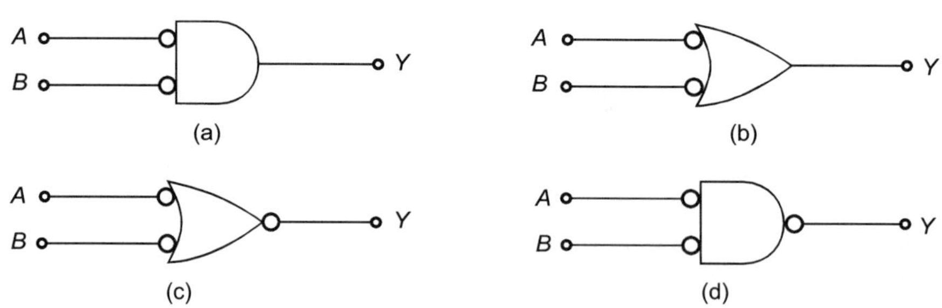

Fig. 1.14
Circuits for Problem 1.4

(c) Output is LOW if and only if all inputs are HIGH.
(d) Output is LOW if and only if all inputs are LOW.

1.7 For the logic expression,

$$Y = A\bar{B} + \bar{A}B$$

(a) Obtain the truth table.
(b) Name the operation performed.
(c) Realize this operation using AND, OR, NOT gates.
(d) Realize this operation using only NAND gates.

1.8 Prove the following:
(a) A positive logic AND operation is equivalent to a negative logic OR operation and vice-versa.
(b) A positive logic NAND operation is equivalent to a negative logic NOR operation and vice-versa.

1.9 Prove the following using the Boolean algebraic theorems:
(a) $A + \overline{A} \cdot B + A \cdot \overline{B} = A + B$
(b) $A \cdot B + \overline{A} \cdot B + \overline{A} \cdot \overline{B} = \overline{A} + B$
(c) $\overline{A}BC + A\overline{B}C + AB\overline{C} + ABC = AB + BC + CA$

1.10 Prove the logic equations of Problem 1.9 using the truth table approach.

1.11 Realize the left hand side and the right hand side of the logic equations of Problem 1.9 using AND, OR, and NOT gates and find the saving in hardware in each case (number of gates and the number of inputs for the gates).

1.12 Prove the following using De Morgan's theorems:
(a) $AB + CD = \overline{\overline{AB} \cdot \overline{CD}}$
(b) $(A + B) \cdot (C + D) = \overline{\overline{(A + B)} + \overline{(C + D)}}$
And hence, prove the following statements:
(*i*) An AND-OR configuration is equivalent to a NAND-NAND configuration.
(*ii*) An OR-AND configuration is equivalent to a NOR-NOR configuration.

1.13 (a) Realize the logic equation (a) of Problem 1.12 using
(*i*) AND and OR gates.
(*ii*) only NAND gates.
(b) Realize the logic equation (b) of Problem 1.12 using
(*i*) OR and AND gates.
(*ii*) only NOR gates.

1.14 Verify that the following operations are commutative and associative
(a) AND (b) OR (c) EX-OR

1.15 Verify that the following operations are commutative but not associative.
(a) NAND (b) NOR

1.16 Realize the logic expression
$Y = A \oplus B \oplus C \oplus D$
using EX-OR gates.

1.17 Consider the expression: $Z = A \oplus B \oplus C \oplus D \oplus$ .... Show that $Z = 1$ if an odd number of variables are 1 and that $Z = 0$ if an even number of variables are 1.

1.18 For a gate with N inputs, how many combinations of inputs are possible? State the general rule to obtain the possible combinations.

1.19 Determine the number of pins in the following ICs.
(a) 7402 (b) 7404 (c) 7408 (d) 7410
(e) 7411 (f) 7420 (g) 7427 (h) 7432
(i) 7486

1.20 Determine the IC chips required for the implementation of each of the circuits of Problem 1.13.

1.21 The logic levels for two typical logic circuits A and B are given below:
A: 0.4 V and 2 V
B: – 0.75 V and –1.55 V
Express these levels in binary form assuming positive logic system.

1.22 Make truth table for a 3-input
(a) AND gate (b) OR gate (c) NAND gate (d) NOR gate

1.23 Is it possible to use a 3-input gate as a 2-input gate for the following gates? If yes, how?
(a) AND (b) OR (c) NAND (d) NOR

1.24 Is it possible to INHIBIT (or DISABLE) AND, OR, NAND, NOR gates? If yes, how?

1.25 One of the inputs of a gate is used to control the operation of the gate and is labelled as ENABLE. Is it active-high or active-low if the gate is
(a) AND ? (b) OR ? (c) NAND? (d) NOR?

1.26 Is the INHIBIT input active-high or active-low in Prob. 1.24.

1.27 Realize a 3-input gate using 2-input gates for the following gates:
(a) AND (b) OR (c) NAND (d) NOR

1.28 Prove the following:
(a) $A \oplus B = \overline{A} \oplus \overline{B}$
(b) $\overline{A \oplus B} = A \oplus \overline{B} = \overline{A} \oplus B$
(c) $B \oplus (B \oplus A \cdot C) = A \cdot C$

2

NUMBER SYSTEMS AND CODES

2.1 INTRODUCTION

We all are familiar with the number system in which an ordered set of ten symbols – 0, 1, 2, 3, 4, 5, 6, 7, 8, and 9, known as *digits*–are used to specify any number. This number system is popularly known as the *decimal number system.* The *radix* or *base* of this number system is 10 (number of distinct digits). Any number is a collection of these digits. For example, 1982.365 signifies a number with an *integer* part equal to 1982 and a *fractional* part equal to 0.365, separated from the integer part with a *radix point* (.) also known as *decimal point.* There are some other systems also, used to represent numbers. Some of the other commonly used number systems are: *binary, octal* and *hexadecimal* number systems. These number systems are widely used in digital systems like microprocessors, logic circuits, computers, etc. and therefore, the knowledge of these number systems is very essential for understanding, analysing and designing digital systems.

As discussed in Chapter 1, computers and other digital circuits use binary signals but are required to handle data which may be numeric, alphabets or special characters. Therefore, the information available in any other form is required to be converted into suitable binary form before it can be processed by digital circuits. This means that the information available in the form of numerals, alphabets and special characters or in any combination of these must be converted into binary format. To achieve this, a process of coding is employed whereby each numeral, alphabet or special character is coded in a unique combination of 0 s and 1s using a coding scheme, known as a *code.* The process of coding is known as *encoding.*

There can be a variety of coding schemes (codes) to serve different purposes, such as arithmetic operations, data entry, error detection and correction, etc. In digital systems, a large number of codes are in use. Selection of a particular code depends on its suitability for the purpose. In one digital system, different codes may be used for different operations and it may be necessary to convert data from one code to another code. For this purpose, code converter circuits are required which will be discussed later.

2.2 NUMBER SYSTEMS

In general, in any number system there is an ordered set of symbols known as digits with rules defined for performing arithmetic operations like addition, multiplication, etc. A collection of these digits makes a number which in general has two parts–integer and fractional, set apart by a radix point (.), that is

$$(N)_b = \underbrace{d_{n-1} d_{n-2} \ldots d_i \ldots d_1 \, d_0}_{\text{Integer portion}} \underset{\substack{\text{Radix} \\ \text{point}}}{\overset{\uparrow}{.}} \underbrace{d_{-1} \, d_{-2} \ldots d_{-f} \ldots d_{-m}}_{\text{Fractional portion}} \tag{2.1}$$

where N = a number

b = radix or base of the number system

n = number of digits in integer portion

m = number of digits in fractional portion

d_{n-1} = most significant digit (msd)

d_{-m} = least significant digit (1sd)

and $0 \leq (d_i \text{ or } d_{-f}) \leq b - 1$

The digits in a number are placed side by side and each position in the number is assigned a *weight* or *index* of importance by some predesigned rule. Table 2.1 gives the details of commonly used number systems.

Table 2.1 Characteristics of commonly used number systems

Number system	*Base or radix (b)*	*Symbols used (d_i or d_{-f})*	*Weight assigned to position*		*Example*
			i	$-f$	
Binary	2	0, 1	2^i	2^{-f}	1011.11
Octal	8	0, 1, 2, 3, 4, 5, 6, 7	8^i	8^{-f}	3567.25
Decimal	10	0, 1, 2, 3, 4, 5, 6, 7, 8, 9	10^i	10^{-f}	3974.57
Hexadecimal	16	0, 1, 2, 3, 4, 5, 6, 7, 8, 9, *A*, *B*, *C*, *D*, *E*, *F*	16^i	16^{-f}	3*FA*9.56

2.3 BINARY NUMBER SYSTEM

The number system with base (or radix) two is known as the *binary number system.* Only two symbols are used to represent numbers in this system and these are 0 and 1. These are known as bits. This system has the minimum base (0 is not possible and 1 is not useful). It is a positional system, that is every position is assigned a specific weight.

Table 2.2 illustrates counting in binary number system. The corresponding decimal numbers are given in the right-hand column. Similar to decimal number system, the left-most bit is known as the *most significant bit* (MSB) and the right-most bit is known as the *least significant bit* (LSB). Any number of 0s can be added to the left of the number without changing the value of the number. In the binary number system, a group of four bits is known as a *nibble*, and a group of eight bits is known as a *byte*.

Table 2.2 4-bit binary numbers and their corresponding decimal numbers

Binary number				*Decimal number*	
B_3	B_2	B_1	B_0	D_1	D_0
0	0	0	0	0	0
0	0	0	1	0	1
0	0	1	0	0	2
0	0	1	1	0	3
0	1	0	0	0	4
0	1	0	1	0	5
0	1	1	0	0	6
0	1	1	1	0	7
1	0	0	0	0	8
1	0	0	1	0	9
1	0	1	0	1	0
1	0	1	1	1	1
1	1	0	0	1	2
1	1	0	1	1	3
1	1	1	0	1	4
1	1	1	1	1	5

2.3.1 Binary-to-Decimal Conversion

Any binary number can be converted into its equivalent decimal number using the weights assigned to each bit position as given in Table 2.1.

Example 2.1 Find the decimal equivalent of the binary number $(1\ 1\ 1\ 1\ 1)_2$.

Solution The equivalent decimal number is

$$= 1 \times 2^4 + 1 \times 2^3 + 1 \times 2^2 + 1 \times 2^1 + 1 \times 2^0$$
$$= 16 + 8 + 4 + 2 + 1$$
$$= (31)_{10}$$

To differentiate between numbers represented in different number systems, either the corresponding number system may be specified along with the number or a small subscript at the end of the number may be added signifying the number system. For example, $(1000)_2$ represents a binary number and is not one thousand.

Example 2.2 Determine the decimal numbers represented by the following binary numbers:
(a) 110101 (b) 101101 (c) 11111111 (d) 00000000

Solution

(a) $(110101)_2 = 1 \times 2^5 + 1 \times 2^4 + 0 \times 2^3 + 1 \times 2^2 + 0 \times 2^1 + 1 \times 2^0$
$= 32 + 16 + 0 + 4 + 0 + 1$
$= (53)_{10}$

(b) $(101101)_2 = 32 + 0 + 8 + 4 + 0 + 1$
$= (45)_{10}$

(c) $(11111111)_2 = 128 + 64 + 32 + 16 + 8 + 4 + 2 + 1$
$= (255)_{10}$

(d) $(00000000)_2 = (0)_{10}$

Example 2.3 Determine the decimal numbers represented by the following binary numbers:
(a) 101101.10101 (b) 1100.1011 (c) 1001.0101 (d) 0.10101

Solution

(a) $(101101.10101)_2 = 1 \times 2^5 + 0 \times 2^4 + 1 \times 2^3 + 1 \times 2^2 + 0 \times 2^1 + 1 \times 2^0 + 1 \times 2^{-1} + 0 \times 2^{-2} + 1 \times 2^{-3} + 0 \times 2^{-4} + 1 \times 2^{-5}$

$$= 32 + 0 + 8 + 4 + 0 + 1 + \frac{1}{2} + 0 + \frac{1}{8} + 0 + \frac{1}{32}$$

$= (45.65625)_{10}$

(b) $(1100.1011)_2 = 8 + 4 + 0 + 0 + 0.5 + 0 + 0.125 + 0.0625$
$= (12.6875)_{10}$

(c) $(1001.0101)_2 = 8 + 0 + 0 + 1 + 0 + 0.25 + 0 + 0.0625$
$= (9.3125)_{10}$

(d) $(0.10101)_2 = 0.5 + 0 + 0.125 + 0 + 0.03125$
$= (0.65625)_{10}$

2.3.2 Decimal-to-Binary Conversion

Any decimal number can be converted into its equivalent binary number. For integers, the conversion is obtained by continuous division by 2 and keeping track of the remainders, while for fractional parts, the conversion is affected by continuous multiplication by 2 and keeping track of the integers generated. The conversion process is illustrated by the following examples.

Example 2.4 Convert $(13)_{10}$ to an equivalent base-2 number.

Solution

	Quotient	Remainder
$\frac{13}{2}$	6	1
$\frac{6}{2}$	3	0
$\frac{3}{2}$	1	1
$\frac{1}{2}$	0	1

1 1 0 1

Thus, $(13)_{10} = (1101)_2$

Example 2.5 Convert $(0.65625)_{10}$ to an equivalent base-2 number.

Solution

0.65625 × 2	0.31250 × 2	0.62500 × 2	0.25000 × 2	0.50000 × 2
1.31250	0.62500	1.25000	0.50000	1.00000
1	0	1	0	1

Thus, $(0.65625)_{10} = (0.10101)_2$

Example 2.6 Express the following decimal numbers in the binary form:

(a) 25.5 (b) 10.625 (c) 0.6875

Solution

(a) *Integer part*

	Quotient	Remainder
$\frac{25}{2}$	12	1
$\frac{12}{2}$	6	0
$\frac{6}{2}$	3	0
$\frac{3}{2}$	1	1
$\frac{1}{2}$	0	1

1 1 0 0 0

Therefore, $(25)_{10} = (11001)_2$

Fractional part

$$\begin{array}{r} 0.5 \\ \times 2 \\ \hline 1.0 \end{array} \rightarrow 1$$

i.e., $(0.5)_{10} = (0.1)_2$

Therefore, $(25.5)_{10} = (11001.1)_2$

(b) *Integer part* $(10)_{10} = (1010)_2$

Fractional part

$$\begin{array}{r} 0.625 \\ \times 2 \\ \hline 1.250 \end{array} \rightarrow 1 \qquad \begin{array}{r} 0.250 \\ \times 2 \\ \hline 0.500 \end{array} \rightarrow 0 \qquad \begin{array}{r} 0.500 \\ \times 2 \\ \hline 1.000 \end{array} \rightarrow 1$$

i.e., $(0.625)_{10} = (0.101)_2$

Therefore, $(10.625)_{10} = (1010.101)_2$

(c)

$$\begin{array}{r} 0.6875 \\ \times 2 \\ \hline 1.3750 \end{array} \rightarrow 1 \qquad \begin{array}{r} 0.3750 \\ \times 2 \\ \hline 0.7500 \end{array} \rightarrow 0 \qquad \begin{array}{r} 0.7500 \\ \times 2 \\ \hline 1.5000 \end{array} \rightarrow 1 \qquad \begin{array}{r} 0.5000 \\ \times 2 \\ \hline 1.0000 \end{array} \rightarrow 1$$

Therefore, $(0.6875)_{10} = (0.1011)_2$

2.4 SIGNED BINARY NUMBERS

2.4.1 Sign-Magnitude Representation

In the decimal number system a plus (+) sign is used to denote a positive number and a minus (–) sign for denoting a negative number. The plus sign is usually dropped, and the absence of any sign means that the number has positive value. This representation of numbers is known as *signed number.* As is well known, digital circuits can understand only two symbols, 0 and 1; therefore, we must use the same symbols to indicate the sign of the number also. Normally, an additional bit is used as the *sign bit* and it is placed as the most significant bit. A 0 is used to represent a positive number and a 1 to represent a negative number. For example, an 8-bit signed number 01000100 represents a positive number and its value (magnitude) is

$(1000100)_2 = (68)_{10}$. The left most 0 (MSB) indicates that the number is positive. On the other hand, in the signed binary form, 11000100 represents a negative number with magnitude $(1000100)_2 = (68)_{10}$. The 1 in the left most position (MSB) indicates that the number is negative and the other seven bits give its magnitude. This kind of representation for signed numbers is known as *sign-magnitude representation.* The user must take care to see the representation used while dealing with the binary numbers.

Example 2.7 Find the decimal equivalent of the following binary numbers assuming sign-magnitude representation of the binary numbers.

(a) 101100 (b) 001000 (c) 0111 (d) 1111

Solution

(a) Sign bit is 1, which means the number is negative.

Magnitude = 01100 = $(12)_{10}$

$\therefore$ $(101100)_2 = (-12)_{10}$

(b) Sign bit is 0, which means the number is positive.

Magnitude = 01000 = 8

$\therefore$ $(001000)_2 = (+8)_{10}$

(c) $(0111)_2 = (+7)_2$

(d) $(1111)_2 = (-7)_2$

2.4.2 One's Complement Representation

In a binary number, if each 1 is replaced by 0 and each 0 by 1, the resulting number is known as the *one's complement* of the first number. In fact, both the numbers are complement of each other. If one of these numbers is positive, then the other number will be negative with the same magnitude and vice-versa. For example, $(0101)_2$ represents $(+5)_{10}$, whereas $(1010)_2$ represents $(-5)_{10}$ in this representation. This method is widely used for representing signed numbers. In this representation also, MSB is 0 for positive numbers and 1 for negative numbers.

Example 2.8 Find the one's complement of the following binary numbers.

(a) 0100111001 (b) 11011010

Solution

(a) 1011000110 (b) 00100101

Example 2.9 Represent the following numbers in one's complement form.

(a) + 7 and –7 (b) + 8 and –8 (c) + 15 and –15

Solution In one's complement representation,

(a) $(+7)_{10} = (0111)_2$

and $(-7)_{10} = (1000)_2$

(b) $(+8)_{10} = (01000)_2$

and $(-8)_{10} = (10111)_2$

(c) $(+15)_{10} = (01111)_2$

and $(-15)_{10} = (10000)_2$

From the above examples, it can be observed that for an n-bit number, the maximum positive number which can be represented in 1's complement representation is $(2^{n-1} - 1)$ and the maximum negative number is $-(2^{n-1} - 1)$.

2.4.3 Two's Complement Representation

If 1 is added to 1's complement of a binary number, the resulting number is known as the *two's complement* of the binary number. For example, 2's complement of 0101 is 1011. Since 0101 represents $(+5)_{10}$, therefore, 1011 represents $(-5)_{10}$ in 2's complement representation. In this representation also, if the MSB is 0 the number is positive, whereas if the MSB is 1 the number is negative. For an n-bit number, the maximum positive number which can be represented in 2's complement form is $(2^{n-1} - 1)$ and the maximum negative number is -2^{n-1}. Table 2.3 gives sign-magnitude, 1's and 2's complement numbers represented by 4-bit binary numbers. From the table, it is observed that the maximum positive number is 0111 = + 7, whereas the maximum negative number is 1000 = – 8 using four bits in 2's complement format.

It is also observed that the 2's complement of the 2's complement of a number is the number itself.

Table 2.3 Sign-magnitude, 1's and 2's complement representation using four bits

Decimal number	*Binary number*		
	Sign-magnitude	*One's complement*	*Two's complement*
0	0000	0000	0000
1	0001	0001	0001
2	0010	0010	0010
3	0011	0011	0011
4	0100	0100	0100
5	0101	0101	0101
6	0110	0110	0110
7	0111	0111	0111
– 8	–	–	1000
– 7	1111	1000	1001
– 6	1110	1001	1010
– 5	1101	1010	1011
– 4	1100	1011	1100
– 3	1011	1100	1101
– 2	1010	1101	1110
– 1	1001	1110	1111
– 0	1000	1111	–

Example 2.10 Find the 2's complement of the numbers:
(i) 01001110 (ii) 00110101

Solution

(i) Number 0 1 0 0 1 1 1 0
1's complement 1 0 1 1 0 0 0 1
Add 1 1

1 0 1 1 0 0 1 0

(ii) Number	0 0 1 1 0 1 0 1
1's complement	1 1 0 0 1 0 1 0
Add 1	1
	1 1 0 0 1 0 1 1

From the above example, we observe the following:

1. If the LSB of the number is 1, its 2's complement is obtained by changing each 0 to 1 and 1 to 0 except the least-significant bit.
2. If the LSB of the number is 0, its 2's complement is obtained by scanning the number from the LSB to MSB bit by bit and retaining the bits as they are up to and including the occurrence of the first 1 and complement all other bits.

Example 2.11 Find two's complement of the numbers:

(i) 01100100 (ii) 10010010
(iii) 11011000 (iv) 01100111

Solution

Using the rules of conversion given above, we obtain

	↓↓↓ (last three bits)
(i) Number	0 1 1 0 0 1 0 0
2's Complement	1 0 0 1 1 1 0 0
	↓↓ (last two bits)
(ii) Number	1 0 0 1 0 0 1 0
2's Complement	0 1 1 0 1 1 1 0
	↓↓↓↓ (last four bits)
(iii) Number	1 1 0 1 1 0 0 0
2's Complement	0 0 1 0 1 0 0 0
	↓ (last bit)
(iv) Number	0 1 1 0 0 1 1 1
2's Complement	1 0 0 1 1 0 0 1

Example 2.12 Represent $(-17)_{10}$ in

(i) Sign-magnitude,
(ii) one's complement,
(iii) two's complement representation.

Solution The minimum number of bits required to represent $(+17)_{10}$ in signed number format is six.

$$\therefore \quad (+17)_{10} = (010001)_2$$

Therefore, $(-17)_{10}$ is represented by
(i) 110001 in sign-magnitude representation.
(ii) 101110 in 1's complement representation.
(iii) 101111 in 2's complement representation.

2.5 BINARY ARITHMETIC

We all are familiar with the arithmetic operations such as addition, subtraction, multiplication, and division of decimal numbers. Similar operations can be performed on binary numbers; infact, binary arithmetic is much simpler than decimal arithmetic because here only two digits, 0 and 1 are involved.

2.5.1 Binary Addition

The rules of binary addition are given in Table 2.4.

Table 2.4 Rules of binary addition

Augend	*Addend*	*Sum*	*Carry*	*Result*
0	0	0	0	0
0	1	1	0	1
1	0	1	0	1
1	1	0	1	10

In the first three rows above, there is no *carry*, that is, carry = 0, whereas in the fourth row a *carry* is produced (since the largest digit possible is 1), that is, carry = 1, and similar to decimal addition it is added to the next higher binary position.

Example 2.13 Add the binary numbers:
(i) 1011 and 1100
(ii) 0101 and 1111

Solution

(i)

		1	0	1	1
	(+)	1	1	0	0
	1	0	1	1	1

↑ carry

(ii)

		(1)	(1)	(1)	← carry
		0	1	0	1
	(+)	1	1	1	1
	1	0	1	0	0

↑ carry

Example 2.14 Add the binary numbers:

0 1 1 0 1 0 1 0
0 0 0 0 1 0 0 0
1 0 0 0 0 0 0 1
1 1 1 1 1 1 1 1

Solution

				(1)					← Two pair of 1's in the previous column
	(1)	(1)	(1)	(1)	(1)	(1)	(1)	←	one pair of 1's in the previous column
	0	1	1	0	1	0	1	0	
	0	0	0	0	1	0	0	0	
	1	0	0	0	0	0	0	1	
	1	1	1	1	1	1	1	1	
1	1	1	1	1	0	0	1	0	
Carry					↑	↑			Even number of 1's in column
	↑	↑	↑	↑			↑		odd number of 1's in column

∴ The sum = 1 1 1 1 1 0 0 1 0

From the above example, we observe the following:

(i) if the number of 1's to be added in a column is even then the sum bit is 0, and if the number of 1's to be added in a column is odd then the sum bit is 1.

(ii) Every pair of 1's in a column produces a carry (1) to be added to the next higher bit column.

2.5.2 Binary Subtraction

The rules of binary subtraction are given in Table 2.5.

Table 2.5 Rules of binary subtraction

Minuend	Subtrahend	Difference	Borrow
0	0	0	0
0	1	1	1
1	0	1	0
1	1	0	0

Except in the second row above, the *borrow* = 0. When the *borrow* = 1, as in the second row, this is to be subtracted from the next higher binary bit as it is done in decimal subtraction.

Example 2.15 Perform the following subtraction:

```
  1 0 1 1
– 0 1 1 0
```

Solution

	Column 4	Column 3	Column 2	Column 1	
	↓	↓	↓	↓	
	1	0	1	1	Minuend
(–)	0	1	1	0	Subtrahend
	0	1	0	1	Difference

Here, in columns 1 and 2, borrow = 0 and in column 3 it is 1. Therefore, in column 4 first subtract 0 from 1 and from this result obtained subtract the borrow bit.

2.5.3 Binary Multiplication

Binary multiplication is similar to decimal multiplication. In binary, each partial product is either zero (multiplication by 0) or exactly same as the multiplicand (multiplication by 1). An example of binary multiplication is given below:

Example 2.16 Multiply 1001 by 1101.

Solution

```
        1 0 0 1              Multiplicand
      × 1 1 0 1              Multiplier
    -------------
        1 0 0 1   }  I
      0 0 0 0     }  II      Partial Products
    1 0 0 1       }  III
  1 0 0 1         }  IV
    -------------
  1 1 1 0 1 0 1              Final Product
```

In a digital circuit, the multiplication operation is performed by repeated additions of all partial products to obtain the full product.

2.5.4 Binary Division

Binary division is obtained using the same procedure as decimal division. An example of binary division is given below:

Example 2.17 Divide 1 1 1 0 1 0 1 by 1 0 0 1.

Solution

```
                  1101        ← Quotient
Divisor → 1001)1110101        ← Dividend
               1001
               ----
                1011
                1001
                ----
                001001
                  1001
                  ----
                  0000
       Ans: 1101
```

2.6 2'S COMPLEMENT ARITHMETIC

Digital circuits are used for performing binary arithmetic operations. It is possible to use the circuits designed for binary addition to perform the binary subtraction also if we can change the problem of subtraction to that of an addition. This concept eliminates the need of additional circuits for subtraction, rather the same adder circuits are used for both the operations. This makes design of arithmetic circuits very convenient and cheaper. For this purpose, 2's complement representation discussed in Section 2.4.3 is used.

2.6.1 Subtraction Using 2's Complement

Binary subtraction can be performed by adding the 2's complement of the subtrahend to the minuend. If a final carry is generated, discard the carry and the answer is given by the remaining bits which is positive (the minuend is greater than the subtrahend). If the final carry is 0, the answer is negative (the minuend is smaller than the subtrahend) and is in 2's complement form.

Example 2.18 Perform binary subtraction using 2's complement representation of negative numbers.

Solution

(i)

Decimal		Binary	
7		0 1 1 1	Minuend
– 5	⟹	(+) 1 0 1 1	2's complement of subtrahend
+ 2		1 0 0 1 0	

↑ Discard final carry

The answer is 0 0 1 0 equivalent to $(+\,2)_{10}$.

(ii)

Decimal		Binary	
5		0 1 0 1	Minuend
– 7	⟹	(+) 1 0 0 1	2's complement of subtrahend
– 2		1 1 1 0	

The final carry = 0. Therefore, the answer is negative and is in 2's complement form.

2's complement of 1110 = 0010

Therefore, the answer is $(-2)_{10}$.

2.6.2 Addition/Subtraction in 2's Complement Representation

The addition/subtraction of signed binary numbers can most conveniently be performed using 2's complement representation of both the operands. This is the method most commonly used when these operations are performed using digital circuits and microprocessors.

Example 2.19 Perform the following operations using 2's complement method:

(i) 48 – 23 (ii) 23 – 48 (iii) 48 – (–23) (iv) – 48 – 23

Use 8-bit representation of numbers.

Solution

(i) 2's complement representation of + 48 = 00110000

2's complement representation of – 23 = 11101001

48 + (– 23)

	Decimal		Binary	
	48		0 0 1 1 0 0 0 0	
⟹	+ (–23)	⟹	(+) 1 1 1 0 1 0 0 1	
	+ 25		1 0 0 0 1 1 0 0 1	= + 25

↑ Discard Carry

(ii) 2's complement representation of + 23 = 0 0 0 1 0 1 1 1
2's complement representation of – 48 = 1 1 0 1 0 0 0 0
23 – 48 = 23 + (– 48)

```
        23          0 0 0 1 0 1 1 1
 ⇒  + (–48)  ⇒  (+) 1 1 0 1 0 0 0 0
    ------       -------------------
      – 25          1 1 1 0 0 1 1 1  ⇒  – 25
```

(iii) 48 – (–23) = 48 + 23

```
        48          0 0 1 1 0 0 0 0
 ⇒   + (23)  ⇒  (+) 0 0 0 1 0 1 1 1
    ------       -------------------
      + 71          0 1 0 0 0 1 1 1  ⇒  + 71
```

(iv) – 48 – 23 = (– 48) + (– 23)

```
      – 48          1 1 0 1 0 0 0 0
 ⇒ + (– 23)  ⇒  (+) 1 1 1 0 1 0 0 1
    ------       -------------------
      – 71        1 1 0 1 1 1 0 0 1  ⇒  – 71
                  ↑
          Carry be ignored
```

From the above example, we observe the following:

(a) If the two operands are of the opposite sign, the result is to be obtained by the rule of subtraction using 2's complement (Sec. 2.6.1)

(b) If the two operands are of the same sign, the sign bit of the result (msb) is to be compared with the sign bit of the operands. In case the sign bits are same, the result is correct and is in 2's complement form. If the sign bits are not same there is a problem of *overflow*, i.e. the result can not be accommodated using eight bits and the result is to be interpreted suitably. The result in this case will consist of nine bits, i.e. carry and eight bits, and the carry bit will give the sign of the number.

2.7 OCTAL NUMBER SYSTEM

The number system with base (or radix) eight is known as the *octal number system.* In this system, eight symbols, 0, 1, 2, 3, 4, 5, 6, and 7 are used to represent numbers. Similar to decimal and binary number systems, it is also a positional system and has, in general, two parts: integer and fractional, set apart by a radix (octal) point (.). Any number can be expressed in the form of Eq. (2.1) with $b = 8$ and $0 \le (d_i \text{ or } d_{-f}) \le 7$. The weights assigned to the various positions are given in Table 2.1. For example, $(6327.4051)_8$ is an octal number.

2.7.1 Octal-to-Decimal Conversion

Any octal number can be converted into its equivalent decimal number using the weights assigned to each octal digit position as given in Table 2.1.

Example 2.20 Convert $(6327.4051)_8$ into its equivalent decimal number.

Solution

Using the weights given in Table 2.1, we obtain

$$\begin{aligned}(6327.4051)_8 &= 6 \times 8^3 + 3 \times 8^2 + 2 \times 8^1 + 7 \times 8^0 + 4 \times 8^{-1} \\ &\quad + 0 \times 8^{-2} + 5 \times 8^{-3} + 1 \times 8^{-4} \\ &= 3072 + 192 + 16 + 7 + \frac{4}{8} + 0 + \frac{5}{512} + \frac{1}{4096} \\ &= (3287.5100098)_{10}\end{aligned}$$

Thus, $(6327.4051)_8 = (3287.5100098)_{10}$

2.7.2 Decimal-to-Octal Conversion

The conversion from decimal to octal (base-10 to base-8) is similar to the conversion procedure for base-10 to base-2 conversion. The only difference is that number 8 is used in place of 2 for division in the case of integers and for multiplication in the case of fractional numbers.

Example 2.21

(a) Convert $(247)_{10}$ into octal
(b) Convert $(0.6875)_{10}$ into octal
(c) Convert $(3287.5100098)_{10}$ into octal

Solution

(a)

	Quotient	Remainder
$\frac{247}{8}$	30	7
$\frac{30}{8}$	3	6
$\frac{3}{8}$	0	3

Remainders read from bottom to top: 3 6 7

Thus, $(247)_{10} = (367)_8$

(b)

$$\begin{array}{r} 0.6875 \\ \times 8 \\ \hline 5.5000 \\ \downarrow \\ 5 \end{array} \qquad \begin{array}{r} 0.5000 \\ \times 8 \\ \hline 4.0000 \\ \downarrow \\ 4 \end{array}$$

Thus, $(0.6875)_{10} = (0.54)_8$

(c) *Integer part*:

	Quotient	Remainder
3287/8	410	7
410/8	51	2
51/8	6	3
6/8	0	6

Remainders read from bottom to top: 6 3 2 7

Thus, $(3287)_{10} = (6327)_8$

Fractional part:

0.5100098	0.0800784	0.6406272	0.1250176
× 8	× 8	× 8	× 8
4.0800784	0.6406272	5.1250176	1.0001408
↓	↓	↓	↓
4	0	5	1

Thus $(0.5100098)_{10} \approx (0.4051)_8$

Therefore, $(3287.5100098)_{10} = (6327.4051)_8$

From the above examples we observe that the conversion for fractional numbers may not be exact. In general, an approximate equivalent can be determined by terminating the process of multiplication by eight at the desired point.

2.7.3 Octal-to-Binary Conversion

Octal numbers can be converted into equivalent binary numbers by replacing each octal digit by its 3-bit equivalent binary. Table 2.6 gives octal numbers and their binary equivalents for decimal numbers 0 to 15.

Table 2.6 Binary and decimal equivalents of octal numbers

Octal	*Decimal*	*Binary*
0	0	000
1	1	001
2	2	010
3	3	011
4	4	100
5	5	101
6	6	110
7	7	111
10	8	001000
11	9	001001
12	10	001010
13	11	001011
14	12	001100
15	13	001101
16	14	001110
17	15	001111

Example 2.22 Convert $(736)_8$ into an equivalent binary number.

Solution From Table 2.6, we observe the binary equivalents of 7, 3 and 6 as 111, 011, and 110, respectively. Therefore, $(736)_8 = (111\ 011\ 110)_2$.

2.7.4 Binary-to-Octal Conversion

Binary numbers can be converted into equivalent octal numbers by making groups of three bits starting from LSB and moving towards MSB for integer part of the number and then replacing each group of three bits by its octal representation. For fractional part, the groupings of three bits are made starting from the binary point.

Example 2.23 Convert $(1001110)_2$ to its octal equivalent.

Solution

$$
\begin{aligned}
(1001110)_2 &= (001\ 001\ 110)_2 \\
&= (1\ 1\ 6)_8 \\
&= (116)_8
\end{aligned}
$$

Example 2.24 Convert $(0.10100110)_2$ to its equivalent octal number.

Solution

$$
\begin{aligned}
(0.10100110)_2 &= (0.101\ 001\ 100)_2 \\
&= (0.5\ 1\ 4)_2 \\
&= (0.514)_2
\end{aligned}
$$

Example 2.25 Convert the following binary numbers to octal numbers

(a) 11001110001.000101111001

(b) 1011011110.11001010011

(c) 111110001.10011001101

Solution

(a) 011 001 110 001.000 101 111 001 = $(3161.0571)_8$

(b) 001 011 011 110.110 010 100 110 = $(1336.6246)_8$

(c) 111 110 001.100 110 011 010 = $(761.4632)_8$

From the above examples we observe that in forming the 3-bit groupings, 0's may be required to complete the first (most significant digit) group in the integer part and the last (least significant digit) group in the fractional part.

2.7.5 Octal Arithmetic

Octal arithmetic rules are similar to the decimal or binary arithmetic. Normally, we are not interested in performing octal arithmetic operations using octal representation of numbers. This number system is normally used to enter long strings of binary data into a digital system like a microcomputer. This makes the task of entering binary data in a microcomputer easier. Arithmetic operations can be performed by converting the octal numbers to binary numbers and then using the rules of binary arithmetic.

Example 2.26 Add $(23)_8$ and $(67)_8$.

Solution

$$
\begin{array}{rcr}
23 & = & 010011 \\
(+)67 & = & 110111 \\
\hline
(112)_8 & = & 1001\ 010
\end{array}
$$

Example 2.27
Subtract (a) $(37)_8$ from $(53)_8$
(b) $(75)_8$ from $(26)_8$

Solution
Using 8-bit representation,

(a)

$(53)_8$	=	00101011	
$-(37)_8$	=	(+) 11100001	Two's complement of $(37)_8$
$(14)_8$	=	100 001 100	

Discard carry (leading 1)

(b)

$(26)_8$	=	00010110	
$-(75)_8$	=	(+) 11000011	Two's complement of $(75)_8$
$-(47)_8$		11011001	Two's complement of result

Two's complement of 11011001 = 00 100 111 = $(47)_8$

Multiplication and division can also be performed using the binary representation of octal numbers and then making use of multiplication and division rules of binary numbers.

2.7.6 Applications of Octal Number System

In digital systems, binary numbers are required to be entered and certain results or status signals are required to be displayed. It is highly inconvenient to handle long strings of binary numbers. It may cause errors also. Therefore, octal numbers are used for entering the binary data and displaying certain informations. Therefore, the knowledge of octal number system is very important for the efficient use of microprocessors and other digital circuits. For example, the binary number 011111110 can easily be remembered as 376 and can be entered as 376 using keys. Since digital circuits can process only zeros and ones, the octal numbers have to be converted into binary form using special circuits known as octal-to-binary converters before being processed by the digital circuits.

2.8 HEXADECIMAL NUMBER SYSTEM

Hexadecimal number system is very popular in computer uses. The base for hexadecimal number system is 16 which requires 16 distinct symbols to represent the numbers. These are numerals 0 through 9 and alphabets A through F. Since numeric digits and alphabets both are used to represent the digits in the hexadecimal number system, therefore, this is an alphanumeric number system. Table 2.7 gives hexadecimal numbers with their binary equivalents for decimal numbers 0 through 15. From the table, it is observed that there are 16 combinations of 4-bit binary numbers and sets of 4-bit binary numbers can be entered in the computer in the form of hexadecimal (hex.) digits. These numbers are required to be converted into binary representation, using hexadecimal-to-binary converter circuits before these can be processed by the digital circuits.

Table 2.7 Binary and decimal equivalents of hexadecimal numbers

Hexadecimal	*Decimal*	*Binary*
0	0	0000
1	1	0001
2	2	0010
3	3	0011
4	4	0100
5	5	0101
6	6	0110
7	7	0111
8	8	1000
9	9	1001
A	10	1010
B	11	1011
C	12	1100
D	13	1101
E	14	1110
F	15	1111

2.8.1 Hexadecimal-to-Decimal Conversion

Using Eq. (2.1), hexadecimal numbers can be converted to their equivalent decimal numbers.

Example 2.28 Obtain decimal equivalent of hexadecimal number $(3A.2F)_{16}$

Solution Using Eq. (2.1),

$$(3A.2F)_{16} = 3 \times 16^1 + 10 \times 16^0 + 2 \times 16^{-1} + 15 \times 16^{-2}$$
$$= 48 + 10 + \frac{2}{16} + \frac{15}{16^2}$$
$$= (58.1836)_{10}$$

The fractional part may not be an exact equivalent and therefore, may give a small error.

2.8.2 Decimal-to-Hexadecimal Conversion

For conversion from decimal to hexadecimal, the procedure used in binary as well as octal systems is applicable, using 16 as the dividing (for integer part) and multiplying (for fractional part) factor.

Example 2.29 Convert the following decimal numbers into hexadecimal numbers.

(a) 95.5 (b) 675.625

Solution

(a) *Integer part*

	Quotient	Remainder
95/16	5	15
5/16	0	5

5 F

Thus, $(95)_{10} = (5F)_{16}$

Fractional part

$$\begin{array}{r} 0.5 \\ \times\ 16 \\ \hline 8.0 \\ \downarrow \\ 8 \end{array} \qquad \text{Thus, } (0.5)_{10} = (0.8)_{16}$$

Therefore, $(95.5)_{10} = (5F.8)_{16}$

(b) *Integer part*

	Quotient	Remainder
675/16	42	3
42/16	2	10
2/16	0	2

2 A 3

Thus, $(675)_{10} = (2A3)_{16}$

Fractional part

$$\begin{array}{r} 0.625 \\ \times\ 16 \\ \hline 10.000 \\ \downarrow \\ A \end{array} \qquad \text{Thus, } (0.625)_{10} = (0.A)_{16}$$

Therefore, $(675.625)_{10} = (2A3.A)_{16}$

2.8.3 Hexadecimal-to-Binary Conversion

Hexadecimal numbers can be converted into equivalent binary numbers by replacing each hex digit by its equivalent 4-bit binary number.

Example 2.30 Convert $(2F9A)_{16}$ to equivalent binary number.

Solution Using Table 2.7, find the binary equivalent of each hex digit.

$$(2F9A)_{16} = (0010\ 1111\ 1001\ 1010)_2$$
$$= (0010111110011010)_2$$

2.8.4 Binary-to-Hexadecimal Conversion

Binary numbers can be converted into the equivalent hexadecimal numbers by making groups of four bits starting from LSB and moving towards MSB for integer part and then replacing each group of four bits by its hexadecimal representation.

For the fractional part, the above procedure is repeated starting from the bit next to the binary point and moving towards the right.

Example 2.31 Convert the following binary numbers to their equivalent hex numbers.

(a) 10100110101111

(b) 0.00011110101101

Solution

(a) $(10100110101111)_2 = \underbrace{0010}_{2}\ \underbrace{1001}_{9}\ \underbrace{1010}_{A}\ \underbrace{1111}_{F}$

$\therefore (10100110101111)_2 = (29AF)_{16}$

(b) $(0.00011110101101)_2 = \underbrace{0.0001}_{1}\ \underbrace{1110}_{E}\ \underbrace{1011}_{B}\ \underbrace{0100}_{4}$

$\therefore (0.00011110101101)_2 = (0.1EB4)_{16}$

Example 2.32 Convert the binary numbers of Example 2.25 to hexadecimal numbers.

Solution

(a) 110 0111 0001.0001 0111 1001 $= (671.179)_{16}$

(b) 10 1101 1110.1100 1010 011 $= (2DE.CA6)_{16}$

(c) 1 1111 0001.1001 1001 101 $= (1F1.99A)_{16}$

From the above examples, we observe that in forming the 4-bit groupings 0's may be required to complete the first (most significant digit) group in the integer part and the last (least significant digit) group in the fractional part.

2.8.5 Conversion from Hex-to-Octal and Vice-Versa

Hexadecimal numbers can be converted to equivalent octal numbers and octal numbers can be converted to equivalent hex numbers by converting the hex/octal number to equivalent binary and then to octal/hex, respectively.

Example 2.33 Convert the following hex numbers to octal numbers.

(a) *A72E* (b) 0.*BF*85

Solution

(a) $(A72E)_{16} = (1010\ 0111\ 0010\ 1110)_2$

$= \underbrace{001}\ \underbrace{010}\ \underbrace{011}\underbrace{100}\ \underbrace{101}\ \underbrace{110}$

$= (123456)_8$

(b) $(0.BF85)_{16} = (0.1011\ 1111\ 1000\ 0101)_2$

$= 0.\underbrace{101}\ \underbrace{111}\ \underbrace{111}\underbrace{000}\ \underbrace{010}\ \underbrace{100}$

$= (0.577024)_8$

Example 2.34 Convert $(247.36)_8$ to equivalent hex number.

Solution

$(247.36)_8 = (010\ 100111.011\ 110)_2$

$= (0\ \underbrace{1010}\ \underbrace{0111}\cdot\underbrace{0111}\underbrace{1000})_2$

$= (A\,7.78)_{16}$

2.8.6 Hexadecimal Arithmetic

The rules for arithmetic operations with hexadecimal numbers are similar to the rules for decimal, octal, and binary systems. The information can be handled only in binary form in a digital circuit and it is easier to enter the information using hexadecimal number system. Since arithmetic operations are performed by the digital circuits on binary numbers, therefore hexadecimal numbers are to be first converted into binary numbers. Arithmetic operations will become clear from the following examples.

Example 2.35 Add $(7F)_{16}$ and $(BA)_{16}$

Solution

$$
\begin{array}{rl}
7F = & 01111111 \\
(+)\ BA = & 10111010 \\
\hline
(139)_{16} = & 100111001
\end{array}
$$

Example 2.36
Subtract (a) $(5C)_{16}$ from $(3F)_{16}$
(b) $(7A)_{16}$ from $(C0)_{16}$

Solution

(a)

$$
\begin{array}{rll}
3F = & 00111111 & \\
-\ 5C = & (+)10100100 & \text{Two's complement of } (5C)_{16} \\
\hline
-\ 1D = & 11100011 & \text{Two's complement of result}
\end{array}
$$

Two's complement of 11100011 = 0001 1101 = $(1D)_{16}$

(b)

$$
\begin{array}{rll}
C0 = & 11000000 & \\
-\ 7A = & (+)\ 10000110 & \text{Two's complement of } (7A)_{16} \\
\hline
46 = & 101000110 &
\end{array}
$$

Discard carry (the leading 1)

Multiplication and division can also be performed using the binary representation of hexadecimal numbers and then making use of multiplication and division rules of binary numbers.

2.9 CODES

Computers and other digital circuits process data in the binary format. Various binary codes are used to represent data which may be numeric, alphabets or special characters. Although, in every code used the information is represented in binary form, the interpretation of this binary information is possible only if the code in which this information is available is known. For example, the binary number 1000001 represents 65 (decimal) in straight binary, 41 (decimal) in BCD and alphabet *A* in ASCII code. A user must be very careful about the code being used while interpreting information available in the binary format. Codes are also used for error detection and error correction in digital systems.

Some of the commonly used codes are given below.

2.9.1 Straight Binary Code

This is used to represent numbers using natural (or straight) binary form as discussed in Section 2.3. Various arithmetic operations can be performed in this form. Binary codes for decimal numbers 0 to 15 are given in Table 2.8. It is a weighted code since a weight is assigned to every position.

2.9.2 Natural BCD Code

In this code, decimal digits 0 through 9 are represented (coded) by their natural binary equivalents using four bits and each decimal digit of a decimal number is represented by this four bit code individually. For example, $(23)_{10}$ is represented by 0010 0011 using BCD code, rather than $(10111)_2$. From this it is observed that it requires more number of bits to code a decimal number using BCD code than using the straight binary code. However, inspite of this disadvantage it is very convenient and useful code for input and output operations in digital systems.

This code is also known as 8-4-2-1 code or simply BCD code. 8, 4, 2, and 1 are the weights of the four bits of the binary code of each decimal digit similar to straight binary number system. Therefore, this is a *weighted* code and arithmetic operations can be performed using this code, which will be discussed in Chapter 6. BCD codes for decimal digits 0 through 9 are given in Table 2.8.

Table 2.8 Various binary codes

Decimal Number	*Binary*				*BCD*				*Excess-3*				*Gray*			
	B_3	B_2	B_1	B_0	D	C	B	A	E_3	E_2	E_1	E_0	G_3	G_2	G_1	G_0
0	0	0	0	0	0	0	0	0	0	0	1	1	0	0	0	0
1	0	0	0	1	0	0	0	1	0	1	0	0	0	0	0	1
2	0	0	1	0	0	0	1	0	0	1	0	1	0	0	1	1
3	0	0	1	1	0	0	1	1	0	1	1	0	0	0	1	0
4	0	1	0	0	0	1	0	0	0	1	1	1	0	1	1	0
5	0	1	0	1	0	1	0	1	1	0	0	0	0	1	1	1
6	0	1	1	0	0	1	1	0	1	0	0	1	0	1	0	1
7	0	1	1	1	0	1	1	1	1	0	1	0	0	1	0	0
8	1	0	0	0	1	0	0	0	1	0	1	1	1	1	0	0
9	1	0	0	1	1	0	0	1	1	1	0	0	1	1	0	1
10	1	0	1	0									1	1	1	1
11	1	0	1	1									1	1	1	0
12	1	1	0	0									1	0	1	0
13	1	1	0	1									1	0	1	1
14	1	1	1	0									1	0	0	1
15	1	1	1	1									1	0	0	0

2.9.3 Excess-3 Code

This is another form of BCD code, in which each decimal digit is coded into a 4-bit binary code. The code for each decimal digit is obtained by adding decimal 3 to the natural BCD code of the digit. For example, decimal 2 is coded as 0010 + 0011 = 0101 in Excess-3 code. It is not a weighted code. This code is a self-complementing code, which means 1's complement of the coded number yields 9's complement of the number itself. For example, Excess-3 code of decimal 2 is 0101, its 1's complement is 1010 which is Excess-3 code for decimal 7, which is 9's complement of 2. The self complementing property of this code helps considerably in performing subtraction operation in digital systems. Excess-3 codes for decimal digits 0 through 9 are given in Table 2.8.

2.9.4 Gray Code

It is a very useful code in which a decimal number is represented in binary form in such a way so that each Gray-code number differs from the preceding and the succeeding number by a single bit. For example, the Gray code for decimal number 5 is 0111 and for 6 is 0101. These two codes differ by only one bit position (third from the left). This code is used extensively for shaft encoders because of this property. It is not a weighted code. The Gray code is a reflected code and can be constructed using this property as given below.

(i) A 1-bit Gray code has two code words 0 and 1 representing decimal numbers 0 and 1 respectively

(ii) An n-bit ($n \geq 2$) Gray code will have first 2^{n-1} Gray codes of (n–1)-bits written in order with a leading 0 appended.

(iii) The last 2^{n-1} Gray codes will be equal to the Gray code words of an (n–1)- bit Gray code, written in reverse order (assuming a mirror placed between first 2^{n-1} and last 2^{n-1} Gray codes) with a leading 1 appended.

Example 2.37 Determine (a) 1-bit (b) 2-bit (c) 3-bit Gray codes and tabulate along with their equivalent decimal numbers.

Solution

(a) 1-bit Gray code is constructed using (i) above.

Decimal number	*Gray code*
0	0
1	1

(b) 2-bit Gray code is constructed using (ii) and (iii) above and Gray code of 1-bit

Decimal number	*Gray code*
0	0 0
1	0 1
2	1 1
3	1 0

(c) 3-bit Gray code is constructed using 2-bit Gray code,

Decimal Number	*Gray code*
0	0 0 0
1	0 0 1
2	0 1 1
3	0 1 0
4	1 1 0
5	1 1 1
6	1 0 1
7	1 0 0

4-bit Gray code is given in Table 2.8.

2.9.5 Octal Code

It is a 3-bit binary code, in which each of the octal digits 0 through 7 is coded into 3-bit straight binary number. For example, code for octal digit 4 is 100. Using this code, octal numbers can be coded into straight binary form or the binary numbers can be represented by octal numbers as discussed in Section 2.7.

This code is used for binary inputs in digital computers, microprocessors, etc.

2.9.6 Hexadecimal Code

It is a 4-bit binary code (Section 2.8) used for input/output in digital computers, microprocessors, etc.

Example 2.38 Represent the decimal number 27 in binary form using

(i) Binary Code (ii) BCD code
(iii) Excess-3 code (iv) Gray code
(v) Octal code (vi) Hexadecimal code

Solution

(i) The decimal number is converted into straight binary form. Its value is 11011
(ii) Each digit of the decimal number is coded using 4-bit BCD code as given below

0010 0111

(iii) Each digit of the decimal number is coded using 4-bit Excess-3 code as given below

0101 1010

(iv) 5 bits are required to represent 27, therefore, 5-bit Gray code is constructed and 27 is represented as 10110.
(v) $(27)_{10} = (33)_8 = 011\ 011$
(vi) $(27)_{10} = (1B)_{16} = 0001\ 1011$

Example 2.39 Represent the decimal numbers (a) 396 and (b) 4096 in binary form in

(i) Binary code (straight binary) (ii) BCD code
(iii) Excess-3 code (iv) Octal code
(v) Hex code

Solution

(a)
(i) 396 = 110001100
(ii) 396 = 001110010110
(iii) 396 = 011011001001
(iv) 396 = $(614)_8$ = 110001100
(v) 396 = $(18C)_{16}$ = 000110001100

(b) (i) 4096 = 1000000000000
(ii) 4096 = 0100000010010110
(iii) 4096 = 01110011 11001001
(iv) 4096 = $(10000)_8$ = 001 000 000 000 000
(v) 4096 = $(1000)_{16}$ = 0001 0000 0000 0000

2.9.7 Alphanumeric Codes

In many situations, digital systems are required to handle data that may consist of numerals, letters, and special symbols. For example, an university with thousands of students may use a digital computer to process the examination results. In this the names of the students, subjects, etc. are to be represented in the binary form. Therefore, it is necessary to have a binary code for alphabets also. If we use an n-bit binary code, we can represent 2^n elements using this code. Therefore, to represent 10 digits 0 through 9 and 26 alphabets A through Z, we need a minimum of 6 bits ($2^6 = 64$, but $2^5 = 32$ is not sufficient). One possible 6-bit alphanumeric code is given in Table 2.9. It is used in many computers to represent alphanumeric characters and symbols internally and therefore can be called *internal code*. Frequently, there is a need to represent more than 64 characters including the lower case letters and special control characters for the transmission of digital information. For this reason the following two codes are normally used:

1. Extended BCD Interchange Code (EBCDIC)
2. American Standard Code for Information Interchange (ASCII)

These are given in Tables 2.9 and 2.10 respectively.

Table 2.9 Some alphanumeric codes

Character	*6-bit Internal code*	*8-bit EBCDIC code*
A	010001	11000001
B	010010	11000010
C	010011	11000011
D	010100	11000100
E	010101	11000101
F	010110	11000110
G	010111	11000111
H	011000	11001000
I	011001	11001001
J	100001	11010001
K	100010	11010010
L	100011	11010011
M	100100	11010100
N	100101	11010101
O	100110	11010110
P	100111	11010111
Q	101000	11011000
R	101001	11011001

(*Contd.*)

Table 2.9 *(Contd.)*

Character	*6-bit Internal code*	*8-bit EBCDIC code*
S	110010	11100010
T	110011	11100011
U	110100	11100100
V	110101	11100101
W	110110	11100110
X	110111	11100111
Y	111000	11101000
Z	111001	11101001
0	000000	11110000
1	000001	11110001
2	000010	11110010
3	000011	11110011
4	000100	11110100
5	000101	11110101
6	000110	11110110
7	000111	11110111
8	001000	11111000
9	001001	11111001
blank	110000	01000000
.	011011	01001011
(	111100	01001101
+	010000	01001110
$	101011	01011011
*	101100	01011100
)	011100	01011101
–	100000	01100000
/	110001	01100001
,	111011	01101011
=	001011	01111110

Table 2.10 The ASCII code

b_6					0	0	0	0	1	1	1	1
	b_5				0	0	1	1	0	0	1	1
		b_4			0	1	0	1	0	1	0	1
b_3	b_2	b_1	b_0		0	1	2	3	4	5	6	7
0	0	0	0	0	NUL	DLE	SP	0	@	P	`	p
0	0	0	1	1	SOH	DC1	!	1	A	Q	a	q
0	0	1	0	2	STX	DC2	"	2	B	R	b	r
0	0	1	1	3	ETX	DC3	#	3	C	S	c	s
0	1	0	0	4	EOT	DC4	$	4	D	T	d	t
0	1	0	1	5	ENQ	NAK	%	5	E	U	e	u
0	1	1	0	6	ACK	SYN	&	6	F	V	f	v
0	1	1	1	7	BEL	ETB	'	7	G	W	g	w

(Contd.)

Table 2.10 *(Contd.)*

				b_6	0	0	0	0	1	1	1	1
				b_5	0	0	1	1	0	0	1	1
				b_4	0	1	0	1	0	1	0	1
1	0	0	0	8	BS	CAN	(	8	H	X	h	x
1	0	0	1	9	HT	EM	)	9	I	Y	i	y
1	0	1	0	*A*	LF	SUB	*	:	J	Z	j	z
1	0	1	1	*B*	VT	ESC	+	;	K	[	k	{
1	1	0	0	*C*	FF	FS	,	<	L	\	1	|
1	1	0	1	*D*	CR	GS	–	=	M	]	m	}
1	1	1	0	*E*	SO	RS	.	>	N	^	n	~
1	1	1	1	*F*	SI	US	/	?	O	–	o	DEL

The code is read from this table as:

	b_6	b_5	b_4	b_3	b_2	b_1	b_0
H =	1	0	0	1	0	0	$0 = (48)_{16}$
n =	1	1	0	1	1	1	$0 = (6E)_{16}$

Example 2.40 Encode the following in ASCII code:

(a) My dear Rajendra,

(b) I am 20 years old.

Solution

Each of the alphabet, numeral, and the special character is represented by 7-bit ASCII code given in Table 2.10.

(a) $\underbrace{1001101}_{M}\ \underbrace{1111001}_{y}\ \square\ \underbrace{1100100}_{d}\ \underbrace{1100101}_{e}\ \underbrace{1100001}_{a}\ \underbrace{1110010}_{r}\ \square$

$\underbrace{1010010}_{R}\ \underbrace{1100001}_{a}\ \underbrace{1101010}_{j}\ \underbrace{1100101}_{e}\ \underbrace{1101110}_{n}\ \underbrace{1100100}_{d}.$

$\underbrace{1110010}_{r}\ \underbrace{1100001}_{a}\ \underbrace{0101100}_{,}$

(b) $\underbrace{1001001}_{I}\ \square\ \underbrace{1100001}_{a}\ \underbrace{1101101}_{m}\ \square\ \underbrace{0110010}_{2}\ \underbrace{0110000}_{0}\ \square$

$\underbrace{1111001}_{y}\ \underbrace{1100101}_{e}\ \underbrace{1100001}_{a}\ \underbrace{1110010}_{r}\ \underbrace{1110011}_{s}\ \square$

$\underbrace{1101111}_{o}\ \underbrace{1101100}_{l}\ \underbrace{1100100}_{d}\ \underbrace{0101110}_{.}$

The code for SPACE has not been added above , which is 0100000

2.10 ERROR DETECTING AND CORRECTING CODES

Digital signals are processed for performing various operations and are transmitted from one circuit or system to another circuit or system. When these binary signals are transmitted from one location (transmitter) to another location (receiver), transmission errors may occur because of electrical noise in the transmission channel. Due to transmission error a signal transmitted as *a* 0 may be received as *a* 1 or vice-versa. In complex digital systems, millions of bits per second are manipulated and it is desired to have high data integrity, or at least a violation of data integrity must be detectable.

A digital data in the form of a code word such as BCD, ASCII etc. is transmitted and there is always a finite probability of occurrence of an error in a single bit position. The probability of occurrence of error in two or more bit positions simultaneously is substantially smaller. It is desired to *detect* the error in the received data word, *locate* its bit position and *correct* it. Various codes are used for the detection and correction of error. Since the probability of simultaneous occurrence of error in two or more bit positions is negligibly small, therefore, we restrict our discussion to the detection and correction of single error, i.e. error in one bit position.

2.10.1 Error-detecting Codes

When a digital information is transmitted, it may not be received correctly by the receiver. At the receiving end it may or may not be possible to detect whether the information has been received correctly or not. Let us consider BCD code given in Table 2.8 is transmitted and the code corresponding to decimal 9, i.e. 1001 is transmitted and is received as 1011. Since 1011 is an invalid BCD code, therefore, it may be detected by the receiver. On the other hand if it is received as 0001 which is a valid BCD code for decimal 1, the receiver will intrepret it as decimal 1 and the error is not detected. In general, the erroneous word received may or may not be a valid code.

To ensure that the occurrence of a single error always results in an invalid code, so as to avoid its incorrect interpretation by the receiver, the code must possess the property that the occurrence of any single error transforms a valid code word into an invalid code word. Since for an n-bit code there are 2^n possible combinations, therefore, if it is desired to make this code an error-detecting code, only half of the possible 2^n combinations should be included to form the code. This means, if an extra bit is attached to the n-bit code word to make the number of bits $n + 1$ in such a way so as to make the number of ones in the resulting $(n + 1)$-bit code even or odd, it will certainly be an error-detecting code.

The criterion of minimum distance of a code can also be used as a useful property of an error-detecting code. The minimum distance of a code is the smallest number of bits in which any two code words differ. A code is an error-detecting code if and only if its minimum distance is two or more.

For detection of error an extra bit known as ***parity bit*** is attached to each code word to make the number of ones in the code even (***even parity***) or odd (***odd parity***).

Table 2.11 shows BCD code with parity bit attached making it even parity code or odd parity code.

Table 2.11 BCD code with parity bit.

BCD code				*BCD code with even parity*					*BCD code with odd parity*				
D	*C*	*B*	*A*	*P*	*D*	*C*	*B*	*A*	*P*	*D*	*C*	*B*	*A*
0	0	0	0	0	0	0	0	0	1	0	0	0	0
0	0	0	1	1	0	0	0	1	0	0	0	0	1
0	0	1	0	1	0	0	1	0	0	0	0	1	0
0	0	1	1	0	0	0	1	1	1	0	0	1	1
0	1	0	0	1	0	1	0	0	0	0	1	0	0
0	1	0	1	0	0	1	0	1	1	0	1	0	1
0	1	1	0	0	0	1	1	0	1	0	1	1	0
0	1	1	1	1	0	1	1	1	0	0	1	1	1
1	0	0	0	1	1	0	0	0	0	1	0	0	0
1	0	0	1	0	1	0	0	1	1	1	0	0	1

From Table 2.11 we observe that a parity bit (P) 0 or 1 is attached to every code word so as to make the number of ones even or odd for even and odd parity respectively. It can be easily verified that the minimum distance between any two code words with parity bit attached is two. The parity bit 1 or 0 is attached to the code to be transmitted at the transmitter end and the parity of the received $(n + 1)$- bit word is checked at the receiving end. If there is only one error, the erroneous code is detected at the receiving end by parity check. If odd number of bits are transmitted erroneously, then also the parity check will detect the incorrect code but if there are even number of bits received incorrectly, this method will not detect error. The parity check method can only detect error in the transmitted word at the receiving end. It can not locate the bit which has changed and, therefore, the question of correction does not arise.

The parity generator and parity checker circuits have been discussed in Chapter 6.

Example 2.41 Formulate 8-bit ASCII code for Example 2.40 and represent it in hexadecimal code with

(i) even parity
(ii) odd parity

Solution

(i) *For Example 2.40(a):*

The 7-bit ASCII code for M is 1001101. It contains four ones. To make an 8-bit even parity code corresponding to this, one 0 is attached to it as the most-significant bit. Therefore, the 8-bit code will be 01001101. Its hexadecimal representation is 4D. Similarly ASCII code for y is 1111001, which contains five 1s and therefore, a 1 as MSB is attached to it for obtaining its 8-bit even parity code, i.e. 11111001. Its hex. representation is F_9.

In the same way each character is formulated. The complete sentence (including the spaces) in hexadecimal code is

4DF9A0E465E172A0D2E16A65EEE472E1AC

For Example 2.40 (b):

Using the above procedure, we get,

C9A0E1EDA0B230A0F965E172F3A06F6CE42E

(ii) For Example 2.40(a):
The 7-bit ASCII code for *M* is 1001101, when *a* 1 is attached to this as MSB, it becomes the 8-bit code for *M*. In hexadecimal code it is *CD*. Similarly, *a* 0 is attached as MSB to the 7-bit ASCII code of *y* to make it 8-bit with odd parity. In the same way each character is formulated. The complete sentence (including the spaces) in hexadecimal code is

*CD*792064*E*561*F*2205261*EAE*56*E*64*F*2612*C*

For Example 2.40 (b)
Using the above procedure, we get,

4920616*D*2032*B*02079*E*561*F*27320*EFEC*64*AE*

Example 2.42 Find out the minimum distance of
(a) ASCII code.
(b) 8-bit ASCII code with even parity.
(c) 8-bit ASCII code with odd parity.

Solution
From the Table 2.10, examine the ASCII codes of various alphabets, numerals, special characters, etc. and find out the distance between any two code words. For example, ASCII code for alphabet *a* is 1100001 and for alphabet *c* is 1100011. These two code words differ in only one bit position (b_1), which shows that the minimum distance of ASCII code is 1.

(b) ASCII code with even parity for
a is 11100001 and for
c is 01100011
These two code words differ in two bit positions (b_1 and b_7). Similarly, it can be verified for any two code words that the distance is 2 or more. Therefore, the minimum distance of ASCII code with even parity is two.

(c) Similar to part – (b) it can be verified that the minimum distance of 8-bit ASCII code with odd parity is two.

2.10.2 Error-correcting Codes

As discussed above, by adding a single parity bit alongwith the information, or message being transmitted, an error in single bit position can be detected. The parity check gives only information that the received message is incorrect. It can not locate the bit position in which error has occurred and, therefore, can not correct the error.

Let us consider a 4-bit binary word 0101 is transmitted alongwith an even parity bit. Due to transmission error in one bit position, the erroneous word received may be 00100, 00111, 00001, or 01101 depending on the error in bit position b_0, b_1, b_2, and b_3 respectively.

Let us examine whether these erroneous words are possible with any other message being transmitted. If the word 01100 is transmitted, it results in 00100 at the receiving end due to an error in bit position b_3. Similarly, the other erroneous words are received due to the transmission error in some other words being transmitted. This indicates that a minimum distance of two can not locate the bit position in the incorrectly received word. Therefore, for a code to be error correcting, its minimum distance must be more than two.

If the minimum distance is three, every error in single bit results in an invalid code word which is at a distance of one from the original code word and at a distance of two from any other valid code word. Therefore, a single bit error can be detected and located using this code. Once the error bit is located it can be inverted to correct the erroneously received message.

In general, a code is said to be error-correcting code if the correct code word can be deduced from the erroneous word. Hamming code is an error-correcting code.

Hamming code is constructed by adding a number of parity bits to each group of n-bit information, or message in such a way so as to be able to locate the bit position in which error occurs. Let us assume that k parity bits $p_1, p_2 \ldots, p_k$ are added to the n-bit message to form an $(n + k)$- bit code. The value of k must be chosen in such a way so as to be able to describe the location of any of the $n + k$ possible error bit locations and also 'no error' condition. Consequently, k must satisfy the inequality

$$2^k \geq n + k + 1. \tag{2.2}$$

The location of each of the $n + k$ bits within a code word is assigned a decimal number, starting from 1 to the *MSB* and $n + k$ to the *LSB*. k parity checks are performed on selected bits of each code word. Each parity check includes one of the parity bits. The result of each parity check is recorded as 1 if error has been detected and as 0 if no error has been detected. Let the results of the parity checks involving the parity bits $p_k, p_{k-1}, \ldots$ are $c_1, c_2, \ldots$ respectively. Bit c_1 is 1 if an error is detected and 0 if there is no error. Similarly $c_2, c_3, \ldots$, etc. The decimal value of the binary word formed $c_1, c_2 \ldots, c_k$ gives the decimal value assigned to the location of the erroneous bit. If there is no error, then the decimal value will be 0. This decimal number is the position or location number. The parity bits $p_1, p_2, \ldots$ are placed in locations 1, 2, 4, ..., 2^{k-1}.

Example 2.43 Find out the value of k for converting BCD code into Hamming code and the bit positions of the resulting Hamming code.

Solution

The value of k must be chosen to satisfy the eqn. (2.2) since $n = 4$, therefore,

$$2^k \geq k + 5$$

The minimum value of k for which it is satisfied is 3. Therefore, three parity bits are attached to each of the BCD code for constructing the Hamming code. It will be a 7-bit code with bit positions

p_1	p_2	n_1	p_3	n_2	n_3	n_4
1	2	3	4	5	6	7

Values (0 or 1) are assigned to the parity bits so as to make the Hamming code have either even parity or odd parity and when an error occurs, the position number will take on the value assigned to the location of the erroneous bit.

In the case of BCD code with three parity bits there are seven error positions. Table 2.12 gives these error positions and the corresponding values of the position number.

Table 2.12

Error position	*Position number*		
	c_1	c_2	c_3
0 (no error)	0	0	0
1	0	0	1
2	0	1	0
3	0	1	1
4	1	0	0
5	1	0	1
6	1	1	0
7	1	1	1

From Table 2.12 we observe, that if an error occurs in positions

1, 3, 5, 7 then $c_3 = 1$
2, 3, 6, 7 then $c_2 = 1$
4, 5, 6, 7 then $c_1 = 1$

Therefore, p_1 is selected so as to establish even (or odd) parity in positions 1, 3, 5, 7
p_2 is selected so as to establish even (or odd) parity in positions 2, 3, 6, 7
p_3 is selected so as to establish even (or odd) parity in positions 4, 5, 6, 7

Example 2.44 Construct Hamming code for BCD 0110. Use even parity.

Solution

For 4-bit code three parity bits p_1, p_2, and p_3 are appended in locations 1, 2, and 4 respectively as shown below:

Position	1	2	3	4	5	6	7
	p_1	p_2	n_1	p_3	n_2	n_3	n_4
Original BCD			0		1	1	0
Even parity in positions 1, 3, 5, 7 requires $p_1 = 1$	1		0		1	1	0
Even parity in positions 2, 3, 6, 7 requires $p_2 = 1$		1	0	0	1	1	0
Even parity in positions 4, 5, 6, 7 requires $p_3 = 0$	1	1	0	0	1	1	0

Therefore, Hamming code for BCD digit 0110 with even parity is 1100110.

Example 2.45 If the Hamming code sequence 1100110 is transmitted and due to error in one position, is received as 1110110, locate the position of the error bit using parity checks and give the method for obtaining the correct sequence.

Solution

Parity check on 4, 5, 6, 7 (0110) positions gives $c_1 = 0$ (even parity)
Parity check on 2, 3, 6, 7 (1110) positions gives $c_2 = 1$ (odd parity)
Parity check on 1, 3, 5, 7 (1110) positions gives $c_3 = 1$ (odd parity)
Therefore, the position number formed is $c_1\,c_2\,c_3 = 011$, which means that the location of the error is in position 3.

To correct the error the bit received in location 3 is complemented and the correct message 1100110 is received.

Example 2.46

(a) Find the distance between the BCD digits 0110 and 0111.

(b) Determine Hamming codes for 0110 and 0111 and find the distance between them. Use even parity.

Solution

(a) The distance between the BCD numbers 0110 and 0111 is 1 since only the LSB is different.

(b) The Hamming codes for these are given below

BCD code				Hamming code						
				p_1	p_2	n_1	p_3	n_2	n_3	n_4
D	C	B	A	1	2	3	4	5	6	7
0	1	1	0	1	1	0	0	1	1	0
0	1	1	1	0	0	0	1	1	1	1

These Hamming codes differ in positions 1, 2, 4, and 7, thus the distance between them is four.

2.11 SUMMARY

Various number systems that are widely used in digital circuits, microprocessors, computers, etc. have been presented. The rules of arithmetic operations like addition, subtraction, multiplication, division, etc. are given.

Different codes are in use in digital systems for representing numerals, alphabets and special symbols and some of the more commonly used codes have been introduced. The ASCII is the most commonly used code in computers. Any program and data are entered into the memory of the computer using key-board. When any key is pressed, its ASCII code is generated which gets stored in the memory.

The knowledge of these number systems and codes is very essential for the effective understanding of various digital systems including microprocessors.

When digital data is transmitted from one location to another location error may occur in transmission due to the presence of electrical noise. To detect and correct the error, error-detection and error-correction codes are used. These codes are based on the principle of partity checking. The concepts of error-detection, error-correction together with the codes used have been discussed.

Glossary

Alphanumeric codes Codes that represent numerals, alphabets (letters) and other usual symbols, for example, ASCII code.

ASCII Code (American Standard Code for Information Interchange) A 7-bit code widely used in computers and related areas for representation of alphanumeric characters and special symbols.

Base (or Radix) of a number system The number of distinct symbols (digits) used in a number system.

BCD (Binary-coded decimal) A code for representing decimal numbers is which each decimal digit is represented by its 4-bit binary code. See also Natural BCD.

Byte A group of eight bits.

Code A system of representation of numeric, alphabets or special characters in a binary form for processing and transmission using digital techniques.

Complement Inversion of the value of a binary number, variable, or expression.

Complementation The process of determining complement of a binary number, variable, or expression.

Decimal number system A number system with base (or radix) 10. The ten digits used to represent any number are: 0, 1, 2, 3, 4, 5, 6, 7, 8, and 9.

Encoding The process of coding alphabets, numerals and symbols in binary format.

Even parity A digital word (string of 0s and 1s) having an even number of ones.

Error correcting code A code used for correction of error in the transmission of digital signals.

Error detecting code A code used for detection of error in the transmission of digital signals.

Excess-3 code A BCD code formed by adding 3(0011) to the binary equivalent of the decimal number.

Gray code A code in which only one bit changes between successive numbers.

HEX Abbreviation for hexadecimal.

Hexadecimal code A method of representing binary numbers in which each group of 4 bits (starting from the right most bit) is represented by its hex digit.

Hexadecimal number system A number system that uses digits 0 through 9 and the alphabets A through F. Its base (or radix) is 16.

Least-significant bit (LSB) The right-most bit of a binary number. It has the least weight.

Least-significant digit (LSD) The right most digit of a number.

Most-significant bit (MSB) The left-most bit of a binary number.

Most-significant digit (MSD) The left-most digit of a number.

Natural BCD BCD representation that uses natural binary numbers.

Nibble A group of four bits.

Nine's complement Nine's complement of an N-digit-decimal number is the number obtained by subtracting it from an N-digit number consisting of all 9's.

Octal code A code in which each group of three bits starting from LSB is represented by its equivalent octal digit.

Octal number system A number system with base (or radix) 8 that uses digits 0, 1, 2, 3, 4, 5, 6, and 7.

Odd parity A digital word having an odd number of ones.

One's complement The number obtained by complementing each bit of a binary number.

One's complement representation A binary representation used for representing positive as well as negative numbers (signed numbers).

Parity A term used to specify the number of ones in a digital word as odd or even.

Parity bit An extra bit attached to a binary word to make the parity of the resultant word even or odd.

Parity checker A logic circuit that checks the parity of a binary word.

Parity generator A logic circuit that generates an additional bit which when appended to a digital word makes its parity as desired (odd or even).

Positional number system A number system in which value of a digit depends upon its position in the number.

Radix Same as the base.

Sign bit The MSB of a signed binary number. If 0, the number is positive, when it is 1, the number is negative.

Signed binary number A binary number that is either positive or negative.

Sign-magnitude representation A representation system for signed binary numbers in which the MSB represents the sign and the remaining bits represent the magnitude of the number.

Two's complement Binary number obtained by adding one to the one's complement of a binary number.

Two's complement representation A method of representation of binary number in which negative numbers are represented by two's complement of their positive equivalents.

Weighted code A binary code in which weight is assigned to each position in the number.

Word A group of bits.

Review Questions

2.1 The radix of a binary number system is ________ and the digits used are ________.

2.2 In ________ number system, 16 distinct symbols are used to specify any number.

2.3 A byte contains ________ bits.

2.4 The weights assigned to an 8-bit binary number to LSB and MSB are ________ and ________ respectively.

2.5 The MSB of a signed-binary number indicates its ________.

2.6 2's complement of a 2's complement is ________.

2.7 The number of bits required to represent 25 in BCD is ________.

2.8 The Excess-3 code for decimal number 8 is ________.

2.9 The number of bits in ASCII code is ________.

2.10 The number of characters represented by ASCII code is ________.

2.11 The parity of 01110010 is ________.

2.12 The minimum distance required for a code to be error detecting code is ________.

2.13 A minimum distance of ________ is required for a code to be error correcting code.

2.14 The process of subtraction gets converted into that of addition by using ________.

2.15 Gray code is ________.

Problems

2.1 Determine the decimal numbers represented by the following binary numbers:
(a) 111001 (b) 101001 (c) 11111110 (d) 1100100
(e) 1101.0011 (f) 1010.1010 (g) 0.11100

2.2 Determine the binary numbers represented by the following decimal numbers:
(a) 37 (b) 255 (c) 15 (d) 26.25
(e) 11.75 (f) 0.1.

2.3 Add the following groups of binary numbers:
(a) 1011 + 1101 (b) 1010.1101 + 101.01

2.4 Perform the following subtractions using 2's complement method:
(a) 01000 – 01001 (b) 01100 – 00011 (c) 0011.1001 – 0001.1110

2.5 Convert the following numbers from decimal to octal and then to binary. Compare the binary numbers obtained with the binary numbers obtained directly from the decimal numbers.
(a) 375 (b) 249 (c) 27.125

2.6 Convert the following binary numbers to octal and then to decimal. Compare the decimal numbers obtained with the decimal numbers obtained directly from the binary numbers.
(a) 11011100.101010 (b) 01010011.010101 (c) 10110011

2.7 Convert the decimal numbers in Problem 2.5 to hexadecimal and then to binary. Compare the binary numbers obtained with the binary numbers obtained directly from the decimal numbers.

2.8 Convert the binary numbers in Problem 2.6 to hexadecimal and then to decimal. Compare the decimal numbers obtained with the decimal numbers obtained directly from the binary numbers.

2.9 Encode the following decimal numbers in BCD code:
(a) 46 (b) 327.89 (c) 20.305

2.10 Encode the decimal numbers in Problem 2.9 to Excess-3 code.

2.11 Encode the decimal number 46 to Gray code.

2.12 Use the 6-bit internal code to represent the statement.

$$P = 3 * Q$$

2.13 Write your full name in
(a) ASCII code (b) EBCDIC code (c) 6-bit internal code
Include blanks wherever necessary.

2.14 Attach an even parity bit as MSB for
(a) ASCII code. (b) EBCDIC code.

2.15 Repeat Problem 2.14 for odd parity.

2.16 Find the number of bits required to encode:
(a) 56 elements of information. (b) 130 elements of information

2.17 Write 8-bit ASCII code (parity and 7-bit code) obtained in Problems 2.14 and 2.15 in hexadecimal format.

2.18 Develop the binary subtraction rules using one's complement representation for negative numbers.

2.19 How many bits of memory are required for storing 100 names of a group of people, assuming that no name occupies more than 20 characters (including spaces)? Assume 7-bit ASCII code with parity bit.

2.20 A line printer is capable of printing 132 characters in a single line and each character is represented by ASCII code. How many bits are required to print each line?

2.21 Determine Hamming Code sequence with odd parity for natural BCD for making it an error correcting code.

3

SEMICONDUCTOR DEVICES—SWITCHING MODE OPERATION

3.1 INTRODUCTION

The basic functional blocks of digital systems have been discussed in Chapter 1. It was also shown there that these functions can be realized using switches. If we make a digital system requiring hundreds of gates using mechanical switches, perhaps we may not be able to operate the switches as desired. Even if a system with a few gates using mechanical switches is used, the operation will be quite complex and slow. The speed of switching can be increased if we use relays. Relays are bulky and generate a lot of electrical noise and hence are not very convenient except for large power handling systems. These difficulties can be overcome if semiconductor devices such as *p-n* junction diode, bipolar junction transistor (BJT) or unipolar transistor like metal-oxide-semiconductor field-effect transistor (MOSFET) are used as switches. These devices are much faster than relays and mechanical switches, and are best suited for digital circuits. Earlier digital systems, including the complex digital computers, used vacuum tubes as switches which were later replaced by semiconductor devices. This resulted in considerable savings in terms of cost, size, weight, and power requirements. These circuits were also faster in comparison to vacuum tube circuits.

Because of the tremendous progress in semiconductor technology, it became possible to fabricate thousands of components like diodes, transistors, resistors, and capacitors on tiny chips of silicon. This made possible integration, that is the fabrication of complete circuits on a small silicon chip. The resulting devices are popularly known as ICs (integrated circuits). ICs have significant advantages over discrete circuits as far as size, weight, power consumption, cost, speed of operation, and reliability are concerned. A large number of circuits can be fabricated on a single chip, requiring processes that are essentially similar to those for discrete transistors resulting in a considerable savings in size and cost. With the developments in semiconductor technology there has been a steady improvement in reliability, production yields, packaging density, and speed of operation. In fact, the number of components per chip has doubled every year since 1959, when the planar technology for the manufacture of transistors was introduced. These developments have made possible electronic calculators, wrist watches, microcomputers, etc. on single chips of a few millimetre square size. Intel's 8086 a 16-bit microprocessor introduced in 1978 contains about 29,000 transistors on a chip of about 6 × 6 mm^2 size and the Pentium II introduced in May 1997 consists of about 7.5 million transistors. Currently ICs in 0.13 micron (micro meter) technology are being manufactured making it possible to fabricate about 16 million transistors per cm^2 of silicon chip. This number is expected to become about 100 million transistors by the year 2012[1].

Since digital logic circuits consist of switches and the switches are implemented (or built) using semiconductor devices, therefore, for understanding, analysing, designing, and practical applications of digital circuits and systems, it is essential to learn the physical principles of working of various semiconductor devices and the circuits constructed using these devices. A clear understanding of the functioning of these devices will help a user to make effective use of these electronic devices and the circuits implemented using these devices in various technologies, such as, biploar technology and MOS technology. The principles of working of biploar and MOS devices have been discussed in this chapter and the technologies have been discussed in Chapter 4.

3.2 SEMICONDUCTORS

The flow of current in a material is due to the flow of electrons and the conductivity of the material is proportional to the concentration (n electrons/m^3) of free charge carriers. For a good conductor (copper, silver, etc.), n is very large ($\sim 10^{28}$) whereas for an insulator (wood, plastic, etc.) n is very small ($\sim 10^7$). There is another class of materials, *semiconductors*, for which the conductivity lies in between the conductivities of conductors and insulators. Silicon and germanium, which are tetravalent elements, are the two most important semiconductors used in electronic devices. These are briefly discussed here.

Silicon and germanium have four valence electrons, which are not free to move about as they are in a conductor. Instead, they combine with the valence electrons of neighbouring atoms and form electron-pairs, known as *covalent bonds*. Due to these bonds, the valence electrons are not free to move but are tightly bound to their nucleus. The outer most orbit of each atom of silicon in the crystal structure has eight electrons, which means it is completely filled. At 0 K, all covalent bonds are intact and the conductivity of the material is zero. If energy in the form of heat or light is supplied to this sample, some of the covalent bonds break away and free electrons and holes are generated. The absence of the electron in the covalent bond is referred to as a *hole*. Holes can be considered as mobile physical entities, the movement of which constitutes a flow of current. It behaves as a free positive charge of value equal to that of an electron. Because of the generation of free electrons and holes, it has conductivity which is

[1] Semiconductor Industry Association, "National Technology Roadmap for Semiconductors," http://www.semichips.org/

very low at room temperature. The energy required to break a covalent bond is about 1.1 eV for silicon and 0.72 eV for germanium at room temperature. The concentration of electrons (n) is same as the concentration of holes (p) in this pure (or *intrinsic*) semiconductor and is very small at room temperature.

The conductivity of silicon is increased by adding a small amount of pentavalent (antimony, phosphorous, or arsenic) or trivalent (boron, gallium or indium) atoms (~ 1 part in 10^8). This process is known as *doping* and the resulting semiconductor is known as *doped* or *extrinsic* semiconductor. The impurity atoms are approximately of the same size as the atoms of silicon and will displace some of the silicon atoms in the crystal lattice, resulting in each impurity atom being surrounded by silicon atoms, because of the small amount of impurity. In the case of the pentavalent impurity, four of the five valence electrons will form covalent bonds with the neighbouring silicon atoms and the fifth will be loosely attached to its nucleus. The energy required to detach the fifth electron from the atom is of the order of only 0.05 eV for silicon and 0.01 eV for germanium. This increases the concentration of electrons (one electron/impurity atom) and decreases the concentration of holes because of the increase in the rate of recombination (when an electron combines with a hole, both get vanished). The resulting semiconductor, which has a very large concentration of electrons and a very small concentration of holes, is referred to as an n-type semiconductor. Since the impurity atoms donate excess electrons, it is referred to as *donor*, or n-type of impurity.

In the case of the trivalent impurity, three valence electrons form covalent bonds with the neighbouring silicon atoms, causing a vacancy (hole) in the fourth covalent bond. This increases the concentration of holes (one hole/impurity atom) and decreases the concentration of electrons because of increase in the recombination rate. The resulting semiconductor, which has a very large concentration of holes and a very small concentration of electrons is referred to as a p-type semiconductor. Since the impurity atoms accept electrons, it is referred to as *acceptor* or p-type impurity.

From the above discussion we conclude that there are two types of extrinsic semiconductors, namely, n-type and p-type. In an n-type of semiconductor, the electrons are the majority charge carriers and the holes are the minority charge carriers, whereas in a p-type of semiconductor the holes are the majority charge carriers and the electrons are the minority charge carriers. Semiconductor devices are fabricated using extrinsic semiconductors.

3.3 *p-n* JUNCTION DIODE

A p-n junction diode is formed by introducing n-type of impurity into one side and p-type impurity into the other side of a single crystal of a semiconductor, as shown in Fig. 3.1. The concentration of charged particles on the two sides of the junction are given in Table 3.1.

Because of the concentration gradient across the junction, the holes from the p side diffuse to the n-side, and the electrons from the n-side to the p-side. This process results in recombination of electrons and holes near the junction on both sides and the region near the junction becomes devoid of charge carriers. This region which is depleted of mobile charge carriers is referred to as the *depletion*, *space charge,* or *transition* region. At the junction, the space-charge density is zero, being negative on the *p*-side and positive on the *n*-side. This gives rise to an electric field intensity and consequent electrostatic potential, which vary with the diffusion of charge carriers across the junction. This constitutes a potential-energy barrier against the further diffusion of charge carriers across the junction and thermal equilibrium is established. Metallic contacts are made with the *p*-type and *n*-type semiconductors for applying a voltage across the junction.

The symbol used for a *p-n* junction diode is shown in Fig. 3-2*a*. Figures 3.2*b* and 3.2*c* show a diode with a battery connected across it.

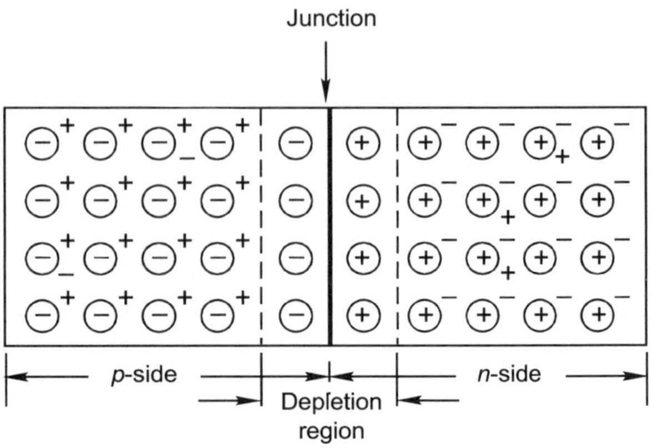

Fig. 3.1
A *p-n* junction diode

Table 3.1 Concentration of charged particles in a *p-n* junction diode

Charged Particles	*Represented by*	*p-side*	*n-side*	*Remarks*
Electrons	–	Negligibly small	Plentiful	Mobile charge carriers
Holes	+	Plentiful	Negligibly small	-do-
Negative ions	⊖	Equal to acceptor impurity atoms	Nil	Immobile
Positive ions	⊕	Nil	Equal to donor impurity atoms	-do-

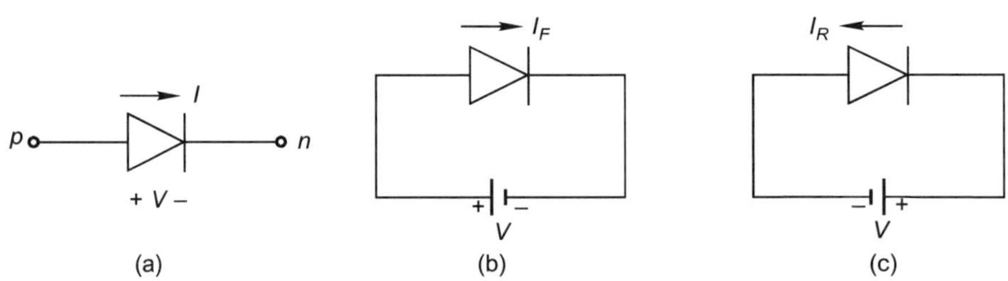

Fig. 3.2
(a) Symbol of *p-n* junction diode
(b) A forward-biased diode
(c) A reverse-biased diode

3.3.1 Forward Bias

In Fig. 3.2*b*, the positive terminal of the battery is connected to the *p*-side and the negative terminal to the *n*-side of the junction. This is referred to as the *forward-biased* junction. This causes both the holes in the p-side and the electrons in the n-side to move closer to the junction. Consequently, the width of the depletion region decreases, the height of the potential-energy barrier at the junction decreases and the equilibrium initially established is disturbed. Hence, the holes cross the junction from the p-side into the n-side, where they are referred to as *injected minority carriers.* Similarly, the electrons cross the junction in the reverse direction and are injected minority carriers in the p-side. The resulting current is I_F in the direction from p-side to n-side as indicated in Fig. 3.2b.

The minority-carrier distribution as a function of the distance x from the junction is illustrated in Fig. 3.3a. $n_p(p_n)$ is the electron (hole) concentration in the $p(n)$ side, n_{p0} and p_{n0} are the corresponding values under thermal equilibrium. The excess minority carrier concentrations are p'_n and n'_p on the n-side and p-side of the junction respectively, which decrease with the distance x, due to recombination and reduce to zero at distances far away from the junction. The minority-carrier concentrations at the junction are maximum and decrease approximately exponentially with the distance x from the junction.

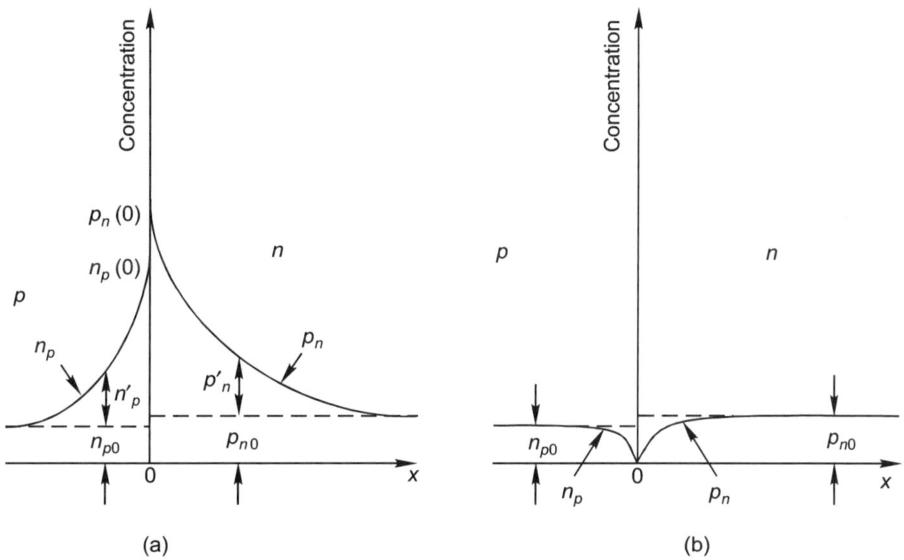

Fig. 3.3

Minority-carrier concentration in a p-n junction diode
(a) Forward-biased (b) Reverse-biased

3.3.2 Reverse Bias

In Fig. 3.2c, the polarity of the battery is opposite to that in Fig. 3.2b and is referred to as the *reverse-biased* junction. This causes both, holes in the p-side and electrons in the n-side, to move away from the junction. Consequently, the depletion layer width increases which prevents the holes from the p-side to cross over to the n-side and electrons from n-side to p-side. However, the minority charge carriers, the electrons from the p-side and the holes from the n-side, can easily cross the junction and constitute a reverse current I_R in the direction indicated in

Fig. 3.2*c*. This current is very small, of the order of a few microamperes for germanium diodes and a few nano-amperes for silicon diodes, and is dependent on the temperature (approximately doubles for every 10 °C rise in temperature).

The minority carriers near the junction cross over to the other side, recombine, and the concentration diminishes to zero at the junction. Far away from the junction, the minority carrier concentrations are equal to their thermal equilibrium values. This is shown in Fig. 3.3*b*.

3.3.3 The Volt-Ampere Characteristic

The volt-ampere (V-I) characteristic of a semiconductor diode is given by

$$I = I_0(\in^{V/\eta V_T} - 1) \tag{3.1}$$

The symbol V_T is the volt equivalent of the temperature and is given by

$$V_T = kT/q \tag{3.2}$$

where k = Boltzmann constant = 1.381×10^{-23} J/K

q = electronic charge = 1.602×10^{-19} C

T = absolute temperature in kelvin. At room temperature ($T \approx 300$ K), $V_T \approx 26$ mV.

The value of η is unity for germanium and is approximately two for silicon. I_0 is known as the *reverse saturation current* which is very small.

As indicated in Fig. 3.2*a*, the current I is positive when it flows in the direction of the arrow, i.e. from p side to n side of the diode. The voltage V is the voltage of p side with respect to n side The form of the V–I characteristic described by Eq. (3.1) is illustrated in Fig. 3.4. From the characteristic, we note that when the junction is forward biased, the current I_F is negligibly small up to the voltage V_γ which is referred to as the *cut-in*, *break-point* or *threshold* voltage and beyond V_γ the current rises very rapidly. The value of V_γ is approximately 0.2 V and 0.6 V for germanium and silicon diodes, respectively.

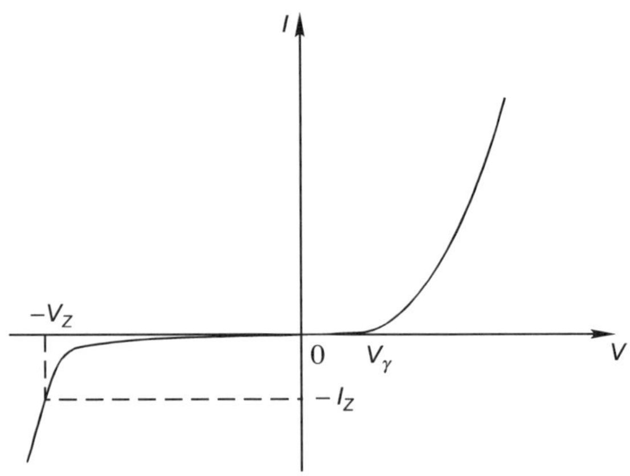

Fig. 3.4
V–I characteristic of a semiconductor diode

Thus, we see that the diode is a unidirectional device, i.e. it allows the current to flow in the forward direction (conducting) and does not allow the current to flow in the reverse direction (non-conducting) and hence acts as a switch.

3.3.4 Zener Diode

From the V–I characteristic of Fig. 3.4, we observe that a large current flows in the reverse direction as the reverse voltage across a diode is increased beyond a voltage known as the *Zener breakdown voltage*. When the diode is operating in the breakdown mode, the voltage across the diode is almost constant, V_Z, and the current in the diode is controlled by an external resistor. These diodes are used as voltage reference or constant-voltage sources. Its symbol is given in Fig. 3.5.

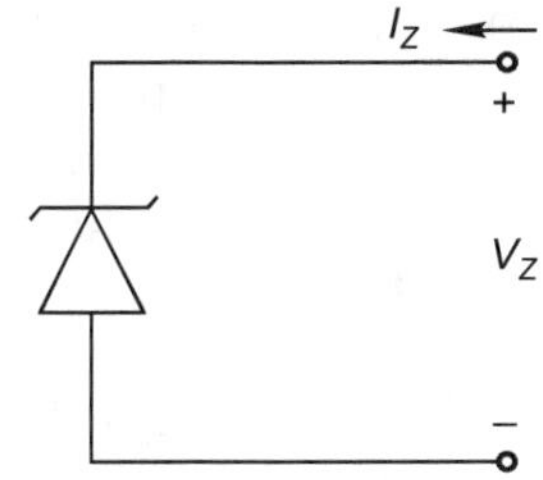

Fig. 3.5
Symbol for a Zener diode

3.3.5 Transition Capacitance of a *p-n* Junction Diode

When a *p-n* junction diode is reverse-biased, there exists a capacitance across it due to the presence of the immobile charges on the two sides of the junction. This is referred to as the *barrier* or *transition* capacitance C_T. It decreases with increasing reverse voltage and is approximately given by

$$C_T = \frac{C_o}{(1+V_R)^n} \tag{3.3}$$

where V_R is the reverse bias voltage

C_o is the capacitance of the open-circuited junction.

n is $\frac{1}{2}$ for an abrupt junction, and $\frac{1}{3}$ for a linearly graded junction.

This capacitance effectively appears across the very high reverse resistance of the diode.

There is another capacitance which is much larger than the transition capacitance C_T and is effective when the junction is forward biased. This appears across the very small, forward resistance of the diode. This capacitance is referred to as *diffusion*, or *storage* capacitance C_D and is proportional to the forward current.

3.3.6 Switching Characteristics of a Semiconductor Diode

When a *p-n* junction diode is forward biased, the voltage across the diode is a few tens of millivolts above the cut-in voltage and is almost constant. The current flowing in the diode, I_F, is limited by an external resistor in series with the diode. The minority-carrier concentrations on both sides of the junction are large as shown in Fig. 3.3*a*. This corresponds to the diode switch in the ON position. On the other hand, when it is reverse-biased the current flowing through the diode ($I_R \approx -I_0$) is negligibly small and is almost constant. The minority-carrier concentrations on both sides of the junction are negligibly small, as shown in Fig. 3.3*b*. This corresponds to the switch in the OFF position.

When the voltage in a diode circuit changes suddenly from the reverse-biased to the forward-biased, then the steady-state condition will be reached when the minority-carrier concentration changes from the one shown in Fig. 3.3*b* to that shown in Fig. 3.3*a*. Similarly, when the voltage changes suddenly from forward to reverse bias, then the minority carrier concentration has to change accordingly. This change of minority-carrier concentration requires a finite time, which is negligibly small when the bias voltage changes from reverse to forward condition, but is significant when the voltage changes from forward to reverse bias condition. This is due to accumulation of excess minority-charge carriers on the two sides of the junction under forward-bias condition, which must be removed before the diode comes to steady-state in the reverse-bias condition. As soon as the voltage changes sign to reverse bias the diode, a large current will flow in the reverse direction, due to the excess charge carriers stored, till all the excess charge has been removed. The time required for the removal of the excess charge is referred to as the *storage time* (t_s). Beyond that, the current changes exponentially and comes to a steady-state value. The time required to reach the steady-state in the reverse bias condition is referred to as the *transition time* (t_r). The sum of storage and transition times is the total time delay and is referred to as the *switching time*. The switching time determines the maximum frequency at which the diode can switch from ON to OFF and from OFF to ON. The concept of minority-carrier storage and its removal is vital in switching mode operation of semiconductor devices and therefore must be understood thoroughly.

Example 3.1 The voltage waveform shown in Fig. 3.6*a* is applied to the diode network of Fig. 3.6*b*. Sketch the waveforms of (a) excess minority charge concentration at the junction, (b) the diode current, and (c) the diode voltage.

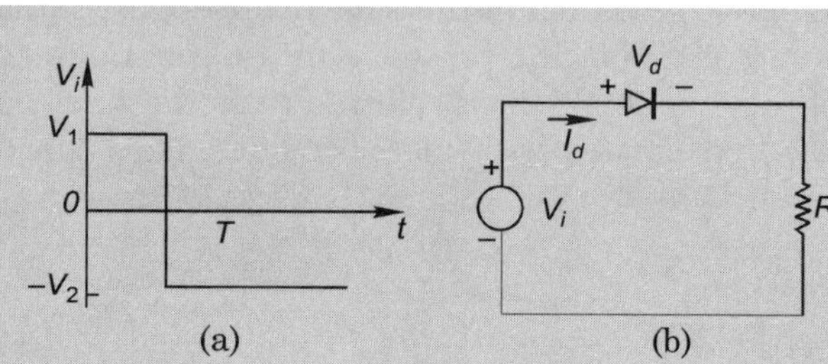

Fig. 3.6
The waveform in (a) is applied to the diode network in (b)

Solution The required waveforms are given in Fig. 3.7.

Upto time T the diode is forward-biased and the current $I_F \approx V_1/R$. At $t = T$, the

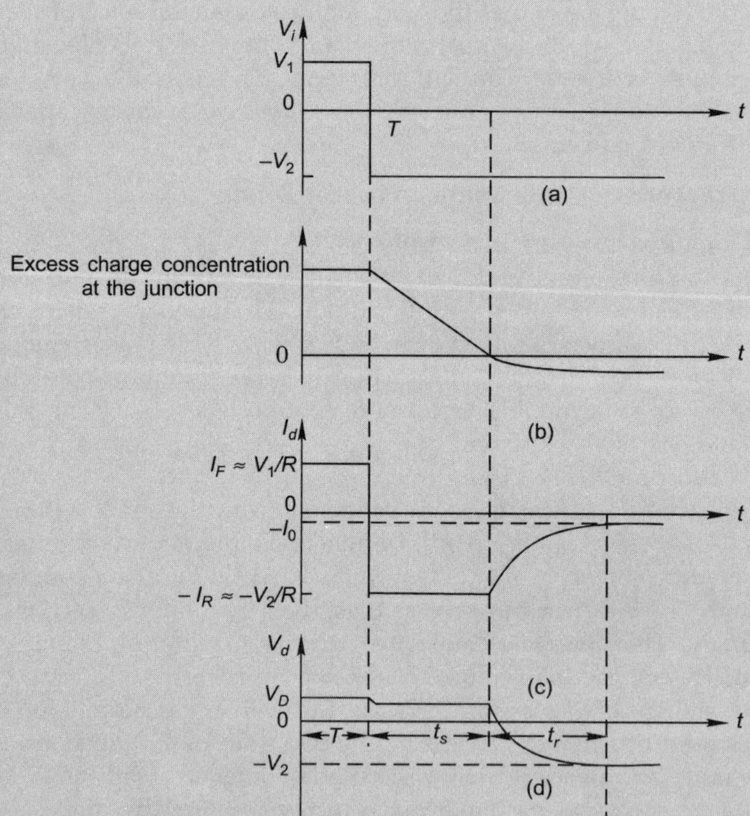

Fig. 3.7
Waveforms of Ex. 3.1.

input voltage changes from V_1 to $-V_2$, consequently the current flows in the reverse direction ($I_d = -I_R \approx -V_2/R$) which removes the excess minority charge. When the excess charge stored is completely removed, the current I_d and the voltage V_d change exponentially and come to their steady-state values $-I_0$ and $-V_2$ respectively. The storage time t_s, and the transition time t_r are indicated in Fig. 3.7.

The storage time increases with increase in I_F and decreases with the increasing magnitude of the reverse current I_R.

3.4 SCHOTTKY DIODE

The speed of operation of a semiconductor diode is reduced due to storage of minority carriers as discussed above. The storage time can be reduced significantly by using the junction formed by a semiconductor and a metal. For example, the junction between an aluminium and *n*-type semiconductor constitutes a diode, known as Schottky diode and is represented by the symbol shown in Fig. 3.8. The *V-I* characteristics of Schottky diodes are similar to those of semiconductor diodes, except for the cut-in voltage, which is in the range of 0.2 to 0.5 V, depending on the metal used. The aluminium *n*-type Schottky diode has a cut-in voltage of about 0.35 V.

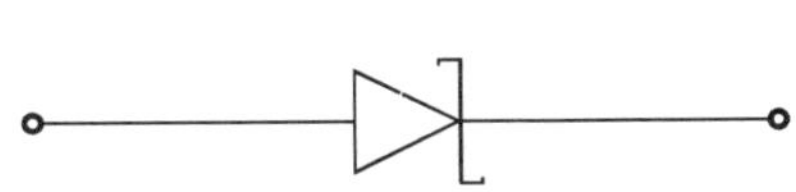

Fig. 3.8
Symbol of a Schottky diode

When a Schottky diode is forward-biased, that is the positive terminal of the battery is connected to the metal and the negative terminal to the semiconductor, current flows across the junction due to the flow of electrons from the semiconductor to the metal. Electrons thus entering the metal cannot be distinguished from the plentiful electrons already present in the metal and hence these do not constitute minority-carriers. Therefore, when the junction voltage is reversed the problem of removal of the excess minority charge does not exist. Hence, Schottky diodes exhibit a negligible storage time.

3.5 BIPOLAR JUNCTION TRANSISTOR

A bipolar junction transistor (BJT) consists of a silicon (or germanium) crystal in which either a thin layer of *p*-type silicon is sandwiched between two layers of *n*-type silicon or a layer of *n*-type is sandwiched between two layers of *p*-type silicon. The former is referred to as an *n-p-n transistor*, and the latter as a *p-n-p transistor*. The two types of transistors, along with their circuit symbols are given in Fig. 3.9. This is a three terminal device and its terminals are designated as the *emitter* (E), *base* (B), and *collector* (C). The arrow on the emitter lead indicates the direction of current flow when the emitter-base junction is forward-biased. The emitter, base, and collector currents, I_E, I_B, and I_C, respectively are assumed as positive when the currents flow into the device, as shown in Fig. 3.9.

The operation of the *n-p-n* transistor is explained below and the operation of the *p-n-p* is identical to that of the *n-p-n*. Consider the transistor circuit of Fig. 3.10 in which the emitter-base (E-B) junction is forward-biased and the collector-base (C-B) junction is reverse biased, the

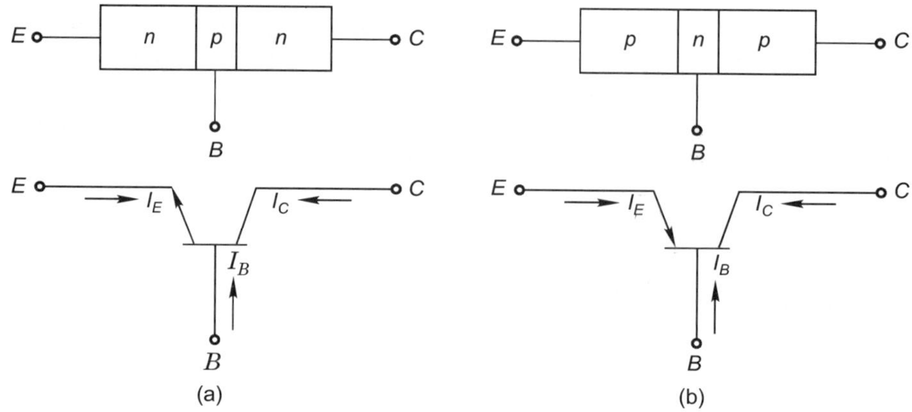

Fig. 3.9
A BJT and its circuit symbol (a) *n-p-n* (b) *p-n-p*

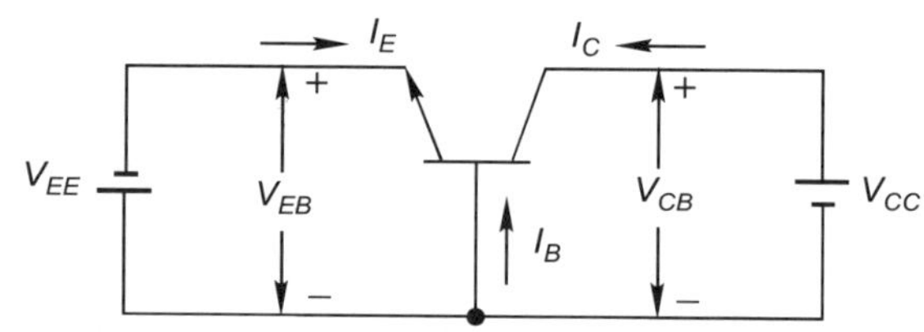

Fig. 3.10
A biased *n-p-n* transistor

forward-bias voltage being much smaller than the reverse-bias voltage. Since E-B junction is forward-biased, electrons from emitter (*n*-type) will diffuse into the base (*p*-type) and similarly the holes from the base will diffuse into the emitter region. This gives rise to an emitter current which is negative (it flows in the direction opposite to the positive direction). Since the collector-base junction is reverse-biased, most of the electrons which have been injected into the base region and behave as minority carriers there will be swept over to the collector region. The collector current is slightly less than the emitter current, because of recombination in the base region which results in a small base current.

The currents I_E, I_C, and I_B are related by

$$I_E = -(I_C + I_B) \tag{3.4}$$

The common-base output characteristics are shown in Fig. 3.11.

The operation of a BJT depends upon the biasing of the two junctions. The range of its operation is divided into three regions: the cut-off, the active, and the saturation regions. When both the junctions are reverse-biased, very small reverse saturation currents will flow across the junctions. This is referred to as the *cut-off region* and corresponds to an open (non-conducting) switch. This region is below the curve for $I_E = 0$ in Fig. 3.11.

When the emitter–base junction is forward-biased and the collector–base junction is reverse-biased, the output (collector) current is almost linearly dependent on the input (emitter) current. This is referred to as the *active region* and is of very little interest when transistor is to be used as a switch. This region is indicated in Fig. 3.11.

The *saturation region* is of great importance in the transistor switch. In this region both the junctions are forward biased. This corresponds to a closed (ON) switch and is indicated in Fig. 3.11.

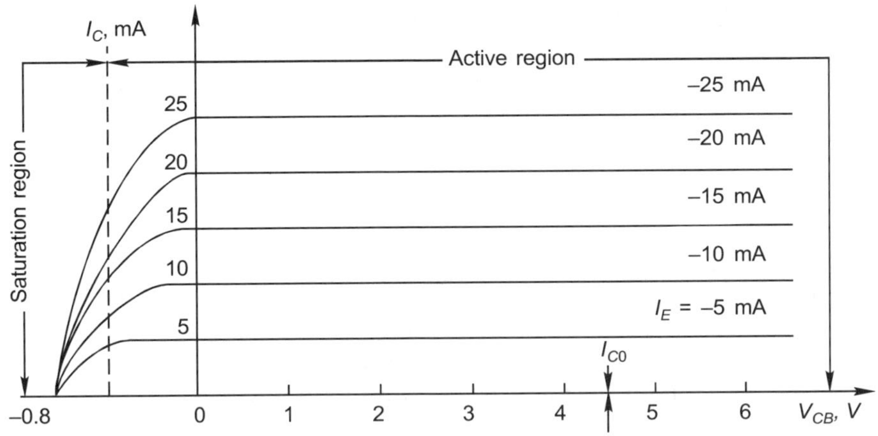

Fig. 3.11

CB output characteristics of an *n-p-n* transistor

3.5.1 Transistor Configurations

A transistor can be used in any one of the three configurations: the common-base (CB), the common-emitter (CE), and the common-collector (CC) configurations. These are illustrated in Fig. 3.12.

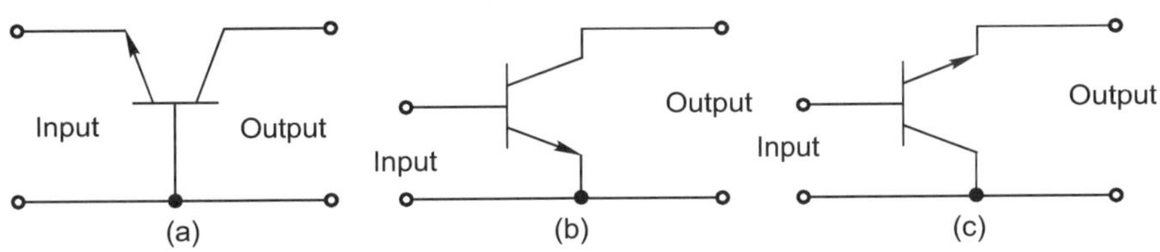

Fig. 3.12

The transistor configurations (a) CB (b) CE (c) CC

Consider CB configuration with E-B junction open-circuited ($I_E = 0$) and C-B junction reverse-biased as shown in Fig. 3.13*a*. In this circuit, the reverse saturation current I_{CO} flows in the output circuit. When the E-B junction is forward biased, currents I_E, I_C, and I_B flow, which are constrained by Eq. (3.4). A parameter α, referred to as d.c. or large signal current gain of a CB transistor configuration is defined as

$$\alpha = -\frac{I_C - I_{CO}}{I_E} \tag{3.5}$$

α is always positive and lies in the range 0.9 to 0.998.
Usually $I_{CO} \ll I_C$, therefore,

$$|I_C| = |\alpha I_E| \tag{3.6}$$

which shows that the output current I_C is almost equal to the input current I_E, and I_B is negligibly small.

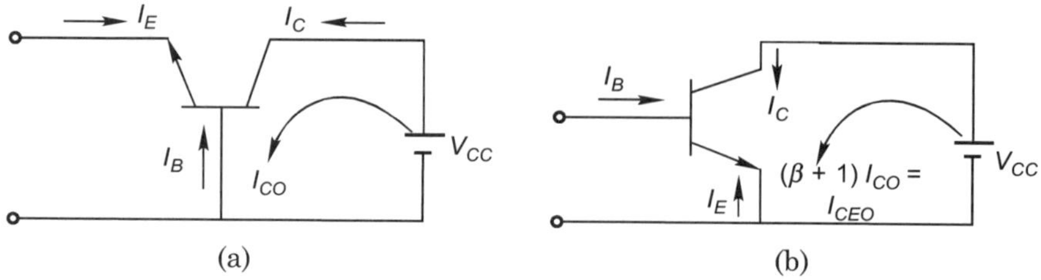

Fig. 3.13
The transistor circuits with input open-circuited (a) CB (b) CE

Similarly in CE configuration, if the input terminals are open-circuited (Fig. 3.13*b*), the collector current I_C can be found using Eqs (3.4) and (3.5) and is given by

$$I_C = \frac{\alpha}{1-\alpha} I_B + \frac{1}{1-\alpha} \cdot I_{CO} \tag{3.7}$$

or

$$I_C = \beta I_B + (\beta + 1) I_{CO} \tag{3.8}$$

where

$$\beta \equiv \frac{\alpha}{1-\alpha} \tag{3.9}$$

β is known as the large-signal current gain of CE configuration. It is also represented by h_{FE}. Since α is close to unity, therefore, β is a large number, for example, for $\alpha = 0.98$, $\beta = 0.98/(1 - 0.98) = 49$.

From Eq. (3.8) we see that

$$I_C \approx \beta I_B \tag{3.10}$$

which means that the collector current is proportional to the base current.

The large signal current gain of a common-collector configuration can be found using Eqs (3.4) and (3.8) and is equal to $\beta + 1$.

3.5.2 Transistor as a Switch

When a transistor is employed as a switch, it connects and disconnects the load R_L from the source V. It is an electrically operated switch with the controlling quantities being applied at the input terminals and the load in series with the source are connected at the output terminals as shown in Fig. 3.14a.

Ideally, the transistor must not allow any current to flow through the load R_L corresponding to the switch in the open position, i.e. the transistor must be biased to cut-off (OFF). The corresponding equivalent circuit is shown in Fig. 3.14b. On the other hand, whole of the voltage V must get connected across the load R_L corresponding to the switch in the closed position, i.e. the transistor must be driven into saturation (ON). Its corresponding equivalent circuit is shown in Fig. 3.14c. For a practical transistor switch, the current through the load R_L is negligibly small in the OFF state and the voltage across the load R_L is slightly less than V because of the very small voltage appearing across the switch.

Let us examine the three configurations from the point of view of using a transistor as a switch. If it is used in CB configuration, the input emitter current required to operate the switch

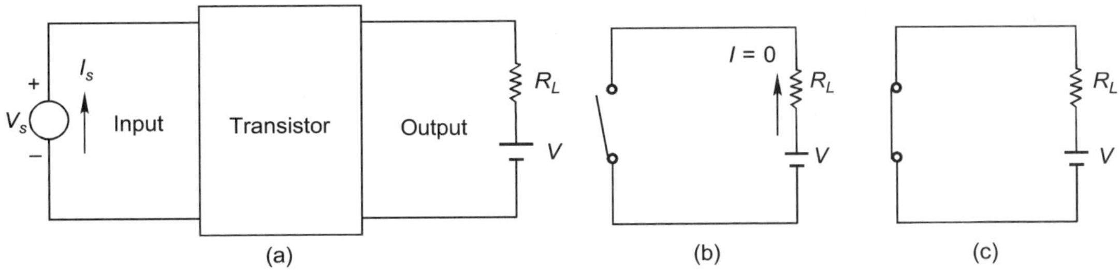

Fig. 3.14
(a) A transistor switch (b) Switch in the open (OFF) position (c) Switch in the closed (ON) position

is nearly as large as the collector current being switched, while in CC configuration, the input voltage required to operate the switch is nearly as large as the supply voltage. In CE configuration, the input switching signal (current or voltage) is very small in comparison with the current or voltage being switched. Hence, the common-emitter configuration is the most useful and therefore most commonly used configuration for a transistor switch.

3.5.3 CE Transistor Switch

A transistor in the common-emitter configuration being used as a switch is shown in Fig. 3.15*a*. In this circuit, the voltage V_i controls the operation of the switch. If the voltage V_i is less than the cut-in voltage of the transistor, the base current $I_B \approx 0$ and hence the collector current $I_C = I_{CEO} \approx 0$. I_{CEO} is the current in the collector-emitter circuit when $I_B = 0$. This corresponds to the switch in the OFF (open) position (Fig. 3.15*b*). As the voltage V_i is increased, the base current I_B increases which causes an increase in the collector current I_C, since $I_C \propto I_B$. The collector current can rise up to a maximum possible value of $I_{C,\,max}$ given by

$$I_{C,\,\max} = \frac{V_{CC} - V_{CE,\text{sat}}}{R_C}$$

where, $V_{CE,\,sat}$ is the value of V_{CE} corresponding to transistor in saturation. It is negligibly small, and therefore, $I_{C,\,max} \approx V_{CC}/R_C$. Any increase in I_B beyond this does not increase I_C. While the collector current I_C is increasing, the voltage V_{CE} decreases and ultimately goes below the voltage V_{BE}, which makes the collector-base junction forward-biased.

Hence, both the junctions are forward-biased and consequently the transistor is driven into saturation. This corresponds to the switch in the ON (closed) position (Fig. 3.15*c*) and the voltage $V_{CE,\,sat}$ will be in the range 0.1 to 0.2 *V* for a silicon transistor, giving a voltage across R_C nearly equal to V_{CC}.

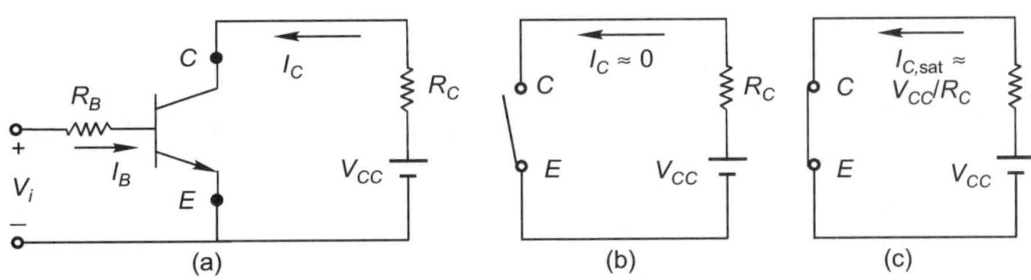

Fig. 3.15
(a) A CE transistor switch (b) The switch in the OFF position (c) The switch in the ON position.

The collector current corresponding to the saturation is $I_{C,sat}$ and the base current required for driving the transistor to saturation is $I_{C,\,sat}/h_{FE}$. $I_{C,\,sat}$ mainly depends upon values of V_{CC} and R_C and is not fixed for a given transistor. However, the values of V_{CC} and R_C must be chosen in such a way so that the $I_{C,\,sat} \approx V_{CC}/R_C$ does not exceed the maximum collector current rating of the transistor.

3.5.4 Switching Speed of BJT

When the transistor in Fig. 3.15*a* is cut-off the minority-carrier concentration in the base region is very small. In the active region, and in the saturation region, the minority-carrier concentration in the base region builds up as illustrated in Fig. 3.16. The slope of the excess minority-carrier concentration at the two edges of the base determine the steady-state emitter and collector currents while the base current is proportional to the total stored excess minority carriers. In Fig. 3.16*a*, the slope of the minority-carrier concentration curve increases as the charge builds up in the base region signifying an increase in the emitter and collector currents.

The area of the shaded region corresponds to the excess minority base charge under steady-state condition. Once the transistor enters saturation, the slope of the minority concentration curve does not increase. For the transistor in saturation, the charge build up in the base region is shown in Fig. 3.16*b*. The shaded area corresponds to the excess charge stored up to the edge of saturation (shown under steady-state condition in saturation region), beyond which it moves in the upward direction. The total excess minority base charge is the sum of the shaded area and the area enclosed by the lines *AA′* and *RR′*.

When a transistor is to be switched from OFF to ON, the charge build up requires time to reach the steady-state condition. Similarly, when it is switched from ON to OFF, the excess charge stored must be removed which again requires some time. Due to these reasons delay is introduced in the operation of the transistor as a switch. The delay times are explained further with the help of the following example.

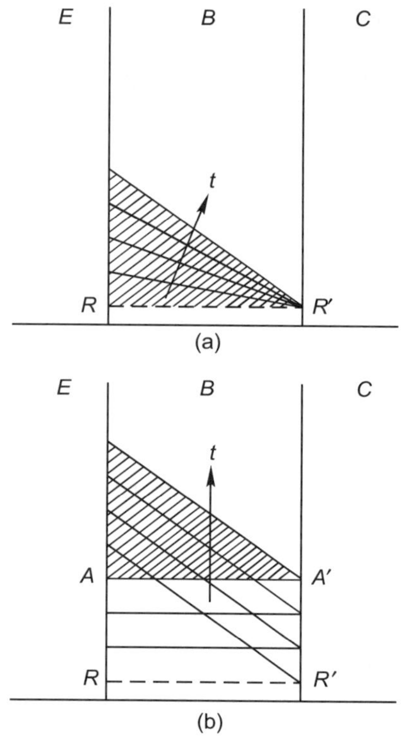

Fig. 3.16

Minority charge build up in the base region for a transistor (a) In active region (b) In saturation region.

Example 3.2 Determine the response of the transistor inverter circuit of Fig. 3.17*b* to the voltage waveform shown in Fig. 3.17*a*.

Solution The base–emitter junction is reverse-biased before the input voltage changes from –10 V to 10 V at $t = 0$. Therefore, the transistor is cut-off for $t < 0$ and consequently $I_B \approx 0$, $I_C \approx 0$, $V_O = V_{CC}$, and the transition capacitance C_{BE} is charged to –10 V.

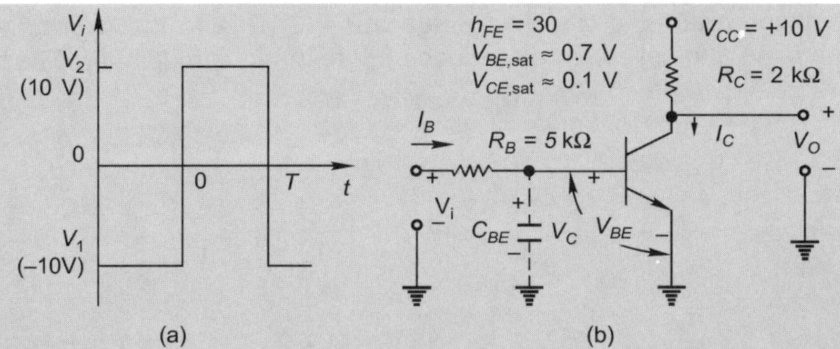

Fig. 3.17

Transistor inverter

When the input makes the transition at $t = 0$, the voltage at the base can not change instantaneously because of the presence of transistor input capacitance C_{BE}. It changes from –10 V exponentially with the time constant (= $R_B\, C_{BE}$) to the steady-state value of +10 V, but as soon as V_{BE} reaches the cut-in voltage of the transistor, it comes to conduction. Further increase in the voltage at the base causes a corresponding increase in the base current and ultimately the transistor is driven into saturation and consequently the base current is

$$I_{B1} = \frac{V_2 - V_{BE}}{R_B} = \frac{10 - 0.7}{5} \text{ mA}$$

$$= 1.86 \text{ mA}$$

the collector current is

$$I_{C,\,sat} = \frac{V_{CC} - V_{CE,\,sat}}{R_C}$$

$$= \frac{10 - 0.1}{2} \text{ mA}$$

$$\approx 5 \text{ mA}$$

The base current required for the transistor to be in saturation is

$$I_{B,\,sat} = \frac{I_{C,\,sat}}{h_{FE}} = \frac{5}{30} \text{ mA}$$

$$= 0.167 \text{ mA.}$$

Since $I_{B1} \gg I_{B,\,sat}$ the transistor is well inside the saturation region.

Also $$V_O = V_{CE,\,sat} \approx 0.1 \text{ V}$$

The time required for the collector current to rise to 10 per cent of $I_{C,\,sat}$ is referred to as the *delay time* (t_d) and the time required for the current to rise through the active region from 10 to 90 per cent of $I_{C,\,sat}$ is referred to as the *rise time* (t_r). The sum of the delay and rise times is referred to as the *turn-on time*, $t_{ON} = t_d + t_r$.

The input voltage returns to the initial value at $t = T$. Due to the excess minority charge stored in the base, the collector current does not respond immediately. The base current is

$$I_{B2} = \frac{V_1 - V_{BE}}{R_B}$$

$$= \frac{-10 - 0.7}{5} \text{ mA}$$

$$= -2.14 \text{ mA}$$

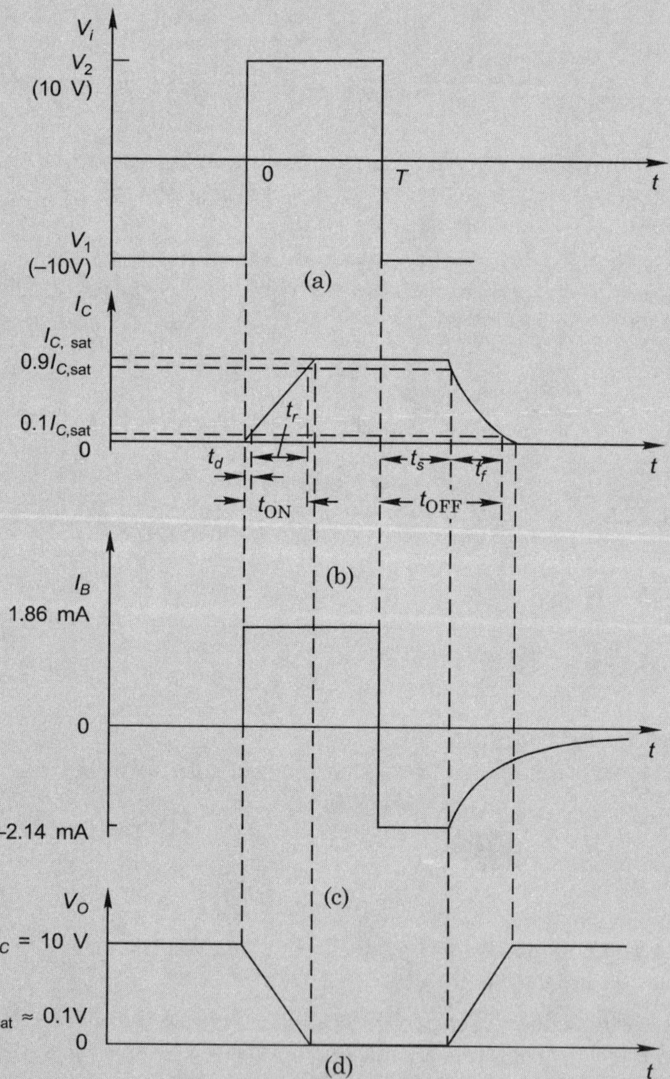

Fig. 3.18

Waveforms of the inverter circuit of Fig. 3.17.

Since this base current is negative, i.e. it is flowing in the opposite direction and thus helps in the removal of the excess base charge (electrons) stored. The time interval which elapses between the transition of the input waveform and the time when I_C has dropped to 90 per cent of $I_{C,\,sat}$ is referred to as the *storage time* (t_s). The time required for I_C to fall from 90 to 10 per cent of $I_{C,\,sat}$ is known as the *fall time* (t_f). The sum of the storage and the fall times is referred to as the *turn off time*, $t_{OFF} = t_s + t_f$.

The actual calculation of the time intervals t_d, t_r, t_s, and t_f is complex and is beyond the scope of this book. The various waveforms are sketched in Fig. 3.18.

The turn-on time can be reduced by increasing the voltage V_2 but this increases the storage time due to an increase in the value of I_{B1}. The turn-off time can be reduced by making V_1 more negative, which causes an increased base current to flow in the reverse direction on transition and hence reduces the time required for the removal of the excess stored charge in the base region.

3.6 SCHOTTKY TRANSISTOR

The storage delay time can be reduced considerably by preventing the transistor from going into saturation. One way of achieving this is to connect a Schottky diode between the base and collector of the transistor as shown in Fig. 3.19. When the transistor is in the active region, diode D is reverse-biased. The diode conducts when the base–collector junction voltage falls to about 0.4 V and does not allow the collector voltage to fall lower than 0.4 V below the base voltage. Hence, the collector junction is not sufficiently forward-biased and the transistor is thus prevented from entering into saturation.

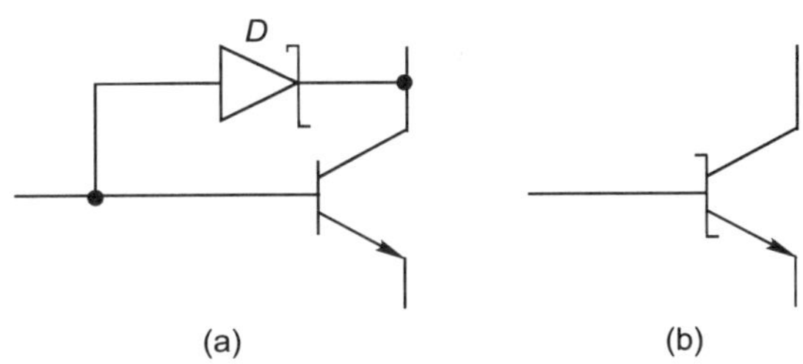

Fig. 3.19

(a) A Schottky diode connected to prevent transistor saturation (b) Symbol for the Schottky transistor.

3.7 FIELD-EFFECT TRANSISTOR

The field-effect transistor is another three-terminal semiconductor device which can be used as a fast operating switch. The operation of this device depends on the control of the flow of majority charge carriers by an electric field. Since only one type of charge carriers contribute to the current flow, it is an unipolar device, in contrast to the BJT in which both types of charge carriers contribute to the flow of current. There are two types of field-effect transistors: the junction field-effect transistor (JFET) and the metal-oxide-semiconductor field-effect transistor (MOSFET).

3.7.1 Junction Field-Effect Transistor

A junction field-effect transistor consists of a bar of n (or p) type of silicon with ohmic contacts at the two ends. Current flows along the length of the bar when a voltage supply is connected between the ends. It is referred to as an *n-channel* or *p-channel* device, depending upon the type of material used. In an *n*-channel JFET, the current flow is due to the flow of electrons (majority carriers) whereas in a *p*-channel JFET, flow of holes constitutes the current. The two ends of the channel are known as the *source* and the *drain*.

In an *n*-channel device, *p*-type impurity is diffused between source and drain which forms a *p-n* junction. A metallic contact is made to the *p*-type material and this terminal is known as the *gate*. An electric field is applied to the channel through this *p-n* junction which controls the flow of current through the channel. The structure of an *n*-channel field-effect transistor is shown in Fig. 3.20*a*. The symbols used for JFETs are shown in Fig. 3.20*b*.

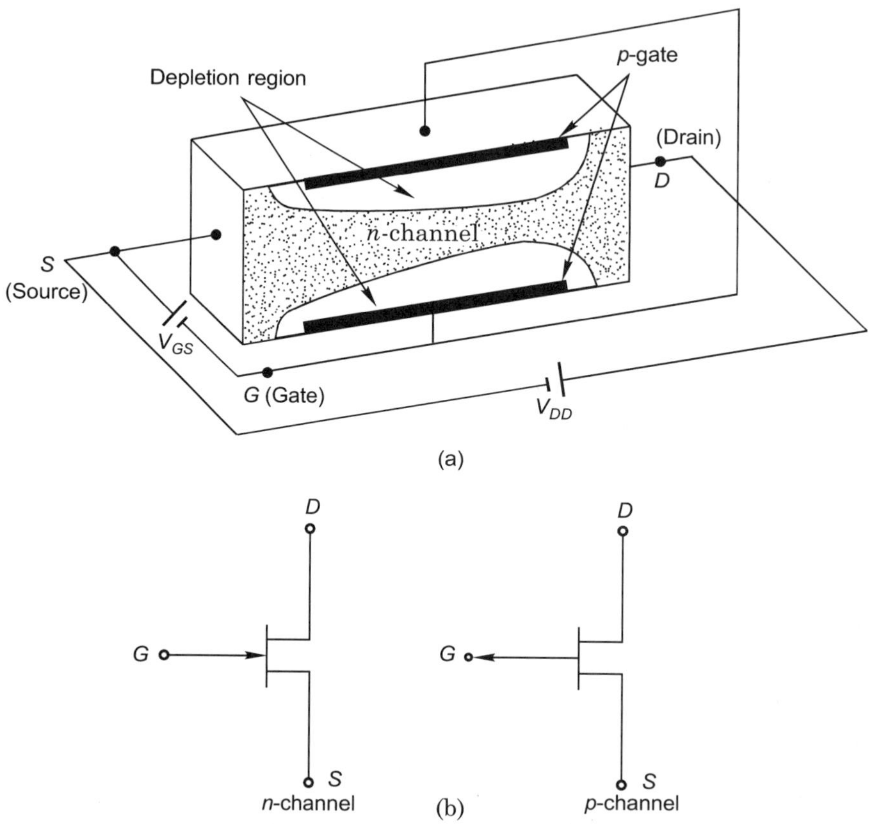

Fig. 3.20
(a) An *n*-channel junction field-effect transistor
(b) Circuit symbols for JFETs

Let us assume that the gate-to-source voltage, $V_{GS} = 0$ and drain-to-source voltage, V_{DS} is low. Corresponding to this, the channel width is almost uniform between the source and drain (depletion region is very small) and the device behaves like a resistor. With increasing drain-to-source voltage, the current in the channel increases and results in the reverse biasing of the gate junction. The reverse biasing is not uniform, rather it increases with the distance from the source end and is maximum at the drain end. Due to the reverse-biased gate junction, the depletion region spreads and constricts the channel which is more pronounced at distances farther from the source. This results in the increase of channel resistivity and the rate of increase of drain current with drain-to-source voltage becomes smaller. Eventually, a voltage V_{DS} is reached at which the channel "pinches-off" and the current begins to level off, i.e. there is hardly any increase in the drain current with further increase in V_{DS}. The drain-to-source voltage at which the saturation of drain current begins when $V_{GS} = 0$ is referred to as the *pinch-off voltage* (V_{P0}). It is not possible for the channel to close completely and thereby reduce the drain current to zero because, in such a case, the ohmic drop along the channel required to provide the necessary reverse-bias would be lacking.

If now a gate voltage V_{GS} is applied to reverse-bias the gate–source *p-n* junction, pinch-off will occur for smaller values of V_{DS} and the saturation drain current will be smaller. The pinch-off voltage V_P for various gate-to-source voltages is given by

$$V_P \approx V_{P0} + V_{GS} \tag{3.11}$$

From Eq. (3.11) we note that $V_P = 0$ when $V_{GS} = -V_{P0}$. At this gate voltage, the drain current will be zero and the FET is cut-off. For $V_{GS} < -V_{P0}$ also the device is cut-off. A typical drain–source volt–ampere characteristics of an *n*-channel JFET is shown in Fig. 3.21.

For the operation of JFET, the gate–channel *p-n* junction is reverse-biased and consequently, the gate current is very small (~ 10^{-9} A). A slight gate bias (less than the cut-in voltage) may be allowed in the forward direction since no appreciable gate current will flow. A plot of $V_{GS} = 0.5$ V is also given in Fig. 3.21.

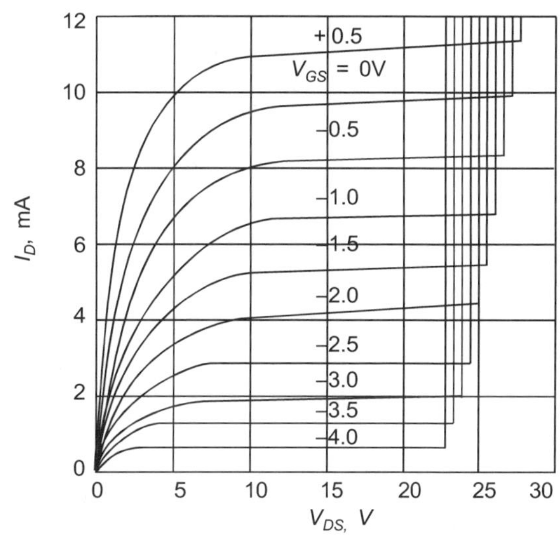

Fig. 3.21
Output characteristics of an *n*-channel JFET

3.7.2 Metal-Oxide–Semiconductor Field-Effect Transistor

A field-effect transistor is made by growing a very thin layer of SiO_2 (0.1 μm thick) over a semiconductor material. A metal such as aluminium is deposited over the dielectric layer of SiO_2. This structure is known as metal-oxide-semiconductor field-effect transistor (MOSFET). The metal gate is insulated from the channel and therefore it is also referred to as an *insulated gate FET* (IGFET).

There are two types of MOSFETs:

1. Enhancement MOSFET, and
2. Depletion MOSFET

Similar to junction field-effect transistors, we can have *n*-channel as well as *p*-channel MOSFETs. Since *n*-channel devices are more popular than *p*-channel devices because of their higher speed, only *n*-channel devices are discussed here.

However, the physical principles of operation of *p*-channel MOSFETs are the same as *n*-channel devices. *p*-channel MOS devices have almost become obsolete as independent devices but have become extremely useful in Complementary-MOS (CMOS) devices which will be discussed later.

Enhancement MOSFET The basic structure of an *n*-channel enhancement mode MOSFET is shown in Fig. 3.22*a*. Two *n*-type regions, known as the source and the drain, are diffused into a *p*-type substrate. A thin layer of silicon dioxide is grown over this and by masking, etching, and metallization processes metallic contacts are taken from the source and the drain and a metal gate is formed above the dielectric layer.

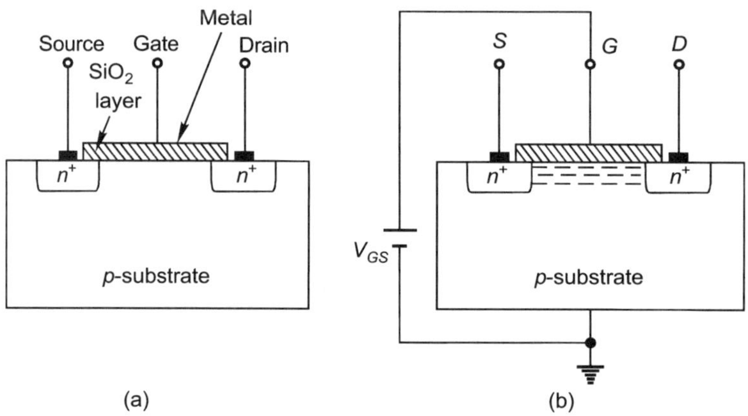

Fig. 3.22

(a) The structure of an *n*-channel enhancement MOSFET
(b) The channel is induced when $V_{GS} > V_T$

In this device, no channel exists between the source and the drain unless a positive voltage greater than the threshold voltage V_T is applied at the gate (Fig. 3.22*b*). Due to this field from the gate, the electrons (minority carriers) from the substrate are attracted towards the gate between the source and the drain regions, and will change this region between the source and drain to *n*-type thereby forming an *n*-channel. This channel makes it possible for the electrons to flow from the source to the drain when a positive voltage is applied at the drain. Since the application of a positive voltage at the gate enhances the channel width, this device is referred to as an *enhancement mode* device or *enhancement* MOSFET. An increase in the drain voltage increases the drain current, producing a resistor type operation for small voltages similar to JFET. Pinch-off occurs when the drain-to-source voltage is sufficiently large, which reduces the field near the drain to zero and makes the drain current relatively constant.

The drain characteristics and the transfer characteristic are shown in Fig. 3.23.

Depletion MOSFET It is fabricated by diffusing *n*-type of impurity between the two *n*-type regions which acts as a channel between the source and the drain. In this, current would flow between drain and source even in the absence of a positive voltage at the gate. The drain current is controlled by the application of negative voltage at the gate, which causes depletion

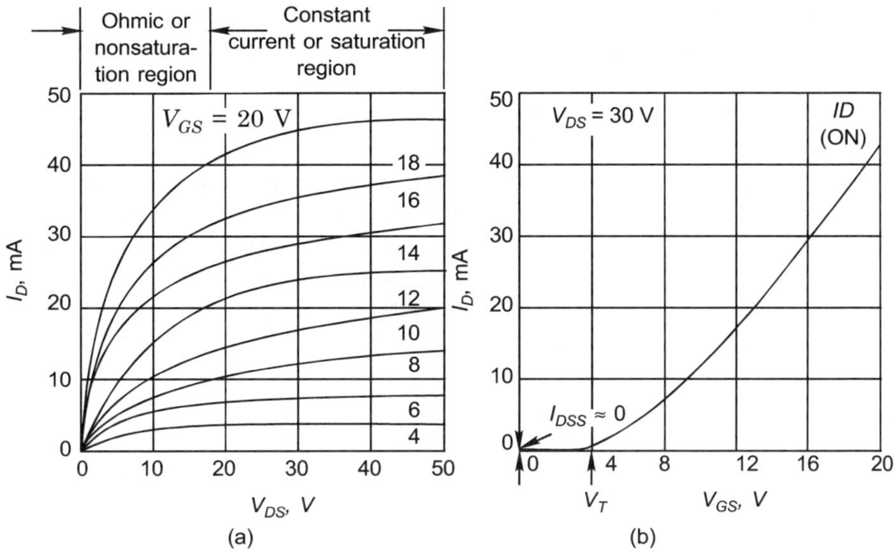

Fig. 3.23
n-channel enhancement MOSFET (a) Drain characteristics (b) Transfer characteristic.

of the channel and hence it is called *depletion* MOSFET. Its operation is similar to a JFET which is also a depletion type of device.

The circuit symbols for an n-channel MOSFET are shown in Fig. 3.24. For a p-channel MOSFET the direction of the arrow is reversed.

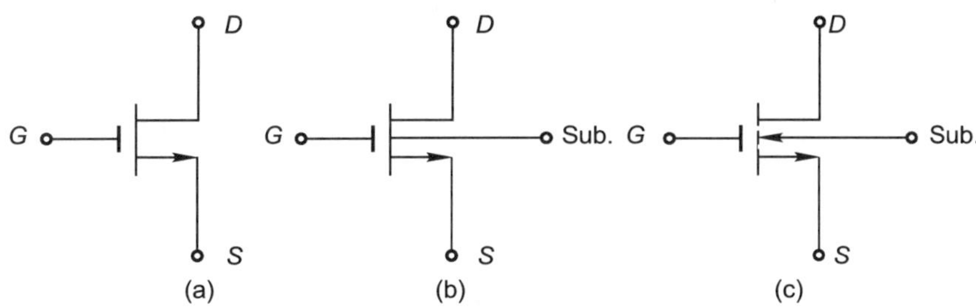

Fig. 3.24
Circuit symbols for an n-channel MOSFET (a) and (b) can be either enhancement or depletion type, whereas (c) represents specifically an enhancement device.

3.7.3 FET Switches

The JFET and MOSFET devices can be used as switches. These devices are unipolar, i.e. the current flows due to the majority charge carriers only, and are voltage controlled devices. A JFET switch uses JFET in common source (CS) configuration with a resistor R_D in the drain circuit (See Example 3.3). Similarly, operation of a MOSFET switch is given in Example 3.4. A

MOSFET can be used as a resistor also (See Example 3.5) and therefore, a switch can be made using two MOSFETs only. One of the MOSFETs in this acts as an active element (driver) and the other one as a non-linear resistor.

FET devices do not have the problem of storage and removal of minority charge carriers when these devices are switched from OFF to ON and ON to OFF. Therefore, the problem of redistribution of charges do not arise in these devices which contributes to delay in bipolar devices. FET devices have significant capacitances contributing to delay in these switches because of the charging and discharging of these capacitors through the resistors in the drain circuit in switching operation. In general, the turn-on and turn-off delay times of unipolar devices are significantly higher than those of bipolar devices but developments in the MOS devices have made it possible for these devices to have speeds comparable to those of bipolar devices.

Example 3.3 For the circuit shown in Fig. 3.25, determine the output voltage V_O for the input voltage V_i of (a) –5 V (b) 0 V. The output characteristics of the JFET are given in Fig. 3.21.

Solution

Load line for V_{DD} = 20 V and R_D = 5 kΩ is drawn on the output characteristics of the JFET as shown in Fig. 3.26.

(a) When the input voltage V_i = –5 V, the JFET is operating at point *A*, where

$$I_D \approx 0 \quad \text{and} \quad V_O \approx V_{DD} = 20 \text{ V}$$

This corresponds to the switch in the OFF state.

(b) When V_i = 0 V, the JFET is operating at point *B*, where $I_D \approx$ 3.8 mA and V_O = 1 V

This corresponds to the switch in the ON state.

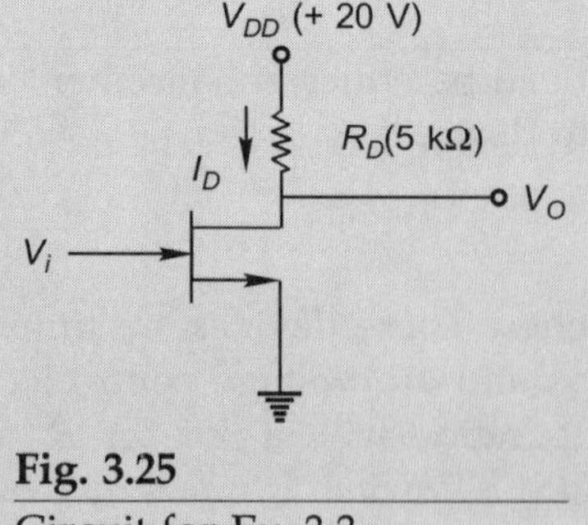

Fig. 3.25
Circuit for Ex. 3.3

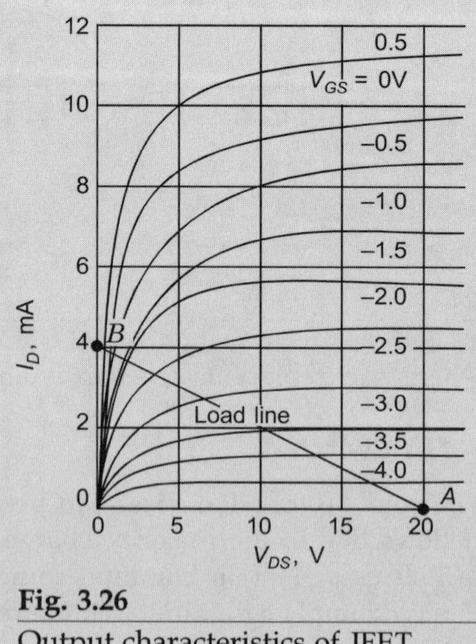

Fig. 3.26
Output characteristics of JFET

Example 3.4 For the circuit shown in Fig. 3.27, determine V_O for V_i (a) 0 V (b) 5 V. The output characteristics of the MOSFET are given in Fig. 3.28.

Solution
The load line is shown in Fig. 3.28.

(a) When V_i = 0, the transistor is cut-off because the voltage between the gate and source is below the threshold voltage. Correspondingly, the output voltage V_O = 5 V (Point *N*).

(b) When V_i = 5 V, the transistor is operating at point *M* and $V_O \approx 0$ V. This corresponds to ON state.

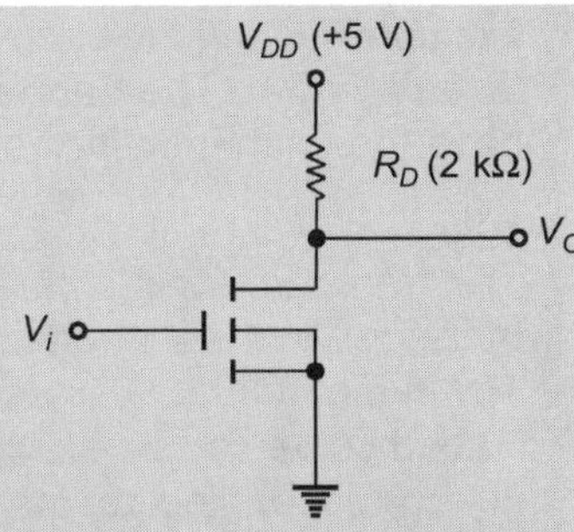

Fig. 3.27
Circuit for Ex. 3.4

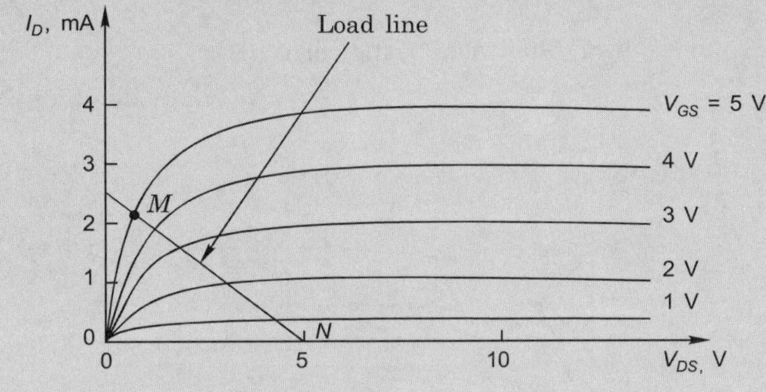

Fig. 3.28

Example 3.5 Obtain the V_{DS} vs. I_D characteristic for the enhancement MOSFET connected as shown in Fig. 3.29. The output characteristics of the MOSFET are given in Fig. 3.28.

Solution Here, $V_{DS} = V_{GS}$
On the output characteristics, the locus of all points (curve *MN*) with $V_{DS} = V_{GS}$ is drawn as shown in Fig. 3.30. Curve *MN* indicates that the transistor in this connection acts like a non-linear resistor.

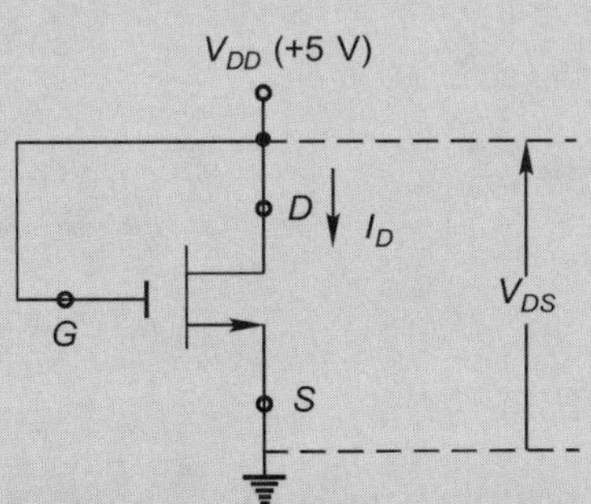

Fig. 3.29
Circuit for Ex. 3.5

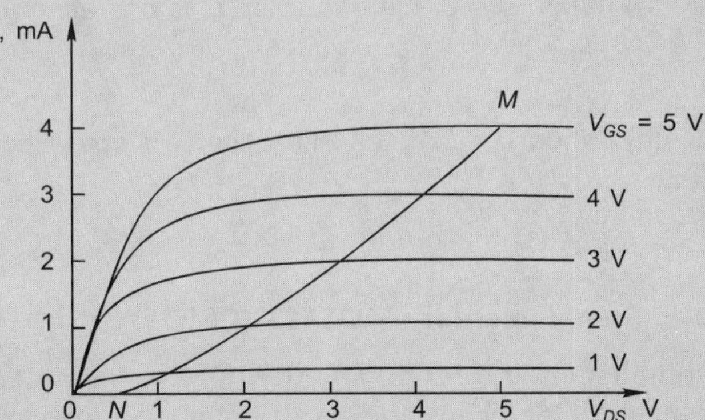

Fig. 3.30

Example 3.6 In the circuit of Fig. 3.31, find V_O for (a) $V_i = 0$, (b) $V_i = 5$ V. The transistors T_1 and T_2 are identical and have the characteristics shown in Fig. 3.28.

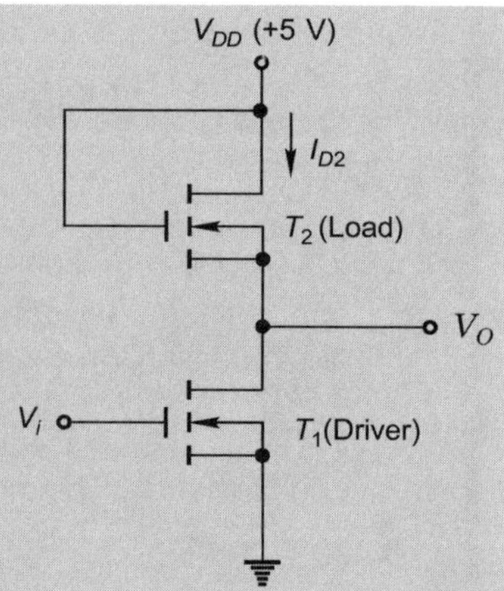

Fig. 3.31
Circuit for Ex. 3.6

Solution

The transistor T_2 is connected as shown in Fig. 3.29 and therefore acts as a non-linear resistor as shown by curve *MN* in Fig. 3.30.

Load curve can be drawn on the characteristics V_{DS1} vs. I_{D1} corresponding to the curve *MN* using the following relations.

$$V_{DS1} = V_{DD} - V_{DS2}$$

and $$I_{D1} = I_{D2}$$

The load curve thus drawn is indicated by the curve *AB* in Fig. 3.32.

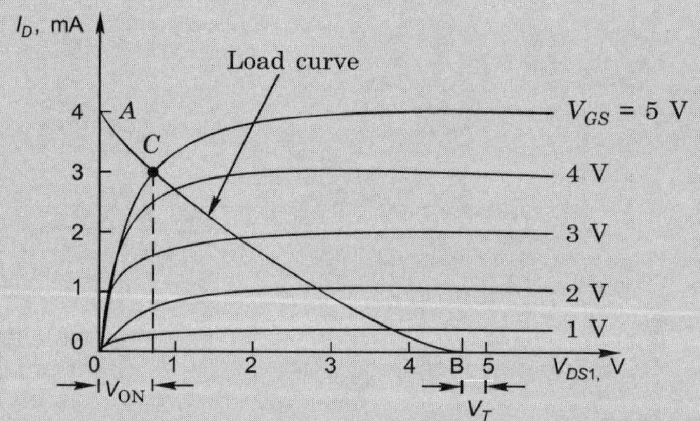

Fig. 3.32

(a) When $V_i = 0$, the transistor T_1 is operating at point *B*.

Here,
$$V_O = V_{DD} - V_T$$
$$\approx V_{DD} = 5 \text{ V}$$

(b) When $V_i = 5$ V, the transistor T_1 is operating at point *C*.

$$V_O = V_{ON}$$
$$\approx 0 \text{ V}$$

3.7.4 Complementary MOSFET (CMOS)

A complementary MOSFET (CMOS) is obtained by connecting a *p*-channel and an *n*-channel MOSFET in series, with drains tied together and the output is taken at the common drain point. Input is applied at the common gate connection formed by connecting the two gates together. A CMOS is shown in Fig. 3.33.

In this circuit, when $V_i = V_{CC}$, T_1 turns ON ($V_{GS1} > V_T$) and T_2 is OFF since $V_{GS2} = 0$ V. Therefore, $V_O \approx 0$ V and since the transistors are connected in series the current I_D will be the drain current of the OFF transistor T_2 which is negligibly small. On the other hand, when $V_i = 0$ V, T_1 turns OFF ($V_{GS1} < V_T$) and T_2 turns ON ($|V_{GS2}| > |V_T|$) giving an output voltage $V_O \approx V_{CC}$, and I_D is again very small being the drain current of the OFF transistor T_1. In either logic state, T_1 or T_2 is OFF and therefore, the quiscent power dissipation, which is the product of the OFF leakage current and V_{CC} is very small. Because of this advantage CMOS switches have become very popular in logic circuits.

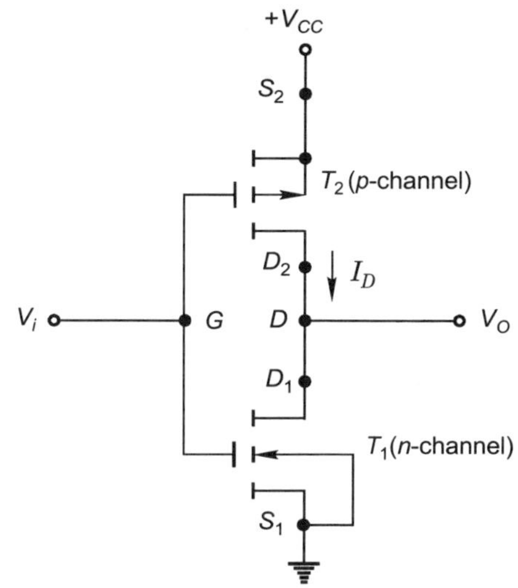

Fig. 3.33
A CMOS switch

3.8 SUMMARY

In this chapter, the principles of operation of semiconductor devices like *p-n* junction, BJT, FET, etc. have been discussed briefly. Their operation in the switching mode has been clearly explained in terms of the physical processes involved in the operation of these devices.

Semiconductor devices in the discrete form are no longer in use for complex digital systems but their study is essential for understanding the operation of ICs. Interfacing between ICs of different logic families usually requires discrete devices.

Glossary

Cut-off State of a transistor when its collector (or drain) current is zero.

Packaging density It is a measure of components that can be fabricated on a chip.

Relay An electromechanical switch.

Saturation State of a transistor when its collector (or drain) current is maximum and V_{CE} or $V_{DS} \approx 0$.

Speed of operation It is a measure of the speed at which a digital circuit operates. It is specified as propagation delay time.

Yield The percentage of acceptable ICs resulting from a manufacturing process.

Review Questions

3.1 ____________ is added in intrinsic semiconductor for making semiconductor devices.

3.2 The principle cause of propagation delay in a *p-n* junction is removal of ____________ charge carriers.

3.3 The speed of switching is ______ in *n-p-n* devices than *p-n-p* devices.

3.4 ______ configuration is the most commonly used configuration when a BJT is used as a switch.

3.5 The time required for a voltage waveform to change from 10% to 90% of its final value is known as ______.

3.6 A BJT operating in saturation has ______ charge stored in the base region.

3.7 A Schottky diode is connected between collector and base terminals of a BJT to prevent BJT from entering into ______ region.

3.8 Gate-to-source voltage must be ______ than the threshold voltage for an enhancement mode MOSFET to be cut-off.

3.9 Power dissipation is negligibly small in ______ devices.

3.10 The most commonly used devices for fabricating very large scale integrated circuits are ______ devices.

Problems

3.1 Give reasons for the following:
(a) The temperature coefficient of resistance of a semiconductor is negative while that of a metal is positive.
(b) A semiconductor behaves as an insulator at 0 K while it has some conductivity at room temperature.

3.2 (a) In a silicon *p-n* junction, calculate the increase in voltage across the diode if the forward current is doubled. Assume $V_T = 26$ mV.
(b) If the diode voltage in part (a) for the lower current is 700 mV, find the per cent change in diode voltage.

3.3 The voltage across a silicon diode at room temperature (300 K) is 0.7 V when 2 mA current flows through it. If the voltage increases to 0.75 V, calculate
(a) the diode current,
(b) per cent change in diode current.

3.4 If the current flowing through a *p-n* junction diode increases ten times, what is the increase in diode voltage? Assume forward-biased silicon diode operating at room temperature.

3.5 The time rate of change of excess minority charge carriers in a volume of semiconductor is given by $\frac{dQ}{dt} + \frac{Q}{\tau} = I$
where Q is the excess minority charge
τ is the storage time constant
I is the current flowing.
In the diode circuit of Fig. 3.6,

$$V_1 = 10 \text{ V}, \; V_2 = 5 \text{ V}, \; R = 10 \text{ k}\Omega,$$
$$V_d = 0.6 \text{ V}, \; \tau = 1 \; \mu\text{s}, \; T = 20 \; \mu\text{s}$$

(a) Calculate the excess minority charge stored in diode at $t = 20\ \mu$s.
(b) Calculate the time at which the diode will be turned-off.

(c) Assume a total capacitance (transition and external shunt capacitance) across the diode to be 10 pF. Draw the waveforms of diode voltage, current, and excess minority-charge.

3.6 (a) Find whether the transistor in Fig. 3.34 is conducting in the active region or in the saturation region. What is the region of operation for V_{CC}= 6 V?

(b) Find the value of R_C in Fig. 3.34 which will be just sufficient to drive the transistor into saturation. What happens if the value of R_C used is greater than the calculated value?

(c) Find the value of R_B which will be just sufficient to drive the transistor into saturation (R_C = 3 kΩ). What happens if R_B of value less than the calculated value is used in the circuit?

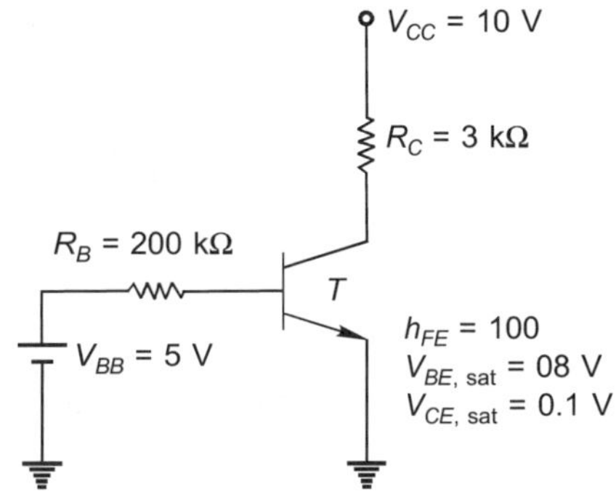

Fig. 3.34
Circuit for Problem 3.6.

3.7 For the transistor of Fig. 3.35 find the range of V_{BB} for the transistor to be

(a) in the cut-off region

(b) in the active region

(c) in the saturation region.

3.8 Find the minimum value of h_{FE} ($h_{FE(\min)}$) for the transistor of Fig. 3.36 to be in the saturation region.

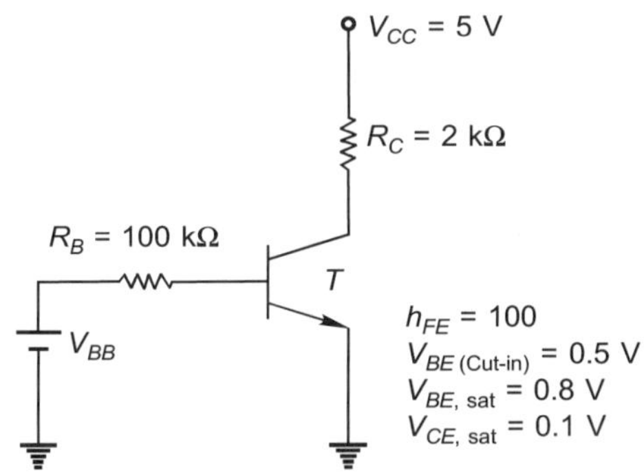

Fig. 3.35
Transistor circuit for Problem 3.7

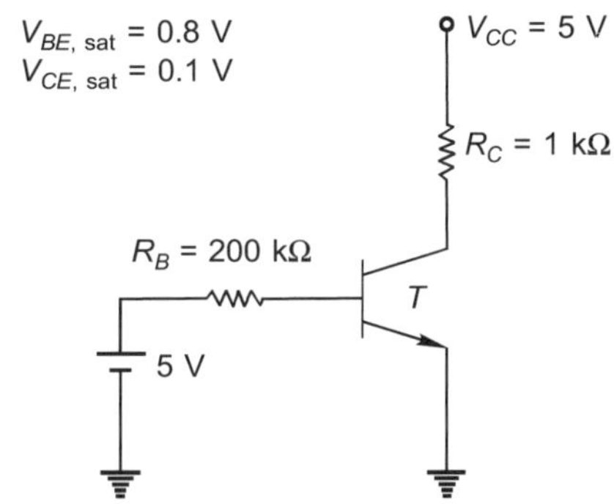

Fig. 3.36
Transistor circuit for Problem 3.8

3.9 For the circuit of Fig. 3.37, determine whether the transistor is in the active or saturation region. Calculate the currents I_C, I_E, and I_B.

3.10 Consider a transistor inverter circuit with a capacitor C connected across the resistor R_B. This capacitor helps in improving the switching speed of the transistor. Explain.

3.11 In the circuit of Fig. 3.38,

(a) determine, the minimum value of voltage V_i required for the load transistors to be in saturation.

(b) Calculate $V_i = V_O$ assuming the load transistors to be in saturation.

(c) Obtain the base current of T'_1 and T'_2.

3.12 In the circuit of Fig. 3.39, the outputs of the two inverters are connected together. Find the voltage at Y when

(a) $V_1 = V_2 = 0$ V (T_1 and T_2 are cut-off)

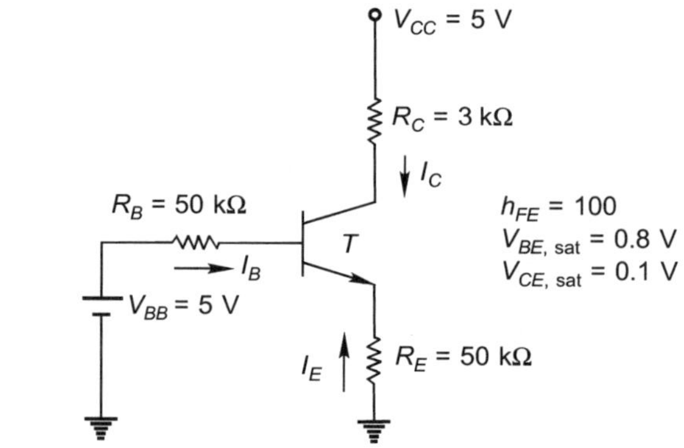

Fig. 3.37
Transistor circuit for Problem 3.9

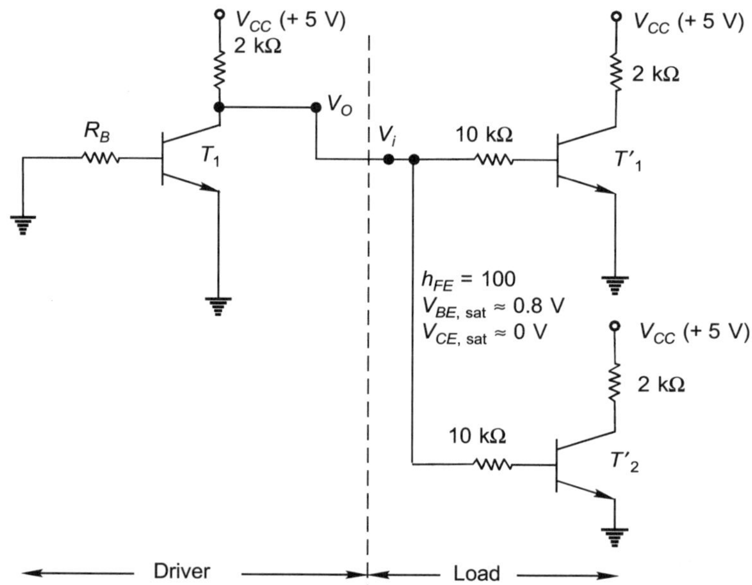

Fig. 3.38
Circuit for Problem 3.11

(b) $V_1 = V_2 = 5$ V (T_1 and T_2 are in saturation)
(c) One of the transistors is cut-off and the other one is in saturation.
(d) Find the function performed by this circuit with V_1, V_2 as inputs and Y as output.

3.13 In the circuit shown in Fig. 3.40, find the state of the transistor, when
(a) S_1 and S_2 are open.
(b) S_1 is closed and S_2 is open.
(c) S_1 and S_2 are closed.

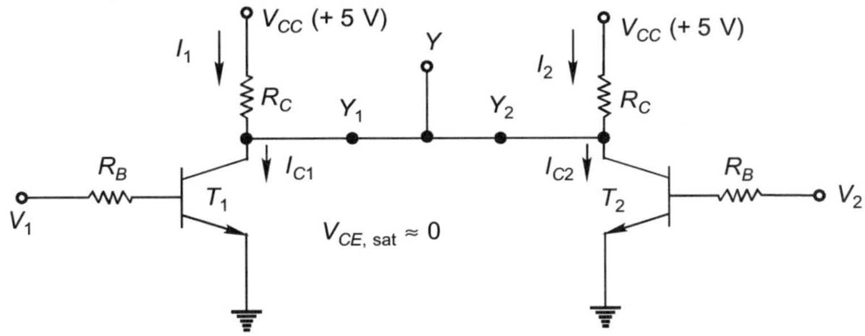

Fig. 3.39
Circuit for Problem 3.12

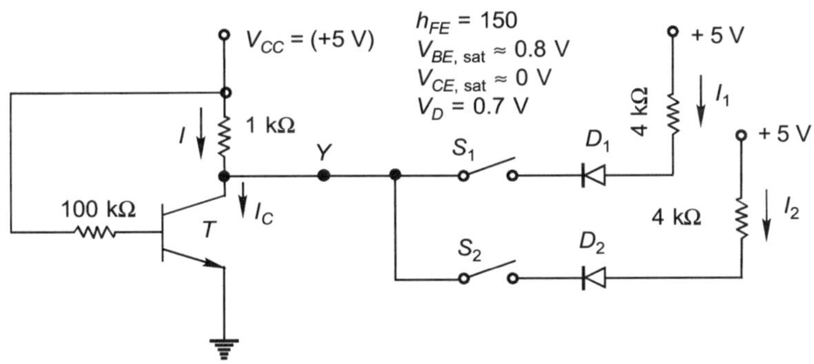

Fig. 3.40
Circuit for Problem 3.13

3.14 In the circuit of Fig. 3.38, will the load transistors be in saturation if the number of load transistors is increased from 2 to 100?

3.15 In the circuit of Fig. 3.39, let Y_1 and Y_2 be not connected. The base current of each transistor is 10 μA which is just sufficient to drive the transistor in saturation. Now, if Y_1 and Y_2 are connected together and T_1 is cut-off, will T_2 be driven into saturation? If not, why?

3.16 In the circuit of Fig. 3.39, determine the time constant with which the voltage at Y will be pulled-up when both the transistors are switched from saturation to cut-off. Assume C_O to be the capacitance at the output.

3.17 (a) Obtain the V_{DS} vs I_D characteristic for the depletion mode MOSFET connected as shown in Fig. 3.41*a*. The output characteristics of the transistor are shown in Fig. 3.41*b*.

(b) Find the operation of the circuit shown in Fig. 3.42. Use the characteristics given in Fig. 3.28 and Fig. 3.41*b* for T_1 and T_2, respectively.

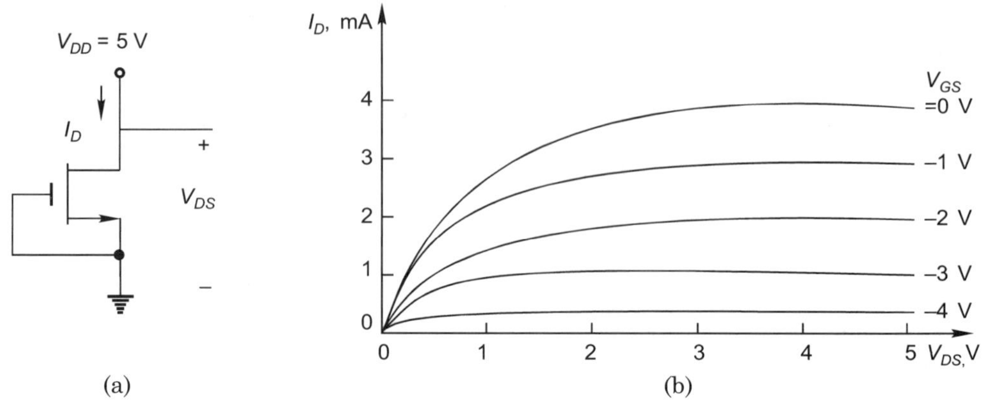

Fig. 3.41

Circuit and characteristics for Problem 3.17(a).

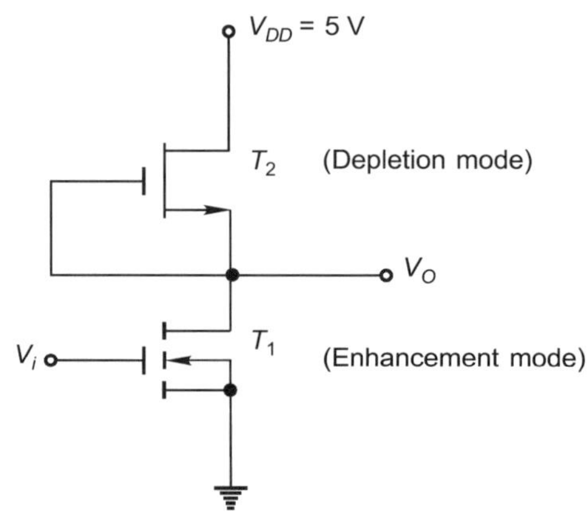

Fig. 3.42

Circuit for Problem 3.17(b).

4

DIGITAL LOGIC FAMILIES

4.1 INTRODUCTION

The switching characteristics of semiconductor devices have been discussed in Chapter 3. Basically, there are two types of semiconductor devices: bipolar and unipolar. Based on these devices, digital integrated circuits have been made which are commercially available. Various digital functions are being fabricated in a variety of forms using bipolar and unipolar technologies. A group of compatible ICs with the same logic levels and supply voltages for performing various logic functions have been fabricated using a specific circuit configuration which is referred to as a *logic family*.

4.1.1 Bipolar Logic Families

The main elements of a bipolar IC are resistors, diodes (which are also capacitors) and transistors. Basically, there are two types of operations in bipolar ICs:

1. Saturated, and
2. Non-saturated.

In saturated logic, the transistors in the IC are driven to saturation, whereas in the case of non-saturated logic, the transistors are not driven into saturation.

The saturated bipolar logic families are:

1. Resistor–transistor logic (RTL),
2. Direct–coupled transistor logic (DCTL),
3. Integrated–injection logic (I^2L),
4. Diode–transistor logic (DTL),
5. High–threshold logic (HTL), and
6. Transistor-transistor logic (TTL).

The non-saturated bipolar logic families are:

1. Schottky TTL, and
2. Emitter-coupled logic (ECL).

4.1.2 Unipolar Logic Families

MOS devices are unipolar devices and only MOSFETs are employed in MOS logic circuits. The MOS logic families are:

1. PMOS,
2. NMOS, and
3. CMOS

While in PMOS only *p*-channel MOSFETs are used and in NMOS only *n*-channel MOSFETs are used, in complementary MOS (CMOS), both *p*- and *n*-channel MOSFETs are employed and are fabricated on the same silicon chip.

All the above logic families are discussed in this chapter.

4.2 CHARACTERISTICS OF DIGITAL ICs

With the widespread use of ICs in digital systems and with the development of various technologies for the fabrication of ICs, it has become necessary to be familiar with the characteristics of IC logic families and their relative advantages and disadvantages. Digital ICs are classified either according to the complexity of the circuit, as the relative number of individual basic gates (2-input NAND gates) it would require to build the circuit to accomplish the same logic function or the number of components fabricated on the chip. The classification of digital ICs is given in Table 4.1.

Table 4.1 Classification of digital ICs

IC Classification	*Equivalent individual basic gates*	*Number of components*
Small-scale integration (SSI)	Less than 12	Up to 99
Medium-scale integration (MSI)	12–99	100–999
Large-scale integration (LSI)	100–999	1,000–9,999
Very large-scale integration (VLSI)	Above 1,000	Above 10,000

The various characteristics of digital ICs used to compare their performances are:

1. Speed of operation,
2. Power dissipation,
3. Figure of merit,
4. Fan-out,
5. Current and voltage parameters,
6. Noise immunity,
7. Operating temperature range,
8. Power supply requirements, and
9. Flexibilities available.

4.2.1 Speed of Operation

The speed of a digital circuit is specified in terms of the propagation delay time. The input and output waveforms of a logic gate are shown in Fig. 4.1. The delay times are measured between the 50 per cent voltage levels of input and output waveforms. There are two delay times: t_{pHL}, when the output goes from the HIGH state to the LOW state and t_{pLH}, corresponding to the output making a transition from the LOW state to the HIGH state. The propagation delay time of the logic gate is taken as the average of these two delay times.

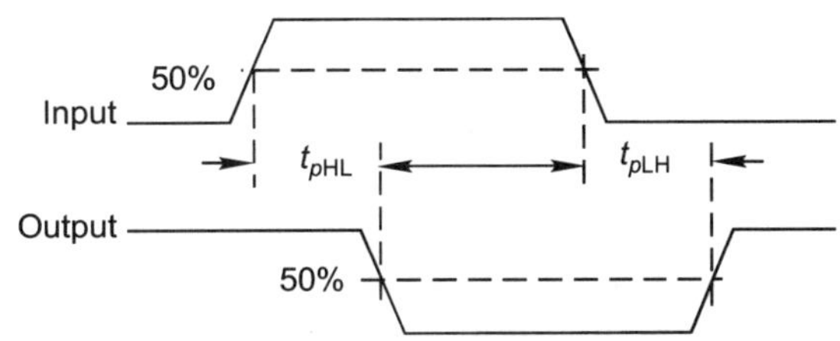

Fig. 4.1
Input and output voltage waveforms to define propagation delay times.

4.2.2 Power Dissipation

This is the amount of power dissipated in an IC. It is determined by the current, I_{CC}, that it draws from the V_{CC} supply, and is given by $V_{CC} \times I_{CC}$. I_{CC} is the average value of $I_{CC}(0)$ and $I_{CC}(1)$. This power is specified in milliwatts.

4.2.3 Figure of Merit

The figure of merit of a digital IC is defined as the product of speed and power. The speed is specified in terms of propagation delay time expressed in nanoseconds.

$$\text{Figure of merit} = \text{propagation delay time } (ns) \times \text{power (mW)}$$

It is specified in pico joules ($ns \times \text{mW} = \text{pJ}$)

A low value of speed-power product is desirable. In a digital circuit, if it is desired to have high speed, i.e. low propagation delay, then there is a corresponding increase in the power dissipation and vice-versa.

4.2.4 Fan-Out

This is the number of similar gates which can be driven by a gate. High fan-out is advantageous because it reduces the need for additional drivers to drive more gates.

4.2.5 Current and Voltage Parameters

The following currents and voltages are specified which are very useful in the design of digital systems.

High-level input voltage, V_{IH}: This is the minimum input voltage which is recognized by the gate as logic 1.

Low-level input voltage, V_{IL}: This is the maximum input voltage which is recognized by the gate as logic 0.

High-level output voltage, V_{OH}: This is the minimum voltage available at the output corresponding to logic 1.

Low-level output voltage, V_{OL}: This is the maximum voltage available at the output corresponding to logic 0.

High-level input current, I_{IH}: This is the minimum current which must be supplied by a driving source corresponding to 1 level voltage.

Low-level input current, I_{IL}: This is the minimum current which must be supplied by a driving source corresponding to 0 level voltage.

High-level output current, I_{OH}: This is the maximum current which the gate can sink in 1 level.

Low-level output current, I_{OL}: This is the maximum current which the gate can sink in 0 level.

High-level supply current, I_{CC} (1): This is the supply current when the output of the gate is at logic 1.

Low-level supply current, I_{CC} (0): This is the supply current when the output of the gate is at logic (0).

The current directions are illustrated in Fig. 4.2.

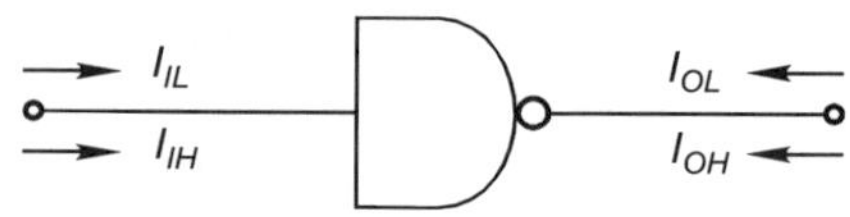

Fig. 4.2
A gate with current directions marked.

4.2.6 Noise Immunity

The input and output voltage levels defined above are shown in Fig. 4.3. Stray electric and magnetic fields may induce unwanted voltages, known as *noise*, on the connecting wires between logic circuits. This may cause the voltage at the input to a logic circuit to drop below V_{IH} or rise above V_{IL} and may produce undesired operation. The circuit's ability to tolerate noise signals is referred to as the *noise immunity*, a quantitative measure of which is called *noise margin*. Noise margins are illustrated in Fig. 4.3.

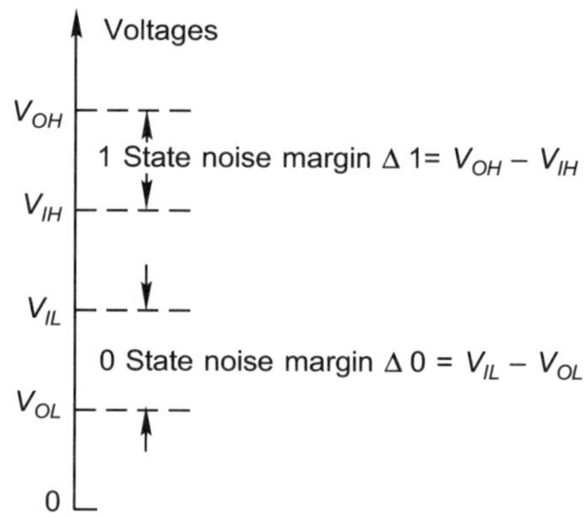

Fig. 4.3
Voltage levels and noise margins of ICs.

The noise margins defined above are referred to as *dc noise margins*. Strictly speaking, the noise is generally thought of as an a.c. signal with amplitude and pulse width. For high speed ICs, a pulse width of a few microseconds is extremely long in comparison to the propagation delay time of the circuit and therefore, may be treated as d.c. as far as the response of the logic circuit is concerned. As the noise pulse width decreases and approaches the propagation delay time of the circuit, the pulse duration is too short for the circuit to respond. Under this condition, a large pulse amplitude would be required to produce a change in the circuit output. This means that a logic circuit can effectively tolerate a large noise amplitude if the noise is of a very short duration. This is referred to as *ac noise margin* and is substantially greater than the dc noise margin. It is generally supplied by the manufacturers in the form of a curve between noise margin and noise pulse width.

4.2.7 Operating Temperature

The temperature range in which an IC functions properly must be known. The accepted temperature ranges are: 0 to + 70 °C for consumer and industrial applications and –55 °C to + 125 °C for military purposes.

4.2.8 Power Supply Requirements

The supply voltage(s) and the amount of power required by an IC are important characteristics required to choose the proper power supply.

4.2.9 Flexibilities Available

Various flexibilities are available in different IC logic families and these must be considered while selecting a logic family for a particular job. Some of the flexibilities available are:

1. *The breadth of the series:* Type of different logic functions available in the series.
2. *Popularity of the series:* The cost of manufacturing depends upon the number of ICs manufactured. When a large number of ICs of one type are manufactured, the cost per function will be very small and it will be easily available because of multiple sources.
3. *Wired-logic capability:* The outputs can be connected together to perform additional logic without any extra hardware.
4. *Availability of complement outputs:* This eliminates the need for additional inverters.
5. *Type of output:* Passive pull-up, active pull-up, open-collector/drain, and tristate. These will be explained in subsequent sections.

4.3 RESISTOR–TRANSISTOR LOGIC (RTL)

The resistor–transistor logic was the most popular form of logic in common use before the development of ICs. RTL circuits consist of resistors and transistors and was the earliest logic family to be integrated. Although RTL has become obsolete now, because of its simplicity and for historical reasons, it is proper to devote some attention to it and introduce some of the important concepts, useful for all types of gates, through this. The basic RTL gate is a NOR gate as shown in Fig. 4.4. For the sake of simplicity, a two-input NOR gate driving N similar gates is shown in the figure, which can be extended to accommodate a larger number of inputs. The number of input terminals is referred to as the *fan-in*.

4.3.1 Logic Operation

Inputs representing the logic levels are applied at A and B terminals. The voltage corresponding to LOW level should be low enough to drive the corresponding transistor to cut-off. Similarly, the input voltage corresponding to HIGH level should be high enough to drive the corresponding transistor to saturation.

If both the inputs are LOW, transistors T_1 and T_2 are cut-off and the output is HIGH. A HIGH level on any input will drive the corresponding transistor to saturation causing the output to go LOW. The LOW (0) level output voltage is $V_{CE,\text{sat}}$ of a transistor (~ 0.2 V) and the HIGH (1) level output voltage depends on the number of gates connected to the output. This causes the output voltage to be variable and is a deciding factor for the fan-out of the gate.

4.3.2 Loading Considerations

If all the inputs to the gate are LOW, the output is HIGH and if the gate is not driving any other gate, i.e. no load is connected, the output voltage will be slightly less than V_{CC} (there is voltage drop across the common collector resistor due to I_{CO} of T_1 and T_2).

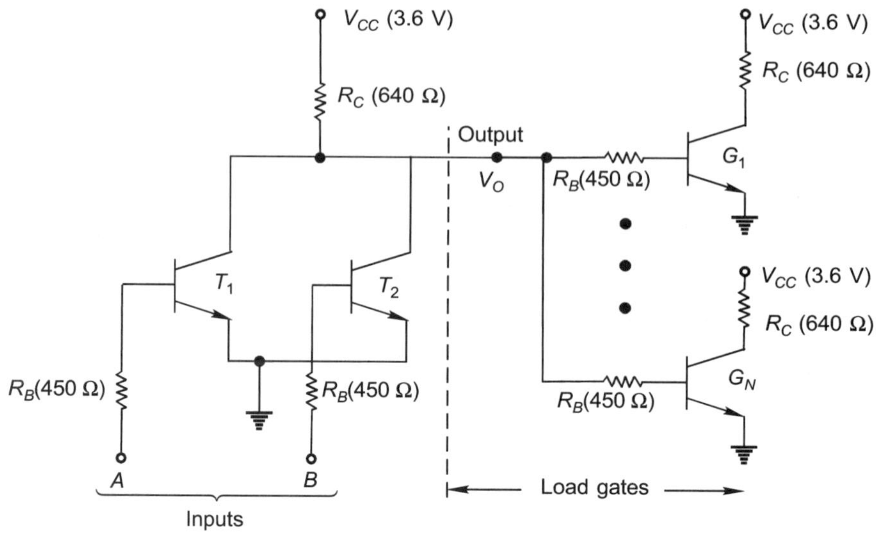

Fig. 4.4
A 2-input RTL NOR gate driving N similar gates.

When N similar gates are being driven, the load will be equivalent to a resistor of value $450/N$ ohms in series with a voltage source of 0.8 V (being the voltage between base and emitter of a transistor in saturation). The relevant portion of the circuit is shown in Fig. 4.5.

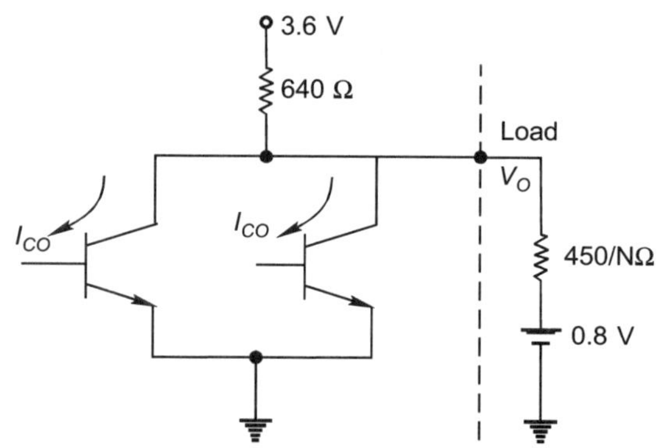

Fig. 4.5
A circuit illustrating the equivalent circuit at the input of the load gates.

The base current for each load transistor is

$$I_B = \left(\frac{3.6 - 0.8}{640 + \dfrac{450}{N}}\right) \cdot \frac{1}{N} = \frac{2.8}{640N + 450} \qquad (4.1)$$

The collector current for the load transistor in saturation is

$$I_{C,\text{sat}} = \frac{3.6 - 0.2}{640} = 5.31 \text{ mA} \qquad (4.2)$$

The value of N must satisfy the following relation,

$$h_{FE} \cdot I_B \geq I_{C,\text{sat}} \qquad (4.3)$$

For $N = 5$, $I_B = 0.767$ mA. Therefore, h_{FE} must be greater than 7.

4.3.3 Noise Margins

When the output is in 0 state $V_O = 0.2$ V. If this voltage becomes about 0.5 V (cut-in voltage of transistor), the load transistor comes to conduction which causes malfunction of the circuit. Hence, the logic 0 noise margin $\Delta 0 \approx 0.3$ V.

The logic 1 noise margin depends upon the number of gates being driven. For $N = 5$,

$$V_O = \frac{90}{90+640}\times(3.6) + \frac{640}{90+640}\times(0.8) = 1.14\text{ V} \tag{4.4}$$

For $h_{FE} = 10$, the total base current required for load transistors to be driven into saturation will be $5 \times \left(\frac{5.31}{10}\right)$ mA and the corresponding V_O must be 1.04 V. Therefore, the noise margin for 1 level is $\Delta 1 = 1.14 - 1.04 = 0.1$ V.

4.3.4 Propagation Delay Time

The propagation delay time is also affected by the number of gates it drives. When the output of the gate is in LOW state all the load transistors are cut-off and the base–emitter junction of each of these transistors appears to be a capacitor, C. When the output has to change from LOW to HIGH level due to changes at the input, it will do so with a time constant given by

$$\left(640 + \frac{450}{N}\right) NC = (640N + 450)\ C \tag{4.5}$$

The resistance in the collector circuit pulls up the output voltage from LOW to HIGH level and hence is known as the *pull-up resistor.* It is passive pull-up in this case in contrast to an *active pull-up* which can be used to decrease the propagation delay time. Active pull-up will be discussed later.

4.3.5 Current Source Logic

The gate supplies current to the load transistors when in 1 level, whereas the leakage-current (reverse-saturation base current) of load transistors flow through T_1 or T_2 in 0 level. Since the source current is much greater than the sink current, it is known as *current source logic.*

4.3.6 Wired-Logic

If the outputs of the gates are connected together as shown in Fig. 4.6, the output Y is given by

$$Y = Y_1 \cdot Y_2$$

$$= \overline{A+B}\cdot\overline{C+D}$$

$$= \overline{A+B+C+D}$$

This shows that fan-in can be increased by this connection which is referred to as *wired-* AND or *implied*-AND. The effect of this connection on fan-out, power dissipation and speed of operation can be seen in Prob. 4.3.

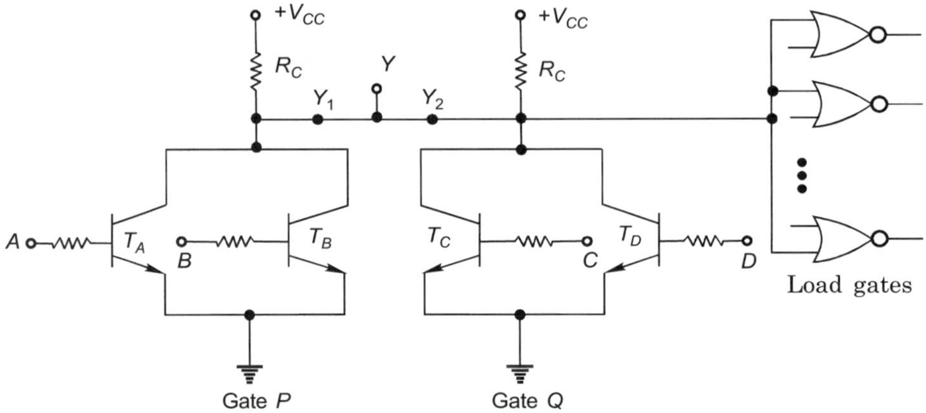

Fig. 4.6
Wired-AND connection of RTL gates driving similar gates

The characteristics of RTL can be summarized as: Poor noise margin, poor fan-out capabilities, low speed, and high power dissipation.

4.4 DIRECT-COUPLED TRANSISTOR LOGIC (DCTL)

In the RTL gate of Fig. 4.4, if the base resistors R_B are omitted, we obtain what is known as the direct-coupled transistor logic (DCTL) gate, in which the inputs are directly coupled to the bases. This circuit performs positive NOR logic and the voltages corresponding to logic 1 and 0 levels are $V_{BE,sat}$ (~ 0.8 V) and $V_{CE,\ sat}$ (~ 0.2 V) respectively. The separation between the logic 1 and 0 level voltages, which is referred to as the *logic swing*, is very small ($V_{BE,sat} - V_{CE,sat}$ = 0.6 V). Therefore, the noise margin of this circuit is very poor.

Although the DCTL is simpler than RTL, it never became popular because of the problem of *current hogging*. The gate should be able to drive the transistors of the load gates to saturation corresponding to logic level 1.

This does not pose any problem if all the transistors have same input characteristics but, unfortunately, the input characteristics differ due to the manufacturing tolerances of different IC packages operating at different temperatures. Owing to these differences, the saturation voltages of the load transistors may be different. Let the base–emitter voltages of the transistors corresponding to saturation be 0.78, 0.79, and 0.80 V. The transistor with the base–emitter voltage of 0.78 V, when it enters saturation, will not allow other transistors to enter saturation and will take whole of the current supplied from the driver gate. This is known as *current hogging*.

4.5 INTEGRATED-INJECTION LOGIC (I²L)

As discussed above, the DCTL suffers from the difficulty of current hogging which makes it unsuitable. However, based on DCTL a new logic referred to as the integrated-injection logic

(I^2L), has been developed. I^2L has the simplicity of DCTL, uses very small silicon chip area, consumes very little power, and requires only four masks and two diffusions (compared to five masks and three diffusions for BJT) and hence, is easier and cheaper to fabricate. Due to these advantages it is eminently suited for medium- and large-scale integration. It is not used for small-scale integration and is the only saturated bipolar logic employed for large-scale integration. Texas Instruments SBP 9900 is a 16-bit microprocessor using I^2L technology.

The genesis of I^2L technology is the concept of merging the components, viz. one semiconductor region is part of two or more devices. Because of this type of merging it is also referred to as the *merged-transistor logic* (MTL). There is considerable saving in the silicon chip area in this process.

4.5.1 I^2L Inverter

The basic operation of I^2L is explained with the help of the inverter circuit shown in Fig. 4.7. If the input V_i is at LOW logic level ($V_i \approx 0$), T_1 is OFF so that $I_{B1} = 0$. The input source acts as a sink for the current I_1. Therefore, I_2 flows through the base of T_2 driving it to saturation. When T_1 is OFF and T_2 is ON, $V_{BE2} = V_{CE1} \approx 0.8$ V.

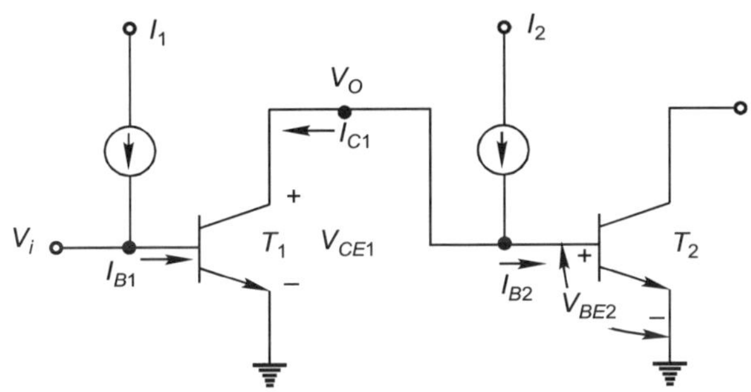

Fig. 4.7
An I^2L inverter directly coupled to the following stage.

On the other hand, if the input is at HIGH logic level ($V_1 \approx 0.8$ V), the base current I_{B1} will have two components, one of them being I_1 and the other is due to the source V_i, and consequently T_1 saturates. Therefore, $V_{CE1} = V_{CE,\text{sat}} \approx 0.2$ V, which drives T_2 to cut-off and T_1 acts as a sink for I_2. This shows that the logic level at V_O is complement to that of V_i, viz. T_1 acts as an inverter. The logic swing is about 0.6 V.

4.5.2 I^2L Configuration

Consider the DCTL gate structure shown in Fig. 4.8 in which there are two logical variables which are assumed to be outputs of similar DCTL gates and we need to generate the functions

$\overline{A+B}, \overline{A+\bar{B}}, \overline{\bar{A}+B}$, and $\overline{\bar{A}+\bar{B}}$.

We observe from the figure that the bases of transistors T_1, T_2, and T_4 are connected together, also their emitters are connected together (grounded). Therefore, the combination of T_1, T_2, and T_4 can be replaced with single transistor having one base, one emitter and three

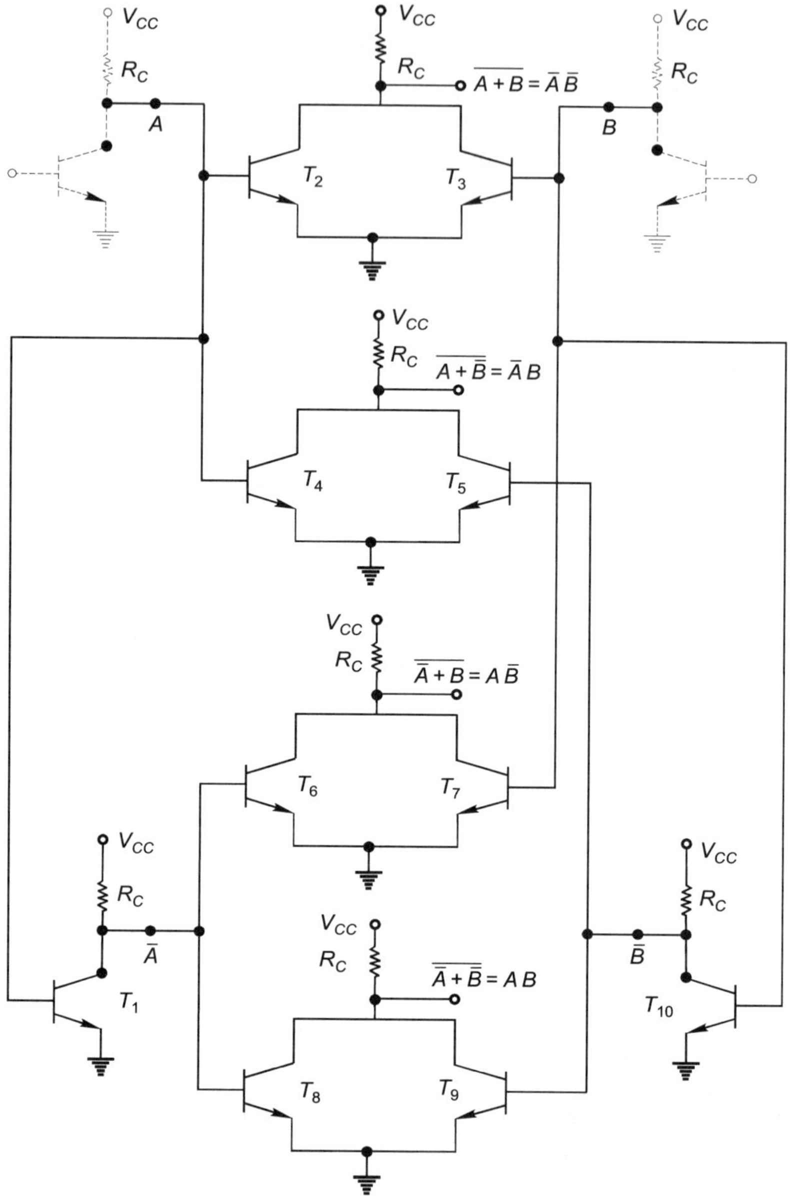

Fig. 4.8
A DCTL gate structure for generating functions of two logical variables.

collectors. Similarly, other transistors with common-bases are replaced with multiple-collector transistors. Using this concept, Fig. 4.8 is redrawn as shown in Fig. 4.9.

As shown in Fig. 4.7, a mechanism for supplying base currents is required. To achieve this, the collector resistors of driving gates (shown dotted in Fig. 4.8) are treated as the base resistors

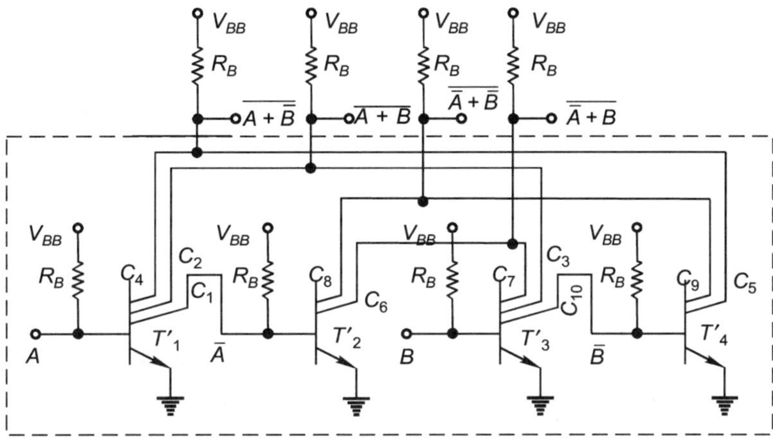

Fig. 4.9
Figure 4.8 redrawn with multiple-collector transistors.

of multiple-collector transistors (T'_1 and T'_3). Similarly, the collector resistors of T_1 and T_{10} are treated as the base resistors of T'_2 and T'_4, respectively. Correspondingly, the supply voltages are indicated as V_{BB}. The portion of the circuit shown outside the dotted box is either a part of other gates driven by the outputs shown or is omitted altogether. This means that an I^{2}L circuit has open-collector outputs, which either feed another I^{2}L circuit or are to be connected to the supply voltage through resistors. Suitable values of supply voltage and resistor are to be used for getting proper output voltage levels, for driving other gates such as TTL.

4.5.3 Fabrication of I^{2}L

The resistor R_B required to inject the base current would require a large silicon area if fabricated on the chip and thus, would render the circuit useless for LSI applications. It can be eliminated by replacing it with a current source. The grounded-base *p-n-p* transistor shown in Fig. 4.10 acts as a current source, which is referred to as a *current injector.* The resistor R_X is external to the chip and the current I_O is given by

$$I_O = \frac{V_{CC} - V_{EB}}{R_X} \tag{4.6}$$

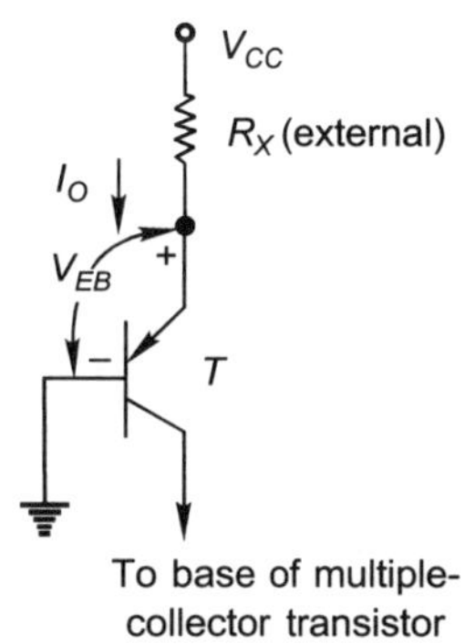

Fig. 4.10
A current injector for I^{2}L

The collector of the current injector transistor T of Fig. 4.10 and the base of the multiple-collector transistor are merged, viz. one *p* region serves both as collector of *p-n-p* transistor and base of *n-p-n* transistor. Similarly, the base of T is merged with the emitter of the multiple-collector transistor. A simplified physical structure of a portion of I^{2}L circuit is shown in Fig. 4.11. This shows the simplicity of I^{2}L structure.

The speed of operation of I^{2}L depends upon the charging current. The propagation delay time is inversely proportional to the charging current, also the power dissipation is proportional

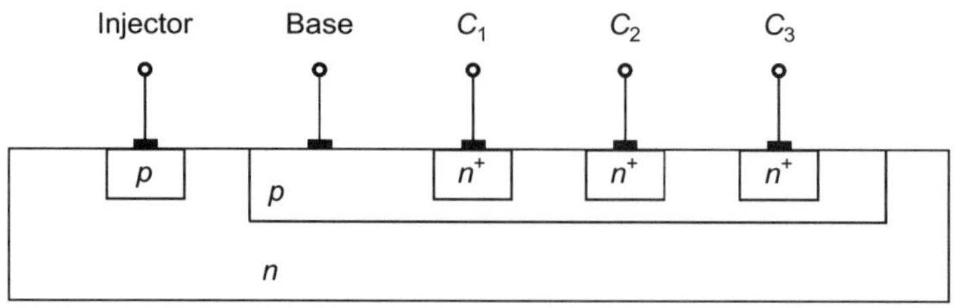

Fig. 4.11
The simplified physical structure of a portion of I^2L.

to the charging current, therefore we have to trade-off between power dissipation and speed. The figure of merit is independent of I_O and it is in the range of 0.1 to 0.7 p*J*. The silicon area required is very small and packing density in the range 120 to 200 gates per square millimetre have been realized.

4.6 DIODE–TRANSISTOR LOGIC (DTL)

The diode–transistor logic is somewhat more complex than RTL but because of its greater fan-out and improved noise margins it has replaced RTL. Its main disadvantage is slower speed and because of this it was modified and emerged as transistor– transistor logic (TTL) which is the most popular logic family today, as far as small- and medium-scale ICs are concerned. Although TTL has completely replaced DTL, for historical reasons as well as for better appreciation of TTL circuit, it is worthwhile discussing the details of DTL.

DTL circuit using discrete components was made using input diodes and a transistor inverter (NOT), which was modified for integrated circuit implementation as shown in Fig. 4.12.

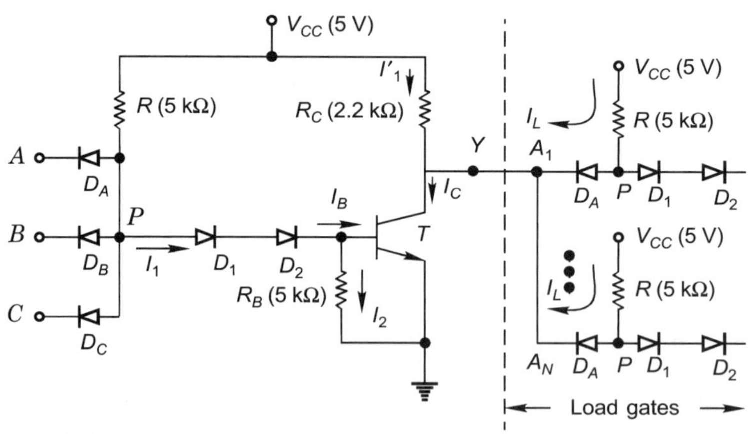

Fig. 4.12
A 3-input DTL NAND gate driving *N* similar gates.

4.6.1 Operation of DTL NAND Gate

The basic DTL gate is a NAND gate. A 3-input NAND gate driving N similar gates is shown in Fig. 4.12. The input diodes D_A, D_B, and D_C conduct through the resistor R, if the corresponding input is in the LOW state, while corresponding to HIGH state the diode is nonconducting. Therefore, if at least one of the inputs is LOW, the diode connected to this input conducts and the voltage V_P at point P is one diode drop above the low level voltage at the input. The voltage V_P should be such as to keep T in cut-off. Therefore, the output of T is V_{CC}. On the other hand, if all the three inputs are in HIGH state, the input diodes are cut-off and consequently current flowing from V_{CC} through R should be sufficient to drive T in saturation. Therefore, the output of T is $V_{CE,sat}$.

If we consider the voltages corresponding to logic 1 and 0 as V_{CC} and $V_{CE,sat}$ respectively, this circuit performs NAND operation. The following example illustrates the loading (fan-out) considerations and the noise-margins.

Example 4.1 For the DTL NAND gate of Fig. 4.12 calculate (*a*) fan-out (*b*) noise-margins, and (*c*) average power, P, dissipated by the gate. The diode and transistor parameters are:

Diode:	Voltage across a conducting diode	= 0.7 V
	Cut-in voltage V_γ	= 0.6 V
Transistor:	Cut-in voltage V_γ	= 0.5 V
	$V_{BE,sat}$	= 0.8 V
	$V_{CE,sat}$	= 0.2 V
	h_{FE}	= 30

Solution

(a) As discussed above, the logic levels are:

$$\text{LOW level} = V(0) = V_{CE,sat} = 0.2 \text{ V}$$
$$\text{HIGH level} = V(1) = V_{CC} = 5 \text{ V}$$

(*i*) If all the inputs are HIGH, the input diodes are reverse-biased. Assuming diodes D_1, D_2 to be conducting and T to be in saturation, the voltage $V_p = 0.7 + 0.7 + 0.8 = 2.2$ V.

Writing Kirchhoff's current law (KCL) equation at the base of T,

$$I_B = I_1 - I_2$$

where
$$I_1 = \frac{V_{CC} - V_P}{R} = \frac{5 - 2.2}{5} = 0.56 \text{ mA}$$

and
$$I_2 = \frac{V_{BE,sat}}{R_B} = \frac{0.8}{5} = 0.16 \text{ mA}$$

which gives a base current $I_B = 0.4$ mA. The collector current (without load gates connected) is

$$I_C = \frac{V_{CC} - V_{CE,sat}}{R_C} = \frac{5 - 0.2}{2.2} = 2.182 \text{ mA}.$$

Since $h_{FE} \times I_B = 30 \times 0.4 = 12$ mA is greater than I_C (2.182 mA), it is confirmed that the transistor is in saturation and the output is in LOW state. Now, if N load gates are fed from this gate, the input diodes of the driven gates will conduct through the output transistor T, i.e. T acts as a sink for the current in the input to the gates it drives. Assuming that all the other inputs to each of the load gates are HIGH except the one driven by T, the current $I_L = \frac{V_{CC} - V_P}{R} = \frac{5 - 0.9}{5}$ = 0.82 mA. This current is referred to as *standard load*. The fan-out is given by I_C £ $h_{FE} I_B$, or $0.82 N + 2.182 \leq 12$ mA or $N < 12$ since N must be an integer. A conservative choice is $N = 10$. The Maximum collector current rating of T must be about 12 mA.

(ii) If at least one of the inputs is LOW, the corresponding input diode conducts and $V_P = 0.2 + 0.7 = 0.9$ V. The minimum voltage required for D_1, D_2, and T to be conducting is 0.6 + 0.6 + 0.5 = 1.7 V, which confirms that D_1, D_2 are nonconducting and hence T is cut-off. Consequently, the output voltage is V_{CC} (5 V) if the load gates are not connected.

If the load gates are connected, the input diodes of the load gates are nonconducting, which means the reverse-saturation current of these diodes must be supplied through the collector resistor R_C, which will produce a voltage drop across R_C and consequently the output voltage corresponding to HIGH state will be a little less than V_{CC}. The maximum current which can be supplied by the gate will depend upon V_{OH}. The fan-out is determined on the basis of maximum current.

(b) (*i*) If all the inputs are HIGH, the output is LOW. Since $V_P = 2.2$ V, the input diodes are reverse-biased by 5 – 2.2 = 2.8 V. Since the cut-in voltage of the diode is 0.6 V, a negative noise spike of at least 3.4 V present at the input will cause malfunction of the circuit, i.e. the 0 level noise margin Δ 0 = 3.4 V.

(*ii*) If at least one input is LOW, the output is HIGH. Since $V_P = 0.9$ V and a voltage of at least 1.7 V [part a (*ii*)] is required for D_1, D_2, and T to conduct, therefore a positive noise spike of at least 0.8 V will cause malfunction of the circuit, i.e. the 1 level noise margin Δ 1 = 0.8 V.

(c) The power $P(0)$ when the output is LOW is given by $P(0) = V_{CC} (I_1 + I'_1) = 5\ (0.56 + 2.182) = 13.71$ mW. When the output is in the HIGH state at least one of the input diodes conduct. Therefore, $I_1 = 0.82$ mA and $I'_1 = 0$. Hence $P(1) = (0.82)\ (5) = 4.1$ mW.

If we assume that the occurrence of LOW and HIGH is equally likely then the average power is

$$P_{av} = \frac{P(0) + P(1)}{2} = \frac{13.71 + 4.1}{2} = 8.905 \text{ mW}$$

4.6.2 Propagation Delays

Delays are associated with the turning-on (*turn-on delay*) and the turning-off (*turn-off delay*) of the output transistor. While turning on, any capacitance shunting the output of the gate discharges rapidly through the low impedance of the output transistor in saturation. On the other hand, at turn-off the shunt capacitor must charge through the pull-up resistor R_C in addition to the storage time delay. The turn-off delay is considerably larger than the turn-on delay, often by a factor of 2 or 3. The propagation delay time of commercially available DTL gates are of the order of 30 to 80 ns.

4.6.3 Current Sink Logic

This gate supplies the reverse-saturation current of input diodes of the load gates in 1 state and sinks the current flowing through the forward-biased input diodes of the load gates in the output transistor of the gate in 0 state. Since the sink current is much greater than the source current, this is known as *current sink logic.*

4.6.4 Wired-Logic

If the outputs of gates are connected together as shown in Fig. 4.13, additional logic is performed without additional hardware. This type of connection is referred to as *wired-logic*, *wired-*AND, or *implied-*AND.

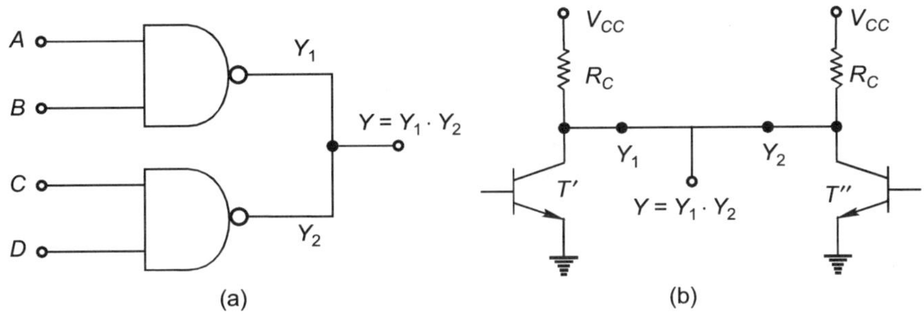

Fig. 4.13
The wired-AND connection of DTL gates

If $Y_1 = Y_2 = 1$, then $Y = 1$, whereas if any one (Y_1 or Y_2) or both are 0, then $Y = 0$. The output is

$$Y = Y_1\ Y_2 = (\overline{AB}) \cdot (\overline{CD}) = \overline{AB + CD} \tag{4.7}$$

Let us consider the effect of wired-AND connection on power dissipation, speed, and fan-out. The power dissipation in LOW output state $P(0)$ increases because of reduction in effective collector resistor ($R_C \parallel R_C = R_C/2$). Consequently, the speed of operation increases due to reduction in charging resistor ($R_C/2$).

There is an effective reduction in the fan-out of the gate in the wired-AND connection. If only one output transistor (say T') is conducting, then this transistor must not only sink the current of the load gates and the current due to its own pull-up resistor but must also sink the current in the pull-up resistor of the other output transistor T''. This situation makes it necessary to reduce the allowable fan-out of each gate in the wired-AND connection.

4.6.5 Modified Integrated DTL NAND Gate

From Ex. 4.1 we note that the fan-out may be increased by increasing the base current of the output transistor. This can be done by replacing D_1 by a transistor T_1, as illustrated in Fig. 4.14.

The circuit can be analysed in a similar way as in Ex. 4.1 (Prob. 4.9). Its fan-out is considerably higher than that of the circuit of Fig. 4.12.

4.7 HIGH-THRESHOLD LOGIC (HTL)

Due to the presence of electric motors, on–off control circuits, high voltage switches, etc. in an industrial environment, the noise level is quite high and the logic families discussed so far do

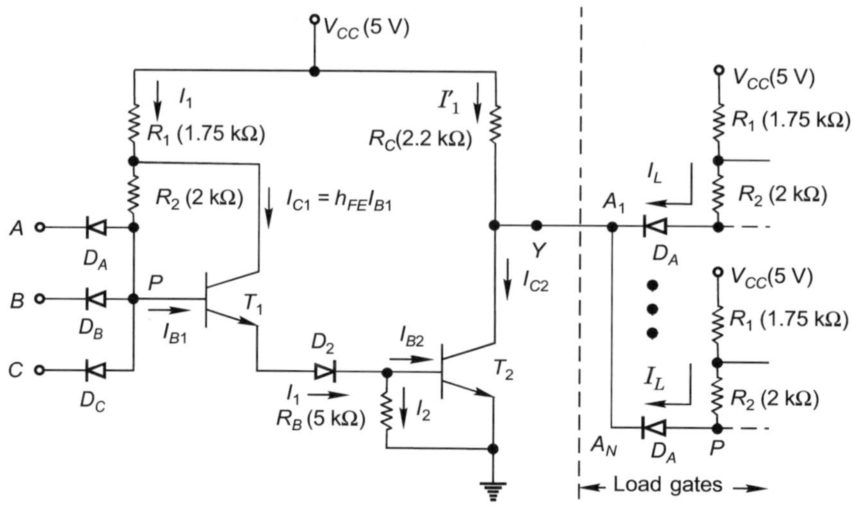

Fig. 4.14
A modified integrated 3-input DTL NAND gate driving N similar gates.

not perform the intended functions. For this purpose, the DTL gate of Fig. 4.14 has been redesigned with a higher supply voltage (15 V instead of 5 V). The diode D_2 has been replaced by a Zener diode with a Zener breakdown voltage of 6.9 V and the resistances have been modified so that approximately the same currents are obtained as in DTL. A 3-input HTL NAND gate with a fan-out of N is shown in Fig. 4.15. The circuit can be analysed to determine the noise-margins, fan-out and power dissipation (Prob. 4.10).

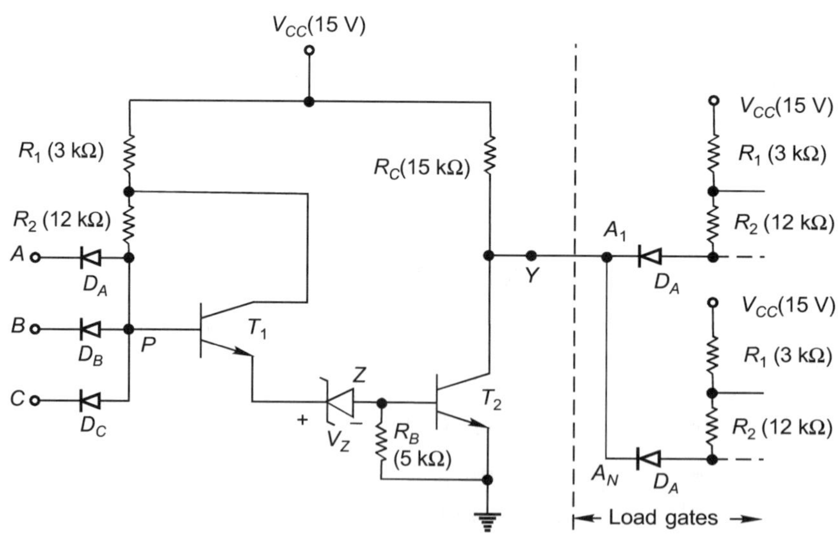

Fig. 4.15
A 3-input HTL NAND gate driving N similar gates.

The propagation delay time is adversely affected due to large resistance values. It is as high as hundreds of nano-seconds. The temperature sensitivity of the HTL gate is considerably less than that of DTL (Prob. 4.12).

4.8 TRANSISTOR – TRANSISTOR LOGIC (TTL)

Because of its speed limitations, DTL has become outdated and is completely replaced by another logic family referred to as transistor – transistor logic (TTL). The main cause for the speed limitation in DTL is the slow process of removal of stored base charge of the output transistor. For example, in the DTL gate of Fig. 4.12, when T goes from saturation to cut-off, the diodes D_1 and D_2 are nonconducting and hence, the base charge must leak-off through the resistor R_B, which is a relatively slow mechanism.

The DTL speed limitation is overcome by making the following modifications in the circuit of Fig. 4.12:

1. The input diodes D_A, D_B, and D_C are replaced by emitter–base junctions of multiple-emitter transistor (T_1), which is easily and economically fabricated in IC.
2. The collector – base junction of T_1 acts as the diode D_1.
3. The diode D_2 is replaced by emitter – base junction of another transistor (T_2).

The modified circuit is known as TTL and is shown in Fig. 4.16.

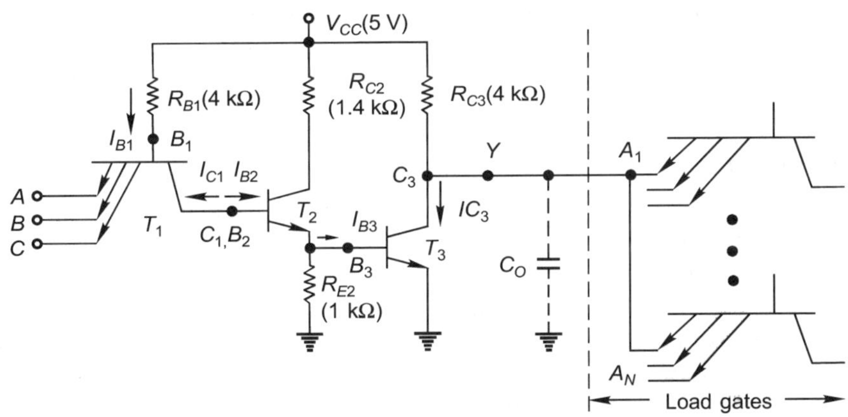

Fig. 4.16
A 3-input TTL NAND gate driving N similar gates.

4.8.1 Operation of TTL NAND Gate

The operation of the TTL gate of Fig. 4.16 is similar to the operation of the DTL gate of Fig. 4.12 as far as the steady-state operation is concerned, as is evident from conditions I and II discussed below. It is condition III that differentiates the operation of TTL from that of DTL and makes it the fastest of all saturating logic families.

For the operation discussed below, we assume that the load gates are not present and the voltages for logic 0 and 1 are $V_{CE,\,sat} \approx 0.2$ V and $V_{CC} = 5$ V respectively.

Condition I At least one input is LOW. The emitter–base junction of T_1 corresponding to the input in the LOW state is forward-biased making voltage at B_1, $V_{B1} = 0.2 + 0.7 = 0.9$ V. For base–collector junction of T_1 to be forward-biased, and for T_2 and T_3 to be conducting, V_{B1} is required to be at least $0.6 + 0.5 + 0.5 = 1.6$ V. Hence, T_2 and T_3 are OFF.

Since T_3 is OFF, therefore $Y = V(1) = V_{CC}$.

Condition II All inputs are HIGH. The emitter–base junctions of T_1 are reverse-biased. If we assume that T_2 and T_3 are ON, then $V_{B2} = V_{C1} = 0.8 + 0.8 = 1.6$ V. Since B_1 is connected to V_{CC} (5 V) through R_{B1}, the collector-base junction of T_1 is forward-biased. The transistor T_1 is operating in the active inverse mode, making I_{C1} flow in the reverse direction. This current flows into the base of T_2 driving T_2 and T_3 into saturation. Therefore, $Y = V(0) \approx 0.2$ V.

From conditions I and II, it appears that T_1 is acting as back-to-back diodes. The importance of T_1 will become clear from condition III.

Condition III Let the circuit be operating under condition II when one of the inputs suddenly goes to $V(0)$. The corresponding emitter–base junction of T_1 starts conducting and V_{B1} drops to 0.9 V. T_2 and T_3 will be turned off when the stored base charge is removed. Since $V_{C1} = V_{B2} = 1.6$ V, therefore the collector–base junction of T_1 is back-biased, making T_1 operate in the normal active region. This large collector current of T_1 is in a direction which helps in the removal of stored base charge in T_2 and T_3 and improves the speed of circuit.

The discussion in Sec. 4.6 regarding loading (fan-out) considerations, noise-margins, average power dissipation, propagation delays, and wired-AND connection, is equally valid for TTL gate of Fig. 4.16 with passive pull-up resistor.

The speed of the circuit can be improved by decreasing R_{C3} which decreases the time constant ($R_{C3} \cdot C_O$) with which the output capacitance charges from 0 to 1 logic level. Such a reduction, however, would increase dissipation and would make it more difficult for T_3 to saturate.

4.8.2 Active Pull-up

It is possible in TTL gates to hasten the charging of output capacitance without corresponding increase in power dissipation with the help of an output circuit arrangement (Fig. 4.17) referred to as an *active pull-up* or *totem-pole* output.

The operation of the circuit can qualitatively be described as: For output Y to be in LOW state, transistor T_4 and diode D are cut-off. When the output makes a transition from LOW to HIGH corresponding to any input going to LOW, transistor T_4 enters saturation and supplies current for the charging of the output capacitor with a small time constant. This current decreases and eventually becomes zero under steady-state condition when $Y = V(1)$.

Diode D is used in the circuit to keep T_4 in cut-off when the output is at logic 0. Corresponding to this, T_2 and T_3 are in saturation, therefore,

$$V_{C2} = V_{B4} = V_{BE3,\,\text{sat}} + V_{CE2,\,\text{sat}} = 0.8 + 0.2 = 1.0 \text{ V} \tag{4.8}$$

Since $V_O = V_{CE3,\,\text{sat}} \approx 0.2$ V, the voltage across the base-emitter junction of T_4 and diode D equals $1.0 - 0.2 = 0.8$ V, which means T_4 and D are cut-off.

If one of the inputs drops to LOW logic level, T_2 and T_3 go to cut-off. The output voltage cannot change instantaneously (being the voltage across C_O) and because of T_2 going to cut-off, the voltage at the base of T_4 rises driving it to saturation.

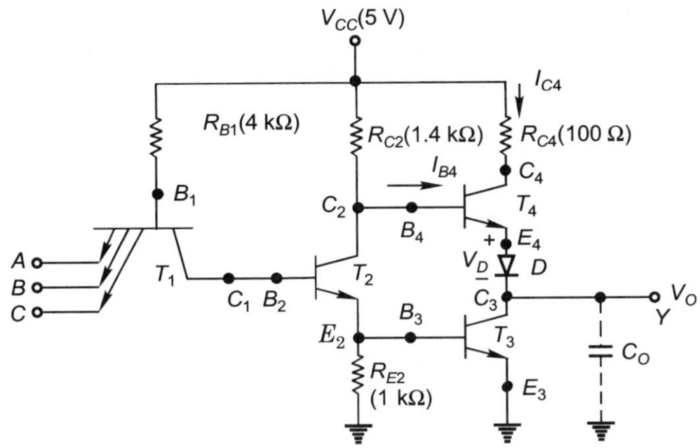

Fig. 4.17
A TTL gate with totem-pole output driver.

As soon as T_2 is cut-off,

$$V_{B4} = V_{BE4,\text{ sat}} + V_D + V_O$$

$$= 0.8 + 0.7 + 0.2 = 1.7 \text{ V} \tag{4.9}$$

Therefore,

$$I_{B4} = \frac{V_{CC} - V_{B4}}{R_{C2}} = \frac{5 - 1.7}{1.4} = 2.36 \text{ mA} \tag{4.10}$$

and

$$I_{C4} = \frac{V_{CC} - V_{CE4,\text{sat}} - V_D - V_O}{R_{C4}}$$

$$= \frac{5 - 0.2 - 0.7 - 0.2}{0.1} = 39 \text{ mA} \tag{4.11}$$

Hence, T_4 is in saturation if h_{FE} exceeds $\frac{39}{2.36} = 16.5$.

The output voltage V_O rises exponentially towards V_{CC} with the time constant = $(R_{C4} + R_{CS4} + R_f)\, C_O$, where R_{CS4} is the saturation resistance of T_4 and R_f is the forward resistance of the diode.

As V_O increases, the base and collector currents of T_4 are decreased and eventually T_4 just comes out of conduction at steady-state. Therefore,

$$V(1) = V_{CC} - V_\gamma\,(T_4) - V_\gamma\,(\text{diode}) = 5 - 0.5 - 0.6 = 3.9 \text{ V}$$

Now, if the output is at $V(1)$ and all the inputs go to HIGH, T_2 goes ON. Consequently T_4 and D go OFF and T_3 conducts. The capacitor C_O discharges through T_3 and as V_O approaches $V(0)$, T_3 enters into saturation.

From the above discussion it is clear that the maximum current is drawn from the supply when the output makes a transition from $V(0)$ to $V(1)$ and equals $I_{C4} + I_{B4} = 39 + 2.4 = 41.4$ mA.

This current spike generates noise in the power supply distribution system and increases power dissipation in the gate, more so when it is operated at high frequencies.

4.8.3 Wired-AND

Wired-AND connection must not be used for totem-pole output circuits because of the current spike problem discussed above (Prob. 4.16). TTL circuits with open-collector outputs are available which can be used for wired-AND connections.

4.8.4 Open-Collector Output

A circuit with open-collector output is same as the circuit of Fig. 4.16 except for the collector–resistor R_{C3} of T_3 which is missing. The collector terminal C_3 is available outside the IC and the passive pull-up is to be connected externally. Naturally, the advantages of active pull-up are not available in this. Gates with open-collector output can be used for wired-AND operation (Prob. 4.18).

4.8.5 Unconnected Inputs

If any input of a TTL gate is left disconnected (open or floating) the corresponding E–B junction of T_1 will not be forward-biased. Hence, it acts exactly in the same way as if a logical 1 is applied to that input. Therefore, in TTL ICs, all unconnected inputs are treated as logical 1s. However, the unused inputs should either be connected to some used input(s) or returned to V_{CC} through a resistor.

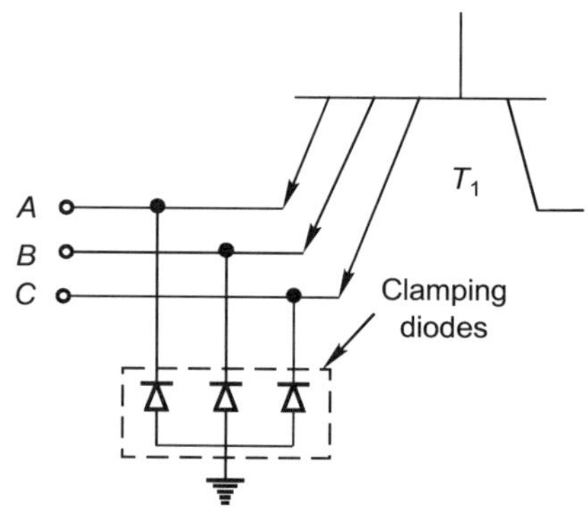

Fig. 4.18
A portion of a TTL gate showing the clamping diodes.

4.8.6 Clamping Diodes

Clamping diodes are commonly used in all TTL gates to suppress the ringing caused from the fast voltage transitions found in TTL. These diodes shown in Fig. 4.18 clamp the negative undershoot at approximately – 0.7 V.

4.9 SCHOTTKY TTL

The speed limitation of TTL is mainly due to the turn-off time delays involved in transistors while making transitions from saturation to cut-off. This can be eliminated by replacing the transistors of TTL gate by Schottky transistors.

With this, the transistors are prevented from entering saturation and hence, there is saving in turn-off time. Schottky TTL gates have propagation delay time of the order of 2 ns which is very small in comparison with the propagation delay time of standard TTL which is of the order of 10 ns. It is a nonsaturating bipolar logic.

4.10 5400/7400 TTL SERIES

TTL 5400/7400 series is the most popular and commonly used series of digital ICs. 7400 devices are used for commercial applications whereas the 5400 devices are used for military applications. The only difference in these two series are in the temperature and the power supply range. The temperature range is 0 °C to 70 °C for the 7400 series and −55 °C to 125 °C for the 5400 series. The supply voltage range is 5 ± 0.25 V for the 7400 series and 5 ± 0.5 V for the 5400 series.

There are seven different series of TTL 54-/74-logic family. These are given in Table 4.2.

Table 4.2 54-/74-TTL ICs with numbering scheme

Series	*Prefix*	*Examples*
Standard TTL	74-	7402, 74193
High Power TTL	74H-	74H02, 74H193
Low Power TTL	74L-	74L02, 74L193
Schottky TTL	74S-	74S02, 74S193
Low Power Schottky TTL	74LS-	74LS02, 74LS193
Advanced Schottky TTL	74AS-	74AS02, 74AS193
Advanced Low Power Schottky TTL	74ALS-	74ALS02, 74ALS193

Table 4.3 Specifications of TTL IC families

Parameter		*5400* *7400*	*54H00* *74H00*	*54L00* *74L00*	*54S00* *74S00*	*54LS00* *74LS00*	*54AS00* *74AS00*	*54ALS00* *74ALS00*	*Units*
V_{IH}		2	2	2	2	2	2	2	Volts
V_{IL}	54 Series	0.8	0.8	0.7	0.8	0.7	0.8	0.8	Volts
	74 Series	0.8	0.8	0.7	0.8	0.8	0.8	0.8	
V_{OH}	54 Series	2.4	2.4	2.4	2.5	2.5	3	3	Volts
	74 Series	2.4	2.4	2.4	2.7	2.7	3	3	
V_{OL}	54 Series	0.4	0.4	0.3	0.5	0.4	0.5	0.4	Volts
	74 Series	0.4	0.4	0.4	0.5	0.5	0.5	0.5	
I_{IH}		40	50	10	50	20	20	20	μA
I_{IL}		−1.6	− 2.0	− 0.18	− 2.0	− 0.36	− 0.5	− 0.1	mA
I_{OH}		− 400	− 500	− 200	− 1000	− 400	− 2000	− 400	μA
I_{OL}	54 Series	16	20	2	20	4	20	4	mA
	74 Series	16	20	3.6	20	8	20	8	mA
$I_{CC}(1)$		8	16.8	0.8	16	1.6	3.2	0.85	mA
$I_{CC}(0)$		22	40	2.04	36	4.4	17.4	3	mA
t_{pHL}		15	10	60	5	15	4	8	ns
t_{pLH}		22	10	60	4.5	15	4.5	11	ns

Table 4.3 summarizes various specifications of 54/74 TTL logic families. Table 4.4 summerizes fan-out capabilities of each series when it drives ICs of the same series or of other series.

Table 4.4 Summary of TTL fan-out capabilities

Source TTL device ↓ / Load TTL device →	54/74	54H/74H	54L/74L	54S/74S	54LS/74LS	54AS/74AS	54ALS/74ALS
54/74	10	8	40	8	20	20	20
54H/74H	12	10	50	10	25	25	25
54L/74L	2	1	20	1	10	7	10
54S/74S	12	10	100	10	50	40	50
54LS/74LS	5	4	40	4	20	16	26
54AS/74AS	12	10	110	10	55	40	100
54ALS/74ALS	5	4	40	4	20	16	20

From Table 4.3, we observe the following:

(i) The input and output voltage specifications are compatible for each of the TTL series, which makes it possible to use any mix of ICs of these series to achieve optimum design from the point of view of propagation delay and power dissipation.

(ii) The input and output current specifications are compatible and the number of gates of each of the series, which can be safely driven from any series can be determined as given in Table 4.4.

(iii) The low power dissipation series L, LS, and ALS have minimum power requirement and are suitable for battery operated circuits. Out of these series ALS series has the minimum propagation delay and therefore it is fast replacing other series.

(iv) H series has low propagation delay (high speed) but requires maximum power.

(v) S and AS series have very low propagation delay. The AS series is fast replacing S series because of its lower dissipation and propagation delay.

4.11 EMITTER-COUPLED LOGIC (ECL)

Emitter-coupled logic (ECL) is the fastest of all logic families and therefore is used in applications where very high speed is essential. High speeds have become possible in ECL because the transistors are used in difference amplifier configuration, in which they are never driven into saturation and thereby the storage time is eliminated. Here, rather than switching the transistors from ON to OFF and vice-versa, they are switched between cut-off and active regions. Propagation delays of less than 1 ns per gate have become possible in ECL.

Basically, ECL is realized using difference amplifier in which the emitters of the two transistors are connected and hence it is referred to as emitter-coupled logic. A 3-input ECL gate is shown in Fig. 4.19, which has three parts: The middle part is the difference amplifier which performs the logic operation.

Emitter follower are used for d.c. level shifting of the outputs, so that $V(0)$ and $V(1)$ are same for the inputs and the outputs. Note that two output Y_1 and Y_2 are available in this circuit which are complementary. Y_1 corresponds to OR logic and Y_2 to NOR logic and hence it is named as an OR/NOR gate.

Additional transistors are used in parallel to T_1 to get the required fan-in. There is a fundamental difference between all other logic families (including MOS logic) and ECL as far as the

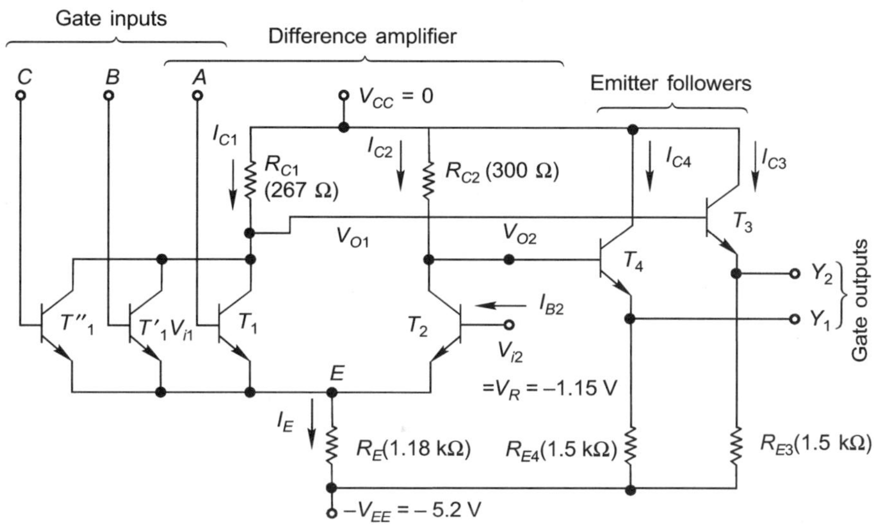

Fig. 4.19
A 3-input ECL OR/NOR gate.

supply voltage is concerned. In ECL, the positive end of the supply is connected to ground in contrast to other logic families in which negative end of the supply is grounded. This is done to minimize the effect of noise induced in the power supply (Prob. 4.22), and protection of the gate from an accidental short circuit developing between the output of a gate and ground (Prob. 4.23). The voltage corresponding to $V(0)$ and $V(1)$ are both negative due to positive end of the supply being connected to ground. The symbol of an ECL OR/NOR gate is shown in Fig. 4.20.

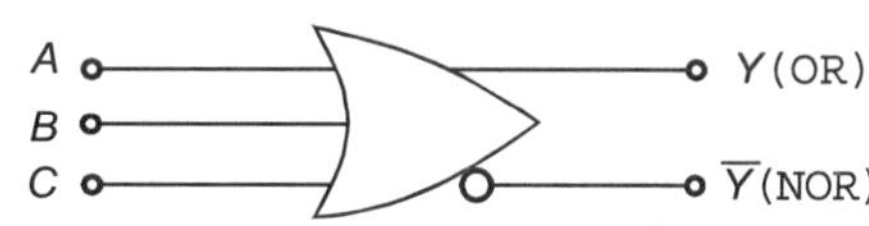

Fig. 4.20
The symbol for a 3-input OR/NOR gate.

Example 4.2 (a) Verify that the circuit of Fig. 4.19 performs OR/NOR operations. (b) Show that the transistors in this circuit operate in the active region and not in saturation. (c) Calculate the noise margins. (d) Find the average power dissipated by the gate.

Assume a base–emitter voltage of 0.7 V for a transistor conducting in active region.

Solution

(a) (i) Assume all inputs to be LOW.

Let us assume that the input transistors T_1, T'_1, T''_1 are cutt-off and T_2 is conducting in the active region. The voltage at the common emitter is $V_E = V_{i2} - V_{BE2} = -1.15 - 0.7 = -1.85$ V. The current

$$I_E = \frac{V_E - (-V_{EE})}{R_E} = \frac{-1.85 + 5.2}{1.18} = 2.84 \text{ mA}$$

Since $I_{B2} << I_{C2}$, therefore $I_{C2} \approx I_E$

$$V_{O2} = -0.3\, I_{C2} = -0.3\,(2.84) = -0.852 \text{ V}$$

Transistor T_4 will be conducting and the output at $Y_1 = V_{O2} - V_{BE4} = -0.852 - 0.7 = -1.55$ V which is assumed to be $V(0)$.

Therefore, if all the inputs are at $V(0) = -1.55$ V, then the base-to-emitter voltage of the input transistor is

$$V_{BE} = V_{i1} - V_E = -1.55 + 1.85 = 0.3 \text{ V}$$

which is less than the cut-in voltage (0.5 V) of the transistor and hence the input transistors are non-conducting, as was assumed above.

The base and collector of T_3 are effectively at the same potential, hence T_3 behaves as a diode. The current flowing through this diode is approximately 3 mA which corresponds to a voltage of about 0.75 V across the diode. Therefore, the voltage at $Y_2 = -0.75$ V which is assumed to be $V(1)$. This shows that Y_1 and Y_2 are complementary, i.e. $Y_2 = \overline{Y_1}$.

(ii) Assume at least one input to be HIGH. Corresponding to this the input transistor T_1 is assumed to be conducting and T_2 to be cut-off.

Then $\qquad V_E = V_{i1} - V_{BE1} = -0.75 - 0.7 = -1.45$ V

Hence, $\qquad V_{BE2} = V_{i2} - V_E = -1.15 + 1.45 = 0.3$ V which verifies the assumption that T_2 is cut-of.

The voltage $\qquad V_{O1} = -R_{C1} \times I_{C1}$

where
$$I_{C1} = \frac{V_E - (-V_{EE})}{R_E}$$

$$= \frac{(-1.45 + 5.2)}{1.18} = 3.18 \text{ mA}$$

Since the collector current of T_1 is higher than the collector current of T_2 when it is conducting, hence $R_{C1} < R_{C2}$ to get the same voltage levels.

This gives voltage at $Y_2 = -1.55 = V(0)$. The voltage at $Y_1 = -0.75 = V(1)$. From (i) and (ii) above, we see that OR function is performed at Y_1 and NOR at Y_2. Hence it is an OR/NOR gate. Its voltages corresponding to logic 0 and 1 are –1.55 V and – 0.75 V respectively. The logic swing is 0.8 V.

(b) From part (a) (i), the voltage between collector and base of T_2 is $V_{CB2} = V_{O2} - V_{i2} = -0.85 + 1.15 = 0.30$ V which shows that the *C–B* junction is reverse-biased and hence T_2 is operating in its active region.

From part (a) (ii), the voltage between the collector and base of T_1 is

$$V_{CB1} = V_{O1} - V_{i1} = -0.85 + 0.75 = -0.1 \text{ V}$$

This shows that the *C–B* junction of T_1 is forward-biased but its magnitude is much less than the cut-in voltage and hence T_1 is operating in its active region.

(c) From part (a)(i), the base–emitter voltage of the input transistors is 0.3 V which is 0.2 V less than the cut-in voltage. Hence the noise margin $\Delta 0 = 0.2$ V.

From part (a) (ii) the base-emitter voltage of T_2 is 0.3 V which again gives a noise margin $\Delta 1 = 0.2$ V. The noise margins are equal and are quite small.

(d) From part (a) (i),

$$I_{C2} = 2.84 \text{ mA}$$

$$I_{C3} = \frac{5.2 - 0.75}{1.5} = 2.97 \text{ mA}$$

and $$I_{C4} = \frac{5.2 - 1.55}{1.5} = 2.43 \text{ mA}$$

From part (a) (ii),

$$I_{C1} = 3.18 \text{ mA}$$
$$I_{C3} = 2.43 \text{ mA}$$
$$I_{C4} = 2.97 \text{ mA}$$

Therefore, average $I_E = \frac{2.84 + 3.18}{2} = 3.01$ mA. The total power supply current drain

$$I_{EE} = 3.01 + 2.97 + 2.43 = 8.41 \text{ mA}$$

Therefore, the power dissipation $= V_{EE} \cdot I_{EE} = (5.2)(8.41)$

$$= 43.7 \text{ mW}$$

4.11.1 Fan-Out

If all the inputs are LOW, the input transistors are cut-off. Therefore the input resistance is very high. On the other hand, if an input is HIGH, the input resistance is that of an emitter follower which is also high. Therefore, the input impedance is always high.

The output resistance is either that of an emitter follower or the forward resistance of a diode (T_3 or T_4 acts as a diode) which is always low. Because of the low output impedance and high input impedance, the fan-out is large.

4.11.2 Wired-OR Logic

The outputs of two or more ECL gates can be connected to obtain additional logic without using additional hardware. The wired-OR configurations are shown in Fig. 4.21.

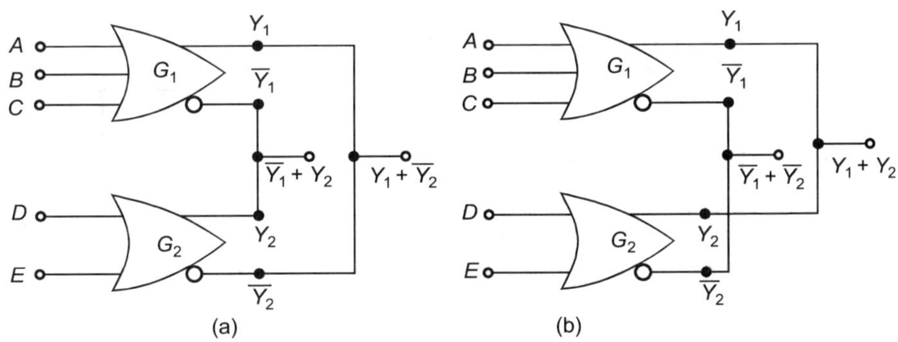

Fig. 4.21
Wired-OR connection of ECL gates.

4.11.3 Open-Emitter Outputs

Similar to open-collector output in TTL, open-emitter outputs are available in ECL which is useful for wired-OR applications.

4.11.4 Unconnected Inputs

If any input of an ECL gate is left unconnected, the corresponding E-B junction of the input transistor will not be conducting. Hence it acts as if a logical 0 level voltage is applied to that input. Therefore, in ECL ICs, all unconnected inputs are treated as logical 0s.

4.11.5 ECL Families

There are two popular ECL families:10xxx (or 10K) series and 100xxx (or 100K) series. The 100K series is the fastest of all logic families and has a propagation delay time less than 1 ns. Their voltage specifications are given in Table 4.5.

Table 4.5 Voltage specifications of ECL series

Series	*Supply voltage* V_{EE}, *V*	V_{OL} *V*	V_{OH} *V*	V_{IL} *V*	V_{IH} *V*
10K	5.2	–1.7	–0.9	–1.4	–1.2
100K	4.5	–1.7	–0.9	–1.4	–1.2

4.12 INTERFACING ECL AND TTL

It is often necessary to mix logic circuits of different families in the design of a digital system to realize the speed and power requirements by choosing the appropriate logic families for different parts of the system. Consider the interfacing between TTL and ECL gates. The logic levels in the two systems are entirely different and therefore level shifting circuits are required to be interposed between TTL and ECL gates. For TTL-to-ECL and ECL-to-TTL interfacing two level translator ICs are available. Interfacing using these ICs are described below.

4.12.1 TTL-to-ECL Translator

The MC10H124 is a quad TTL-to-ECL translator IC. It is a 16-pin IC and its logic diagram is shown in Fig. 4.22. It uses two power supplies; one positive and another negative for the generation of proper logic levels for ECL and TTL.

The logic levels of the translator circuit are:

$V_{IH} = 2\text{V},\ V_{IL} = 0.8\text{ V}$

$V_{OH} = -0.98\text{V},\ V_{OL} = -1.63\text{V}$

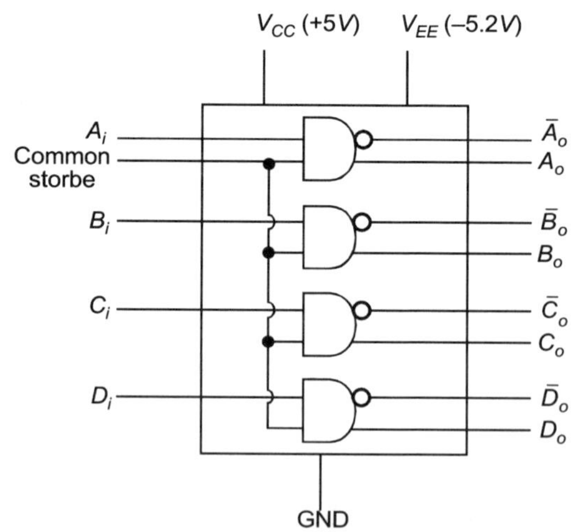

Fig. 4.22
Logic diagram of MC10H124 TTL-to-ECL translator

From Table 4.3, we have

$V_{OH} = 2.4$ V and $V_{OL} = 0.4$ V for TTL ICs. Comparing the output logic levels of TTL and the input logic levels of the translator IC, we observe,

$$V_{IH}\text{ (Translator)} < V_{OH}\text{ (TTL)}$$

and
$$V_{IL}\text{ (Translator)} > V_{OL}\text{(TTL)}$$

which shows that the input logic levels of the translator are compatible with the output logic levels of TTL.

Similarly, comparing the output logic levels of the translator with the input logic levels of ECL (Table 4.5), we obtain

$$V_{IH}\text{ (ECL)} < V_{OH}\text{ (Translator)}$$

and
$$V_{IL}\text{ (ECL)} > V_{OL}\text{ (Translator)}$$

which demonstrates that the output logic levels of the translator are compatible with the input logic levels of ECL.

Figure 4.23 shows a TTL NAND gate driving an ECL NOR gate through a TTL-to-ECL translator gate.

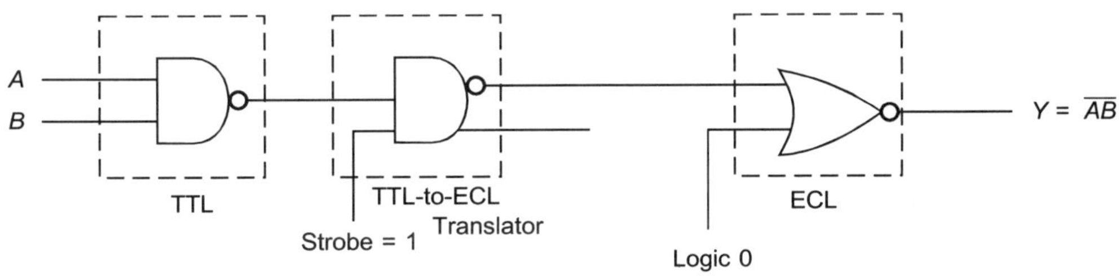

Fig. 4.23

A TTL NAND gate driving an ECL NOR gate through a TTL-to-ECL translator

4.12.2 ECL-to-TTL Translator

The MC10H125 is a quad ECL-to-TTL translator IC. It is a 16-pin IC and its logic diagram is shown in Fig. 4.24. It also uses two power supplies for the generation of proper logic levels for ECL and TTL. Its logic levels are:

$$V_{IH} = -1.13\text{ V},\quad V_{IL} = -1.48\text{ V}$$

$$V_{OH} = 2.5\text{ V},\quad V_{OL} = 0.5\text{ V}$$

Its input logic levels are compatible with ECL and the output logic levels are compatible with TTL (Prob. 4.26).

4.13 MOS LOGIC

MOSFETs have become very popular for logic circuits due to high density of fabrication and low power dissipation. When MOS devices are used in logic circuits, there can be circuits in which either only *p*– or only *n*-channel devices are used. Such circuits are referred to as PMOS and NMOS logic respectively. It is also possible to fabricate enhancement mode *p*-channel and *n*-channel MOS devices on the same chip. Such devices are referred to as complementary

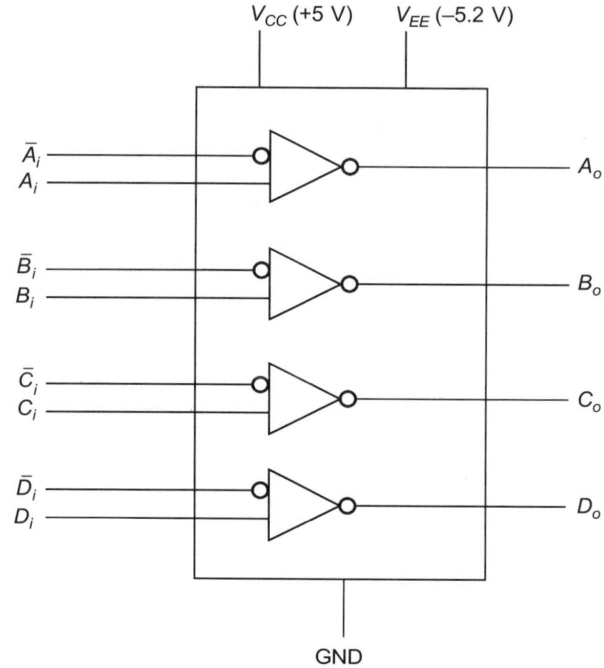

Fig. 4.24

Logic diagram of MC10H125 ECL-to-TTL translator

MOSFETs and logic based on these devices is known as CMOS logic. The power dissipation is extremely small for CMOS and hence CMOS logic has become very popular.

The basic MOS gate is an inverter as shown in Fig. 4.25, in which T_1 is an enhancement MOSFET which acts as driver and T_2 may be an enhancement (Fig. 4.25a) or depletion (Fig. 4.25b) MOSFET, which acts as load. Instead of fabricating diffusion resistor for load,

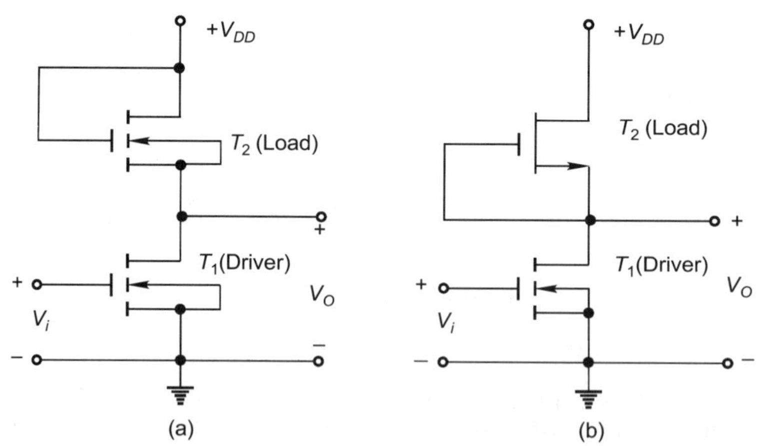

Fig. 4.25

A MOS inverter with (a) Enhancement load (b) Depletion load.

which usually occupies an area about 20 times that of a MOS device, MOSFET itself is used as the load. This makes possible high density of fabrication and therefore MOS logic made large scale integration possible.

The logic levels for the MOS circuits are

$$V(0) \approx 0$$
$$V(1) \approx V_{DD}$$

Although the MOS logic circuits are identical in configuration to bipolar DCTL, the problem of current hogging is not present. The operation of MOSFET switches is given in Section 3.7.

MOS logic is mainly used for LSI and VLSI ICs and not for SSI and MSI ICs. Most of the microprocessors, memories, and peripheral devices are available in NMOS.

4.13.1 MOSFET NAND and NOR Gates

NOR gates can be obtained by using multiple drivers in parallel, whereas for NAND gates the drivers are to be connected in series. A two-input NOR gate is shown in Fig. 4.26*a* and a two-input NAND gate in Fig. 4.26*b*.

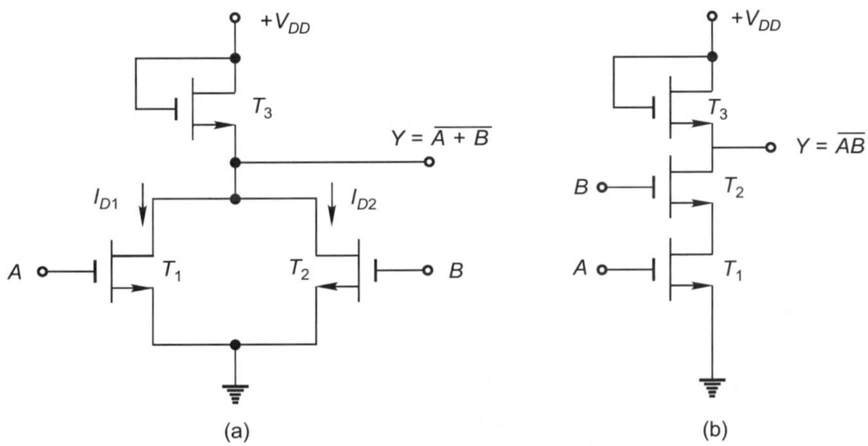

Fig. 4.26
2-input NMOS gates (a) NOR (b) NAND.

In the gate of Fig. 4.26*a*, if both inputs are 0, both transistors T_1 and T_2 are OFF ($I_{D1} = I_{D2} = 0$), hence the output is V_{DD}. If either one or both of the inputs are $V(1) = V_{DD}$, the corresponding FETs will be ON and the output is 0 V. Its truth table is given in Table 4.6, which obviously shows NOR operation.

In the gate of Fig. 4.26*b*, if either one or both the inputs are $V(0) = 0$, the corresponding FETs will be OFF, the voltage across the load FET will be 0, hence the output is V_{DD}. If both inputs are $V(1) = V_{DD}$, both T_1 and T_2 are ON and the output is 0. Its truth table is given in Table 4.7, which shows NAND operation.

Table 4.6 Truth table of Fig. 4.26*a*

Inputs		Output
A	B	Y
0	0	V_{DD}
0	V_{DD}	0
V_{DD}	0	0
V_{DD}	V_{DD}	0

Table 4.7 Truth table of Fig. 4.26*b*

Inputs		Output
A	B	Y
0	0	V_{DD}
0	V_{DD}	V_{DD}
V_{DD}	0	V_{DD}
V_{DD}	V_{DD}	0

4.13.2 Fan-Out

Since MOS devices have very high input impedance, therefore, the fan-out is large. But driving a large number of MOS gates increases the capacitance at the output of the driving gate which reduces, considerably, the speed of MOS gates.

The voltage and current parameters for 8085, 8086 microprocessors and other NMOS devices are:

$$V_{CC} = 5 \text{ V}$$
$$V_{IL} = 0.8 \text{ V}$$
$$V_{IH} = 2 \text{ V}$$
$$V_{OL} = 0.45 \text{ V}, \; I_{OL} = 2 \text{ mA}$$
$$V_{OH} = 2.4 \text{ V}, \; I_{OH} = -400 \; \mu\text{A}$$

The input and output leakage currents are $\pm$ 10 μA.

These voltages are directly compatible with TTL ICs. Usually, NMOS devices are available with higher sink currents which are directly compatible with TTL ICs. This helps in easy interfacing between NMOS devices and TTL devices.

4.13.3 Propagation Delay Time

The propagation delay time is large in MOS devices because of large capacitances present at the input and output of these devices. Also, the resistance through which these capacitors get charged and discharged is high.

In MOS devices, the phenomenon of minority charge storage is not present, and the speed of operation is mainly determined by the speed with which the capacitors get charged and discharged.

Due to the developments in the technology of MOS fabrication, it has become possible to obtain speeds which are comparable to TTL.

4.13.4 Power Dissipation

In the NAND gate of Fig. 4.26*b*, current is drawn from the power supply only during one of the four possible input conditions, whereas in the NOR gate of Fig. 4.23*a* power is drawn during three out of four input conditions. Therefore, the power consumption in MOS circuits is small which is very useful for large-scale integration.

4.13.5 Unconnected Inputs

MOS devices have very high input impedance and even a very small static charge flowing into this high impedance can develop a dangerously high voltage. This may cause damage to the device by rupturing the insulation layer and also to the persons handling such devices. Therefore, MOS ICs inputs must not be left unconnected. Even for storage of such devices, conductive foam or aluminium foil should be used which will ensure shorting of IC pins together so that no voltage can be developed between the pins. Necessary precautions must be taken while handling such devices.

4.14 CMOS LOGIC

A complementary MOSFET (CMOS) is obtained by connecting a *p*-channel and an *n*-channel MOSFET in series, with drains tied together and the output is taken at the common drain. Input is applied at the common gate formed by connecting the two gates together (Fig. 3.33). in a CMOS, *p*-channel and *n*-channel enhancement MOS devices are fabricated on the same chip, which makes its fabrication more complicated and reduces the packing density. But because of negligibly small power consumption, CMOS is ideally suited for battery operated systems.

Its speed is limited by substrate capacitances. To reduce the effect of these substrate capacitances, the latest technology known as silicon on sapphire (SOS) is used in microprocessor fabrication which employs an insulating substrate (sapphire). CMOS has become the most popular in MSI and LSI areas and is the only possible logic for the fabrication of VLSI devices.

4.14.1 CMOS Inverter

The basic CMOS logic circuit is an inverter shown in Fig. 3.33. For this circuit the logic levels are 0 V (logic 0) and V_{CC} (logic 1). When $V_i = V_{CC}$, T_1 turns ON and T_2 turns OFF. Therefore $V_O \approx 0$ V, and since the transistors are connected in series the current I_D is very small. On the other hand, when $V_i = 0$ V, T_1 turns OFF and T_2 turns ON giving an output voltage $V_O \approx V_{CC}$ and I_D is again very small. In either logic state, T_1 or T_2 is OFF and the quiscent power dissipation which is the product of the OFF leakage current and V_{CC} is very low. More complex functions can be realized by combinations of inverters.

4.14.2 CMOS NAND and NOR Gates

A 2-input CMOS NAND gate is shown in Fig. 4.27 and NOR gate in Fig. 4.28. In the NAND gate, the NMOS drivers are connected in series, where as the PMOS loads are connected in parallel. On the other hand, the CMOS NOR gate is obtained by connecting the NMOS drivers in parallel and PMOS loads in series. The operation of NAND gate can be understood from Table 4.8. The operation of the NOR gate can be verified in the similar manner (Prob. 4.29).

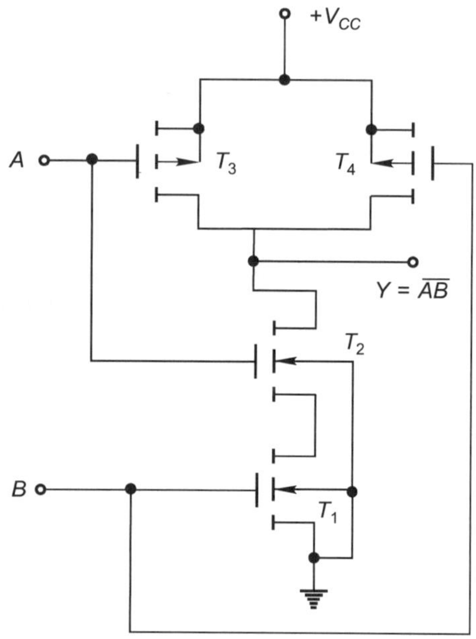

Fig. 4.27
A 2-input CMOS NAND gate.

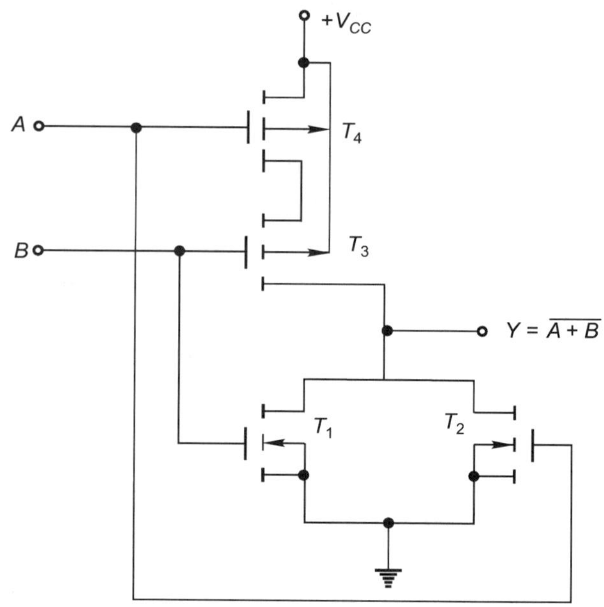

Fig. 4.28
A 2-input CMOS NOR gate.

Table 4.8 Operation of CMOS NAND gate

Inputs		*State of MOS devices*				*Output*
A	B	T_1	T_2	T_3	T_4	Y
0	0	OFF	OFF	ON	ON	V_{CC}
0	V_{CC}	ON	OFF	ON	OFF	V_{CC}
V_{CC}	0	OFF	ON	OFF	ON	V_{CC}
V_{CC}	V_{CC}	ON	ON	OFF	OFF	0

4.14.3 CMOS Transmission Gate

A CMOS transmission gate controlled by gate voltages C and $\bar{C}$ is shown in Fig. 4.29. Assume $C = 1$. If $A = V(1)$, then T_1 is OFF and T_2 conducts in the ohmic region because there is no voltage applied at the drain. Therefore, T_2 behaves as a small resistance connecting the output to the input and $B = A = V(1)$. Similarly, if $A = V(0)$, then T_2 is OFF and T_1 conducts, connecting the output to the input and $B = A = V(0)$. This means the signal is transmitted from A to B when $C = 1$.

In a similar manner, it can be shown that if $C = 0$, transmission is not possible.

In this gate the control C is binary, whereas the input at A may be either digital or analog [the instantaneous value must lie between $V(0)$ and $V(1)$].

4.14.4 Noise Margin

Noise margin of CMOS logic ICs is considerably higher than that of TTL ICs. CMOS devices

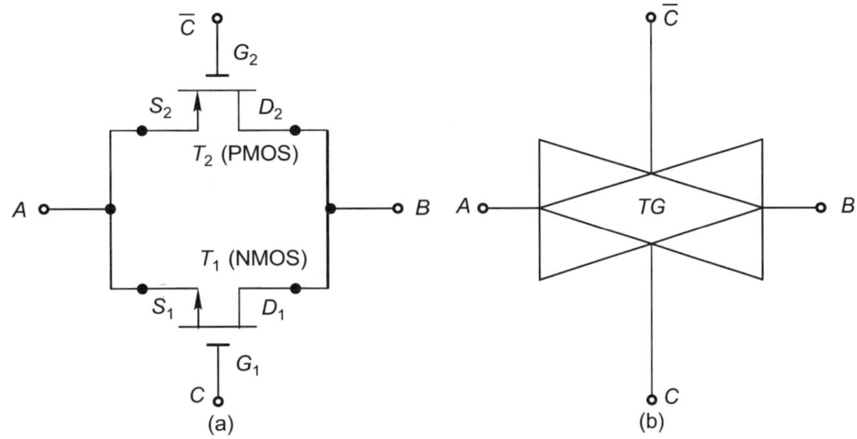

Fig. 4.29
(a) A CMOS transmission gate (b) Its symbol.

have wide supply voltage range and the noise margin increases with the supply voltage V_{CC}. Typically, it is 0.45 V_{CC}.

4.14.5 Unconnected Inputs

The unconnected CMOS ICs inputs behave in a way similar to MOS devices discussed in Section 4.13. Therefore, the unused inputs must be connected to either the supply voltage terminal or one of the used inputs provided that the fan-out of the signal source is not exceeded. This is highly unlikely for CMOS circuits because of their high fan-out.

Some CMOS ICs have Zener diodes connected at the inputs for protection against high input voltages.

4.14.6 Wired-Logic

Figure 4.30 shows two CMOS inverters with their outputs connected together. In this circuit,

(i) When $A = B = V(0)$
 T_1 and T_1' are cut-off and $Y = V(1) = V_{CC}$

(ii) When $A = B = V(1)$
 T_1 and T_1' are ON and $Y = V(0) = 0$

(iii) When $A = V(1)$ and $B = V(0)$
 T_1 and T_2' are ON whereas
 T_1' and T_2 are OFF

Therefore, a large current I will flow as shown in Fig. 4.30.
This will make voltage at Y equal to $V_{CC}/2$ which is neither in the range of logic 0 nor in the range of logic 1. Therefore, the circuit will not operate properly. Also because of large current I, the transistors will be damaged.

Similarly, corresponding to $A = V(0)$ and $B = V(1)$ the operation will not be proper.

Therefore, wired-logic must not be used for CMOS logic circuits.

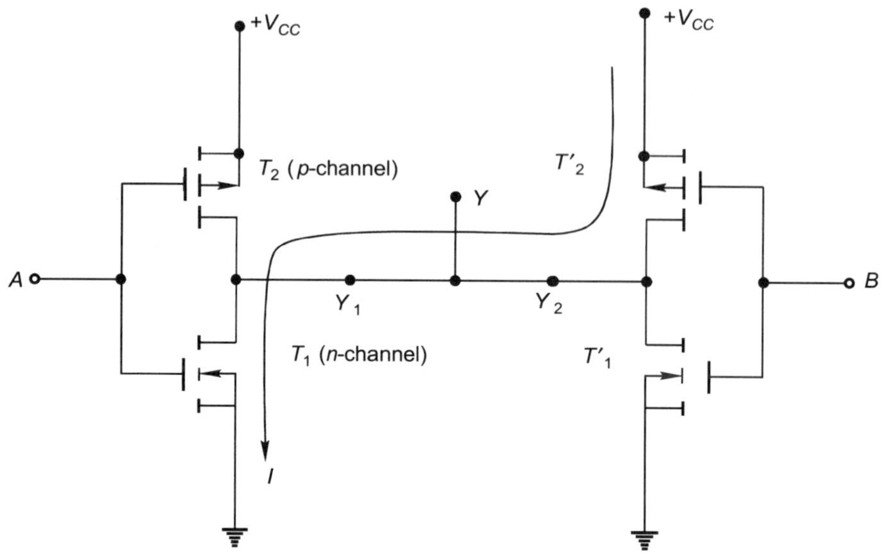

Fig. 4.30
CMOS inverters with outputs connected.

4.14.7 Open-Drain Outputs

CMOS gates with open-drain output are available which are useful for wired-AND operation. In this the drain terminal of the output transistor (*n*-channel) is available outside and the load resistor is to be connected externally since *p*-channel load does not exist.

4.14.8 54C00/74C00 CMOS Series

There are two commonly used CMOS series ICs. These are the 4000 series and 54C/74C series. 54C/74C CMOS series is pin-for-pin, function-for-function equivalent to the 54/74 TTL family and has, therefore, become very popular. The temperature range for 54C series is –55 °C to + 125 °C and for 74C series is –40 °C to 85 °C. It has a wide supply voltage range, 3 V to 15 V. A person can take full advantage of his knowledge of the 54/74 TTL series for the effective use of 54C/74C series.

There have been significant improvements in 54C/74C series. The 74HC/74HCT have higher speed and better current capabilities. 74HC is known as *high-speed* CMOS and 74HCT is known as *high-speed, TTL compatible* CMOS series. 74AC/74ACT are very fast and have very high current sinking capabilities. These are known as *advanced* CMOS and *advanced, TTL compatible* CMOS, respectively. The 74 HC/74HCT/74AC/74ACT series can be operated at supply voltages in the range of 2–6 volts.

The voltage and current parameters of various 74 CMOS series with 5 V supply voltage are given in Table 4.9. From the table, we observe that the output currents and voltages for 74HC/74HCT/74AC/74ACT are different when gates of these series are driving CMOS circuits and TTL circuits. 74 HCT and 74 ACT series are compatible with TTL series for input as well as output and therefore, can easily be used along with TTL ICs for optimum system design from the point of view of speed, power dissipation, noise margins, cost, etc.

The fan-out of 74 HC/74HCT series is 20, whreas for 74AC/74ACT series it is 50 while driving these CMOS series. The fan-out of these gates while driving various TTL series gates can be determined using the specifications of TTL (Table 4.3) and CMOS (Table 4.9).

Table 4.9 Specifications of CMOS IC families

Parameter	*Load*	*74C*	*74HC*	*74HCT*	*74AC*	*74ACT*	*Units*
V_{IH}		3.5	3.85	2.0	3.85	2.0	volts
V_{IL}		1.5	1.35	0.8	1.35	0.8	volts
V_{OH}	CMOS	4.5	4.4	4.4	4.4	4.4	volts
	TTL		3.84	3.84	3.76	3.76	volts
V_{OL}	CMOS	0.5	0.1	0.1	0.1	0.1	volt
	TTL		0.33	0.33	0.37	0.37	volt
I_{IH}		1	1	1	1	1	μA
I_{IL}		– 1	– 1	– 1	– 1	– 1	μA
I_{OH}	CMOS	– 0.1	– 0.02	– 0.02	– 0.05	– 0.05	mA
	TTL		– 4.0	– 4.0	– 24.0	– 24.0	mA
I_{OL}	CMOS	0.36	0.02	0.02	0.05	0.05	mA
	TTL		4.0	4.0	24.0	24.0	mA

4.15 INTERFACING CMOS AND TTL

To achieve optimum performance in a digital system, devices from more than one logic family can be used, taking advantages of the superior characteristics of each family for different parts of the system. For example, CMOS logic ICs can be used in those parts of the system where low power dissipation is required, while TTL can be used in those portions of the system which requires high speed of operation. Also, some functions may be easily available in TTL and others may be available in CMOS. Therefore, it is necessary to examine the interface between TTL and CMOS devices.

The 74C series of CMOS ICs can be operated for any supply voltage in the range of 3 V to 15 V, whereas the 74HC/74HCT/74AC/74ACT series have the supply voltage range of 2 V to 6 V. Since the supply voltage used for all 74 series TTL ICs is 5 V, therefore, it is necessary to operate CMOS devices at + 5 V, to make it compatible with TTL devices.

4.15.1 CMOS Driving TTL

Figure 4.31 shows a CMOS gate driving *N* TTL gates. For such an arrangement to operate properly the following conditions are required to be satisfied,

$$V_{OH}\,(\text{CMOS}) \geq V_{IH}\,(\text{TTL}) \tag{4.12}$$

$$V_{OL}\,(\text{CMOS}) \leq V_{IL}\,(\text{TTL}) \tag{4.13}$$

$$-I_{OH}\,(\text{CMOS}) \geq NI_{IH}\,(\text{TTL}) \tag{4.14}$$

$$I_{OL}\,(\text{CMOS}) \geq -NI_{IL}\,(\text{TTL}) \tag{4.15}$$

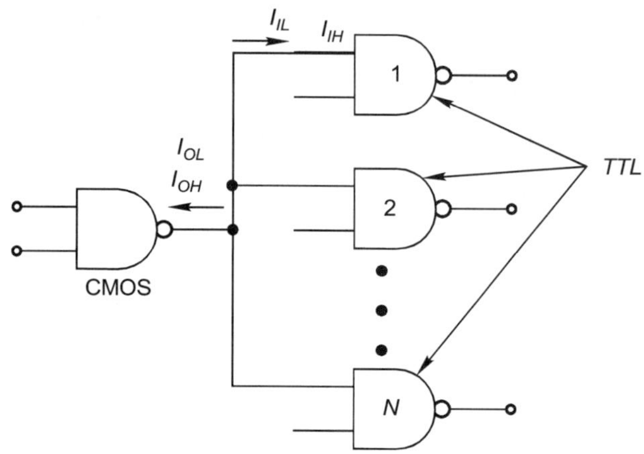

Fig. 4.31
A CMOS gate driving *N* TTL gates.

From the specifications given in Tables 4.3 and 4.9, we observe the following:

(i) The conditions of Eqs (4.12) and (4.13) are always satisfied. The noise margins when 74ACT is driving 74ALS gates are

$$\Delta\, 1 = 3.76 - 2.0 = 1.76 \text{ V}$$
$$\Delta\, 0 = 0.8 - 0.37 = 0.43 \text{ V}$$

(ii) The conditions of Eqs (4.14) and (4.15) are always satisfied for 74 HC/74 HCT/74 AC/ 74 ACT series. The value of *N* is different for different series. The value of *N* when 74 ACT is driving 74ALS gates is 240.

In case of 74 C series, the condition of Eq. (4.14) is satisfied for small values of *N* but the condition of Eq. (4.15) is not satisfied even for $N = 1$, except in case of 74L and 74ALS TTL series. This difficulty can be overcome by using CMOS buffers having an adequate available output current.

If 74C series gate is driving 74L series gates, the condition of Eq. (4.15) is satisfied for $N = 2$ and in case of 74ALS gates for $N = 3$.

4.15.2 TTL Driving CMOS

Figure 4.32 shows a TTL gate driving *N* CMOS gates. For such an arrangement to operate properly, the following conditions are required to be satisfied:

$$V_{OH}\,(\text{TTL}) \geq V_{IH}\,(\text{CMOS}) \tag{4.16}$$
$$V_{OL}\,(\text{TTL}) \leq V_{IL}\,(\text{CMOS}) \tag{4.17}$$
$$-I_{OH}\,(\text{TTL}) \geq NI_{IH}\,(\text{CMOS}) \tag{4.18}$$
$$I_{OL}\,(\text{TTL}) \geq -NI_{IL}\,(\text{CMOS}) \tag{4.19}$$

All the above conditions are always satisfied in case of 74 HCT and 74 ACT series for high values of *N*. This shows that these two CMOS series are TTL compatible. In the case of 74C/ 74HC/74AC series, the condition of Eq. (4.16) is not satisfied. A circuit modification used to

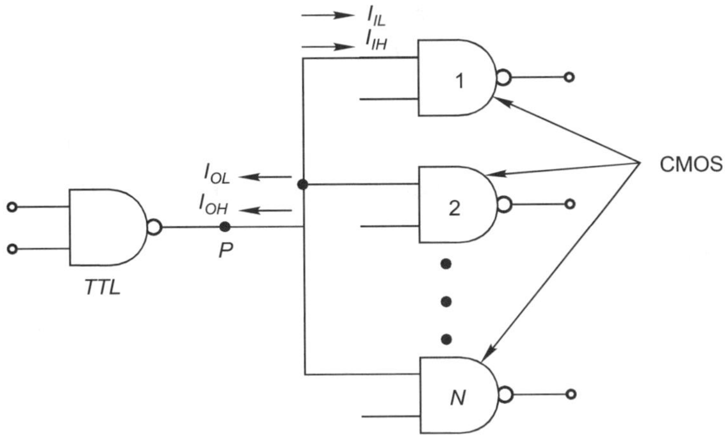

Fig. 4.32
A TTL gate driving *N* CMOS gates.

raise V_{OH} (TTL) above 3.5 V is obtained by connecting a resistance (≈ 2 K Ω) between points *P* and V_{CC} as shown in Fig. 4.33. This acts as a passive pull-up, which pulls up the voltage at *P*, by charging the capacitor C_O present between *P* and the ground terminal, to a higher value (≈ V_{CC}) after the transistor T_4 of the TTL becomes non-conducting.

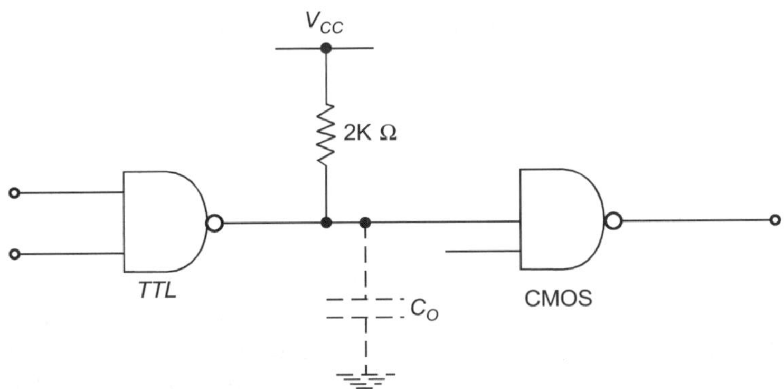

Fig. 4.33
Circuit to pull up the output voltage of TTL.

4.16 INTERFACING CMOS AND ECL

Using MC10H124 TTL-to-ECL translator and MC10H125 ECL-to-TTL translator ICs, it is possible to interface CMOS and ECL logic families. The input of MC10H124 translator is compatible to the output logic voltages of CMOS and therefore, this can be used for CMOs-to-ECL interfacing (Prob. 4.33). Similarly, the output of MC10H125 translator is compatible to the input logic voltages of CMOS (74 HCT and 74 ACT) families which makes it possible to be used it for ECL-to-CMOS interfacing. Other CMOS logic families can also be interfaced using pull-up resistor similar to Fig. 4.33 (Prob. 4.34).

4.17 TRI-STATE LOGIC

In normal logic circuits there are two states of the output, LOW and HIGH. If the output is not in the LOW state, it is definitely in the other state (HIGH). Similarly, if the output is not in the HIGH state, it is definitely in the LOW state. In complex digital systems like microcomputers and microprocessors, a number of gate outputs may be required to be connected to a common line which is referred to as a *bus* which, in turn, may be required to drive a number of gate inputs. When a number of gate outputs are connected to the bus, we encounter some difficulties. These are:

1. Totem-pole outputs cannot be connected together because of very large current drain from the supply and consequent heating of the ICs which may get damaged.
2. Open-collector outputs can be connected together with a common collector-resistor connected externally. This causes the problems of loading and speed of operation.

To overcome these difficulties, special circuits have been developed in which there is one more state of the output, referred to as the *third state* or *high-impedance state*, in addition to the LOW and HIGH states. These circuits are known as *TRI-STATE, tri-state logic* (TSL) or *three-state logic.* TRI-STATE, is a registered trade mark of National Semiconductor Corporation of USA.

There is a basic functional difference between wired-OR and the TSL. For the wired-OR connection of two functions Y_1 and Y_2 is

$$Y = Y_1 + Y_2 \tag{4.20}$$

whereas for TSL, the result is not a Boolean function but an ability to multiplex many functions economically.

4.17.1 TSL Inverter

A TSL inverter circuit with tri-state output is shown in Fig. 4.34. When the control input is LOW, the drive is removed from T_3 and T_4. Hence, both T_3 and T_4 are cut-off and the output is in the third state. When the control input is HIGH, the output Y is logic 1 or 0 depending on the data input. The logic symbol of a TSL inverter is shown in Fig. 4.35 and its truth table is given in Table 4.10.

Table 4.10 Truth table of a TSL inverter

Data input	Control	Data output
0	0	HIGH – Z
1	0	HIGH – Z
0	1	1
1	1	0

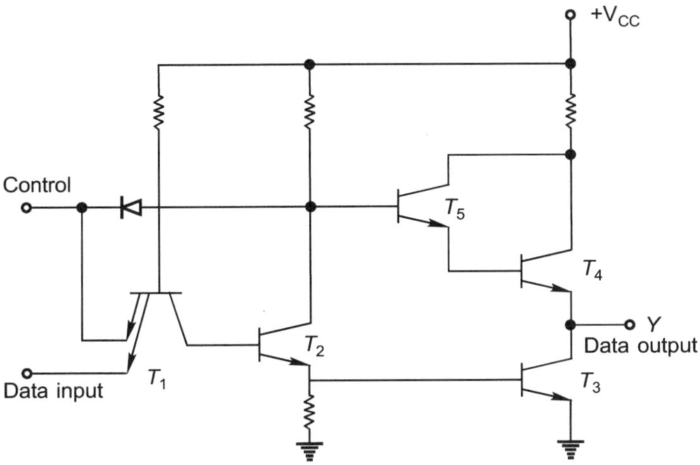

Fig. 4.34
A TSL inverter.

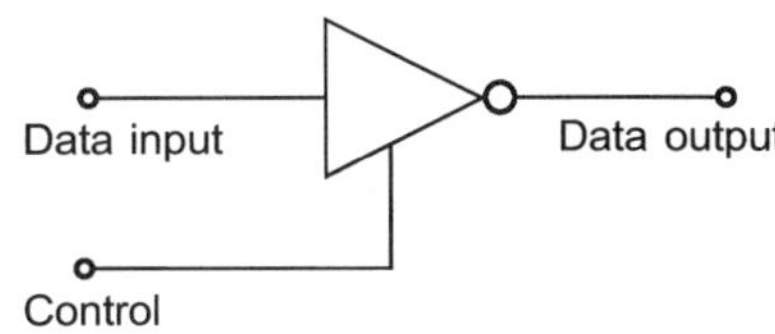

Fig. 4.35
Logic symbol of a TSL inverter.

The output and input current specifications of TSL family are given in Table 4.11.

Table 4.11 Current specifications of TSL family

Parameter	*Control input*	
	LOW (DISABLE)	*HIGH (ENABLE)*
I_{IH}	40 μA	40 μA
I_{IL}	– 1.6 mA	– 1.6 mA
I_{OH}	40 μA	– 5.2 mA
I_{OL}	– 40 μA	16 mA

Example 4.3 Consider the arrangement shown in Fig. 4.36. At any time one of the gates drives the bus line. Calculate the maximum possible value of N.

Solution
Let I_1 be driving the bus line. All other gates, I_2 through I_N must be tristated.

If the output of I_1 is in logic 1 state, it has to supply leakage current (40 μA) to each of the tri-stated gates and input current to G_1 and G_2 (40 μA). From Table 4.11, we have the maximum possible output current of TSL in logic 1 state as 5.2 mA. Therefore,

$$40(N - 1) + 40(2) \leq 5.2 \times 10^3$$

or

$$N \leq 129$$

which means 129 TSL outputs can be connected to the bus line.

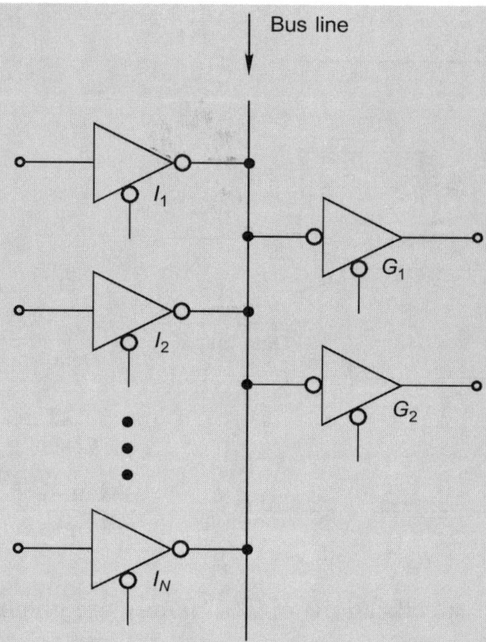

Fig. 4.36
N TSL gates driving a bus line.

4.18 SUMMARY

Essential features of all the major logic families have been discussed and the important conclusion are given below:

1. RTL and DTL families are no more used for new systems because of their low speed, high power dissipation, and low fan-out.
2. TTL is the most popular general purpose logic family. It is available in seven different series with a wide range of operating speed, power dissipation, and fan-out. There are a large number of functions in SSI and MSI available in TTL.

   TTL ICs are available with totem-pole output (which decreases speed-power product), open-collector output (which makes possible wired-AND connection and bus operation), and tri-state (TSL) outputs (which are ideally suited for bus operation).
3. HTL are best suited for an industrial environment where electrical noise level is high.
4. ECL is the fastest logic family. Its main disadvantages are low noise-margins and high power dissipation. For interfacing with other logic families, level-shifting networks are required.
5. I^2L is the only saturated bipolar logic suitable for LSI because of small silicon chip area required, and low power consumption. The supply voltage required is low hence it is highly suitable for battery operated systems.

I^2L circuits can drive TTL circuits if a resistive load is connected to the output stage of I^2L with a higher supply voltage (5 V).

6. MOS devices occupy a very small fraction of silicon chip area in comparison to bipolar devices and require very small power. Therefore, MOS logic is the most popular logic for LSI. The main drawback of MOS logic is slow speed, which is being improved upon by improvements in the technology of MOS fabrication. HMOS, a variety of NMOS has speeds comparable to bipolar logic families.
7. CMOS has the lowest speed power product and requires very small power.

   It is the most popular logic family and has led to the VLSI chips.
8. Corresponding to TTL 54/74 series, 54C/74C, 54HC/74HC, 54HCT/74HCT, 54AC/74AC and 54ACT/74ACT series have been developed which are directly compatible with various 54/74TTL series and have the same numbering scheme and pin-outs.

   A comparison of various digital IC logic families is given in Table 4.12.

Glossary

Active pull-up A circuit with active devices used to pull up the output voltage of a logic circuit from LOW to HIGH in response to the appropriate inputs.

Bipolar logic Logic circuits using bipolar junction semiconductor devices.

Breadth of logic family The number of different types of gates and other functions available in an IC logic family.

Buffer A circuit or gate that can drive a substantially higher number of gates or other loads. Also known as Buffer driver.

Bus A group of conductors carrying a related set of signals.

CMOS (Complementary metal-oxide semiconductor) A MOS device that uses one *p*-channel one *n*-channel device to make an inverter circuit.

Current sink logic A logic circuit in which the output sink current corresponding to logic 0 state is appreciably higher than the output source current corresponding to logic 1 state.

Current source logic A logic circuit in which the output source current corresponding to logic 1 state is appreciably higher than the output sink current corresponding to logic 0 state.

DCTL (Direct-coupled transistor logic) A form of bipolar logic that uses direct coupling.

Depletion mode MOSFET A MOS device in which channel width gets depleted when the voltage of proper polarity is applied at the gate.

DTL (Diode transistor logic) A form of bipolar logic circuit that uses diodes and bipolar junction transistors to realize a logic operation.

ECL (Emitter-coupled logic) A form of bipolar logic circuit that uses emitter-coupled configuration.

Enhancement mode MOSFET A MOS device in which the channel is formed only when a proper voltage is applied at the gate. The channel width enhances with the increased voltage at the gate.

Fan-in The number of inputs of a logic gate.

Table 4.12 Comparison of digital IC logic families

Parameter \ Logic family	*RTL*	*I²L*	*DTL*	*HTL*	*TTL*				
					Standard	*High-Power high-speed H*	*Low-power low-speed L*	*Schottky low-power LS*	*Schottky S*
Basic gate	NOR	NOR	NAND	NAND	NAND	NAND	NAND	NAND	NAND
Fan-out	5	Depends on injector current	8	10	10	10	20	20	10
Power dissipation in mW per gate	12	6nW-70 μW	8-12	55	10	22	1	2	19
Noise immunity	Nominal	Poor	Good	Excel-lent	Very good	Very good	Very good	Very good	Very good
Propagation delay in ns per gate	12	25-250	30	90	10	6	33	9.5	3
Speed-power product (pJ)	144	<1	300	4950	100	132	33	19	57
Clock rate (MHz) for FFs	8	–	72	4	35	50	3	45	125
Available functions	High	LSI only	Fairly high	Nomi-nal	Very high	Very high	Very high	Very high	Very high

(*Contd.*)

Table 4.12 (*Contd.*)

		ECL		*MOS*	*CMOS*				
Advanced Schottky AS	*Advanced low-power Schottky ALS*	*10K*	*100K*		*74C*	*74HC*	*74HCT*	*74AC*	*74ACT*
NAND	NAND	OR−NOR	OR−NOR	NAND	NOR or NAND	NOR or NAND	NOR or NAND	NOR or NAND	NOR or NAND
40	20	25		20	50	20	20	50	50
10	1	40-55	40-55	0.2-10	Static 0.01	0.0025	0.0025	0.005	0.005
					Dynamic 1 MHz 1	0.6	0.6	0.75	0.75
					Total 1.01	0.6025	0.6025	0.755	0.755
Very good	Very good	Poor	Poor	Good	Very good	Very good	Very good	Very good	Very good
1.5	4	2	0.75	300	70.0	18	18	5.25	4.75
15	4	100	40	60	d.c. 0.7	0.045	0.045	0.026	0.024
					dynamic (at 1 MHz)70.7	10.8	10.8	4.0	3.6
175	50	>60	600 MHz	2	10	60	60	100	100
Very high	Very high	High	High	Low	High	High	High	High	High

Fan-out The maximum number of similar logic gates which can be driven by a logic gate.

Field-effect transistor (FET) A three terminal semiconductor device in which the current flow is due to the flow of one type of charge carriers only. The current in the channel is controlled by the field produced due to the applied voltage at the gate.

Figure of merit (of digital ICs) It is the product of speed expressed as propagation delay time and power dissipation. It is also known as the speed power product.

High-impedance state The third state of a tristate logic (TSL) in which the device is inactive and is effectively disconnected from the circuit.

HTL (High-threshold logic) A form of bipolar logic circuit which is identical to DTL but has appreciably higher noise margins.

I^2L (Integrated-injection logic) A form of bipolar logic circuit that uses only bipolar transistors. It is an alternative form of DCTL.

Logic Swing The difference between the voltages corresponding to HIGH and LOW levels.

LSI (Large-scale integration) An IC chip containing logic circuits equivalent of 100 to 1000 gates or containing 1000 –10,000 transistors.

MOSFET (Metal-oxide-semiconductor field-effect transistor) A field-effect transistor consisting of a semiconductor substrate over which an oxide layer is grown and above the oxide layer a metalic layer is deposited which acts as the gate. If is also known as the insulated-gate FET (IGFET).

Merged-transistor logic (MTL) Same as the I^2L.

MSI (Medium-scale integration) An IC chip containing logic circuits equivalent of 13 to 99 gates or containing 100–1000 transistors.

Noise immunity A circuits ability to tolerate noise.

Noise margin A measure of the noise which can be tolerated by a logic circuit.

Noise-margin, high-level For a logic circuit, the difference between the minimum voltage that is produced at the output corresponding to logic 1 and the minimum voltage that is recognised as logic 1 of the input.

Noise-margin, low-level For a logic circuit the difference between the maximum voltage that is recognised as logic 0 at the input and the maximum voltage that is produced corresponding to logic 0 at the output.

Non-saturated logic A logic circuit in which the BJTs are not driven to saturation corresponding to ON state.

Open-collector output An output of a digital IC which is the collector terminal of a BJT not connected to any other point inside the IC.

Open-drain output An output of a MOS IC which is the drain terminal of a MOS device not connected to any other point in the IC.

Open-emitter output An output of an ECL IC which is the emitter terminal of a BJT not connected to any other point in the IC.

Passive pull-up A resistance used to pull-up the output voltage of a logic circuit from LOW to HIGH in response to appropriate inputs.

Pull-up resistor A resistor connected between the output (collector or drain of a transistor) and the supply voltage (V_{CC} or V_{DD}).
Saturated logic A logic circuit in which the BJTs are driven to saturation corresponding to ON state.
Schottky TTL The TTL circuit in which each BJT is replaced by a Schottky transistor.
SSI (small-scale integration) An IC chip containing circuits equivalent of upto 12 gates or 100 transistors.
Three-state gate (tristate gate) A gate having a 1, 0, or high-impedance output states.
Tristate output An output of a logic circuit having 1, 0, or high-impedance states.
Totem-pole output Same as the active pull-up.
TSL (Tristate logic) Same as tristate output.
TTL (Transistor-transistor logic) A form of bipolar logic circuit that uses transistors to realize the logic operations.
Unipolar logic Logic circuits using only MOS devices.
VLSI (Very large-scale integration) An IC chip containing logic circuits equivalent of above 1000 gates or above 10,000 transistors.
Wire-ANDing Tying the outputs of two or more gates together to perform additional logic. Also known as Wired-Logic.

Review Questions

4.1 A logic family using BJTs is known as __________ logic family.
4.2 A unipolar logic family uses only __________ devices.
4.3 Figure of merit of a digital IC is given by __________.
4.4 The number of similar gates which a gate can drive is known as its __________ .
4.5 Fan-in signifies the __________ of a gate.
4.6 A TTL gate is driving another TTL gate. The output transistor of the driver gate is driven __________ into saturation when its output is at low level.
4.7 For interfacing logic gates V_{OH} must be __________ than V_{IH}.
4.8 Outputs of TTL gates with active pull-up must __________ connected together.
4.9 Unconnected input terminal of a TTL gate behaves as __________ .
4.10 The input terminal of a CMOS circuit must __________ .
4.11 Schottky TTL has __________ propagation delay time than TTL.
4.12 The temperature range for 74-series ICs is __________ .
4.13 The states of a TSL are __________ .
4.14 TTL gates with __________ output can be used for wired-logic operation.
4.15 __________ is the fastest logic family.

Problems

4.1 In the RTL NOR gate of Fig. 4.4, calculate the average power supplied by V_{CC} to the driver gate when it is driving 5 gates. Assume $V_{BE,\text{ sat}} \approx 0.8$ V, $V_{CE,\text{ sat}} \approx 0.2$ V, $h_{FE} = 10$. Neglect leakage currents.

4.2 In the circuit of Fig. 4.4, calculate
- (a) Output voltage V_O and noise margin $\Delta 1$ for $N = 5, 6, 7, 8, 9, 10$. Assume $h_{FE} = 10$.
- (b) Repeat (a) for $h_{FE} = 20$.
- (c) Comment on the effect of h_{FE} on the fan-out and noise margin of circuit.
- (d) Comment on the effect of N on the noise margin for a given h_{FE}.

4.3 In the circuit of Fig. 4.6, the fan-out of RTL NOR gates P and Q is 5 each.
- (a) Calculate the fan-out of the combined gate.
- (b) Evaluate the propagation delay time constant and power dissipation, and comment on the effect of wired-logic on these.

4.4 A buffer is used to increase the output drive capability of a logic circuit. An RTL buffer inverter is shown in Fig. 4.37.
- (a) Explain the operation of this circuit.
- (b) Calculate the fan-out. Assume $h_{FE} = 30$.
- (c) Consider outputs of two such buffers A and B connected in parallel. Let the input to buffer A be logic 1 and the input to buffer B be logic 0. Calculate the current flowing in T_3 of buffer A.

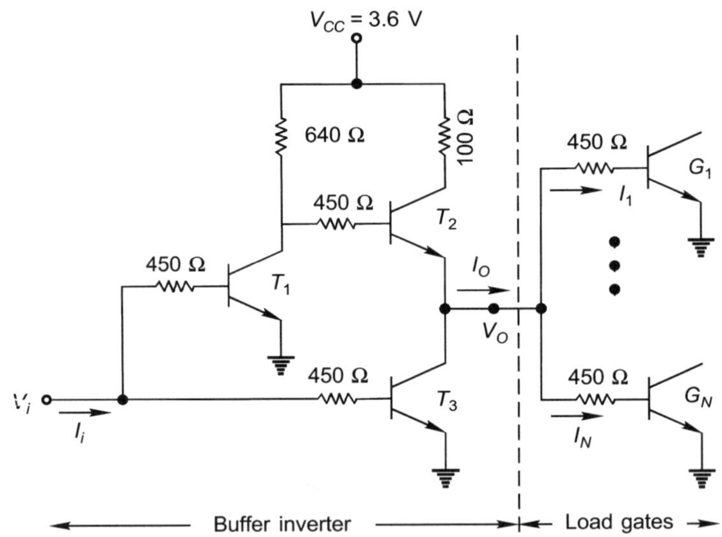

Fig. 4.37
An RTL buffer inverter driving N RTL gates.

4.5 What will happen in the DTL circuit of Fig. 4.12 if
- (a) one of the diodes D_1 or D_2 is removed,
- (b) one more diode D_3 is inserted in series with D_1 and D_2.

4.6 Calculate the value of h_{FE} required for a fan-out of 10 in the DTL gate of Fig. 4.12.

4.7 M DTL gates (Fig 4.12) each with a fan-out of N are connected in a wired-AND connection. Determine the fan-out of this combination as a function of number M.

4.8 In the modified DTL NAND gate of Fig. 4.14, show that when T_1 is conducting it is in its active region and not in the saturation region.

4.9 Explain the operation of the modified DTL gate of Fig. 4.14 and calculate its (a) fan-out (b) noise-margins, and (c) average power dissipation. Assume $h_{FE} = 30$.

4.10 Calculate (a) noise-margins, (b) fan-out, and (c) power dissipation of HTL gate of Fig. 4.15. Assume $h_{FE} = 40$.

4.11 Repeat Problem 4.7 for the HTL gate shown in Fig. 4.15.

4.12 Explain why the temperature sensitivity of HTL is significantly better than that of DTL.

4.13 In the TTL NAND gate of Fig. 4.17 determine the current drawn from the supply, when the output
(a) is LOW
(b) is HIGH
(c) makes a transition from LOW to HIGH.

4.14 Consider the circuit shown in Fig. 4.38 which uses TTL gates. The current I is 1.6 mA when terminal B is left unconnected. Find the value of I when B is connected to A. Comment on the effect of this connection on the fan-out of gate G_1.

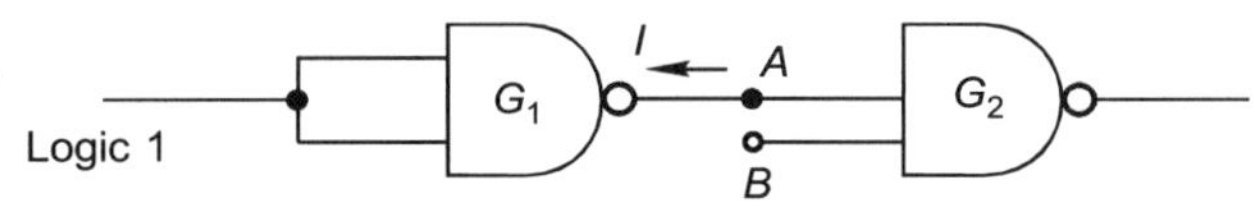

Fig. 4.38

Circuit for Problem 4.14.

4.15 In the TTL gate of Fig. 4.17, what happens if
(a) $R_{C4} = 0$,
(b) diode D is not present,
(c) the output accidently gets shorted to ground?

4.16 The outputs of two totem-pole TTL gates (Fig. 4.17) are connected as shown in Fig. 4.39. Obtain the current drawn from the supply for all the possible combinations of the inputs to the two gates.

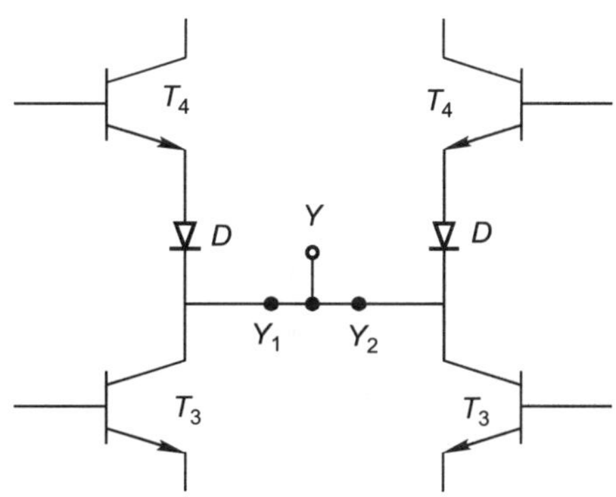

Fig. 4.39

Circuit for Problem 4.16.

4.17 For an open-collector TTL gate, the specifications are:

$$V_{OH} = 2.4 \text{ V}$$
$$V_{OL} = 0.4 \text{ V}$$
$$I_{OH} = 250\ \mu\text{A}$$
$$I_{OL} = 16 \text{ mA}$$
$$I_{IH} = 40\ \mu\text{A}$$
$$I_{IL} = -1.6 \text{ mA}$$

Calculate the value of R_C required for the open-collector gate. Assume $V_{CC} = 5$ V and a fan-out of 8.

4.18 If 5 open-collector gates of Problem 4.17 are wire-ANDed, and are loaded by similar 6 gates, calculate the value of collector resistor R_C required.

4.19 For an open-collector TTL, non-inverting buffer (7407) the specifications are:

$V_{OH} = 30$ V (maximum), $V_{IH} = 2.0$ V
$V_{OL} = 0.4$ V, $V_{IL} = 0.8$ V
$I_{OH} = 250\ \mu$A, $I_{IH} = 40\ \mu$A
$I_{OL} = 40$ mA, $I_{IL} = -1.6$ mA

If 7 such gates are wire-ANDed and drive 7 standard TTL gates of 74-series, determine the value of supply voltage V_{CC} and the collector resistor R_C to be used.

4.20 If it is desired to use a 10 V, 30 mA lamp as load in a digital circuit, can you use a 74-series TTL gate with (a) totem-pole output (b) with passive pull-up (c) open-collector output (specifications given in Problem 4.17), (d) open-collector buffer 7407?
In case your answer is yes, give the circuit arrangement and explain its operation.

4.21 Verify Table 4.4 using the specifications given in Table 4.3.

4.22 Consider the ECL circuit shown in Fig. 4.40. Here, V_n represents the noise. Calculate the noise component in the output taken between.
(a) Y and P terminals (b) Y and Q terminals
Hence justify the grounding of positive end of the supply voltage. Assume $h_{FE} = 100$.

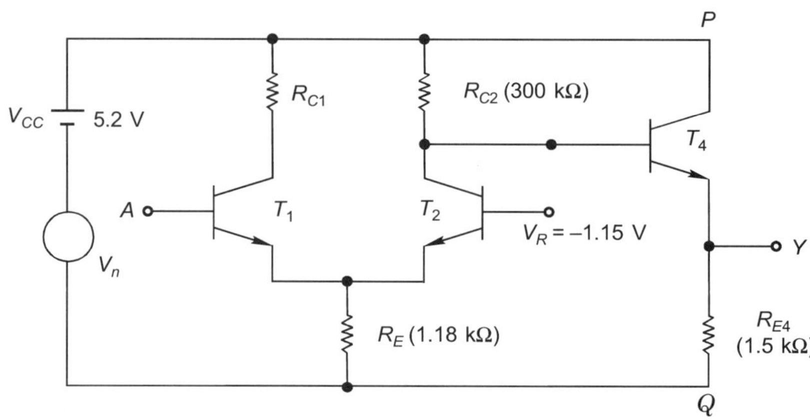

Fig. 4.40
Circuit for Problem 4.22.

4.23 (a) What will happen in the ECL gate of Fig. 4.19 if Y_1 or Y_2 accidently gets shorted to ground.
(b) Repeat part (a) if negative end of the supply is grounded.

4.24 Compare the current spikes in ECL and TTL gates.

4.25 Verify the operation of wired-OR connections of ECL gates shown in Fig. 4.21.

4.26 Prove that the input of MC10H125 IC is ECL compatible and its output is TTL compatible.

4.27 Design a circuit for interfacing an ECL 2-input NOR gate with a TTL inverter to obtain NOR function of the combined circuit.

4.28 What happens if output accidently gets shorted to ground in
(i) NMOS? (ii) CMOS?

4.29 Explain the operation of CMOS NOR gate of Fig. 4.28.

4.30 Find the fan-out of each of the 74 TTL series driving 74 HC/74 HCT/74 AC/74 ACT gates using the specifications given in Tables 4.3 and 4.9.

4.31 Consider a CMOS gate driving TTL gates. Find the fan-out when
(a) 74HC/74HCT gate is driving each of TTL series gates.
(b) 74AC/74ACT gate is driving each of TTL series gates.

4.32 A 74AC/74ACT gate is driving twenty 74AS gates. It is desired to drive some 74ALS gates in addition to this. Find the maximum possible number of 74ALS gates which can be connected.

4.33 Is it possible to use TTL-to-ECL translator for CMOS-to-ECL interfacing? Justify your answer.

4.34 Is it possible to use ECL-to-TTL translator for ECL-to-CMOS interfacing? Justify your answer.

5

COMBINATIONAL LOGIC DESIGN

5.1 INTRODUCTION

Logic operations and Boolean algebra have already been discussed in Chapter 1. Boolean algebraic theorems are used for the manipulations of logical expressions. It has also been demonstrated that a logical expression can be realized using the logic gates. The number of gates and the number of input terminals for the gates required for the realization of a logical expression, in general, get reduced considerably if the expression can be simplified. Therefore, the simplification of logical expression is very important as it saves the hardware required to design a specific system. A large number of functions are available in IC form and therefore, we should be able to make optimum use of these ICs in the design of digital systems. That is, our aim should be to minimize the number of IC packages. Some of the logical gates available in ICs have been discussed in Section 1.7. We will be discussing in Chapter 6, some of the MSI digital circuits available in IC form and design of digital systems using these ICs. Programmable logic devices (PLDs), field-programmable gate arrays (FPGAs) and the design using these devices have been discussed in Chapter 12.

Basically, digital circuits are divided into two broad categories:

1. Combinational circuits, and
2. Sequential circuits.

In *combinational circuits*, the outputs at any instant of time depend upon the inputs present at that instant of time. This means there is no memory in these circuits. There are other types of circuits in which the outputs at any instant of time depend upon the present inputs as well as past inputs/outputs. This means that there are elements used to store past information. These elements are known as *memory*. Such circuits are known as *sequential circuits*. A sequential logic system may have combinational logic sub-systems. The design of combinational circuits will be discussed here. Sequential circuits design will be discussed later.

The design requirements of combinational circuits may be specified in one of the following ways:

1. A set of statements,
2. Boolean expression, and
3. Truth table.

The aim is to design a circuit using the gates already discussed or some other circuits which are in fact derived from the basic gates. As is usual in any engineering design, the number of components used should be minimum to ensure low cost, saving in space, power requirements, etc. There can be two different approaches to the design of combinational circuits. One of these is the traditional method, wherein the given Boolean expression or the truth table is simplified by using standard methods and the simplified expression is realized using the gates. The other method normally does not require any simplification of the logical expression or truth table, instead the complex logic functions available in medium scale integrated circuits (MSI) can be directly used. Computer-aided design (CAD) tools are used for the design using PLDs and FPGAs. Combinational circuit design using the traditional design methods have been discussed below, however, the concepts used in these methods will help in the understanding of design using MSIs, PLDs, and FPGAs, etc.

The following methods can be used to simplify the Boolean functions:

1. Algebraic method,
2. Karnaugh-map technique,
3. Quine–McCluskey method, and
4. Variable entered mapping (VEM) technique.

The algebraic method, the Karnaugh map (K-map) technique, and the Quine-McCluskey method have been discussed here. The K-map is the simplest and most commonly used method. It is a manual method and depends to a great extent upon human intuition. This method can be used conveniently up to six variables beyond which it is very cumbersome.

The Quine–McCluskey method is suitable for computer mechanization and is seldom used by logic designers manually. The variable entered mapping (VEM) has been discussed in Fletcher[2] and the interested reader can refer to it.

5.2 STANDARD REPRESENTATIONS FOR LOGICAL FUNCTIONS

Logical functions are expressed in terms of logical variables. The values assumed by the logical functions as well as the logical variables are in the binary form. Any arbitrary logic function can be expressed in the following forms:

1. Sum-of-products form (SOP), and
2. Product-of-sums form (POS)

This does not mean that the logic function cannot be written in any other form. It can be written in various forms but the above two forms are conveniently suited in arriving at the standard methods for designing the circuits which will become clear from the following discussions.

[2] Fletcher, W.L., *An Engineering Approach to Digital Design*, Prentice-Hall, Englewood Cliffs, New Jersey, 1980.

Example 5.1 Given the logical equation

$$Y = (A + BC)(B + \bar{C}A) \tag{5.1}$$

(a) Design a circuit using gates to realize this function.
(b) Find out whether it is possible to design the circuit with only one type of gates (NAND or NOR). If yes, design the circuits.
(c) Find out whether it is possible to simplify this equation. If yes, simplify it.
(d) Now design the circuit using the simplified expression obtained in part (c).
(e) Compare the circuits obtained in parts (a), (b), and (d) from the point of view of number of gates, number of inputs for the gates, types of gates, and propagation delay.

Solution

In Eq. (5.1), there are three input logical variables A, B, and C, and Y is the output. The variable C appears as C in one term and as $\bar{C}$ in the other term. A variable in *uncomplemented* or *complemented* form is known as a *literal.*

(a) A circuit using gates can simply be designed by looking at the expression and finding out the basic gates which can be used to realize the various terms and then connect these gates appropriately. We assume that the signals corresponding to each literal are available, that is the variables are available in uncomplemented as well as the complemented form.
   (i) The first term (A) has only one literal A and the second term (BC) has two literals B and C. The second term is recognized as an AND operation and can be realized by using a 2-input AND gate. The combination of these two terms is realized by using a 2-input OR gate. The complete realization of first two terms is shown in Fig. 5.1*a*.
   (ii) The third term (B) is again a single literal term and the fourth term ($\bar{C}A$) has two literals. The combination of these two terms is similar to the combination of the first two terms and is realized in a similar way. This realization is shown in Fig. 5.1*b*.
   (iii) Now, the complete realization is obtained by using a 2-input AND gate with Y_1 and Y_2 as the inputs and the output of this gate will be the required output Y. This realization is given in Fig. 5.1*c*. The above design requires three 2-input AND gates and two 2-input OR gates.

(b) (i) Sum-of-Products Form

Equation (5.1) can be written as

$$Y = A(B + \bar{C}A) + (BC)(B + \bar{C}A) \qquad \text{(Theorem 1.9)}$$

$$= AB + A\bar{C}A + BCB + BC\bar{C}A \qquad \text{(Theorem 1.9)} \tag{5.2}$$

$$= AB + A\bar{C} + BC \tag{5.3}$$

Equation (5.3) is obtained from Eq. (5.2) in the following way:

$$A\bar{C}A = (A \cdot A) \cdot \bar{C} = A\bar{C} \qquad \text{(Theorem 1.6)}$$

$$BCB = (B \cdot B) \cdot C = BC \qquad \text{(Theorem 1.6)}$$

$$BC\bar{C}A = B \cdot (C \cdot \bar{C}) \cdot A = B \cdot O \cdot A \qquad \text{(Theorem 1.8)}$$

$$= 0 \qquad \text{(Theorem 1.4)}$$

This term does not appear in Eq. (5.3), because of Theorem 1.1.

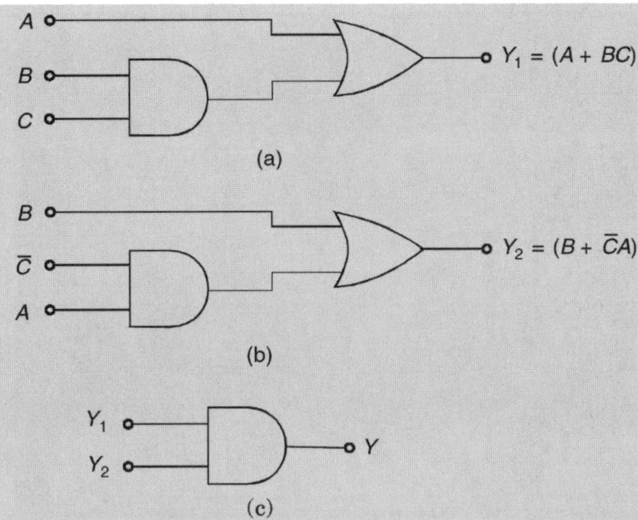

Fig. 5.1
Logic circuits for the realization of Eq. (5.1).

The representation of Eq. (5.3) is known as *sum-of-products* (SOP) form. This can be realized using AND–OR configuration as shown in Fig. 5.2*a*. This realization is known as *two-level realization*. The first level consists of AND gates and the second level consists of the OR gate. By making use of the De-Morgans theorem (Theorem 1.22) we can write Eq. (5.3) as

$$\bar{Y} = \overline{AB + A\bar{C} + BC}$$

$$= \overline{AB} \cdot \overline{A\bar{C}} \cdot \overline{BC}$$

or $$Y = \overline{Y_1 \cdot Y_2 \cdot Y_3} \quad (5.4)$$

Where $$Y_1 = \overline{AB}$$

$$Y_2 = \overline{A\bar{C}}$$

$$Y_3 = \overline{BC}$$

Equation (5.4) can be realized using NAND gates only. The realization is given in Fig. 5.2*b*. This is also a two-level realization and in this only NAND gates are used. Therefore, if we express the equation in the SOP form we can always design the circuit using only one type of gates (NAND).

(ii) Product-of-Sums Form

It is possible to represent Eq. (5.1) in another form.

$$Y = (A + B)(A + C)(B + \bar{C})(B + A) \quad \text{(Theorem 1.10)}$$

$$= (A + B)(A + C)(B + \bar{C}) \quad \text{(Theorem 1.6)} \quad (5.5)$$

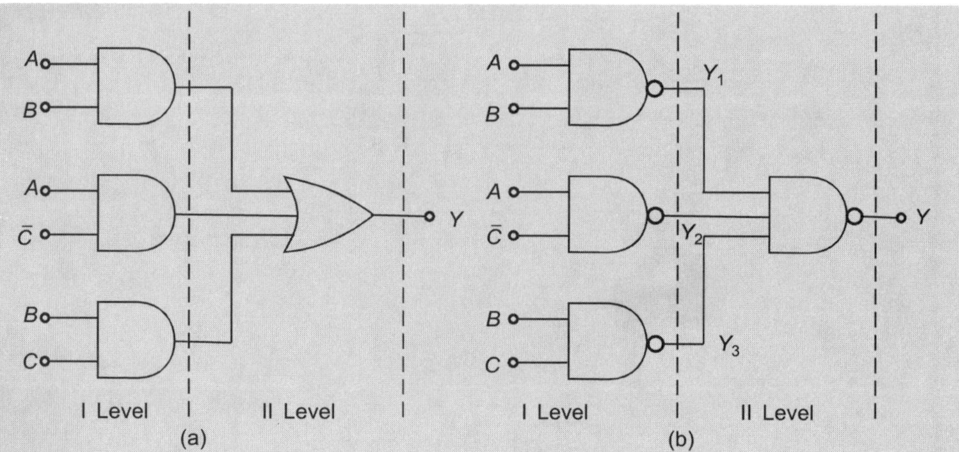

Fig. 5.2

Realization of Eq. (5.3) using (a) AND–OR (b) NAND–NAND.

The representation of Eq. (5.5) is known as *Product-of-sums* (POS) form. This can be realized using OR–AND gates as shown in Fig. 5.3*a*. This is also a two-level realization.

Using De-Morgans theorem (Theorem 1.21), we can write Eq. (5.5) as

$$\bar{Y} = \overline{(A+B)(A+C)(B+\bar{C})}$$

$$= (\overline{A+B}) + (\overline{A+C}) + (\overline{B+\bar{C}})$$

or $$Y = \overline{Y_A + Y_B + Y_C} \tag{5.6}$$

Where $$Y_A = \overline{A+B}$$

$$Y_B = \overline{A+C}$$

$$Y_C = \overline{B+\bar{C}}$$

Equation (5.6) can be realized using NOR gates only. This realization is given in Fig. 5.3*b* which is a two-level realization. Hence, if we express the equation in POS form we can always design the circuit using only one type of gates (NOR).

(c) (i) Simplification of Eq. (5.3),

$$Y = AB + A\bar{C} + BC$$

$$= BC + A\bar{C} \qquad \text{(Theorem 1.19)} \tag{5.7}$$

(ii) Simplification of Eq. (5.5),

$$Y = (A + B)(A + C)(B + \bar{C})$$

$$= (A + C)(B + \bar{C}) \qquad \text{(Theorem 1.20)} \tag{5.8}$$

(d) Realization of Eqs (5.7) and (5.8) are given in Figs 5.4*a* and 5.4*b*, respectively.

(e) The gate requirements corresponding to parts (a), (b), and (d) are given in Table 5.1. The realization of part (a) needs maximum number of gates. Also, it is a three-level

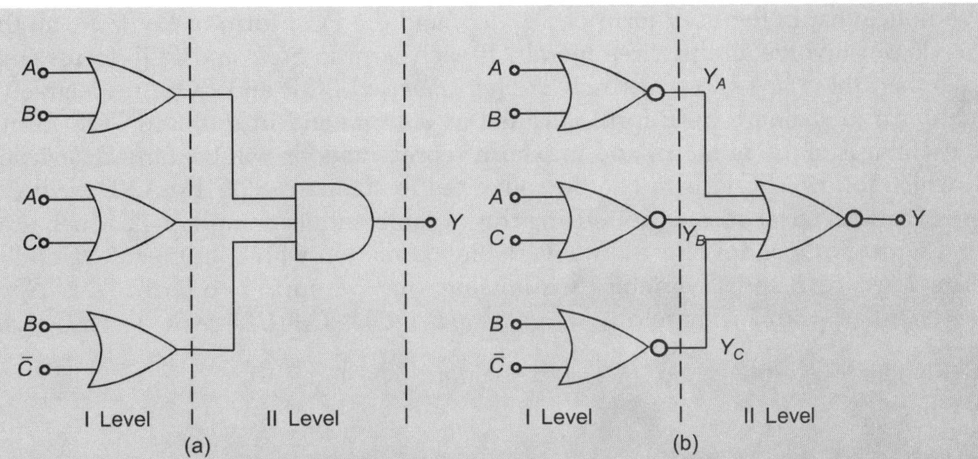

Fig. 5.3
Realization of Eq. (5.5) using (a) OR–AND (b) NOR–NOR

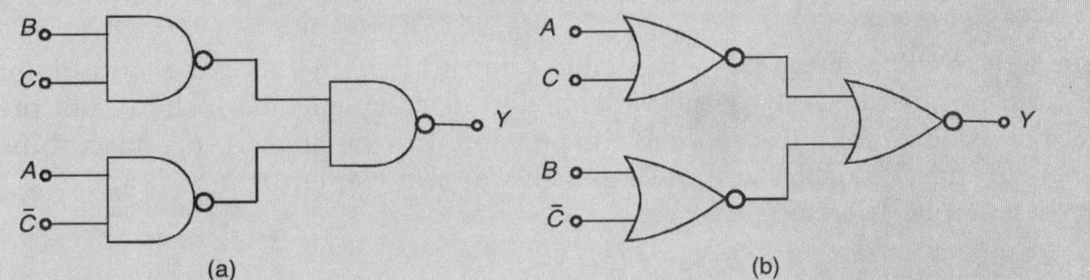

Fig. 5.4
Realization of (a) Equation (5.7) (b) Equation (5.8).

Table 5.1 Gate requirements

Part (a)	*Part (b)*		*Part (d)*
	AND–OR/NAND–NAND	*OR–AND/NOR–NOR*	
3, 2-input AND	3, 2-input AND	3, 2-input OR	3, 2-input NAND or 3, 2-input NOR
2, 2-input OR	1, 3-input OR or 3, 2-input NAND 1, 3-input NAND	1, 3-input AND or 3, 2-input NOR 1, 3-input NOR	

realization which increases the propagation delay time which is same as decrease in speed of operation. Realizations corresponding to parts (b) and (d) are very useful since only one type of gates (NAND / NOR) are required which is very convenient to use when we use ICs because a number of similar gates are available in the same package. Realization corresponding to part (d) requires minimum number of gates. Therefore, the simplification of logical expressions is very useful.

We notice that in the SOP form of Eq. (5.3) and the POS form of Eq. (5.5), all the individual terms do not involve all the three literals. If each term in SOP and POS forms contains all the literals then these are known as *standard* (*or canonical*) SOP and POS, respectively. Each individual term in standard SOP form is called as *minterm* and in standard POS form as *maxterm.* The usefulness of the minterm and maxterm representations will become clear from the discussion which follows. SOP form can be converted to standard SOP by ANDing the terms in the expression with terms formed by ORing the variable and its complement which are not present in that term. For example for a three-variable expression with variables A, B, and C, if there is a term A, where B and C variables are missing, then we form two terms $(B + \overline{B})$ and $(C + \overline{C})$ and AND them with A. Therefore, we get $A \cdot (B + \overline{C}) \cdot (C + \overline{C}) = ABC + A\overline{B}C + AB\overline{C} + A\overline{B}\overline{C}$.

Example 5.2 Convert Eq. (5.3) into standard SOP form.

Solution

$$\begin{aligned} Y &= AB(C + \overline{C}) + A\overline{C}(B + \overline{B}) + BC(A + \overline{A}) \\ &= ABC + AB\overline{C} + AB\overline{C} + A\overline{B}\overline{C} + ABC + \overline{A}BC \\ &= ABC + AB\overline{C} + A\overline{B}\overline{C} + \overline{A}BC \qquad \text{(Theorem 1.5)} \qquad (5.9) \end{aligned}$$

Similarly, POS form can be converted to standard POS by ORing the terms in the expression with terms formed by ANDing the variable and its complement which are not present in that term. For example for a three-variable expression with variables A, B, and C, if there is a term A where B and C variables are missing, then we form two terms $B \times \overline{B}$ and $C \times \overline{C}$ and OR A with these terms. Therefore, we get

$$\begin{aligned} & A + B\overline{B} + C\overline{C} \\ &= (A + B\overline{B} + C)(A + B\overline{B} + \overline{C}) \qquad \text{(Theorem 1.10)} \\ &= (A + B + C)(A + \overline{B} + C)(A + B + \overline{C})(A + \overline{B} + \overline{C}) \qquad (5.10) \end{aligned}$$

Example 5.3 Convert Eq. (5.5) into standard POS form.

Solution

$$\begin{aligned} Y &= (A + B + C\overline{C})(A + B\overline{B} + C)(A\overline{A} + B + \overline{C}) \\ &= (A + B + C)(A + B + \overline{C})(A + B + C)(A + \overline{B} + C) \\ &\quad (A + B + \overline{C})(\overline{A} + B + \overline{C}) \qquad \text{(Theorem 1.10)} \\ &= (A + B + C) \cdot (A + B + \overline{C}) \cdot (A + \overline{B} + C) \cdot (\overline{A} + B + \overline{C}) \\ &\qquad \text{(Theorem 1.6)} \qquad (5.11) \end{aligned}$$

The concept of minterm and maxterm introduced above allows us to introduce a very convenient shorthand notation to express logical functions. Table 5.2 gives the minterms and maxterms for a four variable logical function where the number of minterms as well as maxterms is $2^4 = 16$. In general, for an n-variable logical function there are 2^n minterms and an equal number of maxterms. While writing a particular minterm or maxterm, we shall always write it with the variables in an orderly way as can be seen from Table 5.2. Each minterm is represented by m_i where the subscript i is the decimal equivalent of the natural binary number corresponding to the minterm with normal (uncomplemented) variables taken as 1's and the complemented variables taken as 0's.

Table 5.2 Minterms/maxterms for four variables

Variable A B C D	*Minterm* m_i	*Maxterm* M_i
0 0 0 0	$\bar{A}\,\bar{B}\,\bar{C}\bar{D} = m_0$	$A + B + C + D = M_0$
0 0 0 1	$\bar{A}\,\bar{B}\,\bar{C}D = m_1$	$A + B + C + \bar{D} = M_1$
0 0 1 0	$\bar{A}\,\bar{B}\,C\bar{D} = m_2$	$A + B + \bar{C} + D = M_2$
0 0 1 1	$\bar{A}\,\bar{B}\,C\,D = m_3$	$A + B + \bar{C} + \bar{D} = M_3$
0 1 0 0	$\bar{A}\,B\bar{C}\bar{D} = m_4$	$A + \bar{B} + C + D = M_4$
0 1 0 1	$\bar{A}\,B\bar{C}D = m_5$	$A + \bar{B} + C + \bar{D} = M_5$
0 1 1 0	$\bar{A}\,B\,C\bar{D} = m_6$	$A + \bar{B} + \bar{C} + D = M_6$
0 1 1 1	$\bar{A}\,B\,C\,D = m_7$	$A + \bar{B} + \bar{C} + \bar{D} = M_7$
1 0 0 0	$A\,\bar{B}\,\bar{C}\bar{D} = m_8$	$\bar{A} + B + C + D = M_8$
1 0 0 1	$A\,\bar{B}\,\bar{C}D = m_9$	$\bar{A} + B + C + \bar{D} = M_9$
1 0 1 0	$A\,\bar{B}\;C\,\bar{D} = m_{10}$	$\bar{A} + B + \bar{C} + D = M_{10}$
1 0 1 1	$A\,\bar{B}\;C\,D = m_{11}$	$\bar{A} + B + \bar{C} + \bar{D} = M_{11}$
1 1 0 0	$A\,B\;\bar{C}\bar{D} = m_{12}$	$\bar{A} + \bar{B} + C + D = M_{12}$
1 1 0 1	$A\,B\bar{C}D = m_{13}$	$\bar{A} + \bar{B} + C + \bar{D} = M_{13}$
1 1 1 0	$A\,B\,C\,\bar{D} = m_{14}$	$\bar{A} + \bar{B} + \bar{C} + D = M_{14}$
1 1 1 1	$A\,B\,C\,D = m_{15}$	$\bar{A} + \bar{B} + \bar{C} + \bar{D} = M_{15}$

Similar to minterms, each maxterm is represented by M_i, where the subscript i is the decimal equivalent of the natural binary number corresponding to the maxterm with uncomplemented variables taken as 0's and complemented variables taken as 1's. Using these notations the standard SOP (Eq. (5.9)) can be written as:

$$Y = m_3 + m_4 + m_6 + m_7$$
$$= \Sigma m\,(3, 4, 6, 7) \tag{5.12}$$

Where $\bar{A}BC = m_3$

$A\bar{B}\bar{C} = m_4$

$AB\bar{C} = m_6$

and $ABC = m_7$

Similarly, the standard POS (Eq. (5.11)) can be written as

$$Y = M_0 \cdot M_1 \cdot M_2 \cdot M_5$$
$$= \Pi\, M\,(0, 1, 2, 5) \tag{5.13}$$

Where $A + B + C = M_0$

$A + B + \bar{C} = M_1$

$A + \bar{B} + C = M_2$

and $\bar{A} + B + \bar{C} = M_5$

Equations (5.12) and (5.13) are the shorthand forms of standard SOP and POS respectively. Since these two equations represent the same logical function (Eq. (5.1)), therefore we notice that there is a complementary type of relationship between a function expressed in terms of minterms and in terms of maxterms. Here, we are dealing with a three-variable function where the number of minterms and maxterms are $2^3 = 8$ and the corresponding decimal numbers are 0 through 7. Out of this the terms corresponding to decimal numbers 3, 4, 6, and 7 are minterms, and the terms corresponding to decimal numbers 0, 1, 2, and 5 are maxterms. Hence, if a logical function is specified in terms of minterm/maxterm, its maxterm/minterm representation can be determined by using this complementary property. For example, for a four-variable case if

$$Y = \Sigma\, m\,(0, 3, 6, 7, 10, 12, 15)$$

then

$$Y = \Pi\, M\,(1, 2, 4, 5, 8, 9, 11, 13, 14).$$

5.3 KARNAUGH MAP REPRESENTATION OF LOGICAL FUNCTIONS

We have discussed the two standard forms of logical functions and their realizations using gates. We have also established the need for the simplification of the Boolean expressions and introduced the algebraic method of simplification using Boolean algebraic theorems. Sometimes it is difficult to be sure that a logical expression can be simplified. There is another technique, which is graphical, known as the Karnaugh map technique which provides a systematic method for simplifying and manipulating Boolean expressions. In this technique, the information contained in a truth table or available in POS or SOP form is represented on Karnaugh map (K-map). This is perhaps the most extensively used tool for simplification of Boolean functions. Although the technique may be used for any number of variables, it is generally used up to six variables beyond which it becomes very cumbersome.

Figure 5.5 shows the K-maps for two, three and four variables. In an n-variable K-map there are 2^n cells. Each cell corresponds to one of the combinations of n variables, since there are 2^n combinations of n variables. Therefore, we see that for each row of the truth table, for each minterm and for each maxterm there is one specific cell in the K-map. The variables have been designated as A, B, C, and D, and the binary numbers formed by them are taken as AB, ABC, and $ABCD$ for two, three, and four variables respectively. In each map the variables and all possible values of the variables are indicated (the first bit corresponds to the first variable and the second bit

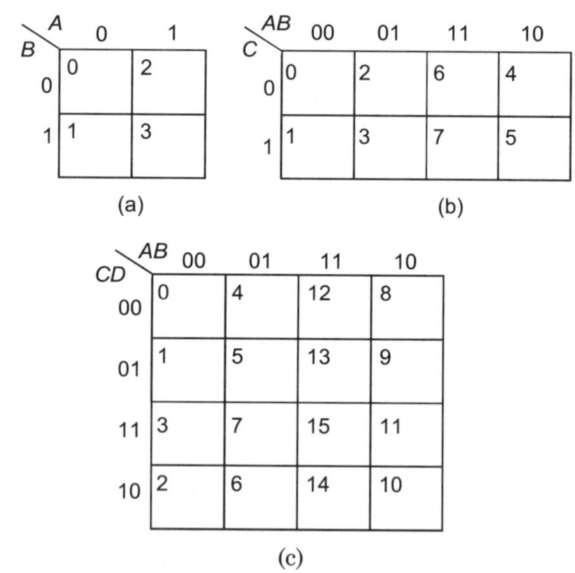

Fig. 5.5 Karnaugh-maps (a) two-variable (b) three-variable (c) four-variable.

corresponds to the second variable) to identify the cells. Gray code has been used for the identification of cells. The reason for using Gray code will become clear when we discuss the application of K-map. You can verify the decimal number corresponding to each cell which is written in the top left corner of the cell as shown in Fig. 5.5.

Figure 5.6 shows the minterm/maxterm corresponding to each cell and the term is written inside the cell for clear understanding.

B \ A	0	1
0	$\bar{A}\bar{B}$	$A\bar{B}$
1	$\bar{A}B$	AB

(a)

B \ A	0	1
0	$A + B$	$\bar{A} + B$
1	$A + \bar{B}$	$\bar{A} + \bar{B}$

(b)

C \ AB	00	01	11	10
0	$\bar{A}\bar{B}\bar{C}$	$\bar{A}B\bar{C}$	$AB\bar{C}$	$A\bar{B}\bar{C}$
1	$\bar{A}\bar{B}C$	$\bar{A}BC$	ABC	$A\bar{B}C$

(c)

CD \ AB	00	01	11	10
0	$A + B + C$	$A + \bar{B} + C$	$\bar{A} + \bar{B} + C$	$\bar{A} + B + C$
1	$A + B + \bar{C}$	$A + \bar{B} + \bar{C}$	$\bar{A} + \bar{B} + \bar{C}$	$\bar{A} + B + \bar{C}$

(d)

CD \ AB	00	01	11	10
00	$\bar{A}\bar{B}\bar{C}\bar{D}$	$\bar{A}B\bar{C}\bar{D}$	$AB\bar{C}\bar{D}$	$A\bar{B}\bar{C}\bar{D}$
01	$\bar{A}\bar{B}\bar{C}D$	$\bar{A}B\bar{C}D$	$AB\bar{C}D$	$A\bar{B}\bar{C}D$
11	$\bar{A}\bar{B}CD$	$\bar{A}BCD$	$ABCD$	$A\bar{B}CD$
10	$\bar{A}\bar{B}C\bar{D}$	$\bar{A}BC\bar{D}$	$ABC\bar{D}$	$A\bar{B}C\bar{D}$

(e)

CD \ AB	00	01	11	10
00	$A+B+C+D$	$A+\bar{B}+C+D$	$\bar{A}+\bar{B}+C+D$	$\bar{A}+B+C+D$
01	$A+B+C+\bar{D}$	$A+\bar{B}+C+\bar{D}$	$\bar{A}+\bar{B}+C+\bar{D}$	$\bar{A}+B+C+\bar{D}$
11	$A+B+\bar{C}+\bar{D}$	$A+\bar{B}+\bar{C}+\bar{D}$	$\bar{A}+\bar{B}+\bar{C}+\bar{D}$	$\bar{A}+B+\bar{C}+\bar{D}$
10	$A+B+\bar{C}+D$	$A+\bar{B}+\bar{C}+D$	$\bar{A}+\bar{B}+\bar{C}+D$	$\bar{A}+B+\bar{C}+D$

(f)

Fig. 5.6

Maxterm/minterm corresponding to each cell of K-maps.

5.3.1 Representation of Truth Table on K-Map

Consider the truth table of the 3-variable logic function given in Table 5.3. The output Y is logical 1 corresponding to the rows 1, 2, 4, and 7. Corresponding to this we can write the equation in terms of standard SOP as given below:

$$Y = \bar{A}\bar{B}C + \bar{A}B\bar{C} + A\bar{B}\bar{C} + ABC \tag{5.14}$$

Equation (5.14) represents the complete truth table in standard SOP form. Similarly, we note that the output Y is logical 0 corresponding to the rows 0, 3, 5, and 6 and the output Y can be represented in terms of standard POS form as given below:

$$Y = (A + B + C)(A + \bar{B} + \bar{C})(\bar{A} + B + \bar{C})(\bar{A} + \bar{B} + C) \tag{5.15}$$

Equation (5.15) also represents the complete truth table, and Eqs (5.14) and (5.15) are equivalent. We shall make use of the 3-variable K-map of Fig. 5.5*b* and enter the value of the output variable Y (0 or 1) in each cell corresponding to its decimal or minterm or maxterm identification. Figure 5.7 gives the complete K-map of the truth table given in Table 5.3.

C \ AB	00	01	11	10
0	0: 0	2: 1	6: 0	4: 1
1	1: 1	3: 0	7: 1	5: 0

Fig. 5.7
K-map for Table 5.3.

The procedure used above is general and is used to represent a truth table on the K-map. On the other hand, if a K-map is given we can make the truth table corresponding to this by following the reverse process. That is, the output Y is logical 1 corresponding to the decimal numbers/minterms represented by cells with entries 1. In all other rows, the output Y is logical 0.

Table 5.3 Truth table of a 3-variable function

	Inputs			Output
Row No.	*A*	*B*	*C*	*Y*
0	0	0	0	0
1	0	0	1	1
2	0	1	0	1
3	0	1	1	0
4	1	0	0	1
5	1	0	1	0
6	1	1	0	0
7	1	1	1	1

Example 5.4 Prepare the truth table for K-map of Fig. 5.8.

Solution The truth table is given in Table 5.4.
Note that the 0's are not written in the K-map of Fig. 5.8. Actually, we need enter either 0's or 1's only in the K-map. If only 1's are entered the empty cells are 0's and if only 0's are entered then the empty cells are 1's.

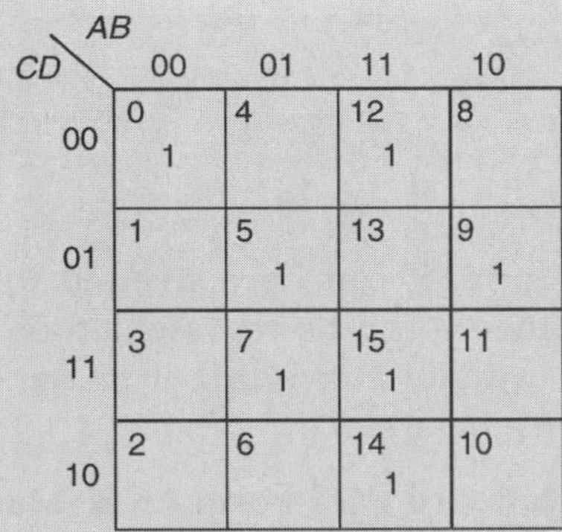

Fig. 5.8
K-map of Ex. 5.4.

Table 5.4 Truth table for K-map of Fig. 5.8

	Inputs				*Output*
Row No.	*A*	*B*	*C*	*D*	*Y*
0	0	0	0	0	1
1	0	0	0	1	0
2	0	0	1	0	0
3	0	0	1	1	0
4	0	1	0	0	0
5	0	1	0	1	1
6	0	1	1	0	0
7	0	1	1	1	1
8	1	0	0	0	0
9	1	0	0	1	1
10	1	0	1	0	0
11	1	0	1	1	0
12	1	1	0	0	1
13	1	1	0	1	0
14	1	1	1	0	1
15	1	1	1	1	1

5.3.2 Representation of Standard SOP Form on K-Map

A logical equation in standard SOP form can be represented on a K-map by simply entering 1's in the cells of the K-map corresponding to each minterm present in the equation.

Example 5.5 Represent Eq. (5.14) on K-map.

Solution
Corresponding to each minterm in the equation, there is a cell in the K-map and a 1 is entered in each one of these cells. The K-map will be as shown in Fig. 5.7.

Similarly, from the K-map, we can write the corresponding logical equation in standard SOP form by ORing the minterms corresponding to each 1 entry in the K-map.

Example 5.6 Write the logical equation in the standard SOP form for the K-map of Fig. 5.8.

Solution

$$Y = \bar{A}\bar{B}\bar{C}\bar{D} + AB\bar{C}\bar{D} + \bar{A}B\bar{C}D + A\bar{B}\bar{C}D + \bar{A}BCD + ABCD + ABC\bar{D}$$
$$= \Sigma\, m\,(0, 5, 7, 9, 12, 14, 15) \qquad (5.16)$$

If the equation is in SOP form, it can be converted to standard SOP form by using the method discussed earlier and then it can be represented on K-map. Another method of representing SOP form on K-map without converting it to standard SOP form will become clear from the discussions of Sections 5.4 and 5.6.

5.3.3 Representation of Standard POS Form on K-Map

Logical equation in standard POS form can be represented on K-map by entering 0's in the cells of K-map corresponding to each maxterm present in the equation.

Example 5.7 Represent Eq. (5.15) on K-map.

Solution Corresponding to each maxterm in the equation, there is a cell in the K-map and a 0 is entered in each one of these cells. The K-map will be as shown in Fig. 5.7.

From a given K-map, we can write the logical equation in the standard POS form by ANDing the maxterms corresponding to each 0 entry in the K-map.

Example 5.8 Write the logical equation in the standard POS form for the K-map of Fig. 5.8.

Solution

$$Y = (A + B + C + \bar{D})\,(A + B + \bar{C} + D)\,(A + B + \bar{C} + \bar{D})$$
$$(A + \bar{B} + C + D)\,(A + \bar{B} + \bar{C} + D)\,(\bar{A} + B + C + D)$$
$$(\bar{A} + B + \bar{C} + D)\,(\bar{A} + B + \bar{C} + \bar{D})\,(\bar{A} + \bar{B} + C + \bar{D})$$
$$= \Pi\, M\,(1, 2, 3, 4, 6, 8, 10, 11, 13) \qquad (5.17)$$

If the equation is in POS form, it can be converted into standard POS form by the method discussed earlier and then it can be represented on K-map. Another method of representing POS form on K-map without converting it to standard POS form will become clear from the discussion of Sections 5.4 and 5.6.

Equation (5.16) represents the K-map of Fig. 5.8 in standard SOP form and Eq. (5.17) represents the same K-map in standard POS form. Therefore, these two equations are equivalent. Alternatively stated, an equation in SOP form can be converted into an equivalent POS form and vice-versa.

5.4 SIMPLIFICATION OF LOGICAL FUNCTIONS USING K-MAP

Simplification of logical functions with K-map is based on the principle of combining terms in adjacent cells. Two cells are said to be adjacent if they differ in only one variable. For example, in the two-variable K-maps of Figs 5.6*a* and *b*, the top two cells are adjacent and the bottom two cells are adjacent. Also, the left two cells and the right two cells are adjacent. It can be verified that in adjacent cells one of the literals is same, whereas the other literal appears in uncomplemented form in one and in the complemented form in the other cell.

Similarly, we observe adjacent cells in the 3-variable and 4-variable K-maps. Table 5.5 gives the adjacent cells of each cell in 2-, 3-, and 4-variable K-maps. From this it becomes clear that if the Gray code is used for the identification of cells in K-map, physically adjacent (horizontal and vertical but not diagonal) cells differ in only one variable. Also, the left-most cells are adjacent to their corresponding right-most cells and similarly the top cells are adjacent to their corresponding bottom cells. The simplification of logical function is achieved by grouping adjacent 1's or 0's in groups of 2^i, where $i = 1, 2, \ldots n$ and n is the number of variables.

The process of simplification involves grouping of minterms and identifying *prime-implicants* (PI) and *essential prime-implicants* (EPI).

A *prime-implicant* is a group of minterms that cannot be combined with any other minterm or groups. An *essential prime-implicant* is a *prime-implicant* in which one or more minterms are unique; i.e. it contains at least one minterm which is not contained in any other prime-implicant.

Table 5.5 Adjacent cells in K-maps

Cell with decimal number	*Decimal numbers of adjacent cells*		
	2-variable	*3-variable*	*4-variable*
0	1, 2	1, 2, 4	1, 2, 4, 8
1	0, 3	0, 3, 5	0, 3, 5, 9
2	0, 3	0, 3, 6	0, 3, 6, 10
3	1, 2	1, 2, 7	1, 2, 7, 11
4		0, 5, 6	0, 5, 6, 12
5		1, 4, 7	1, 4, 7, 13
6		2, 4, 7	2, 4, 7, 14
7		3, 5, 6	3, 5, 6, 15
8			0, 9, 10,12
9			1, 8, 11,13
10			2, 8, 11,14
11			3, 9, 10,15
12			4, 8, 13,14
13			5, 9, 12,15
14			6,10, 12,15
15			7,11, 13,14

5.4.1 Grouping Two Adjacent Ones

If there are two adjacent ones on the map, these can be grouped together and the resulting term will have one less literal than the original two terms. It can be verified for each of the groupings of two ones as given in Table 5.5.

Example 5.9 Simplify the K-map of Fig. 5.9.

Solution The standard SOP form of equation can be written by inspection as

$$Y = \bar{A}\bar{B}\bar{C} + \bar{A}BC + ABC + A\bar{B}\bar{C} \quad (5.18)$$

If we combine the ones in adjacent cells (0, 4) and (3, 7), Eq. (5.18) can be written as

$$Y = (\bar{A} + A)\,\bar{B}\bar{C} + (\bar{A} + A)\,BC \quad (5.19)$$

$$= \bar{B}\bar{C} + BC \text{ (Theorems 1.7 and 1.2)} \quad (5.20)$$

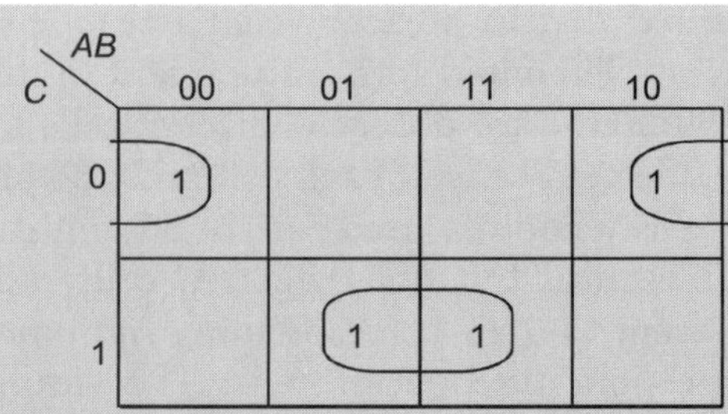

Fig. 5.9
K-map of Ex. 5.9.

Equation (5.20) can be directly obtained from the K-map by using the following procedure:

1. Identify adjacent ones, then see the values of the variables associated with these cells. Only one variable will be different and it gets eliminated. Other variables will appear in ANDed form in the term, it will be in the uncomplemented form if it is 1 and in the complemented form if it is 0.
2. Determine the term corresponding to each group of adjacent ones. These terms are ORed to get the simplified equation in SOP form.

Here, $\bar{B}\bar{C}$ and BC are the two prime-implicants, since grouping of the two minterms m_0 and m_4 cannot be combined with anyone of the remaining minterms m_3 and m_7, or the combination of the minterms m_3 and m_7. Similarly, the grouping of m_3 and m_7 is a prime-implicant. We also observe that both the groups are also essential prime-implicants, since each one of them contains unique minterms which are not contained in the other group.

5.4.2 Grouping Four Adjacent Ones

Four cells form a group of four adjacent ones if two of the literals associated with the minterms/maxterms are not same and the other literals are same. Table 5.6 gives all possible groups of four adjacent ones for each cell in a 3-variable map. In case of 2-variable map, there is only one possibility corresponding to entry 1 in all the four cells, and the simplified expression will be $Y = 1$. That is, Y always equals 1 (independent of the variables).

On the basis of groupings of 4 adjacent ones given in Table 5.6, we can find the groupings in K-maps of four or more variables. In the case of a four-variable K-map, there are six possible groupings of 4-variables involving any cell. It is left to the reader to verify this fact.

Table 5.6 Groups of four adjacent ones in a 3-variable K-map

Cell with decimal number	*Decimal numbers of cells forming groups of adjacent fours*		
0	(0, 2, 6, 4),	(0, 1, 2, 3),	(0, 1, 4, 5)
1	(1, 0, 2, 3),	(1, 3, 7, 5),	(1, 0, 4, 5)
2	(2, 0, 6, 4),	(2, 3, 1, 0),	(2, 3, 6, 7)
3	(3, 1, 7, 5),	(3, 2, 1, 0),	(3, 2, 6, 7)
4	(4, 6, 2, 0),	(4, 5, 6, 7),	(4, 5, 0, 1)
5	(5, 1, 3, 7),	(5, 4, 6, 7),	(5, 4, 0, 1)
6	(6, 0, 2, 4),	(6, 7, 4, 5),	(6, 7, 2, 3)
7	(7, 1, 3, 5),	(7, 6, 4, 5),	(7, 6, 2, 3)

Example 5.10 Simplify the K-map of Fig. 5.10.

Solution The standard SOP form of equation can be written by inspection as

$$Y = m_0 + m_1 + m_3 + m_7 + m_8 + m_9 + m_{11} + m_{15}$$
$$= (m_0 + m_1 + m_8 + m_9) + (m_3 + m_7 + m_{15} + m_{11}) \quad (5.21)$$

In the K-map of Fig. 5.10, there are two groups of four adjacent ones. One corresponding to cells 0, 1, 8 and 9, and the other one corresponding to 3, 7, 15 and 11. In Eq. (5.21), the minterms corresponding to each group are combined. The first term can be written as

CD \ AB	00	01	11	10
00	0: 1	4	12	8: 1
01	1: 1	5	13	9: 1
11	3: 1	7: 1	15: 1	11: 1
10	2	6	14	10

Fig. 5.10 K-map of Ex. 5.10

$$m_0 + m_1 + m_8 + m_9 = \bar{A}\bar{B}\bar{C}\bar{D} + \bar{A}\bar{B}\bar{C}D + A\bar{B}\bar{C}\bar{D} + A\bar{B}\bar{C}D$$
$$= \bar{B}\bar{C}(\bar{A}\bar{D} + \bar{A}D + A\bar{D} + AD)$$
$$= \bar{B}\bar{C}[\bar{A}(\bar{D} + D) + A(\bar{D} + D)]$$
$$= \bar{B}\bar{C}[\bar{A}\cdot 1 + A\cdot 1]$$
$$= \bar{B}\bar{C}(\bar{A} + A)$$
$$= \bar{B}\bar{C}\cdot 1 = \bar{B}\bar{C}$$

In the first term or Eq. (5.21) we observe the following:

1. In this group of four minterms, two of the variables appear as $\bar{B}$ and $\bar{C}$ in all the four terms.
2. The variable A appears as A in two and as $\bar{A}$ in the other two minterms.
3. The variable D appears as D in two and as $\bar{D}$ in the other two minterms.
4. The combination of these four minterms results in one term with two literals which are present in all the four terms. Similarly, the second term of Eq. (5.21) is simplified to CD. Therefore, the K-map is simplified to

$$Y = \bar{B}\bar{C} + CD \quad (5.22)$$

Example 5.11 In Fig. 5.10, show that the following groups of minterms are not prime-implicants:

(a) m_0, m_1 (b) m_1, m_3 (c) m_3, m_7 (d) m_7, m_{15}
(e) m_{11}, m_{15} (f) m_8, m_9 (g) m_9, m_{11} (h) m_0, m_8
(i) m_1, m_9 (j) m_3, m_{11}

Solution

(a) m_0, m_1 group can be combined with m_8, m_9 group to form a group of four adjacent 1s. Hence, m_0, m_1 group is not a prime-implicant.
(b) m_1, m_3 group can be combined with m_9, m_{11} group, therefore, it is not a prime-implicant.
(c) m_3, m_7 group can be combined with m_{15}, m_{11} group.
(d) m_7, m_{15} group can be combined with m_3, m_{11} group.
(e) m_{11}, m_{15} group can be combined with m_3, m_7 group.
(f) See (a)
(g) See (b)

(h) m_0, m_8 group can be combined with m_1, m_9 group
(i) See (h)
(j) See (d)

Example 5.12 Show that the following groups of minterms in Fig. 5.10 are essential prime-implicants:

(a) *m* (0, 1, 8, 9) (b) *m* (3, 7, 11, 15)

Solution

(a) The group formed by the minterms m_0, m_1, m_8, m_9 is a prime-implicant, since it can not be combined with any other minterm or group. Also, it includes minterms which can not be included in the other group, therefore, it is an essential prime-implicant.
(b) The group formed by the minterms m_3, m_7, m_{11}, m_{15} have m_7 and m_{15} which can not be included in any other prime-implicant, hence this is an essential prime-implicant.

5.4.3 Grouping Eight Adjacent Ones

Eight cells form a group of eight adjacent ones if three of the literals associated with the minterms/maxterms are not same and the other literals are same. In case of 3-variable K-map, there is only one possibility of eight ones appearing in the K-map and this corresponds to output equal to 1, irrespective of the values of the input variables. Table 5.7 gives all possible groups of eight adjacent ones in a 4-variable K-map. From an understanding of this, we can easily find out such combinations for 5- and 6-variable K-maps. When eight adjacent ones are combined, the resulting equation will have only one term with the number of literals three less than the number of literals in the original minterms. Similar to the groupings of adjacent two and four ones, the literals which are common in all the eight minterms will be present and the literals which are not same get eliminated in the resulting term.

Table 5.7 Groups of eight adjacent ones in 4-variable K-map

Decimal numbers of cells forming groups of adjacent eights in a 4-variable K-map
0, 4, 12, 8, 1, 5, 13, 9,
0, 4, 12, 8, 2, 6, 14, 10
0, 1, 3, 2, 4, 5, 7, 6
0, 1, 3, 2, 8, 9, 11, 10
1, 5, 13, 9, 3, 7, 15, 11
4, 5, 7, 6, 12, 13, 15, 14
12, 13, 15, 14, 8, 9, 11, 10
3, 7, 15, 11, 2, 6, 14, 10

The reader is advised to verify the simplification of eight adjacent ones into a single term with three variables eliminated. For example, let us take the first group of eight adjacent ones in Table 5.7. For all these eight cells, the variable *C* appears as $\bar{C}$ in the minterms and the other three variables are not same. Therefore, the grouping of these eight cells results in a term $\bar{C}$. Figure 5.11 shows the simplified expression for each of the groupings of eight ones for a 4 variable K-map.

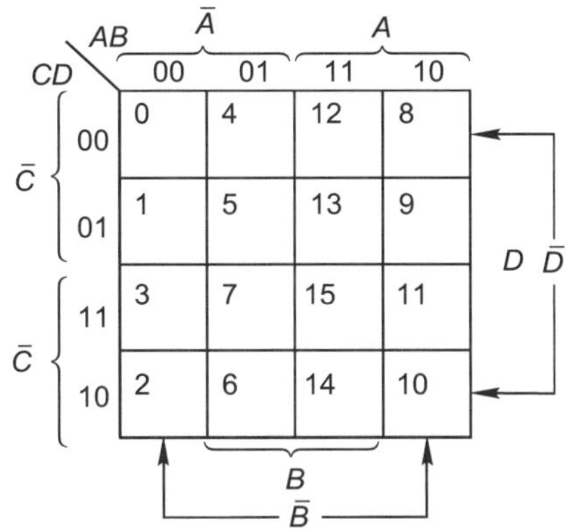

Fig. 5.11

Four-variable K-map illustrating the groupings of eight adjacent ones.

5.4.4 Grouping 2, 4, and 8 Adjacent Zeros

In the above discussion, we have considered groups of 2, 4, and 8 adjacent ones. Instead of making the groups of ones, we can also make groups of zeros. The procedure is similar to the one used above and is as follows:

1. Group of two adjacent zeros result in a term with one literal less than the number of variables. The literal which is not same in the two maxterms gets eliminated.
2. Group of four adjacent zeros result in a term with two literals less than the number of variables. The two literals which are not same in all the four maxterms get eliminated.
3. Group of eight adjacent zeros result in a term with three literals less than the number of variables. The three literals which are not same in all the eight maxterms get eliminated.

We have considered groups of 2, 4, and 8 adjacent ones and zeros. The same logic can be extended to 16, 32, and 64 adjacent ones and zeros which occur in K-maps with more than 4 variables.

5.5 MINIMIZATION OF LOGICAL FUNCTIONS SPECIFIED IN MINTERMS/MAXTERMS OR TRUTH TABLE

5.5.1 Minimization of SOP Form

We have seen the advantages of simplifying a logical expression. If the expression is simplified to a stage beyond which it can not be further simplified, it will require minimum number of gates with minimum number of inputs to the gates. Such an expression is referred to as the *minimized expression.*

For minimizing a given expression in SOP form or for a given truth table, we have to prepare the K-map first and then look for combinations of ones on the K-map. We have to combine the ones in such a way that the resulting expression is minimum. To achieve this, the following algorithm can be used which will definitely lead to minimized expression:

1. Identify the ones which can not be combined with any other ones and encircle them. These are *essential prime implicants.*
2. Identify the ones that can be combined in groups of two in only one way. Encircle such groups of ones.
3. Identify the ones that can be combined with three other ones, to make a group of four adjacent ones, in only one way. Encircle such groups of ones.
4. Identify the ones that can be combined with seven other ones, to make a group of eight adjacent ones, in only one way. Encircle such groups of ones.
5. After identifying the *essential groups* of 2, 4, and 8 ones, if there still remains some ones which have not been encircled then these are to be combined with each other or with other already encircled ones. Of course, however, we should combine the left-over ones in largest possible groups and in as few groupings as possible. In this, the groupings may not be unique and we should make the groupings in an optimum manner. You can verify that any one can be included any number of times without affecting the expression.

The logic function consisting of the essential prime-implicants obtained in steps 1 to 4 and the prime-implicants obtained in step 5 will be the minimized function. The above algorithm will be used to minimize the logical functions in the examples given.

Example 5.13 Minimize the four-variable logic function using K-map.

$$f(A, B, C, D) = \Sigma m\ (0, 1, 2, 3, 5, 7, 8, 9, 11, 14) \qquad (5.23)$$

Solution The K-map of Eq. (5.23) is shown in Fig. 5.12. The equation is minimized in the following steps:

1. Encircle 1 in cell 14 which can not be combined with any other 1. The term corresponding to this is $ABC\bar{D}$.

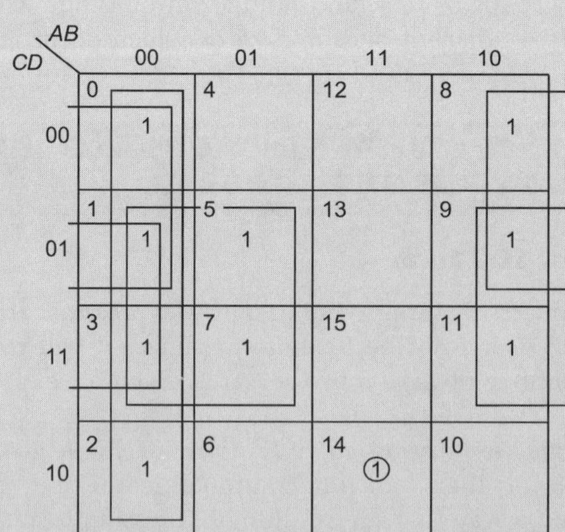

Fig. 5.12
K-map for Eq. (5.23).

2. There are at least two possible ways for every 1 forming groups of two adjacent ones. Therefore, we ignore it for the time being and go to the next step.
3. There is only one possible group of four adjacent ones involving each of the cells 8, 11, 5 or 7 and 2, and these are (8, 9, 0, 1), (11, 9, 1, 3), (5, 7, 3, 1) and (2, 3, 1, 0), respectively. Encircle these groups. The terms corresponding to these groups are $\bar{B}\bar{C}$, $\bar{B}D$, $\bar{A}D$, and $\bar{B}\bar{C}$, respectively.

Since all the ones have been encircled, therefore, the minimized equation is

$$f(A, B, C, D) = ABC\bar{D} + \bar{B}\bar{C} + \bar{B}D + \bar{A}D + \bar{A}\bar{B} \tag{5.24}$$

Example 5.14 Determine the minimized expression in SOP form for the truth table given in Table 5.8.

Solution The K-map for the truth table of Table 5.8 is shown in Fig. 5.13.

Using the minimization steps, we obtain the minimized expression.

$$Y = \bar{B} + A\bar{C} + \bar{A}CD \tag{5.25}$$

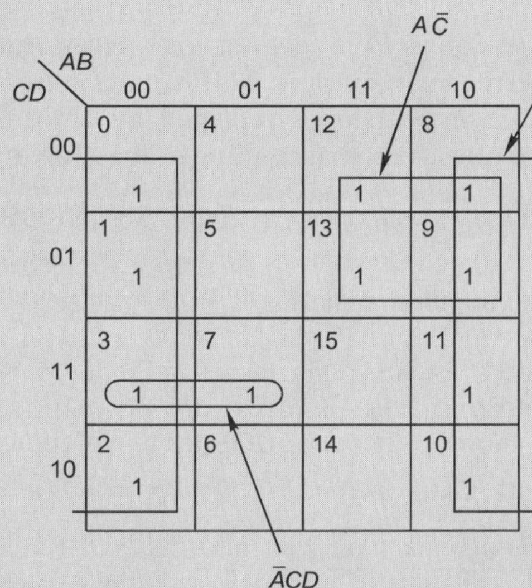

Fig. 5.13
K-map of Table 5.8.

Table 5.8

Inputs				*Output Y*
A	*B*	*C*	*D*	
0	0	0	0	1
0	0	0	1	1
0	0	1	0	1
0	0	1	1	1
0	1	0	0	0

(*Contd.*)

Table 5.8 *(Contd.)*

Inputs				*Output Y*
A	*B*	*C*	*D*	
0	1	0	1	0
0	1	1	0	0
0	1	1	1	1
1	0	0	0	1
1	0	0	1	1
1	0	1	0	1
1	0	1	1	1
1	1	0	0	1
1	1	0	1	1
1	1	1	0	0
1	1	1	1	0

5.5.2 Minimization of POS Form

For minimizing a given expression in POS form or for a given truth table we write zeros in the cells corresponding to maxterms for 0 outputs. The K-map is simplified by following the same procedure as used for SOP form with ones replaced by zeros. In this, groups of zeros are formed rather than groups of ones. We shall minimize the above two examples in POS form.

Example 5.15 Minimize the logic function of Eq. (5.23) in POS form.

Solution

Equation (5.23) can be expressed in standard POS form as

$$f(A, B, C, D) = \Pi M(4, 6, 10, 12, 13, 15) \tag{5.26}$$

The K-map corresponding to Eq. (5.26) is shown in Fig. 5.14. Note that the K-map can also be obtained directly from Eq. (5.23).

Using steps similar to those outlined for SOP form, we obtain the minimized expression,

$$f = (\bar{A} + B + \bar{C} + D) \cdot (\bar{A} + \bar{B} + C) \cdot (\bar{A} + \bar{B} + \bar{D}) \cdot (A + \bar{B} + D) \tag{5.27}$$

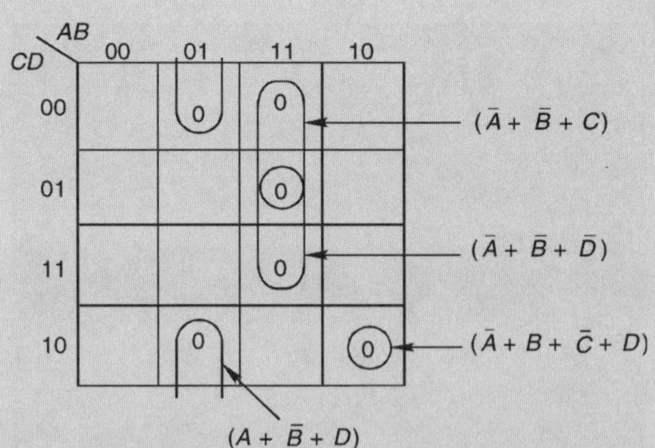

Fig. 5.14
K-map of Eq. (5.26).

If we compare Eqs (5.24) and (5.27), we observe that the number of terms are not same in the two minimizations. In fact, in general the two minimizations will not have the same number of terms and will require different quantities of hardware. Therefore, one can obtain both minimizations and select the one which requires minimum hardware. In some situations there may not be any choice to the designer because of non-availability of certain ICs.

Example 5.16 Minimize the truth table given in Table 5.8 using maxterms.

Solution

The K-map is given in Fig. 5.15.

The simplified expression is

$$Y = (A + \bar{B} + C)(\bar{A} + \bar{B} + \bar{C})(\bar{B} + \bar{C} + D) \tag{5.28}$$

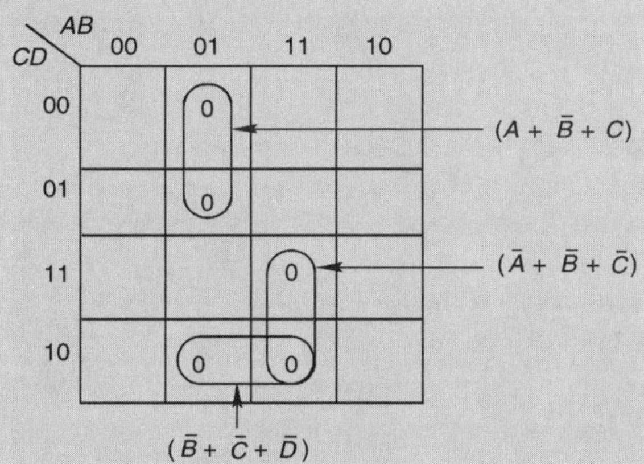

Fig. 5.15
K-map of Table 5.8.

Comparison of Eqs (5.25) and (5.28) confirms our generalizations made in Ex. 5.15 regarding the hardware requirements in the two methods.

5.6 MINIMIZATION OF LOGICAL FUNCTIONS NOT SPECIFIED IN MINTERMS/MAXTERMS

If the function is specified in one of the two standard forms, its K-map can be prepared and the function can be minimized. Now we consider the cases where the functions are not specified in standard forms. In such cases, the equations can be converted into standard forms using the techniques given in Section 5.2, the K-maps obtained and minimized. Alternately, we can directly prepare K-map using the following algorithm:

1. Enter ones for minterms and zeros for maxterms.
2. Enter a pair of ones/zeros for each of the terms with one variable less than the total number of variables.

3. Enter four adjacent ones/zeros for terms with two variables less than the total number of variables.
4. Repeat for other terms in the similar way.

Once the K-map is prepared the minimization procedure is same as discussed earlier. The following examples will help in understanding the above procedure:

Example 5.17 Minimize the four variable logic function

$$f(A, B, C, D) = AB\bar{C}D + \bar{A}BCD + \bar{A}\bar{B}\bar{C} + \bar{A}\bar{B}\bar{D} + A\bar{C} + A\bar{B}C + \bar{B} \tag{5.29}$$

Solution

The method for obtaining K-map is

1. Enter 1 in the cell with $A = 1$, $B = 1$, $C = 0$, $D = 1$ corresponding to the minterm $AB\bar{C}D$.
2. Enter 1 in the cell with $A = 0$, $B = 1$, $C = 1$, $D = 1$ corresponding to the minterm $\bar{A}BCD$.
3. Enter 1's in the two cells with $A = 0$, $B = 0$, $C = 0$ corresponding to the term $\bar{A}\bar{B}\bar{C}$.
4. Enter 1's in the two cells with $A = 0$, $B = 0$, $D = 0$ (one of these is already entered) corresponding to the term $\bar{A}\bar{B}\bar{D}$.
5. Enter 1's in the two cells with $A = 1$, $B = 0$, $C = 1$ corresponding to the term $A\bar{B}C$.
6. Enter 1's in the four cells with $A = 1$, $C = 0$ (one of them is already entered) corresponding to the term $A\bar{C}$.
7. Enter 1's in the eight cells with $B = 0$ (all of them except one have already been entered) corresponding to the term $\bar{B}$.

The K-map is given in Fig. 5.16 which is same as the K-map of Fig. 5.13 and the minimized expression is given by Eq. (5.25).

CD \ AB	00	01	11	10
00	1		1	1
01	1		1	1
11	1	1		1
10	1			1

Fig. 5.16 K-map of Eq. (5.29).

Example 5.18 Minimize the four variable logic function

$$f(A, B, C, D) = (A + B + \bar{C} + \bar{D}) \cdot (\bar{A} + C + \bar{D}) \cdot (\bar{A} + B + \bar{C} + \bar{D}) \cdot (\bar{B} + C) \cdot (\bar{B} + \bar{C}) \cdot (A + \bar{B}) \cdot (\bar{B} + \bar{D}) \tag{5.30}$$

Solution

The K-map cells in which 0's are to be entered corresponding to each term are given in Table 5.9. Even if a cell is involved in more than one terms, a 0 is to be entered only once.

Table 5.9 K-map cells with 0 entries

Term	Cell(s) with 0's
$A + B + \bar{C} + \bar{D}$	$A = 0$, $B = 0$, $C = 1$, $D = 1$
$\bar{A} + C + \bar{D}$	$A = 1$, $C = 0$, $D = 1$
$\bar{A} + B + \bar{C} + \bar{D}$	$A = 1$, $B = 0$, $C = 1$, $D = 1$
$\bar{B} + C$	$B = 1$, $C = 0$

(*Contd.*)

Table 5.9 (*Contd.*)

Term	Cell(s) with 0's
$\bar{B} + \bar{C}$	$B = 1, C = 1$
$A + \bar{B}$	$A = 0, B = 1$
$\bar{B} + \bar{D}$	$B = 1, D = 1$

The K-map is given in Fig. 5.17. The minimized expression is

$$f(A, B, C, D) = \bar{B} \cdot (\bar{A} + \bar{D})(\bar{C} + \bar{D}) \tag{5.31}$$

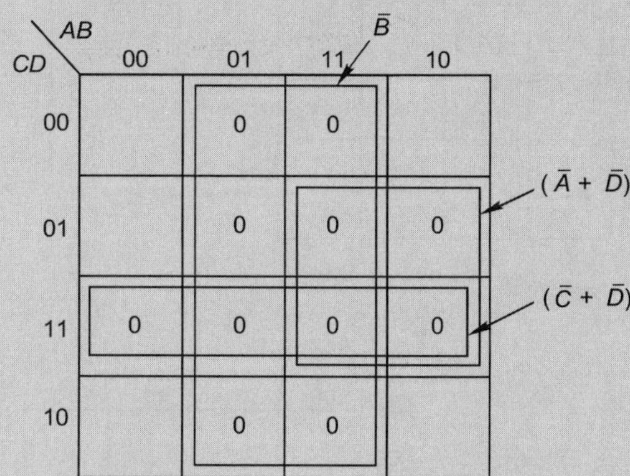

Fig. 5.17
K-map of Eq. (5.30).

5.7 DON'T-CARE CONDITIONS

We enter 1's and 0's in the map corresponding to input variables that make the function equal to 1 or 0, respectively. The maps are simplified using either 1's or 0's. Therefore, we make the entries in the map for either 1's or 0's. The cells which do not contain 1 are assumed to contain 0 and vice-versa. This is not always true since there are cases in which certain combinations of input variables do not occur. Also, for some functions the outputs corresponding to certain combinations of input variables do not matter. In such situations the designer has a flexibility and it is left to him whether to assume a 0 or a 1 as output for each of these combinations. This condition is known as *don't-care* condition and can be represented on the K-map as a × mark in the corresponding cell. The × mark in a cell may be assumed to be a 1 or a 0 depending upon which one leads to a simpler expression. The function can be specified in one of the following ways:

1. In terms of minterms and don't-care conditions. For example,

$$f(A, B, C, D) = \Sigma m(1, 3, 7, 11, 15) + d(0, 2, 5) \tag{5.32}$$

Its K-map and the minimized expression are given in Fig. 5.18*a*.

2. In terms of maxterms and don't-care conditions. For example,

$$f(A, B, C, D) = \Pi m\ (4, 5, 6, 7, 8, 12) \cdot d(1, 2, 3, 9, 11, 14) \tag{5.33}$$

Its K-map and the minimized expression are given in Fig. 5.18*b*.

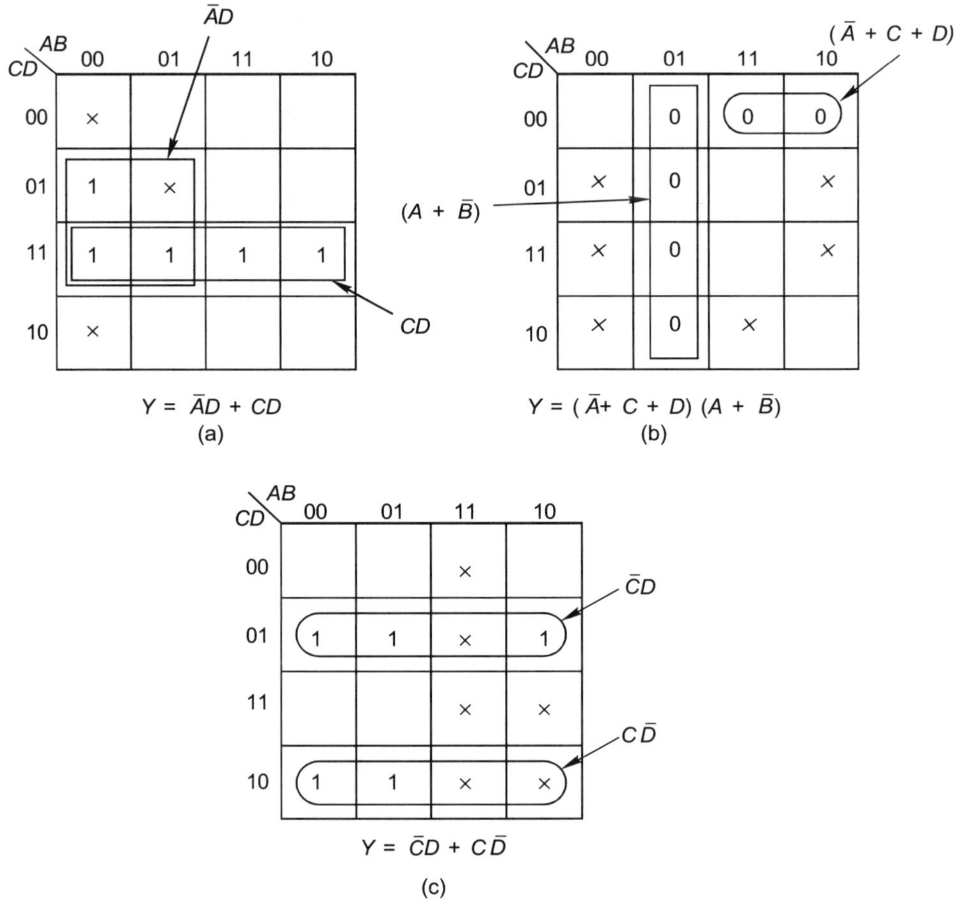

Fig. 5.18
K-maps with don't-care conditions.

3. In terms of truth table. For example, consider the truth table of Table 5.10.

Table 5.10

Inputs				*Output*
A	*B*	*C*	*D*	*Y*
0	0	0	0	0
0	0	0	1	1
0	0	1	0	1

(*Contd.*)

Table 5.10 *(Contd.)*

Inputs				*Output*
A	*B*	*C*	*D*	*Y*
0	0	1	1	0
0	1	0	0	0
0	1	0	1	1
0	1	1	0	1
0	1	1	1	0
1	0	0	0	0
1	0	0	1	1
1	0	1	0	×
1	0	1	1	×
1	1	0	0	×
1	1	0	1	×
1	1	1	0	×
1	1	1	1	×

Its K-map and the minimized expression in SOP form are given in Fig. 5.18*c*.

5.8 DESIGN EXAMPLES

5.8.1 Arithmetic Circuits

1. Half-adder A logic circuit for the addition of two one-bit numbers is referred to as an *half-adder.* The addition process is illustrated in Section 2.5 and is reproduced in truth table form in Table 5.11. Here, *A* and *B* are the two inputs and *S* (SUM) and *C* (CARRY) are the two outputs.

Table 5.11 Truth table of an half-adder

Inputs		*Outputs*	
A	*B*	*S*	*C*
0	0	0	0
0	1	1	0
1	0	1	0
1	1	0	1

From the truth table, we obtain the logical expressions for *S* and *C* outputs as

$$S = \bar{A}B + A\bar{B} = A \oplus B \tag{5.34a}$$

$$C = AB \tag{5.34b}$$

The realization of an half-adder using gates is shown in Fig. 5.19.

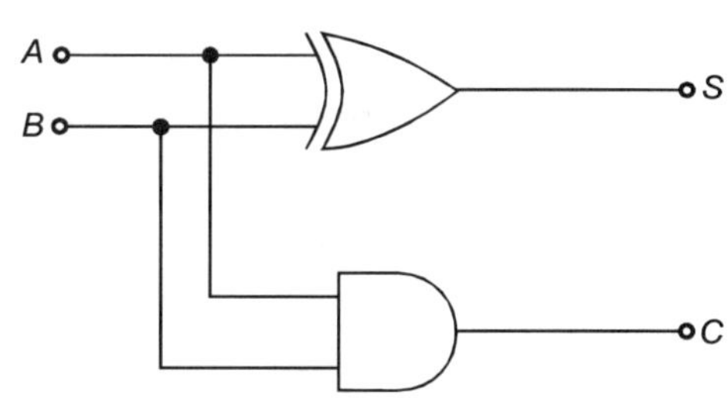

Fig. 5.19
Realization of an half-adder.

2. Full-adder An half-adder has only two inputs and there is no provision to add a carry coming from the lower order bits when multibit addition is performed.

For this purpose, a third input terminal is added and this circuit is used to add A_n, B_n, and C_{n-1}, where A_n and B_n are the nth order bits of the numbers A and B respectively and C_{n-1} is the carry generated from the addition of (n–1)th order bits. This circuit is referred to as *full-adder* and its truth table is given in Table 5.12.

Table 5.12 Truth table of a full-adder

Inputs			*Outputs*	
A_n	B_n	C_{n-1}	S_n	C_n
0	0	0	0	0
0	0	1	1	0
0	1	0	1	0
0	1	1	0	1
1	0	0	1	0
1	0	1	0	1
1	1	0	0	1
1	1	1	1	1

The K-maps for the outputs S_n and C_n are given in Fig. 5.20 and the minimized expressions are given by Eq. (5.35).

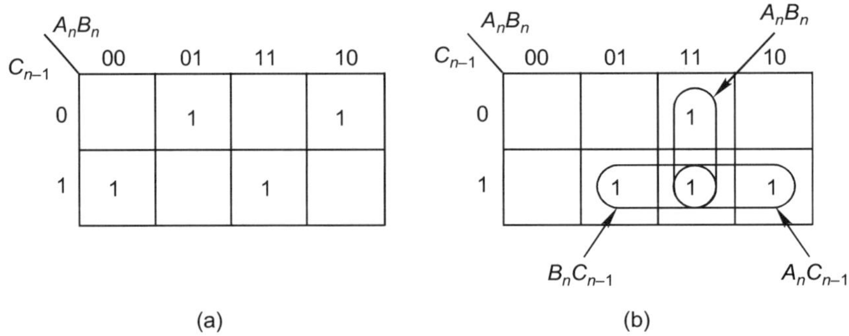

Fig. 5.20
K-maps for (a) S_n (b) C_n.

$$S_n = \bar{A}_n B_n \bar{C}_{n-1} + \bar{A}_n \bar{B}_n C_{n-1} + A_n \bar{B}_n \bar{C}_{n-1} + A_n B_n C_{n-1} \quad (5.35a)$$

$$C_n = A_n B_n + B_n C_{n-1} + A_n C_{n-1} \quad (5.35b)$$

The NAND-NAND realizations are given in Fig. 5.21.

3. *Half-subtractor* A logic circuit for the subtraction of B (subtrahend) from A (minuend) where A and B are 1-bit numbers is referred to as a *half-subtractor*. The subtraction process is illustrated in Section 2.5 and is reproduced in truth table form in Table 5.13. Here, A and B are the two inputs and D (difference) and C (borrow) are the two outputs.

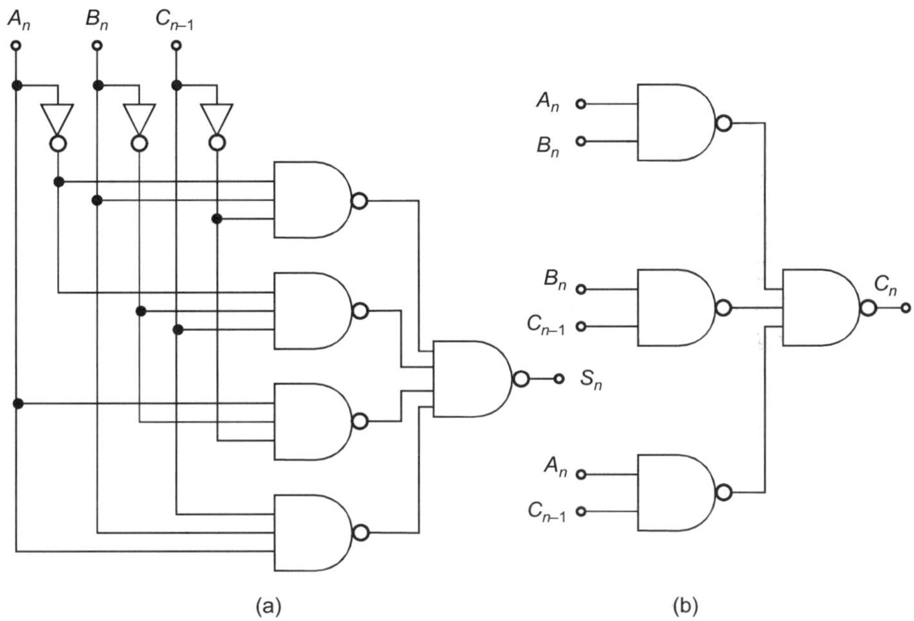

Fig. 5.21

NAND-NAND realization of (a) S_n (b) C_n.

Table 5.13 Truth table of a half-subtractor

Inputs		*Outputs*	
A	*B*	*D*	*C*
0	0	0	0
0	1	1	1
1	0	1	0
1	1	0	0

From the truth table, the logical expressions for D and C are obtained as

$$D = \bar{A}B + A\bar{B} = A \oplus B \tag{5.36a}$$

$$C = \bar{A}B \tag{5.36b}$$

The realization of a half-subtractor using gates is shown in Fig. 5.22.

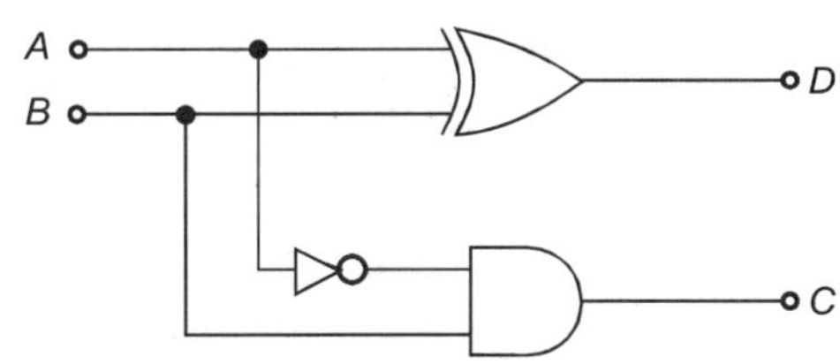

Fig. 5.22

Realization of a half-subtractor.

4. Full-subtractor Just like a full-adder, we require a *full-subtractor* circuit for performing multibit subtraction wherein a borrow from the previous bit position may also be there. A full-subtractor will have three inputs, A_n (minuend), B_n (subtrahend) and C_{n-1} (borrow from the previous stage) and two outputs, D_n (difference) and C_n (borrow). Its truth table is given in Table 5.14. The K-map for the output D_n is exactly same as the K-map for S_n of the adder circuit and therefore, its realization is same as given in Fig. 5.21*a*.

Table 5.14 Truth table of a full-subtractor

Inputs			*Outputs*	
A_n	B_n	C_{n-1}	D_n	C_n
0	0	0	0	0
0	0	1	1	1
0	1	0	1	1
0	1	1	0	1
1	0	0	1	0
1	0	1	0	0
1	1	0	0	0
1	1	1	1	1

The K-map for C_n is given in Fig. 5.23*a* and its realization is given in Fig. 5.23*b*. The simplified expression for C_n is

$$C_n = \bar{A}_n B_n + \bar{A}_n C_{n-1} + B_n C_{n-1} \tag{5.37}$$

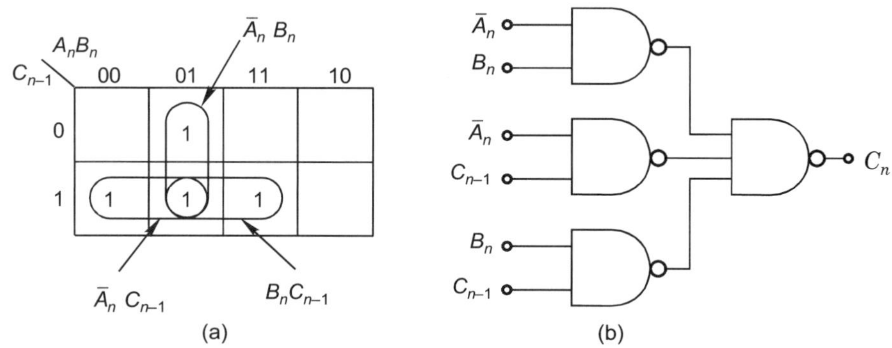

Fig. 5.23
(a) Truth table for C_n (b) Realization of C_n.

5.8.2 BCD-to-7-Segment Decoder

A digital display that consists of seven LED segments is commonly used to display decimal numerals in digital systems. Most familiar examples are electronic calculators and watches where one 7-segment display device is used for displaying one numeral 0 through 9. For using this display device, the data has to be converted from some binary code to the code required for the display. Usually, the binary code used is natural BCD. Figure 5.24*a* shows the display device, Fig. 5.24*b* shows the segments which must be illuminated for each of the numerals and Fig. 5.24*c* gives the display system.

Table 5.15 gives the truth table of BCD-to-7-segment decoder. Here *ABCD* is the natural BCD code for numerals 0 through 9. The K-maps for each of the outputs *a* through *g* are given in Fig. 5.25. The entries in the K-map corresponding to six binary combinations not used in the

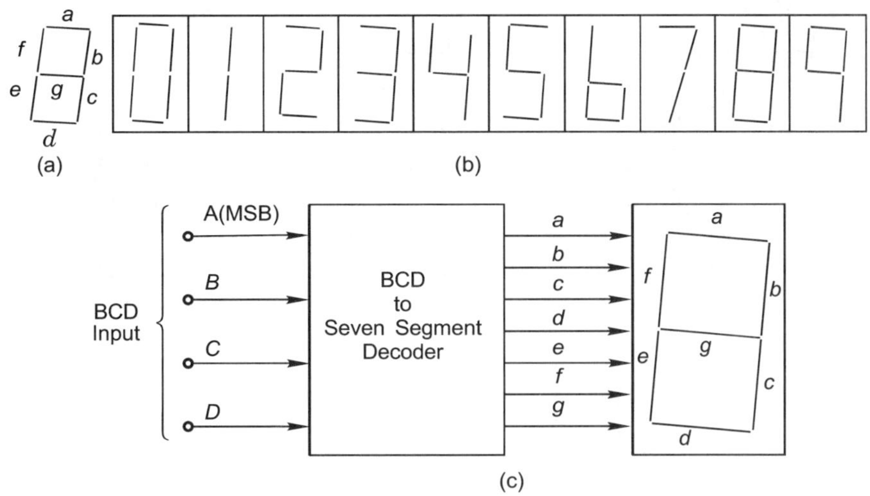

Fig. 5.24

(a) 7-segment display (b) display of numerals (c) display system.

truth table are × – don't-care. The K-maps are simplified and the minimum expressions are given by:

$$a = \bar{B}\bar{D} + BD + CD + A \tag{5.38}$$

$$b = \bar{B} + \bar{C}\bar{D} + CD \tag{5.39}$$

$$c = B + \bar{C} + D = \overline{\bar{B}C\bar{D}} \tag{5.40}$$

$$d = \bar{B}\bar{D} + C\bar{D} + \bar{B}C + B\bar{C}D \tag{5.41}$$

$$e = \bar{B}\bar{D} + C\bar{D} \tag{5.42}$$

$$f = A + \bar{C}\bar{D} + B\bar{C} + B\bar{D} \tag{5.43}$$

$$g = A + B\bar{C} + \bar{B}C + C\bar{D} \tag{5.44}$$

Table 5.15 Truth table of BCD-to-7 segment decoder

Decimal digit displayed	*Inputs*				*Outputs*						
	A	*B*	*C*	*D*	*a*	*b*	*c*	*d*	*e*	*f*	*g*
0	0	0	0	0	1	1	1	1	1	1	0
1	0	0	0	1	0	1	1	0	0	0	0
2	0	0	1	0	1	1	0	1	1	0	1
3	0	0	1	1	1	1	1	1	0	0	1
4	0	1	0	0	0	1	1	0	0	1	1
5	0	1	0	1	1	0	1	1	0	1	1
6	0	1	1	0	0	0	1	1	1	1	1
7	0	1	1	1	1	1	1	0	0	0	0
8	1	0	0	0	1	1	1	1	1	1	1
9	1	0	0	1	1	1	1	0	0	1	1

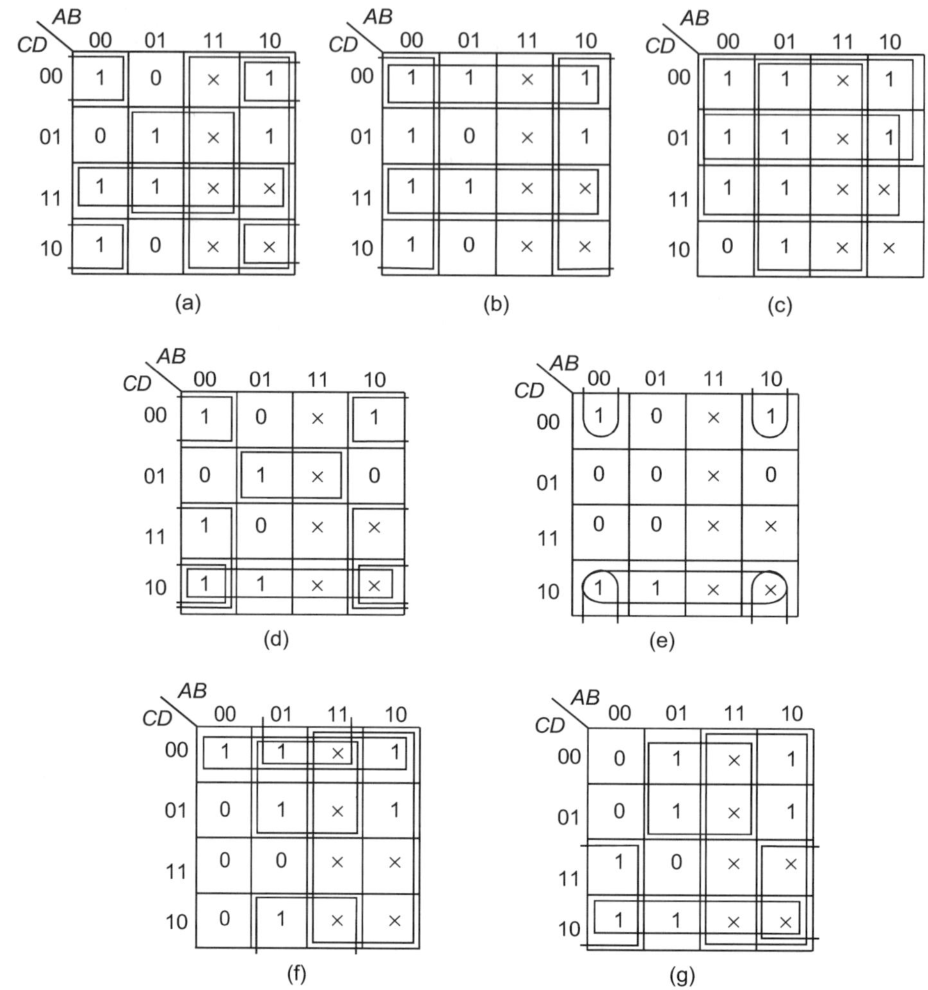

Fig. 5.25 K-maps of Table 5.15.

The NAND gate realizations are shown in Fig. 5.26.

We see from the realizations of Fig. 5 26 that a term with single literal must be inverted and then applied to the second level NAND gate.

5.9 EX-OR AND EX-NOR SIMPLIFICATION OF K-MAPS

There are many situations in logic design in which simplification of logic expression is possible in terms of EX–OR and EX–NOR operations. These functions are widely used in digital design and therefore are available in IC form. This section deals with recognizing K-map patterns indicating EX–OR and EX–NOR functions. Ring map has been proposed by Fronek, Donald K in his paper entitled 'Ring Map Minimizes Logic Circuit', *Electronic Design* 17 (1972) for simplify-

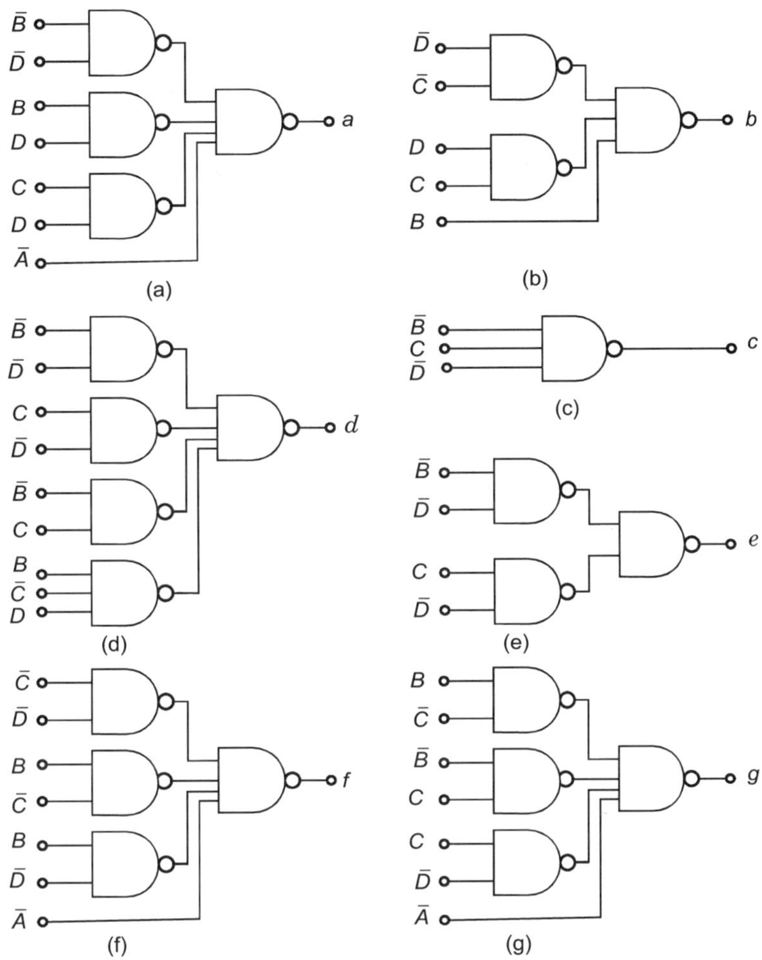

Fig. 5.26

NAND gate realizations of Eqs (5.38) to (5.44).

ing EX–OR functions and their complements. This is an unnecessary modification of K-map and hence not adopted in practice. The interested reader may refer to the original paper. In AND–OR or OR–AND simplification, the adjacent ones or zeros are grouped. The adjacency used is horizontal or vertical. In the case of EX–OR /EX–NOR simplification we have to look for:

1. Diagonal adjacencies, and
2. Offset adjacencies.

Examples of diagonal and offset adjacencies for single ones are given in Fig. 5.27. All the possible diagonal and offset adjacencies are marked in the figures and the terms corresponding to each group of such adjacency involving EX–OR or EX–NOR operation are given below.

Two-variable K-maps

$$F_1 = \bar{A}\bar{B} + AB = \overline{A \oplus B} = A \odot B$$
$$F_2 = A\bar{B} + \bar{A}B = A \oplus B$$

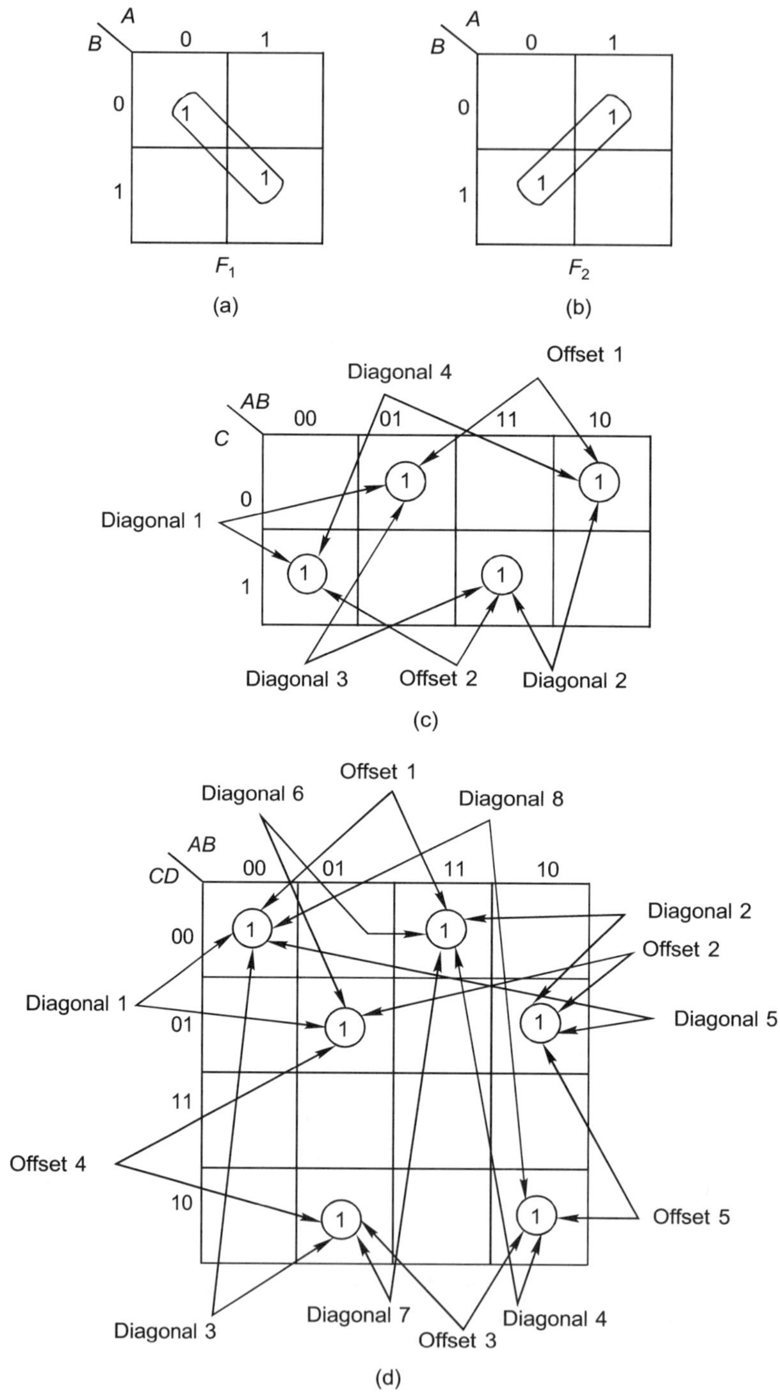

Fig. 5.27

K-maps illustrating diagonal and offset adjacencies groups of two ones.

Three-variable K-map

Offset 1: $F_3 = (\bar{A}B + A\bar{B})\ \bar{C}$

$= (A \oplus B)\ \bar{C}$

Offset 2: $F_4 = (\bar{A}\,\bar{B} + AB)\ C$

$= (\overline{A \oplus B})C$

$= (A \odot B)\ C$

Diagonal 1: $F_5 = \bar{A}B\bar{C} + \bar{A}\,\bar{B}C$

$= \bar{A}\ (B \oplus C)$

Diagonal 2: $F_6 = A\bar{B}\bar{C} + ABC$

$= A(B \odot C)$

Diagonal 3: $F_7 = B(A \odot C)$

Diagonal 4: $F_8 = \bar{B}(A \oplus C)$

Four-variable K-map

Offset 1: $F_9 = \bar{A}\,\bar{B}\bar{C}\bar{D} + AB\bar{C}\bar{D} = \bar{C}\bar{D}\ (A \odot B)$

Offset 2: $F_{10} = \bar{C}D\,(A \oplus B)$

Offset 3: $F_{11} = C\bar{D}(A \oplus B)$

Offset 4: $F_{12} = \bar{A}B(C \oplus D)$

Offset 5: $F_{13} = A\bar{B}(C \oplus D)$

Diagonal 1: $F_{14} = \bar{A}\,\bar{B}\bar{C}\bar{D}\ + \bar{A}B\bar{C}D\ = \bar{A}\bar{C}(B \odot D)$

Diagonal 2: $F_{15} = A\bar{C}\ (B \oplus D)$

Diagonal 3: $F_{16} = \bar{A}\bar{D}(B \odot C)$

Diagonal 4: $F_{17} = A\bar{D}(B \oplus C)$

Diagonal 5: $F_{18} = \bar{B}\bar{C}(A \odot D)$

Diagonal 6: $F_{19} = B\bar{C}(A \oplus D)$

Diagonal 7: $F_{20} = B\bar{D}(A \oplus C)$

Diagonal 8: $F_{21} = \bar{B}\bar{D}(A \odot C)$

From this we can find the way to identify these adjacencies and the method of obtaining the term corresponding to each grouping.

5.9.1 Diagonal and Offset Adjacencies of Groups of Ones

Figure 5.28 illustrates the diagonal and offset adjacencies of standard groups of two ones. Their simplified terms are also given in the figures.

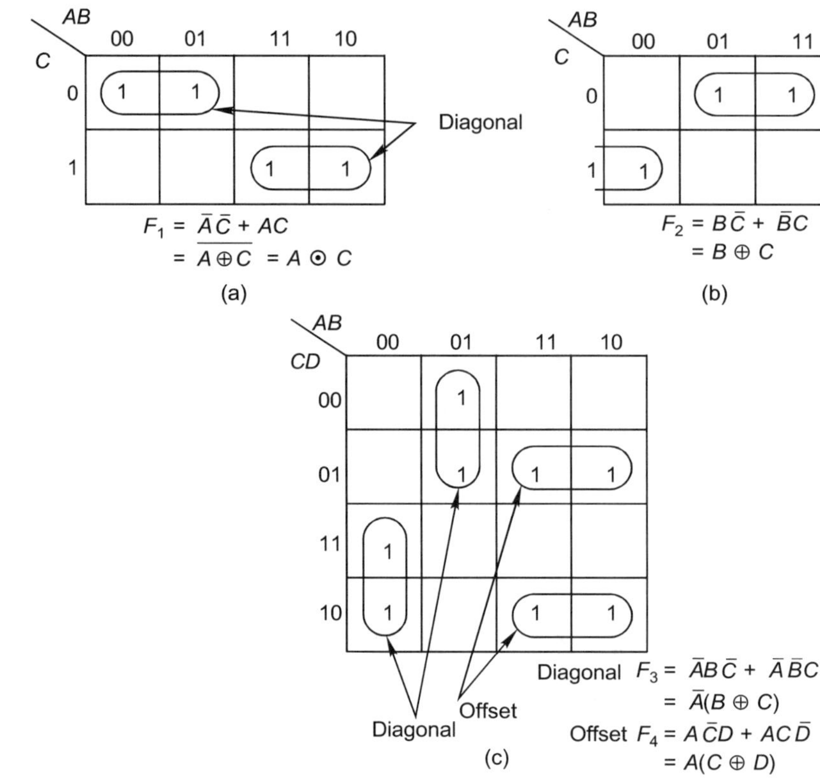

Fig. 5.28

K-maps illustrating diagonal and offset adjacencies of groups of two ones.

From this, we observe that if standard K-map groupings of two ones occur in a diagonal or offset adjacent pattern, these can be recognized as EX-OR or EX–NOR functions and the function can be simplified in terms of EX–OR/EX–NOR operations.

Figure 5.29 illustrates the diagonal and offset adjacencies of standard groups of four ones. Their simplified terms are also given in the figures.

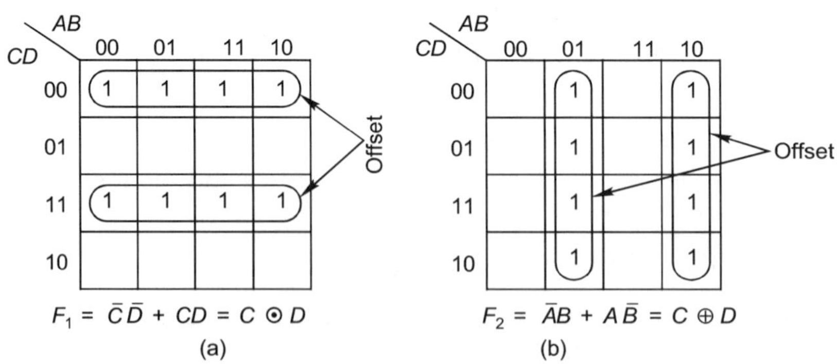

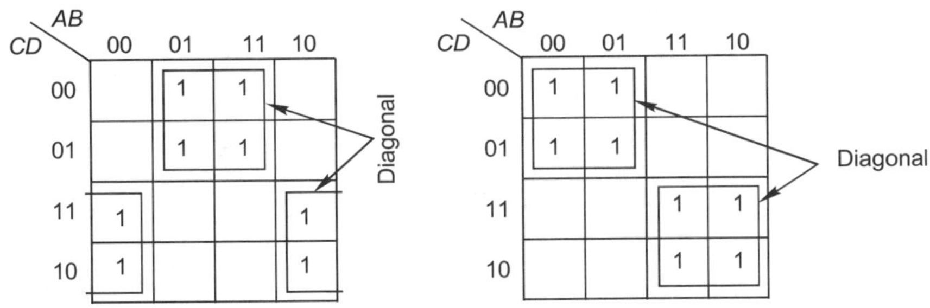

$F_3 = B\bar{C} + \bar{B}C = B \oplus C$

(c)

$F_4 = \bar{A}\bar{C} + AC == A \odot C$

(d)

Fig. 5.29
K-maps illustrating diagonal and offset adjacencies of groups of four ones

From this we observe the EX–OR/EX–NOR function patterns for standard K-map groupings of four ones.

Although it is possible to write a set of rules for K-map simplification in terms of EX–OR/OR–NOR operations, it will unnecessarily complicate the matter. The K-map should be simplified first using standard methods and then the diagonal and offset adjacencies must be identified and the expression simplified partially by using Boolean theorems and EX–OR/EX–NOR operations. The following examples will clarify the procedure further.

Example 5.19 Design a Binary-to-Gray code converter.

Solution
The truth table of Binary-to-Gray code converter is given in Table 2.8. For each of the four outputs, K-maps are prepared and simplified. The K-maps are given in Fig. 5.30 and the simplified expressions are given by Eqs (5.45). The circuit is given in Fig. 5.31.

$$G_3 = B_3 \quad (5.45a)$$
$$G_2 = B_2 \oplus B_3 \quad (5.45b)$$
$$G_1 = B_1 \oplus B_2 \quad (5.45c)$$
$$G_0 = B_0 \oplus B_1 \quad (5.45d)$$

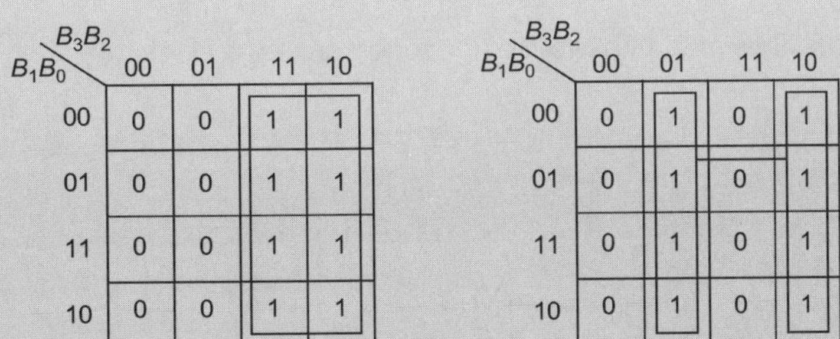

G_3
(a)

G_2
(b)

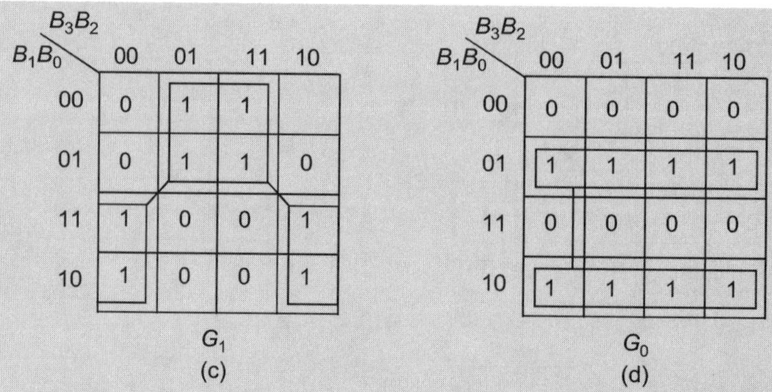

Fig. 5.30

K-maps of Ex. 5.19.

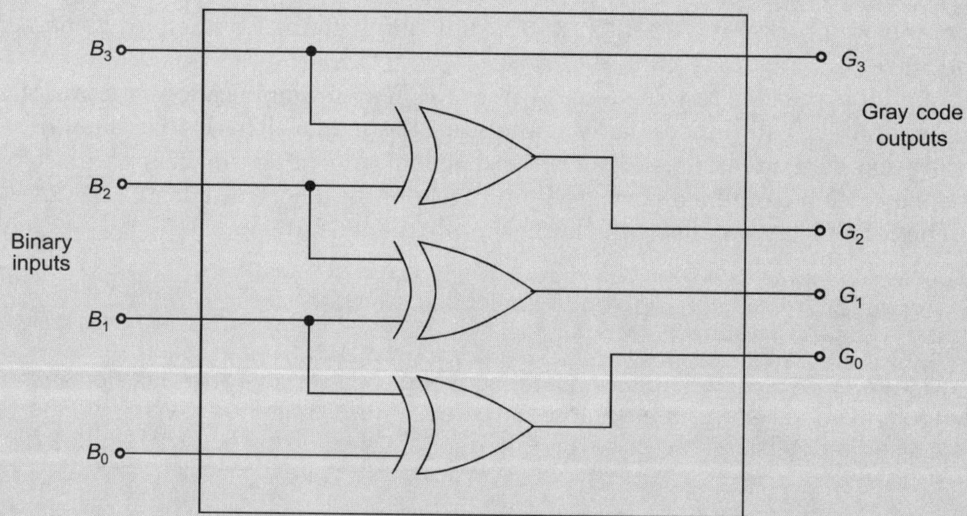

Fig. 5.31

Binary-to-Gray code converter.

Example 5.20 Design a Gray-to-Binary code converter.

Solution

The truth table for this is given in Table 2.8. The K-maps are given in Fig. 5.32 and the simplified expressions are given by Eqs (5.46).

The converter circuit is given in Fig. 5.33.

$$B_3 = G_3 \quad (5.46a)$$

$$B_2 = G_2 \oplus G_3 \quad (5.46b)$$

$$B_1 = G_1 \oplus G_2 \oplus G_3 \quad (5.46c)$$

$$B_0 = G_0 \oplus G_1 \oplus G_2 \oplus G_3 \quad (5.46d)$$

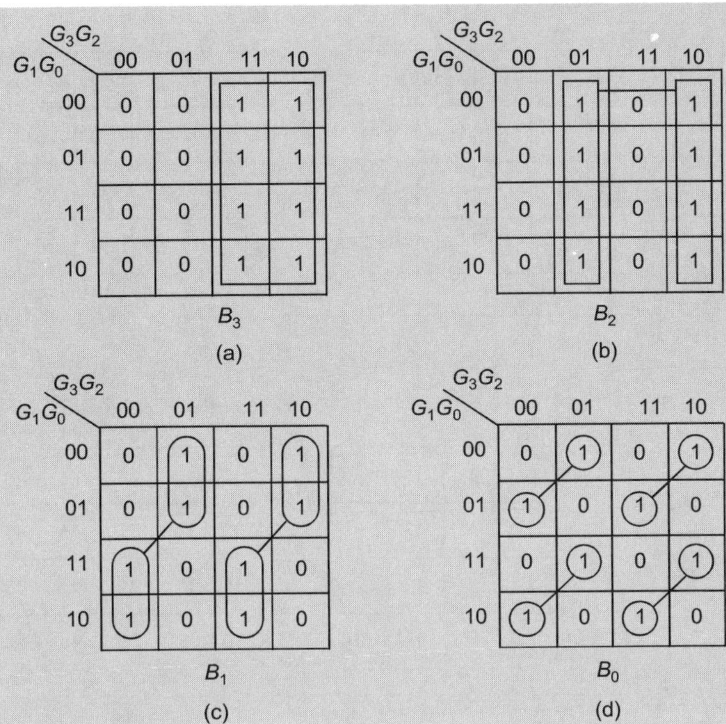

Fig. 5.32
K-maps of Ex. 5.18.

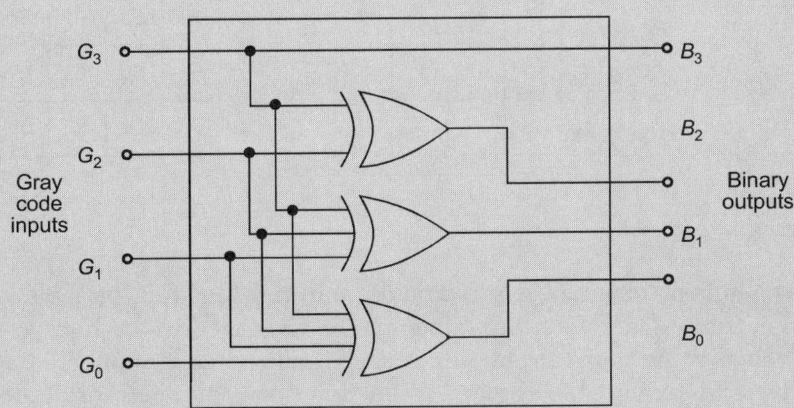

Fig. 5.33 Gray-to-Binary code converter.

5.10 FIVE- AND SIX-VARIABLE K-MAPS

A five and a six variable K-maps are shown in Figs 5.34 and 5.35, respectively. In a five variable map, we have to consider the two four-variable maps superimposed on one another;

A = 0

DE \ BC	00	01	11	10
00	0	4	12	8
01	1	5	13	9
11	3	7	15	11
10	2	6	14	10

A = 1

DE \ BC	00	01	11	10
00	16	20	28	24
01	17	21	29	25
11	19	23	31	27
10	18	22	30	26

Fig. 5.34
A five-variable K-map.

A = 0, B = 0

EF \ CD	00	01	11	10
00	0	4	12	8
01	1	5	13	9
11	3	7	15	11
10	2	6	14	10

A = 1, B = 0

EF \ CD	00	01	11	10
00	32	36	44	40
01	33	37	45	41
11	35	39	47	43
10	34	38	46	42

A = 0, B = 1

EF \ CD	00	01	11	10
00	16	20	28	24
01	17	21	29	25
11	19	23	31	27
10	18	22	30	26

A = 1, B = 1

EF \ CD	00	01	11	10
00	48	52	60	56
01	49	53	61	57
11	51	55	63	59
10	50	54	62	58

Fig. 5.35
A six-variable K-map

not 'hinged' or 'mirror imaged'. A six-variable map has four four-variable maps. The adjacencies between entries in each four-variable map are visualized in the normal way. However, the adjacencies between the four-variable maps are visualized as groupings in a two variable map. The use of five and six variable K-maps is illustrated with the help of the following examples.

Example 5.21 Simplify the logic expression

$$F(A, B, C, D, E) = \Sigma m\,(0, 5, 6, 8, 9, 10, 11, 16, 20, 24, 25, 26, 27, 29, 31) \qquad (5.47)$$

Solution

The K-map is shown in Fig. 5.36 and the simplified expression is,

$$F = \bar{A}\bar{B}C\bar{D}E + \bar{A}\bar{B}CD\bar{E} + AB\bar{D}\bar{E} + \bar{C}\bar{D}\bar{E} + ABE + B\bar{C} \qquad (5.48)$$

Equation (5.48) can be realized using NAND–NAND configuration. Try this.

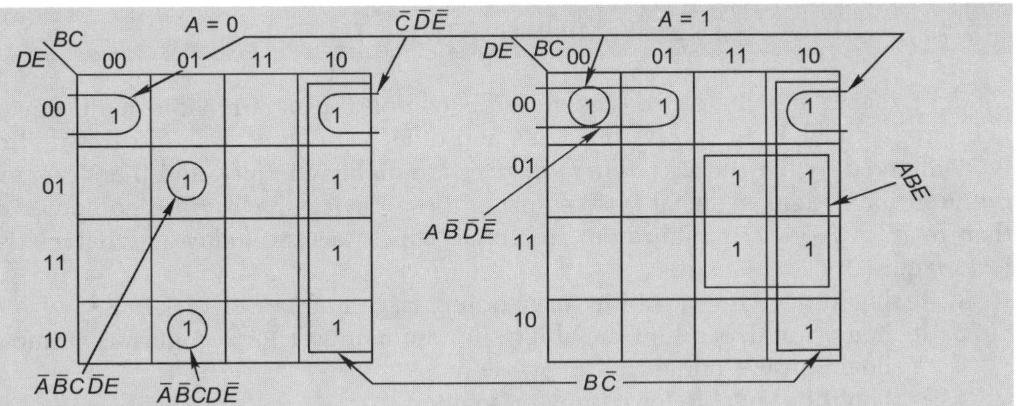

Fig. 5.36
K-map of Eq. (5.47).

Example 5.22 Simplify the six-variable logic expression

$$f(A, B, C, D, E, F) = \Sigma m\ (0, 5, 7, 8, 9, 12, 13, 23, 24, 25, 28, 29, 37, 40, 42, 44, 46, 55, 56, 57, 60, 61) \quad (5.49)$$

Solution

The K-map of Eq. (5.49) is shown in Fig. 5.37 and the simplified expression is

$$F = \bar{A}\bar{B}\bar{D}\bar{E}\bar{F} + \bar{A}\bar{B}\bar{C}DF + \bar{B}\bar{C}D\bar{E}F + B\bar{C}DEF + A\bar{B}C\bar{F} + \bar{A}C\bar{E} + BC\bar{E} \quad (5.50)$$

Equation (5.50) can be realized using NAND–NAND configuration. Try this.

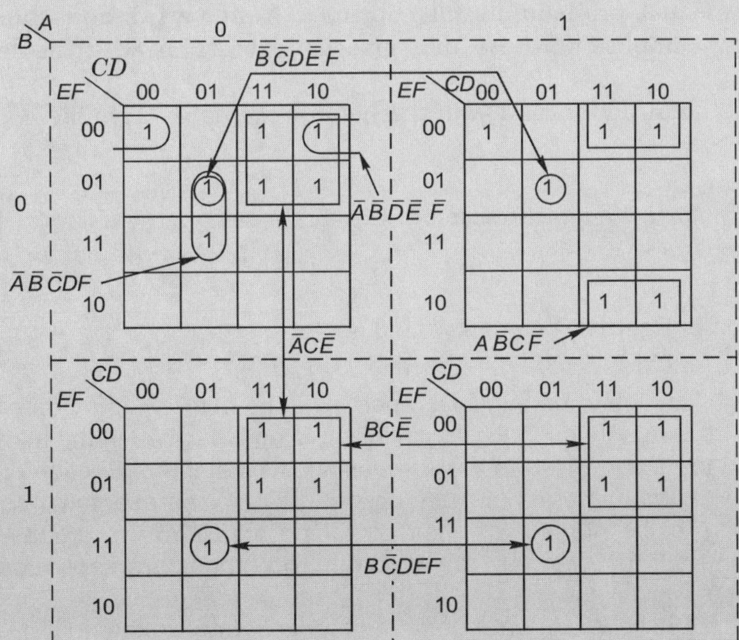

Fig. 5.37
K-map of Eq. (5.49).

5.11 QUINE-McCLUSKEY MINIMIZATION TECHNIQUE

Modern digital systems are designed using complex programmable logic devices (CPLDs), field-programmable gate arrays (FPGAs), and other very large scale integrated circuits that can be configured by the end user. These devices are highly complex and therefore, the techniques required for designing digital systems using these devices have to be computer driven rather than manual. A logic minimization technique which has the following characteristics is therefore, required:

1. It should have the capability of handling large number of variables.
2. It should not depend on the ability of a human user for recognising prime-implicants.
3. It should ensure minimized expression.
4. It should be suitable for computer solution.

The Quine-McCluskey minimization technique satisfies the above requirements and hence can be effectively used for the design of logic circuits. The K-map technique is not suitable for handling the design of complex digital systems because of the following disadvantages:

1. Minimization of logic functions involving more than six variables is unwieldy.
2. Recognition of prime-implicants that may form part of the simplified function relies on the ability of the human user making it difficult to be sure whether the best selection has been made.

The Quine-McCluskey method consists of two parts:

1. To find by an exhaustive search all the prime-implicants that may form part of the simplified function.
2. To identify essential prime-implicants obtained from part 1 and choose among the remaining prime-implicants those that give an expression with the least number of literals.

The method can be best understood with the help of examples. This method is also known as Tabular method.

Example 5.23 Simplify the logic function of Eq. (5.21) using the Quine-McCluskey minimization technique.

Solution

The logic function of Eq. (5.21) can be written as

$$Y(A, B, C, D) = \Sigma m\ (0, 1, 3, 7, 8, 9, 11, 15) \tag{5.51}$$

Step 1. The logic function to be minimized here is in the minterm form, therefore, we go to step 2. In case a function to be minimized is not specified in the minterm form, it is required to be converted to this form using the method of Section 5.2.

Step 2. Arrange all the minterms of the function in binary representation form in a Table according to the number of ones contained and form the groups containing no ones, one 1, two 1s, three 1s and so on. The groups are separated by horizontal lines.

Table 5.16 shows the arrangement of groups, minterms, and the variables. Here group 0 contains no 1s {0}; group 1 contains minterms having a single 1 {1,8}; group

2 contains minterms with two 1s {3,9}; group 3 contains minterms with three 1s {7,11}; and group 4 contains minterms with four 1s {15}.

Table 5.16 Grouping minterms according to the number of 1s

Group	*Minterm*	*Variables* A B C D	*See Step 3*
0	0	0 0 0 0	✓
1	1	0 0 0 1	✓
	8	1 0 0 0	✓
2	3	0 0 1 1	✓
	9	1 0 0 1	✓
3	7	0 1 1 1	✓
	11	1 0 1 1	✓
4	15	1 1 1 1	✓

Step 3. The Boolean algebraic theorem $A + \overline{A} = 1$ (Theorem 1.7) is applied to pairs of minterms in which only one variable is different and all the other variables are same. This kind of relationship will be applicable only to the minterms belonging to adjacent groups of minterms. For this search, compare each minterm in group n with each minterm in group $(n + 1)$ and identify the matched pairs. Put a check (✓) on each matched pair as shown in Table 5.16.

The detailed procedure for comparison and identification of matched pairs for Table 5.16 is given below and another Table 5.17 is prepared.

(i) The minterm 0 from group 0 is compared with the minterm 1 in the adjacent group 1. The three variables A, B, C are same in both with value 0 and the variable D is 0 in minterm 0 and 1 in minterm 1. Check (✓) marks are placed on both the terms in Table 5.16 and a new term is generated as a result of the matching of these two terms which will contain $A = 0$, $B = 0$, $C = 0$ and a dash (–) mark is placed in D. Since the variable D differs and hence it gets eliminated and the resulting combination of these two terms is $\overline{A}\overline{B}\overline{C}$.

(ii) Next the minterm 0 is compared with the minterm 8. The minterms match and their combination results in a term $\overline{B}\overline{C}\overline{D}$. Check mark is placed on minterm 8 in Table 5.16, check mark on minterm 0 has already been placed. A – is placed under variable A.

(iii) Similarly, comparison of minterm 1 with 3 results in
$A = 0$, $B = 0$, $C = –$, and $D = 1$
The minterm 3 is checked in Table 5.16.

(iv) Comparison of minterm 1 with minterm 9 yields
$A = –$, $B = 0$, $C = 0$, and $D = 1$
and the minterm 9 is checked

(v) Now compare 8 with 3 and 9. The minterms 8 and 3 do not match. The comparison of minterms 8 and 9 results in
$A = 1$, $B = 0$, $C = 0$, and $D = –$

(vi) Next compare the minterms 3 and 7, it results in
$A = 0$, $B = –$, $C = 1$, and $D = 1$
and the minterm 7 is checked in Table 5.16.

(vii) Comparison of the minterms 3 and 11 results in
$A = –, B = 0, C = 1$, and $D = 1$
and the minterm 11 is checked in Table 5.16

(viii) The minterms 9 and 7 do not match.

(ix) Comparison of the minterms 9 and 11 results in
$A = 1, B = 0, C = –$, and $D = 1$

(x) Next compare the minterms 7 and 15, the result is
$A = –, B = 1, C = 1$, and $D = 1$
and the minterm 15 is checked in Table 5.16

(xi) Comparison of the minterms 11 and 15 results in
$A = 1, B = –, C = 1$, and $D = 1$

Table 5.17 lists the results of all the above matchings and all of the minterms in each group have been compared to those in the next higher group.

Table 5.17 Combination of minterm groups of two

Group	*Minterms*	*Variables* A B C D	*See Step 4*
0	0, 1	0 0 0 –	✓
	0, 8	– 0 0 0	✓
1	1, 3	0 0 – 1	✓
	1, 9	– 0 0 1	✓
	8, 9	1 0 0 –	✓
2	3, 7	0 – 1 1	✓
	3, 11	– 0 1 1	✓
	9, 11	1 0 – 1	✓
3	7, 15	– 1 1 1	✓
	11, 15	1 – 1 1	✓

Step 4. Next all the minterms in the adjacent groups in Table 5.17 are compared to see if groups of four can be made by matching. For this the dashes must be in the same bit position in the groups of two and only one variable must differ (0 in one group and 1 in the other). The matched pairs of minterms are checked in Table 5.17 and a new Table 5.18 is created.

In Table 5.18 we observe that in each group the two terms are same, therefore, only one is to be taken in each group.

Table 5.18 Combination of minterm groups of four

Group	*Minterms*	*Variables* A B C D
0	0, 1, 8, 9	– 0 0 –
	0, 8, 1, 9	– 0 0 –
1	1, 3, 9, 11	– 0 – 1
	1, 9, 3, 11	– 0 – 1
2	3, 7, 11, 15	– – 1 1
	3, 11, 7, 15	– – 1 1

Step 5. Repeat the process of grouping of eight minterms. In this case both dashes (–) must be in the same bit position and only one other variable must be different for matching. Since, there is no matching possible here, therefore, the process is complete. In general, this same process is repeated until no further combinations of minterm groups is possible.

Step 6. All nonchecked minterm groups in Tables 5.16, 5.17, and 5.18 are the prime-implicants of the function. The function can now be written as

$$Y(A, B, C, D) = \bar{B}\bar{C} + \bar{B}D + CD \tag{5.52}$$

Step 7. Next a prime-implicant table is prepared listing each of the minterms contained in the original function, PI terms, and the decimal numbers of minterms that make up the PI. Put cross (×) marks in the table in each row under the minterms contained in that PI. Table 5.19 is the PI table for the logic equation Eq. (5.21). Find the minterms that contain only one × in its column. These ×'s are encircled and the corresponding PI terms are checked (✓).

Table 5.19 PI table

PI terms	*Decimal numbers*	*Minterms*							
		0	*1*	*3*	*7*	*8*	*9*	*11*	*15*
$\bar{B}\bar{C}$ ✓	0, 1, 8, 9	⊗	×			⊗	×		
$\bar{B}D$	1, 3, 9, 11		×	×			×	×	
CD ✓	3,7, 11, 15			×	⊗			×	⊗

The minterms 0 and 8 are contained in only one PI $\bar{B}\bar{C}$, the minterms 7 and 15 are contained in only PI CD. Therefore, the prime-implicants $\bar{B}\bar{C}$ and CD are essential prime-implicants. Now observe the other minterms and see whether these are contained in EPIs or not. Here, the minterms 1 and 9 are contained in $\bar{B}\bar{C}$ and 7 and 11 are contained in CD. Therefore, all the minterms of the original function are included in the two EPIs and the minimized expression will be

$$Y(A, B, C, D) = \bar{B}\bar{C} + CD \tag{5.53}$$

This is same as the minimized function obtained using K-map technique Eq. (5.22).

Example 5.24 Simplify the logic function of Eq. (5.23) using the Quine-McCluskey method.

Solution

Table 5.20 is prepared grouping the minterms according to the number of 1s contained.

Table 5.20

Group	*Minterm*	*Variables*				
		A	*B*	*C*	*D*	
0	0	0	0	0	0	✓
1	1	0	0	0	1	✓
	2	0	0	1	0	✓
	8	1	0	0	0	✓

(*Contd.*)

Table 5.20 *(Contd.)*

Group	Minterms	A	B	C	D	
		Variables				
2	3	0	0	1	1	✓
	5	0	1	0	1	✓
	9	1	0	0	1	✓
3	7	0	1	1	1	✓
	11	1	0	1	1	✓
	14	1	1	1	0	

Table 5.21 is created by comparing and matching the minterms in adjacent groups. In Table 5.20 all the minterms except m_{14} are checked (✓).

Table 5.21

Group	Minterms	A	B	C	D	
		Variables				
0	0, 1	0	0	0	–	✓
	0, 2	0	0	–	0	✓
	0, 8	–	0	0	0	✓
1	1, 3	0	0	–	1	✓
	1, 5	0	–	0	1	✓
	1, 9	–	0	0	1	✓
	2, 3	0	0	1	–	✓
	8, 9	1	0	0	–	✓
2	3, 7	0	–	1	1	✓
	3, 11	–	0	1	1	✓
	5, 7	0	1	–	1	✓
	9, 11	1	0	–	1	✓

Next Table 5.22 is created by comparing and matching minterms is Table 5.21. All the minterms in Table 5.21 are checked (✓).

Table 5.22

Group	Minterms	A	B	C	D
		Variables			
0	0, 1, 2, 3	0	0	–	–
	0, 1, 8, 9	–	0	0	–
	0, 2, 1, 3	0	0	–	–
	0, 8, 1, 9	–	0	0	–
1	1, 3, 5, 7	0	–	–	1
	1, 3, 9, 11	–	0	–	1
	1, 5, 3, 7	0	–	–	1
	1, 9, 3, 11	–	0	–	1

There is no matching possible in Table 5.22. The function f consists of the prime-implicant m_{14} (unchecked minterm in Table 5.20) and the terms corresponding to the prime-implicants of Table 5.22. It is given as

$$f(A, B, C, D) = ABC\bar{D} + \bar{A}\bar{B} + \bar{B}\bar{C} + \bar{A}D + \bar{B}D \tag{5.54}$$

The PI table corresponding to Eq. (5.54) is prepared as Table 5.23.

Table 5.23

PI terms		*Decimal numbers*	*Minterms*									
			0	1	2	3	5	7	8	9	11	14
$ABC\bar{D}$	✓	14										$\otimes$
$\bar{A}\bar{B}$	✓	0, 1, 2, 3	$\times$	$\times$	$\otimes$	$\times$						
$\bar{B}\bar{C}$	✓	0, 8, 1, 9	$\times$	$\times$					$\otimes$	$\times$		
$\bar{A}D$	✓	1, 3, 5, 7		$\times$		$\times$	$\otimes$	$\otimes$				
$\bar{B}D$	✓	1, 9, 3,11		$\times$		$\times$				$\times$	$\otimes$	
					✓		✓	✓	✓		✓	✓

Observe the PI table and find the columns containing only a single cross (×). There are six minterms whose columns have a single cross; 2, 5, 7, 8, 11, and 14. Prime-implicants that cover minterms with a single cross in their column are essential prime–implicants. The single crosses are encircled. Check marks are placed below these columns. A check mark is also placed in the table on the essential prime-implicants.

Since all the prime-implicants are essential prime-implicants, therefore, Eq. (5.54) is the minimized function. This is same as Eq.(5.24) obtained by K-map method.

In case of incompletely specified function, i.e., for functions containing don't-care terms, the don't-care terms are considered as if they are the required minterms. The prime implicants are determined using these minterms alongwith the other specified minterms. To differentiate the minterms corresponding to the don't-care conditions, an asterisk (*) mark is put on the decimal numbers corresponding to the don't-care terms. The PIs are thus determined. When forming the PI chart, the don't-care terms are not listed at the top. The PI chart is solved and the resulting expression will contain all the required don't-care minterms. The following example illustrates the complete procedure.

Example 5.25 Simplify the logic function of Eq. (5.32) using the Quine-McCluskey method

Solution

Using the procedure discussed above, Tables 5.24(a), (b), and (c) are prepared.

Table 5.24(a)

Group	*Minterm/ Don't care term**	*Variables* A B C D	
0	0*	0 0 0 0	✓
1	1	0 0 0 1	✓
	2*	0 0 1 0	✓

(*Contd.*)

Table 5.24(a) *(Contd.)*

Group	*Minterm/ Don't care term**	*A*	*B*	*C*	*D*	
2	3	0	0	1	1	✓
	5 *	0	1	0	1	✓
3	7	0	1	1	1	✓
	11	1	0	1	1	✓
4	15	1	1	1	1	✓

Table 5.24(b)

Group	*Minterms/ Don't care term**	*A*	*B*	*C*	*D*	
0	0*, 1	0	0	0	–	✓
	0*, 2*	0	0	–	0	✓
1	1, 3	0	0	–	1	✓
	1, 5*	0	–	0	1	✓
	2*, 3	0	0	1	–	✓
2	3, 7	0	–	1	1	✓
	3, 11	–	0	1	1	✓
	5*, 7	0	1	–	1	✓
3	7, 15	–	1	1	1	✓
	11, 15	1	–	1	1	✓

Table 5.24(c)

Group	*Minterms/ Don't care term**	*A*	*B*	*C*	*D*
0	0*, 1, 2*, 3	0	0	–	–
	0*, 2*, 1, 3	0	0	–	–
1	1, 3, 5*, 7	0	–	–	1
	1, 5*, 3, 7	0	–	–	1
2	3, 7, 11, 15	–	–	1	1
	3, 11, 7, 15	–	–	1	1

Table 5.25 gives the PI table.

Table 5.25

PI terms	*Decimal numbers (Minterms)*	*1*	*3*	*7*	*11*	*15*
$\bar{A}\bar{B}$	0*, 1, 2*, 3	×	×			
$\bar{A}D$	1, 3, 5*, 7	×	×	×		
CD ✓	3, 7, 11, 15		×	×	⊗	⊗
					✓	✓

From the PI table, we observe only one × mark in columns 11 and 15, therefore, the PI term CD is an essential prime implicant. Check marks are placed at the EPI and in columns 11 and 15. The remaining three minterms 1,3, and 7 are covered by the PI term $\bar{A}D$. Therefore, the minimal function is given by

$$Y = \bar{A}D + CD$$

This is same as obtained by K-map method given in Fig. 5.18(a).

From the above result, we see that the minterm $5(m_5)$ has been included in the simplification.

5.12 SUMMARY

For any logic design, it is always essential to design a product which meets the following requirements:

- Minimum cost.
- Minimum space requirement.
- Maximum speed of operation.
- Easy availability of components.
- Ease of interconnecting the components.
- Easy to design.

To achieve most of the above requirements, it is necessary to have some tools for simplification of logic expressions. Some of the traditional and widely used methods of logic design, such as, Boolean algebraic method, Karnaugh map technique, and Quine-McCluskey methods have been discussed. The logic expressions can be expressed in one of the standard forms: SOP or POS and then simplified using, K-maps or Quine-McCluskey method. This results in saving in required hardware in terms of number of gates required and the number of input terminals (fan- in) of the gates.

As discussed in Chapter 1, a large number of gates are available in IC form. The number of gates which can be packaged in an IC chip decreases with the number of input terminals of the gates. Therefore, the simplification results in lesser number of IC chips required, thereby resulting in reduced cost and space requirement.

Since the simplified boolean expressions are in SOP or POS form, therefore, only one type (NAND or NOR) of gates are required which again is useful in reducing the number of IC chips required. The propagation delay time is also minimum because of two-level realisation. Because of the availability of a number of logic functions in a logic family which are compatible, it is very convenient to interconnect them to obtain the complete circuit.

The simplification methods discussed are very convenient to use. However, K-maps can conveniently be used upto six variables beyond which it becomes very cumbersome also it can not be mechanised. The Quine-McCluskey method can be used for larger number of variables and can be mechanised.

In addition to minimization of logical functions for their realization using NAND/NOR gates, simplification techniques have also been discussed which enables us to design logic circuits using EX-OR and EX-NOR gates, since EX-OR is a very useful and commonly used gate.

Many complex digital functions have been designed which are directly available in IC form. Such circuits have been discussed in Chapter 6 and their use reduces our dependence on simplification procedures. However, discrete gates are employed for interfacing MSI and LSI devices.

A large number of design examples have been included to illustrate the design procedures discussed.

Glossary

Adder A logic circuit that can add two numbers.

AND-OR realization A two-level realization of logic functions that has AND gates in the first level and an OR gate in the second level.

Code converter A logic circuits that converts data from one binary code to another binary code.

Combinational logic The logic in which the outputs at any instant of time are dependent only on the inputs present at that time.

Decoder A logic circuit used to decode a coded binary word.

Don't care A minterm/maxterm in a logic function which may or may not be included.

Essential prime implicant A term in SOP form in a logic function that must be present in the minimal expression.

Full-adder A logic circuit that accepts two one-bit signals and a carry-in as inputs and produces their sum and carry bits as outputs.

Full-subtactor A digital circuit that accepts two one-bit signals and a borrow-in as inputs and produces their difference and borrow as outputs.

Half-adder A logic circuit that accepts two single bit inputs and produces their sum bit and carry bit as outputs.

Half-subtractor A logic circuit that accepts two bits as inputs and produces their difference bit and borrow bit as outputs.

Karnaugh-map (K-map) A mapping technique used to minimize logical expressions.

LED (Light-emitting diode) A diode that emits visible radiant energy when conducting.

Literal A logical variable in either inverted or non-inverted form.

Maxterm A logical term consisting of all the literals in the ORed form in logic function.

Minimization The process of reducing a logical expression to a minimum form to make it realizable using minimum number of gates with minimum fan-in.

Minterm A logical term consisting of all the literals in the ANDed form.

OR-AND realization A two-level realization of logic functions that has OR gates in the first level and an AND gate in the second level.

Product-of-sums (POS) A logic expression in the form of ORed terms ANDed together.

Sequential logic Logic circuits whose outputs are determined by the sequence in which input signals are applied.

Standard POS A POS form of logic expression consisting of only maxterms.
Standard SOP A SOP form of logic expression consisting of only minterms.
Sum-of-products (SOP) A logic expression in the form of ANDed terms ORed together.
Two-level realization Realization of logic functions in which signals pass through two gates, such as AND-OR, OR-AND, NAND-NAND, NOR-NOR realizations.

Review Questions

5.1 A logical variable in complemented or uncomplemented form is known as a ________ .

5.2 A combinational circuit does not have ________ .

5.3 A standard SOP logic expression of 4 variables contains each term with ________ literals.

5.4 A logic expression form most suitable for realization using only NAND gates is ________ .

5.5 AND-OR realization is equivalent to ________ realization.

5.6 NOR-NOR realization is preferred over OR-AND realization because of reduced ________ count.

5.7 A logic term containing literals corresponding to all the variables in ANDed form is known as a ________ .

5.8 A logic term containing literals corresponding to all the variables in ORed form is known as a ________ .

5.9 A K-map of n-variables contains ________ cells.

5.10 ________ code is used for labelling the cells of K-map.

5.11 A minterm corresponding to don't care condition may have a value ________ .

5.12 Number of inputs for a full adder is ________ .

5.13 A NOR-NOR realization is a ________ level realization.

5.14 A 1 in a cell of K-map can be combined with three other 1's in only one combination. The resulting term of these four 1's is ________ .

5.15 In a PI chart of Quine-McCluskey simplification a column containing only one × contributes a term ________ .

Problems

5.1 A staircase light is controlled by two switches, one at the top of the stairs and another at the bottom of stairs.
(a) Make a truth table for this system.
(b) Write the logic equation in SOP form.
(c) Realize the circuit using AND–OR gates.
(d) Realize the circuit using NAND gates only.

5.2 Given the logic equation

$$f = ABC + B\bar{C}D + \bar{A}BC$$

(a) Make a truth table.
(b) Simplify using K-map.
(c) Realize f using NAND gates only.

5.3 For the truth table given in Table 5.26
(a) Write logic equations for f_1 and f_2 in POS form.
(b) Use K-map for minimization and obtain the minimized expressions.
(c) Realize these using OR–AND gates.
(d) Realize these using only NOR gates.
(e) Compare the IC packages required in parts (c) and (d).

Table 5.26 Truth table for Problem 5.3

Inputs				*Outputs*	
A	*B*	*C*	*D*	f_1	f_2
0	0	0	0	1	0
0	0	0	1	0	0
1	0	0	0	0	1
1	1	0	0	0	1
1	1	1	0	1	1
1	1	1	1	1	1
0	1	1	1	0	0
0	0	1	1	0	0
0	0	1	0	1	0
0	1	0	0	0	1
0	1	0	1	1	1
0	1	1	0	0	0
1	0	0	1	1	0
1	0	1	0	0	1
1	0	1	1	1	0
1	1	0	1	0	0

5.4 (a) Realize Eq. (5.24) using only NAND gates.
(b) Realize Eq. (5.27) using only NOR gates.
(c) Compare the IC packages required in parts (a) and (b). See Table 1.10 for IC packages.

5.5 (a) Realize Eq. (5.25) using minimum number of available NAND gate ICs.
(b) Realize Eq. (5.28) using minimum number of available NOR gate ICs.
(c) Compare the IC packages required in parts (a) and (b).

5.6 Realize Eq. (5.31) using minimum number of available NOR gate chips.

5.7 (a) Make a K-map for the function

$$f = AB + A\bar{C} + C + AD + A\bar{B}C + ABC$$

(b) Express f in standard SOP form.
(c) Minimize it and realize the minimized expression using NAND gates only.

5.8 Minimize the following functions and realize using minimum number of gates.
(a) $f_1 = \Sigma\, m\,(0, 3, 5, 6, 9, 10, 12, 15)$
(b) $f_2 = \Sigma\, m\,(0, 1, 2, 3, 11, 12, 14, 15)$

5.9 Design a BCD-to-Excess-3-code converter using minimum number of NAND gates.

5.10 Design a Excess-3-to-BCD-code converter using minimum number of NAND gates.

5.11 Minimize the following expressions using K-maps and realize using NOR gates only.
(a) $f_1(A, B, C, D) = \Pi\, M(1, 2, 3, 5, 6, 7, 9, 10, 11, 13, 14, 15)$
(b) $f_2(A, B, C, D) = \Pi\, M(1, 4, 6, 9, 10, 11, 14, 15)$
(c) $f_3(A, B, C, D) = \Pi\, M(2, 7, 8, 9, 10, 12)$.

5.12 Minimize the following expressions using K-maps and realize with NAND gates.
$f_1(A, B, C, D, E) = \Sigma\, m(8, 9, 10, 11, 13, 15, 16, 18, 21, 24, 25, 26, 27, 30, 31)$
$f_2(A, B, C, D, E) = \Pi\, M(6, 9, 11, 13, 14, 17, 20, 25, 28, 29, 30)$

5.13 Realize the logical function of Truth Table 5.10 using minimum number of NAND gates
(a) assuming 0s for all the don't-care conditions.
(b) assuming a 0 or 1 for the don't-care conditions leading to a simpler expression.
(c) comment on the effect of don't-care conditions on the hardware requirement.

5.14 Minimize the following logic functions and realize using NAND/NOR gates.
(a) $f_1(A, B, C, D) = \Sigma\, m(1, 3, 5, 8, 9, 11, 15) + d(2, 13)$
(b) $f_2(A, B, C, D) = \Pi\, M(1, 2, 3, 8, 9, 10, 11, 14) \cdot d(7, 15)$

5.15 Realize the following expressions using EX–OR and EX–NOR gates.
(a) $f_1 = \bar{A}B\bar{C}\,\bar{D} + A\bar{B}\,\bar{C}\,\bar{D} + \bar{A}\,\bar{B}\,\bar{C}D + AB\bar{C}D$
(b) $f_2 = \bar{A}B\bar{C} + \bar{A}\,\bar{B}C + A\bar{C}D + AC\bar{D}$
(c) $f_3 = AB\bar{C}\,\bar{D} + A\bar{B}\,\bar{C}D + \bar{A}\,\bar{B}CD + \bar{A}BC\bar{D}$

5.16 Design a parity generator to generate an odd parity bit for a 4-bit word. Use EX–OR and $\overline{\text{EX–OR}}$ gates.

5.17 Repeat Prob. 5.16 for an even parity bit.

5.18 Simplify the following logic expressions and realize using NAND / NOR gates.
(a) $f_1(A, B, C, D, E, F) = \Sigma\, m(6, 9, 13, 18, 19, 25, 27, 29, 41, 45, 57, 61)$
(b) $f_2(A, B, C, D, E, F) = \Pi\, M(4, 5, 6, 7, 8, 12, 13, 16, 17, 18, 19, 21, 22, 25, 28, 32, 35, 37, 38, 39, 40)$.

5.19 Design a full-adder circuit using two half-adders. Use
(a) Block diagram of half-adder.
(b) Circuit of Fig. 5.19 for half-adder.

5.20 For the full-adder circuit of Prob. 5.19(b) determine the propagation delay time for Sum output (S_n) and carry output (C_n), assuming the propagation delay time of gates as

EX-OR	20 ns
AND	10 ns
OR	10 ns

and presence of data inputs A_n, B_n and carry-in (C_{n-1}) simultaneously.

5.21 Realize the logic function in SOP form using Quine-McCluskey method
$$f(A, B, C, D) = \Pi M(2, 7, 8, 9, 10, 12)$$

5.22 Minimize the logic function using Quine-McCluskey method

$$f(A, B, C, D) = \Sigma m(1, 3, 5, 8, 9, 11, 15) + d\,(2, 13)$$

5.23 Minimize the logic function using Quine-McCluskey method

$$f(A, B, C, D, E) = \Sigma m\,(8, 9, 10, 11, 13, 15, 16, 18, 21, 24, 25, 26, 27, 30, 31)$$

6

COMBINATIONAL LOGIC DESIGN USING MSI CIRCUITS

6.1 INTRODUCTION

The traditional methods of combinational circuit design have been discussed in Chapter 5. These methods involve simplification and realization using gates. Using these methods, complex functions have been integrated (MSI) and are easily available in IC form. There is an attractive array of devices such as multiplexers, demultiplexers, adders, parity generators/checkers, priority encoders, decoders, comparators, etc. This chapter presents these complex integrated circuits and their applications in combinational system design. These devices significantly reduce IC package count thereby reducing the system cost. The system design is greatly simplified because the laborious and time consuming simplification methods are generally not required. This also improves the reliability of the system by reducing the number of external wired connections. Therefore, a designer must become familiar with the functions performed, the options available, and the limitations of these devices in order to make an effective and optimum use of such devices.

6.2 MULTIPLEXERS AND THEIR USE IN COMBINATIONAL LOGIC DESIGN

6.2.1 Multiplexer

The multiplexer is a special combinational circuit that is one of the most widely used standard circuits in digital design. The multiplexer (or data selector) is a logic circuit that gates one out of several inputs to a single output. The input selected is controlled by a set of *select* inputs. Figure 6.1 shows the block diagram of a multiplexer with n input lines and one output line. For selecting one out of n inputs for connection to the output, a set of m select inputs is required,

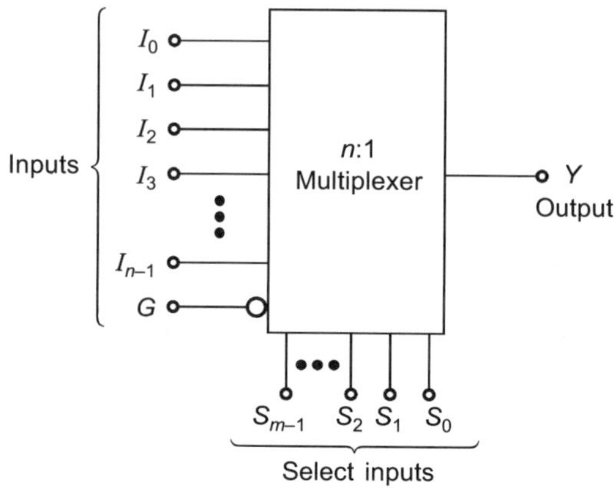

Fig. 6.1
Block diagram of a digital multiplexer.

where $2^m = n$. Depending upon the digital code applied at the select inputs one out of n data sources is selected and transmitted to a single output channel. Normally, a *strobe* (or *enable*) input (G) is incorporated which helps in cascading and it is generally *active-low*, which means it performs its intended operation when it is LOW.

Table 6.1 Truth table of a 4:1 multiplexer

Select inputs		*Output*
S_1	S_0	Y
0	0	I_0
0	1	I_1
1	0	I_2
1	1	I_3

Table 6.1 gives the truth table of a 4:1 multiplexer. With LOW input at G, the output Y can be expressed as

$$Y = \bar{S}_1 \bar{S}_0 I_0 + \bar{S}_1 S_0 I_1 + S_1 \bar{S}_0 I_2 + S_1 S_0 I_3 \tag{6.1}$$

Equation (6.1) can be realized using NAND gates and the realization is given in Fig. 6.2.

6.2.2 Combinational Logic Design Using Multiplexers

The multiplexing function discussed above can conveniently be used as a logic element in the design of combinational circuits. Standard ICs are available (Table 6.2) for 2:1, 4:1, 8:1, and 16:1 multiplexers.

Use of multiplexers offers the following advantages:

1. Simplification of logic expression is not required.
2. It minimizes the IC package count.
3. Logic design is simplified.

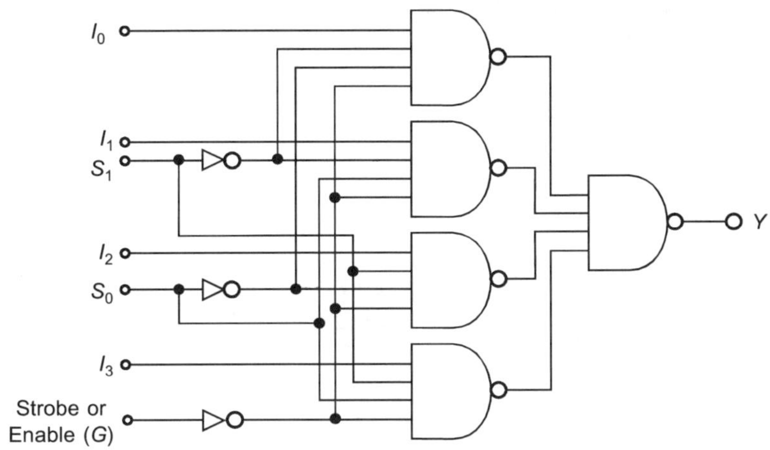

Fig. 6.2
A 4:1 multiplexer with strobe input using NAND gates.

Table 6.2 Available multiplexer ICs

IC No.	*Description*	*Output*
74157	Quad 2:1 Multiplexer	Same as input
74158	Quad 2:1 Multiplexer	Inverted input
74153	Dual 4:1 Multiplexer	Same as input
74352	Dual 4:1 Multiplexer	Inverted input
74151A	8:1 Multiplexer	Complementary outputs
74152	8:1 Multiplexer	Inverted input
74150	16:1 Multiplexer	Inverted input

For using the multiplexer as a logic element, either the truth table or one of the standard forms of logic expression must be available. The design procedure is given below:

1. Identify the decimal number corresponding to each minterm in the expression. The input lines corresponding to these numbers are to be connected to logic 1 level.
2. All other input lines are to be connected to logic 0 level.
3. The inputs are to be applied to select inputs.

The following examples illustrate the above procedure.

Example 6.1 Implement the expression using a multiplexer.

$$f(A, B, C, D) = \Sigma m(0, 2, 3, 6, 8, 9, 12, 14)$$

Solution

Since there are four variables, therefore, a multiplexer with four select inputs is required. The circuit of 16:1 multiplexer connected to implement the above expression is shown in Fig. 6.3. This implementation requires only one IC package. In case the output of the multiplexer is active-low, the logic 0 and logic 1 inputs of Fig. 6.3 are to be interchanged. The reader should verify the validity of this statement.

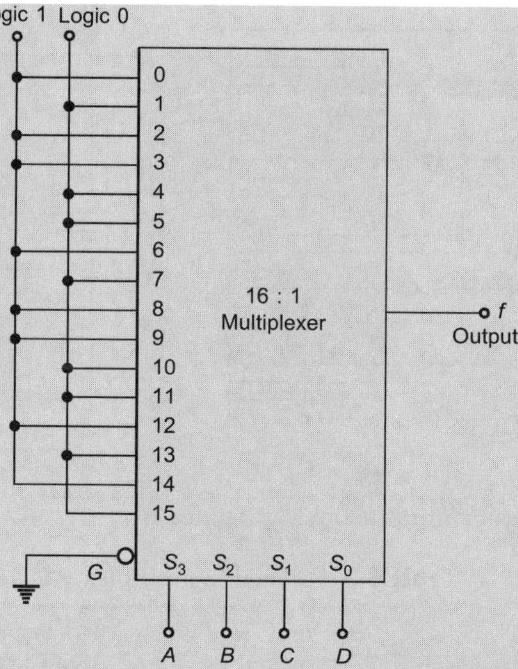

Fig.6.3
Implementation of logic expression of Ex. 6.1.

Example 6.2 Realize the logic function of the truth table given in Table 6.3.

Table 6.3 Truth table of Example 6.2

Inputs				*Output*
A	*B*	*C*	*D*	*Y*
0	0	0	0	0
0	0	0	1	0
0	0	1	0	1
0	0	1	1	0
0	1	0	0	1
0	1	0	1	0
0	1	1	0	1
0	1	1	1	1
1	0	0	0	0
1	0	0	1	1
1	0	1	0	1
1	0	1	1	1
1	1	0	0	1
1	1	0	1	0
1	1	1	0	0
1	1	1	1	1

Solution

(i) *First Method*: This can be realized using the method used in Example 6.1. Here, the input lines 2, 4, 6, 7, 9, 10, 11, 12, and 15 are to be connected to logic 1 and the input lines 0, 1, 3, 5, 8, 13, and 14 are to be connected to logic 0.

(ii) *Second Method:* A four variable truth table or logic expression can be realized by using an 8:1 multiplexer instead of a 16:1 multiplexer. For this, partition the truth table as shown by dotted lines. Here the inputs A, B, and C are to be connected to S_2, S_1, and S_0 select inputs respectively. Now, we observe the relationship between input D and output Y for each group of two rows. There are four possible values of Y and these are 0, 1, D, and $\bar{D}$. These are given in Table 6.4. From this table, we note the output Y for each of the combinations of A, B, and C, and then make the connections accordingly. The implementation of this function using an 8:1 multiplexer is shown in Fig. 6.4.

Table 6.4

Inputs			*Output*
A	B	C	Y
0	0	0	0
0	0	1	$\bar{D}$
0	1	0	$\bar{D}$
0	1	1	1
1	0	0	D
1	0	1	1
1	1	0	$\bar{D}$
1	1	1	D

The second method can also be used if the logic expression is specified.

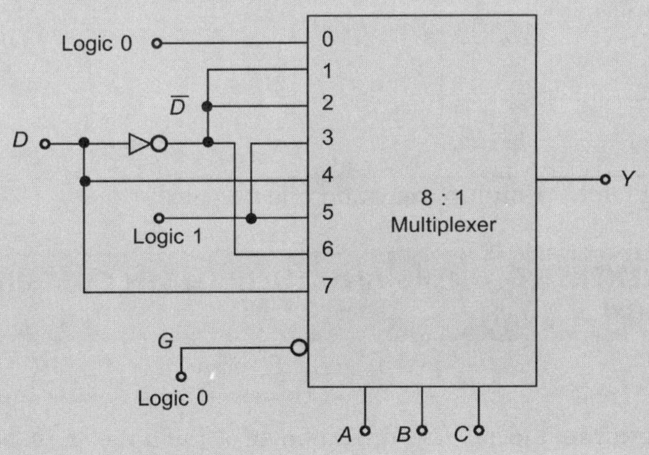

Fig. 6.4

Realization of 4-variable truth table using 8:1 multiplexer.

6.2.3 Multiplexer Tree

Since 16-to-1 multiplexers are the largest available ICs, therefore, to meet the larger input needs there should be a provision for expansion. This is achieved with the help of enable/strobe inputs and multiplexers stacks or trees are designed. Two commonly used methods for this purpose are illustrated in Figs 6.5 and 6.6.

These two general techniques can be used to expand to an n input multiplexer without any difficulty.

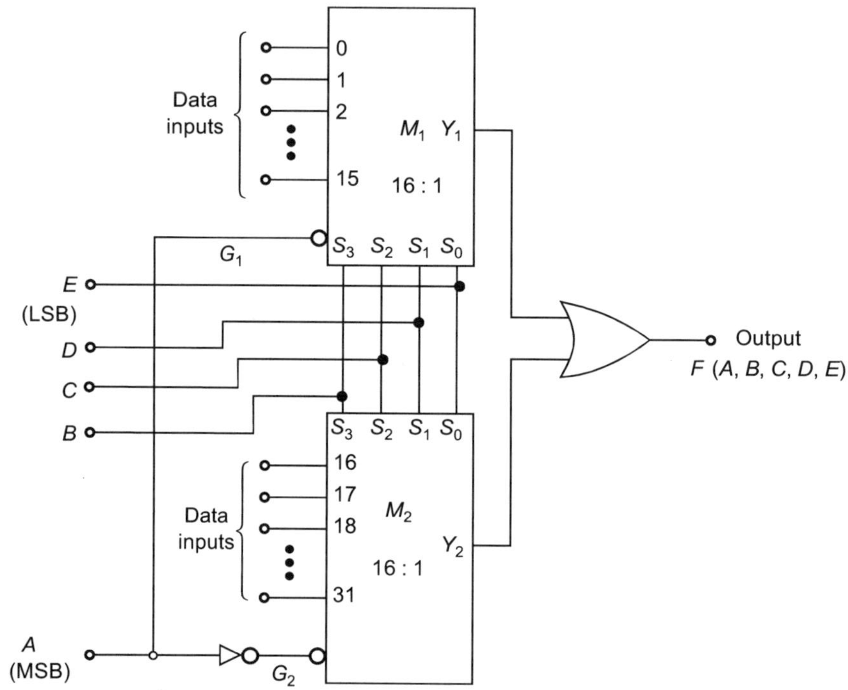

Fig. 6.5

32:1 multiplexer using two 16:1 multiplexers and one OR gate.

6.3 DEMULTIPLEXERS/DECODERS AND THEIR USE IN COMBINATIONAL LOGIC DESIGN

6.3.1 Demultiplexer

The demultiplexer performs the reverse operation of a multiplexer. It accepts a single input and distributes it over several outputs. Figure 6.7 gives the block diagram of a demultiplexer. The select input code determines to which output the data input will be transmitted.

The number of output lines is n and the number of select lines is m, where $n = 2^m$. This circuit can also be used as binary-to-decimal decoder with binary inputs applied at the select input lines and the output will be obtained on the corresponding line. The data input line is to

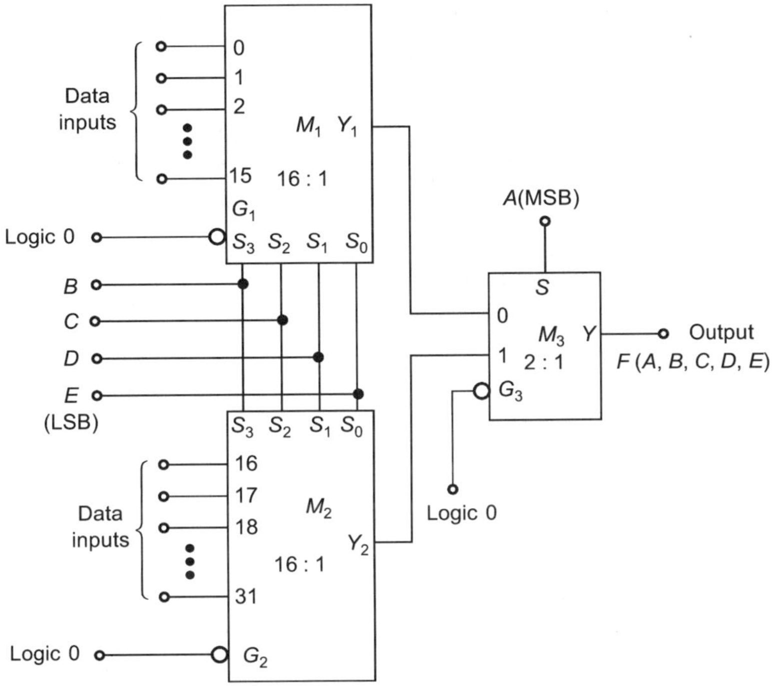

Fig. 6.6
32:1 multiplexer using two 16:1 multiplexers and one 2:1 multiplexer.

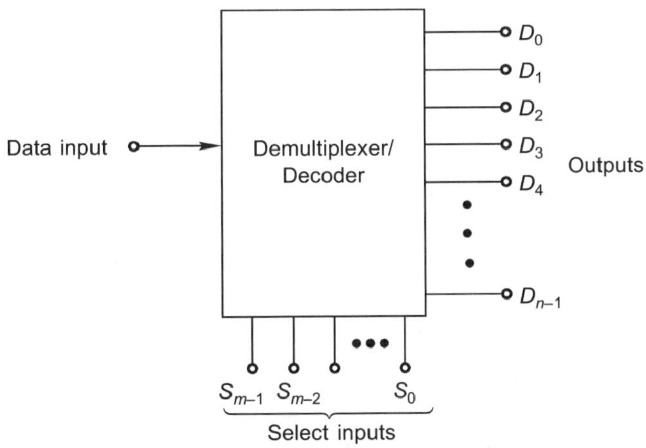

Fig. 6.7
Block diagram of a demultiplexer.

be connected to logic 1 level. This circuit can be designed using gates and it is left as an exercise for the reader. However, this device is available as an MSI IC and can conveniently be used for the design of combinational circuits. The device is very useful if multiple-output combinational circuit is to be designed, because this needs minimum package count. These devices are available (Table 6.5) as 2-line-to-4-line, 3-line-to-8-line, and 4-line-to-16-line decoders. The outputs of most of these devices are active-low, also there is an active-low enable/data input terminal available.

Table 6.5 Available demultiplexer ICs

IC No.	*Description*	*Output*
74139	Dual 1:4 Demultiplexer (2-line-to-4-line decoder)	Inverted input
74155	Dual 1:4 Demultiplexer (2-line-to-4-line decoder)	$1Y-$ Inverted input $2Y-$ Same as input
74156	-do-	Open-collector $1Y$–Inverted input $2Y$–Same as input
74138	1:8 Demultiplexer (3-line-to-8-line decoder)	Inverted input
74154	1:16 Demultiplexer (4-line-to-16-line decoder)	Same as input
74159	-do-	Same as input Open–collector

Unlike the multiplexer, the decoder does require some gates in order to realize Boolean expressions in the standard SOP form. The following example illustrates its use in combinational logic design.

Example 6.3 Implement the following multi-output combinational logic circuit using a 4-to-16-line decoder.

$$F_1 = \Sigma m\ (1, 2, 4, 7, 8, 11, 12, 13)$$
$$F_2 = \Sigma m\ (2, 3, 9, 11)$$
$$F_3 = \Sigma m\ (10, 12, 13, 14)$$
$$F_4 = \Sigma m\ (2, 4, 8)$$

Solution

The realization is shown in Fig. 6.8.

The four-bit input *ABCD* is applied at the Select input terminals S_3, S_2, S_1, and S_0. The output F_1 is required to be 1 for minterms 1, 2, 4, 7, 8, 11, 12, and 13. Therefore, a NAND gate is connected as shown. Similarly NAND gates are used for the outputs F_2, F_3, and F_4. Here, the decoder's outputs are active-low, therefore a NAND gate is required for every output of the combinational circuit.

In the combinational logic design using multiplexer, additional gates are not required, whereas design using demultiplexer requires additional gates. However, even with this disadvantage, the decoder is more economical in cases where nontrivial, multiple-output expressions of the same input variables are required. In such cases, one multiplexer is required for each output whereas it is likely that only one decoder will be required, supported with a few gates. Therefore, using a decoder could have advantages over using a multiplexer.

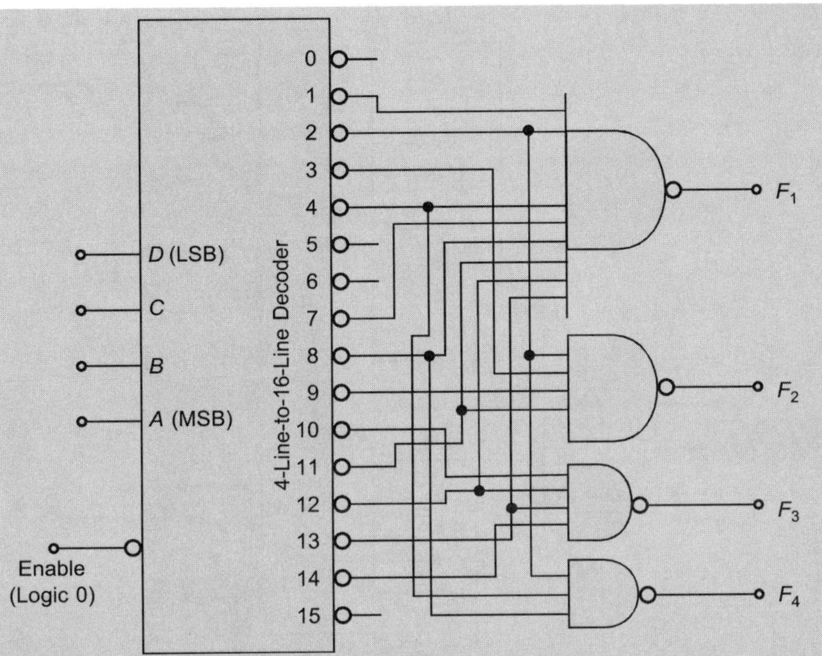

Fig. 6.8
Implementation of combinational logic circuit of Ex. 6.3.

6.3.2 Demultiplexer Tree

Since 4-line-to-16-line decoders are the largest available circuits in ICs, to meet the larger inputs need there should be a provision for expansion. This is made possible by using *enable* input terminal. Figure 6.9 shows a 5-line-to-32-line decoder and Fig. 6.10 shows a 8-line-to-256-line decoder using 4-line-to-16-line decoders. In a similar way, any *m*-line-to-*n*-line decoder can be implemented. However, if only a few codes of a large number need be recognized, the alternative approach, such as the one shown in Fig. 6.11, can be used. This is connected to detect the digital number 00011111. The most-significant 4-bits are applied at *A B C D* inputs and the least-significant bits are applied at *E F G H* inputs. The output goes low when the most-significant bits are 0 0 0 1 and the least-significant bits are 1 1 1 1. The circuit can be expanded to detect other 8-bit codes.

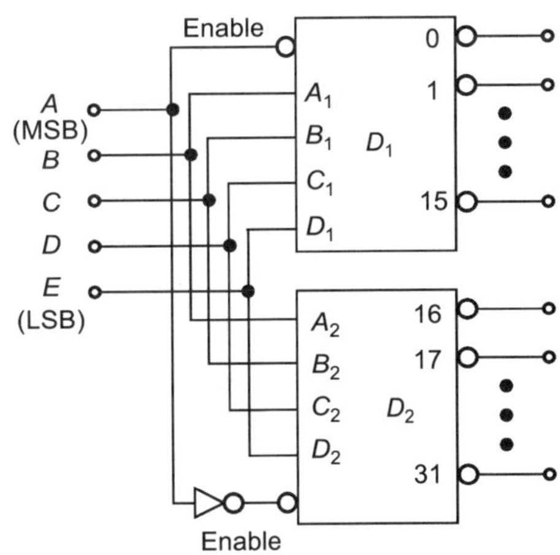

Fig. 6.9
5-line-to-32-line decoder using two 4-line-to-16-line decoders.

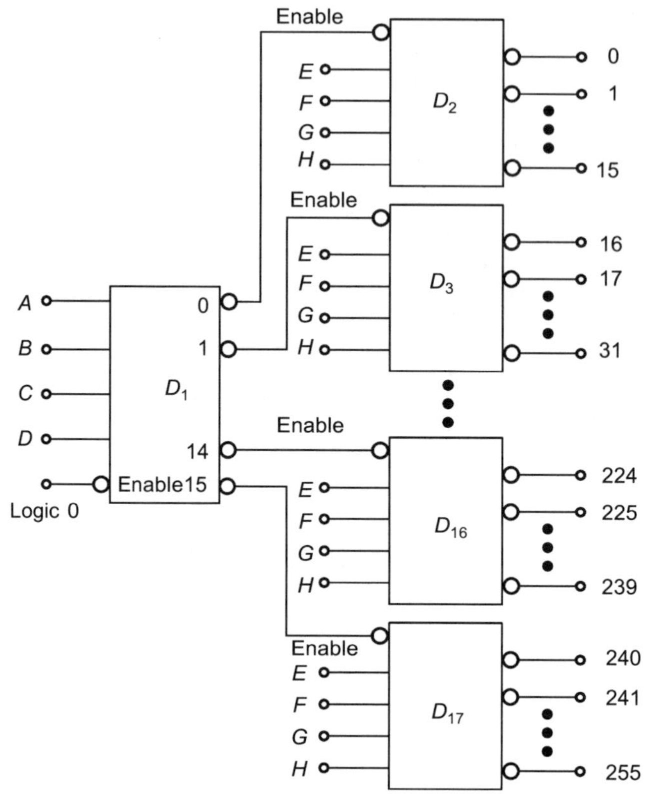

Fig. 6.10

8-line-to-256-line decoder using 4-line-to-16-line decoders tree.

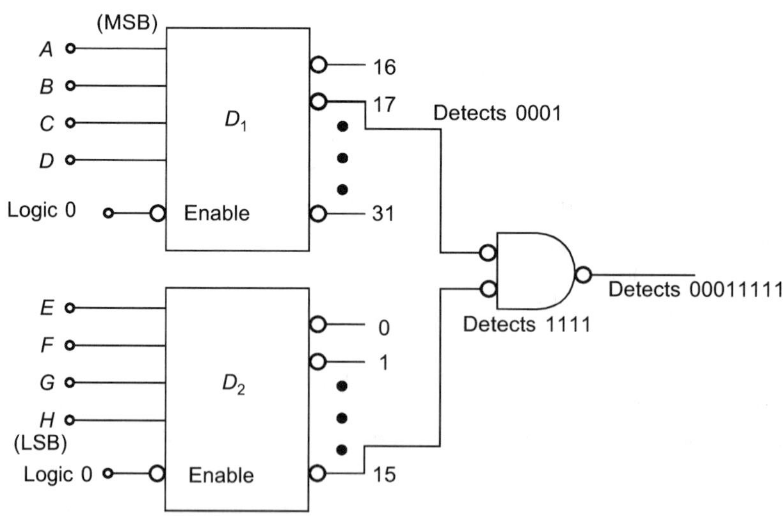

Fig. 6.11

An alternative approach to decode some combinations.

6.4 ADDERS AND THEIR USE AS SUBTRACTORS

Half- and full-adders and subtractors for two one-bit binary numbers have been discussed in Chapter 5. Using these circuits, we can design adders and subtractors for n-bit numbers. However, we can perform both addition and subtraction using only adders because the problem of subtraction becomes that of an addition when we use 1's and 2's complement representation of negative numbers. Therefore, adders have become widely-used standard circuits available in MSI.

An adder circuit for addition of two n-bit binary numbers consists of n full-adder circuits. It accepts two n-bit binary numbers as inputs and produces an $(n + 1)$-bit binary number as the sum. Figure 6.12*a* shows an n-bit adder using full-adders and Fig. 6.12*b* shows its block diagram.

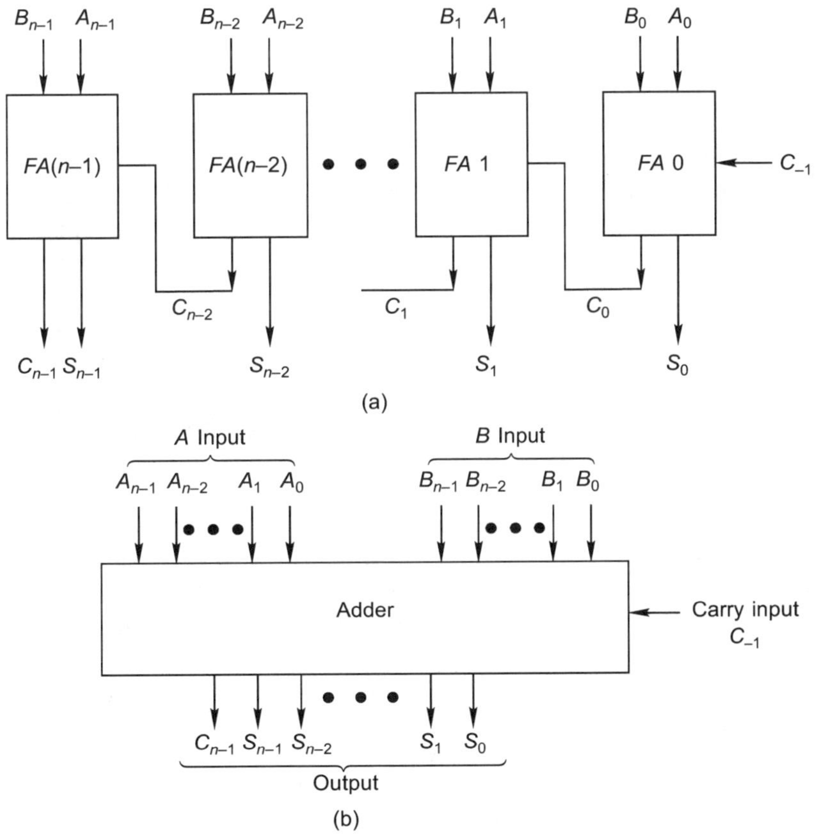

Fig. 6.12
(a) n-bit binary adder using n full-adders.
(b) block diagram of an n-bit binary adder.

Here, A and B are the two n-bit inputs to be added and $C_{n-1}\ S_{n-1}\ S_{n-2}\ ..\ S_2\ S_1\ S_0$ is their sum. In this, a half-adder may be used to add the least-significant bits A_0 and B_0. However, for cascading these adders to increase the number of bits to be added, the CARRY input terminal is required for the adder to add the least-significant bits. Therefore, all the adders used are full-adders. 2-bit and 4-bit adder circuits are available in ICs.

If the n-bit adder is implemented using the scheme of Fig. 6.12*a*, the carry has to ripple down the line of cascaded adders from the LSB to MSB position. This decreases the operating speed of the adder. The time required for addition operation to be completed is limited by the amount of time required to complete the ripple-carry operation (Prob. 6.11). A technique known as the *look-ahead carry* generation process is used for increasing the speed of operation. This technique anticipates the carry bits for each stage before the actual summing operation.

Adder ICs are designed incorporating the look-ahead carry generator circuit.

6.4.1 Adder with Look-Ahead Carry

In the half-adder circuit of Fig. 5.19 and full-adder circuits of Fig. 5.21 and Prob. 5.19(b), it was assumed that the inputs (augend and addend) and carry input are simultaneously present and the sum and carry outputs are generated instantaneously. However, in Prob. 5.20 it has been shown that even if the augend, addend, and carry inputs are present simultaneously, then the sum and carry outputs will be delayed due to the propagation delays of gates through which the signals are passing. Now, if we consider the n-bit parallel adder of Fig. 6.12(a), we observe that the sum (S_0) and carry (C_0) outputs are delayed because of the propagation delays of the gates involved in FA0. The S_0 output, however, is the LSB of the final result and it is not required to be passed through other gates and hence does not get further delayed. The carry output C_0 acts as carry input of the full-adder FA1 and therefore, the outputs of FA1 S_1 and C_1 will reach steady-state only after arrival of C_0 and propagation delay introduced by FA1. This means the total delay upto FA1 will be the sum of delays introduced by FA0 and FA1. Similarly going further in the chain of adders towards MSB, the carry has to ripple through all the stages, thereby reducing the speed of the adder as the number of adder stages are increased (Prob. 6.11).

To design fast operating parallel adders, we can use gates with lower propagation delay time. Even with this, the delay time of the adder will increase with increasing number of bits to be added. Another approach most commonly used is the concept of *look-ahead carry.* Although it requires additional circuitry but the speed of the adder becomes independent of the number of bits.

Let us consider a full-adder circuit in block diagram form and its EX-OR realization (Prob. 5.19(b)) shown in Fig. 6.13. Here,

$$P_i = A_i \oplus B_i \tag{6.2a}$$

$$G_i = A_i B_i \tag{6.2b}$$

$$S_i = P_i \oplus C_{i-1} = A_i \oplus B_i \oplus C_{i-1} \tag{6.2c}$$

$$C_i = G_i + P_i C_{i-1} \tag{6.2d}$$

The output G_i of the first half-adder is 1 if A_i and B_i both are 1 and a carry is generated. The variable G_i is known as a *carry generate.* Its value is independent of the input carry. The variable P_i is known as a *carry propagate* because this is the term associated with the propagation of the

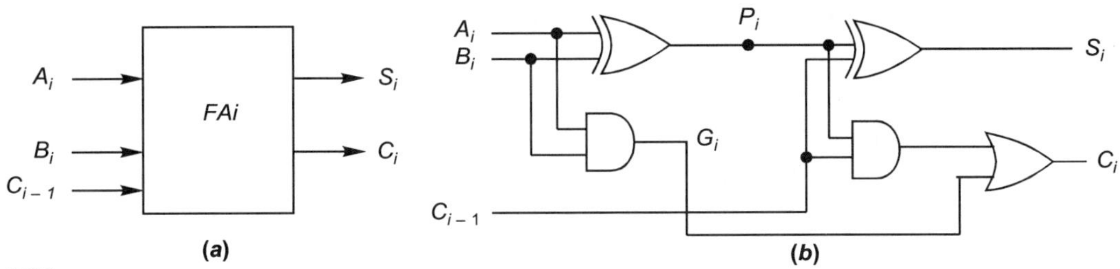

Fig. 6.13
(a) Block diagram of ith stage of a full-adder
(b) EX-OR implementation of full-adder.

carry from C_{i-1} to C_i. Using Eq. (6.2d), we write Boolean expression for carry output of each stage:

$$C_{i+1} = G_{i+1} + P_{i+1} \cdot C_i$$

Substituting the value of C_i from Eq. (6.2d), we obtain,

$$C_{i+1} = G_{i+1} + P_{i+1}G_i + P_{i+1}\, P_iC_{i-1} \qquad (6.3)$$

Similarly, for an n-stage adder, the final carry C_{n-1} can be determined. For clear understanding of the advantage of this approach, we consider a 4-bit adder and formulate Boolean expressions for the carry outputs C_0, C_1, C_2, and C_3. We obtain,

$$P_0 = A_0 \oplus B_0 \text{ and } G_0 = A_0B_0$$

$$C_0 = G_0 + P_0C_{-1} \qquad (6.4a)$$

$$C_1 = G_1 + P_1G_0 + P_1P_0C_{-1} \qquad (6.4b)$$

$$C_2 = G_2 + P_2G_1 + P_2P_1G_0 + P_2P_1P_0\, C_{-1} \qquad (6.4c)$$

$$C_3 = G_3 + P_3G_2 + P_3P_2G_1 + P_3P_2P_1G_0 + P_3P_2\, P_1\, P_0\, C_{-1} \qquad (6.4d)$$

Now, let us observe the logic variables involved in Eq. (6.4). these are,

$$G_0, G_1, G_2, G_3, P_0, P_1, P_2, P_3, \text{ and } C_{-1}.$$

The G variables can be generated directly from A and B inputs using AND gates (Eq. 6.2b). the P variables are obtained again directly from A and B inputs using EX-OR gates (Eq. 6.2a). C_{-1} is the carry input. If Gs, Ps, and C_{-1} are available simultaneously, the carry outputs C_0, C_1, C_2, and C_3 are produced by using 2-level realization (AND-OR or NAND-NAND) since Eq. 6.4 gives these outputs in SOP form. Therefore, for the generation of these carry outputs, propagation delay time of two gates only will be there. These carry outputs are connected to the carry inputs of the succeeding stages, thereby eliminating the problem of carry rippling through all the stages.

Logic circuit of a look-ahead carry generator can be prepared using Eqs. 6.4. A 4-bit adder with look-ahead carry is shown in Fig. 6.14. Thus we see that by generating all the individual carry terms needed by each full-adder in a 2-level circuit, the propagation delay time through the adder is considerably reduced and it becomes independent of the number of full-adders in the circuit.

6.4.2 Cascading of Adders

The bit lengths of the numbers to be added can be increased by cascading the adders. Figure 6.15 shows an 8-bit adder using two 4-bit adders. In a similar way, we can form an adder tree for making an n-bit adder.

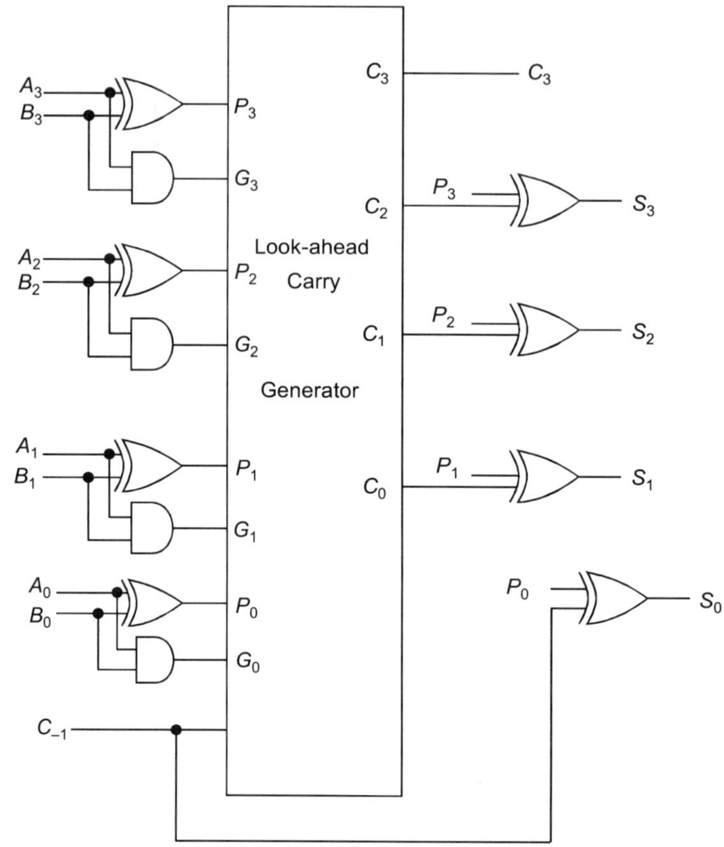

Fig. 6.14

A 4-bit adder with look-ahead carry.

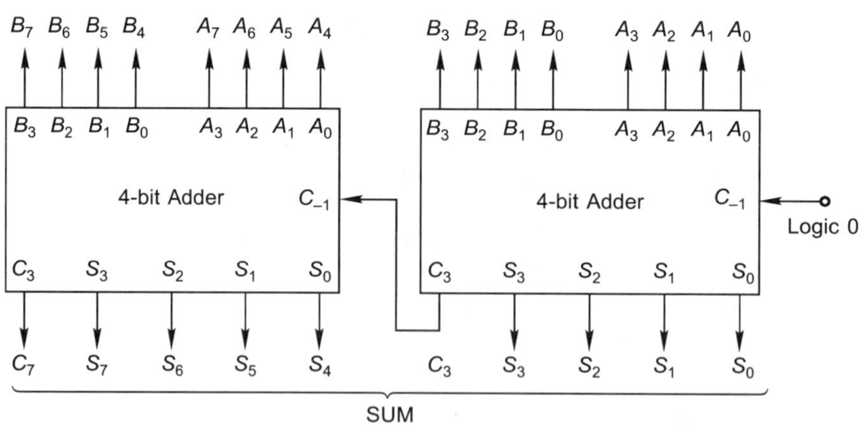

Fig. 6.15

An 8-bit adder as a cascade of two 4-bit adders.

6.4.3 Subtraction Using Adder

As discussed in Chapter 2, the problem of subtraction gets converted into that of addition if 1's and 2's complement representation are used for representing negative numbers. The algorithm for a subtractor using adder is given in Fig. 6.16.

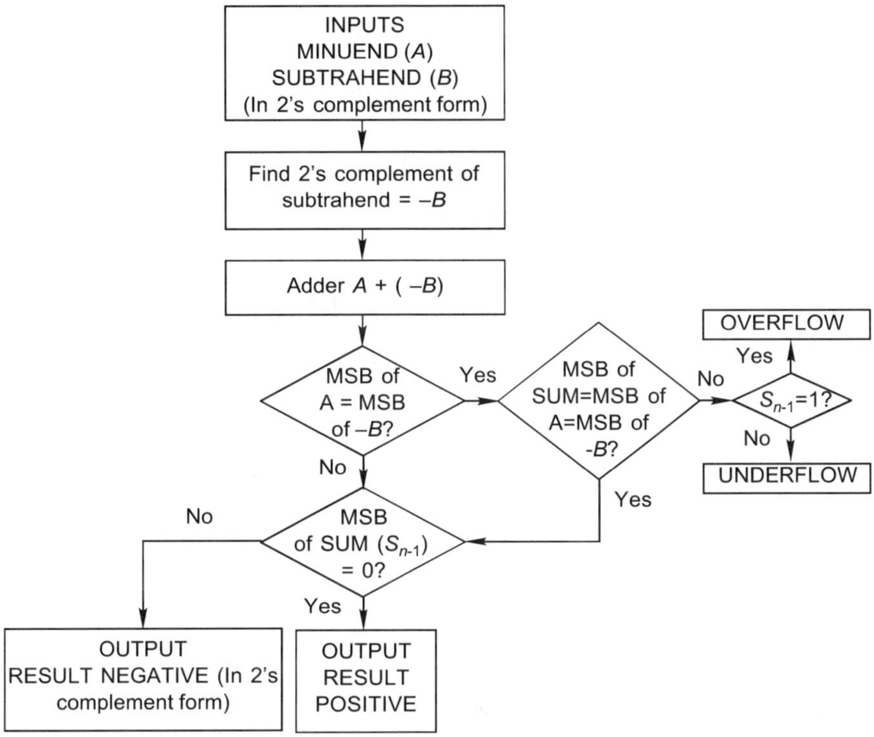

Fig. 6.16
Algorithm for performing subtraction using adder.

The two numbers A and B can be of the same sign or of opposite signs. If the two numbers are of unlike sign, we may come across the problem of *overflow* or *underflow*. Overflow occurs when the subtraction operation produces a number larger than the largest possible number which can be represented by n-bits. On the other hand, underflow occurs when the result produced is smaller than the smallest number which can be represented by n-bits. The overflow and underflow logic is illustrated in Fig. 6.16. The subtractor circuit is given in Fig. 6.17. The reader can verify the operation of this circuit. If overflow or underflow occurs then the result is wrong. This circuit can be converted into an ADDER/SUBTRACTOR circuit with ADD/SUB control (Problem 6.4).

6.5 BCD ARITHMETIC

Quite often, BCD code is used to represent decimal numbers, for example, in a calculator. Therefore, addition and subtraction are required to be performed in BCD code.

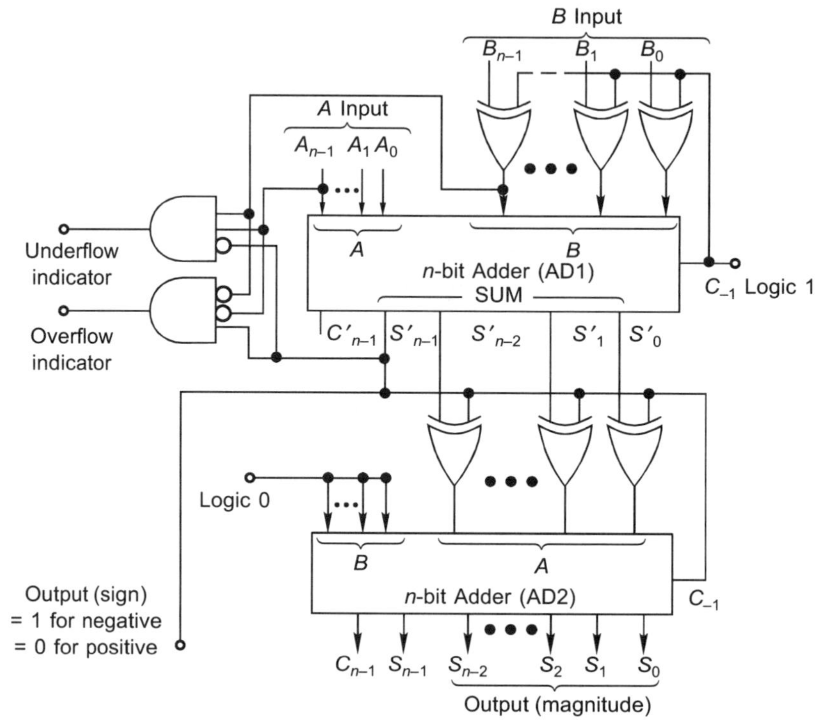

Fig. 6.17
n-bit subtractor circuit using n-bit adders.

6.5.1 BCD Adder

The 4-bit binary adder IC (7483) can be used to perform addition of BCD numbers. In this, if the four-bit sum output is not a valid BCD digit, or if a carry C_3 is generated, then decimal 6 (0 1 1 0 binary) is to be added to the sum to get the correct result. Figure 6.18 shows a 1-digit BCD adder. BCD adders can be cascaded to add numbers several digits long by connecting the carry-out of a stage to the carry-in of the next stage.

6.5.2 BCD Subtractor

For BCD subtraction, nine's complement of the subtrahend is added to the minuend. The nine's complement of a BCD number is given by nine minus that number. For example, nine's complement of 7 = 9 – 7 = 2. In BCD code we say that the nine's complement of 0 1 1 1 is 0 0 1 0. Examples of BCD subtraction using nine's complement to represent negative numbers are given below:

Example 6.4
(a) Subtract 5 from 9
(b) Subtract 1 from 8
(c) Subtract 8 from 4

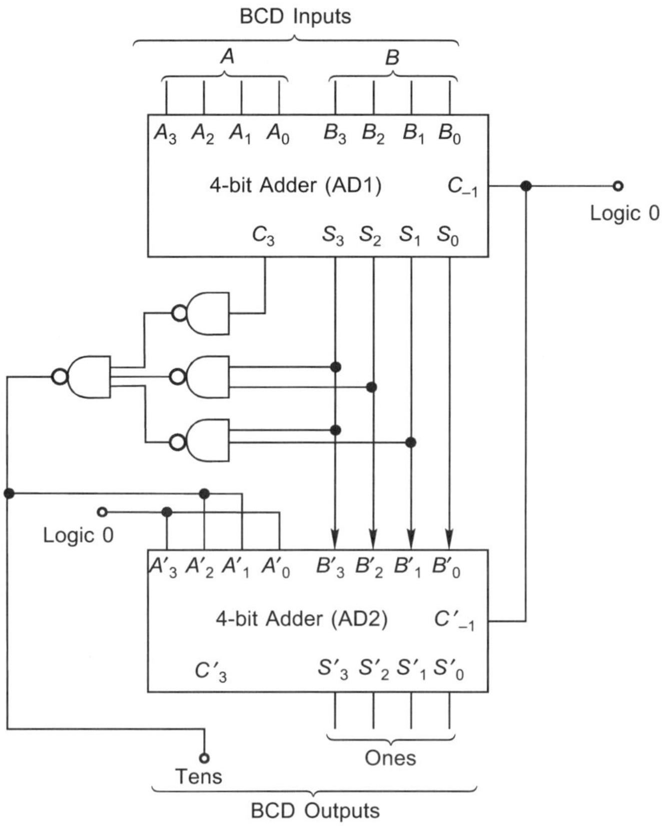

Fig. 6.18
One digit BCD adder.

Solution

(a)	9 =	1001	
	– 5 =	(+) 0100	(nine's complement of 5)
		1101	(Invalid)
	Add	0110	
		10011	
		1	Add End Around Carry (EAC)
		0100	= + 4
(b)	8 =	1000	
	– 1 =	(+) 1000	(nine's complement of 1)
		10000	(Invalid)
	Add	0110	
		10110	
		1	Add (EAC)
		0111	= + 7

(c) 4 = 0100
− 8 = (+) 0001 (nine's complement of 8)
0101

Nine's complement of 0101 = 4
Therefore, the answer is –4.

From the above example, we conclude the following:

1. If the sum of minuend and the nine's complement of subtrahend is an invalid BCD code (Ex. 6.4(a)) or if a carry is produced from the MSB (Ex. 6.4(b)), add decimal 6 (binary 0110) and the end around carry (EAC) to this sum. The result is positive number represented by this sum.
2. If the sum of the minuend and the nine's complement of the subtrahend is a valid BCD code, the result is negative and is in the nine's complement form (Ex. 6.4(c)).

Nine's complement of a number can be determined by adding 1010 to the one's complement of the number. Figure 6.19 shows a nine's complementer circuit using a 4-bit adder and EX-OR gates.

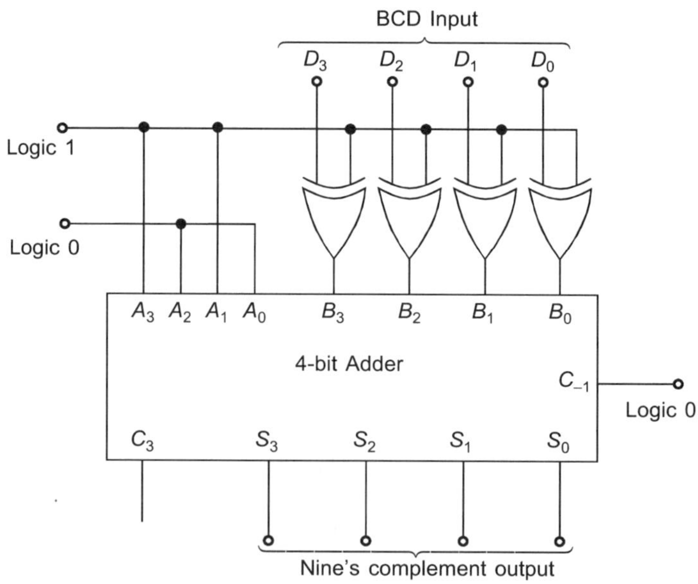

Fig. 6.19
A nine's complementer circuit.

Figure 6.20 shows a one-digit BCD subtractor, using the nine's complementer circuit of Fig. 6.19.

We observe that the BCD arithmetic is rather complex in comparison to the straight binary arithmetic. Therefore, BCD arithmetic requires more hardware and results in reduction of speed. Also, more number of bits are required to represent a given number in BCD, therefore, BCD is normally not used in computers. However, digital calculators use BCD because the

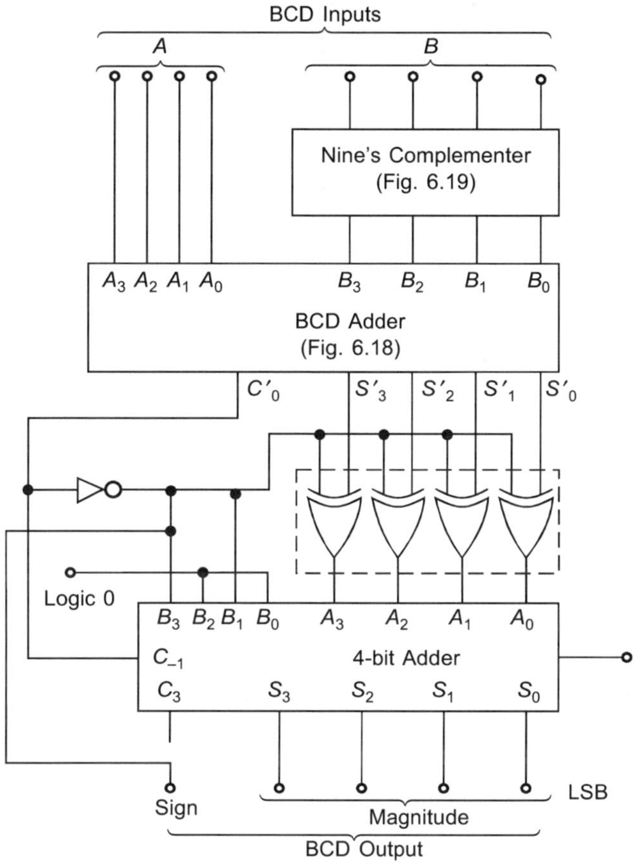

Fig. 6.20
A one-digit BCD subtractor.

input data from the keyboard as well as the output display are decimal. Although the process of encoding BCD from decimal is simple and changing from decimal to binary is complex, even then the calculators have rather complex circuitry which are costly and considerably slower than computers.

6.6 ARITHMETIC LOGIC UNIT (ALU)

A very popular and widely used combinational circuit is ALU which is capable of performing arithmetic as well as logical operations. This is the heart of any microprocessor. Figure 6.21 shows the block diagram of 74181 ALU and Table 6.6 gives its function table.

The functions of various input, output and control lines are given below:

A and B: 4-bit binary data inputs and Table 6.6 gives its function table.

$\overline{C}_n$: Carry input (active-low)

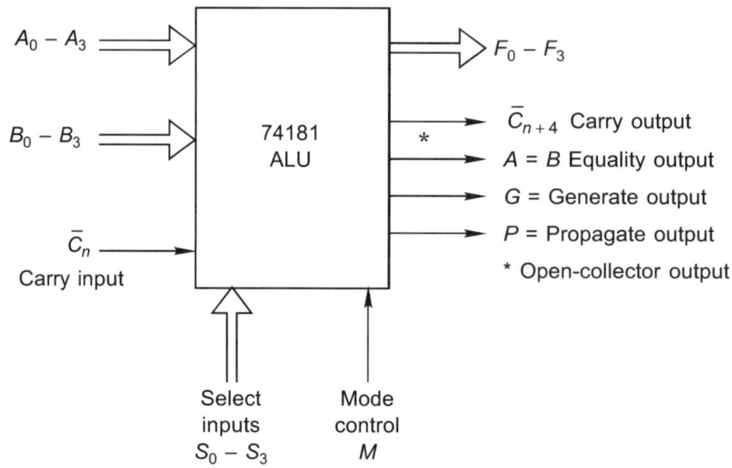

Fig. 6.21
Block diagram of 74181 ALU.

F:4-bit binary data output.

$\overline{C}_{n+4}$: Carry output (active-low)

For subtraction operation, it indicates the sign of the output. Logic 0 indicates positive result and logic 1 indicates negative result expressed in 2's complement form.

$A = B$: Logic 1 on this line indicates $A = B$

G : Carry generate output

P : Carry propagate output

G and P outputs are used when a number of 74181 circuits are used in cascade along with 74182 Look-ahead Carry-generator circuit to make the arithmetic operations faster.

Select input (S) : Used to select any operation (Table 6.6)

Mode Control (M) : $M = 0$ Arithmetic operations

$M = 1$ Logic operations

The 74181 can be cascaded by connecting the carry-out of a stage to the carry-in of the succeeding stage.

Example 6.5 Design an 8-bit adder/subtractor using 74181s in cascade. Show how it works if (a) A = 97 and B = 29 (b) A = 24 and B = 58.

Solution

For designing an 8-bit adder/subtractor, two 74181 ICs are to be cascaded. The least-significant four bits of A and B are applied at the A and B inputs of the least-significant 74181 and the most-significant four bits of A and B are applied at the A and B inputs of the most-significant 74181. The carry-out of the least-significant 74181 is connected to the

Table 6.6 Function table of 74181 ALU

Line	Selection $S_3S_2S_1S_0$	Active high data: $M = 1$ Logic Functions	$M = 0$; Arithmetic operations: $\bar{C}_n = 1$ (no carry)	$\bar{C}_n = 0$ (with carry)
0	0 0 0 0	$F = \bar{A}$	$F = A$	$F = A$ PLUS 1
1	0 0 0 1	$F = \overline{A + B}$	$F = A + B$	$F = (A + B)$ PLUS 1
2	0 0 1 0	$F = \bar{A} \cdot B$	$F = A + B$	$F = (A + \bar{B})$ PLUS 1
3	0 0 1 1	$F = 0$	$F =$ MINUS 1 (2's COMPL)	$F =$ ZERO
4	0 1 0 0	$F = \overline{AB}$	$F = A$ PLUS $A\bar{B}$	$F = A$ PLUS $A\bar{B}$ PLUS 1
5	0 1 0 1	$F = \bar{B}$	$F = (A + B)$ PLUS $A\bar{B}$	$F = (A + B)$ PLUS $A\bar{B}$ PLUS 1
6	0 1 1 0	$F = A \oplus B$	$F = A$ MINUS B MINUS 1	$F = A$ MINUS B
7	0 1 1 1	$F = A\bar{B}$	$F = A\bar{B}$ MINUS 1	$F = A\bar{B}$
8	1 0 0 0	$F = \bar{A} + B$	$F = A$ PLUS AB	$F = A$ PLUS AB PLUS 1
9	1 0 0 1	$F = \overline{A \oplus B}$	$F = A$ PLUS B	$F = A$ PLUS B PLUS 1
10	1 0 1 0	$F = B$	$F = (A + \bar{B})$ PLUS AB	$F = (A + \bar{B})$ PLUS AB PLUS 1
11	1 0 1 1	$F = AB$	$F = AB$ MINUS 1	$F = AB$
12	1 1 0 0	$F = 1$	$F = A$ PLUS A*	$F = A$ PLUS A PLUS 1
13	1 1 0 1	$F = A + \bar{B}$	$F = (A + B)$ PLUS A	$F = (A + B)$ PLUS A PLUS 1
14	1 1 1 0	$F = A + B$	$F = (A + \bar{B})$ PLUS A	$F = (A + \bar{B})$ PLUS A PLUS 1
15	1 1 1 1	$F = A$	$F = A$ MINUS 1	$F = A$

* Each bit is shifted to the next more significant position.

carry-in of the most-significant 74181. The 8-bit output will be available on F outputs. The select lines of both the 74181s are connected together.

Addition is performed with $M = 0$ and $S = 1001$, and subtraction is performed with $M = 0$ and $S = 0110$. Connect $\bar{C}_n$ of least-significant 74181 to 1 for addition and 0 for subtraction.

(a) $A = 97 = 01100001$
$B = 29 = 00011101$

(b) $A = 24 = 00011000$
$B = 58 = 00111010$

The various inputs and outputs are given in Table 6.7.

Table 6.7 Table for Example 6.5

Part	Mode control M	Select inputs $S_3S_2S_1S_0$	Least-significant ALU Inputs A_L	B_L	$\bar{C}_n$	Outputs S_L	$\bar{C}_{n+4}$	Most-significant ALU Inputs A_H	B_H	$\bar{C}_n$	Outputs S_H	$\bar{C}_{n+4}$	Output	Remarks
(a)	0	1 0 0 1	0001	1101	1	1110	1	0110	0001	1	0111	1	01111110	$(126)_{10}$
	0	0 1 1 0	0001	1101	0	0100	1	0110	0001	1	0100	0	01000100	$(68)_{10}$
(b)	0	1 0 0 1	1000	1010	1	0010	0	0001	0011	0	0101	1	01010010	$(82)_{10}$
	0	0 1 1 0	1000	1010	0	1110	1	0001	0011	1	1101	1	11011110	$-(34)_{10}$

6.7 DIGITAL COMPARATORS

The principle of comparing digital signals has been given in Section 1.5. Comparators can be designed for comparing multibit numbers. Figure 6.22 shows the block diagram of an *n*-bit comparator. It receives two *n*-bit numbers *A* and *B* as inputs and the outputs are $A > B$, $A = B$, and $A < B$. Depending upon the relative magnitude of the two numbers, one of the outputs will be HIGH. Table 6.8 gives the truth table of a 2-bit comparator. The reader is advised to simplify the expressions for $A > B$, $A = B$, and $A < B$ outputs using K-map and design the circuit using gates. However, 4-bit comparators are available in MSI (7485) which can compare straight binary and natural BCD codes. These ICs can be cascaded to compare words of greater lengths without external gates. The $A > B$, $A = B$, and $A < B$ outputs of a stage handling less-significant bits are connected to the corresponding $A > B$, $A = B$, and $A < B$ cascading inputs of the next stage handling more-significant bits. The stage handling the least-significant bits must have $A = B$ input connected to logic 1 level and $A > B$ and $A < B$ inputs connected to logic 0 or 1 level.

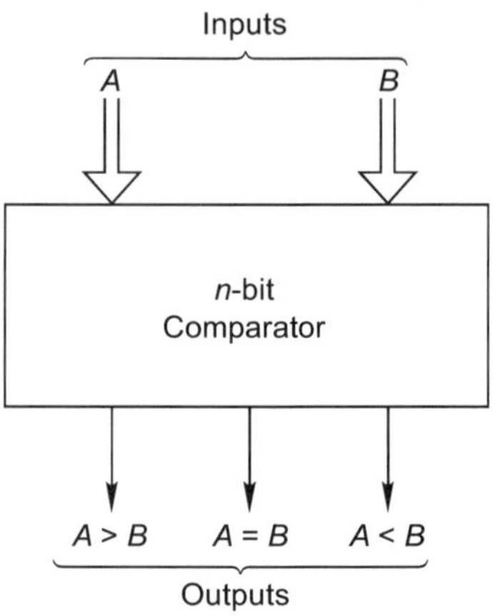

Fig. 6.22
Block diagram of *n*-bit comparator.

Table 6.8 Truth table of a 2-bit comparator

Inputs				*Outputs*		
A_1	A_0	B_1	B_0	$A > B$	$A = B$	$A < B$
0	0	0	0	0	1	0
0	0	0	1	0	0	1
0	0	1	0	0	0	1
0	0	1	1	0	0	1
0	1	0	0	1	0	0
0	1	0	1	0	1	0
0	1	1	0	0	0	1
0	1	1	1	0	0	1
1	0	0	0	1	0	0
1	0	0	1	1	0	0
1	0	1	0	0	1	0
1	0	1	1	0	0	1
1	1	0	0	1	0	0
1	1	0	1	1	0	0
1	1	1	0	1	0	0
1	1	1	1	0	1	0

The function table of 7485 comparator is given in Table 6.9.

Table 6.9 Function table of 7485 4-bit comparator

Comparing inputs	*Cascading inputs*			*Outputs*		
A, B	$A > B$	$A = B$	$A < B$	$A > B$	$A = B$	$A < B$
$A > B$	×	×	×	1	0	0
$A = B$	1	0	0	1	0	0
	×	1	×	0	1	0
	0	0	1	0	0	1
	0	0	0	1	0	1
	1	0	1	0	0	0
$A < B$	×	×	×	0	0	1

Example 6.6 Design a 5-bit comparator using a single 7485 and one gate.

Solution

The circuit is shown in Fig. 6.23. The two 5-bit numbers to be compared are $A_4\,A_3\,A_2\,A_1\,A_0$ and $B_4\,B_3\,B_2\,B_1\,B_0$.

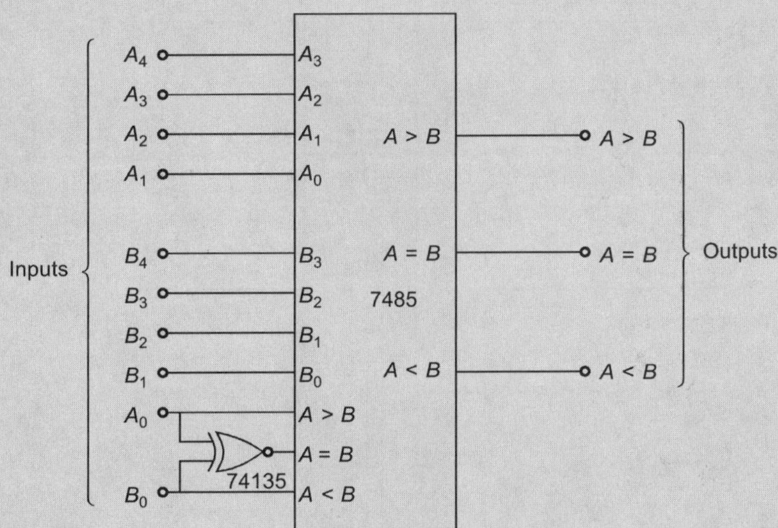

Fig. 6.23
A 5-bit comparator.

The operation of this circuit can be understood from the function table of 7485.

Example 6.7 Design a 24-bit comparator using six 7485 comparators in two levels.

Solution

Based on the approach used in Example 6.6, the circuit is given in Fig. 6.24. The reader is advised to verify the circuit assuming any two arbitrary 24-bit binary numbers. This method of cascading reduces comparison time in comparison to the straight method of cascading.

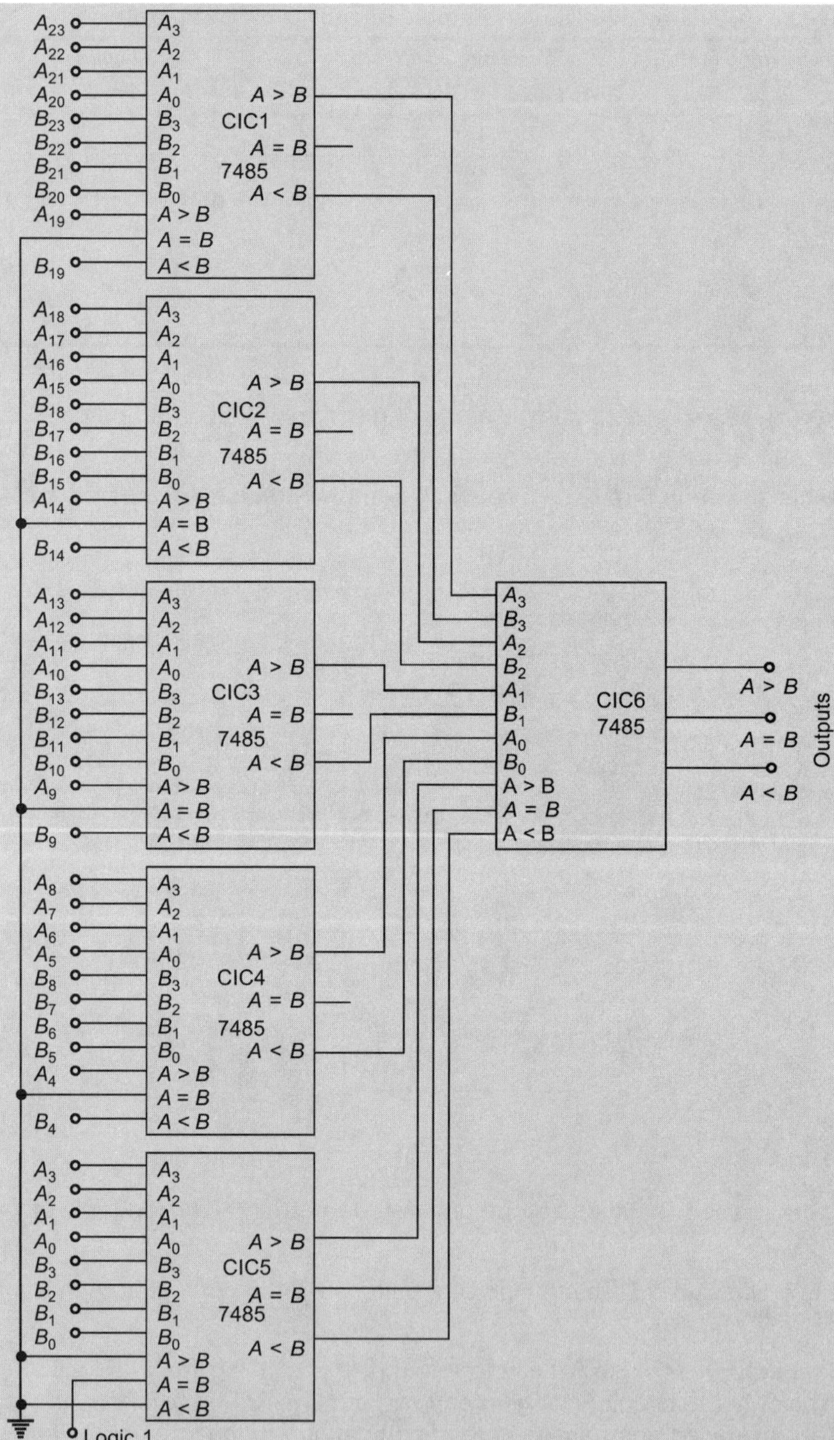

Fig. 6.24
A 24-bit comparator

6.8 PARITY GENERATORS/CHECKERS

The concept of parity, wherein an additional bit known as the *parity-bit* is added to a binary word to make the number of 1's, in the new word formed, even (*even parity*) or odd (*odd parity*), has been discussed in Section 2.10. The circuits for the generation of parity bits and checking the parity of a given word can be designed using gates (Problems 5.17 and 5.18). Because of its wide use, an 8-bit parity generator/checker circuit has been designed and is available as a MSI chip (74180).

Figure 6.25 gives the block diagram of 74180 in which there are eight parity inputs A through H and two cascading inputs. There are two outputs Σ EVEN and Σ ODD. Its function table is given in Table 6.10.

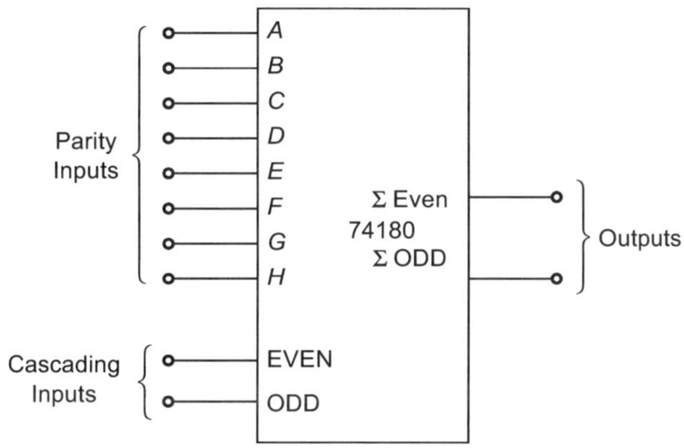

Fig. 6.25
Block diagram of 74180 parity generator/checker.

Table 6.10 Function table of 74180

Parity of inputs A through H	*Cascading inputs*		*Outputs*	
	EVEN	*ODD*	*Σ EVEN*	*Σ ODD*
EVEN	1	0	1	0
ODD	1	0	0	1
EVEN	0	1	0	1
ODD	0	1	1	0
×	1	1	0	0
×	0	0	1	1

The 74180 can be used as a parity generator as well as parity checker. Its function as a parity checker can be understood from the function table. Its function as a parity generator is given in Table 6.11.

Table 6.11 Function table of 74180 as a 9-bit parity generator

Parity of inputs A through H	*Cascading inputs EVEN*	*ODD*	*Parity of A through H and Σ EVEN*	*Parity of A through H and Σ ODD*
ODD	1	0	ODD	EVEN
EVEN	1	0	ODD	EVEN
ODD	0	1	EVEN	ODD
EVEN	0	1	EVEN	ODD

The cascading inputs must not be equal and the unused parity inputs must be tied to logic 0 level.

Example 6.8

(a) Make a 9-bit odd parity checker using single 74180 and an inverter.
(b) Make a 10-bit even parity generator using single 74180 and an inverter.
(c) Make a 16-bit even parity checker using two 74180s.

Solution

(a) Eight of the nine bits are applied at *A–H* inputs and the ninth bit, *I*, is applied to the ODD input. Its operation can be verified with the help of Table 6.10. The circuit is given in Fig. 6.26*a*.

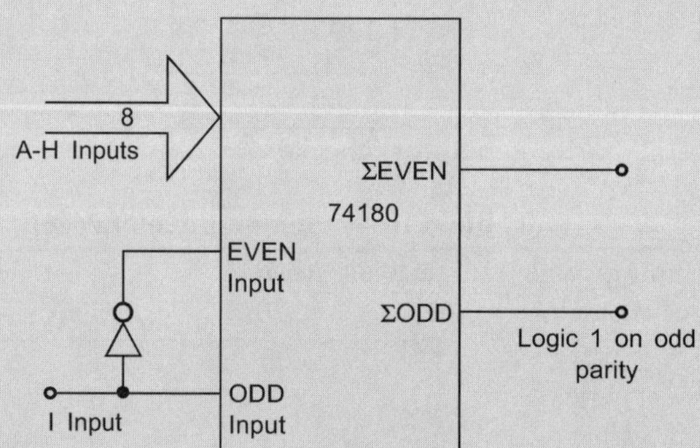

Fig. 6.26
(a) A 9-bit odd parity checker.

(b) Here the 9-bit word consisting of *A* through *I* is converted to a 10-bit word with even parity. The circuit is given in Fig. 6.26*b*.
(c) The 16-bit even parity checker is shown in Fig. 6.26*c*.

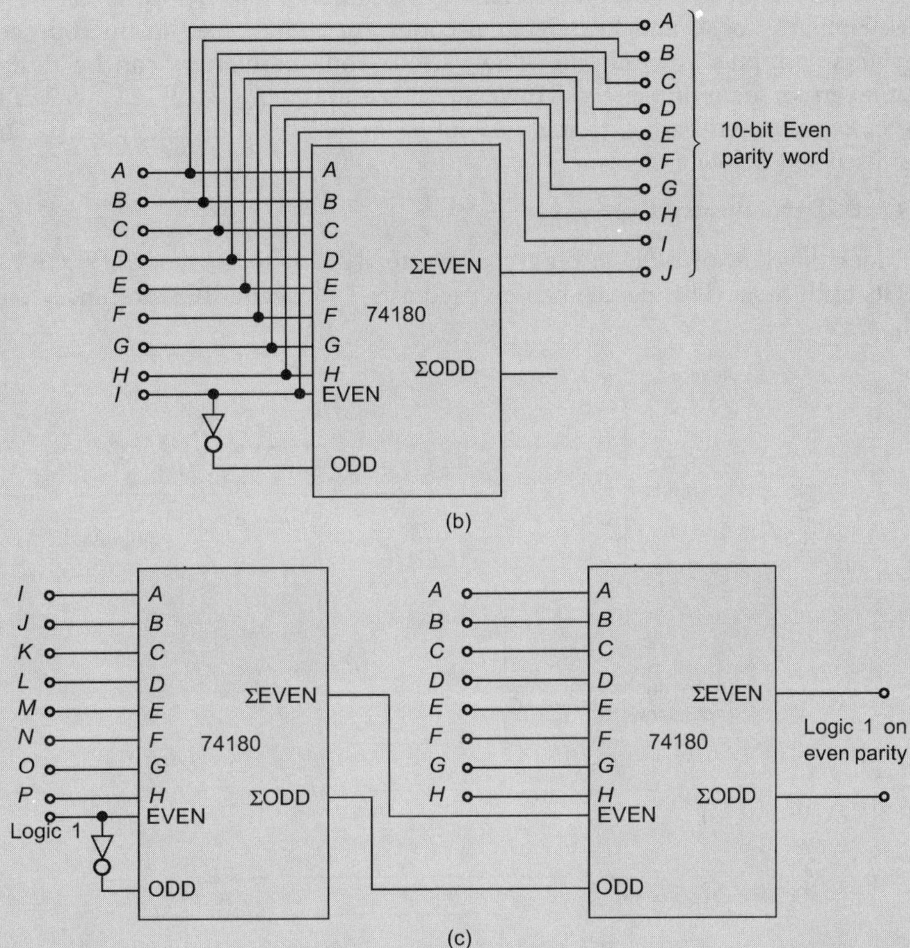

Fig. 6.26
(b) A 10-bit even parity generator.
(c) A 16-bit even parity checker.

6.9 CODE CONVERTERS

There is a wide variety of binary codes used in digital systems. Some of these codes are binary-coded-decimal (BCD), Excess-3, Gray, octal, hexadecimal, etc. Often, it is required to convert from one code to another. For example the input to a digital system may be in natural BCD and the output may be 7-segment LEDs. The digital system used may be capable of processing the data in straight binary format. Therefore, the data has to be converted from BCD to binary at the output.

The BCD output has to be converted to 7-segment code before it can be used to drive the LEDs. Similarly, octal and hexadecimal codes are widely used in microprocessors and digital computers as inputs and outputs. The various code converters can be designed using gates, multiplexers or demultiplexers. However, there are some MSI ICs available for performing these conversions and are extremely useful in the design of digital systems. These devices have been discussed below.

6.9.1 BCD-to-Binary Converter

The block diagram of BCD-to-binary converter IC 74184 is given in Fig. 6.27 and Table 6.12 gives its truth table. This device can be used as a $1\frac{1}{2}$ decade BCD-to-binary converter as shown in Fig. 6.28.

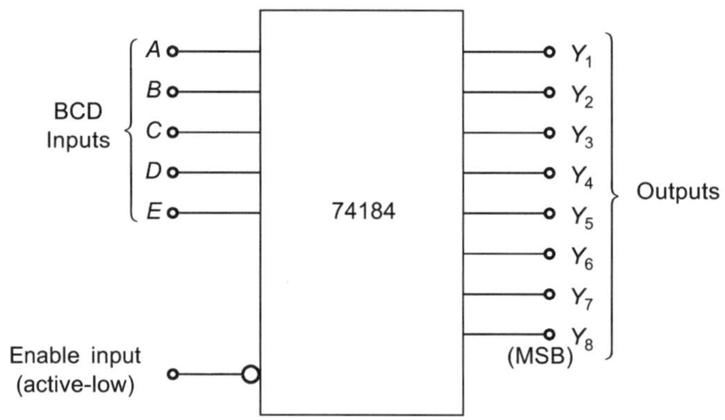

Fig. 6.27
Block diagram of 74184.

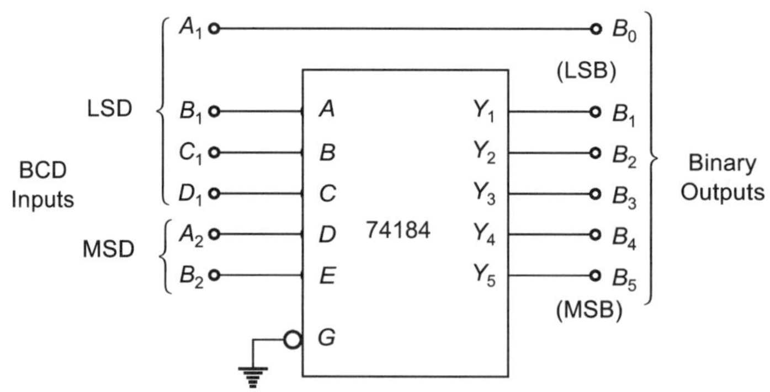

Fig. 6.28
A $1\frac{1}{2}$ decade BCD-to-binary converter.

The BCD inputs are applied at the input terminals *A* through *E* and the LSB of the least-significant BCD digit bypasses the converter and appears as the LSB of the binary output. It accepts two BCD digits–a full digit D_1 C_1 B_1 A_1 and the two least-significant bits of a second

Table 6.12 Truth table of 74184 BCD-to-binary converter

BCD Words	*Inputs*						*Outputs*				
	E	D	C	B	A	G	Y_5	Y_4	Y_3	Y_2	Y_1
0 – 1	0	0	0	0	0	0	0	0	0	0	0
2 – 3	0	0	0	0	1	0	0	0	0	0	1
4 – 5	0	0	0	1	0	0	0	0	0	1	0
6 – 7	0	0	0	1	1	0	0	0	0	1	1
8 – 9	0	0	1	0	0	0	0	0	1	0	0
10 – 11	0	1	0	0	0	0	0	0	1	0	1
12 – 13	0	1	0	0	1	0	0	0	1	1	0
14 – 15	0	1	0	1	0	0	0	0	1	1	1
16 – 17	0	1	0	1	1	0	0	1	0	0	0
18 – 19	0	1	1	0	0	0	0	1	0	0	1
20 – 21	1	0	0	0	0	0	0	1	0	1	0
22 – 23	1	0	0	0	1	0	0	1	0	1	1
24 – 25	1	0	0	1	0	0	0	1	1	0	0
26 – 27	1	0	0	1	1	0	0	1	1	0	1
28 – 29	1	0	1	0	0	0	0	1	1	1	0
30 – 31	1	1	0	0	0	0	0	1	1	1	1
32 – 33	1	1	0	0	1	0	1	0	0	0	0
34 – 35	1	1	0	1	0	0	1	0	0	0	1
36 – 37	1	1	0	1	1	0	1	0	0	1	0
38 – 39	1	1	1	0	0	0	1	0	0	1	1
Any	×	×	×	×	×	1	1	1	1	1	1

digit B_2 A_2. This means that the BCD inputs 00 through 39 can be converted to corresponding binary output by this circuit. Terminals Y_6, Y_7, and Y_8 are not used for BCD-to-binary conversion. These terminals are used to obtain the 9's complement and the 10's complement of BCD numbers, useful for BCD arithmetic operations. Figure 6.29 gives the block diagram of BCD 9's complement converter and Table 6.13 gives its truth table.

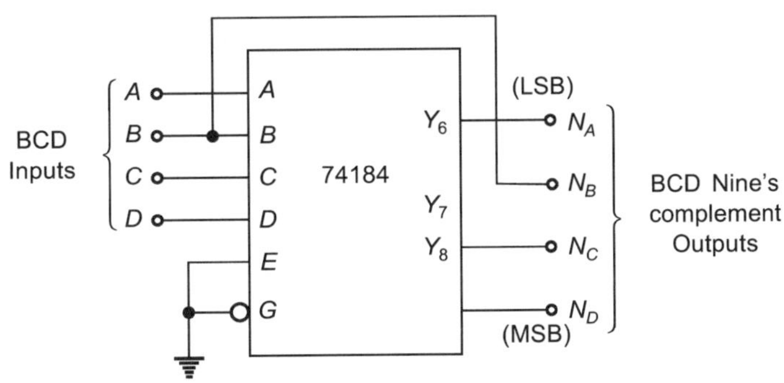

Fig. 6.29
Block diagram of BCD 9's complement converter using 74184.

Table 6.13 Truth table of 74184 as BCD 9's complement converter

BCD Word	*Inputs*						*Outputs*			*BCD 9's complement*			
	E	*D*	*C*	*B*	*A*	*G*	Y_8	Y_7	Y_6	N_D	N_C	N_B	N_A
0	0	0	0	0	0	0	1	0	1	1	0	0	1
1	0	0	0	0	1	0	1	0	0	1	0	0	0
2	0	0	0	1	0	0	0	1	1	0	1	1	1
3	0	0	0	1	1	0	0	1	0	0	1	1	0
4	0	0	1	0	0	0	0	1	1	0	1	0	1
5	0	0	1	0	1	0	0	1	0	0	1	0	0
6	0	0	1	1	0	0	0	0	1	0	0	1	1
7	0	0	1	1	1	0	0	0	0	0	0	1	0
8	0	1	0	0	0	0	0	0	1	0	0	0	1
9	0	1	0	0	1	0	0	0	0	0	0	0	0
Any	×	×	×	×	×	1	1	1	1				

BCD input is applied at DCBA input terminals and its 9's complement appears at $N_D N_C N_B N_A$ terminals.

Figure 6.30 gives the block diagram of BCD 10's complement converter and Table 6.14 gives its truth table.

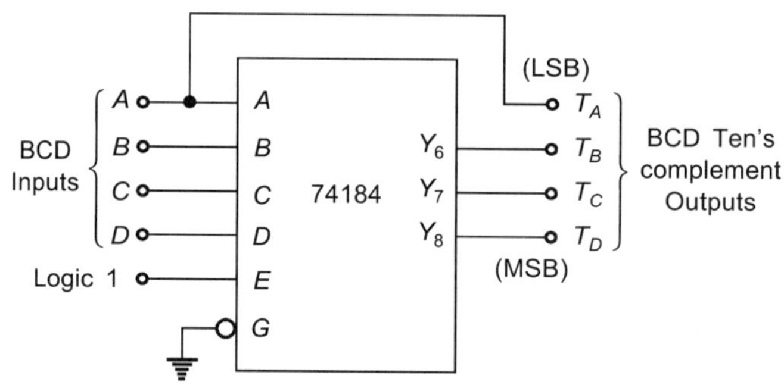

Fig. 6.30
Block diagram of BCD 10's complement converter using 74184.

Table 6.14 Truth table of 74184 as BCD 10's complement converter

BCD Word	*Inputs*						*Outputs*			*BCD 10's complement*			
	E	*D*	*C*	*B*	*A*	*G*	Y_8	Y_7	Y_6	T_D	T_C	T_B	T_A
0	1	0	0	0	0	0	0	0	0	0	0	0	0
1	1	0	0	0	1	0	1	0	0	1	0	0	1
2	1	0	0	1	0	0	1	0	0	1	0	0	0
3	1	0	0	1	1	0	0	1	1	0	1	1	1
4	1	0	1	0	0	0	0	1	1	0	1	1	0
5	1	0	1	0	1	0	0	1	0	0	1	0	1
6	1	0	1	1	0	0	0	1	0	0	1	0	0
7	1	0	1	1	1	0	0	0	1	0	0	1	1
8	1	1	0	0	0	0	0	0	1	0	0	1	0
9	1	1	0	0	1	0	0	0	0	0	0	0	1
Any	×	×	×	×	×	1	1	1	1				

The design of circuits for the conversion of 2 or more decades of BCD is quite involved. The circuit for 2-decade BCD-to-binary converter is given in Fig. 6.31. The number of 74184 ICs required increases tremendously as the number of input BCD decades increases. Table 6.15 gives the number of ICs required for a given number of input BCD decades. From the table we observe that the circuits become unmanageable for longer BCD words and it may be worthwhile to examine an alternative approach.

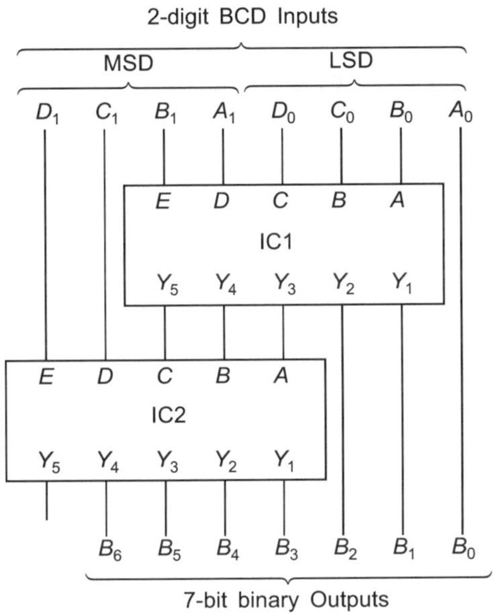

Fig. 6.31
A 2-decade BCD-to-binary converter.

Table 6.15 Number of 74184 ICs required

BCD decades	Number of 74184 ICs required
2	2
3	6
4	11
5	19
6	28

A read-only-memory (ROM) may be a convenient alternative, which will be discussed later.

Example 6.9 Verify the operation of a 2-decade BCD-to-binary converter of Fig. 6.31 for an input of 29.

Solution

The input is

$D_1\ C_1\ B_1\ A_1\ D_0\ C_0\ B_0\ A_0 = 00101001$

The binary output $B_0 = A_0 = 1$

The inputs of IC1 are $E = B_1 = 1$

$D = A_1 = 0$

$C = D_0 = 1$

$B = C_0 = 0$

$A = B_0 = 0$

The outputs of IC1 are $Y_5 = 0$

$Y_4 = 1$

$Y_3 = 1$

$Y_2 = 1$

$Y_1 = 0$

The binary outputs B_1 and B_2 are

$B_1 = Y_1 = 0$

$B_2 = Y_2 = 1$

The inputs of IC2 are $E = D_1 = 0$

$D = C_1 = 0$

$C = Y_5$ of IC1 $= 0$

$B = Y_4$ of IC1 $= 0$

$A = Y_3$ of IC1 $= 1$

The outputs of IC2 are $Y_5 = 0$

$Y_4 = 0$

$Y_3 = 0$

$Y_2 = 1$

$Y_1 = 1$

The binary outputs B_3, B_4, B_5, and B_6 are

$B_3 = Y_1 = 1$

$B_4 = Y_2 = 1$

$B_5 = Y_3 = 0$

$B_6 = Y_4 = 0$

Therefore the full binary output is 0011101.

6.9.2 Binary-to-BCD Converter

The block diagram of binary-to-BCD converter IC 74185A is given in Fig. 6.30 and Table 6.16 gives its truth table. It has the same pin out as the 74184. The binary inputs are applied at terminals A through E and the BCD outputs are available at terminals Y_1 through Y_6 (Y_7 and Y_8 are not used and these are always at logic 1). Similar to the BCD-to-binary converter, the least-significant bit bypass the circuit as shown in Fig. 6.32.

Table 6.16 Truth table of 74185A binary-to-BCD converter

Binary Words	*Inputs*						*Outputs*					
	E	*D*	*C*	*B*	*A*	*G*	Y_6	Y_5	Y_4	Y_3	Y_2	Y_1
0 – 1	0	0	0	0	0	0	0	0	0	0	0	0
2 – 3	0	0	0	0	1	0	0	0	0	0	0	1
4 – 5	0	0	0	1	0	0	0	0	0	0	1	0
6 – 7	0	0	0	1	1	0	0	0	0	0	1	1
8 – 9	0	0	1	0	0	0	0	0	0	1	0	0
10 – 11	0	0	1	0	1	0	0	0	1	0	0	0
12 – 13	0	0	1	1	0	0	0	0	1	0	0	1
14 – 15	0	0	1	1	1	0	0	0	1	0	1	0
16 – 17	0	1	0	0	0	0	0	0	1	0	1	1
18 – 19	0	1	0	0	1	0	0	0	1	1	0	0
20 – 21	0	1	0	1	0	0	0	1	0	0	0	0
22 – 23	0	1	0	1	1	0	0	1	0	0	0	1
24 – 25	0	1	1	0	0	0	0	1	0	0	1	0
26 – 27	0	1	1	0	1	0	0	1	0	0	1	1
28 – 29	0	1	1	1	0	0	0	1	0	1	0	0
30 – 31	0	1	1	1	1	0	0	1	1	0	0	0
32 – 33	1	0	0	0	0	0	0	1	1	0	0	1
34 – 35	1	0	0	0	1	0	0	1	1	0	1	0
36 – 37	1	0	0	1	0	0	0	1	1	0	1	1
38 – 39	1	0	0	1	1	0	0	1	1	1	0	0
40 – 41	1	0	1	0	0	0	1	0	0	0	0	0
42 – 43	1	0	1	0	1	0	1	0	0	0	0	1
44 – 45	1	0	1	1	0	0	1	0	0	0	1	0
46 – 47	1	0	1	1	1	0	1	0	0	0	1	1
48 – 49	1	1	0	0	0	0	1	0	0	1	0	0
50 – 51	1	1	0	0	1	0	1	0	1	0	0	0
52 – 53	1	1	0	1	0	0	1	0	1	0	0	1
54 – 55	1	1	0	1	1	0	1	0	1	0	1	0
56 – 57	1	1	1	0	0	0	1	0	1	0	1	1
58 – 59	1	1	1	0	1	0	1	0	1	1	0	0
60 – 61	1	1	1	1	0	0	1	1	0	0	0	0
62 – 63	1	1	1	1	1	0	1	1	0	0	0	1
All	×	×	×	×	×	1	1	1	1	1	1	1

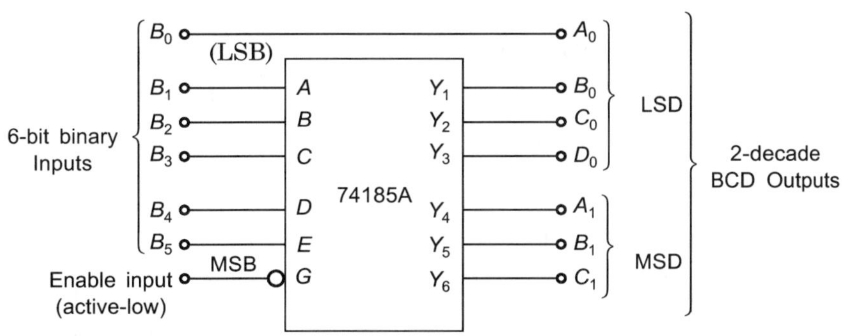

Fig. 6.32
A 6-bit binary-to-BCD converter.

The method of expansion to accommodate larger number of binary inputs is quite complicated. The circuit for conversion of 8-bit binary number to BCD is given in Fig. 6.33. This requires three ICs. Table 6.17 gives the number of ICs required for a given number of input binary bits. Similar to the BCD-to-binary converters we may have to resort to ROM for conversion of longer binary words.

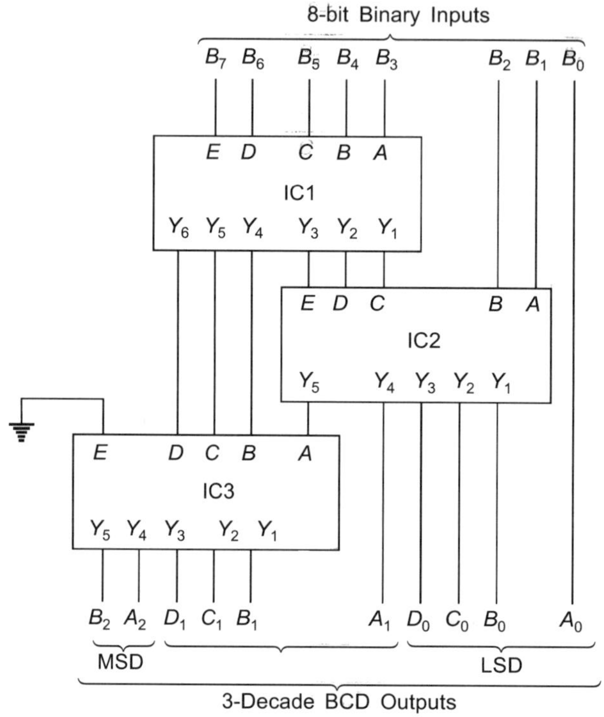

Fig. 6.33
An 8-bit binary-to-BCD converter.

Table 6.17 Number of 74185A ICs required

Binary bits	Number of 74185 ICs required
4 – 6	1
7, 8	3
9	4
10	6
11	7
12	8
13	10
14	12
15	14
16	16

Example 6.10 Verify the operation of 8-bit binary-to BCD converter of Fig. 6.33 for the binary input of 11010011.

Solution

The inputs of IC1 are $E = B_7 = 1$
$D = B_6 = 1$
$C = B_5 = 0$
$B = B_4 = 1$
$A = B_3 = 0$

The outputs of IC1 are $Y_6 = 1$
$Y_5 = 0$
$Y_4 = 1$
$Y_3 = 0$
$Y_2 = 0$
$Y_1 = 1$

The inputs of IC2 are $E = Y_3$ of IC1 = 0
$D = Y_2$ of IC1 = 0
$C = Y_1$ of IC1 = 1
$B = B_2 = 0$
$A = B_1 = 1$

The outputs of IC2 are $Y_6 = 0$
$Y_5 = 0$
$Y_4 = 1$
$Y_3 = 0$
$Y_2 = 0$
$Y_1 = 0$

The inputs of IC3 are $E = 0$
$D = Y_6$ of IC1 = 1
$C = Y_5$ of IC1 = 0
$B = Y_4$ of IC1 = 1
$A = Y_5$ of IC2 = 0

The outputs of IC3 are $Y_6 = 0$
$Y_5 = 1$
$Y_4 = 0$
$Y_3 = 0$
$Y_2 = 0$
$Y_1 = 0$

The BCD outputs are $B_2 = Y_5$ of IC3 = 1
$A_2 = Y_4$ of IC3 = 0
$D_1 = Y_3$ of IC3 = 0
$C_1 = Y_2$ of IC3 = 0
$B_1 = Y_1$ of IC3 = 0
$A_1 = Y_4$ of IC2 = 1
$D_0 = Y_3$ of IC2 = 0
$C_0 = Y_2$ of IC2 = 0

$$B_0 = Y_1 \text{ of IC2} = 0$$
$$A_0 = B_0 = 1$$

Therefore, the BCD output is
10 0001 0001 = 211.

6.10 PRIORITY ENCODERS

6.10.1 Decimal-to-BCD Encoder

One of the most commonly used input device for a digital system is a set of ten switches, one for each numeral between 0 and 9. These switches generate 1 or 0 logic levels in response to turning them OFF or ON. When a particular number is to be fed to the digital circuit in BCD code, the switch corresponding to that number is pressed. There is an IC available for performing this function (74147) which is a priority encoder. The block diagram of 74147 IC is given in Fig. 6.34 and Table 6.18 gives its truth table. It has active-low inputs and outputs. The meaning of the word priority can be seen from the truth table, for example, if inputs 2 and 5 are LOW, the output will be corresponding to 5 which has a higher priority than 2, i.e. the highest numbered input has priority over lower numbered inputs.

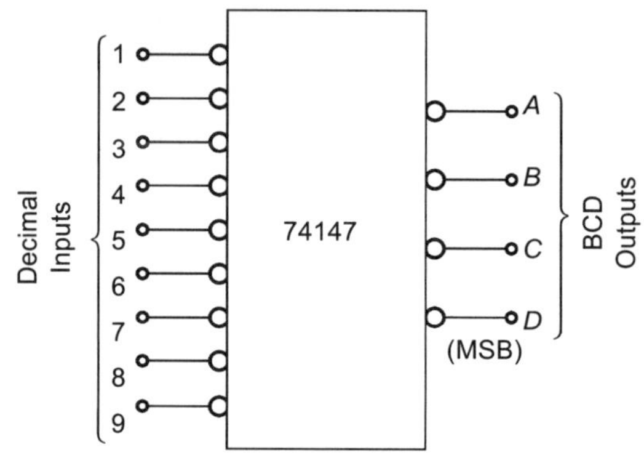

Fig. 6.34 Block diagram of 74147 decimal-to-BCD priority encoder.

Table 6.18 Truth table of 74147

Active-low decimal inputs									*Active-low BCD outputs*			
1	2	3	4	5	6	7	8	9	*D*	*C*	*B*	*A*
1	1	1	1	1	1	1	1	1	1	1	1	1
0	1	1	1	1	1	1	1	1	1	1	1	0
×	0	1	1	1	1	1	1	1	1	1	0	1
×	×	0	1	1	1	1	1	1	1	1	0	0
×	×	×	0	1	1	1	1	1	1	0	1	1
×	×	×	×	0	1	1	1	1	1	0	1	0
×	×	×	×	×	0	1	1	1	1	0	0	1
×	×	×	×	×	×	0	1	1	1	0	0	0
×	×	×	×	×	×	×	0	1	0	1	1	1
×	×	×	×	×	×	×	×	0	0	1	1	0

6.10.2 Octal-to-Binary Encoder

The octal code is often used at the inputs of digital circuits that require manual entering of long binary words. Priority encoder 74148 IC has been designed to achieve this operation. Its block diagram is given in Fig. 6.35 and Table 6.19 gives its truth table. This circuit also has active-low inputs and active-low outputs. The enable input and carry outputs, which are also active-low, are used to cascade circuits to handle more inputs. A hexadecimal-to-binary encoder, which is also a very useful circuit because of widespread use of hexadecimal code in computers, microprocessors, etc. can be designed using this facility.

The priority encoders can conveniently be used for handling priority interrupts in computers, microprocessors, etc.

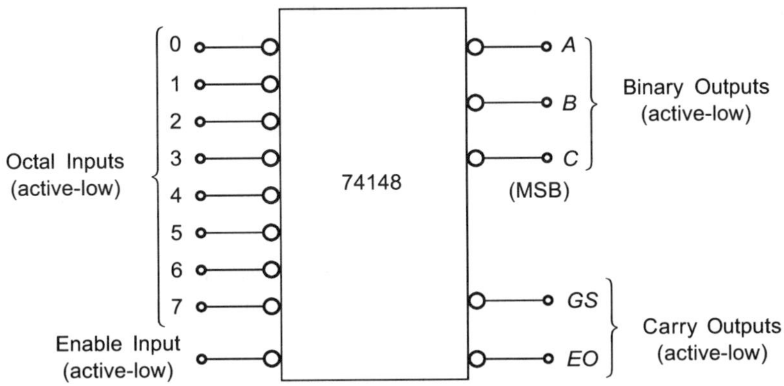

Fig. 6.35
Block diagram of 74148 octal-to-binary priority encoder.

Table 6.19 Truth table of 74148

EI	*Inputs*								*Outputs*				
	0	1	2	3	4	5	6	7	*C*	*B*	*A*	*GS*	*EO*
1	×	×	×	×	×	×	×	×	1	1	1	1	1
0	0	1	1	1	1	1	1	1	1	1	1	0	1
0	×	0	1	1	1	1	1	1	1	1	0	0	1
0	×	×	0	1	1	1	1	1	1	0	1	0	1
0	×	×	×	0	1	1	1	1	1	0	0	0	1
0	×	×	×	×	0	1	1	1	0	1	1	0	1
0	×	×	×	×	×	0	1	1	0	1	0	0	1
0	×	×	×	×	×	×	0	1	0	0	1	0	1
0	×	×	×	×	×	×	×	0	0	0	0	0	1
0	1	1	1	1	1	1	1	1	1	1	1	1	0

Example 6.11 Design a hexadecimal-to binary encoder using 74148 encoders and 74157 multiplexer.

Solution Since there are sixteen symbols (0 – *F*) in hexadecimal number system, two 74148 encoders are required. Hexadecimal inputs 0 through 7 are applied to IC1 input lines and inputs 8 through *F* to IC2 input lines. Whenever one of the inputs of IC2 is active (LOW), IC1 must be disabled, on the other hand, if all the inputs of IC2 are HIGH then IC1 must be enabled. This is achieved by connecting the *EO* line of IC2 to the *EI* line of IC1. A quad 2:1 multiplexer is required to get the proper 4-bit binary outputs. The complete circuit is shown in Fig. 6.36. The *GS* output of 74148 goes LOW whenever one of its inputs is active. Therefore, *GS* of IC2 is connected to select input of 74157, which selects *A* inputs if it is LOW, otherwise *B* inputs are selected. The outputs of the multiplexer are the required binary outputs and are active-low. This circuit is also a priority encoder.

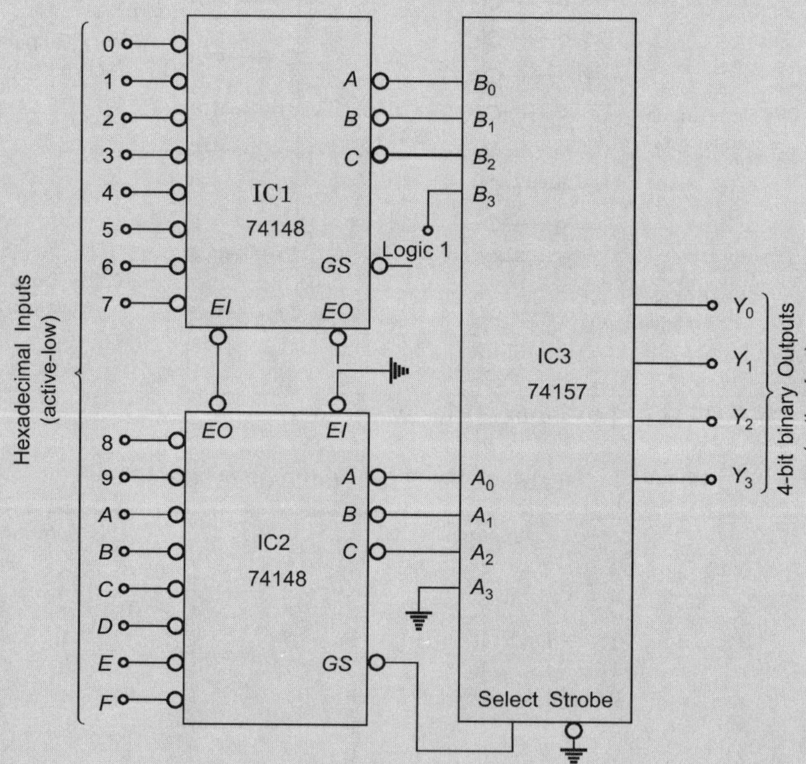

Fig. 6.36
Hexadecimal-to-binary priority encoder.

6.11 DECODER/DRIVERS FOR DISPLAY DEVICES

6.11.1 BCD-to Decimal Decoder/Driver

In many digital systems, we prefer to see the output in a decimal format. The outputs can either be displayed using display devices like LEDs, Nixie tubes, etc. or can be used to actuate some indicators or relays. Table 6.20 gives the available BCD-to-decimal decoder/driver ICs. All these ICs have active-high inputs and active-low outputs.

Table 6.20 Available BCD-to-decimal decoder/driver ICs

IC No.	*Output circuit*	*Application*
7441	Open-collector	Nixie tube driver
7442	Totem-pole	LED driver
7445	Open-collector	Indicator/relay driver
74141	Open-collector	Nixie tube driver
74145	Open-collector	Indicator/relay driver
74445	Open-collector	Indicator/relay driver

6.11.2 BCD-to-7-Segment Decoder/Driver

Seven segment display is the most popular display device used in digital systems. For displaying data using this device, the data have to be converted from BCD to 7-segment code. A number of MSI ICs are available (Table 6.21) for performing this function. The decoder/driver circuit has 4 input lines for BCD data and 7 output lines to drive a 7-segment display. Output terminals *a* through *g* of the decoder are to be connected to *a* through *g* terminals of the display respectively. If the outputs are active-low, then the 7-segment LED must be of the common anode type, whereas if the outputs are active-high then the 7-segment LED must be of the common cathode type. The function of LT, RBI, RBO and BI are given in Table 6.22 and are explained below.

Table 6.21 Available BCD-to-7-segment decoder/driver ICs

IC No.	*Output*	*Rating (Max voltage, Sink current)*	*Facilities available*			
			Lamp Test LT	*Ripple blanking Input RBI*	*Ripple blanking Output RBO*	*Blanking Input BI*
7446, 74246	Active-low; Open-collector	30 V, 40 mA	Yes	Yes	Yes	Yes
7447, 74247	Active-low; Open-collector	15 V, 40 mA	Yes	Yes	Yes	Yes
7448, 74248	Active-high; Pull-up resistor = 2 kΩ	5.5 V, 6.4 mA	Yes	Yes	Yes	Yes
7449, 74249	Active-high; Open-collector	5.5 V, 8 mA	No	No	No	Yes

Table 6.22 Summary of BCD-to-7-segment decoder functions

LT	*RBI*	*BI/RBO*	*BCD inputs*	*Display mode*
0	×	1 output	×	Lamp test
×	×	0 input	×	Display blank
1	1	1 output	Any number	Normal decoding
1	0	Normally at logic 1 output. Goes to logic 0 during zero blanking interval	Any number	Normal decoding with zero blanking

LT This is used to check the segments of LED. If it is connected to logic 0 level, all the segments of the display connected to the decoder will be ON. For normal decoding operation, this terminal is to be connected to logic 1 level.

RBI It is to be connected to logic 1 for normal decoding operation. If it is connected to 0 level, the segment outputs will generate data for normal 7-segment decoding for all BCD inputs except zero. Whenever the BCD inputs correspond to zero, the 7-segment display switches off. This is used for blanking out leading zeros in multi-digit displays.

BI If it is connected to logic 0 level, the display is switched off irrespective of BCD inputs. This is used for conserving power in multiplexed displays.

RBO This output, which is normally at logic 1 goes to logic 0 during zero blanking interval. This is used for cascading purposes and is connected to RBI of the succeeding stage.

Example 6.12

(a) Set up a single 7-segment LED display using 7447 BCD-to-7-segment decoder/driver.
(b) Design a 4-digit 7-segment LED display system with leading zero blanking.
(c) Design a 4-digit multiplexed 7-segment LED display system with leading-zero blanking.

Solution

(a) The set up is given in Fig. 6.37.
(b) A 4-digit display system can display numbers from 0000 to 9999. If the number to be displayed is smaller than a 4-digit number then there is a problem of leading zeros. It is very often desirable to design a display system so that any number less than 1000 will not show leading zeros. For example, the number 0203 should appear as 203. The block diagram of a 4-digit display system with leading zero blanking is shown in Fig. 6.38. The reader should verify the operation of this circuit.
(c) When a multi-digit, 7-segment LED display system is used, it requires power which is *n*-times the power required for a single 7-segment LED, where *n* is the number of such LEDs in the system. Typically, the current requirement of one segment is 20 mA at full brightness which makes the maximum current requirement for a single 7-segment LED to be 140 mA (corresponding to zero). If there are four such LEDs in the display system, then the maximum current required would be 560 mA. It is pos-

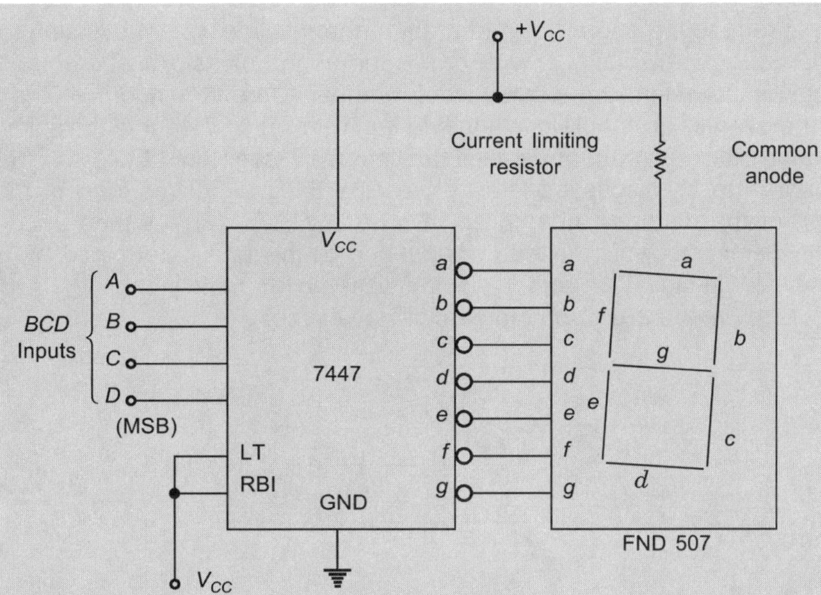

Fig. 6.37
A 7447 driving a 7-segment LED display.

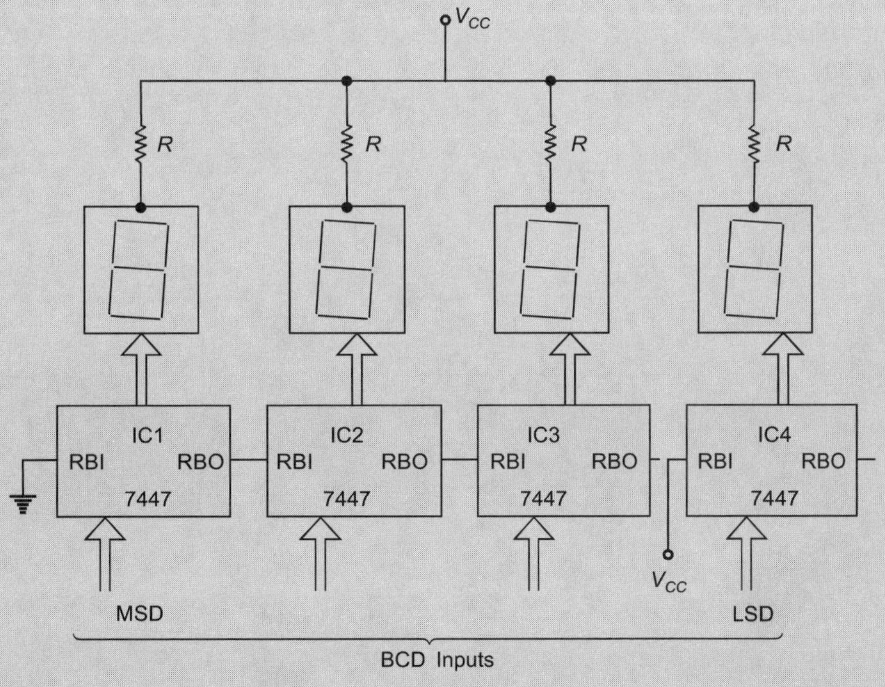

Fig. 6.38
A 4-digit LED display system with leading zero blanking.

sible to reduce the overall current drain of the display system by using the concept of multiplexing the display, i.e. to energize the output digits one at a time. If this is done in rapid succession, the display would appear to be continuous to the human eyes and the overall current drain is equal to that of a single 7-segment display. Figure 6.39 gives the block diagram of a 4-digit multiplexed 7-segment LED display system. Here, a counter (to be discussed later) drives the BCD-to-decimal decoder 7442 and the output of the decoder through an inverter pulls up the common anode terminal to HIGH. Each anode is excited in sequence at the rate determined by the clock frequency. Multi-digit 7-segment display assemblies which can be used only in the multiplexed mode are also commercially available.

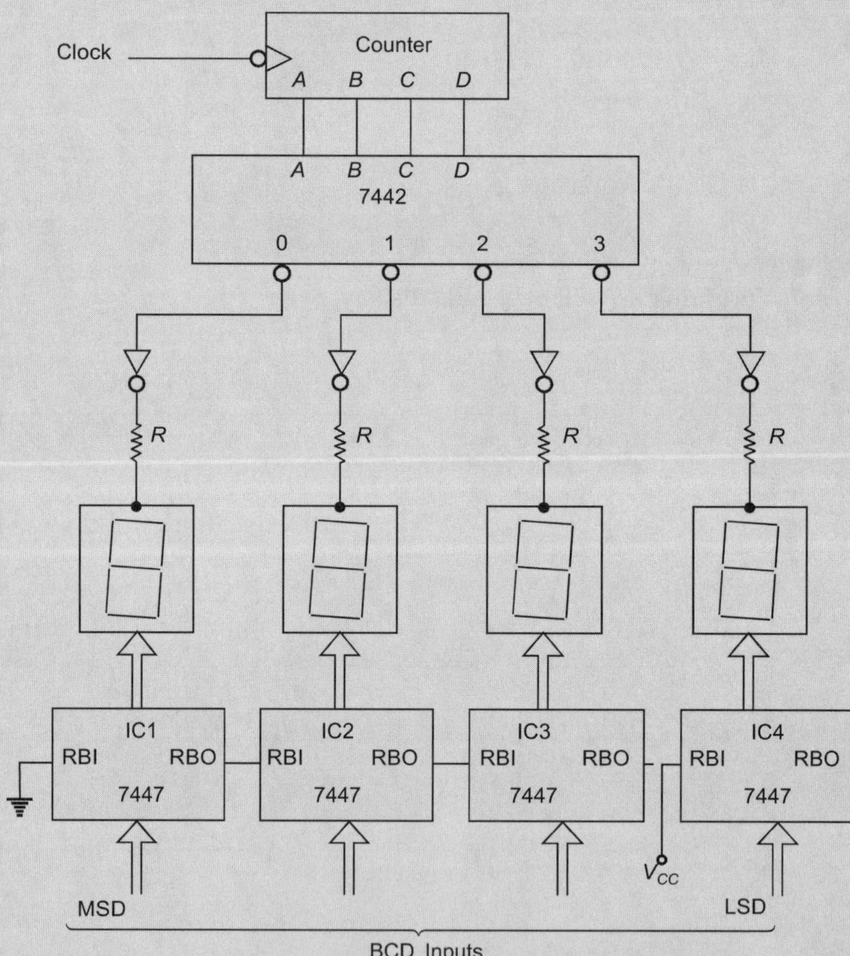

Fig. 6.39
A 4-digit multiplexed display system with leading zero blanking.

6.12 SUMMARY

The most popular and commonly available MSI ICs and their applications have been discussed in this chapter. Since the logic equations for these circuits are very complex, therefore, their applicability must be recognized at the system level. Quite often, the system specifications are affected by the initial knowledge of available standard circuits. Each technique discussed can be thought of as a tool and each tool has its place. The designer has to make a proper choice of the tool for the job. Though more than one tool may work for a given job, the key is to select the right one. Although system design has been oriented around making use of higher levels of integration, a lot of little jobs of interfacing the MSI devices are still best done with discrete gates.

Active-low inputs and outputs have been indicated by small circles throughout. Recently, many authors have started using an alternative symbol (small right triangle) to represent the active-low input/output to avoid confusion (of inversion) associated with the small circle. Figure 6.35 is redrawn in Fig. 6.40 to indicate this new system.

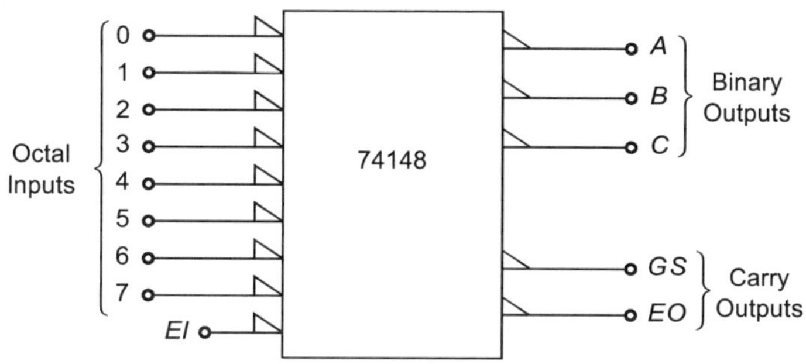

Fig. 6.40

An alternative representation of active-low input/output.

Glossary

Arithmetic-logic unit (ALU) A logic circuit that performs arithmetic and logical operations.

Comparator A logic circuit that compares two binary numbers and produces an output indicating whether they are equal or which is greater if they are unequal.

Demultiplexer A logic circuit that connects a single input line to one of the several output lines.

Dot matrix display A display device consisting of elements in the form of dots arranged in a matrix form. It is used to display alphanumeric and special characters.

Driver A circuit capable of supplying high load current.

Encoder A logic circuit that produces coded binary outputs from unencoded inputs.

Hardware The physical parts forming a system.

Look-ahead carry A method of generating carry for fast addition.

Multiplexed display A display system used in multiplexed mode.

Multiplexer A logic circuit that selects one of several input lines and connects it to a single output line.

Priorty encoder An encoder that responds to the highest number when two or more numbers are applied simultaneously.

Strobe An input signal to a digital circuit that can enable or disable the circuit.

Subtractor A logic circuit used for subtraction.

Ten's complement It is 9's complement of a decimal number plus one.

Review Questions

6.1 A logic circuit that gates one out of several inputs to single output is known as a ______________ .

6.2 A ____________ is a logic circuit that accepts one data input and distributes it over several outputs.

6.3 The minimum number of selection inputs required for selecting one out of 32 inputs is ____________ .

6.4 A logic circuit required for converting BCD code to 7-segment code is known as ______________ .

6.5 In a BCD adder ________________ is added in case the sum is not a valid BCD number.

6.6 A logic circuit used to compare binary numbers is ________________ .

6.7 ________________ is used in display systems to save ________________ .

6.8 A 4-variable logic expression can be realized using single ________________ multiplexer.

6.9 A decoder with 64 output lines has ________________ data inputs.

6.10 ALU stands for ________________ .

6.11 A bubble at the output of a multiplexer indicates ________________ output.

6.12 Chip select input of a digital IC has ______◺ symbol. This indicates ____________ .

6.13 A 4-digit display system with leading zero blanking displays 47 as ________________ .

6.14 ____________________ complementation is used for performing BCD subtraction.

6.15 Subtractors are designed using ____________________ ICs.

Problems

6.1 Realize the logic function of Table 6.3 using
   (a) A 16:1 multiplexer IC 74150, and
   (b) An 8:1 multiplexer IC 74152.

6.2 Design a 32:1 multiplexer using two 16:1 multiplexer ICs.

6.3 Design a full adder using 8:1 multiplexer ICs. Compare the IC package count with the NAND–NAND realization.

6.4 Design a 4-bit ADDER/SUBTRACTOR circuit with ADD/SUB control line.

6.5 Design a Gray-to-BCD code converter using
   (a) Two dual 4:1 multiplexer ICs (74153) and some gates, and
   (b) One 1:16 demultiplexer IC 74154 and NAND gates.

6.6 Design a BCD-to-7-segment decoder with active-low outputs using
(a) Dual 4:1 multiplexers and some gates,
(b) 1:16 demultiplexer and some gates, and
(c) A BCD-to-decimal decoder and NAND gates.
(d) Compare the IC package count.

6.7 Design a BCD-to-Gray code converter using
(a) 8:1 multiplexers,
(b) Dual 4:1 multiplexers and some gates,
(c) Quad 2:1 multiplexers and some gates,
(d) A BCD-to-decimal decoder and NAND gates, and
(e) NAND gates only.
(f) Compare package count for the various approaches.

6.8 Realize the following functions of four variables using
(a) 8:1 multiplexers,
(b) 16:1 multiplexers, and
(c) 4-to-16-line decoder with active-low outputs.
(i) $f_1 = \Sigma m(0, 3, 5, 6, 9, 10, 12, 15)$
(ii) $f_2 = \Sigma m\ (0, 1, 2, 3, 11, 12, 14, 15)$
(iii) $f_3 = \Pi M\ (0, 1, 3, 7, 9, 10, 11, 13, 14, 15)$

6.9 Design a 40:1 multiplexer using 8:1 multiplexers.

6.10 Design a 1:40 demultiplexer using BCD-to-decimal decoders.

6.11 In the n-bit binary adder shown in Fig. 6.12(a), determine the time required for the final carry C_{n-1} to reach steady state. Assume each full-adder to be consisting of half-adder circuits (Prob. 5.19(b)) and propagation delays of gates as given in Prob. 5.20.

6.12 Design a 4-digit BCD adder using 7483 adders.

6.13 Design a 2-bit comparator using gates.

6.14 Design an 8-bit comparator using only two 7485s.

6.15 Verify the operation of the 24-bit comparator of Fig. 6.24 for the following numbers:
$A = 100110000111011001010010$
$B = 101110000111011000100011$

6.16 Design a one digit-BCD-to-binary converter using 74184.

6.17 Show that the hexadecimal-to-binary encoder of Fig. 6.36 is a priority encoder.

6.18 Design a 6-bit odd/even parity checker using 74180.

6.19 Design a parity generator circuit using 74180 to add an odd parity bit to a 7-bit word.

6.20 Design a parity generator circuit to add an even parity bit to a 14-bit word. Use two 74180 packages.

6.21 Design a 10-bit parity checker using one 74180 and an EX-OR gate (7486).

6.22 Design a two level parity checker tree using ten packages of 74180 to check the parity of 81 bits.

6.23 Design a two level parity checker tree using three packages of 74180 to check the parity of 25 bits.

6.24 Design the following circuits:
(a) 7442 BCD-to-decimal decoder driving ten LEDs.
(b) 74141 BCD-to-decimal decoder driving a Nixie tube.

6.25 Suggest a suitable alternative circuit for Fig. 6.11 which does not require an OR gate.

6.26 Consider the four digit 7-segment display system shown in Fig. 6.41. Modify the circuit of Fig. 6.39 if this display system is to be used.

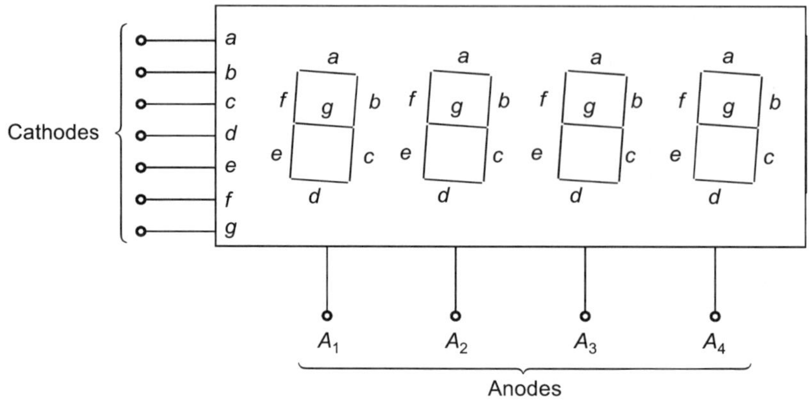

Fig. 6.41
A 4-digit 7-segment display system.

6.27 A commonly used display system known as 7 × 5 dot matrix is shown in Fig. 6.42*a*. This is used to display alphanumeric characters and some special symbols. It is desired to display *R* using this display system whose pattern of glowing of the dots is shown in Fig. 6.42*b*. Design a suitable circuit for this.

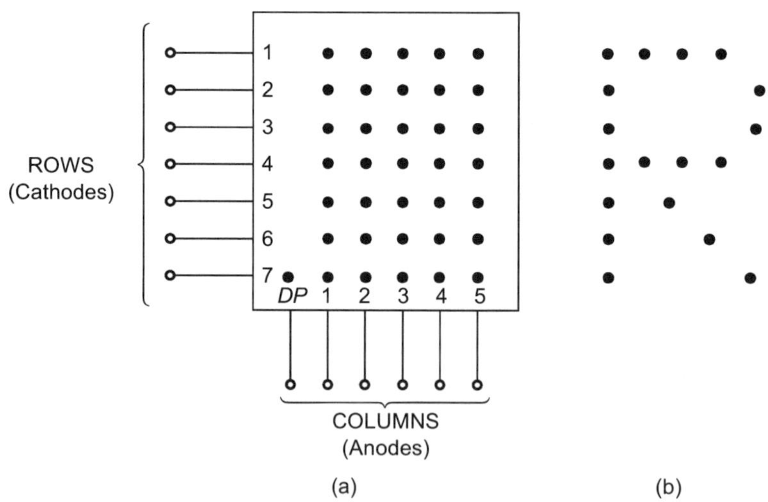

Fig. 6.42
A 7 × 5 dot matrix display.

7

FLIP-FLOPs

7.1 INTRODUCTION

So far we have directed our studies towards the analysis and design of combinational digital circuits. Though very important, it constitutes only a part of digital systems. The other major aspect of digital systems is analysis and design of sequential circuits. However, sequential circuit design depends, to a large extent, on the combinational circuit design discussed earlier.

There are many applications in which digital outputs are required to be generated in accordance with the sequence in which the input signals are received. This requirement can not be satisfied using a combinational logic system. These applications require outputs to be generated that are not only dependent on the present input conditions but they also depend upon the past history of these inputs. The past history is provided by feedback from the output back to the input.

A block diagram of a sequential circuit is shown in Fig. 7.1. It consists of combinational circuits which accept digital signals from external inputs and from outputs of memory elements and generates signals for external outputs and for inputs to memory elements referred to as excitation.

A memory element is some medium in which one bit of information (1 or 0) can be stored or retained until necessary, and thereafter its contents can be replaced by a new value. The contents of memory elements in Fig. 7.1 can be changed by the outputs of the combinational circuit which are connected to its input.

The combinational circuit performs certain operations, some of which are used to determine the digital signals to be stored in memory elements. The other operations are performed on external inputs and memory outputs to generate the external outputs.

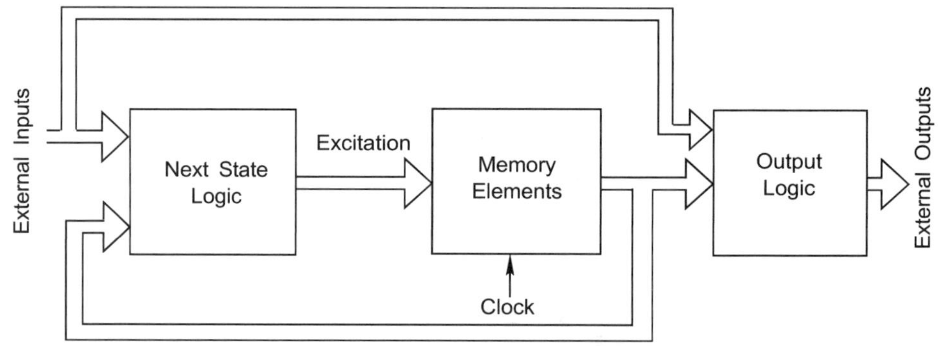

Fig. 7.1
Block diagram of a sequential circuit.

The above process demonstrates the dependence of the external outputs of a sequential circuit on the external inputs and the present contents of the memory elements (referred to as the *present state* of memory elements). The new contents of the memory elements, referred to as the *next state*; depend on the external inputs and the present state. Hence, the output of a sequential circuit is a function of the time sequence of inputs and the *internal states.*

Sequential circuits are classified in two main categories, known as *asynchronous* and *synchronous* sequential circuits depending on timing of their signals.

A sequential circuit whose behaviour depends upon the sequence in which the input signals change is referred to as an asynchronous sequential circuit. The outputs will be affected whenever the inputs change. The commonly used memory elements in these circuits are time delay devices. These can be regarded as combinational circuits with feedback.

A sequential circuit whose behaviour can be defined from the knowledge of its signal at discrete instants of time is referred to as a synchronous sequential circuit. In these systems, the memory elements are affected only at discrete instants of time. The synchronization is achieved by a timing device known as a *system clock* which generates a periodic train of clock pulses as shown in Fig. 7.2. The outputs are affected only with the application of a clock pulse.

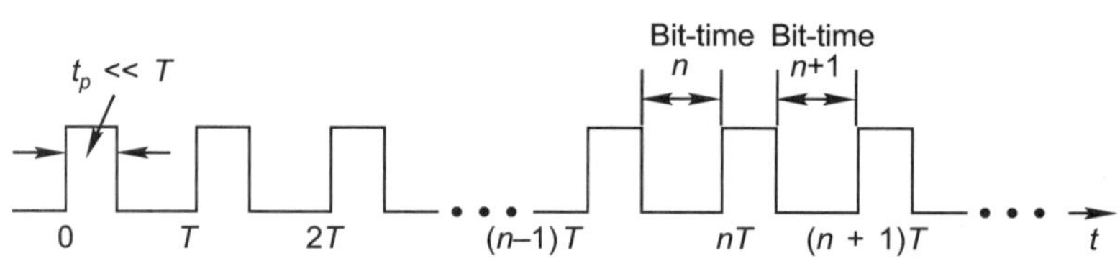

Fig. 7.2
A train of pulses.

Since the design of asynchronous circuits is more-tedious and difficult, therefore their uses are rather limited.

Synchronous circuits have gained considerable domination and wide popularity and are also known as *clocked-sequential circuits.* The memory elements used are FLIP-FLOPs which are capable of storing binary information.

7.2 A 1-BIT MEMORY CELL

The basic digital memory circuit is known as FLIP-FLOP. It has two stable states which are known as the *1 state* and the *0 state.* It can be obtained by using NAND or NOR gates. We shall be systematically developing a FLIP-FLOP circuit starting from the fundamental circuit shown in Fig. 7.3. It consists of two inverters G_1 and G_2 (NAND gates used as inverters). The output of G_1 is connected to the input of G_2 (A_2) and the output of G_2 is connected to the input of G_1 (A_1).

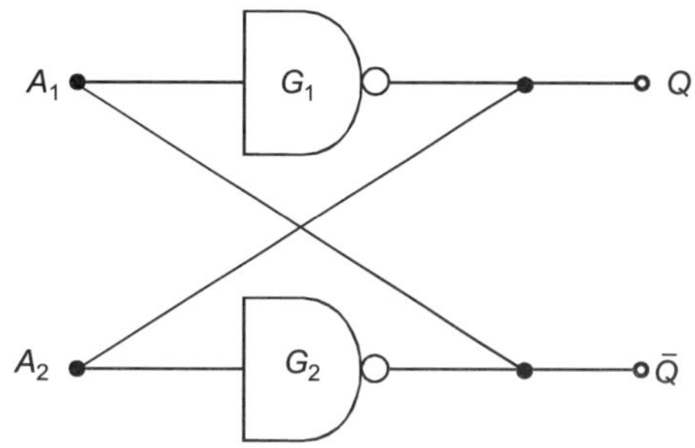

Fig. 7.3

Cross-coupled inverters as a memory element.

Let us assume the output of G_1 to be $Q = 1$, which is also the input of $G_2(A_2 = 1)$. Therefore, the output of G_2 will be $\bar{Q} = 0$, which makes $A_1 = 0$ and consequently $Q = 1$ which confirms our assumption.

In a similar manner, it can be demonstrated that if $Q = 0$, then $\bar{Q} = 1$ and this is also consistent with the circuit connections.

From the above discussion we note the following:

1. The outputs Q and $\bar{Q}$ are always complementary.
2. The circuit has two stable states; in one of the stable state $Q = 1$ which is referred to as the 1 state (*or set state*) whereas in the other stable state $Q = 0$ which is referred to as the 0 state (*or reset state*).
3. If the circuit is in 1 state, it continues to remain in this state and similarly if it is in 0 state, it continues to remain in this state. This property of the circuit is referred to as *memory,* i.e. it can store 1-bit of digital information.

Since this information is locked or latched in this circuit, therefore, this circuit is also referred to as a *latch.*

In the latch of Fig. 7.3, there is no way of entering the desired digital information to be stored in it. In fact, when the power is switched on, the circuit switches to one of the stable states ($Q = 1$ or 0) and it is not possible to predict the state. If we replace the inverters G_1 and G_2 with 2-input NAND gates, the other input terminals of the NAND gates can be used to enter the desired digital information. The modified circuit is shown in Fig. 7.4. Two additional inverters G_3 and G_4 have been added for reasons which will become clear from the following discussion.

If $S = R = 0$, the circuit is exactly the same as that of Fig. 7.3 (Prob. 7.1). If $S = 1$ and $R = 0$, the output of G_3 will be 0 and the output of G_4 will be 1. Since one of the inputs of G_1 is 0, its output will certainly be 1. Consequently, both the inputs of G_2 will be 1 giving an output $\bar{Q} = 0$. Hence, for this input condition, $Q = 1$ and $\bar{Q} = 0$. Similarly, if $S = 0$ and $R = 1$ then the outputs will be $Q = 0$ and $\bar{Q} = 1$. The first of these two input conditions ($S = 1$, $R = 0$) makes $Q = 1$ which is referred to as the *set state,* whereas the second input condition ($S = 0$, $R = 1$) makes

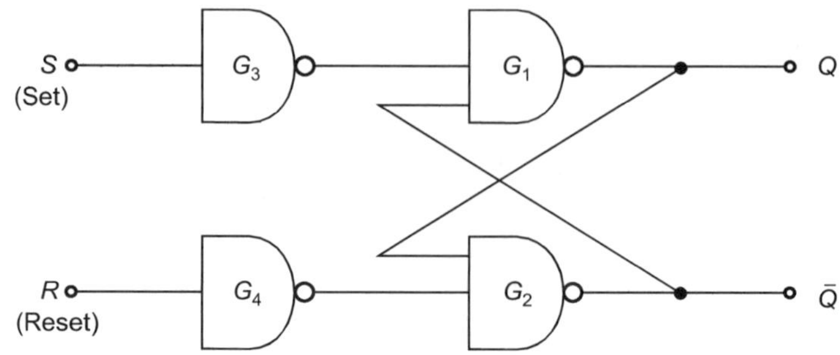

Fig. 7.4
The memory cell with provision for entering data.

$Q = 0$ which is referred to as the *reset state* or *clear state.* This gives us the means for entering the desired bit in the latch.

Now we see what happens if the input conditions are changed from $S = 1, R = 0$ to $S = R = 0$ or from $S = 0, R = 1$ to $S = R = 0$. The output remains unaltered (Prob. 7.2). This shows the basic difference between a combinational circuit and a sequential circuit, even though the sequential circuit is made up of combinational circuits.

The two input terminals are designated as set (S) and reset (R) because $S = 1$ brings the circuit in set state and $R = 1$ brings it to reset or clear state.

If $S = R = 1$, both the outputs Q and $\bar{Q}$ will try to become 1 which is not allowed and therefore, this input condition is prohibited.

7.3 CLOCKED *S–R* FLIP-FLOP

It is often required to set or reset the memory cell (Fig. 7.4) in synchronism with a train of pulses (Fig. 7.2) known as clock (abbreviated as *CK*). Such a circuit is shown in Fig. 7.5, and is referred to as a *clocked set-reset* (*S–R*) FLIP-FLOP.

In this circuit, if a clock pulse is present ($CK = 1$), its operation is exactly the same as that of Fig. 7.4. On the other hand, when the clock pulse is not present ($CK = 0$), the gates G_3 and G_4 are inhibited, i.e. their outputs are 1 irrespective of the values of S or R. In other words, the circuit responds to the inputs S and R only when the clock is present.

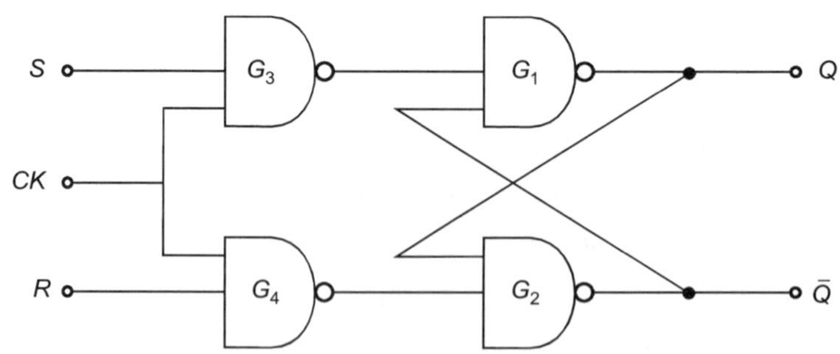

Fig. 7.5
A clocked S–R FLIP-FLOP.

Assuming that the inputs do not change during the presence of the clock pulse, we can express the operation of a FLIP-FLOP in the form of the truth table in Table 7.1 for the *S–R* FLIP-FLOP. Here S_n and R_n denote the inputs and Q_n the output during the bit time *n* (Fig. 7.2). Q_{n+1} denotes the output *Q* after the pulse passes, i.e. in the bit time $n + 1$.

Table 7.1 Truth table of *S–R* FLIP-FLOP

Inputs		*Output*
S_n	R_n	Q_{n+1}
0	0	Q_n
1	0	1
0	1	0
1	1	?

If $S_n = R_n = 0$, and the clock pulse is applied, the output at the end of the clock pulse is same as the output before the clock pulse, i.e. $Q_{n+1} = Q_n$. This is indicated in the first row of the truth table.

If $S_n = 1$ and $R_n = 0$, the output at the end of the clock pulse will be 1, whereas if $S_n = 0$ and $R_n = 1$, then $Q_{n+1} = 0$. These are indicated in the second and third rows of the truth table respectively.

In the circuit of Fig. 7.4, it was mentioned that $S = R = 1$ is not allowed. Let us see what happens in the *S–R* FLIP-FLOP of Fig. 7.5 if $S_n = R_n = 1$. When the clock is present the outputs of gates G_3 and G_4 are both 0, making one of the inputs of G_1 and G_2 NAND gates 0. Consequently, *Q* and $\bar{Q}$ both will attain logic 1 which is inconsistent with our assumption of complementary outputs. Now, when the clock pulse has passed away ($CK = 0$), the outputs of G_3 and G_4 will rise from 0 to 1. Depending upon the propagation delays of the gates, either the stable state $Q_{n+1} = 1$ ($\bar{Q}_{n+1} = 0$) or $Q_{n+1} = 0$ ($\bar{Q}_{n+1} = 1$) will result. That means the state of the circuit is undefined, indeterminate or ambiguous and therefore is indicated by a question mark (fourth row of the truth table).

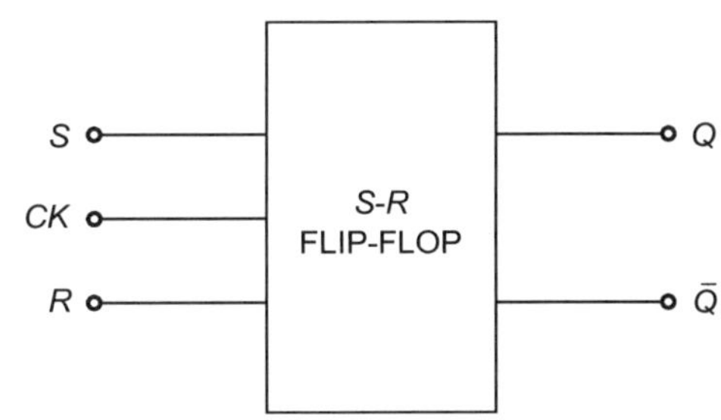

Fig. 7.6 Logic symbol of clocked *S–R* FLIP-FLOP.

The condition $S_n = R_n = 1$ is forbidden and it must not be allowed to occur.

The logic symbol of clocked *S–R* FLIP-FLOP is given in Fig. 7.6.

7.3.1 Preset and Clear

In the FLIP-FLOP of Fig. 7.5, when the power is switched on, the state of the circuit is uncertain. It may come to set ($Q = 1$) or reset ($Q = 0$) state. In many applications it is desired to initially set or reset the FLIP-FLOP, i.e. the initial state of the FLIP-FLOP is to be assigned. This is accomplished by using the direct, or *asynchronous inputs*, referred to as *preset* (*Pr*) and *clear* (*Cr*) inputs. These inputs may be applied at any time between clock pulses and are not in synchronism with the clock. An *S–R* FLIP-FLOP with preset and clear is shown in Fig. 7.7. If

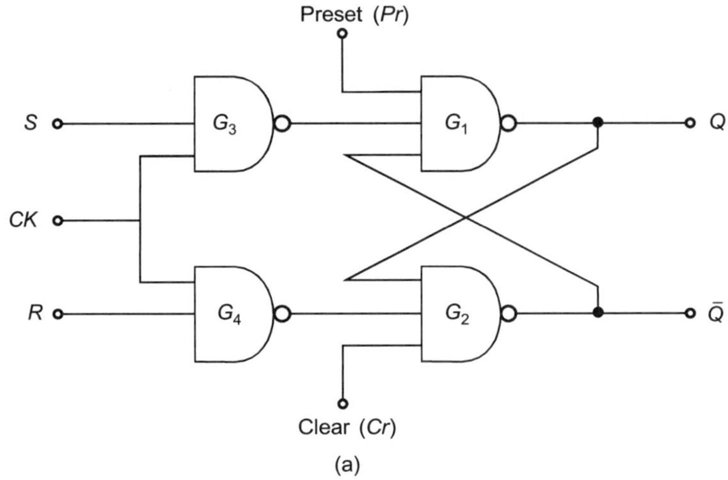

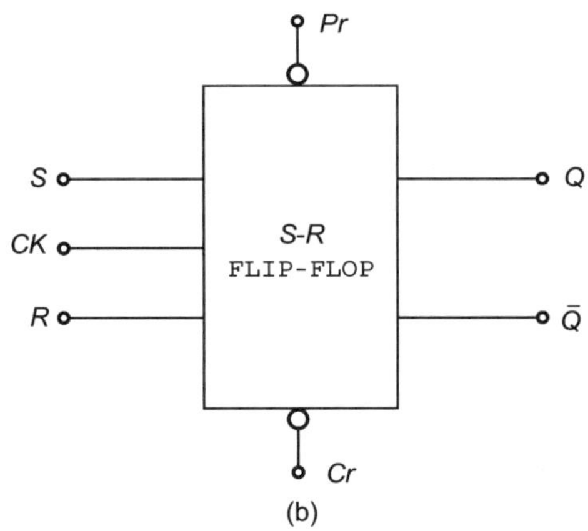

Fig. 7.7
(a) An *S–R* FLIP-FLOP with preset and clear,
(b) Its logic symbol.

$Pr = Cr = 1$ the circuit operates in accordance with the truth table of *S–R* FLIP-FLOP given in Table 7.1.

If $Pr = 0$ and $Cr = 1$, the output of $G_1(Q)$ will certainly be 1. Consequently, all the three inputs to G_2 will be 1 which will make $\bar{Q} = 0$, Hence, making $Pr = 0$ sets the FLIP-FLOP.

Similarly, if $Pr = 1$ and $Cr = 0$, the FLIP-FLOP is reset. Once the state of the FLIP-FLOP is established asynchronously, the asynchronous inputs *Pr* and *Cr* must be connected to logic 1 before the next clock is applied.

The condition $Pr = Cr = 0$ must not be used, since this leads to an uncertain state.

In the logic symbol of Fig. 7.7*b*, bubbles are used for *Pr* and *Cr* inputs, which means these are active-low, i.e. the intended function is performed when the signal applied to *Pr* or *Cr* is LOW. The operation of Fig. 7.7 is summarized in Table 7.2.

Table 7.2 Summary of operation of *S*–R FLIP-FLOP

Inputs			*Output*	*Operation performed*
CK	*Cr*	*Pr*	*Q*	
1	1	1	Q_{n+1}(Table 7.1)	Normal FLIP-FLOP
0	0	1	0	Clear
0	1	0	1	Preset

The circuit can be designed such that the asynchronous inputs override the clock, i.e. the circuit can be set or reset even in the presence of the clock pulse (Prob. 7.4).

7.4 *J–K* FLIP-FLOP

The uncertainty in the state of an *S–R* FLIP-FLOP when $S_n = R_n = 1$ (fourth row of the truth table) can be eliminated by converting it into a *J–K* FLIP-FLOP. The data inputs are *J* and *K* which are ANDed with $\bar{Q}$ and *Q*, respectively, to obtain *S* and *R* inputs, i.e.

$$S = J \cdot \bar{Q} \tag{7.1a}$$

$$R = K \cdot Q \tag{7.1b}$$

A *J–K* FLIP-FLOP thus obtained is shown in Fig. 7.8. Its truth table is given in Table 7.3a which is reduced to Table 7.3b for convenience. Table 7.3a has been prepared for all the possible combinations of *J* and *K* inputs, and for each combination both the states of the output have been considered. The reader can verify this (Prob. 7.5).

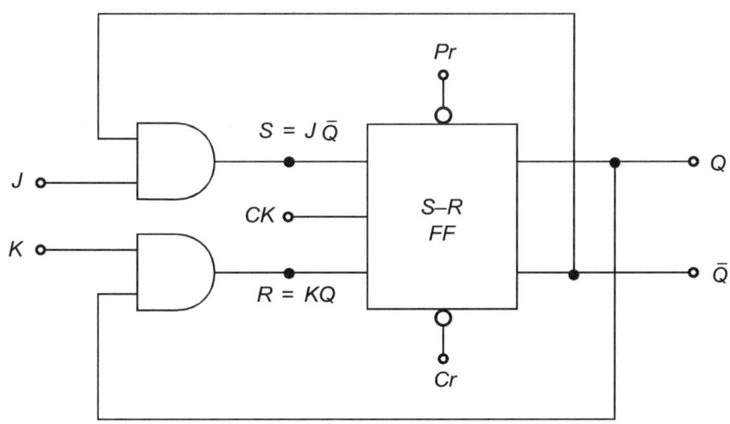

Fig. 7.8
An *S–R* FLIP-FLOP converted into *J–K* FLIP-FLOP.

It is not necessary to use the AND gates of Fig. 7.8, since the same function can be performed by adding an extra input terminal to each NAND gate G_3 and G_4 of Fig. 7.7 (Prob. 7.6). With this modification incorporated in Fig. 7.7, we obtain the *J–K* FLIP-FLOP using NAND gates as shown in Fig. 7.9. The logic symbol of *J–K* FLIP-FLOP is given in Fig. 7.10.

Table 7.3a Truth table for Fig. 7.8

Data inputs		*Outputs*		*Inputs to S–R FF*		*Output* Q_{n+1}	
J_n	K_n	Q_n	$\bar{Q}_n$	S_n	R_n		
0	0	0	1	0	0	0	$= Q_n$
0	0	1	0	0	0	1	
1	0	0	1	1	0	1	$= 1$
1	0	1	0	0	0	1	
0	1	0	1	0	0	0	$= 0$
0	1	1	0	0	1	0	
1	1	0	1	1	0	1	$= \bar{Q}_n$
1	1	1	0	0	1	0	

Table 7.3b Truth table of *J–K* FLIP-FLOP

Inputs		*Output*
J_n	K_n	Q_{n+1}
0	0	Q_n
1	0	1
0	1	0
1	1	$\bar{Q}_n$

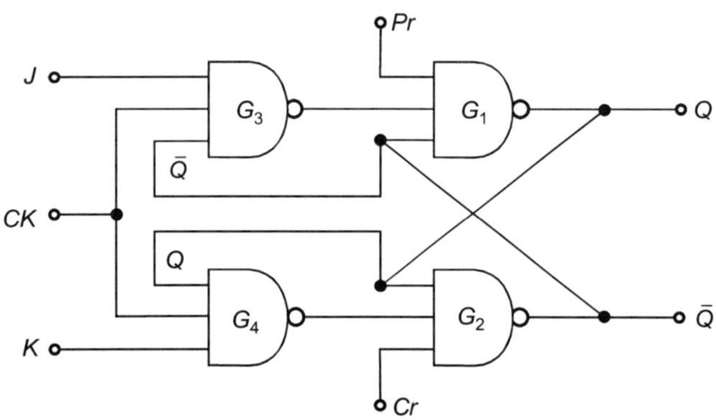

Fig. 7.9
A *J–K* FLIP-FLOP using NAND gates.

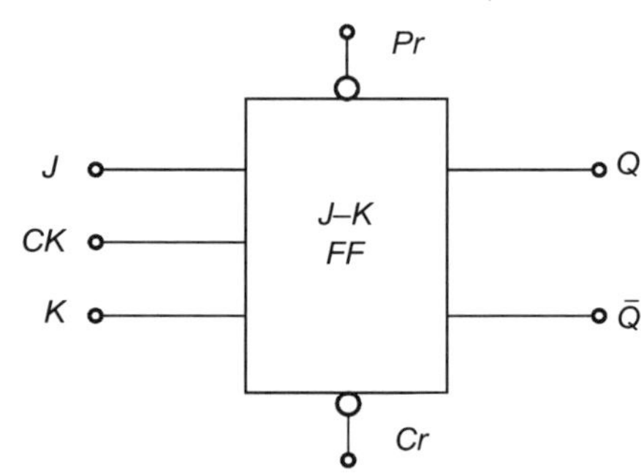

Fig. 7.10
Logic symbols of *J–K* FLIP-FLOP.

7.4.1 The Race-Around Condition

The difficulty of both inputs $1(S = R = 1)$ being not allowed in an S–R FLIP-FLOP is eliminated in a *J–K* FLIP-FLOP by using the feedback connection from outputs to the inputs of the gates G_3 and G_4 (Fig. 7.9). Table 7.3 assumes that the inputs do not change during the clock pulse $(CK = 1)$, which is not true because of the feedback connections. Consider, for example, that the inputs are $J = K = 1$ and $Q = 0$, and a pulse as shown in Fig. 7.11 is applied at the clock input. After a time interval Δt equal to the propagation delay through two NAND

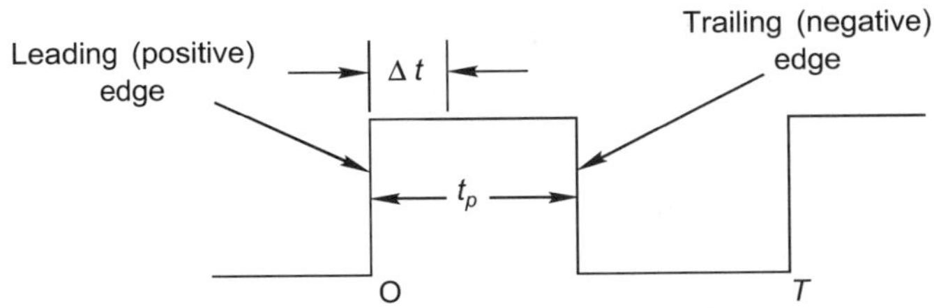

Fig. 7.11
A clock pulse.

gates in series, the output will change to $Q = 1$ (see fourth row of Table 7.3b). Now we have $J = K = 1$ and $Q = 1$ and after another time interval of Δt the output will change back to $Q = 0$. Hence, we conclude that for the duration t_p of the clock pulse, the output will oscillate back and forth between 0 and 1. At the end of the clock pulse, the value of Q is uncertain. This situation is referred to as the *race-around condition.*

The race-around condition can be avoided if $t_p < \Delta t < T$. However, it may be difficult to satisfy this inequality because of very small propagation delays in ICs. A more practical method for overcoming this difficulty is the use of the master–slave (*M–S*) configuration discussed below.

7.4.2 The Master–Slave *J–K* `FLIP-FLOP`

A master–slave *J–K* `FLIP-FLOP` is a cascade of two *S–R* `FLIP-FLOPs`, with feedback from the outputs of the second to the inputs of the first as illustrated in Fig. 7.12. Positive clock pulses are applied to the first `FLIP-FLOP` and the clock pulses are inverted before these are applied to the second `FLIP-FLOP`.

When $CK = 1$, the first `FLIP-FLOP` is enabled and the outputs Q_M and $\bar{Q}_M$ respond to the inputs J and K according to Table 7.3. At this time, the second `FLIP-FLOP` is inhibited

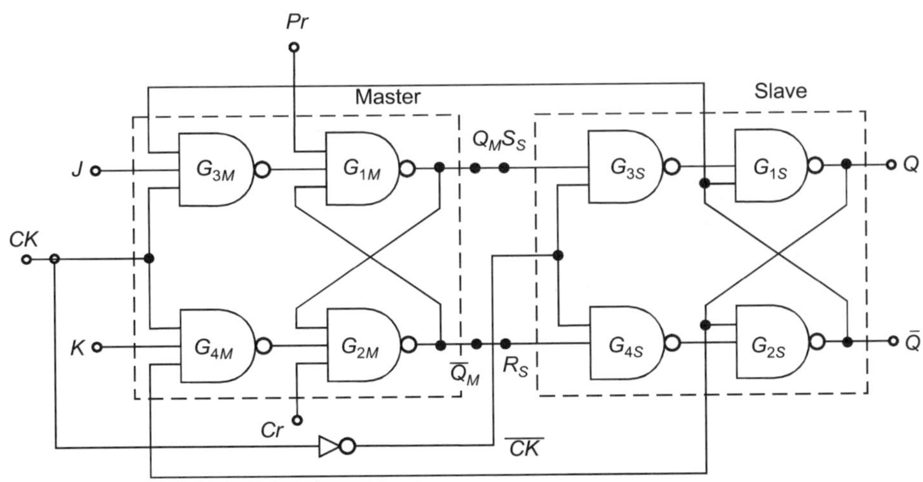

Fig. 7.12
A master-slave *J–K* `FLIP-FLOP`.

because its clock is LOW ($\overline{CK} = 0$). When CK goes LOW ($\overline{CK} = 1$), the first FLIP-FLOP is inhibited and the second FLIP-FLOP is enabled, because now its clock is HIGH ($\overline{CK} = 1$). Therefore, the outputs Q and $\bar{Q}$ follow the outputs Q_M and $\bar{Q}_M$, respectively (second and third rows of Table 7.3b). Since the second FLIP-FLOP simply follows the first one, it is referred to as the *slave* and the first one as the *master*. Hence, this configuration is referred to as *master–slave* (*M–S*) FLIP-FLOP.

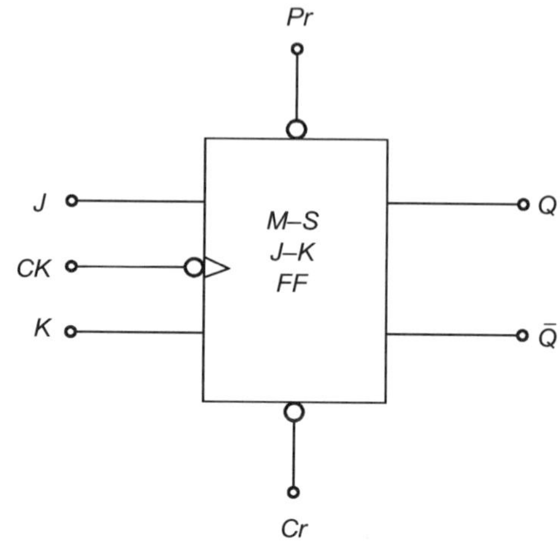

Fig. 7.13
A master–slave *J-K* FLIP-FLOP logic symbol.

In this circuit, the inputs to the gates G_{3M} and G_{4M} do not change during the clock pulse, therefore the race-around condition does not exist. The state of the master–slave FLIP-FLOP changes at the negative transition (trailing edge) of the clock pulse. The logic symbol of a *M–S* FLIP-FLOP is given in Fig. 7.13. At the clock input terminal, the symbol > is used to illustrate that the output changes when the clock makes a transition and the accompanying bubble signifies negative transition (change in CK from 1 to 0).

7.5 D-TYPE FLIP-FLOP

If we use only the middle two rows of the truth table of the *S-R* (Table 7.1) or *J–K* (Table 7.3b) FLIP-FLOP, we obtain a *D*-type FLIP-FLOP as shown in Fig. 7.14. It has only one input referred to as *D*-input or data input. Its truth table is given in Table 7.4 from which it is clear that the output Q_{n+1} at the end of the clock pulse equals the input D_n before the clock pulse.

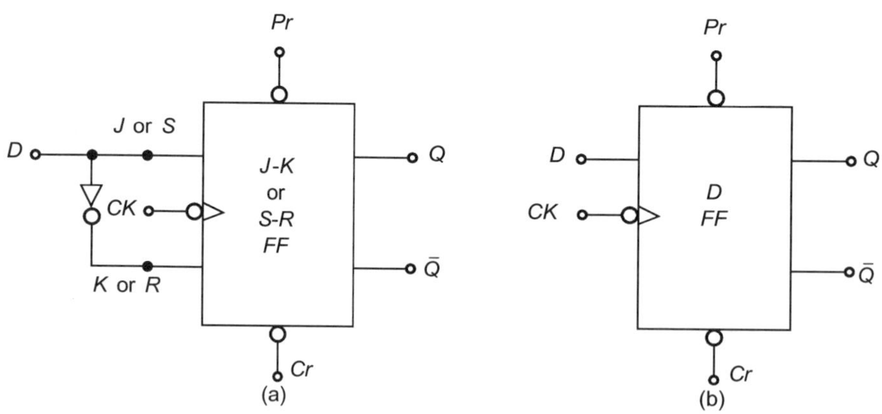

Fig. 7.14
(a) A *J-K* or *S–R* FLIP-FLOP converted into a *D*-type FLIP-FLOP (b) its logic symbol.

Table 7.4 Truth table of a *D*-type FLIP-FLOP

Input D_n	*Output* Q_{n+1}
0	0
1	1

This is equivalent to saying that the input data appears at the output at the end of the clock pulse. Thus, the transfer of data from the input to the output is delayed and hence the name *delay* (*D*) FLIP-FLOP. The *D*-type FLIP-FLOP is either used as a delay device or as a latch to store 1-bit of binary information.

7.6 *T*-TYPE FLIP-FLOP

In a *J–K* FLIP-FLOP , if $J = K$, the resulting FLIP-FLOP is referred to as a *T*-type FLIP-FLOP and is shown in Fig. 7.15. It has only one input, referred to as *T*-input. Its truth table is given in Table 7.5 from which it is clear that if $T = 1$ it acts as a toggle switch. For every clock pulse, the output Q changes.

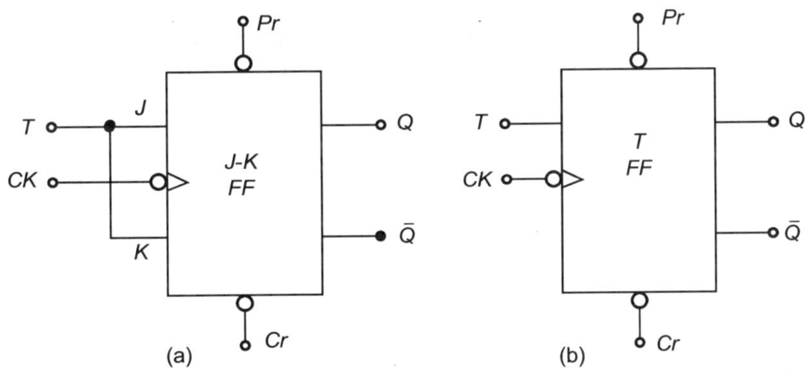

Fig. 7.15
(a) A *J–K* FLIP-FLOP converted into a *T*-type FLIP-FLOP, (b) its logic symbol.

Table 7.5 Truth table of *T*-type FLIP-FLOP

Input T_n	*Output* Q_{n+1}
0	Q_n
1	$\bar{Q}_n$

An *S–R* FLIP-FLOP cannot be converted into a *T*-type FLIP-FLOP since $S = R = 1$ is not allowed. However, the circuit of Fig. 7.16 acts as a toggle switch, i.e. the output Q changes with every clock pulse (Prob. 7.11).

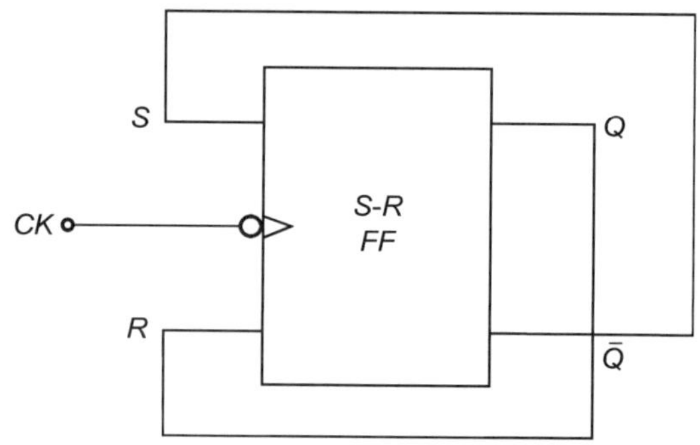

Fig. 7.16
An S–R FLIP-FLOP as a toggle switch.

7.7 EXCITATION TABLE OF FLIP-FLOP

The truth table of a FLIP-FLOP is also referred to as the *characteristic table* and specifies the operational characteristic of the FLIP-FLOP.

In the design of sequential circuits, we usually come across situations in which the *present state* and the *next state* of the circuit are specified, and we have to find the input conditions that must prevail to cause the desired transition of the state. By the present state and the next state we mean the state of the circuit prior to and after the clock pulse respectively. For example, the output of an S–R FLIP-FLOP before the clock pulse is $Q_n = 0$ and it is desired that the output does not change when the clock pulse is applied. What input conditions (S_n and R_n values) must exist to achieve this?

From the truth table (or the characteristic table) of an S–R FLIP-FLOP (Table 7.1) we obtain the following conditions:

1. $S_n = R_n = 0$ (first row)
2. $S_n = 0$, $R_n = 1$ (third-row)

We conclude from the above conditions that the S_n input must be 0, whereas the R_n input may be either 0 or 1 (don't-care). Similarly, input conditions can be found for all possible situations. A tabulation of these conditions is known as the *excitation table*. It is a very important and useful design aid for sequential circuits. Table 7.6 gives the excitation tables of S–R, J–K, T, and D FLIP-FLOPs. This is derived from the characteristic table of the FLIP-FLOP.

Table 7.6 Excitation table of FLIP-FLOPs

Present State	*Next State*	*S–R*	*FF*	*J–K*	*FF*	*T-FF*	*D-FF*
		S_n	R_n	J_n	K_n	T_n	D_n
0	0	0	×	0	×	0	0
0	1	1	0	1	×	1	1
1	0	0	1	×	1	1	0
1	1	×	0	×	0	0	1

7.8 CLOCKED FLIP-FLOP DESIGN

In earlier sections, we defined or specified the operation of different FLIP-FLOPs assuming a circuit without regard to where the circuit came from or how it was designed. In this section, the design of a FLIP-FLOP is given. The design philosophy illustrated is, in fact, a general approach for the design of sequential circuits and systems.

Consider the general model of the FLIP-FLOP shown in Fig. 7.17. Basically, a clocked FLIP-FLOP is a sequential circuit which stores the bits 0 and 1. This operation is accomplished by using a binary cell coupled with some combinational set/reset decoding logic to allow some input control over the set and reset operations of the cell. The steps for the design of FLIP-FLOP are given below.

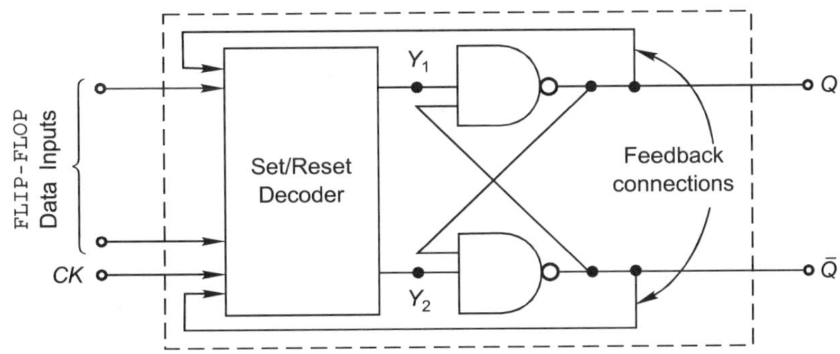

Fig. 7.17
The general model of the FLIP-FLOP.

Step I Examine each row of the given characteristic table, specifying the desired inputs and outputs, and answer the following questions and make a truth table with Y_1 and Y_2 as output variables.

1. Does the cell need to be set $(Q_{n+1} = 1)$ for this condition?
2. Does the cell need to be reset $(Q_{n+1} = 0)$ for this condition?
3. Does the cell need to be left as it is?

Step II Prepare the *K*-map for Y_1 and Y_2 output variables, minimize it and determine the logic for Y_1 and Y_2, respectively. Draw the complete circuit using gates.

The above design steps are illustrated in Example 7.1.

Example 7.1 Using the technique described above, design a clocked *S–R* FLIP-FLOP whose characteristic table is given in Table 7.7.

Solution

Step I Determine the values of Y_1 and Y_2 for each row. For example, for the first row $Q_n = 0$ and $Q_{n+1} = 0$. To obtain $Q_{n+1} = 0$, Y_1 must be equal to 1, since $\bar{Q}_n = 1$, Y_2 can be 0 or 1 since $Q_n = 0$. In a similar manner, complete the truth table (Table 7.7).

Table 7.7 Truth table

Characteristic table					*Truth table for decoder*		
CK	*S*	*R*	Q_n	Q_{n+1}	Y_1	Y_2	
0	0	0	0	0	1	×	
0	0	0	1	1	×	1	
0	0	1	0	0	1	×	
0	0	1	1	1	×	1	
0	1	0	0	0	1	×	
0	1	0	1	1	×	1	
0	1	1	0	0	1	×	} *
0	1	1	1	1	×	1	
1	0	0	0	0	1	×	
1	0	0	1	1	×	1	
1	0	1	0	0	1	×	
1	0	1	1	0	1	0	
1	1	0	0	1	0	1	
1	1	0	1	1	×	1	
1	1	1	0	×	×	×	} **
1	1	1	1	×	×	×	

*$S = R = 1$ can happen with no clock.
**$S = R = 1$ must not happen.

Step II The *K*-maps for Y_1 and Y_2 are given in Fig. 7.18 which give

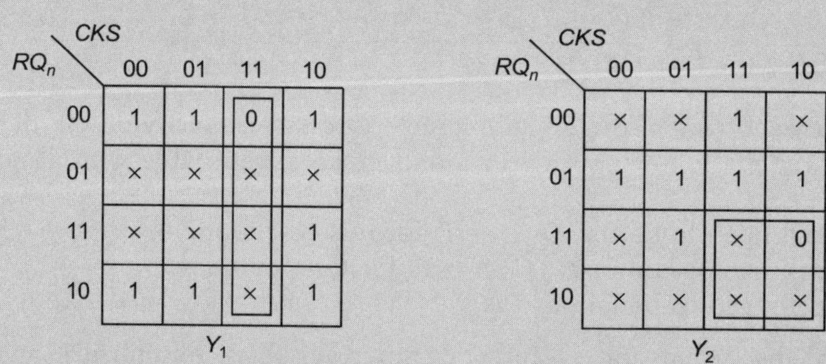

Fig. 7.18
K-maps for Ex. 7.1.

$$Y_1 = \overline{CK} + \bar{S} = \overline{CK \cdot S}$$
$$Y_2 = \bar{R} + \overline{CK} = \overline{CK \cdot R}$$

Thus, we see that the circuit resulting from this design is the same as that shown in Fig. 7.5.

7.8.1 Conversion From One Type of FLIP-FLOP to Another Type

In earlier sections, we have discussed conversion from *S–R* to *J–K*, *S–R* (or *J–K*) to *D*-type, and *J–K* to *T*-type FLIP-FLOPs. Now, we shall effect the conversion from one type of FLIP-FLOP to another type by using a formal technique which is similar to the one used above and will be useful in the design of clocked sequential circuits.

Consider the general model for conversion from one type of FLIP-FLOP to another type (Fig. 7.19). In this, we are required to design the combinational logic decoder (conversion logic) for converting new input definitions into input codes which will cause the given FLIP-FLOP to perform as desired.

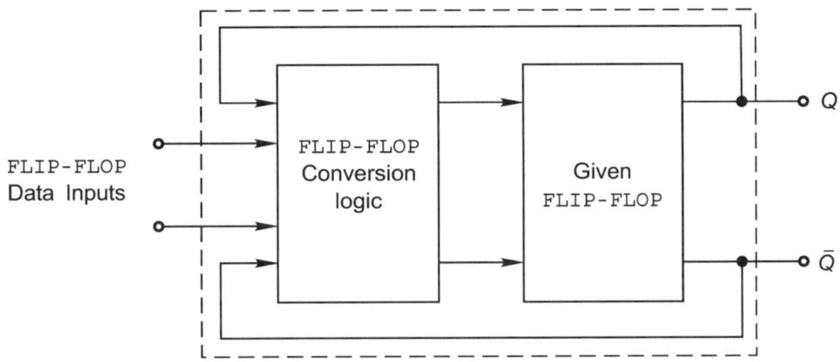

Fig. 7.19

The general model used to convert one type of FLIP-FLOP to another type.

To design the conversion logic we need to combine the excitation tables for both FLIP-FLOPs and make a truth table with data input(s) and Q as the inputs and the input(s) of the given FLIP-FLOP as the output(s). The conventional method of combinational logic design then follows as usual. The conversion is illustrated in Example 7.2.

Example 7.2 Convert an *S–R* FLIP-FLOP to a *J–K* FLIP-FLOP.

Solution

The excitation tables of *S–R* and *J–K* FLIP-FLOPs are given in Table 7.6 from which we make the truth table given in Table 7.8.

Table 7.8 Truth table of conversion logic

Row	*FF data inputs*		*Output*	*S–R FF inputs*	
	J	*K*	*Q*	*S*	*R*
1	0	0	0	0	×
2	0	1	0	0	×
3	1	0	0	1	0
4	1	1	0	1	0
5	0	1	1	0	1
6	1	1	1	0	1
7	0	0	1	×	0
8	1	0	1	×	0

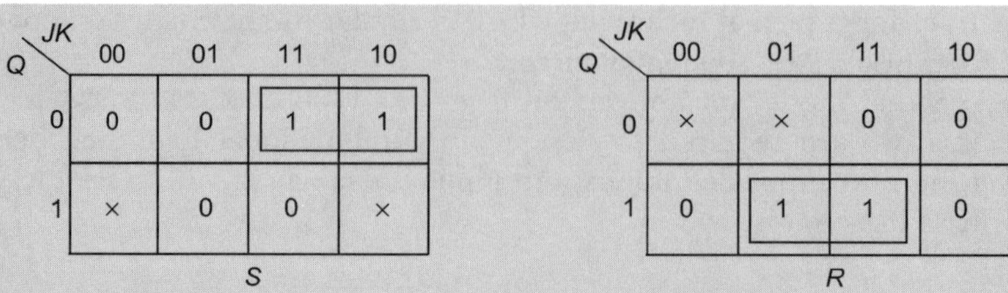

Fig. 7.20
K-maps for Ex. 7.2.

The *K*-maps are given in Fig. 7.20, which give

$$S = J \cdot \bar{Q} \qquad \text{and} \qquad R = K \cdot Q$$

Thus, we see that the circuit resulting from this design is the same as that shown in Fig. 7.8.

7.9 EDGE-TRIGGERED FLIP-FLOPs

All FLIP-FLOPs other than the master–slave type discussed in earlier sections are level triggered, i.e. the outputs respond to the inputs (according to the truth table) as long as the clock is present. The only ICs available in this category are latches, for example 7475 and 74100 are transparent latches.

The master–slave FLIP-FLOPs are also referred to as the *pulse-triggered* FLIP-FLOPs, i.e. the outputs respond to the inputs when a pulse is applied at the clock input. The master FLIP-FLOP responds when the clock is present ($CK = 1$) and the output of the slave will be available at the falling edge of the clock pulse ($CK = 0$). As discussed in Section 7.4, this eliminates the problem of race-around condition. In this the data is *locked-out* at the falling edge of the clock pulse, i.e. the changes occurring at the inputs once *CK* goes to 0 will not affect the operation of the FLIP-FLOP .

The inputs to the FLIP-FLOP may change during the presence of the clock pulse due to certain operations in the system. This causes uncertainty in the outputs of the FLIP-FLOP which is eliminated by using edge-triggered FLIP-FLOPs.

In the case of an edge-triggered FLIP-FLOP, the transfer of information from data input(s) to the output of the FLIP-FLOP occurs at the positive (or negative) edge of the clock pulse. The only time the outputs can change state is during the brief interval of time when the clock signal is making a transition from the 0 to 1 (↑), or in some circuits, from the 1 to 0 (↓) states. A FLIP-FLOP which responds only to rising (or falling) edge is referred to as *positive-edge-triggered* (or *negative-edge-triggered*). The data lock-out occurs at the end of the edge.

The logic symbol used for an edge-triggered FLIP-FLOP is the same as that of a master–slave FLIP-FLOP .

The timing specifications of an edge-triggered FLIP-FLOP are illustrated in Fig. 7.21 and are explained below.

Set-up time It is the time required for the input data to settle in before the triggering edge of the clock.

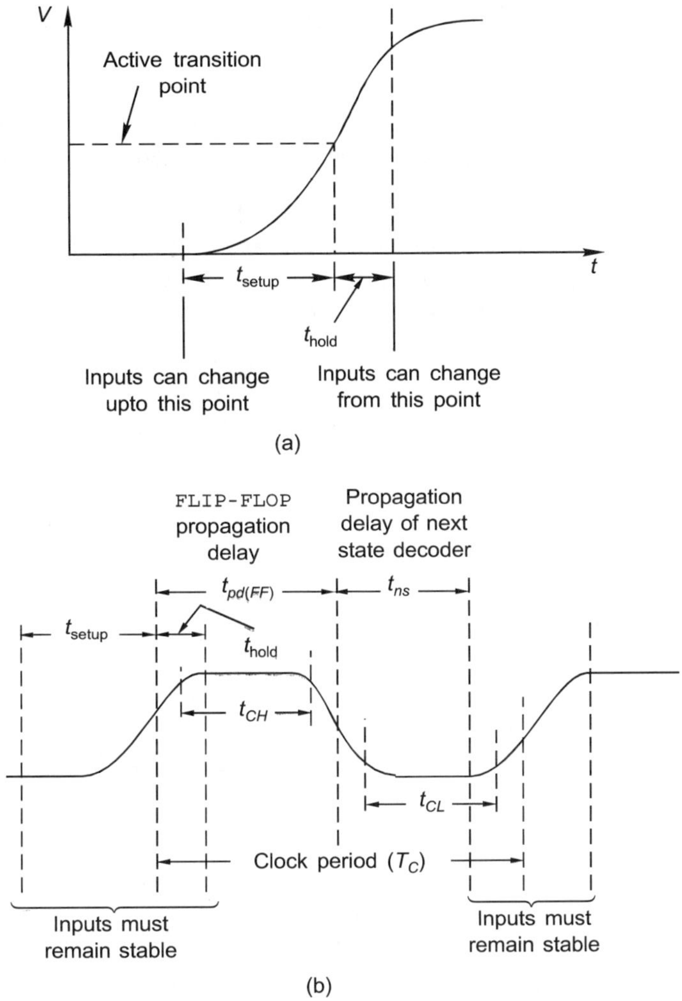

Fig. 7.21

Typical clock waveforms to illustrate timing specifications of FLIP-FLOPs.

Hold time It is the time for which the data must remain stable after the triggering edge of the clock.

Maximum frequency This is the highest frequency of clock pulses for proper and stable operation. It is given by

$$f_{\max} = \frac{1}{T_c} \leq \frac{1}{t_{\text{setup}} + t_{\text{pd(FF)}} + t_{\text{ns}}} \tag{7.2}$$

Clock HIGH pulse width (t_{CH}) This is the minimum time for which the clock must remain HIGH.

Clock LOW pulse width (t_{CL}) This is the minimum time for which the clock must remain LOW.

Example 7.3 The clock (Fig. 7.22*a*) and input (Fig. 7.22*b*) waveforms are applied to *D* or *J* input of each of the following type of FLIP-FLOPs. Sketch the output waveform in each case.

(a) Positive-edge-triggered *D*-type FLIP-FLOP 7474.
(b) Positive-level-triggered *D*-type FLIP-FLOP 7475 (transparent latch).
(c) Negative-edge-triggered *J–K* FLIP-FLOP ($K = 1$) 74112.
(d) Master–slave *J–K* FLIP-FLOP ($K = 1$) 7476.

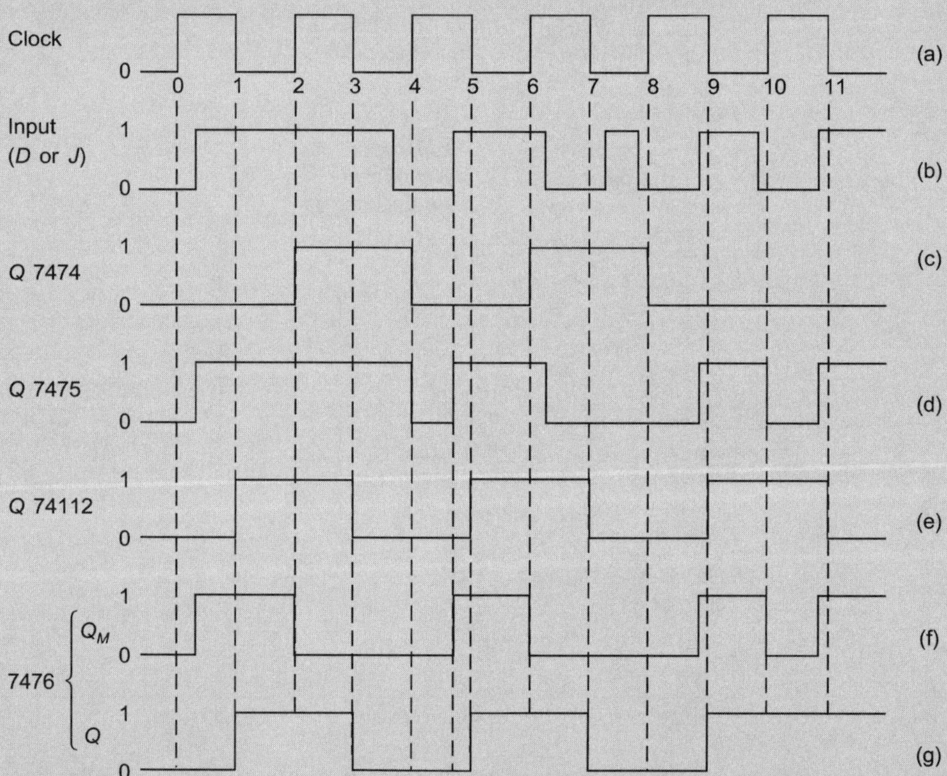

Fig. 7.22
Waveforms of Ex. 7.3.

Solution

(a) In the case of positive-edge-triggered *D*-type FLIP-FLOP the output *Q* is same as the input *D* at the positive edge of the clock pulse. The output does not change till the next positive edge arrives. The output (*Q*) waveform is shown in Fig. 7.22*c*.
(b) 7475 is a transparent latch, i.e. the output (*Q*) follows the input (*D*) as long as $CK = 1$. The output does not change when $CK = 0$. The output (*Q*) waveform is shown in Fig. 7.22*d*.

(c) In the case of negative-edge-triggered *J–K* FLIP-FLOP the output (*Q*) responds to the *J* and *K* inputs present at the negative edge of the clock pulse (according to *J–K* truth table). The output does not change till the arrival of the next negative edge. The output (*Q*) waveform is shown in Fig. 7.22*e*.

(d) In the case of the master–slave *J–K* FLIP-FLOP, the output of the master responds to the *J* and *K* inputs present when $CK = 1$ (according to *J–K* truth table). The output of master (Q_M) is shown in Fig. 7.22*f*.

The output of the slave (*Q*) follows Q_M at the negative edge of the clock-pulse. The output (*Q*) waveform is shown in Fig. 7.22*g*.

7.10 APPLICATIONS OF FLIP-FLOPs

Some of the common uses of FLIP-FLOPs are as:

1. Bounce elimination switch,
2. Latch,
3. Registers,
4. Counters,
5. Memory, etc.

Some examples of the uses of FLIP-FLOPs are given below.

7.10.1 Bounce-Elimination Switch

Mechanical switches are employed in digital systems as input devices by which digital information (0 or 1) is entered into the system. There is a very serious problem associated with these switches, viz. *switch-bouncing* (or *chattering*). When the arm of the switch is thrown from one position to another, it chatters or bounces several times before finally coming to rest in the position of contact. The bounce is the result of the spring-loaded impact of the switch throw contact and the pole contacts.

In a sequential circuit, if a 1 is to be entered through a switch, then the switch is thrown to the corresponding position. As soon as it is thrown to this position, the output is 1 but the output oscillates between 0 and 1 for some time due to make and break (bouncing) of the switch at the point of contact before coming to rest. This changes the output of the sequential circuit and creates difficulties in the operation of the system. This problem is eliminated by using bounce-elimination switches.

Example 7.4 Show that the circuit of Fig. 7.23*a* acts as a bounce-elimination (chatterless) switch.

Solution

The waveforms at $\bar{S}$, $\bar{R}$, *Q*, and $\bar{Q}$ are illustrated in Fig. 7.23*b*. The switch (SW) is thrown from position *A* to *B* at $t = 0$. Therefore, at $t = 0^+$ the voltage at $\bar{S}$ will be V_{CC} (logic 1) and will continue to remain so as long as the switch is not thrown to position *A* again.

At $\bar{R}$(*B*), the voltage at $t = 0^-$ is V_{CC} (logic 1) and goes to 0 *V* (logic 0) at t_1 (t_1 being the time delay of the switch). The switch arm makes contact at *B* at $t = t_1$, and then

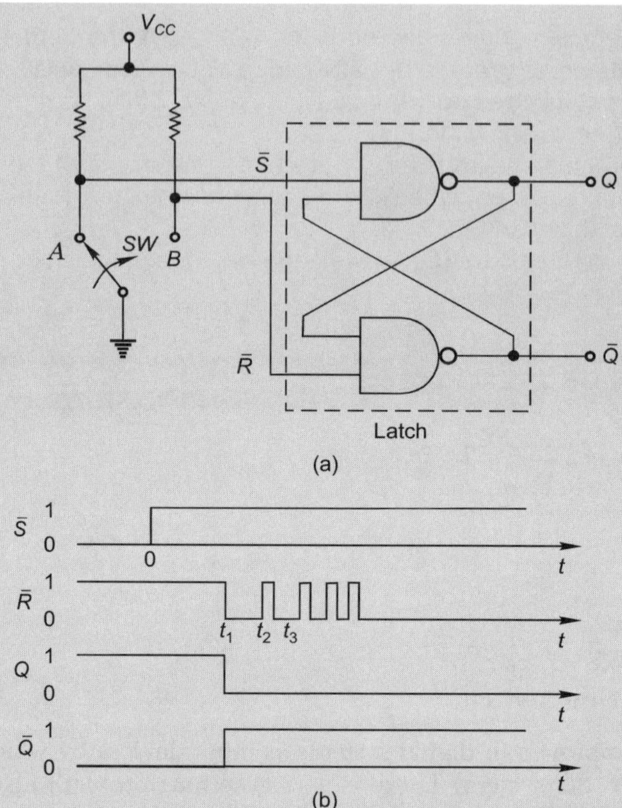

Fig. 7.23

(a) A bounce-elimination switch (b) Waveforms of $\overline{S}$, $\overline{R}$, Q, and $\overline{Q}$.

bounces off. Therefore, the level at $\bar{R}$ changes from 0 to 1 and vice-versa. This is illustrated in Fig. 7.23*b*.

Between $t = 0$ and $t = t_1$, both the inputs $\bar{S}$ and $\bar{R}$ are at logic 1 and therefore, Q does not change. The output Q changes at t_1 and becomes 0. Now, even when $\bar{R}$ is changing at t_2, t_3, etc. Q does not change. This shows that it is a chatterless switch.

The latch used in Fig. 7.23*a* can be replaced by the IC 74279 which is a quad $\bar{S} - \bar{R}$ latch.

7.10.2 Registers

A register is composed of a group of FLIP-FLOPs to store a group of bits (word). For storing an *N*-bit word, the number of FLIP-FLOPs required is *N* (one FLIP-FLOP for each bit). A 3-bit register using 7474 positive-edge-triggered FLIP-FLOPs is shown in Fig. 7.24. The bits to be stored are applied at the *D*-inputs which are clocked in at the leading-edge of the clock pulse. In this register, the data to be entered must be available in parallel form. Other types of registers will be discussed in Chapter 8.

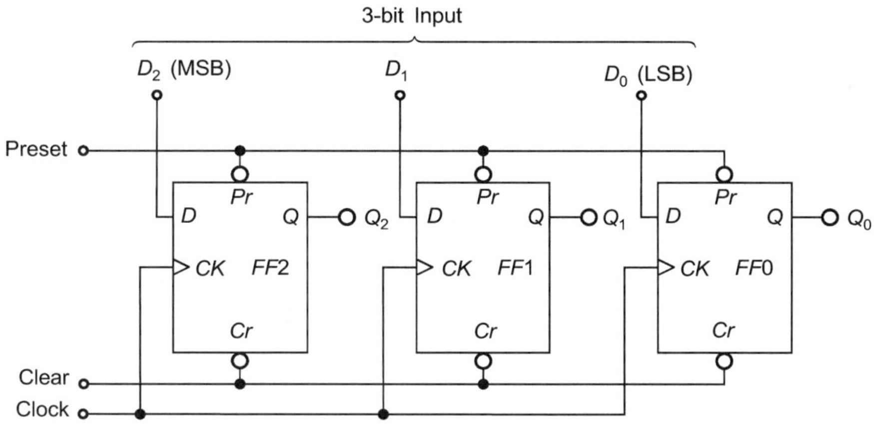

Fig. 7.24

A 3-bit register using FLIP-FLOPs.

7.10.3 Counters

Digital counters are often needed to count events. For example, counting the number of tablets filled in a vial. Electrical pulses corresponding to the event are produced using a transducer and these pulses are counted using a counter.

The counters are composed of FLIP-FLOPs. A 3-bit counter consisting of three FLIP-FLOPs is shown in Fig. 7.25. A circuit with *n*- FLIP-FLOPs has 2^n possible states. Therefore, the 3-bit counter can count from decimal 0 to 7.

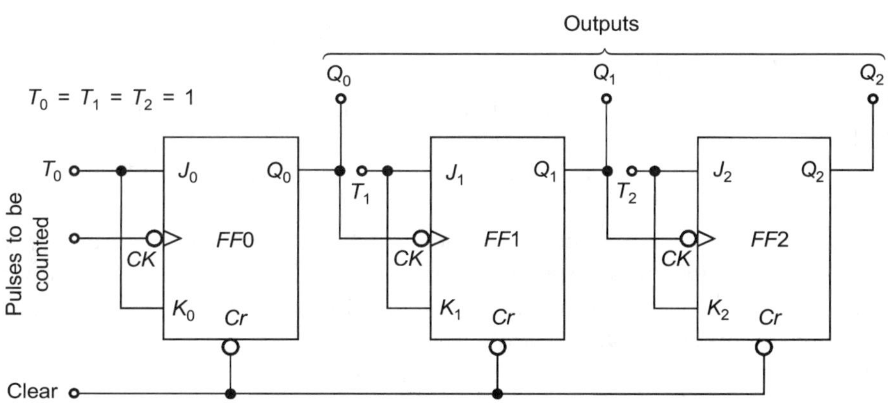

Fig. 7.25

A 3-bit counter using FLIP-FLOPs

The FLIP-FLOPs used are 74107 *J–K* master–slave FLIP-FLOPs, used as *T*-type. The pulses to be counted are connected at the clock input of FF0. The Q_0 output of FF0 is connected to the clock input of FF1 and similarly Q_1 is connected to the clock input of FF2.

The FLIP-FLOPs are cleared by applying logic 0 at the clear input terminal momentarily. For normal counting operation, it is to be maintained at logic 1. The pulses and the output waveforms are illustrated in Fig. 7.26.

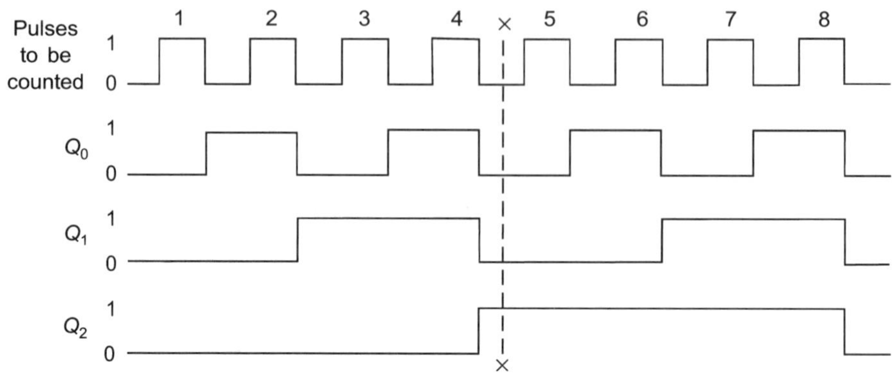

Fig. 7.26
Waveforms of counter of Fig. 7.25.

The output Q_0 of the least-significant stage changes at the negative edge of each pulse (since $T_0 = 1$). The output Q_1 changes at the negative edge of each Q_0 pulse (since Q_0 acts as *CK* for FF1 and $T_1 = 1$) and the output Q_2 changes at the negative edge of each Q_1 pulse (since Q_1 acts as *CK* for FF2 and $T_2 = 1$).

At any time, the decimal equivalent of the binary number $Q_2Q_1Q_0$ is the number of pulses counted till that time. For example, at × the count is 100 (decimal 4). The circuit resets after counting eight pulses.

There are various types of counters, some of which will be discussed in the next chapter.

7.10.4 Random-Access Memory

In computers, digital control systems, information processing systems, etc. it is necessary to *store* digital data and *retrieve* the data as desired. For this purpose, earlier only magnetic memory devices were possible, whereas these days it has become possible to make memory devices using semiconductor devices. Semiconductor memories have become very popular because of their small size (available in ICs) and convenience to use. Chapter 11 deals with various semiconductor memories.

FLIP-FLOPs can be used for making memories in which data can be stored for any desired length of time and then read out whenever required. In such a memory, data can be put into (*writing* into the memory) or retrieved from (*reading* from the memory) the memory in a random fashion and is known as *random-access memory.*

A 1-bit read/write memory is shown in Fig. 7.27 which is the basic memory element and memory ICs are built around a system of basic 1-bit cell.

In this memory cell, a level *D* FLIP-FLOP is used which has *Q* output that follows the *D* input as long as *CK* terminal is at logic 1. The moment the *CK* input changes to logic 0, the *Q* output does not change and it retains the *D* input level that existed just before the transition from 1 to 0 at input *CK*. This input is used to select the memory cell. In the 1-bit cell shown there are three inputs – D_i (data input), A_n (address select) and $R/\overline{W}$ (read/write control) and one output D_0 (data output). $A_n = 1$ enables the cell for reading or writing operation, $R/\overline{W}$ at logic 1 is for reading from the cell and logic 0 for writing into the cell. As long as $A_n = 0$, all

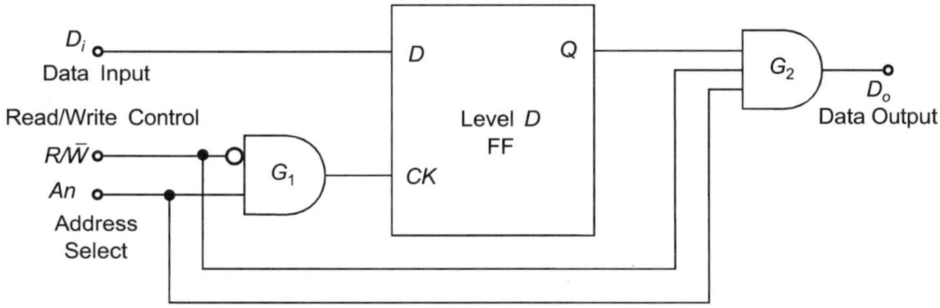

Fig. 7.27

A 1-bit read/write memory cell.

input and output activities are blocked, and the cell is in the hold mode where its stored output is protected.

The complete function of this cell can be understood from the function table of Table 7.9. The read operation is nondestructive, that is, the stored bit can be read out any number of times without disturbing it. The stored bit will be protected as long as power is on. Therefore, this type of memory is known as *volatile memory*.

Table 7.9 Function table of 1-bit memory cell

Inputs			*Mode*
A_n	$R/\overline{W}$	D_i	
0	×	×	Hold, $D_0 = 0$
1	0	0	Write 0 into memory, $D_0 = 0$
1	0	1	Write 1 into memory, $D_0 = 0$
1	1	×	Read, D_0 = stored D_i bit.

As far as writing into the cell is concerned, it is not required to be cleared before entering the new bit. Whenever a new bit is entered the earlier one gets destroyed automatically.

7.11 SUMMARY

The basic element of sequential circuits, FLIP-FLOP has been introduced here. This is a basic memory element which is used to store 1-bit of digital information. The four types of commonly used FLIP-FLOPs *S–R*, *J–K*, *T*-type, and *D*-type have been covered in detail, including their design using gates.

The various triggering systems have been discussed which will be very helpful in understanding the detailed operation of the FLIP-FLOP and other circuits containing these FLIP-FLOPs.

Simple examples of the uses of FLIP-FLOPs in registers, counters, memory elements, etc. have been included. These topics will be covered in greater detail in later chapters.

Glossary

Asynchronous sequential circuit A sequential circuit whose behaviour depends upon the sequence in which the input signals change. It is event driven and does not require clock pulses.

Bit time Amount of time to transmit a single bit.

Bouncing Moving back and forth between two states before reaching a final state. Same as chattering.

Bounce-elimination circuit A circuit that eliminates the effect of switch bouncing.

Characteristic table A table describing operation of a FLIP-FLOP.

Chatterless switch A switch in which the bouncing effect has been eliminated.

Clear Setting the contents of a FLIP-FLOP or a circuit containing FLIP-FLOPs or a memory to zero.

Clear input The input used for clearing a digital circuit.

Clock A train of pulses of usually constant frequency that synchronize the operation of any synchronous sequential circuit including a microprocessor based system.

Clock cycle The interval between successive positive or negative transitions in a clock.

Clocked FLIP-FLOP A FLIP-FLOP that responds to the data inputs only on the occurrence of the appropriate clock signal.

Clock frequency The number of clock cycles per second.

Clocked sequential circuit The sequential circuits whose operation is synchronized with the application of clock pulses, between which no changes of state occur.

Counter A digital circuit that can count the number of pulses.

Data Information in digital (binary) form.

Debouncing switch Same as chatterless switch.

D-type FLIP-FLOP A FLOP-FLOP whose output follows the input when triggered.

Edge-triggered FLIP-FLOP A FLIP-FLOP whose state changes on the rising (positive) or falling (negative) edge of a clock pulse.

Excitation table A tabular representation of the operation of a FLIP-FLOP.

Falling edge A transition from high to low voltage.

Frequeucy The repetition rate of a cyclic signal. It is expressed in Hertz (Hz).

Hold time In FLIP-FLOPs and memories, a minimum amount of time after the application of a clock (or Read/write) signal, when data(s) or address inputs must remain stable.

J-K FLIP-FLOP A FLIP-FLOP with inputs J and K. The state of the FLIP-FLOP does not change when $J = K = 0$, whereas it changes with every clock pulse when $J = K = 1$. The FLIP-FLOP is set when $J = 1$ and $K = 0$ and reset when $J = 0$ and $K = 1$. All the above operations are performed in synchronism with the clock.

Latch A temporary storage device consisting of D-type FLIP-FLOPs. Its contents are fixed at their current values by a transition of the clock and remain fixed until the next clock transition occurs.

Leading edge A transition from low to high voltage.

Level-triggereed FLIP-FLOP A FLIP-FLOP that is triggered (i.e. the outputs respond to the inputs) When the level of the clock signal is appropriate.

Master-slave FLIP-FLOP A FLIP-FLOP consisting of the cascade of two FLIP-FLOPs; the first FF is the master and the second FF is the slave. The master FF is triggered when the clock is HIGH and the slave FF follows the master FF when the clock goes LOW.

Preset The state of a FLIP-FLOP when $Q = 1$ and $\bar{Q} = 0$

Preset input The input used for setting a FLIP-FLOP.

Pulse-triggered FLIP-FLOP A FLIP-FLOP that requires a clock pulse for triggering. It is same as the master-slave FF.

Random-access memory RAM The ability to address a semiconductor memory for read-and-write operation in which any memory location can be accessed at random. It is same as Read and Write memory.

Read (memory) Process of getting (retrieving) a word from a memory.

Register An array of FLIP-FLOPs used to store binary information.

Reset Same as clear.

Reset input Same as the clear input

Rising edge Same as the leading edge.

Set $Q = 1$ state of a FLIP-FLOP.

Setup time The time required for the input data to settle in before the triggering edge of the clock pulse.

S-R FLIP-FLOP A FLIP-FLOP with inputs S and R. The state of the FF does not change when $S = R = 0$. It is set when $S = 1$ and $R = 0$; reset when $S = 0$ and $R = 1$. $S = R = 1$ is not allowed.

Stable state A state in which a digital circuit remains forever unless changed by a triggering signal.

State The value of FLIP-FLOP(s) output in a digital circuit.

Switch bouncing Fluctuations in the switch positions between ON and OFF when the switch position is changed.

Synchronous sequential circuit A sequential circuit whose operation is synchronized with the application of clock pulses, between which no change of state can occur.

Toggling Causing as FF to change its state.

Trailing edge Same as the falling edge.

Trigger To cause change of state.

T-type FLIP-FLOP A FLIP-FLOP with one data input (T) which changes state for every clock pulse if $T = 1$ and does not change the state if $T = 0$.

Unstable state The state which is not a stable state. The circuit comes out of this state without applying any triggering signal.

Write (memory) Process of placing (storing) a word into a specific memory location.

Review Questions

7.1 FLIP-FLOP is a __________ element.

7.2 Number of FLIP-FLOPs required for storing *n*-bit of information is __________ .

7.3 In an S–R FLIP-FLOP $S = R = 1$ __________ permitted.

7.4 'Preset' and 'Clear' inputs are used in a FLIP-FLOP for making $Q =$ __________ and __________ respectively.

7.5 In a J-K FLIP-FLOP if $J = K = 1$, its Q output will be __________ when a clock pulse is applied.

7.6 Master-slave configuration is used in a J-K FLIP-FLOP to eliminate __________ .

7.7 In a T FLIP-FLOP, the Q output __________ when $T = 0$ and clock pulse is applied.

7.8 A FLIP-FLOP has __________ states.

7.9 A latch is used to store 1 __________ of data.

7.10 A negative edge-triggered FLIP-FLOP changes state at the __________ of the clock pulse.

7.11 An active low 'Clear' input clears the FLIP-FLOP when it is __________ .

7.12 A FLIP-FLOP with active-low 'preset' input will have $\bar{Q} =$ __________ when preset is connected to LOW.

7.13 A chatterless switch can be implemented using a __________ .

7.14 A tabulation specifying inputs required for a FLIP-FLOP to change from a present state to a specified next state is known as __________ .

7.15 Registers and Counters can be designed using __________ .

Problems

7.1 Show that the circuit of Fig. 7.4 with $S = R = 0$ is the same as that of Fig. 7.3.

7.2 In a circuit of Fig. 7.4 if the inputs change from
(a) $S = 1, R = 0$ to $S = R = 0$, and
(b) $S = 0, R = 1$ to $S = R = 0$,
show that the outputs do not change.

7.3 Design an S–R latch using two 2-input NOR gates

7.4 In the FLIP-FLOP circuit of Fig. 7.28 show that if
(a) $Pr = 0$ and $Cr = 1$, then $Q = 1$ (independent of S, R, and CK).
(b) $Pr = 1$ and $Cr = 0$, then $Q = 0$ (independent of S, R, and CK).
(c) $Pr = Cr = 1$, then it functions as a clocked S–R FLIP-FLOP.

7.5 Verify Table 7.3a.

7.6 Determine the output Y_1 of Fig. 7.29*a* and Y_2 of Fig. 7.29*b* and show that $Y_1 = Y_2$.

7.7 Identify Q and $\bar{Q}$ outputs of the clocked J–K FLIP-FLOP shown in Fig. 7.30.

7.8 For the circuit shown in Fig. 7.31 the clock and input waveforms shown in Fig. 7.32 are applied. Sketch the waveforms of Q and $\bar{Q}$ if the FLIP-FLOP is edge-triggered.

7.9 Repeat Problem 7.8 if the FLIP-FLOP is master–slave.

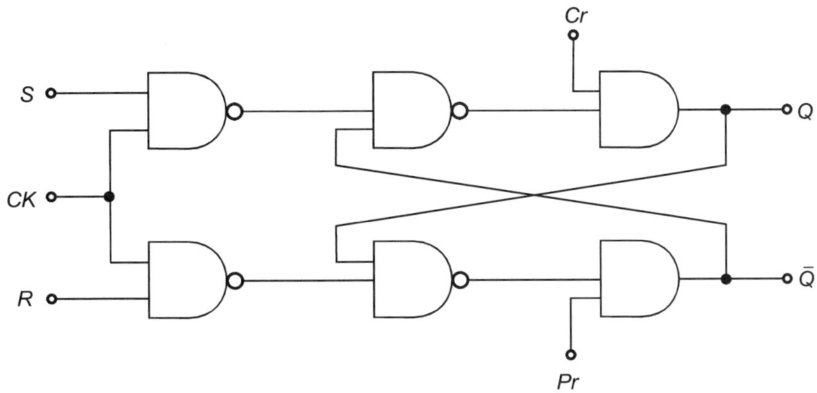

Fig. 7.28

FLIP-FLOP circuit for Problem 7.4.

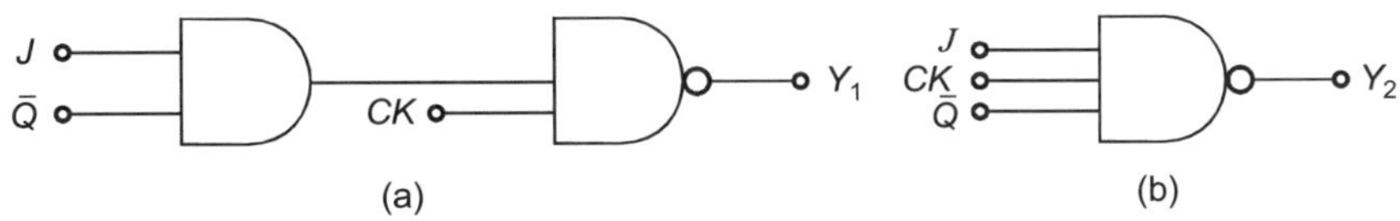

Fig. 7.29

Figures for Problem 7.6.

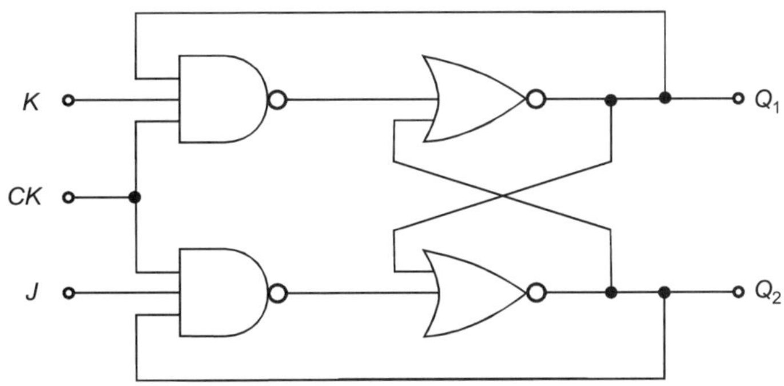

Fig. 7.30

Figure for Problem 7.7.

7.10 If the waveforms shown in Fig. 7.33 are applied to the circuit of Fig. 7.31, sketch the output (Q) waveform. Assume *M–S* FLIP-FLOP.

7.11 Verify that the circuit of Fig. 7.16 acts as a toggle switch.

7.12 Prepare the truth table for the circuit of Fig. 7.34 and show that it acts as a *T*-type FLIP-FLOP.

7.13 If $\bar{Q}$ output of a *D*-type FLIP-FLOP is connected to *D* input, it acts as a toggle switch. Verify.

7.14 Using the method outlined in Section 7.8, design the following FLIP-FLOPs:
(a) *J–K* (b) *D*-type (c) *T*-type

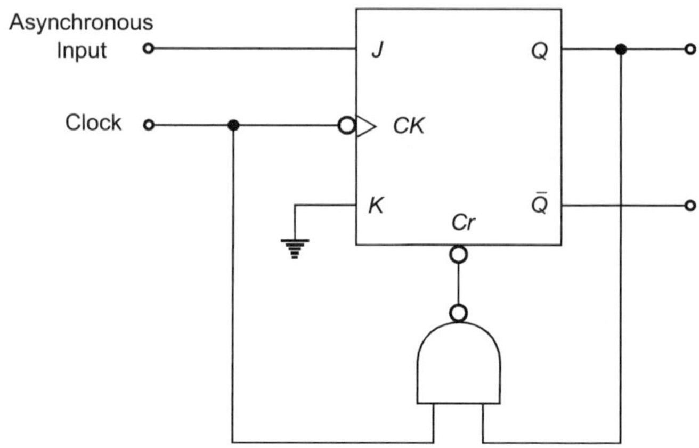

Fig. 7.31

Figure for Problem 7.8.

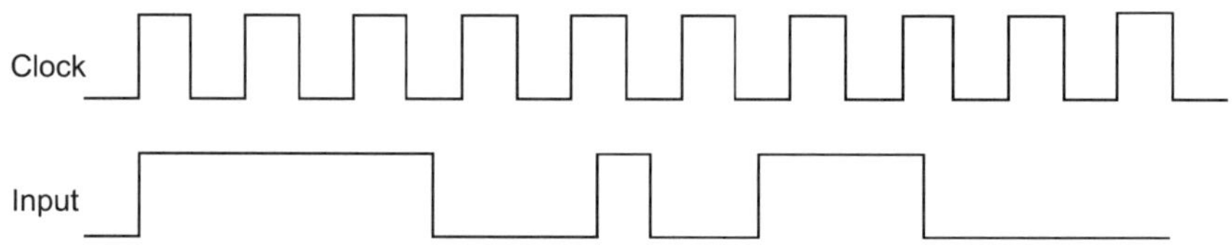

Fig. 7.32

Waveforms for Problem 7.8.

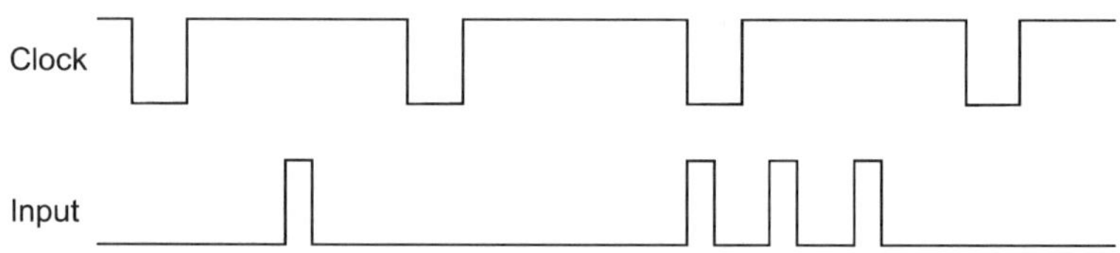

Fig. 7.33

Waveforms for Problem 7.10.

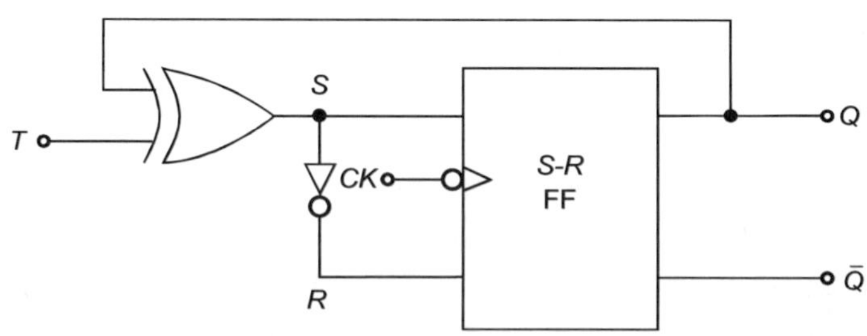

Fig. 7.34

Circuit for Problem 7.12.

7.15 Using the conversion method outlined in Section 7.8, carry out the following conversions:
(a) *S–R* to *D* (b) *J–K* to *D* (c) *D* to *J–K* (d) *S–R* to *T* (e) *J–K* to *T*
(f) *T* to *J–K* (g) *T* to *D* (h) *D* to *S–R* (i) *D* to *T* (j) *T* to *S–R*
(k) *J–K* to *S–R*

7.16 Figure 7.35 shows a positive-edge-triggered *D* FLIP-FLOP. Verify its operation.

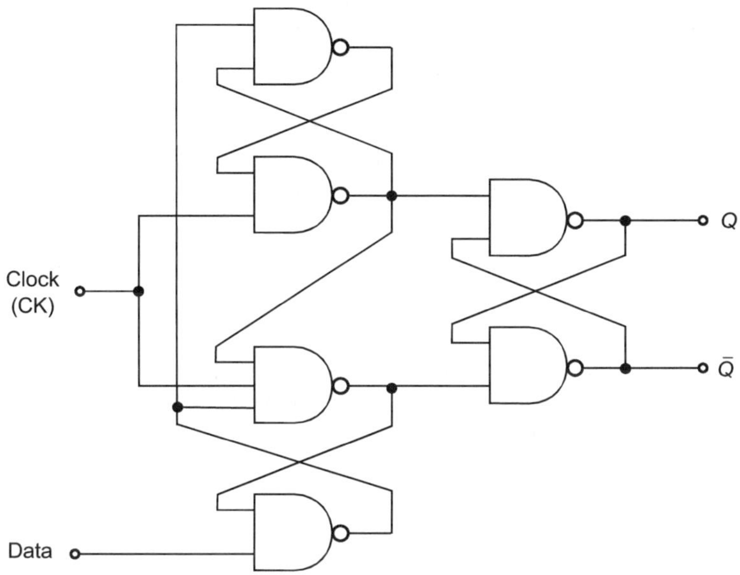

Fig. 7.35

Positive-edge-triggered *D* FLIP-FLOP.

7.17 IC 74111 is a master–slave *J–K* FLIP-FLOP with data lock out at the positive edge of the clock (the state changes at the negative edge of the clock).
If the waveform shown in Fig. 7.22*a* is applied at the clock input and the waveform of 7.22*b* is applied at the *J* input, sketch the output (Q) waveform when
(a) *K* is connected to logic 0.
(b) *K* is connected to logic 1.

7.18 Verify the operation of the debounce switches shown in Fig. 7.36.

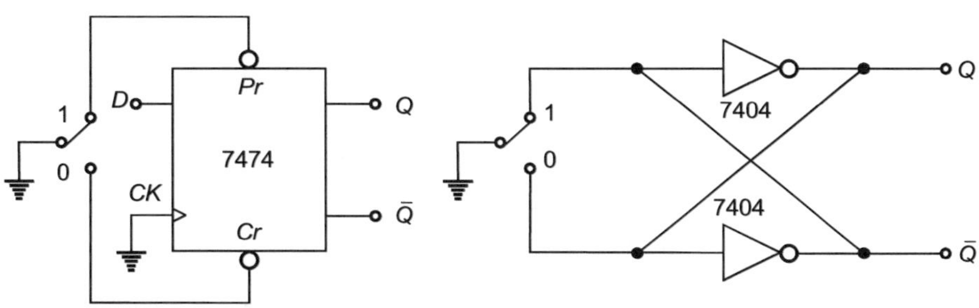

Fig. 7.36

Debounce switches for Problem 7.18.

7.19 Consider the circuit of Fig. 7.37 consisting of positive-edge-triggered FLIP-FLOPs. Δt_1 and Δt_2 are the time delays introduced in the clock (CLOCK SKEW) by buffer devices and the propagation delay of wires. At the rising edge of the clock, new data enters the source FF and the content of source FF enters the destination FF. Show the effect of clock skew ($\Delta t_2 > \Delta t_1$) on the operation of the circuit. Discuss the problems created by clock skew for data transmission.

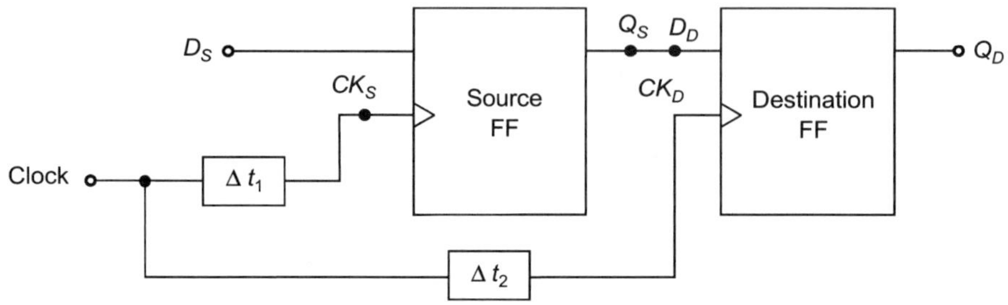

Fig. 7.37
Circuit for Problem 7.19.

7.20 A mod-3 counter (resets after every three pulses) is shown in Fig. 7.38. The FLIP-FLOPs used are master–slave *J–K*. Sketch the waveforms of Q_0 and Q_1 when clock pulses are applied and verify its operation. Assume $Q_0 = Q_1 = 0$ initially.

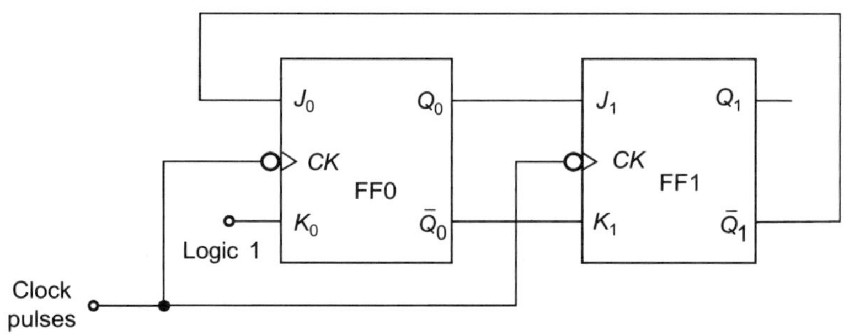

Fig. 7.38
A mod-3 counter.

7.21 In the circuit shown in Fig. 7.39, the time constant (RC) is very small. Explain the operation of this circuit.

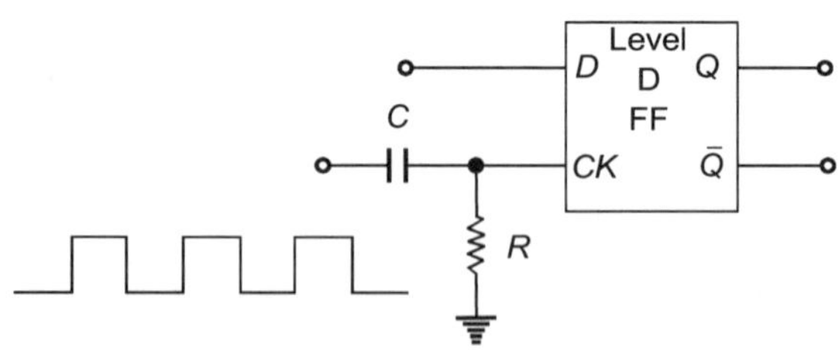

Fig. 7.39
Circuit for Problem 7.21.

8

SEQUENTIAL LOGIC DESIGN

8.1 INTRODUCTION

The FLIP-FLOP is a basic element of sequential logic system. Using FLIP-FLOPs and combinational logic circuits, any sequential logic circuit can be designed. The most important sequential circuits that are widely used in digital systems are registers and counters. These are discussed in detail in the following sections.

The design of registers and counters using FLIP-FLOPs has been introduced followed by standard MSI devices. Some of the common applications of registers and counters have also been discussed.

The design methods for general clocked sequential circuits have also been discussed. Basically, the system may be specified in terms of input–output relationship or/and the sequence of states to be followed when clock pulses are applied. The usual design steps are: reduction of states, state assignment and next-state decoder design.

The synchronous counter is a special case of the general clocked sequential circuit and can be designed using the design methods developed for clocked sequential circuits.

Asynchronous sequential circuits do not use clock pulses and their response depends upon the sequence in which the input signal change. These circuits use delay circuits as memory elements. The delay is inherent in logic circuits and the feedback introduced in combinational circuits provides memory capability to these circuits. The S-R FLIP-FLOPs fall under this category of delay circuits which are extensively used in asynchronous sequential circuits. The analysis and design of asynchronous sequential circuits have been discussed in detail.

8.2 REGISTERS

As discussed in Chapter 7, a FLIP-FLOP can store (or remember) 1-bit of digital information (1 or 0). It is also referred to as a 1-bit *register.* An array of FLIP-FLOPs is required to store binary information, the number of FLIP-FLOPs required being equal to the number of bits in the binary word (one FLIP-FLOP for each bit) and is referred to as a register. Registers find application in a variety of digital systems including microprocessors. For example, Intel's 8085 microprocessor chip contains seven 8-bit registers, referred to as *general purpose registers* and five 1-bit registers, referred to as *flags.*

The data can be entered in *serial* (one-bit at a time) or in *parallel* form (all the bits simultaneously) and can be retrieved in the serial or parallel form. Data in serial form is also referred to as a *temporal code* (a time arrangement of bits) and in parallel form as *spacial code.* A 4-bit data 1010 in serial form is shown in Fig. 8.1*a* and in parallel form in Fig. 8.1*b*. For serial input/output, only one line is required for data input and one line for data output, whereas the number of lines required is equal to the number of bits for parallel input/output.

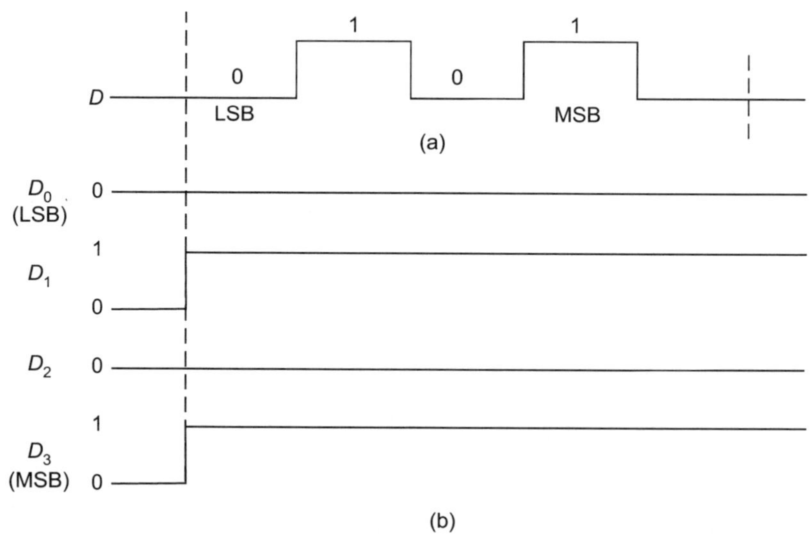

Fig. 8.1

Data representation in (*a*) Serial form (*b*) Parallel form.

Registers are classified depending upon the way in which data are entered and retrieved. There are four possible modes of operation:

1. Serial-in, serial-out (SISO),
2. Serial-in, parallel-out (SIPO),
3. Parallel-in, serial-out (PISO), and
4. Parallel-in, parallel-out (PIPO).

Registers can be designed using discrete FLIP-FLOPs (*S-R* or *J-K* as *D*-type) and are also available as MSI devices. The registers available in 54/74 TTL series are given in Table 8.1.

Table 8.1 Shift registers available in 54/74 TTL series

IC No.	*Description*
7491, 7491A	8-bit serial-in, serial-out
7494	4-bit parallel-in, serial-out
7495	4-bit serial/parallel-in, parallel-out (right-shift, left-shift)
7496	5-bit parallel-in/parallel-out, serial-in/serial-out
7499	4-bit bi-directional (universal)
74164	8-bit serial-in, parallel-out
74165	8-bit serial/parallel-in, serial-out
74166	8-bit serial/parallel-in, serial-out
74178, 74179	4-bit bi-directional (universal)
74194	4-bit bi-directional (universal)
74195	4-bit serial/parallel-in, parallel-out
74198	8-bit bi-directional (universal)
74199	8-bit serial/parallel-in, parallel-out
74295A	4-bit TRI-STATE serial/parallel-in, parallel-out bi-directional
74395	4-bit TRI-STATE cascadable serial/parallel-in, serial/parallel-out.

Registers in which data are entered or/and taken out in serial form are referred to as *shift registers*, since bits are shifted in the FLIP-FLOPs with the occurrence of clock pulses either in the right direction (*right-shift register*) or in the left direction (*left-shift register*). In the *bi-directional shift register*, data can be shifted from left to right as well as in the reverse direction, using the mode control. For example, IC 74295A is a bi-directional shift register.

A register is referred to as a *universal register* if it can be operated in all the four possible modes and also as a bi-directional register. For example, 74194 is a universal register.

8.2.1 Shift Register

A 5-bit shift register using five master-slave *S-R* (or *J-K*) FLIP-FLOPs is shown in Fig. 8.2. This circuit can be used in any of the four modes. The operation of this circuit is explained by

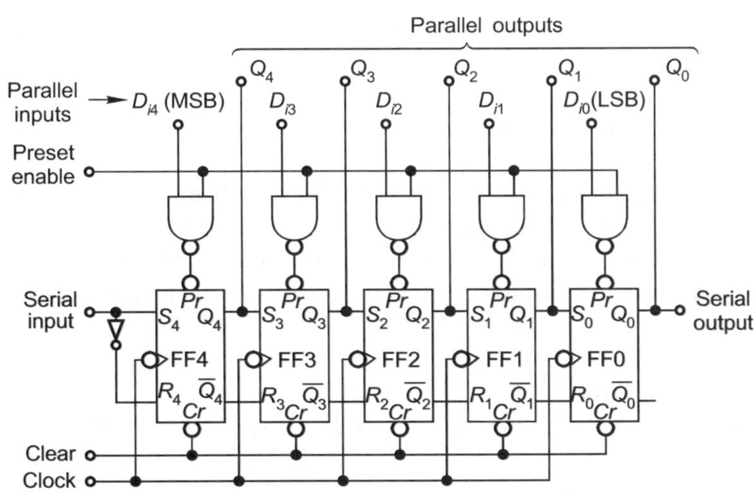

Fig. 8.2

A 5-bit shift register (7496).

assuming the 5-bit data 10110. For any other 5-bit data, the operation will be similar to the one explained.

Serial Input

The data word in the serial form (Fig. 8.1*a*) is applied at the serial input after clearing the FLIP-FLOPs using the clear line. The preset enable is to be held at 0 so that *Pr* for every FLIP-FLOP is 1. The input and output waveforms are illustrated in Fig. 8.3.

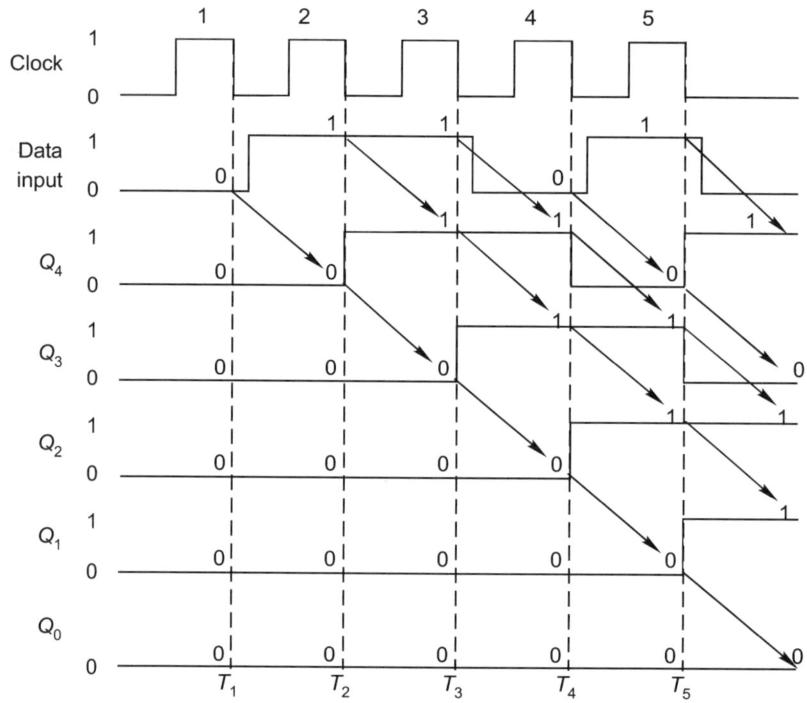

Fig. 8.3
Waveforms of shift register for serial input.

The process of entering the digital word starts with the data input corresponding to the least-significant bit (0) at the serial input and first clock pulse. At the falling edge (T_1) of the first clock pulse the output of FF4 (Q_4) will be 0 and the outputs of all other FLIP-FLOPs are 0 since their inputs are 0. Next, the input corresponding to next bit is applied and, at the falling edge (T_2) of the second clock pulse, the FLIP-FLOP outputs will be

$$Q_4 = 1$$
$$Q_3 = Q_2 = Q_1 = Q_0 = 0$$

Similarly, the input corresponding to each bit is applied till the MSB and the bits go on shifting from left to right at the falling edge of each clock pulse as illustrated in Fig. 8.3. At the end of fifth clock pulse, the outputs of FLIP-FLOPs are

$$Q_4 = 1$$

$$Q_3 = 0$$
$$Q_2 = 1$$
$$Q_1 = 1$$
$$Q_0 = 0$$

which is the same as the number to be stored. The number of clock pulses required for entering the data, is the same as the number of bits. The process of entering the data is also referred to as *writing* into the register.

The data stored can be retrieved (also referred to as *reading*) in two ways: serial-out and parallel-out. The data in the serial form is obtained at Q_0 when clock pulses are applied. The number of clock pulses required will be same as the number of bits (five in this case).

In the parallel form, the data is available at Q_4 Q_3 Q_2 Q_1 Q_0 and clock is not required for reading. In the case of serial output, after the *n*th clock pulse, for an *n*-bit word, each FLIP-FLOP output is 0. This means that once the data is retrieved the register is empty. On the other hand, in the case of parallel output, the contents of the register can be *read* any number of times until new data is stored in the register.

The clock rate may be different for the input data and the output data in case of a serial-in, serial-out shift register. Hence, this method can be used for changing the spacing in time of a binary code which is referred to as *buffering*.

Parallel Input

Data can be entered in the parallel form making use of the preset inputs. After clearing the FLIP-FLOPs, if the data lines are connected to the parallel inputs (D_{i4}, D_{i3}, D_{i2}, D_{i1}, and D_{i0}) and a 1 is applied at the preset input, the data are written into the register. This is referred to as *asynchronous loading*.

As explained above, the stored word may be read in the serial form at Q_0 by applying five clock pulses or in the parallel form at Q outputs.

The data can also be entered in parallel form by using *D*-type FLIP-FLOPs connected as shown in Fig. 7.24. In this, the data is loaded when a clock pulse is applied and hence, it is referred to as *synchronous loading*.

Bi-directional Register

There are applications in which shifting data to the right and/or to the left is required. For example, a binary number can be divided by two by shifting it one stage to the right. In this process the least-significant bit is lost (unless additional circuitry is used to preserve it) causing an error of 0.5 if the number is odd. Similarly, a number stored in a shift register can be multiplied by two by shifting it one stage to the left, provided a 1 is not shifted out of the most-significant stage. A 4-bit bi-directional shift register is shown in Fig. 8.4.

When the mode control $M = 1$, all the A AND gates are enabled and the data at D_R is shifted to the right when clock pulses are applied. On the other hand, when $M = 0$, the A gates are inhibited and B gates are enabled allowing the data at D_L to be shifted to the left. M should be changed only when $CK = 0$, otherwise the data stored in the register may be altered.

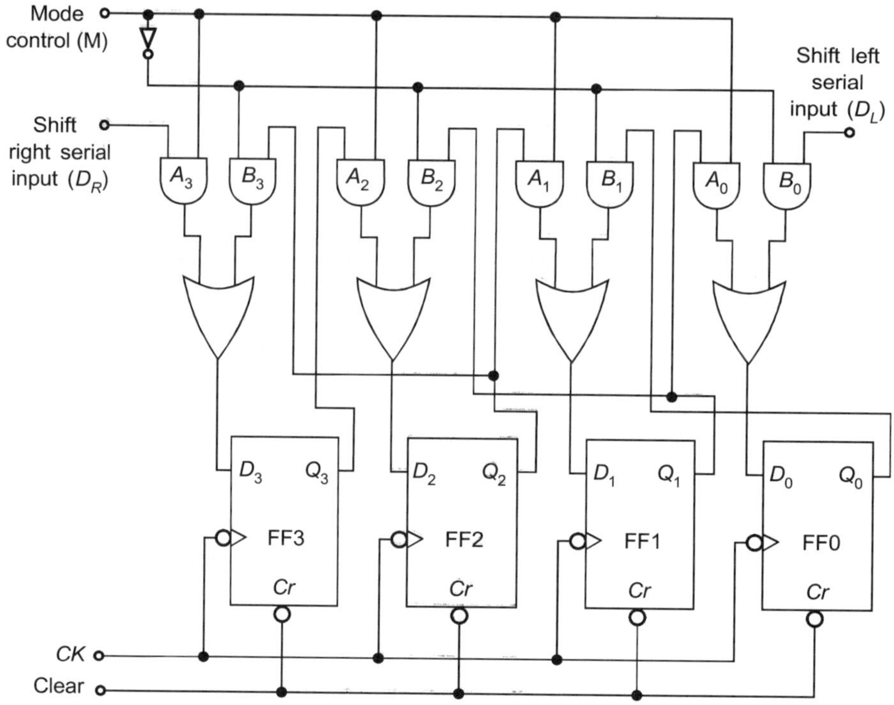

Fig. 8.4
A 4-bit bi-directional shift register.

8.3 APPLICATIONS OF SHIFT REGISTERS

The primary uses of shift registers are temporary data storage and bit manipulations. Some of the common applications of shift registers are discussed below:

8.3.1 Delay Line

A SISO shift register may be used to introduce time delay Δt in digital signals given by

$$\Delta t = N \times \frac{1}{f_C} \tag{8.1}$$

where N is the number of stages and f_C is the clock frequency.

Thus, an input pulse train appears at the output delayed by Δt. The amount of delay can be controlled by the clock frequency or the number of FLIP-FLOPs in the shift register.

8.3.2 Serial-to-Parallel Converter

Data in the serial form can be converted into parallel form by using a SIPO shift register.

8.3.3 Parallel-to-Serial Converter

Data in the parallel form can be converted into serial form by using the PISO shift register.

8.3.4 Ring Counter

If the serial output Q_0 of the shift register of Fig. 8.2 is connected back to the serial input, then an injected pulse will keep circulating. This circuit is referred to as a *ring counter.* The pulse is injected by entering 00001 in the parallel form after clearing the FLIP-FLOPs. When the clock pulses are applied, this 1 circulates around the circuit. The waveforms at the Q outputs are shown in Fig. 8.5. The outputs are sequential non-overlapping pulses which are useful for control-state counters, for stepper motor (which rotates in steps) which require sequential pulses to rotate it from one position to the next, etc.

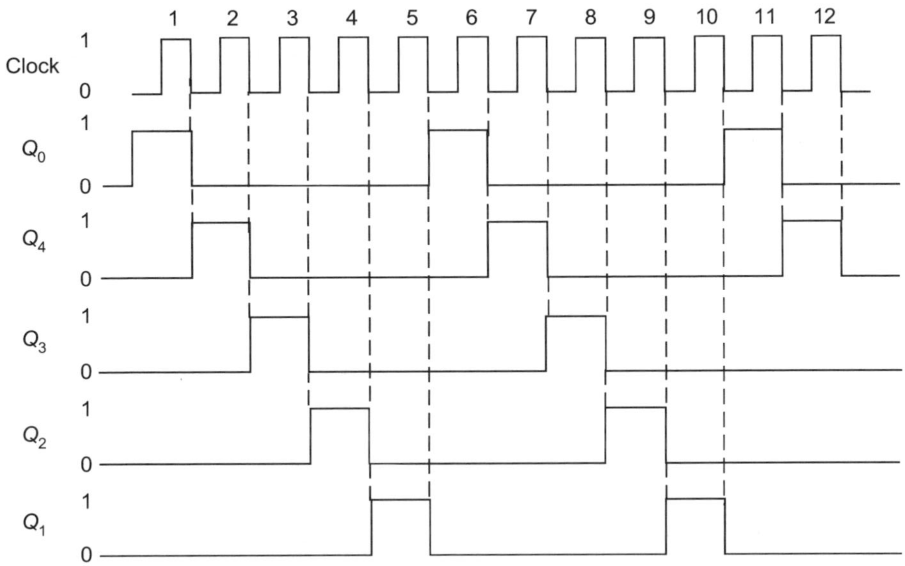

Fig. 8.5

Output waveforms of ring counter.

This circuit can also be used for counting the number of pulses. The number of pulses counted is read by noting which FLIP-FLOP is in 1 state. No decoding circuitry is required. Since there is one pulse at the output for each of the N clock pulses, this circuit is referred to as a *divide-by-N counter* or an *N : 1 scalar.*

8.3.5 Twisted-Ring Counter

In the shift register of Fig. 8.2, if $\bar{Q}_0$ is connected to the serial input, the resulting circuit is referred to as a *twisted-ring, Johnson,* or *moebius* counter. If the clock pulses are applied after clearing the FLIP-FLOPs, square waveform is obtained at the Q outputs as shown in Fig. 8.6.

Similar to ring-counter sequence, the moebius sequence is also useful for control-state counters. It is also useful for the generation of multiphase clock.

The moebius counter is a divide-by-$2N$ counter. For decoding the count, two- input AND gates are required. The decoder circuit for a five-stage counter is shown in Fig. 8.7.

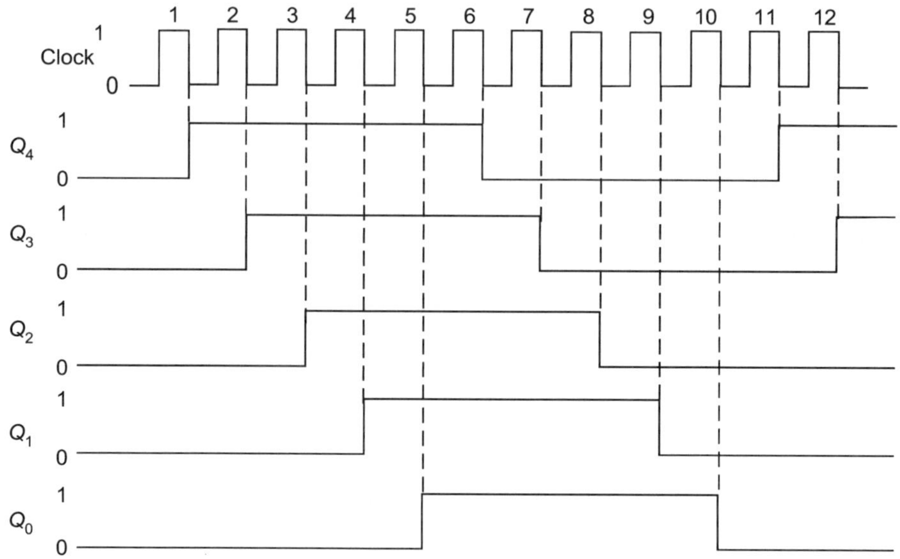

Fig. 8.6
Output waveforms of twisted-ring counter.

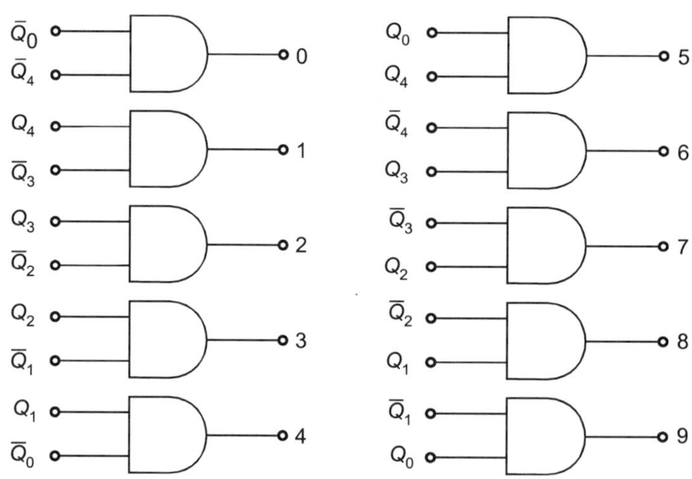

Fig. 8.7
The decoding logic for a 5-stage twisted-ring counter.

8.3.6 Sequence Generator

A circuit which generates a prescribed sequence of bits, in synchronism with a clock, is referred to as a sequence generator. Such generators can be used as

1. Counters,
2. Random bit generators,

3. Prescribed period and sequence generators, and
4. Code generators.

The basic structure of a sequence generator is shown in Fig. 8.8.

The output Y of the next state decoder is a function of $Q_{N-1}\ Q_{N-2}\ \cdots\ Q_1 Q_0$. This system is similar to a ring counter ($Y = Q_0$) or a twisted-ring counter ($Y = \bar{Q}_0$) which are special cases of sequence generators. The design of the decoder will be clear from Example 8.1.

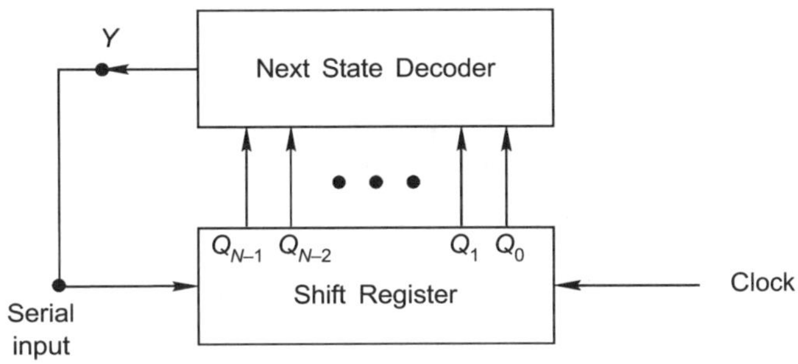

Fig. 8.8

Basic structure of a sequence generator.

Example 8.1 Design a sequence generator to generate the sequence ... 1101011...

Solution

The minimum number of FLIP-FLOPs, N, required to generate a sequence of length S is given by

$$S \le 2^N - 1 \tag{8.2}$$

In this case $S = 7$, therefore, the minimum value of N, which may generate this sequence is 3. However, it is not guaranteed to lead to a solution. If the given sequence leads to seven distinct states, then only three FLIP-FLOPs are sufficient otherwise we have to increase the number of FLIP-FLOPs. We write the states of the circuit as given in Table 8.2. The prescribed sequence is listed under Q_2 and the sequence listed under Q_1 and Q_0 are the same sequence delayed by one and two clock pulses respectively. From the table we observe that all the states are not distinct, which means $N = 3$ is not sufficient. Next we assume $N = 4$ and prepare Table 8.3 in a similar manner as Table 8.2. The last column gives the required serial input for getting the desired change of state when a clock pulse is applied. This is obtained by assuming D-type FLIP-FLOPs and looking at the Q_3 output. For example, at the falling edge of the first clock pulse, $Q_3 = 1$. The second clock pulse must result in $Q_3 = 1$ which requires its D input to be 1. In the same manner, all the entries in column Y are determined. The K-map of Table 8.3 is given in Fig. 8.9 and the simplified expression is given by

$$Y = \bar{Q}_3 + \bar{Q}_1 + \bar{Q}_0 \tag{8.3}$$

Table 8.2 State table of sequence generator ($N = 3$)

Number of clock pulses	*FLIP-FLOP outputs*		
	Q_2	Q_1	Q_0
1	1	1	1
2	1	1	1
3	0	1	1
4	1	0	1
5	0	1	0
6	1	0	1
7	1	1	0

Table 8.3 Truth table of sequence generator ($N = 4$)

Number of clock pulses	*FLIP-FLOP outputs*				*Serial input*
	Q_3	Q_2	Q_1	Q_0	Y
1	1	1	1	0	1
2	1	1	1	1	0
3	0	1	1	1	1
4	1	0	1	1	0
5	0	1	0	1	1
6	1	0	1	0	1
7	1	1	0	1	1
1	1	1	1	0	1
2	1	1	1	1	0
3	0	1	1	1	1
*	1	0	1	1	0
*	0	1	0	1	1
*	1	0	1	0	1

The sequence generator circuit can be designed using a 4-stage shift register using *D*-type FLIP-FLOPs and the decoder circuit given by Eq. (8.3).

8.4 RIPPLE OR ASYNCHRONOUS COUNTERS

A circuit used for counting the pulses is known as a *counter*. In Sec. 8.3, two types of counters have been discussed. The number of states in an N-stage ring counter is N, whereas it is $2N$ in the case of moebius counter. These counters are referred to as modulo N (or divide-by-N) and modulo $2N$ (or divide-by-$2N$) counters respectively, where modulo indicates the number of states in the counter. When the pulses to be counted are applied to a counter, it goes from state to state and the output of the FLIP-FLOPs in the counter is decoded to read the count. The circuit comes back to its starting state after counting N pulses in the case of modulo N counter.

The ring counter and the twisted-ring counter do not make efficient use of FLIP-FLOPs. A FLIP-FLOP has two states, therefore a group of N FLIP-FLOPs will have 2^N states. This

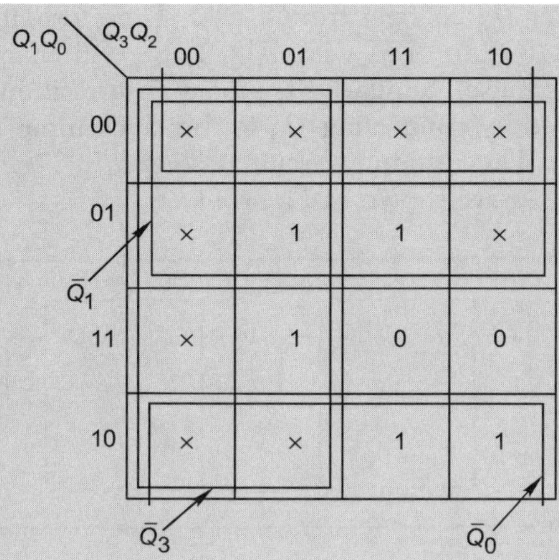

Fig. 8.9
K-map of Table 8.3.

means it is possible to make a modulo 2^N counter using N FLIP-FLOPs. Basically there are two types of such counters:

1. Asynchronous counter (ripple counter), and
2. Synchronous counter.

In the case of an asynchronous counter, all the FLIP-FLOPs are not clocked simultaneously, whereas in a synchronous counter all the FLIP-FLOPs are clocked simultaneously. The ring- and the twisted-ring counters are examples of synchronous counters.

Consider the count sequence shown in Table 8.4. The number of states in this sequence is 8 which requires 3 FLIP-FLOPs ($2^3 = 8$) and Q_2, Q_1, and Q_0 are the outputs of these FLIP-FLOPs. Assume master-slave FLIP-FLOPs.

Table 8.4 Counting sequence of a 3-bit binary counter

Counter state	*Count*		
	Q_2	Q_1	Q_0
0	0	0	0
1	0	0	1
2	0	1	0
3	0	1	1
4	1	0	0
5	1	0	1
6	1	1	0
7	1	1	1

The output Q_0 of the least-significant FLIP-FLOP, changes for every clock pulse. This can be achieved by using T-type FLIP-FLOP with $T_0 = 1$. The output Q_1 makes a transition (from

0 to 1 or 1 to 0) whenever Q_0 changes from 1 to 0. Therefore, if Q_0 is connected to the clock input of next T-type FLIP-FLOP, FF1 with $T_1 = 1$, Q_1 will change whenever Q_0 goes from 1 to 0 (falling edge of clock pulse). Similarly, Q_2 makes a transition whenever Q_1 goes from 1 to 0 and this can be achieved by connecting Q_1 to the clock input of the most-significant FLIP-FLOP, FF2 (and $T_2 = 1$). The resulting circuit is shown in Fig. 8. 10. The waveforms of the outputs of the FLIP-FLOPs are shown in Fig. 8.11.

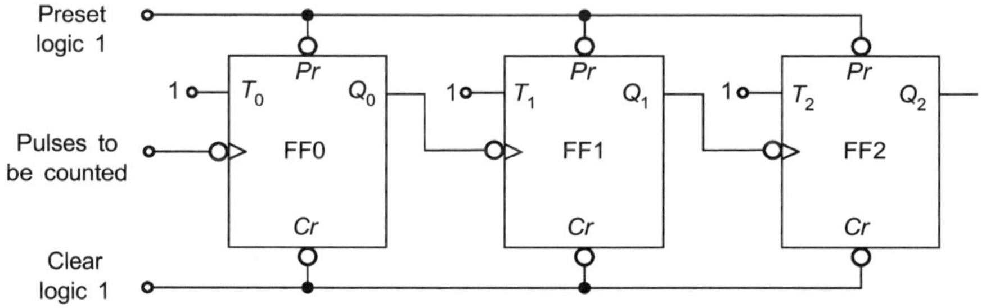

Fig. 8.10

A 3-bit binary counter.

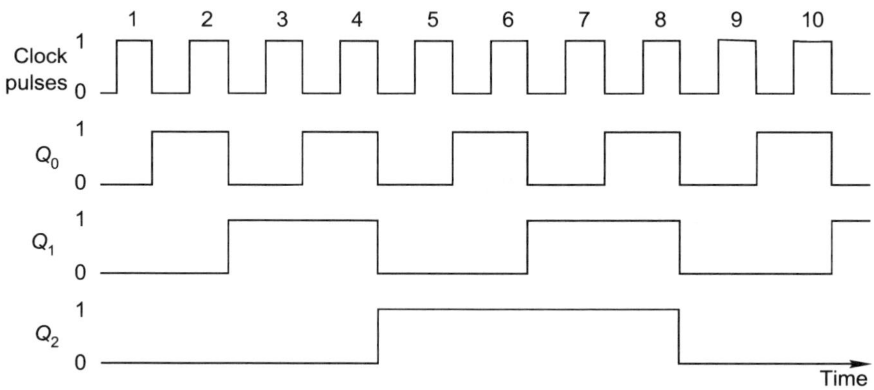

Fig. 8.11

Output waveforms of counter of Fig. 8.10.

A decoder circuit for decoding the count is shown in Fig. 8.12. In this, the output corresponding to the number counted goes low (active-low).

At the decoder outputs, false pulses of a short duration, known as *spikes*, occur as counter FLIP-FLOPs change state. This is because of the propagation delay of the FLIP-FLOPs due to which either all the FLIP-FLOPs do not change state at exactly the same time or only one FLIP-FLOP changes state for any clock pulse.

The problems of spikes at the decoder outputs is eliminated by using a *strobe* pulse input. With this, the decoding will occur only when all the FLIP-FLOPs have come to steady state.

The frequency, f, of clock pulses for reliable operation of the counter is given by

$$\frac{1}{f} \geq N \cdot (t_d) + T_s \tag{8.4}$$

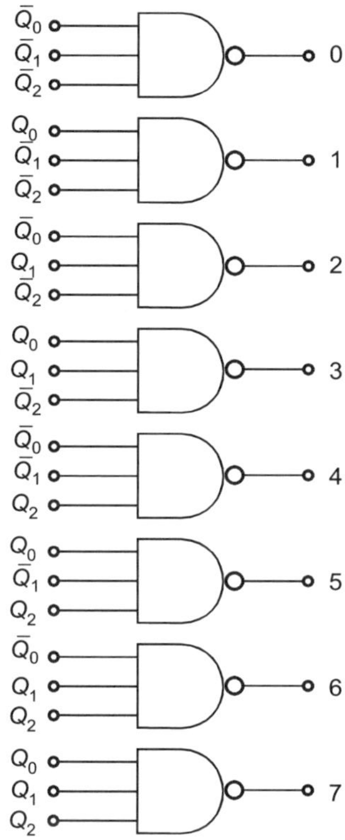

Fig. 8.12

A decoder circuit for a 3-bit binary counter.

where N = number of FLIP-FLOPs
t_d = propagation delay of one FLIP-FLOP
T_s = strobe pulse width.

Example 8.2 In a 4-stage ripple counter, the propagation delay of a FLIP-FLOP is 50 ns. If the pulse width of the strobe is 30 ns, find the maximum frequency at which the counter operates reliably.

Solution

The maximum frequency is

$$f_{max} = \frac{10^3}{4 \times 50 + 30} \text{ MHz}$$
$$= 4.35 \text{ MHz}$$

8.4.1 UP/DOWN Counters

The counter of Fig. 8.10 counts in the UP direction, i.e. the decimal equivalent of the counter output increases with successive clock pulses. It is also possible to make a counter in which the decimal equivalent of the counter output decreases with the application of successive clock pulses, i.e. the counting proceeds in the DOWN direction (Prob. 8.5). The former is referred to as an *UP counter* and the latter as a *DOWN counter.*

An UP/DOWN counter can also be designed which can count in any direction depending upon the direction control input (Prob. 8.6).

8.4.2 Modulus of the Counter

The counter discussed above is referred to as a *ripple counter,* since the pulses applied ripple from stage to stage. It is a modulo n counter where $n = 2^N$.

If it is desired to have a modulo m counter, the number of FLIP-FLOPs required is determined using Eq. (8.5) as the minimum value of N which satisfies the equation

$$m \leq 2^N \tag{8.5}$$

For example, the value of N is 4 for any value of m from 9 to 16. If $m = 16$, then the circuit can be designed as discussed above but if it is less than 16, say 10, then out of 16 states only 10 states are used and the remaining six states are unused. The counter is required to be reset (i.e. the normal counting is to be terminated) at the end of the tenth clock pulse. This can be achieved by generating a logic 0 signal immediately after the tenth pulse and applying it to the clear input of all the FLIP-FLOPs.

For a modulo 10 counter (also referred to as a *decade counter*) the circuit for resetting the counter after the tenth pulse is shown in Fig. 8.13.

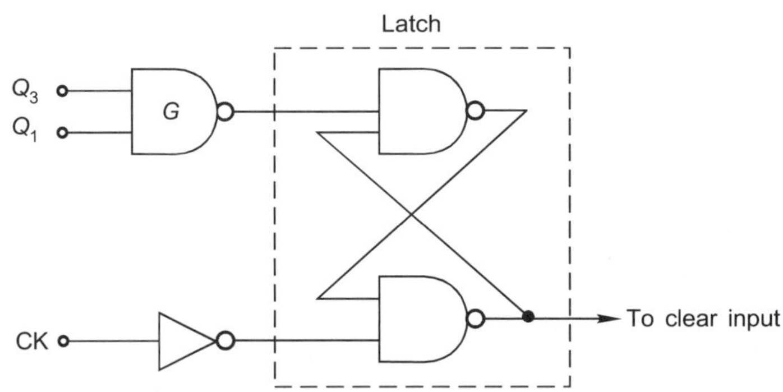

Fig. 8.13
A circuit to be used to obtain a decade counter from modulo-16 counter.

At the end of the tenth pulse, $Q_3 = Q_1 = 1$, therefore, the output of the NAND gate G will be 0, making the output of the latch 0. This will reset the counter. The latch is used in order to keep the clear line at 0 till both the FLIP-FLOPs are cleared.

A counter in which the starting state is not 0 can be designed by making use of the preset inputs of the FLIP-FLOPs, similar to Fig. 8.2. This is referred to as *loading* the counter asynchronously (not synchronous with the clock pulse). This is referred to as a *presettable counter.*

8.4.3 54/74 Series Asynchronous Counter ICs

The design of asynchronous counters using FLIP-FLOPs has been discussed above. Some asynchronous counters are available in MSI and are given in Table 8.5 along with some of their features. Depending upon these features, these ICs are divided into three groups A, B, and C. The group to which a particular IC belongs is also indicated in the table. All these ICs consist of four master-slave FLIP-FLOPs. The load, set, and reset (clear) operations are asynchronous, i.e. independent of the clock pulse.

Table 8.5 Available asynchronous counter ICs

IC No.	*Description*	*Features*	*Group*
7490, 74290	BCD counter	Set, reset	A
7492	Divide-by-12 counter	Reset	B
7493, 74293	4-bit binary counter	Reset	B
74176, 74196	Presettable BCD counter	Reset, load	C
74177, 74197	Presettable 4-bit binary counter	Reset, load	C
74390	Dual decade counters	Reset	B
74393	Dual 4-bit binary counters	Reset	B
74490	Dual BCD counters	Set, reset	A

Group A asynchronous counter ICs

Figure 8.14 shows the basic internal structure of 7490. It consists of four FLIP-FLOPs internally connected to provide a mod-2 counter and a mod-5 counter. The mod-2 and mod-5 counters can be used independently or in combination. FLIP-FLOP FFA operates as a mod-2 counter whereas the combination of FLIP-FLOPs FFB, FFC, and FFD form a mod-5 counter. There are two reset inputs R_1 and R_2 both of which are to be connected to logic 1 level for clearing all the FLIP-FLOPs. The two set inputs S_1 and S_2 when connected to logic 1 level, are used for setting the counter to 1001.

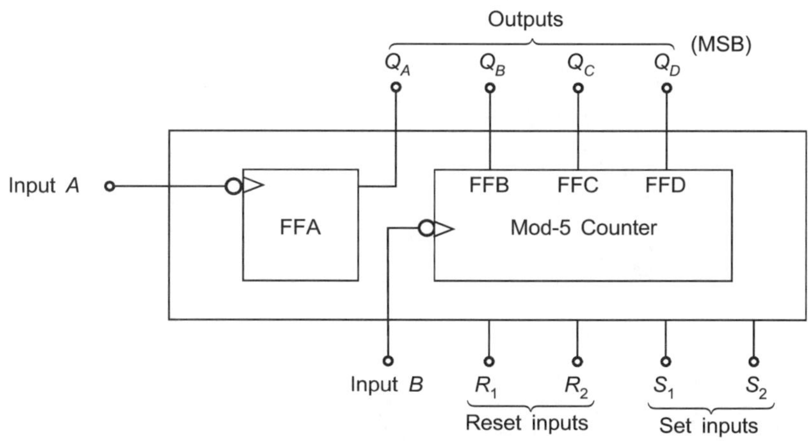

Fig. 8.14

Basic internal structure of 7490 ripple counter IC.

IC 74490 is a dual BCD counter consisting of two independent BCD counters. Each section consists of four FLIP-FLOPs, all connected internally to form a decade counter. For each section there is a reset (*R*) and a set (*S*) input which are active-high.

Example 8.3 In a 7490 if Q_A output is connected to *B* input and the pulses are applied at *A* input, find the count sequence and the waveforms at *Q* outputs.

Solution

When Q_A output is connected to *B* input, we have the mod-2 counter followed by the mod-5 counter. The count sequence obtained is given in Table 8.6.

Table 8.6

Counter state	FLIP-FLOP *outputs*			
	Q_D	Q_C	Q_B	Q_A
0	0	0	0	0
1	0	0	0	1
2	0	0	1	0
3	0	0	1	1
4	0	1	0	0
5	0	1	0	1
6	0	1	1	0
7	0	1	1	1
8	1	0	0	0
9	1	0	0	1
10	0	0	0	0

It may be noted that when Q_A changes from 0 to 1, the state of the mod-5 counter does not change, whereas when Q_A changes from 1 to 0, the mod-5 counter goes to the next state.

The waveforms at *Q* outputs are illustrated in Fig. 8.15.

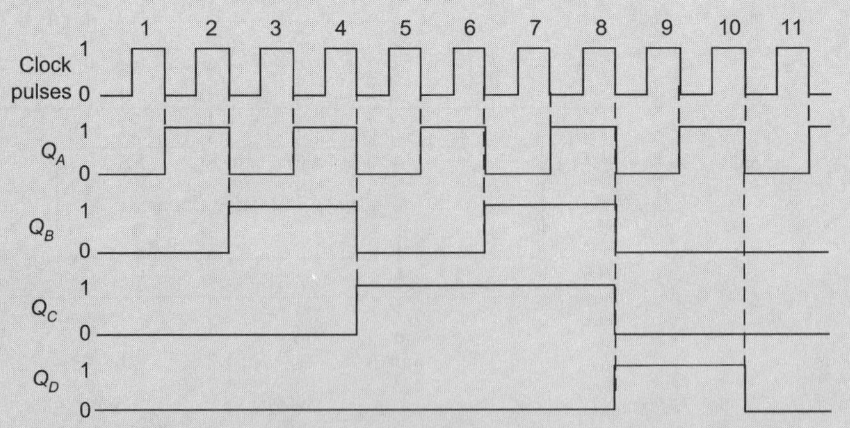

Fig. 8.15

Waveforms at *Q* outputs for Ex. 8.3.

Example 8.4 In a 7490, if Q_D output is connected to A input and the pulses are applied at B input, find the count sequence and the waveform of output Q_A. Compare its count sequence with that of Table 8.6.

Solution

When Q_D output is connected to A input and the pulses are applied at B input, we have the mod-5 counter followed by the mod-2 counter. The count sequence obtained is given in Table 8.7. Here the states of the mod-5 counter change in a normal binary sequence and Q_A changes whenever Q_D goes from 1 to 0.

Table 8.7

Counter State	*FLIP-FLOP outputs*			
	Q_D	Q_C	Q_B	Q_A
0	0	0	0	0
1	0	0	1	0
2	0	1	0	0
3	0	1	1	0
4	1	0	0	0
5	0	0	0	1
6	0	0	1	1
7	0	1	0	1
8	0	1	1	1
9	1	0	0	1
10	0	0	0	0

The waveforms of Q_A output is illustrated in Fig. 8.16 which is a square wave.

The count sequence of this counter is different from that of Table 8.6 although both are divide-by-10 counters. The waveform of Q_A output is square wave. Hence this circuit can be used for the generation of square waves.

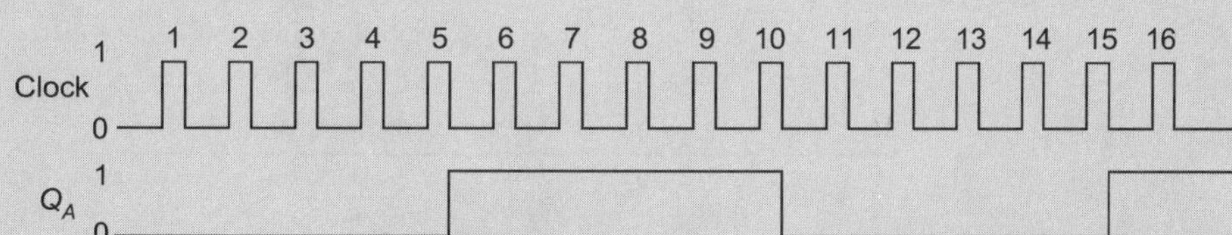

Fig. 8.16

Waveform of output Q_A.

Example 8.5 Design a divide-by-6 counter using 7490.

Solution

Connect the counter as divide-by-10 for normal binary sequence (Ex. 8.3). Outputs Q_B and Q_C are connected to the reset inputs. Therefore as soon as Q_B and Q_C both become 1, the counter is reset to 0000. Figure 8.17 shows the divide-by-6 ripple counter.

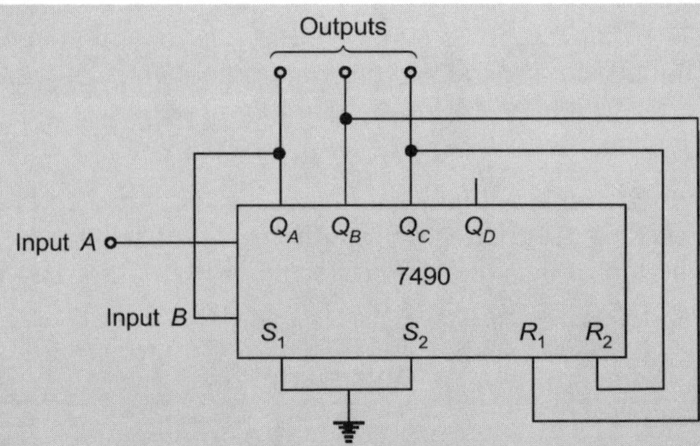

Fig. 8.17
A divide-by-6 ripple counter using 7490.

Group B asynchronous counter ICs

The basic internal structure of 7492, 7493, and 74293 asynchronous counter ICs is shown in Fig. 8.18. The operation of these ICs is identical to the operation of IC 7490 except that the set inputs are not present and mod-6 counter does not count in straight binary sequence. The sequence of mod-6 counter is given in Table 8.8. These ICs are not used as counters but are used for frequency division.

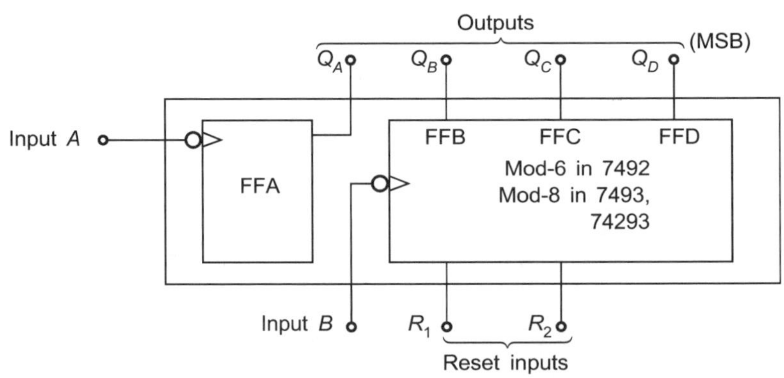

Fig. 8.18
Basic internal structure of 7492, 7493 and 74293 counter ICs.

Table 8.8 Count sequence of Mod-6 counter

Q_D	Q_C	Q_B
0	0	0
0	0	1
0	1	0
1	0	0
1	0	1
1	1	0

74390 IC is a dual BCD counter consisting of two independent BCD counters similar to 7490. There is one reset (R) input for each section. 74393 is a dual 4-bit binary counter with one reset (R) input for each section which is active-high.

Example 8.6 If output Q_A of a divide-by-12 ripple counter 7492 is connected to the B input and the pulses are applied at the A input, find the count sequence.

Solution

The count sequence is given in Table 8.9. It may be noted from it that simultaneous divisions of 2, 6, and 12 are performed at the Q_A, Q_C, and Q_D outputs, respectively.

Table 8.9

Q_D	Q_C	Q_B	Q_A
0	0	0	0
0	0	0	1
0	0	1	0
0	0	1	1
0	1	0	0
0	1	0	1
1	0	0	0
1	0	0	1
1	0	1	0
1	0	1	1
1	1	0	0
1	1	0	1

Group C asynchronous counter ICs

The basic internal structure of group C counter ICs is shown in Fig. 8.19. 74176 and 74196 are both BCD counters with a difference in only maximum clock frequency specification. Similarly, 74177 and 74197 are both 4-bit binary counters with the same difference.

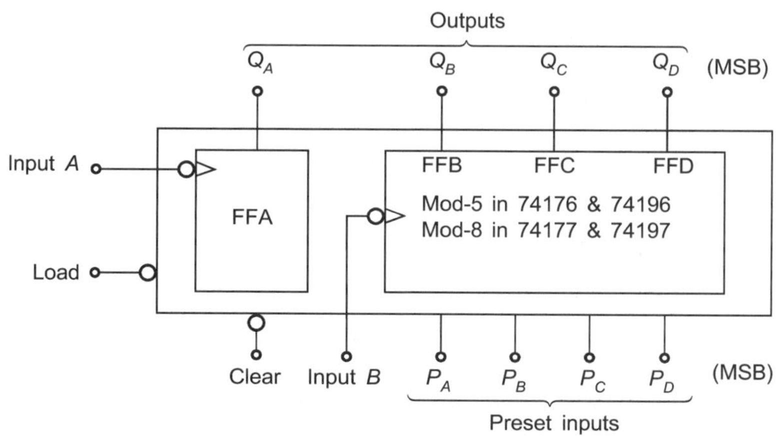

Fig. 8.19

Basic internal structure of group C asynchronous counter ICs

These counters are presettable versions of 7490 and 7493 counters respectively. The counter is cleared by connecting logic 0 to clear input (active-low). Setting load input to logic 0 (while clear is at logic 1) stops the count and loads any binary number present at the preset inputs into the counter. For normal UP counting operation, both the load and clear inputs should be connected to logic 1.

The presettable 4-bit binary counters can be used as variable mod-n counter in which the counter modulus is equal to 15 – P, where P is the binary number connected at the preset inputs. In other words, for designing a mod-n counter, the value of P is 15 – n. When the counter output reaches the count 1111, the counter must be loaded again with P. This is made possible by using a 4-input NAND gate between the Q outputs of the counter and the load input.

Example 8.7 Design a divide-by-12 counter using 74177.

Solution

The circuit of a divide-by-12 counter is shown in Fig. 8.20. The counter is loaded with P = 1111 – 1100 = 0011 as soon as the output becomes 1111.

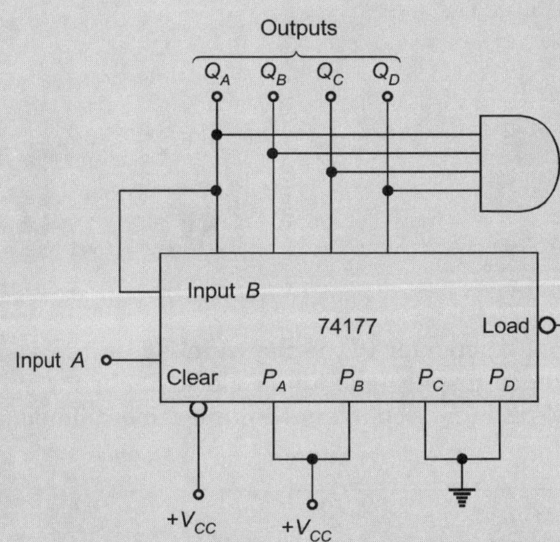

Fig. 8.20
A divide-by-12 counter using 74177.

Cascading of ripple counter ICs

Ripple counters of any cycle length can be obtained by cascading the ICs discussed above. The desired cycle length is decoded and used to reset all the counters to 0. The strobe should be used to eliminate false data.

The cascading arrangement for all the asynchronous counter ICs is same where Q_D of preceding stage goes to the clock input terminal of the succeeding stage. The load and clear inputs of all ICs are to be connected together.

Example 8.8 Design a 2-decade BCD counter using the IC 74390.

Solution

The 74390 IC is a dual BCD counter, therefore only one IC is required to design a 2-decade BCD counter. The 2-decade counter is shown in Fig. 8.21.

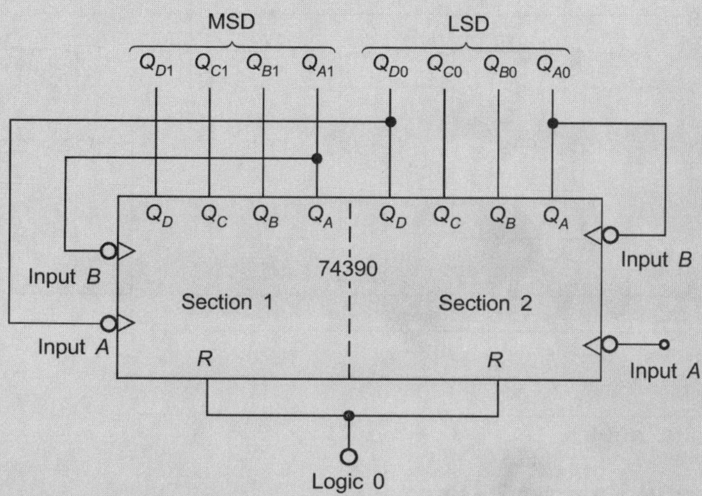

Fig. 8.21
A 2-decade BCD counter using 74390 IC.

8.5 SYNCHRONOUS COUNTERS

The ripple counters have the advantage of simplicity (only FLIP-FLOPs are required) but their speed is low because of ripple action. The maximum time is required when the output changes from 111 ... 1 to 000 ... 0 and this limits the frequency of operation of ripple counters.

The speed of operation improves significantly if all the FLIP-FLOPs are clocked simultaneously. The resulting circuit is known as a *synchronous counter.* Synchronous counters can be designed for any count sequence (need not be straight binary). These counters can be designed following a systematic approach. Before we discuss the formal method of design for such counters, we shall consider an intuitive method.

Consider the count sequence of Table 8.4. The output Q_0 of the least-significant FLIP-FLOP changes for every clock pulse. This can be achieved by using a *T*-type FLIP-FLOP with $T_0 = 1$. The output Q_1 changes whenever Q_0 changes from 1 to 0. Therefore, if Q_0 is connected to *T* input (T_1) of the next FLIP-FLOP, Q_1 will change from 1 to 0 (or 0 to 1) when $Q_0 = 1$ ($T_1 = 1$) and will remain unaffected when $Q_0 = T_1 = 0$. Similarly, we observe from Table 8.4 that Q_2 changes whenever Q_1 and Q_0 are both 1. This can be achieved by making the *T*-input (T_2) of the most-significant FLIP-FLOP equal to $Q_1 \cdot Q_0$. The circuit thus obtained is shown in Fig. 8.22.

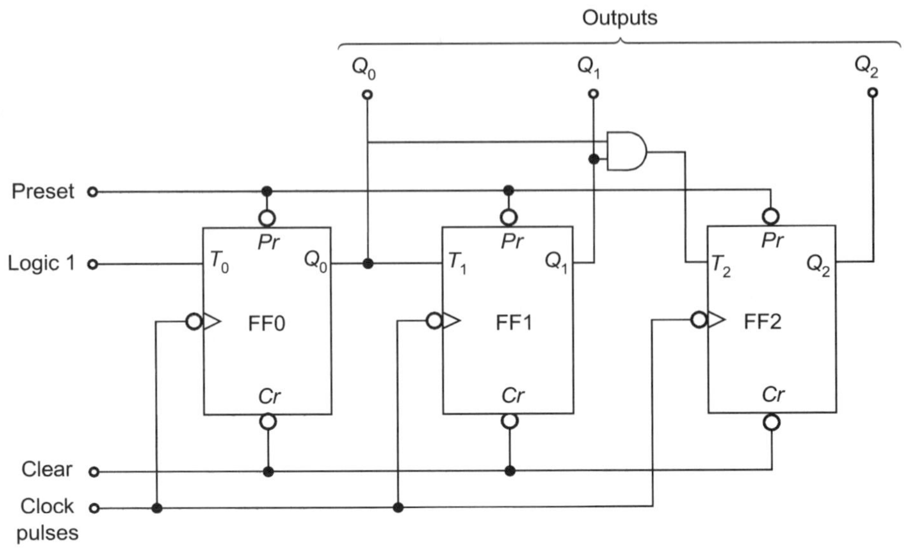

Fig. 8.22
A 3-bit synchronous counter.

8.5.1 Synchronous Counter Design

Synchronous counters for any given count sequence and modulus can be designed in the following way:

1. Find the number of FLIP-FLOPs required using Eq. (8.5).
2. Write the count sequence in the tabular form similar to Table 8.4.
3. Determine the FLIP-FLOP inputs which must be present for the desired next state from the present state using the excitation table of the FLIP-FLOPs.
4. Prepare *K*-map for each FLIP-FLOP input in terms of FLIP-FLOP outputs as the input variables. Simplify the *K*-maps and obtain the minimized expressions.
5. Connect the circuit using FLIP-FLOPs and other gates corresponding to the minimized expressions.

The above design steps can be clearly understood from the following examples.

Example 8.9 Design a 3-bit synchronous counter using *J-K* FLIP-FLOPs.

Solution

The number of FLIP-FLOPs required is 3. Let the FLIP-FLOPs be FF0, FF1, and FF2 and their inputs and outputs are given below:

FLIP-FLOP	*Inputs*	*Output*
FF0	J_0, K_0	Q_0
FF1	J_1, K_1	Q_1
FF2	J_2, K_2	Q_2

Table 8.10

FLIP-FLOP inputs			Counter state FF0		FF1		FF2	
Q_2	Q_1	Q_0	J_0	K_0	J_1	K_1	J_2	K_2
0	0	0	1	×	0	×	0	×
0	0	1	×	1	1	×	0	×
0	1	0	1	×	×	0	0	×
0	1	1	×	1	×	1	1	×
1	0	0	1	×	0	×	×	0
1	0	1	×	1	1	×	×	0
1	1	0	1	×	×	0	×	0
1	1	1	×	1	×	1	×	1
0	0	0						

The count sequence and the required inputs of FLIP-FLOPs are given in Table 8.10. The inputs to the FLIP-FLOPs are determined in the following manner:

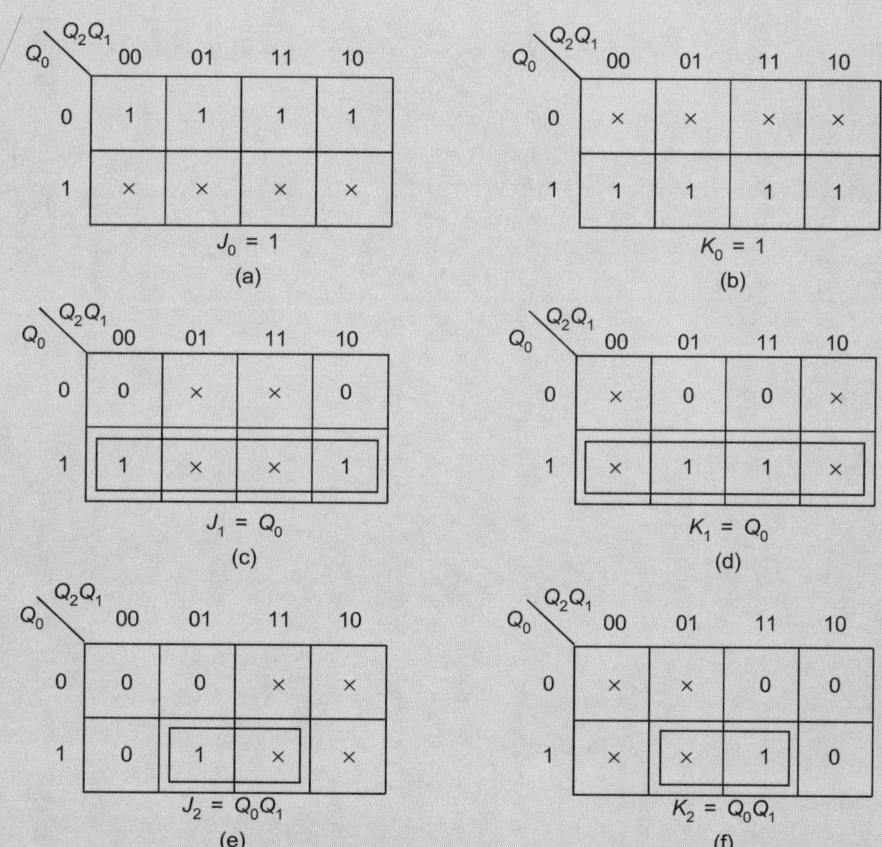

Fig. 8.23

K-maps of Ex. 8.9.

Consider one column of the counter state at a time and start from the first row, for example, consider Q_0. Before the first pulse is applied, $Q_0 = 0$ and it is required to be 1 at the end of the first clock pulse. Therefore, to achieve this condition, the values of J_0 and K_0 are 1 and × respectively (from the excitation Table 7.6). These are entered in the table in the row corresponding to 0 pulse. When the second clock pulse is applied Q_0 is to change from 1 to 0, therefore, the required inputs are

$$J_0 = \times, K_0 = 1$$

In a similar manner inputs of each FLIP-FLOP are determined.

Now, we prepare the *K*-maps (Fig. 8.23) with Q_2, Q_1 and Q_0 as input variables and FLIP-FLOP inputs as output variables. We then minimize the *K*-maps and the resulting minimized expressions are:

$$J_0 = 1, \quad K_0 = 1$$
$$J_1 = Q_0, \quad K_1 = Q_0$$
$$J_2 = Q_0\, Q_1, \; K_2 = Q_0\, Q_1$$

The resulting counter circuit is same as the circuit of Fig. 8.22.

Example 8.10 Design a 3-bit binary UP/DOWN counter with a direction control *M*. Use *J-K* FLIP-FLOPs.

Solution

The count sequence is given in Table 8.11. For *M* = 0, it acts as an UP counter and for *M* = 1 as a DOWN counter. The number of FLIP-FLOPs required is 3. The inputs of the FLIP-FLOPs are determined in a manner similar to the one employed in Ex. 8.9.

Table 8.11

Direction control	*Counter state*			FLIP-FLOP *inputs*					
M	Q_2	Q_1	Q_0	J_0	K_0	J_1	K_1	J_2	K_2
0	0	0	0	1	×	0	×	0	×
0	0	0	1	×	1	1	×	0	×
0	0	1	0	1	×	×	0	0	×
0	0	1	1	×	1	×	1	1	×
0	1	0	0	1	×	0	×	×	0
0	1	0	1	×	1	1	×	×	0
0	1	1	0	1	×	×	0	×	0
0	1	1	1	×	1	×	1	×	1
1	0	0	0	1	×	1	×	1	×
1	1	1	1	×	1	×	0	×	0
1	1	1	0	1	×	×	1	×	0
1	1	0	1	×	1	0	×	×	0
1	1	0	0	1	×	1	×	×	1
1	0	1	1	×	1	×	0	0	×
1	0	1	0	1	×	×	1	0	×
1	0	0	1	×	1	0	×	0	×
	0	0	0						

From Table 8.11, we obtain

$$J_0 = K_0 = 1$$

The K-maps for J_1, K_1, J_2, and K_2 are shown in Fig. 8.24. From the K-maps, the minimized expressions are obtained as

$$J_1 = K_1 = Q_0\bar{M} + \bar{Q}_0 M$$

$$J_2 = K_2 = \bar{M}\,Q_1 Q_0 + M\,\bar{Q}_1\,\bar{Q}_0$$

The counter circuit can be drawn using the above expressions

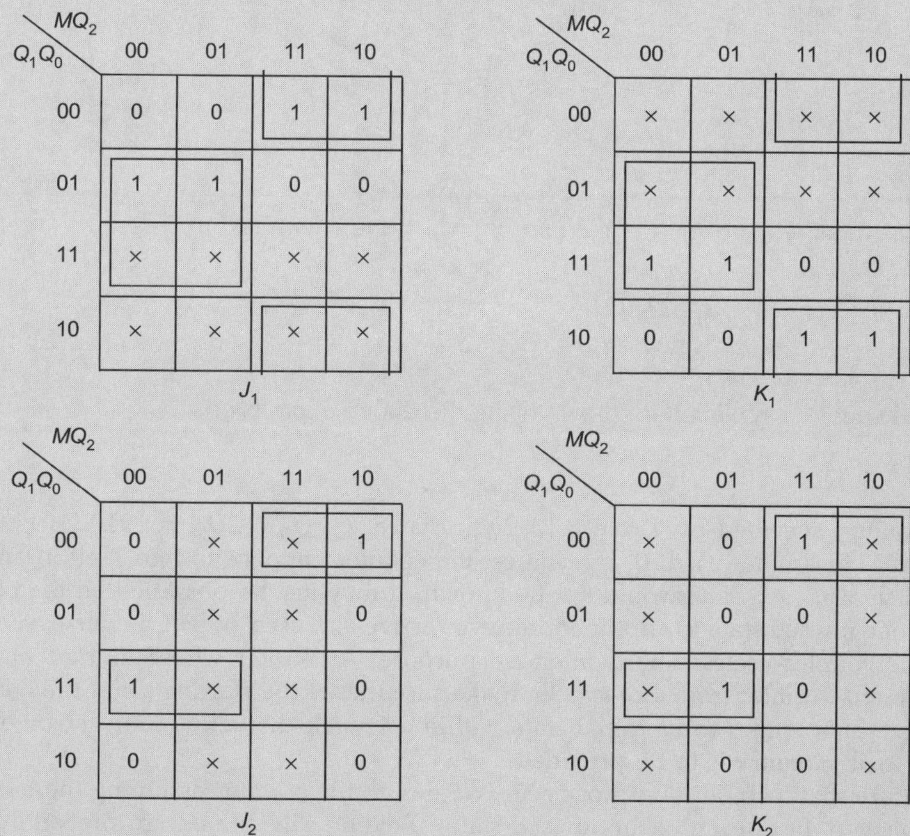

Fig. 8.24

K-maps for Ex. 8.10.

Example 8.11 Design a decade UP counter. Use *J-K* FLIP-FLOPs.

Solution

There are ten states in a decade counter, which requires four FLIP-FLOPs. The remaining six states are unused states. The count sequence and the FLIP-FLOP inputs are given in Table 8.12.

Table 8.12

Counter state				FLIP-FLOP inputs							
Q_3	Q_2	Q_1	Q_0	J_0	K_0	J_1	K_1	J_2	K_2	J_3	K_3
0	0	0	0	1	×	0	×	0	×	0	×
0	0	0	1	×	1	1	×	0	×	0	×
0	0	1	0	1	×	×	0	0	×	0	×
0	0	1	1	×	1	×	1	1	×	0	×
0	1	0	0	1	×	0	×	×	0	0	×
0	1	0	1	×	1	1	×	×	0	0	×
0	1	1	0	1	×	×	0	×	0	0	×
0	1	1	1	×	1	×	1	×	1	1	×
1	0	0	0	1	×	0	×	0	×	×	0
1	0	0	1	×	1	0	×	0	×	×	1
0	0	0	0								

The *K*-maps are shown in Fig. 8.25 from which the minimized expressions are obtained as

$$J_0 = 1, \qquad K_0 = 1$$
$$J_1 = Q_0 \bar{Q}_3, \qquad K_1 = Q_0$$
$$J_2 = Q_0 Q_1, \qquad K_2 = Q_0 Q_1$$
$$J_3 = Q_0 Q_1 Q_2, \qquad K_3 = Q_0$$

The counter circuit can be drawn using the above expressions.

8.5.2 Lock Out

In the counter specified by Table 8.12, logic states, $Q_3 Q_2 Q_1 Q_0$ = 1010, 1011, 1100, 1101, 1110, and 1111 are not used. If, by chance, the counter happens to find itself in any one of the unused states, its next state would not be known. It may just be possible that the counter might go from one unused state to another and never arrive at a used state. Of course, such a situation makes the counter useless for its intended purpose. A counter whose unused states have this feature is said to suffer from *lock out*. To make sure that at the starting point the counter is in its initial state or it comes to its initial state within a few clock cycles (count error due to noise), external logic circuitry is to be provided.

To ensure that lock out does not occur, we design the counter assuming the next state to be the initial state, from each of the unused states. Beyond this, the design procedure is the same as discussed earlier.

8.5.3 54/74 Series Synchronous Counter ICs

The design of synchronous counters using FLIP-FLOPs has been discussed above. Counters for any count sequence and modulus can be designed using these methods. Some synchronous counters are available in MSI and are given in Table 8.13 along with some of their features. All theses ICs are positive-edge-triggered, i.e. the change of state, synchronous loading, and clearing take place on the positive going edge of the input clock pulse. Basically these ICs can be divided into four groups–A, B, C, and D. A brief description of each group is given below:

J_0

$Q_1Q_0 \backslash Q_3Q_2$	00	01	11	10
00	1	1	×	1
01	×	×	×	×
11	×	×	×	×
10	1	1	×	×

K_0

$Q_1Q_0 \backslash Q_3Q_2$	00	01	11	10
00	×	×	×	×
01	1	1	×	1
11	1	1	×	×
10	×	×	×	×

J_1

$Q_1Q_0 \backslash Q_3Q_2$	00	01	11	10
00	0	0	×	0
01	1	1	0	0
11	×	×	×	×
10	×	×	×	×

K_1

$Q_1Q_0 \backslash Q_3Q_2$	00	01	11	10
00	×	×	×	×
01	×	×	×	×
11	1	1	×	×
10	0	0	×	×

J_2

$Q_1Q_0 \backslash Q_3Q_2$	00	01	11	10
00	0	×	×	0
01	0	×	×	0
11	1	×	×	×
10	0	×	×	×

K_2

$Q_1Q_0 \backslash Q_3Q_2$	00	01	11	10
00	×	0	×	×
01	×	0	×	×
11	×	1	×	×
10	×	0	×	×

J_3

$Q_1Q_0 \backslash Q_3Q_2$	00	01	11	10
00	0	0	×	×
01	0	0	×	×
11	0	1	×	×
10	0	0	×	×

K_3

$Q_1Q_0 \backslash Q_3Q_2$	00	01	11	10
00	×	×	×	0
01	×	×	×	1
11	×	×	×	×
10	×	×	×	×

Fig. 8.25
K-maps for Ex. 8.11.

Table 8.13 Available synchronous counter ICs

IC No.	*Description*	*Features*	*Group*
74160	Decade UP counter	Synchronous preset and asynchronous clear	A
74161	4-bit binary UP counter	–do–	A
74162	Decade UP counter	Synchronous preset and clear	A
74163	4-bit binary UP counter	–do–	A
74168	Decade UP/DOWN counter	Synchronous preset and no clear	B
74169	4-bit binary UP/DOWN counter	–do–	B
74190	Decade UP/DOWN counter	Asynchronous preset and no clear	C
74191	4-bit binary UP/DOWN counter	–do–	C
74192	Decade UP/DOWN counter	Asynchronous preset and clear	D
74193	4-bit binary UP/DOWN counter	–do–	D

Group A synchronous counter ICs

The block diagram and the function table of these ICs are given in Fig. 8.26. In these ICs there are two separate enable inputs, *ENT* and *ENP*. Setting either of these inputs to logic 0 stops counting asynchronously. Ripple carry (RC) output is normally at logic 0 and goes to logic 1 whenever the counter reaches its highest count (binary 9 for BCD counters and binary 15 for 4-bit binary counters). Setting *ENT* to logic 0 also inhibits *RC* changing from logic 0 to logic 1.

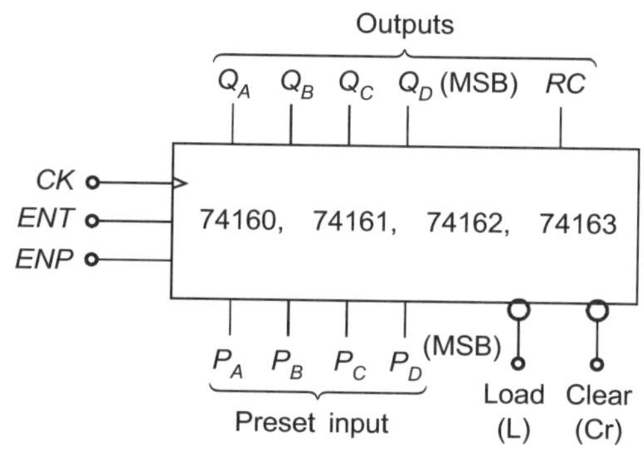

Fig. 8.26
Group A synchronous counter ICs (a) block diagram.

Load L	*ENP*	*ENT*	*Cr*	*CK*	*Mode*
0	×	×	1	↑	Preset
1	0	1	1	×	Stop count
1	×	0	1	×	Stop count, disable RC
×	×	×	0	*	Reset to zero
1	1	1	1	↑	UP count

* × for 74160 and 74161
↑ for 74162 and 74163

Fig. 8.26
(*b*) Function table.

Example 8.12 Design a normal mod-12 counter using 74161.

Solution

The circuit is designed for the normal UP counting (last row of Fig. 8.26*b*). The Q_D and Q_C outputs through a NAND gate are connected to the *Cr* terminal which clears the counter as

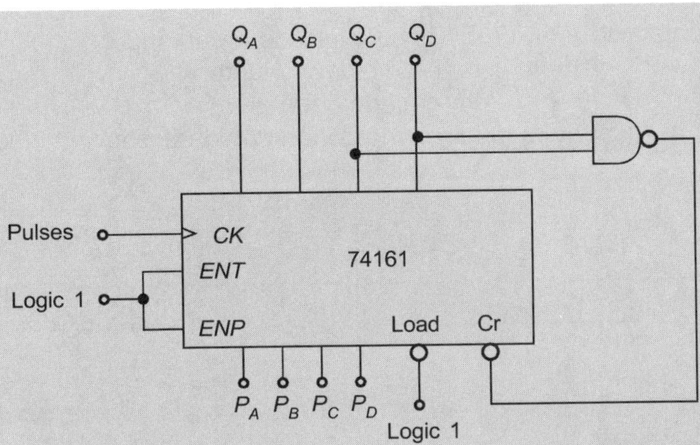

Fig. 8.27
Figure for Ex. 8.12.

soon as the output is 1100. The states of the counter are from 0000 through 1011. The mod-12 counter is shown in Fig. 8.27.

Using the approach followed in Ex. 8.12, the count can be terminated at any desired value and a counter with any modulus (less than 16 for binary and less than 10 for decade counter) can be obtained.

Example 8.13 Design a divide-by-11 counter using 74163. Make use of the *RC* output and preset inputs.

Solution

For obtaining a divide-by-11 counter, the counter is preset at binary 0101 (decimal 5). When the count reaches 1111, *RC* output goes to 1, which is used to load the data present at the preset inputs into the counter. The circuit of the counter is shown in Fig. 8.28.

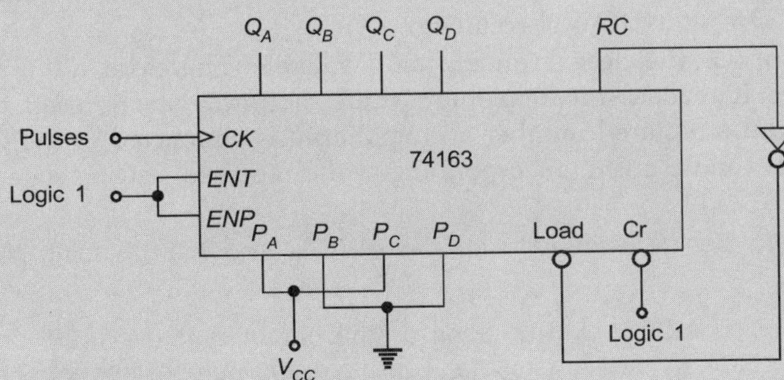

Fig. 8.28
Counter for Ex. 8.13.

In general, for obtaining a divide-by-m counter, the preset input, P, is given by

$$P = 16 - m \quad \text{for 4-bit binary counter}$$
$$\;\; = 10 - m \quad \text{for decade counter}$$

Cascading of group A counters in fully synchronous mode is shown in Fig. 8.29.

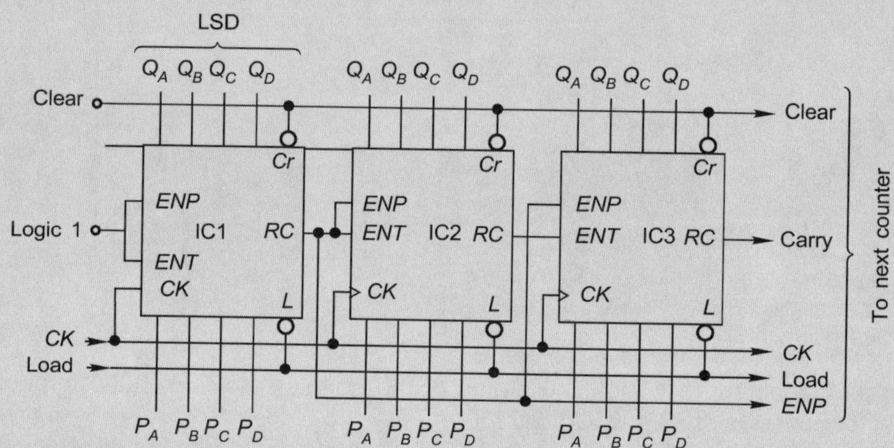

Fig. 8.29
Cascading arrangement of Group A synchronous counter ICs.

Group B synchronous counter ICs

The block diagram and the function table of these ICs are given in Fig. 8.30.

The functions of *ENT* and *ENP* are same as in group A ICs except that these are active-low. Ripple carry (*RC*) output is normally held at logic 1 and goes to logic 0 (i) when the count reaches maximum during UP counting, and (ii) when the count reaches minimum during DOWN counting. Signal at $U/\bar{D}$ terminal decides the direction of counting, $U/\bar{D} = 1$ for UP counting and $U/\bar{D} = 0$ for DOWN counting.

In this group of ICs, the clear terminal is not available. Therefore, if it is desired to terminate the count before it reaches the maximum value, a NAND gate is used to detect the count corresponding to the required number and its output is connected to the load input terminal. The preset inputs can be given corresponding to the required starting state of the counter.

Example 8.14 Design a counter with states 0011 through 1100 using 74169 counter.

Solution

The preset input is 0011 and as soon as the output reaches 1100, on the next pulse it should come back to its original state. Therefore, the number corresponding to the highest required state is to be detected for loading the counter. The counter is shown in Fig. 8.31.

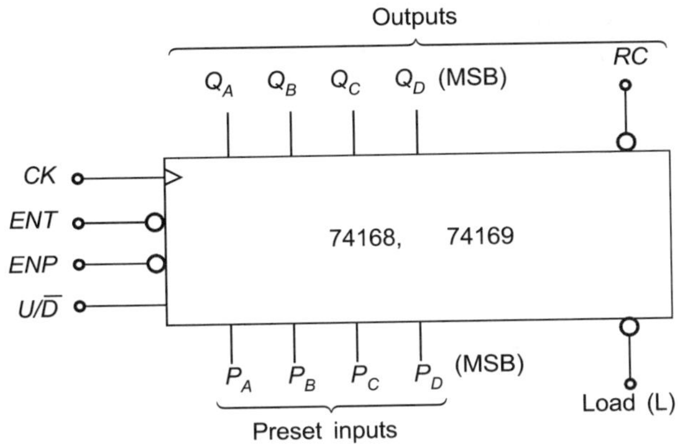

(a)

Load L	*ENP*	*ENT*	$U/\bar{D}$	*CK*	*Mode*
0	×	×	×	↑	Preset
1	1	0	×	×	Stop count
1	×	1	×	×	Stop count, disable RC
1	0	0	1	↑	UP count
1	0	0	0	↑	DOWN count

(b)

Fig. 8.30
Group B synchronous counter ICs (a) block diagram (b) function table.

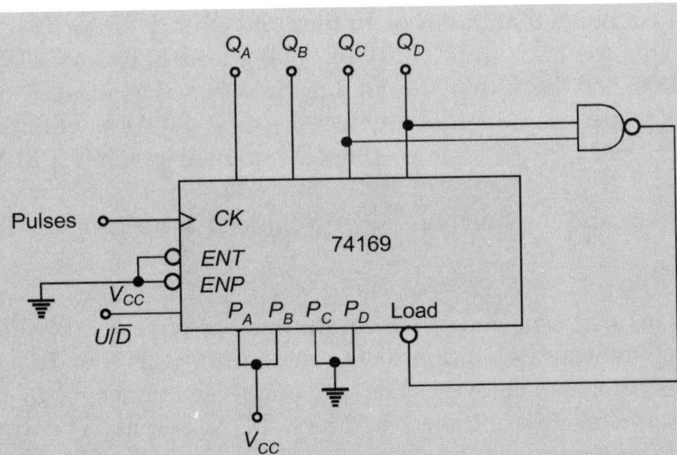

Fig. 8.31
Counter for Ex. 8.14.

If the counter is required to count up to the maximum/minimum value then the RC output is to be connected to the load input, for loading the initial count at the next pulse after maximum/minimum count has been reached. The frequency of the output waveform at RC (f_{out}) is related to the input clock frequency (f_{in}) as follows:

Binary counter 74169

$$f_{out} = \frac{f_{in}}{N+1}, 1 \le N \le 15 \text{ (for DOWN counting)}$$

$$= \frac{f_{in}}{16-N}, 0 \le N \le 14 \text{ (for UP counting)}$$

Decade counter 74168

$$f_{out} = \frac{f_{in}}{N+1}, 1 \le N \le 9 \text{ (for DOWN counting)}$$

$$= \frac{f_{in}}{10-N}, 0 \le N \le 8 \text{ (for UP counting)}$$

where N is the decimal equivalent of the preset input.

The cascading of group B counter ICs is similar to that of group A counter ICs.

Group C synchronous counter ICs

These ICs have only one enable input terminal $ENAB$ which is active-low. MAX/MIN output is used to detect the counter's maximum or minimum count. It is normally at logic 0 and goes to logic 1 when the count is maximum (1001 for 74190 and 1111 for 74191) for the UP counting and minimum (0000) for the DOWN counting. It serves the purpose of an overflow detector while UP counting and an underflow detector while DOWN counting. The RC output is normally at logic 1, and goes to logic 0 when the counter reaches a MAX/MIN point and the CK input is low.

The block diagram and the function table of these ICs are given in Fig. 8.32.

Frequency dividers

These counters can also be used as programmable frequency dividers. By presetting any number into the counter and counting to the maximum (UP counting) or minimum (DOWN counting) count, division is achieved. The RC output is connected to the load input and the required output waveform is obtained at MAX/MIN output. The frequency of the output waveform f_{out} and the input clock frequency (f_{in}) are related as follows:

Binary Counter 74191

$$f_{out} = \frac{f_{in}}{N}, 1 \le N \le 15 \text{ (for DOWN counting)}$$

$$= \frac{f_{in}}{15-N}, 0 \le N \le 14 \text{ (for UP counting)}$$

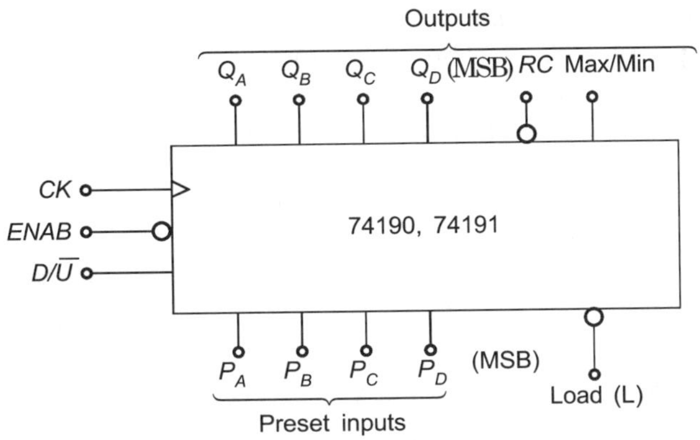

(a)

Load L	*ENAB*	$D/\overline{U}$	*CK*	*Mode*
×	1	×	×	Stop count
0	0	×	×	Preset
1	0	0	↑	UP count
1	0	1	↑	DOWN count

(b)

Fig. 8.32
Group C synchronous counter ICs (a) Black diagram (b) Function table.

Decade Counter 74190

$$f_{\text{out}} = \frac{f_{\text{in}}}{N} \text{ for } 1 \le N \le 9 \text{ (for DOWN counting)}$$

$$= \frac{f_{\text{in}}}{9 - N} \text{ for } 1 \le N \le 8 \text{ (for UP counting)}$$

where N is the decimal equivalent of preset input. For example, in case of 74191, if the preset input is 1000(8), and the clock frequency is 560 Hz, the frequency of output waveform will be 80 Hz for *UP* counting and 70 Hz for DOWN counting.

Cascading of group C counters

These counters can be cascaded in three different ways:

1. *Synchronous counter ICs cascaded as an asynchronous counter:* The *RC* output of each stage is connected to the *CK* input of the succeeding stage and the clock pulses are applied at the *CK* input of the first stage. In this, each IC is synchronous within itself, but between stages the overall system is a ripple counter.
2. *Synchronous counter ICs cascaded with ripple carry between stages:* The *RC* output of each stage is connected to the *ENAB* input of the succeeding stage. All *CK* inputs are connected together and the clock pulses are applied at this common clock terminal.

3. *Synchronous counter ICs cascaded with parallel carry:* Figure 8.33 shows a 3-decade synchronous counter with parallel carry. The speed of operation is maximum in this type of cascading. The number of stages that can be cascaded in this manner may be restricted due to loading of MAX/MIN output by the external gating.

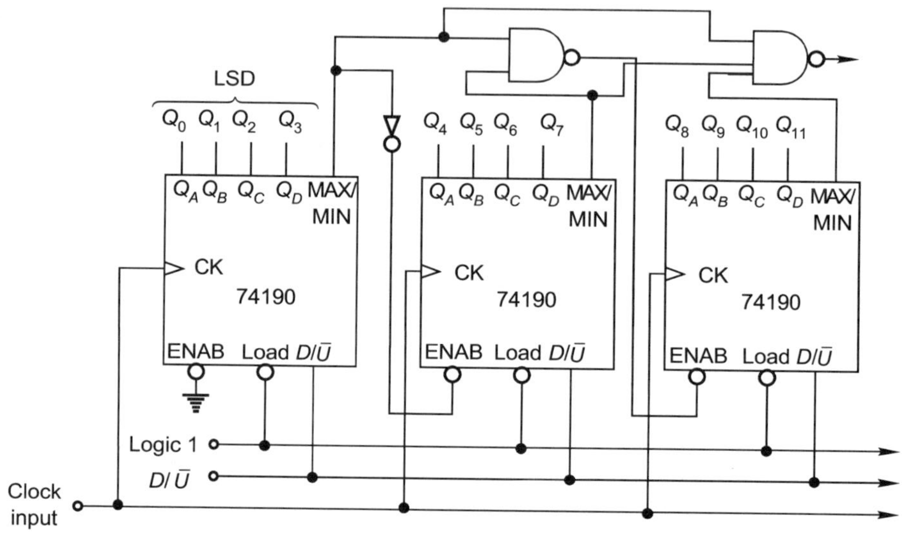

Fig. 8.33

A 3-decade synchronous counter using 74190 counter ICs with parallel carry.

Group D synchronous counter ICs

In these counters, for UP counting the clock is applied at *CK*-UP terminal with *CK*-DOWN connected to logic 1 and for DOWN counting at the *CK*-DOWN terminal with *CK*-UP connected to logic 1.

The carry and borrow outputs are normally at logic 1. The carry output drops to logic 0 when the counter shows its maximum count while UP counting and the *CK*-UP input is at logic 0. The borrow output remains at logic 1 as long as the circuit is operating from the *CK*-UP input.

The function of borrow in DOWN counting, is the same as the function of carry in UP counting.

The block diagram and function table of these ICs are given in Fig. 8.34.

Frequency dividers

These counters can be used as programmable frequency dividers in a manner similar to the one used for group C counters except that carry (or borrow) output is to be connected to the load input for UP (or DOWN) counting.

For example, in the DOWN count mode, if the binary equivalent of decimal number 11 is applied to the preset inputs of a 74193 4-bit binary counter, the frequency of the pulses at borrow output will be one-eleventh of the clock frequency.

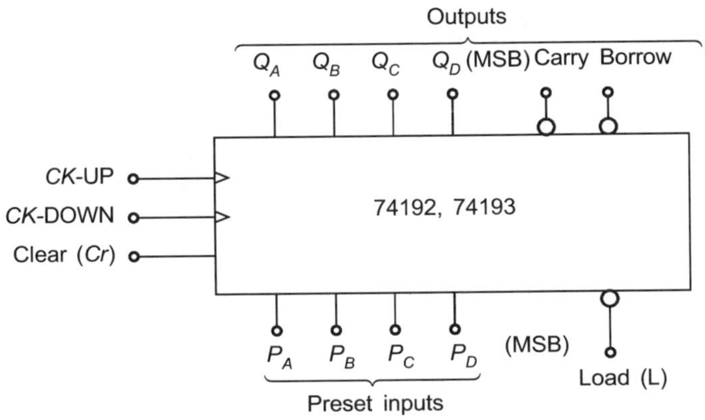

(a)

Fig. 8.34

Group D synchronous counter ICs (a) Block diagram.

Load *L*	*Clear* *Cr*	*CK-UP*	*CK-DOWN*	*Mode*
×	1	×	×	Reset to Zero
1	0	↑	1	UP count
1	0	1	↑	DOWN count
0	0	×	×	Preset
1	0	1	1	Stop count

(b)

Fig. 8.34

(b) Function table.

Cascading of group D counters

For cascading these counter ICs, the carry and borrow outputs of each stage are to be connected to the *CK*-UP and *CK*-DOWN inputs of the succeeding stage respectively. For steering the clock to *CK*-UP or *CK*-DOWN input the circuit shown in Fig. 8.35 can be used.

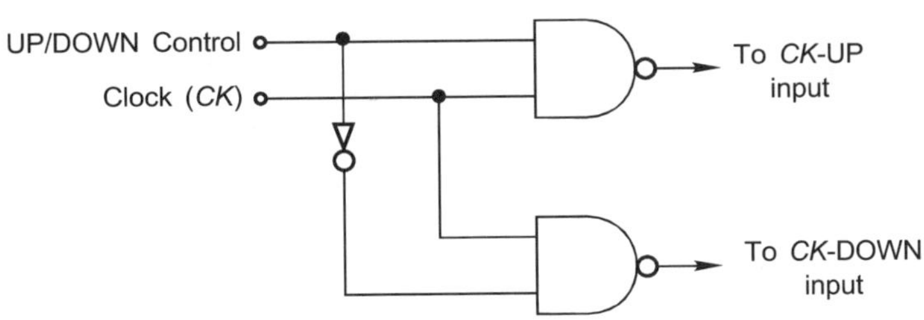

Fig. 8.35

Circuit for steering clock pulses to *CK*-UP (for UP count) or *CK*-DOWN (for DOWN count).

8.6 CLOCKED SEQUENTIAL CIRCUIT DESIGN

The block diagram of a sequential circuit as shown in Fig. 7.1 uses clock signal input for the memory elements which are FLIP-FLOPs. All the FLIP-FLOPs are clocked simultaneously. The outputs and the next state of a clocked sequential circuit depend upon the external inputs and the present state of the circuit. There are two commonly used models of such circuits. These are:

Mealy Model and *Moore Model* The next state and the output for these models are expressed as follows:

For Mealy model

Next State = F_1 (Present state, inputs)
Outputs = F_2 (Present state, inputs)
Whereas for Moore Model
Next state = F_1 (Present state, inputs)
Outputs = F_2 (Present state)

Figure 7.1 represents Mealy model. For the Moore model the signal path from the external inputs to output logic block is not present.

For these circuits, the sequence of inputs, present and next states, and output can be represented by a *state table* or a *state diagram.*

The design of any clocked sequential circuit, also referred to as *state machine* is similar to the design methods used for FLIP-FLOPs (Chapter 7), and counters, etc.

Let us consider a clocked sequential circuit represented by the state table given in Table 8.14. Figure 8.36 represents its state diagram. The state table gives information about the present state of the circuit (in terms of the outputs of the FLIP-FLOPs), the next state of the circuit for all the combinations of the external inputs (X is an external input) and the outputs for all the combinations of the inputs. The column for the output is necessary only if the outputs are taken from the logic gates. In case the outputs are taken directly from the FLIP-FLOPs, then the output column is not required and this information is available in the present state column.

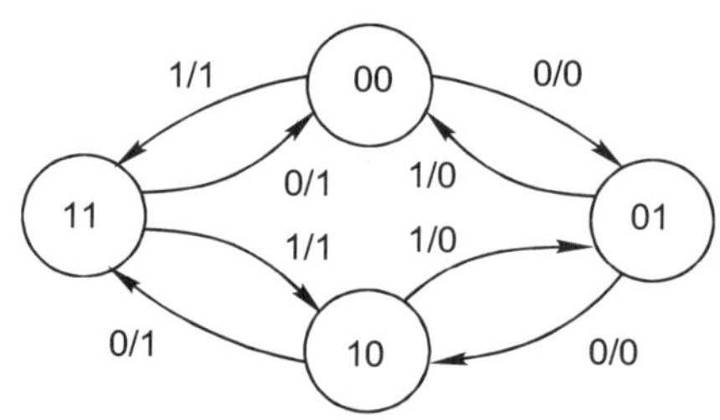

Fig. 8.36
State diagram for state Table 8.14.

Table 8.14 State Table of a sequential circuit

Present state		*Next state*				*Output*	
		$X = 0$		$X = 1$			
A	*B*	*A*	*B*	*A*	*B*	$X = 0$	$X = 1$
0	0	0	1	1	1	0	1
0	1	1	0	0	0	0	0
1	0	1	1	0	1	1	0
1	1	0	0	1	0	1	1

The state diagram represents the information of state table in graphical form. In this diagram, a state is represented by a circle (with the state indicated inside the circle) and the directed lines connecting the states indicate the transition between the states when the inputs are applied and the circuit is clocked. The input and output conditions are labelled with two binary numbers separated by the symbol '/' along with the directed line. The first number indicates the input value that causes the state transition, and the second number (after the symbol /) gives the value of the output.

In case an input condition does not cause a change of state, the fact is indicated by a directed line terminating on the same circle from which it originated.

The design procedure starts with the determination of the number of FLIP-FLOPs required, which depends upon the number of states. Sometimes it may be possible to reduce the number of states, referred to *state reduction*, which may result in the reduction of FLIP-FLOPs. For the purpose of state reduction, we have to identify the *equivalent states*. Two states are said to be equivalent if, for each input condition, they give exactly the same output and the same next state. When two states are equivalent, one of them may be eliminated without altering the input–output relations.

Once the number of states is minimized, the next step in the design is to assign the binary numbers to the states, referred to as *state assignment*. State assignment is required in case we are interested in only obtaining a prescribed sequence of outputs for a given sequence of inputs. In such a situation, the actual binary states are not important and we can assign suitable binary states (on the basis of minimum hardware requirement).

Using the FLIP-FLOPs and the states as obtained above, the inputs to the FLIP-FLOPs are required to be determined using the characteristic table of the FLIP-FLOP.

The expressions for the FLIP-FLOP inputs and external outputs are minimized using *K*-maps.

The synchronous counter is an example of a clocked sequential circuit in which the FLIP-FLOP outputs are used as outputs and external inputs and outputs are not present.

Example 8.15 Design a clocked sequential circuit for the state diagram shown in Fig. 8.37.

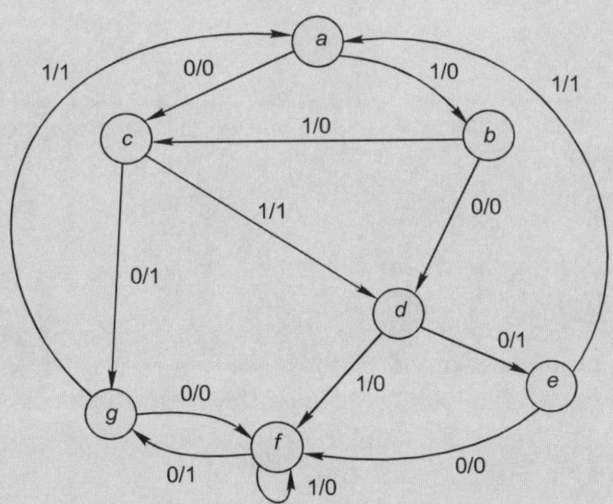

Fig. 8.37.

Solution

From the state diagram, a state table is prepared as shown in Table 8.15. If we are interested only in the sequence of output for a given input sequence, then we should reduce the state diagram if possible. The reduced state diagram will have fewer states and hence will need lesser hardware.

Table 8.15

Present state	Next state		Output	
	$X = 0$	$X = 1$	$X = 0$	$X = 1$
a	*c*	*b*	0	0
b	(*d*)*f*	*c*	0	0
c	(*g*)*e*	(*d*)*f*	1	1
d	*e*	*f*	1	0
e	*f*	*a*	0	1
f	(*g*)*e*	*f*	1	0
g	*f*	*a*	0	1

States *e* and *g* are equivalent because the next states and the outputs are same. We can eliminate any one of them, say *g*. Thus, *g* is replaced by *e* wherever it occurs. Now we notice that the states *d* and *f* are equivalent. Thus, one of them say *d*, can be eliminated. The reduced state table is given in Table 8.16.

Table 8.16

Present state	Next state		Output	
	$X = 0$	$X = 1$	$X = 0$	$X = 1$
a	*c*	*b*	0	0
b	*f*	*c*	0	0
c	*e*	*f*	1	1
e	*f*	*a*	0	1
f	*e*	*f*	1	0

Now each state can be assigned binary values. Since there are five states, therefore the number of FLIP-FLOPs required is 3 and 3-bit binary numbers are to be assigned to the states. For example, the state assignment may be

$$a = 000,\ b = 001,\ c = 010,\ e = 011, \text{ and } f = 100$$

The unused states can be treated as don't-care conditions. The circuit can be designed using the methods discussed earlier.

The reduced state table incorporating the assigned states is given in Table 8.17. In this $Q_2\ Q_1\ Q_0$ are the outputs of the three edge-triggered D FFs and give the present state whereas $Q_2^*\ Q_1^*\ Q_0^*$ gives the next state. The excitation inputs of the D FFs will be

$$D_2 = Q_2^*$$
$$D_1 = Q_1^*$$
$$D_0 = Q_0^*$$

Table 8.17

Present state			*Input*	*Next state*			*Output*
Q_2	Q_1	Q_0	X	Q_2^*	Q_1^*	Q_0^*	Y
0	0	0	0	0	1	0	0
0	0	0	1	0	0	1	0
0	0	1	0	1	0	0	0
0	0	1	1	0	1	0	0
0	1	0	0	0	1	1	1
0	1	0	1	1	0	0	1
0	1	1	0	1	0	0	0
0	1	1	1	0	0	0	1
1	0	0	0	0	1	1	1
1	0	0	1	1	0	0	0

The *K*-maps for D_2, D_1, D_0, and Y as outputs and Q_2, Q_1, Q_0, and X as inputs are given in Fig. 8.38. The unused states have been taken as don't-care conditions.

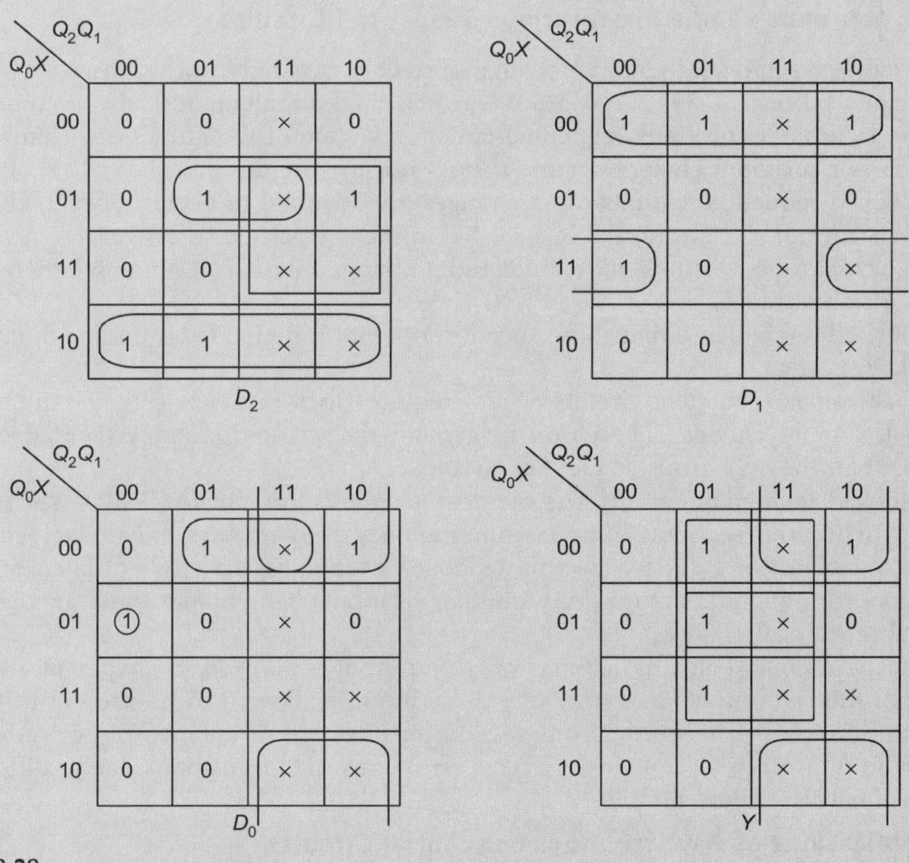

Fig. 8.38
K-maps for Table 8.17.

The minimized expressions are:

$$D_2 = Q_0 \cdot \bar{x} + Q_2 \cdot x + Q_1 \cdot \bar{Q}_0 \cdot x$$
$$D_1 = \bar{Q}_0 \cdot \bar{x} + \bar{Q}_1 \cdot Q_0 \cdot x$$
$$D_0 = Q_2 \cdot \bar{x} + Q_1 \cdot \bar{Q}_0 \cdot \bar{x} + \bar{Q}_2 \cdot \bar{Q}_1 \cdot \bar{Q}_0 \cdot x$$
$$Y = Q_1 \cdot \bar{Q}_0 + Q_1 \cdot x + Q_2 \cdot \bar{x}$$

The complete circuit is given in Fig. 8.39.
The circuit can be designed using *J-K* FFs also (Prob. 8.30).

8.7 ASYNCHRONOUS SEQUENTIAL CIRCUITS

The two types of sequential circuits were introduced in Section 7.1. The design of clocked sequential circuits have been discussed in section 8.6. Another important class of sequential circuits, i.e. asynchronous sequential circuits have been discussed here.

8.7.1 Asynchronus versus Synchronous Sequential Circuits

- In a clocked sequential circuit a change of state occurs only in response to a synchronizing clock pulse. All the FLIP-FLOPs are clocked simultaneously by a common clock pulse. In an asynchronous sequential circuit, the state of the circuit can change immediately when an input change occurs. It does not use a clock.
- In clocked sequential circuits input changes are assumed to occur between clock pulses. The circuit must be in the stable state before next clock pulse arrives.

  In asynchronous sequential circuits input changes should occur only when the circuit is in a stable state.
- In clocked sequential circuits, the speed of operation depends on the maximum allowed clock frequency.

  Asynchronous sequential circuits do not require clock pulses and they can change state with the input change. Therefore, in general the asynchronous sequential circuits are faster than the synchrous sequential circuits.
- In clocked sequential circuits, the memory elements are clocked FLIP-FLOPs.

  In asynchronous sequential circuits, the memory elements are either unclocked FLIP-FLOPs (latches) or gate circuits with feedback producing the effect of latch operation.
- In clocked sequential circuits, any number of inputs can change simultaneously (during the absence of the clock).

  In asynchronous sequential circuits only one input is allowed to change at a time in the case of the level inputs and only one pulse input is allowed to be present in the case of the pulse inputs. If more than one level inputs change simultaneously or more that one pulse input is present, the circuit makes erroneous state transitions due to different delay paths for each input variable.

8.7.2 Applications of Asynchronous Sequential Circuits

- An asynchronous circuit is preferred over synchronous circuit when high speed of operation is required. Since asynchronous sequential circuits respond immediately whenever there is a change in any input variable without having to wait for a clock pulse.

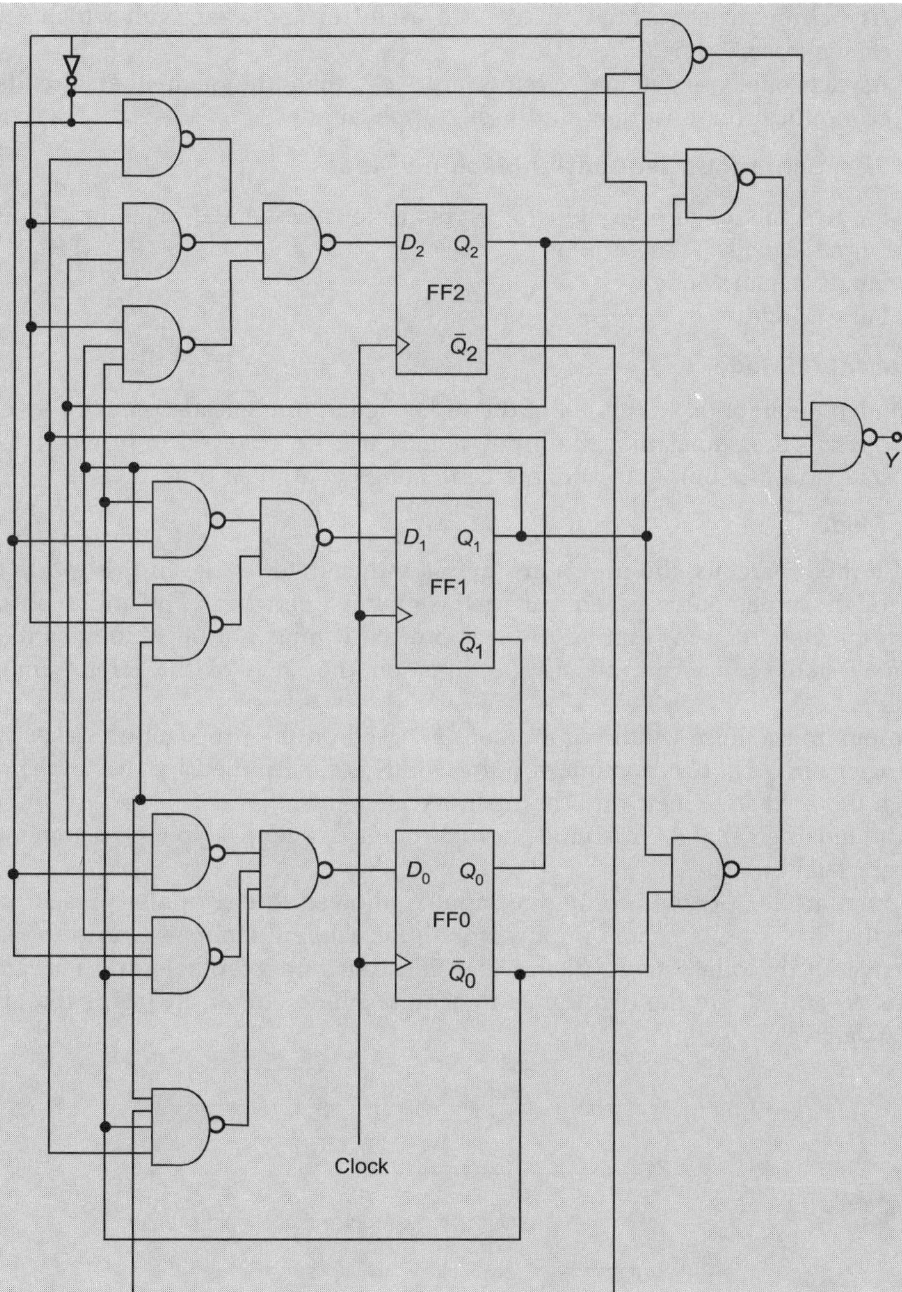

Fig. 8.39

- Asynchronous sequential circuits are useful in applications in which input signals may change at any time.
- Asynchronous sequential circuits cost less than the sequential circuits, therefore, for economical reasons, they find useful applications.

8.7.3 Asynchronous Sequential Machine Modes

There are two modes of operations of asynchronous sequential machines depending upon the type of input signals. These are:

- Fundamental Mode
- Pulse Mode

Fundamental Mode

In a fundamental mode circuit, all of the input signals are considered to be levels. Fundamental mode operation assumes that the input signals will be changed only when the circuit is in a stable state and that only one variable can change at a given time.

Pulse Mode

In pulse mode circuits, the inputs are pulses rather than levels. In this mode of operation the width of the input pulses is critical to the circuit operation. The input pulse must be long enough for the circuit to respond to the input but it must not be so long as to be present even after new state is reached. In such a situation the state of the circuit may make another transition.

The minimum pulse width requirement is based on the propagation delay through the next-state logic (Fig 7.1). The maximum pulse width is determined by the total propagation delay through the next state logic and the memory elements.

Both fundamental and pulse mode asynchronous sequential circuits use unclocked S-R FLIP-FLOPs or latches.

In pulse-mode operation, only one input is allowed to have pulse present at any time. This means that when pulse occurs on any one input, while the circuit is in stable state, pulse must not arrive at any other input. Figure 8.40 illustrates unacceptable and acceptable input pulse change. X_1 and X_2 are the two inputs to a pulse mode circuit. In Fig. 8.40a at time t_3 pulse at input X_2 arrives.

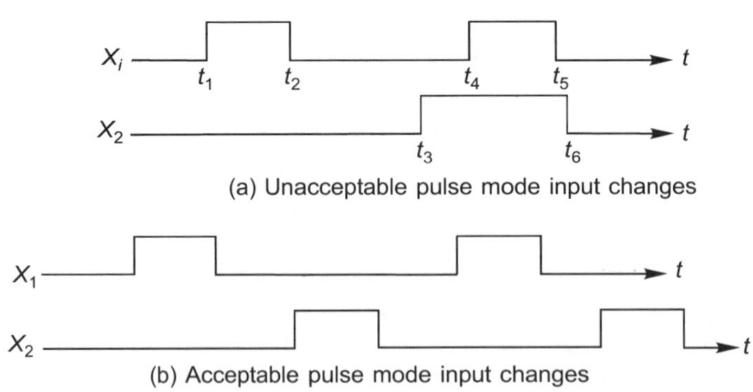

Fig. 8.40

While this pulse is still present, another pulse at X_1 input arrives at t_4. Therefore, this kind of the presence of pulse inputs is not allowed.

8.7.4 Analysis of Asynchronous Sequential Machines

Analysis of asynchronous sequential circuits operation in fundamental mode and pulse mode will help in clearly understanding the asynchronous sequential circuits.

Fundamental Mode Circuits

Fundamental mode circuits are of two types:

- Circuits without latches
- Circuits with latches

Circuits without Latches Consider a fundamental mode circuit shown in Fig. 8.41. This circuit has only gates and no explicit memory elements are present. There are two feedback paths from Q_1 and Q_2 to the next-state logic circuit. This feedback creates the latching effect due to delays, necessary to produce a sequential circuit. It may be noted that a memory element latch is created due to feedback in gate circuit (Fig. 7.4).

The first step in the analysis is to identify the *states* and the *state variables.* The combination of level signals from external sources X_1, X_2 is referred to as the *input state* and X_1, X_2 are the *input state variables.* The combination of the outputs of memory elements are known as *secondary*, or *internal states* and these variables are known as *internal* or *secondary state variables.* Here, Q_1 and Q_2 are the internal variables since no explicit elements are present. The combination of both, input state and the secondary state (Q_1, Q_2, X_1, X_2) is known as the *total state.* Y is the output variable.

The next secondary state and output logic equations are derived from the logic circuit in the next-state logic block. The next-secondary state variables are denoted by Q_1^+ and Q_2^+ these are given by

$$Q_1^+ = X_1 \bar{X}_2 + \bar{X}_1 X_2 Q_2 + X_2 Q_1 \bar{Q}_2 \tag{8.6}$$

$$Q_2^+ = \bar{X}_1 X_2 \bar{Q}_1 + X_1 Q_2 + X_2 Q_2 \tag{8.7}$$

$$Y = X_1 \oplus Q_1 \oplus Q_2 \tag{8.8}$$

Here, Q_1 and Q_2 are the present secondary state variables when X_1 X_2 input-state variables occur, the circuit goes to next secondary state. A state table shown in Table 8.18 is constructed using these logic equations. If the resulting next secondary state is same as the present state, i.e. $Q_1^+ = Q_1$ and $Q_2^+ = Q_2$, the total state Q_1 Q_2 X_1 X_2 is said to be *stable.* Otherwise it is unstable. The stability of the next total state is also shown in Table 8.18.

Transition Table A state table can be represented in another form known as *transition table.* The transition table for the state table of Table 8.18 is shown in Fig. 8.42.

In a transition table, columns represent input states (one column for each input state) and rows represent secondary states (one row for each secondary state). The next secondary state values are written into the squares, each indicating a total state. The stable states are circled. For any given present secondary state (Q_1 Q_2), the next secondary state is located in the square corresponding to row for the present secondary state and the column for the input state (X_1 X_2). For example, for $Q_1Q_2 = 11$ and $X_1X_2 = 00$, the next secondary state is 00 (third row, second column) which is an unstable state.

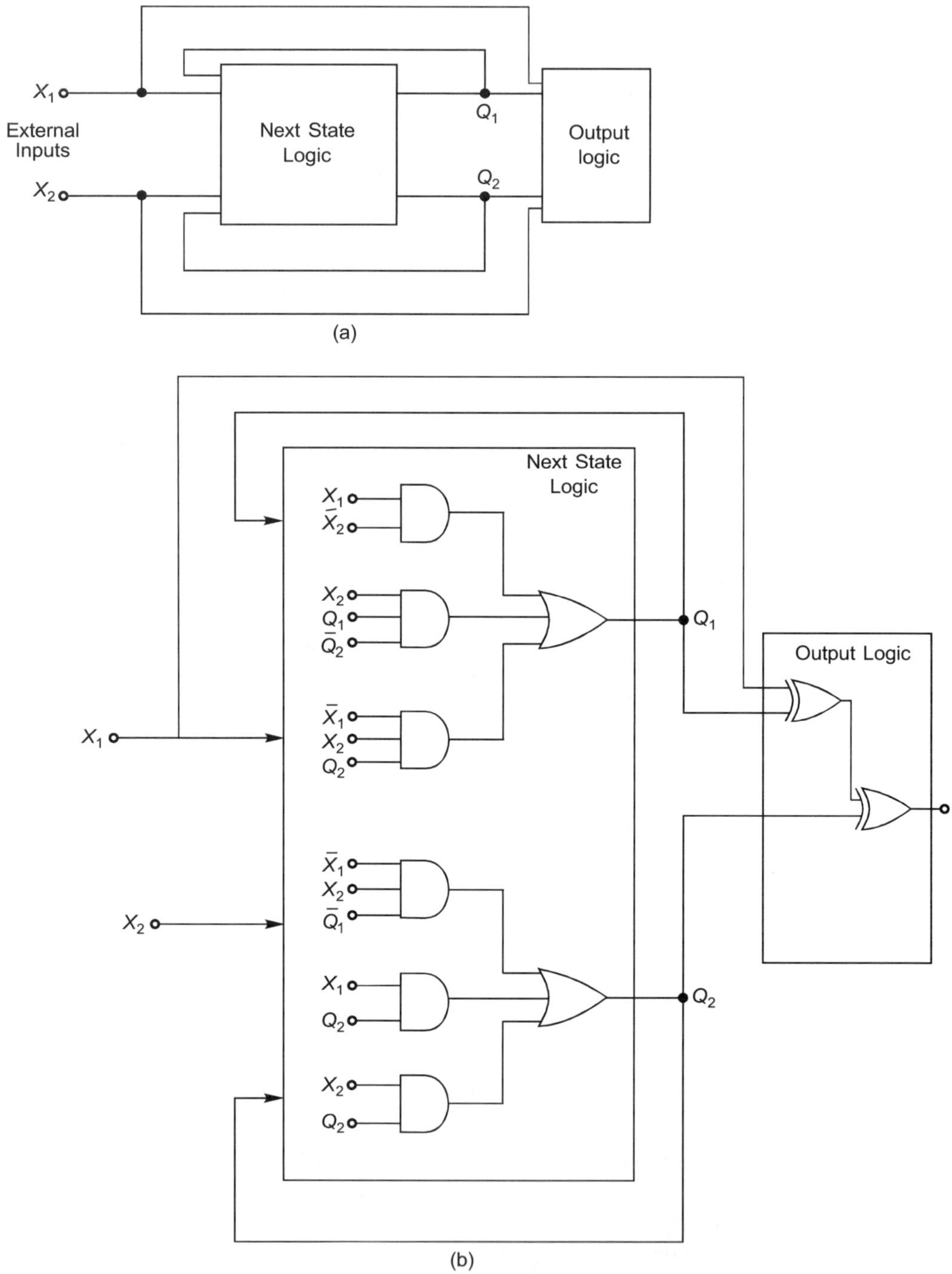

Fig. 8.41

Fundamental mode asynchronous sequential circuit without latch
(a) block diagram (b) circuit diagram

Table 8.18 State Table

Present total state				*Next total state*				*Stable total state*	*Output*
Q_1	Q_2	X_1	X_2	Q_1^+	Q_2^+	X_1	X_2	*Yes/No*	Y
0	0	0	0	0	0	0	0	Yes	0
0	0	0	1	0	1	0	1	No	0
0	0	1	1	0	0	1	1	Yes	1
0	0	1	0	1	0	1	0	No	1
0	1	0	0	0	0	0	0	No	1
0	1	0	1	1	1	0	1	No	1
0	1	1	1	0	1	1	1	Yes	0
0	1	1	0	1	1	1	0	No	0
1	1	0	0	0	0	0	0	No	0
1	1	0	1	1	1	0	1	Yes	0
1	1	1	1	0	1	1	1	No	1
1	1	1	0	1	1	1	0	Yes	1
1	0	0	0	0	0	0	0	No	1
1	0	0	1	1	0	0	1	Yes	1
1	0	1	1	1	0	1	1	Yes	0
1	0	1	0	1	0	1	0	Yes	0

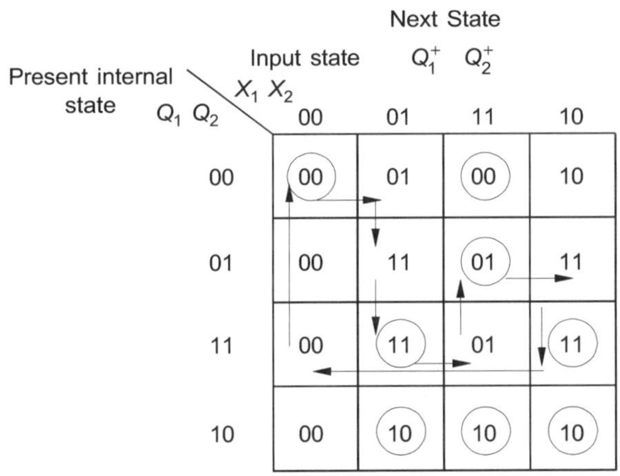

Fig. 8.42
Transition table for Table 8.18.

For a given input sequence, the total state sequence can be determined from the transition table.

Example 8.16 For the transition table shown in Fig 8.42, the initial total state is $Q_1Q_2\,X_1X_2$ = 0000. Find the total state sequence for an input sequence $X_1\,X_2$ = 00, 01, 11, 10, 00.

Solution
For a given internal state of the circuit, a change in the value of the circuit input causes a *horizontal* move in the transition table to the column corresponding to the new input value. A

change in the internal state of the circuit is reflected by a vertical move. Since a change in the input can occur only when the circuit is in a stable state, a horizontal move can emanate only from a circled entry.

The initial total state is 0000 (first row, first column) which is a stable state. When the input state changes from 00 to 01, the circuit makes a transition (horizontal move) from the present total state to the next total state 0101 (first row, second column) which is unstable. Next, the circuit makes another transition from 0101 to 1101 (vertical move) (second row, second column) which is also an unstable state. Finally in the next transition (vertical move) it comes to stable state 1101 (third row, second column). All these transitions are indicated by arrows. Thus we see that a single input change produces two secondary state changes before a stable total state is reached. If the input is next changed to 11 the circuit goes to total state 0111 (horizontal move) which is unstable and then to stable total state 0111 (vertical move). Similarly, the next input change to 10 will take the circuit to unstable total state 1110 (horizontal move) and finally to stable total state 1110 (vertical move). A change in input state from 10 to 00 causes a transition to unstable total state 0000 (horizontal move) and then to stable total state 0000 (vertical move), completing the state transitions for the input sequence. All the state transitions are indicated by arrows.

The total state sequence is

(0000) → 0101 → (1101) → (0111) → (1110) → (0000)

From the above discussions we see that from the logic diagram of an asynchronous sequential circuit, logic equations, state table, and transition table can be determined. Similarly, from the transition table, logic equations can be written and the logic circuit can be designed.

Flow table In asynchronous sequential circuits design, it is more convenient to use *flow table* rather than transition table. A flow table is basically similar to a transition table except that the internal states are represented symbolically rather than by binary states. The column headings are the input combinations and the entries are the next states, and outputs. The state changes occur with change of inputs (one input change at a time) and logic propagation delay.

The flow of states from one to another is clearly understood from the flow table. The transition table of Fig. 8.42 constructed as a flow table is shown in Fig. 8.43. Here, *a*, *b*, *c*, and *d* are the states. The binary value of the output variable is indicated inside the square next to the state symbol and is separated by a comma. A stable state is circled.

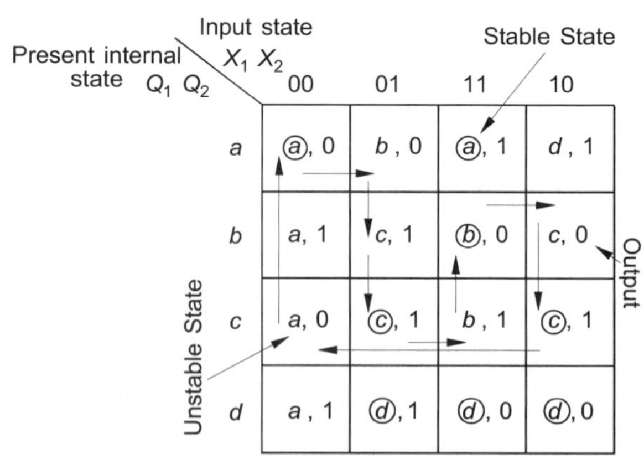

Fig. 8.43
Flow table

From the flow table, we observe the following behaviour of the circuit.

When $X_1 X_2 = 00$, the circuit is in state ⓐ. It is a stable state. If X_2 changes to 1 while $X_1 = 0$, the circuit goes to state *b* (horizontal move) which is an unstable state. Since *b* is an unstable state, the circuit goes to *c* (vertical move), which is again an unstable state. This causes another vertical move and finally the circuit reaches a stable state ⓒ. Now consider X_1 changing to 1 while $X_2 = 1$, there is a horizontal movement to the next column. Here b is an unstable state and therefore, there is a vertical move and the circuit comes to a stable state ⓑ. Next change in X_2 from 1 to 0 while X_1 remaining 1 will cause horizontal move to state c (unstable state) and finally to stable state ⓒ due to the vertical move. Similarly changing X_1 from 1 to 0 while $X_2 = 0$ will cause the circuit to go to the unstable state a and finally to stable state ⓐ. The flow of circuit states are shown by arrows.

In the flow table of Fig. 8.43 there are more than one stable states in rows. For example, the first row contains stable states in two columns. If every row in a flow table has only one stable state, the flow table is known as a *primitive flow table.*

From a flow table, transition table can be constructed by assigning binary values to each state and from the transition table logic circuit can be designed by constructing *K*-maps for Q_1^+ and Q_2^+.

Circuits with Latches

In Section 7.2 latch was introduced using NAND gates. Latch circuits using NAND and NOR gates are shown in Fig. 8.44.

For the circuit of Fig 8.44a, the next-state equation is

$$Q^+ = \overline{\bar{S} \cdot \bar{Q}} = \overline{\bar{S} \cdot \overline{(Q\bar{R})}}$$

$$= S + \bar{R}\,Q \qquad (8.9)$$

Similarly, for the circuit of Fig. 8.44 b, the next-state equation is

$$Q^+ = \overline{R + \bar{Q}} = \overline{R + \overline{(S + Q)}}$$

$$= \bar{R} \cdot (S + Q)$$

$$= S\bar{R} + \bar{R}Q$$

Since, $S = R = 1$ is not allowed, which means $SR = 0$, therefore,

$$S\bar{R} = S\bar{R} + SR = S(\bar{R} + R) = S$$

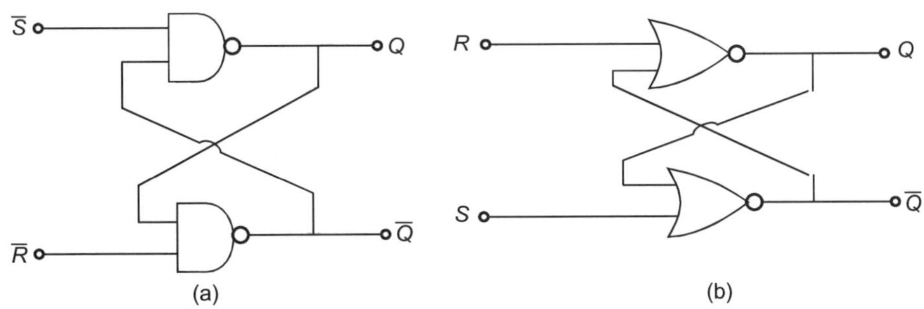

Fig. 8.44

(a) $\bar{S}$-$\bar{R}$ latch using NAND gates.

(b) S-R latch using NOR gates.

which gives,

$$Q^+ = S + \bar{R}Q$$

It is same as Eq. (8.9)

The transition table of *S-R* latch is shown in Fig. 8.45.

Q \ SR (Q^+)	00	01	11	10
0	⓪	⓪	⓪	1
1	①	0	0	①

$$Q^+ = S\bar{R} + \bar{R}Q = S + \bar{R}Q$$

Fig. 8.45

Transition table of *S-R* latch

From the transition table of *S-R* FLIP-FLOP, we observe that when *SR* changes from 11 to 00 the circuit will attain either the stable state ⓪ (first row, first column) or ① (second row, first column) depending upon whether *S* goes to 0 first or *R* goes to 0 first respectively. Therefore, $S = R = 1$ must not be applied.

Consider an asynchronous sequential circuit with latches shown in Fig. 8.46.

For FF-1, $R_1 = 0$ and the excitation equation for S_1 is

$$S_1 = X_1 \bar{X}_2 \bar{Q}_2 + \bar{X}_1 Q_2 \tag{8.10}$$

The next-state equation is

$$Q_1^+ = S_1 + \bar{R}_1 Q_1 \tag{8.11}$$

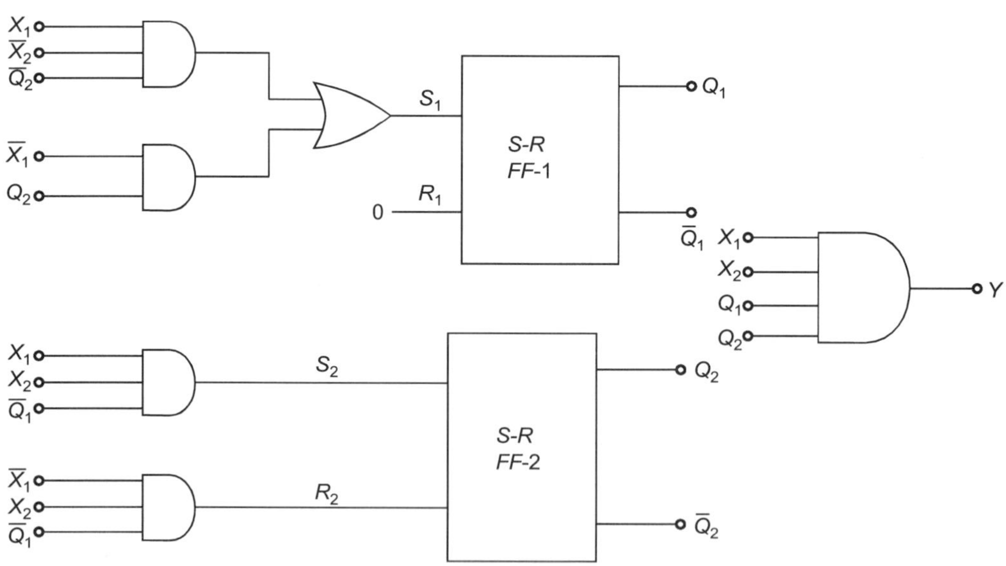

Fig. 8.46

Asynchronous sequential circuit with latches

Substituting the value of S_1 from Eq. (8.10) in Eq.(8.11) we obtain,

$$Q_1^+ = X_1 \bar{X}_2 \bar{Q}_2 + \bar{X}_1 Q_2 + Q_1 \tag{8.12}$$

Similarly, the excitation equations for FF-2 are

$$S_2 = X_1 X_2 \bar{Q}_1, \; R_2 = \bar{X}_1 X_2 \bar{Q}_1 \tag{8.13}$$

The next-state equation is

$$\begin{aligned} Q_2^+ &= S_2 + \bar{R}_2 Q_2 \\ &= X_1 X_2 \bar{Q}_1 + \overline{\bar{X}_1 X_2 \bar{Q}_1} \cdot Q_2 \end{aligned} \tag{8.14}$$

Using Eqs. (8.12) and (8.14), transition table is obtained as shown in Fig. 8.47.
The output function is

$$Y = X_1 X_2 Q_1 Q_2 \tag{8.15}$$

Its flow table is shown in Fig. 8.48.

$Q_1^+ Q_2^+$

$Q_1 Q_2$ \ $X_1 X_2$	00	01	11	10
00	(00)	(00)	01	10
01	11	11	(01)	(01)
11	(11)	10	(11)	(11)
10	(10)	(10)	(10)	(10)

Fig. 8.47
Transition table for the circuit of Fig. 8.46

$Q_1^+ Q_2^+$

$Q_1 Q_2$ \ $X_1 X_2$	00	01	11	10
a	ⓐ, 0	ⓐ, 0	*b*, 0	*d*, 0
b	*c*, 0	*c*, 0	ⓑ, 0	ⓑ, 0
c	ⓒ, 0	*d*, 0	ⓒ, 1	ⓒ, 0
d	ⓓ, 0	ⓓ, 0	ⓓ, 0	ⓓ, 0

Fig. 8.48
Flow table for the circuit of Fig. 8.46

From a flow table , transition table can be obtained by assigning binary values to the states. From the transition table, logic equations can be obtained by constructing *K*-maps for *S* and *R* inputs of every latch. For this, the excitation table of *S-R* latch will be used. Logic circuit can then be designed using the logic equation for *S*, *R* inputs of every latch.

Example 8.17 Design logic circuit using *S-R* latches for the transition table of Fig. 8.42.

Solution

Since, there are two internal states Q_1 and Q_2, therefore, two *S-R* latches are required for the design of logic circuit. Let the two latches be L_1 and L_2. The inputs and outputs of these latches are given below.

Latch	*Inputs*	*Outputs*
L_1	S_1, R_1	$Q_1, \bar{Q}_1$
L_2	S_2, R_2	$Q_2, \bar{Q}_2$

The excitation table of an *S-R* latch is given in Table 8.19. This is same as Table 7.1 for *S-R* FLIP-FLOP.

Table 8.19 Excitation table of *S-R* latch

Present state	*Next state*	*Inputs*	
Q	Q^+	*S*	*R*
0	0	0	×
0	1	1	0
1	0	0	1
1	1	×	0

To determine S_1 and R_1 for different values of $X_1 X_2$, we make use of Q_1 and Q_1^+ values for every square of transition table. For example, this square in the first row and first column gives $Q_1 = 0$ and $Q_1^+ = 0$. This means, for the present state 0 the circuit gives next state as 0 for Q_1. Corresponding to this we find the value of S_1 and R_1 using the Table 8.19, which are

$$S_1 = 0 \quad \text{and} \quad R_1 = X$$

Thus the entry in the cell corresponding to $X_1 X_2 = 00$ and $Q_1 Q_2 = 00$ for *K*-map of S_1 will be 0 and for *K*-map of R_1 it will be *X*. Similarly, *K*-map entries are determined for S_1 and R_1.

Following similar procedure, *K*-maps for S_2 and R_2 are constructed. The *K*-maps are given in Fig. 8.49.

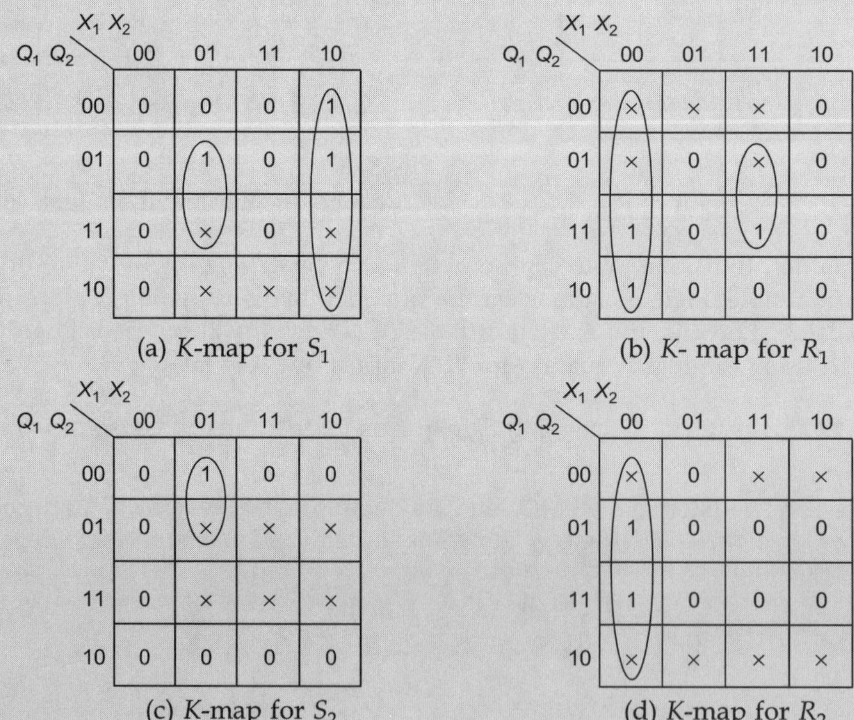

(a) *K*-map for S_1

(b) *K*- map for R_1

(c) *K*-map for S_2

(d) *K*-map for R_2

Fig. 8.49

From the K-map of Fig. 8.49, we obtain logic equations for S_1, R_1 S_2, and R_2.

$$S_1 = X_1\bar{X}_2 + \bar{X}_1 X_2 Q_2$$

$$R_1 = \bar{X}_1 + X_1 X_2 Q_2$$

$$S_2 = \bar{X}_1 X_2 \bar{Q}_2$$

$$R_2 = \bar{X}_1 \bar{X}_2$$

The logic circuit is shown in Fig. 8.50.

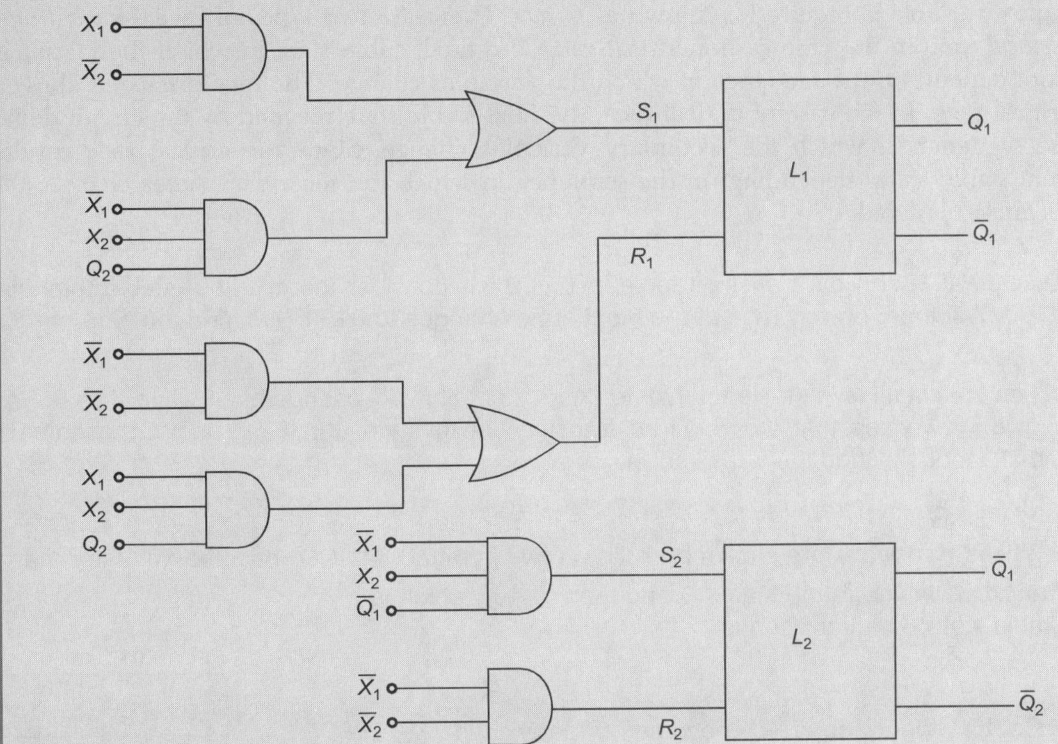

Fig. 8.50
Logic circuit for Ex. 8.17

Races and Cycles

A *race* condition exists in an asynchronous sequential circuit when more than one state variable change value in response to a change in an input variable. This is caused because of unequal propagation delays in the path of different secondary variables in any practical electronic circuit. Consider a transition table shown in Fig. 8.51. When both the inputs X_1 and X_2 are 0 and the present state is $Q_1\ Q_2 = 00$, the resulting next-state $Q_1^+\ Q_2^+$ will have $Q_1^+ = 1$ and $Q_2^+ = 1$ simultaneously if the propagation delays in the paths of Q_1 and Q_2 are equal.

Q_1Q_2 \ X_1X_2 ($Q_1^+\ Q_2^+$)	00	01	11	10
00	11	(00)	10	01
01	11	00	11	(01)
11	(11)	00	10	(11)
10	11	(10)	(10)	11

Fig. 8.51

Since Q_1 and Q_2 both are to change and in general the propagation delays in the paths of Q_1 and Q_2 are not same, therefore, either Q_1 or Q_2 may change first instead of both changing simultaneously. As a consequence of this the circuit will go to either state 01 or to state 10.

If Q_2^+ changes faster than Q_1^+, the next state will be 01, then 11 (first column, second row) and then to the stable state (11) (first column, third row) will be reached. On the other hand, if Q_1^+ changes faster than Q_2^+, the next-state will be 10, then 11 (first column, fourth row) and then to the stable state (11) (first column, third row) will be reached. In both the situations, the circuit goes to the same final stable state (11). This situation, where a change of more than one secondary variable is required is known as a *race*. There are two types of races: *noncritical race* and *critical race*. In the case of noncritical race, the final stable state in which the circuit goes does not depend on the sequence in which the variables change. The race discussed above is a noncritical race. In the case of critical race, the final stable state reached by the circuit depends on the sequence in which the secondary variables change. Since the critical race results in different stable states depending on the sequence in which the secondary states change, therefore, it must be avoided.

Example 8.18 In the transition table of Fig. 8.51, consider the circuit in stable total state 1100. Will there be any race, if the input state changes to 01? If yes, find the type of race.

Solution

When the circuit is in stable total state, X_1X_2 = 00. Now X_2 changes to 1 while X_1 = 0. From Fig. 8.51 we ses that the required transition is to state 00. If Q_1^+ and Q_2^+ become 00 simultaneously, then the transition will be

(11) → 00 → (00)

These transitions are shown by solid arrows in Fig. 8.52. If Q_2^+ becomes 0 faster than Q_1^+, the circuit will go to the state 10 and then to (10), which is a stable state. The transition is

(11) → 10 → (10)

On the other hand, if Q_1^+ becomes 0 faster than Q_2^+, the transition will be

(11) → 01 → 00 → (00)

It is shown by dotted arrow in Fig. 8.52. Thus, we see that the circuit attains different stable states (00) or (10) depending up on the sequence in which the secondary variables change. Therefore, the race condition exists in this circuit and it is critical race.

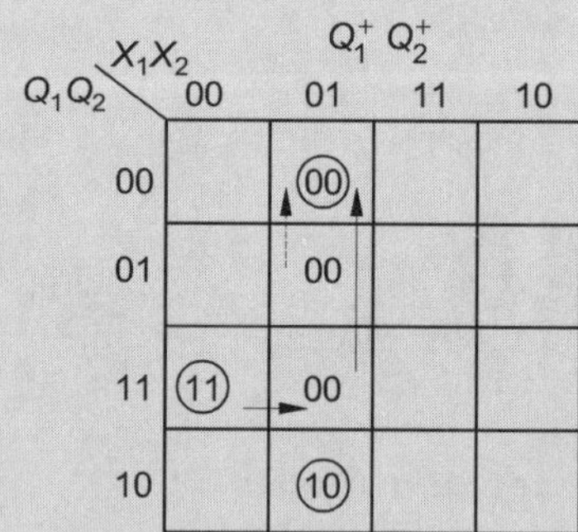

Fig. 8.52

Races can be avoided by making a proper binary assignment to the state variables in a flow table. The state variables must be assigned binary numbers in such a way so that only one state variable can change at any one time when a state transition occurs in the flow table. The state transition is directed through a unique sequence of unstable state variable change. This is referred to as a *cycle*. This unique sequence must terminate in a stable state, otherwise the circuit will go from one unstable state to another unstable state making the entire circuit unstable.

Example 8.19 In the state transition table of Fig. 8.51, if X_1X_2 = 10 and the circuit is in stable state ⓪①, find the cycle when X_2 is changed to 1 while X_1 remaining 1.

Solution

The circuit is in stable state (01) (fourth column, second row). When X_2 changes to 1, the circuit will go to the state 11 (third column, second row), then to state 10 (third column, third row) and finally to the stable state (10) (third column, fourth row). Thus the cycle is

$$\textcircled{01} \rightarrow 11 \rightarrow 10 \rightarrow \textcircled{10}$$

Pulse-mode circuits

In a pulse, mode asynchronous sequential circuit, an input pulse is permitted to occur only when the circuit is in stable state and there is no pulse present on any other input. When an input pulse arrives, it triggers the circuit and causes a transition from one stable state to another stable state so as to enable the circuit to receive another input pulse. In this mode of operation critical race can not occur. To keep the circuit stable between two pulses, FLIP-FLOPs whose outputs are levels must be used as memory elements.

For the analysis of pulse-mode circuits, the model used for the fundamental-mode circuits is not valid since the circuit is stable when there are no inputs and the absence of a pulse conveys no information. For this a model similar to the one used for synchronous sequential circuits will be convenient to use.

In pulse-mode asynchronous circuits the number of columns in the next-state table is equal to the number of input terminals.

Consider a pulse-mode circuit logic diagram shown in Fig. 8.53. In this circuit there are four input variables X_1, X_2, X_3, and X_4, and Y is the output variable. It has two states Q_1 and Q_2.

The excitation and output equations are:

$$\bar{S_1} = \overline{(X_2 + X_3)} \quad \text{or} \quad S_1 = X_2 + X_3 \tag{8.16a}$$

$$\bar{R_1} = \bar{X_4} \quad \text{or} \quad R_1 = X_4 \tag{8.16b}$$

$$\bar{S_2} = \overline{\bar{Q_1} X_1} \quad \text{or} \quad S_2 = \bar{Q_1} X_1 \tag{8.16c}$$

$$\bar{R_2} = \overline{(X_4 + \bar{Q_1}\, X_3)} \quad \text{or} \quad R_2 = X_4 + \bar{Q_1} X_3 \tag{8.16d}$$

$$Y = X_4 \bar{Q_2} \tag{8.17}$$

The next-state equations are obtained by using the excitation equations (Eqs. (8.16)) and the characteristic equation of latch (Eq. (8.9)). These are:

$$\begin{aligned} Q_1^+ &= S_1 + \bar{R_1} Q_1 \\ &= X_2 + X_3 + \bar{X_4}\, Q_1 \end{aligned} \tag{8.18}$$

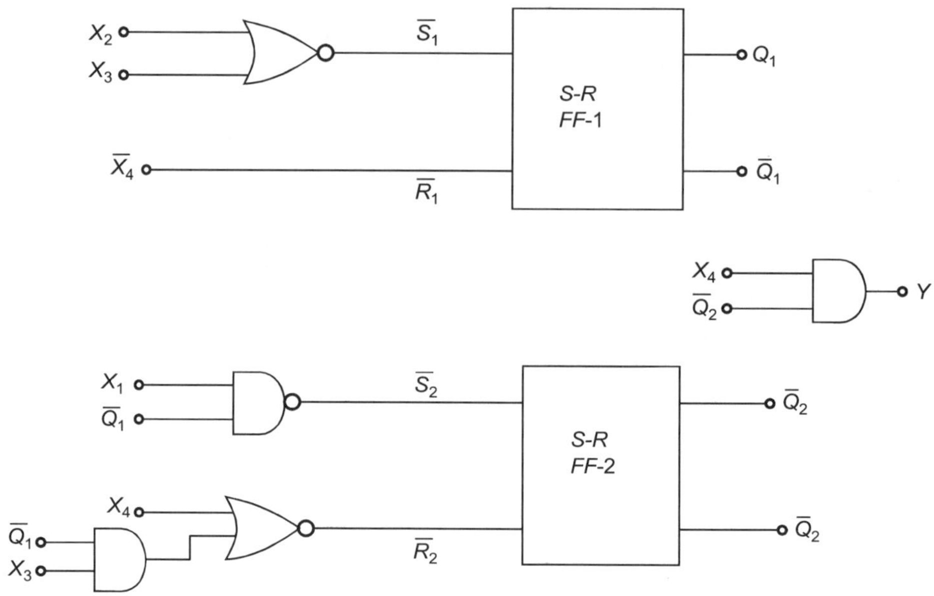

Fig. 8.53
A pulse-mode asynchronous circuit

and

$$\begin{aligned} Q_2^+ &= S_2 + \bar{R}_2 Q_2 \\ &= \bar{Q}_1 X_1 + \overline{(X_4 + \bar{Q}_1 X_3)} \cdot Q_2 \\ &= \bar{Q}_1 X_1 + \bar{X}_4 \cdot \overline{(\bar{Q}_1 X_3)} \cdot Q_2 \\ &= \bar{Q}_1 X_1 + \bar{X}_4 \cdot (Q_1 + \bar{X}_3) \cdot Q_2 \\ &= \bar{Q}_1 X_1 + Q_1 Q_2 \bar{X}_4 + Q_2 \bar{X}_3 \bar{X}_4 \end{aligned} \tag{8.19}$$

The transition table is constructed by evaluating the next-state and output for each present state and input value using Eqs. (8.17), (8.18), and (8.19). The transition table is shown in Fig. 8.54. It has four rows (one row for each combination of state variables) and four columns (one column for each input variable). Since in pulse-mode circuits only one input variable is permitted to be present at a time, therefore, the columns are for each input variable only and not for the combinations of input variables.

Flowtable can be constructed from the transition table and is shown in Fig. 8.55. Here, S_0, S_1, S_2, and S_3 are the four state variables.

From a flow table a transition table can be constructed by assigning binary values to the states. From a transition table next-state equations can be obtained and the logic diagram can then be obtained.

Present State Q_1Q_2 \ Input variables	X_1	X_2	X_3	X_4
00	01, 0	10, 0	10, 0	00, 1
01	01, 0	11, 0	10, 0	00, 0
11	11, 0	11, 0	11, 0	00, 0
10	10, 0	10, 0	10, 0	10, 0

(Next-state value, Output value)

Fig. 8.54

Transition table of Fig. 8.53

Present-State Q_1Q_2 \ Input variables	X_1	X_2	X_3	X_4
S_0	S_1, 0	S_3, 0	S_3, 0	S_0, 1
S_1	S_1, 0	S_2, 0	S_3, 0	S_0, 0
S_2	S_2, 0	S_2, 0	S_2, 0	S_0, 0
S_3	S_3, 0	S_3, 0	S_3, 0	S_0, 1

Fig. 8.55

Flow table of pulse-mode circuit

8.7.5 Asynchronous Sequential Circuit Design

Design of asynchronous sequential circuits is more difficult than that of synchronous sequential circuits because of the timing problems involved in these circuits. Designing an asynchronous sequential circuit requires obtaining logic diagram for the given design specifications. Usually the design problem is specified in the form of statements of the desired circuit performance precisely specifying the circuit operation for every applicable input sequence.

Design Steps

1. Primitive flow table is obtained from the design specifications. When setting up a primitive flow table it is not necessary to be concerned about adding states which may ultimately turn out to be redundant. A sufficient number of states are to be included to completely specify the circuit performance for every allowable input sequence. Outputs are specified only for stable states.
2. Reduce the primitive flow table by eliminating the redundant states, which are likely to be present. These redundant states are eliminated by merging the states. *Merger diagram* is used for this purpose.
3. Binary numbers are assigned to the states in the reduced flow table. The binary state assignment must be made to ensure that the circuit will be free of critical races. The output values are to be chosen for the unstable states with unspecified output entries. These must be chosen in such a way so that momentary false outputs do not occur when the circuit switches from one stable state to another stable state.
4. Transition table is obtained next.
5. From the transition table logic diagram is designed by using the combinational design methods. The logic circuit may be a combinational circuit with feedback or a circuit with *S-R* latches.

The above design steps will be illustrated through an example.

Example 8.20 The output (Y) of an asynchronous sequential circuit must remain 0 as long as one of its two inputs X_1 is 0. While X_1 = 1, the occurrence of first change in another input X_2, should give Y = 1 as long as X_1 = 1 and becomes 0 where X_1 returns to 0. Construct a primitive flow table.

Solution

This circuit has two inputs X_1, X_2 and one output Y. For the construction of flow table, the next-state and output are required to be obtained. The flow table is shown in Fig. 8.56.

Present-state \ $X_1 X_2$	00	01	11	10
a	ⓐ, 0	*b*, 0	–, –	*c*, 0
b	*a*, 0	ⓑ, 0	*d*, 0	–, –
c	*a*, 0	–, –	*e*, –	ⓒ, 0
d	–, –	*b*, 0	ⓓ, 0	*f*, –
e	–, –	*b*, –	ⓔ, 1	*f*, 1
f	*a*, –	–, –	*e*, 1	ⓕ, 1

Fig. 8.56
Flow table of Ex. 8.20

For X_1X_2 = 00, let us take state *a*. When the circuit has X_1X_2 = 00 the output is 0 (since X_1 = 0) and the circuit is in stable state ⓐ. The next-state and output are shown in the first column, first row of Fig. 8.56. Since only one input is allowed to change at a time, therefore, the next input may be X_1X_2 = 01 or 10.

If X_1X_2 = 01, let us take another state *b*, correspondingly the second row of the flow table corresponds to state *b*. when the inputs change from X_1X_2 = 00 to 01, the circuit is required to go to stable state ⓑ and output is 0 (since X_1 = 0). Therefore, the entry in the second column, first row will be *b*, 0 and in the second column, second row will be ⓑ, 0. The output corresponding to unstable state *b* is taken as 0 so that no momentary false outputs occur when the circuit switches between stable states. On the other hand if X_1X_2 = 10, the circuit is required to go to another stable state ⓒ with output 0. Therefore, the entries in the fourth column, first row anf fourth column, third row will be respectively *c*, 0 and ⓒ, 0.

Since both the inputs cannot change simultaneously, therefore, from stable state ⓐ, the circuit cannot go to any specific state corresponding to X_1X_2 = 11 and accordingly the entry in the third column, first row will be –, –. The dashes represent the unspecified state, output.

Now consider the stable state ⓑ. The inputs X_1X_2 can chage to 00 or 11.

If X_1X_2 = 00, the circuit will go to state *a*. Therefore, the entry in the first column, second row will be *a*, 0. From this unstable state the circuit goes to stable state ⓐ. On the other hand if X_1X_2 = 11, then the circuit goes to *a* new state *d*. The output corresponding to X_1X_2 = 11 will be 0 since, there is no change in X_2, which is already 1. Therefore, the entry in the third column, second row will be *d*, 0. The fourth row corresponds to state *d*, and the entry in the third column, fourth row, will be ⓓ, 0. From ⓑ, the circuit is not required to go to any specific state and therefore, the entry in the fourth column, second row will be –,–.

Similarly, now consider stable state ⓒ. The inputs can change to X_1X_2 = 11 or 00. If $X_1 X_2$ = 11, the circuit goes to a new stable state ⓔ and the output will be 1, since X_2 changes from 0 to 1 while X_1 = 1. The entry in the third column, third row will be *c*, –. Output has to change from 0 to 1 from stable state ⓒ to stable state ⓔ, which may or may not change to 1 for unstable *e*. The entry in the third column, fifth row will be ⓔ, 1. The entry in the second column third row will be –,– and the entry in the first column, third row will be *a*, 0 (for X_1X_2 = 00).

In the same manner, we consider the stable ⓓ and obtain the entries *f*, – (fourth column, fourth row); ⓕ, 1 (fourth column, sixth row); *b*, 0 (second column, fourth row) and –,– (first column, fourth row).

Similar procedure applied to ⓔ and ⓕ yields the remaining entries of the flow table (Problem. 8.36).

Since, every row in the flow table of Fig. 8.56 contains only one stable state, therefore, this flow table is a primitive flow table.

Reduction of States

The necessity of reducing the number of states has been discussed in Section 8.6 and the equivalent states have been defined. When asynchronous sequential circuits are designed, the design process starts from the construction of primitive flow table. A primitive flow table is never completely specified. Some states and outputs are not specified in it as shown in Fig. 8.56 by dashes. Therefore, the concept of equivalent states can not be used for reduction of states. However, incompletely specified states can be combined to reduce the number of states in the flow table. Two incompletely specified states can be combined if they are *compatible.*

Two states are *compatible* if and only if, for every possible input sequence both produce the same output sequence whenever both outputs are specified and their next states are compatible whenever they are specified. The unspecified outputs and states shown as dashes in the flow table have no effect for compatible states.

Example 8.21 In the primitive flow table of Fig. 8.56, find whether the states *a* and *b* are compatible or not. If compatible, find out the merged state.

Solution

The rows corresponding to the states *a* and *b* are shown in Fig. 8.57. Each column of these two states is examined.

X_1X_2	00	01	11	10
a	ⓐ, 0	*b*, 0	–, –	*c*, 0
b	*a*, 0	ⓑ, 0	*d*, 0	–, –

Fig. 8.57

Column –1 Both the rows have the same state *a* and the same output 0. *a* in first row is stable state and in the second row is unstable state.

Since for the same input both the states *a* and *b* have the same specified next-state *a* and the same specified output 0. Therefore, this input condition satisfies the requirements of compatibility.

Column –2 The input condition $X_1\ X_2$ = 01 satisfies the requirements of compatibility as discussed for column–1.

Column –3 The first row has unspecified next-state and output and the second row has specified state and output. The unspecified state and output may be assigned any desired state and output and therefore, for this input condition also the requirements of compatibility are satisfied.

Column –4 The requirements of compatibility are satisfied for the reasons same as applicable to column-3.

Therefore, we conclude that since the next-states and the outputs for all the input combinations are compatible for the two states *a* and *b*, the two states are compatible.

The merged state will be as shown in Fig. 8.58

X_1X_2	00	01	11	10
a	ⓐ, 0	ⓐ, 0	*d*, 0	*c*, 0

Fig. 8.58

When the merged state entries are determined a circled entry and an uncircled entry results in a circled entry, since the corresponding state must be stable as shown in Fig. 8.58.

Example 8.22 In the primitive flow table of Fig. 8.56 find whether the states *a* and *e* are compatible or not. Examine their compatibility if the entries in the fourth column for the states *a* and *e* have same output.

Solution

The partial flowtable for states *a* and *e* of Fig. 8.56 is shown in Fig. 8.59.

X_1X_2	00	01	11	10
a	ⓐ, 0	*b*, 0	–, –	*c*, 0
e	–, –	*b*, –	ⓔ, 1	*f*, 1

Fig. 8.59

From this we observe the following

Column-1 compatible
Column-2 compatible
Column-3 compatible
Column-4 not compatible, since the outputs are different.

Therefore, the states *a* and *e* are not compatible.

In case of same output in column-4, the outputs are said to be *not conflicting* and the states *a* and *e* are compatible if and only if the states *c* and *f* are compatible. This is referred to as *c*, *f* is *implied* by *a*, *b* or *a*, *b implies c*, *f*.

Merger Diagram

A *merger diagram (or graph)* is prepared for a primitive flow table to determine all the possible compatible states (maximal compatible states) and from this a minimal collection of compatibles covering all the states.

A merger graph is constructed following the steps outlined below:

- Each state is represented by a vertex, which means it consists of *n* vertices, each of which corresponds to a state of the circuit for an *n*- state primitive flow table. Each vertex is labelled with the state name.
- For each pair of compatible states an undirected arc is drawn between the vertices of the two states. No arc is drawn for incompatible states.
- For compatible states implied by other states a broken arc is drawn between the states and the implied pairs are entered in the broken space.

The flow table is required to be examined for all the possible pairs of states. All the pairs are checked and the merger graph is obtained. Thus we see that the merger graph displays all possible pairs of compatible states and their implied pairs. Next it is necessary to check whether the incompatible pair(s) does not invalidate any other implied pair. If any implied pair is invalidated it is neglected. All the remaining valid compatible pairs form a group of *maximal* compatibles.

The maximal compatible set can be used to construct the reduced flow table by assigning one row to each member of the group. However, the maximal compatibles do not necessarily constitute the set of *minimal* compatibles. The set of minimal compatibles is a smaller collection of compatibles that will satisfy the row merging.

The conditions that must be satisfied for row merging are:

- the set of chosen compatibles must *cover* all the states, and
- the set of chosen compatibles must be *closed.*

The condition of covering requires inclusion of all the states of the primitive flow graph in the set of chosen compatibles. This condition only defines a *lower bound* on the number of states in the minimal set. However, if none of their implied pairs are contained in the set, the set is not sufficient and this is referred to as *closed* condition not being satisfied. Therefore, condition of *closed covering* is essentially required for row merging.

Example 8.23 Construct merger diagram for the primitive flow table of Fig. 8.56. Determine maximal compatibles and the minimal set of compatibles.

Solution

For construction of merger diagram, every row of the primitive flow table is checked with every other row to determine compatibility of states.

Consider row-1 (state *a*)

a, *b* are compatible
a, *c* are compatible
a, *d* are compatible if *c*, *f* are compatible
a, *e* are compatible if *c*, *f* are compatible
a, *f* are compatible if *c*, *f* are compatible

row-2 (state *b*)

b, *c* are compatible if *d*, *e* are compatible
b, *d* are compatible
b, *e* are not compatible (outputs are conflicting)
b, *f* are not compatible (outputs are conflicting)

row-3 (state *c*)

c, *d* are compatible if *e*,*d* and *c*, *f* are compatible
c, *e* are not compatible (outputs are conflicting)
c, *f* are not compatible (outputs are conflicting)

row-4 (state *d*)

d, *e* are not compatible (outputs are conflicting)
d, *f* are not compatible (outputs are conflicting)

row-5 (state *e*)

e, *f* are compatible

The primitive flow table has six states therefore, there are six vertices in the merger diagram as shown in Fig 8.60. Solid arcs are drawn between (*a*, *b*), (*a*, *c*), (*b*, *d*) and (*f*, *e*) vertices. Corresponding to these states being compatibles. Since (*c*, *f*) and (*d*, *e*) are not compatible, therefore, there are no implied pairs available.

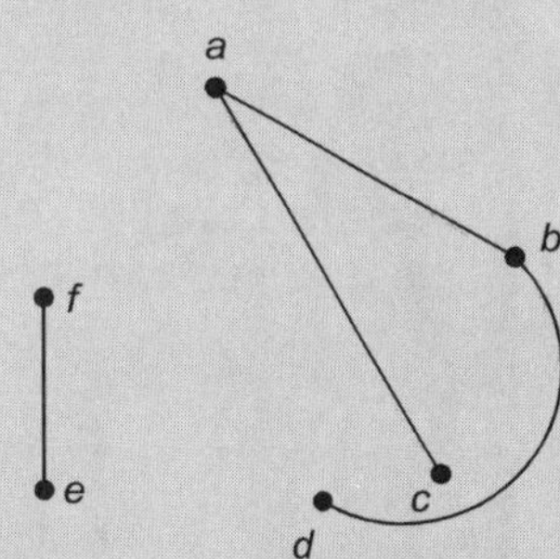

Fig. 8.60 Merger diagram

From the merger diagram, we get the maximal compatibles:
(*a*, *b*), (*a*, *c*), (*b*, *d*), (*e*, *f*)

Since (*a*, *b*) is covered by (*a*, *c*) and (*b*, *d*), therefore, the minimal set is
(*a*, *c*), (*b*, *d*), (*e*, *f*)

Example 8.24 Determine the reduced flow table of Fig. 8.57.

Solution

From the merger diagram, we have obtained three pairs of compatible states: These compatibles are merged and are represented by

$a, c : S_0$
$b, d : S_1$
$e, f : S_2$

The reduced flow table is shown in Fig. 8.61

Present-state \ X_1X_2	00	01	11	10
S_0	(S_0), 0	S_1, 0	S_2, –	(S_0), 0
S_1	S_0, 0	(S_1), 0	(S_1), 0	S_2, –
S_2	S_0, –	S_1, –	(S_2), 1	(S_2), 1

Fig. 8.61
Reduced flow table

Example 8.25 Assign binary states to the reduced flow table of Fig. 8.61. Avoid critical race.

Solution

Let us assign the following binary states to S_0, S_1, and S_2 for the reduced flow table of Fig. 8.61

$S_0 \rightarrow 00$
$S_1 \rightarrow 01$
$S_2 \rightarrow 11$

The transition table will be as shown in Fig. 8.62

In the transition table of Fig. 8.62, we observe that race condition occurs in the following cases:

(i) From stable state (00) to unstable state 11 when $X_1 X_2$ changes from 10 to 11.

(ii) From stable state (11) to unstable state 00 when $X_1 X_2$ changes from 10 to 00.

To avoid critical race, one unstable state 10 is added with the entries

00, –; –, –; 11, –; –, –

and the entries in third column, first row is changed from 11, – to 10, – and in first column, third row from 00, – to 10, –.

The modified transition table is given in Fig. 8.63.

$Q_1 Q_2$ \ $X_1 X_2$	00	01	11	10
00	(00), 0	01, 0	11, –	(00), 0
01	00, 0	(01), 0	(01), 0	11, –
11	00, –	01, –	(11), 1	(11), 1
10				

Fig. 8.62
Transition table

$Q_1 Q_2$ \ $X_1 X_2$	00	01	11	10
00	(00), 0	01, 0	10, –	(00), 0
01	00, 0	(01), 0	(01), 0	11, –
11	10, –	01, –	(11), 1	(11), 1
10	00, –	–, –	11, –	–, –

Fig. 8.63
Modified transition table

Example 8.26 Design logic circuit with feedback for the transition table of Fig. 8.63

Solution

The *K*-maps for Q_1^+, Q_2^+, and Y determined from the transition table are given in Fig.8.64.

From the *K*-maps, we obtain,

$$Q_1^+ = X_1 X_2 \bar{Q}_2 + \bar{X}_2 Q_1 Q_2 + X_1 \bar{X}_2 Q_2 + X_1 Q_1$$

$$Q_2^+ = \bar{X}_1 X_2 + X_1 Q_2 + X_1 Q_1$$

$$Y = Q_1$$

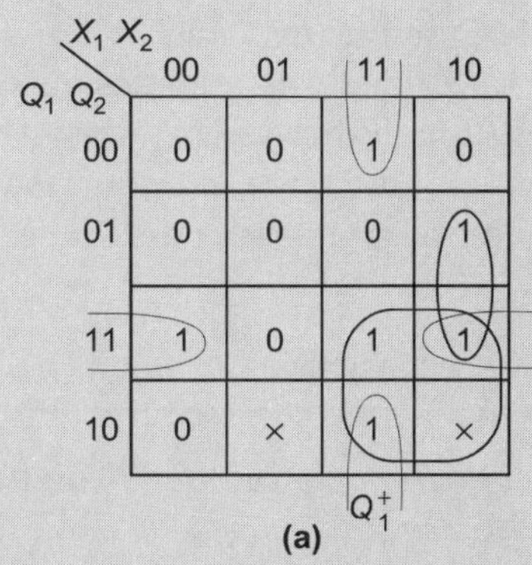

Logic circuit using gates can be obtained from the above logic equations.

Thus we see that the design steps outlined above can be used to design an asynchronous sequential circuit.

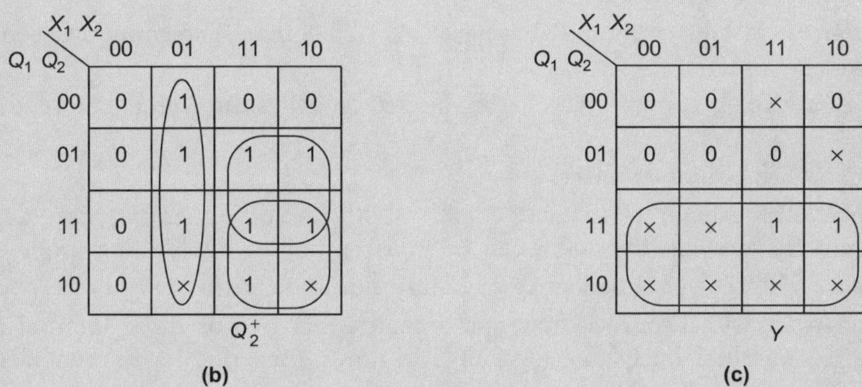

Fig. 8.64

K-maps for (a) Q_1^+ (b) Q_2^+ (c) Y

8.8 SUMMARY

A very important and useful class of digital circuits, sequential circuits, have been discussed in detail in this chapter. Clocked sequential circuits, such as, counters, shift registers, useful in many applications of digital systems have been discussed. Their design using FLIP-FLOPs and standard MSI devices have been included. Design of general clocked sequential circuits using the two models, Mealy and More, will be helpful in designing any clocked sequential circuit.

Standard Counter ICs are available for counting in normal binary sequence and natural BCD for different modulus. In many applications counters can be designed using these available ICs. In case any other sequence of count or modulus is required, then the circuit can be designed using FLIP-FLOPs and gates using the methods discussed here.

The analysis and design of asynchronous sequential circuits have been discussed in detail. Although the design of asynchronous sequential circuits is much more difficult than the design of clocked (synchronous) sequential circuits, nonetheless they find widespread applications requiring high speeds, low cost systems in which input signals may change at any time.

Glossary

Asynchronous counter A digital counter in which all the FLIP-FLOPs are not triggered simultaneously.

BCD counter A modulo-10 counter with count sequence from 0000 to 1001, i.e 0 to 9 decimal digitis.

Bidirectional register A shift register in which data can be shifted in both the directions, i.e from left to right and right to left.

Binary counter An N-stage counter that recycles after 2^N counts. The count proceeds in specificed binary sequence.

Counter, Pressetable A counter which can be set to a desired value before the start of the counting.

Counter, Ripple See Ripple counter.

Counter, UP-DOWN A counter that can count in both up and down direction, depending upon a control input. This means the count can be in ascending or descending order.

Count-sequence The repeating sequence of binary numbers at the output of a counter.

Critical race In an asynchronous sequential circuit when two or more internal state variables have to change simultaneously due to change in input, then due to different delay paths, it is very difficult to predict the state variable which will change first. Due to this race condition is said to exist and the circuit may go to different next-states depending upon the delays involved in different paths.

Cycle A cycle occurs is an asynchronous sequential circuit when multiple state transitions occur without changing the inputs, taking the circuit from one unstable state to another unstable state and finally reaching a stable state.

Decade counter A counter that recycles after every 10 pulses. Same as a BCD counter.

Down-counter A counter in which the count decreases with every clock pulse.

First-in-first-out (FIFO) A register organisation in which the data which is stored first is the first data available for taking out. (See bidirectional shift register)

Flowtable A tabular method for showing state transitions in asynchronous sequential circuits.

Fundamental mode An asynchronous sequential circuit with level inputs.

Frequency divider A sequential logic circuit that can divide the frequency of a pulse source by an integer.

Internal state A state in asynchronous sequential circuit. Same as secondary state.

Left shift register A shift register in which data gets shifted in the left direction in response to clock pulses.

Last-in-first-out (LIFO) A register organisation in which the data which is stored last is the first data available for taking out. (See bidirectional shift register).

Loading of a counter Presetting (or loading) the counter with the desired initial count.

Lock out A condition which may exist in a counter taking the counter from one unused state to another unused state and not allowing it to come to any of the used states.

Merger diagram A graph used for the possible merging of states in an asynchronous sequential circuit.

Modulus (MOD) The number of clock pulses after which a counter reaches its initial state.

Non-critical race. A race condition in which an asynchronous sequential circuit will reach the same stable state.

Parallel-data All the data bits are available simultaneously on different data lines.

Parallel-loading Simultaneous loading of all the bits in a register or counter.

Parallel-to-serial converter A logic circuit that converts data from parallel to serial form.

Primitive flow table A flow table in which every row has only one stable state.

Pulse mode An asynchronous sequential circuit with pulse inputs.

Race A condition occuring in an asynchronous sequential circuit due to different path delays when transition takes place requiring more than one state to change simultaneously.

Right-shift register A shift register in which the data gets shifted in the right direction in response to clock pulses.

Ring counter A shift register whose output is fedback to the input, i.e. the data circulates around the FLIP-FLOPs in response to clock pulses.

Ripple counter A counter in which all the FLIP-FLOPs are not clocked simultaneously. A particular FLIP-FLOP will get triggered only if the state of the previous FLIP-FLOP changes suitably.

Serial data Data arranged as one bit at a time at a regular interval starting from the LSB.

Serial-to-Parallel converter A logic circuit that converts data from serial to parallel form.

Shift register A cascade of FLIP-FLOPs in which the data gets shifted in response to clock pulses.

State assignment Assigning binary values to states in a sequential circuit.

State diagram It is a graphical representation of a sequential circuit indicating the states, input values, output values, and directed arcs for state transitions.

State reduction Process of reducing the states by eliminating the redundant states in sequential circuits.

State table (or State transition table) A tabular form of representing the behaviour of a sequential circuit.

Synchronous counter A digital counter in which all the FLIP-FLOPs are clocked simultaneously.

Timing diagram Graphical representation of various signals with reference to time.
Transition table A tabular form of representing the behaviour of a sequential circuit.
Twisted-ring counter A shift register with its complement output ($\bar{Q}$) of the last stage connected to the *D*-input of the first stage.
UP-counter A counter that increments on every clock pulse.

Review Questions

8.1 The minimum number of FLIP-FLOPs required for a decade counter is ________ .

8.2 Race condition may exist in ________ sequential circuits.

8.3 The modulo of a 4-bit binary counter is ________ .

8.4 The count of a 4-bit binary DOWN Counter is 0000. When a clock pulse is applied its count will be ________ .

8.5 The flow table of an asynchronous sequential circuit with 6 states and 2 inputs will have ________ rows and ________ columns.

8.6 The cascade of divide-by 5 counter followed by divide-by 2 counter is in state 0000. When a clock pulse is applied its state will be ________ . Assume negative edge-triggered circuits.

8.7 The modulo of a 4-stage twisted-ring counter is ________ .

8.8 A ripple counter is ________ sequential circuit.

8.9 A 4-bit binary code 1010 will be represented as ________ in temporal code.

8.10 The speed of an asynchronous counter is ________ than that of a synchronous counter.

8.11 The number of inputs which can change simultaneously in a fundamental mode asynchronous circuit is ________ .

8.12 The number of pulse inputs which are allowed to be present simultaneously in a pulse-mode asynchronous sequential circuit is ________ .

8.13 A ________ shift register can be used for SISO, SIPO, PISO, and PIPO of data.

8.14 The minimum number of FLIP-FLOPs required to generate a sequence of 9 bits is ________ .

8.15 The maximum possible number of states in a clocked sequential circuit having 5 FLIP-FLOPs is ________ .

Problems

8.1 Explain the operation of the bi-directional shift register of Fig. 8.4.

8.2 Explain the operation of a twisted-ring counter and give its state diagram.

8.3 Verify the decoder logic of Fig. 8.7.

8.4 A stepper-motor drive circuit requires four signal waveforms given in Fig. 8.65. Design a sequence generator to provide the necessary signals for this stepper-motor drive.

8.5 Write the count sequence of a 3-bit binary DOWN counter. Design a ripple counter using FLIP-FLOPs for this sequence.

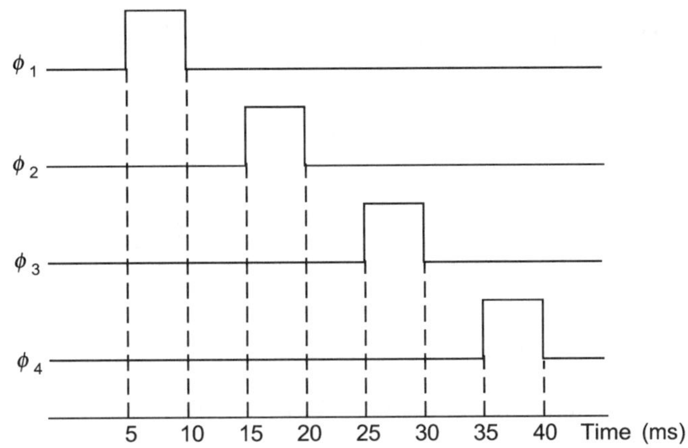

Fig. 8.65
Waveforms for Prob. 8.4.

8.6 Design a 4-bit binary UP/DOWN ripple counter with a control for UP/DOWN counting.

8.7 Design a 4-bit asynchronous counter with provision for asynchronous loading.

8.8 The latch used for eliminating resetting difficulties due to unequal internal delays of FLIP-FLOPs is shown in Fig. 8.13. Explain the operation of this latch and show how the resetting difficulties are eliminated.

8.9 Design the following ripple counters using FLIP-FLOPs:
(a) Divide-by-5, (b) Divide-by-7.

8.10 In a 7492 divide-by-12 ripple counter if Q_D output is connected to A input and the pulses are applied at the B input, find the count sequence.
Show that the frequency divisions of 3, 6, and 12 are obtained at the Q_C, Q_D, and Q_A outputs respectively.

8.11 Design the following counters using 7492 used as in Problem 8.10:
(a) Divide-by-7, (b) Divide-by-9, (c) Divide-by-11.

8.12 Convert 7493 into a 4-bit DOWN counter.

8.13 Design a divide-by-128 counter using 7493 ICs.

8.14 Design a divide-by-96 counter using 7490 ICs.

8.15 Design a divide-by-78 counter using 7493 and 7492 (as divide-by-6) counter ICs.

8.16 Sketch the output waveforms of the counter circuit of Fig. 8.27.

8.17 Sketch the output waveforms of the counter circuit of Fig. 8.28.

8.18 If the counter 74161 is replaced by 74163 in the Ex. 8.12, what will be its modulus? Also draw the output waveforms.

8.19 Explain the operation of the circuit of Fig. 8.29.

8.20 Design a mod-12 UP counter using 74193 IC and compare it with the circuit of Fig. 8.27.

8.21 Design a divide-by-5 DOWN counter using 74192 IC. Make use of the borrow output. Sketch the output waveforms.

8.22 Modify the state diagram of Fig. 8.36 to represent a modulo-4 UP/DOWN counter. Design its circuit using J-K FLIP-FLOPs.

8.23 Design a circuit that gives the input–output relationship shown in Fig. 8.66.

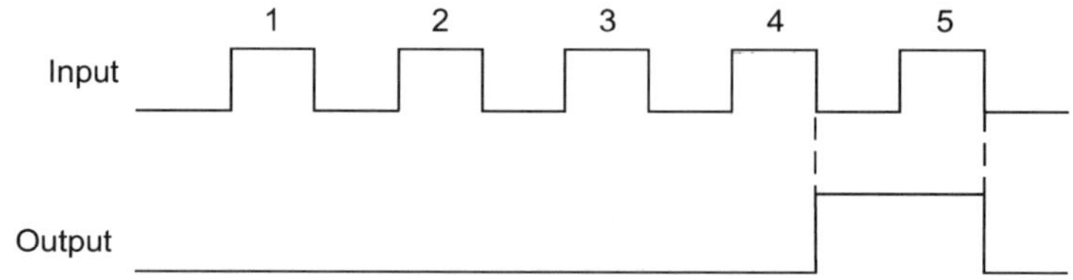

Fig. 8.66
Waveform for Prob. 8.23.

8.24 Design a decade counter to count in the Excess-3 code sequence. Use minimum number of *J-K* FFs.

8.25 Design a mod-12 normal binary counter for the count sequence 0000 to 1011 using 74163 IC.

8.26 A clocked synchronous sequential circuit using positive-edge-triggered D FFs has an input X and an output Y.
The excitation equations are:

$$D_1 = Q_1 \cdot \bar{X} + \bar{Q}_1 \cdot Q_0 \cdot X + Q_1 \cdot \bar{Q}_0 \cdot X$$

$$D_0 = Q_0 \cdot \bar{X} + \bar{Q}_0 \cdot X$$

and the output equation is

$$Y = Q_1 \cdot Q_0 \cdot X$$

(a) Draw its circuit diagram,
(b) Obtain its state diagram,
(c) Redesign the circuit using *J-K* FFs.

8.27 For a clocked sequential circuit shown in Fig. 8.67, obtain
(a) Excitation and output equations,
(b) The output sequence for an input sequence of 101001 assuming initial state to be $Q_1 Q_0 = 00$.

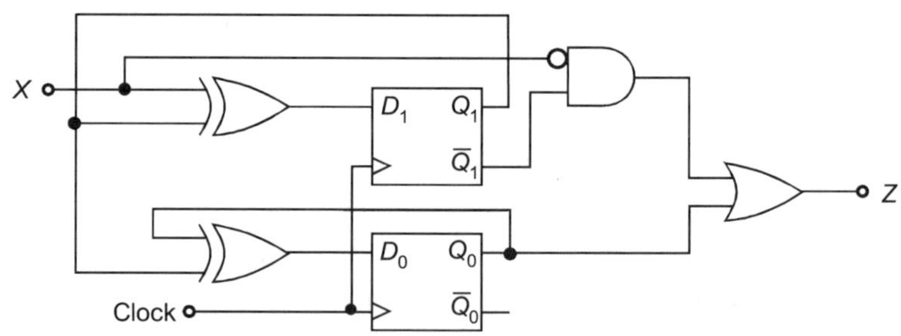

Fig. 8.67
Figure for Prob. 8.27.

8.28 For the state diagram shown in Fig. 8.68, obtain the state table and design the circuit using minimum number of *J-K* FFs.

8.29 For the state diagram shown in Fig. 8.69, design the circuit using
(a) *T* FFs, (b) *S-R* FFs, (c) *J-K* FFs.

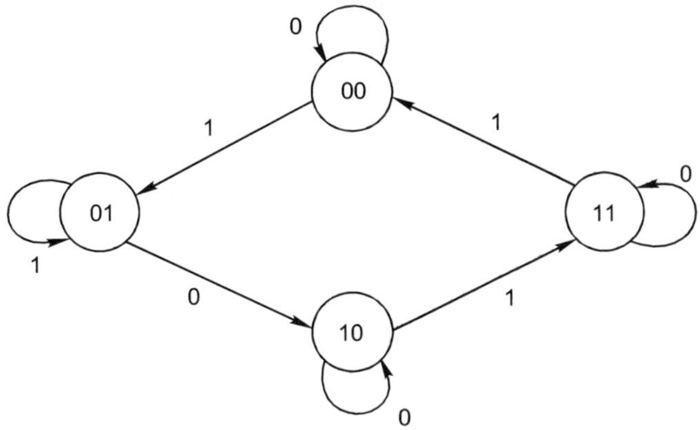

Fig. 8.68
State diagram for Prob. 8.28.

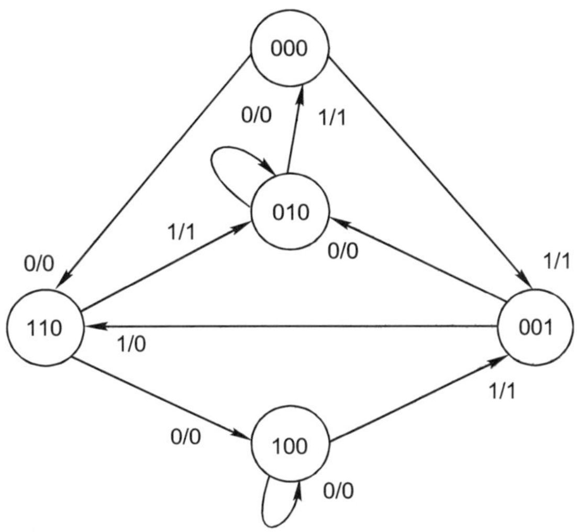

Fig. 8.69
State diagram for Prob. 8.29.

8.30 Design the clocked sequential circuit for Example 8.15 using minimum number of *J-K* FFs.

8.31 For the state table given in Table 8.20, obtain the state diagram. If possible, determine the reduced state table.

8.32 Design the circuit for Prob. 8.31 using minimum number of FFs of
(a) Edge-triggered *D* type,
(b) *J-K* type.

8.33 Obtain the output sequence for Prob. 8.31 for an input sequence of 01110010011. Assume initial state $Q_2Q_1Q_0 = 000$.

Table 8.20

Present State			*Next State*						*Output*	
			$x = 0$			$x = 1$			Y	
Q_2	Q_1	Q_0	Q_2	Q_1	Q_0	Q_2	Q_1	Q_0	$x = 0$	$x = 1$
0	0	0	1	0	1	0	0	1	0	0
0	0	1	0	1	1	0	1	0	0	0
0	1	0	1	0	1	1	0	0	0	0
0	1	1	1	1	0	0	0	0	1	0
1	0	0	0	1	1	0	1	0	0	0
1	0	1	1	0	1	0	0	1	1	1
1	1	0	1	1	0	1	1	1	0	1
1	1	1	1	1	0	0	0	0	1	0

8.34 (a) Explain the operation of the transition table shown in Fig. 8.70 from initial total state Q_1, Q_2, X_1 X_2 of

(i) 0001 when X_1X_2 becomes 11.

(ii) 1111 when X_1X_2 becomes 01.

(b) Determine whether there are race conditions in (a)–(i) & (ii) above and whether they are critical or noncritical.

Q_1Q_2 \ X_1X_2	00	01	11	10
00	10	(00)	11	10
01	(01)	00	10	10
11	01	00	(11)	(11)
10	11	00	(10)	(10)

Fig. 8.70
Transition table

8.35 For an asynchronous sequential circuit, the sequence of internal states Q_1 Q_2 obtained is 00, 00, 01, 11, 11, 01, 00 for an input sequence of X_1 X_2 : 00, 10, 11, 01, 11, 10, 00.

(a) Determine transition table

(b) Next-state logic equations

(c) Logic circuit

8.36. In the flow table of Fig. 8.56, verify the entries in the rows '*e*' and '*f*'.

9

TIMING CIRCUITS

9.1 INTRODUCTION

Most digital systems require some kind of a timing waveform, for example, a source of trigger pulses is required for all clocked sequential systems. In digital systems, a rectangular waveform is most desirable (unlike analog systems where sinusoidal signals are often used). The generators of rectangular waveforms are referred to as *multivibrators.*

There are three types of multivibrators. These are:

1. Astable (or free-running) multivibrator,
2. Monostable multivibrator (or one-shot), and
3. Bistable multivibrator (or `FLIP-FLOP`).

An astable multivibrator is nothing but an oscillator which generates rectangular waveform. It has two quasi-stable states and does not require any triggering, hence it is referred to as a *free-running* multivibrator. This is used as a source of clock pulses in sequential circuits.

A monostable multivibrator has one stable state, i.e. under steady-state condition its output is fixed; it is in either the LOW or the HIGH state. When the circuit is triggered with an externally applied pulse it goes into the other state, i.e. if it was in LOW state, it goes to the HIGH state and if it was in the HIGH state, it goes to the LOW state. The circuit remains in this state for a time duration depending upon the values of the elements used in the circuit. This state of the circuit is referred to as *quasi-stable state* since it recovers back to the stable state without any external triggering pulse. The width of the trigger pulse is very small and the width of the output pulse depends upon the time duration for which the circuit remains in the quasi-stable state. This circuit is also referred to as a *one-shot* since one trigger pulse produces only one pulse but of a different pulse width. This circuit is very useful because it can generate a relatively long pulse (tens of ms) from a narrow pulse, therefore, it is also known as *pulse stretcher.*

For example, a microprocessor may signal an output device to print something by transmitting a pulse. The electromechanical output device in general is slower than the microprocessor and hence would require the signal pulse for a longer duration. This is achieved by the interface circuitry consisting of a monostable multivibrator.

A multivibrator circuit in which both the states are stable is referred to as a *bistable* multivibrator or FLIP-FLOP. This circuit makes transitions from one stable state to another only when a triggering pulse is applied. These are often used as memory elements in digital systems and have been discussed extensively in Chapters 7 and 8.

Until a few years ago, multivibrators were designed using discrete devices like vacuum triodes, BJTs, FETs, etc. which have become obsolete now because of the availability of various ICs. Therefore, we shall be dealing with multivibrator circuits using various ICs only. The ICs used are:

1. Logic gates,
2. OP-AMPs,
3. Monostable multivibrators, and
4. Timers.

9.2 APPLICATIONS OF LOGIC GATES IN TIMING CIRCUITS

Logic gates can be used for generating the pulses required in digital systems. These circuits are easy to design and analyse but due to the lack of precision, their applications are rather restricted.

9.2.1 Free-Running Multivibrator

A simple, free-running multivibrator is shown in Fig. 9.1*a*. This circuit works on the same principle as an *RC* phase-shift oscillator. In this circuit, the phase shift is provided by the propagation delay time of the inverters.

Suppose, for example, the input V_1 to inverter I_1 makes a transition from logic 0 to logic 1 at $t = 0$. At t_1 (after a propagation delay time Δt) its output changes from logic 1 to logic 0. This will cause the output of inverter I_2 to change from logic 0 to logic 1 at t_2. At t_3, the output V_O and the voltage V_1 will change from logic 1 to logic 0. This process will go on indefinitely. The various voltage waveforms are illustrated in Fig. 9.1*b*, *c*, and *d*. The time period T of the output square wave is given by

$$T = 6\,\Delta t \tag{9.1}$$

For TTL gates, Δt is of the order of 10 ns, therefore, the circuit's operating frequency is of the order of 16 MHz.

In this circuit, we have no control over the frequency of the square wave and it is difficult to determine the exact propagation delay time of the logic gate and hence the frequency of the square wave. Therefore, it cannot be used in a system calling for precise and stable operating frequencies. However, because of its simplicity, it is useful whenever it is necessary to get high frequency trigger pulses at a very low cost.

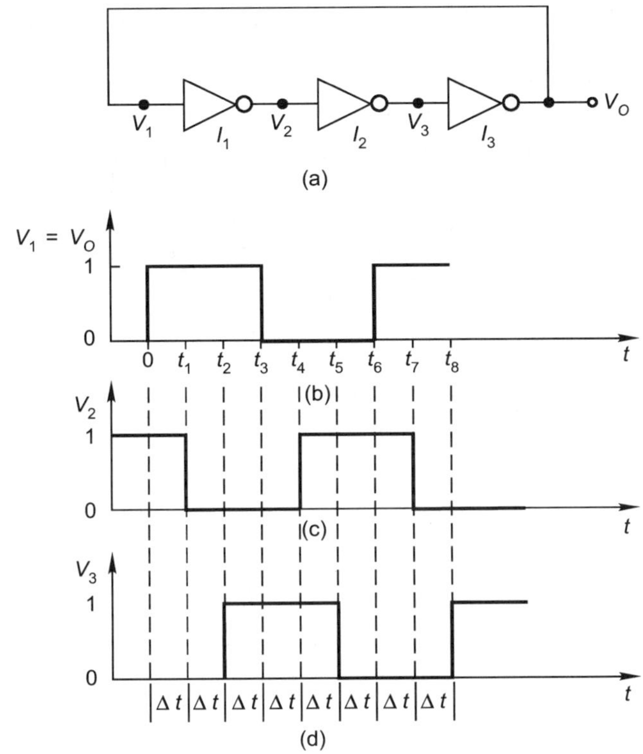

Fig. 9.1
(a) Free-running multivibrator using inverters. Waveforms of (b) $V_1 = V_O$ (c) V_2, and (d) V_3.

It is possible to have some control over the frequency of the square wave by using timing elements–a resistance and a capacitance (Problem 9.1).

9.2.2 Monostable Multivibrator

A monostable multivibrator circuit using gates is shown in Fig. 9.2. Under steady-state condition, the voltage across the resistance R, v_R is 0, i.e. the input to the inverter is at logic 0, hence its output v_O is at logic 1. Now, if the trigger input is at logic 1, then both the inputs of the NAND gate are at logic 1 and its output is at logic 0. Hence, v_O in logic 1 state is the stable state of this circuit. It can be verified that v_O in 0 state is not a stable state (Problem 9.2). When the trigger

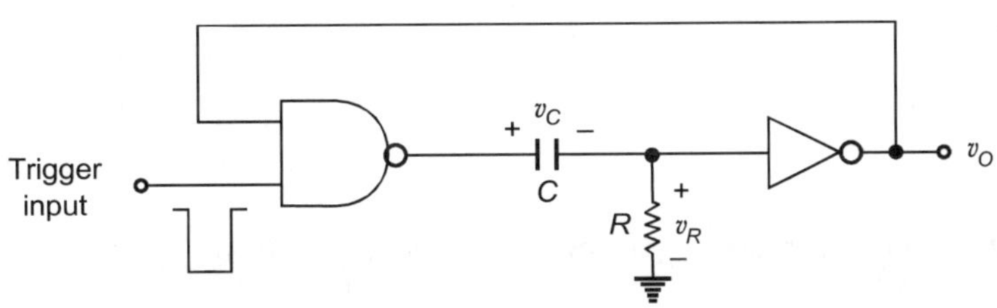

Fig. 9.2
Monostable multivibrator.

input goes from logic 1 to logic 0, the output of the NAND gate goes to logic 1 and hence, v_R corresponds to logic 1.

Therefore, the output v_O of the inverter goes to logic 0. The 0 at v_O which is returned to one input of NAND gate holds the output of the NAND gate at logic 1. As voltage v_C increases exponentially with the time constant $\tau = RC$, the voltage v_R decreases and goes to logic 0 driving the output v_O to logic 1. As soon as both the inputs of the NAND gate become 1, its output is 0. The capacitor now discharges through the output circuit of the NAND gate and the circuit comes to the stable state. The circuit should not be triggered during this recovery interval. The various waveforms are shown in Fig. 9.3.

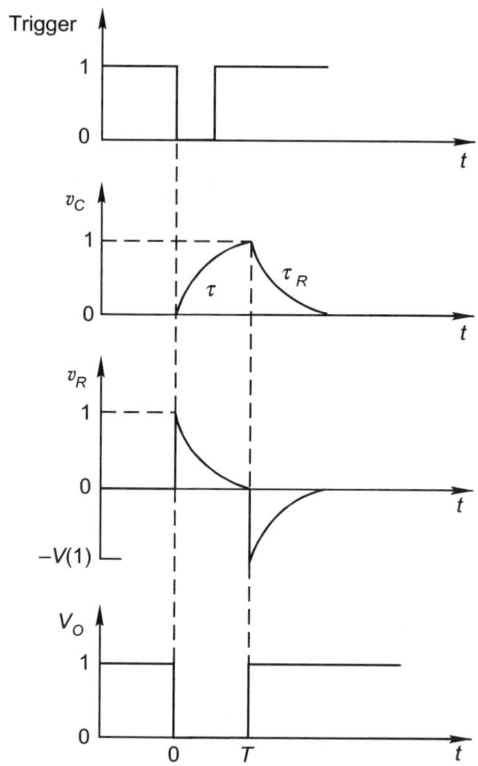

Fig. 9.3
Waveforms of monostable multivibrator.

It is not possible to obtain the exact value of the pulse duration T because of the somewhat uncertain values of the input/output voltages corresponding to 0/1 level and the propagation delay time of the gates. However, the circuit is useful because of its simplicity and low cost.

9.3 OP AMP AND ITS APPLICATIONS IN TIMING CIRCUITS

An *operational amplifier*, popularly known as OP AMP is a very high gain d.c. amplifier. Originally, it was designed using vacuum tubes to perform mathematical operations such as addition, multiplication by a constant, differentiation, integration, etc. and it was a basic building block of analog computers. Now-a-days, it is available as a linear IC and, because of its low cost, versatility, and reliability, has become very popular and finds applications in the generation of waveforms like square, triangular, pulse, sweep, staircase, etc.

An OP AMP is shown in Fig. 9.4. It has two inputs, labelled – and +, and are referred to as *inverting*, and *non-inverting* inputs, respectively. The input V_1 is applied between the inverting input terminal and ground, and input V_2 is applied between the non-inverting input terminal and ground. The ground terminal is not there in this circuit but is the ground terminal of the supplies $+V$ and $-V$. The output, V_O, is obtained between the output terminal and ground. The input terminals are called *differential inputs* and the output is *single-ended*. The output voltage depends on the difference voltage $V_i = V_1 - V_2$ and is given by

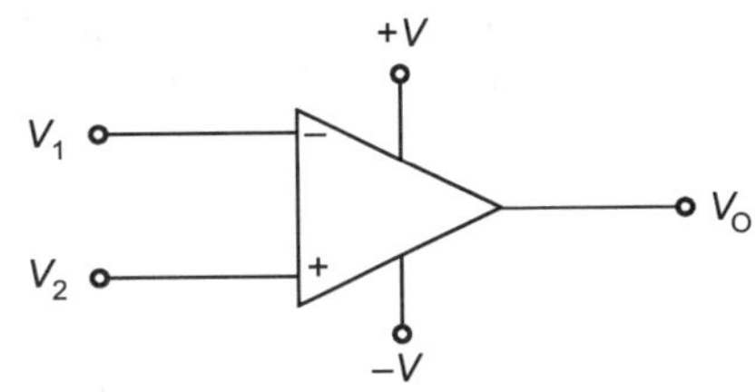

Fig. 9.4
Block diagram of an OP AMP.

$$V_O = A_v \cdot V_i \tag{9.2}$$

where A_v is the voltage gain of the amplifier, which has a large negative value (ideally, $A_v \rightarrow -\infty$).

From Eq. (9.2), we observe that the polarity of the output voltage is the same as the polarity of the non-inverting input voltage and is opposite (or inverted) to the polarity of the inverting input voltage.

Since the value of A_v is extremely large, often 200,000 or more, therefore, the input voltage V_i is extremely small ($V_i \approx 0$) and the output voltage V_O can never exceed positive or negative saturation voltages $+V_{sat}$ and $-V_{sat}$, respectively. These voltages are normally within about 2 V of $+V$ and $-V$. For example, if supply voltages are ±15 V, then $+V_{sat} = +13$ V and $-V_{sat} = -13$ V. Thus, the output is restricted to a peak-to-peak swing of ±13 V.

9.3.1 OP AMP Comparator

The OP AMP can be used as an analog comparator to compare two analog signals. The analog signals to be compared are applied at the two inputs and the polarity of the output voltage indicates the comparison. The magnitude of the output voltage is V_{sat}. This is the basic building block required for non-sinusoidal waveform generators.

Example 9.1 Find the output waveform of an OP AMP under the following conditions:

(*a*) Inverted input terminal connected to ground and a sinusoidal signal of 4 V peak at the non-inverting input.

(*b*) Non-inverting input terminal connected to ground and a sinusoidal signal of 4 V peak at the inverting input.

(*c*) +3 V at the inverting input and a sinusoidal signal of 5 V peak at the non-inverting input.

Solution

(*a*) The circuit with inverting input connected to ground and the voltage source connected at the non-inverting input is shown in Fig. 9.5*a*. The input and the output waveforms are shown in Fig. 9.5*b* and *c*, respectively. When the input passes through 0 V while changing from negative to positive, the output changes from $-V_{sat}$ to + V_{sat}. Similarly, when the input passes through 0 V while changing from positive to negative, the output changes from $+V_{sat}$ to $-V_{sat}$.

(*b*) With the source connected at the inverting input and the non-inverting input connected to ground, the circuit, and the input and output waveforms are shown in Fig. 9.6.

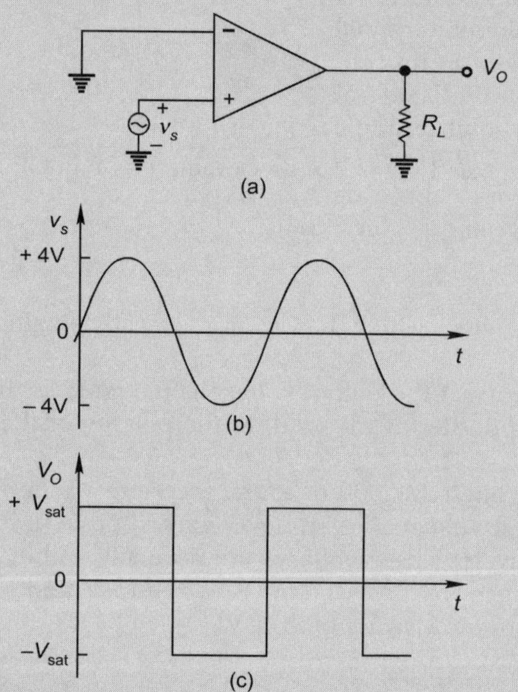

Fig. 9.5
(*a*) Comparator for Ex. 9.1*a* (*b*) Input waveform (*c*) Output waveform.

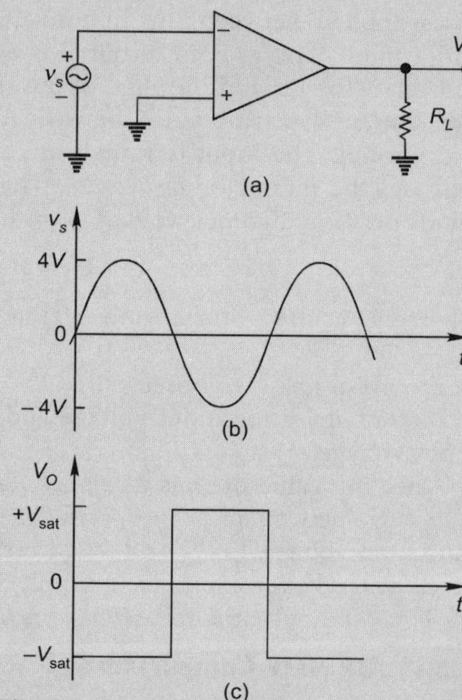

Fig. 9.6
(*a*) Comparator for Ex. 9.1*b* (*b*) Input waveform (*c*) Output waveform.

(*c*) In this case a reference voltage, V_{ref} = 3 V is maintained at the inverting input terminal. Whenever the voltage v_s is greater than V_{ref}, the output is $+V_{sat}$ and whenever it is less than V_{ref}, the output is $-V_{sat}$. The circuit diagram, and the input and output waveforms are shown in Fig. 9.7.

In some practical applications, the input voltage to the comparator may approach the reference voltage very slowly and/or may oscillate around V_{ref}. In such a situation, V_O either would not switch quickly from one saturation voltage to the other or would oscillate between $+V_{sat}$ and $-V_{sat}$. This oscillation may also occur due to ringing caused from the fast voltage transitions or due to the presence of noise on wires leading to the OP AMP's input terminals.

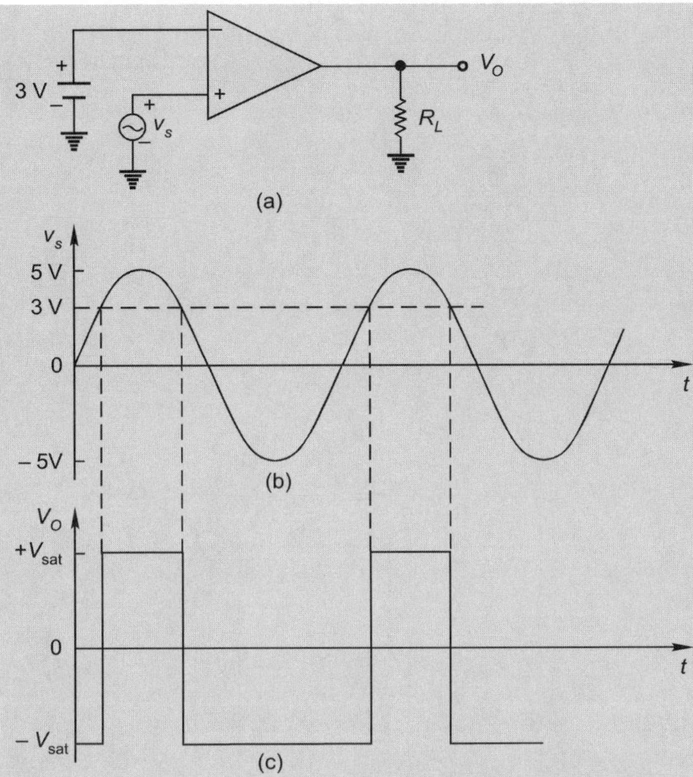

Fig. 9.7
(*a*) Comparator for Ex. 9.1*c* (*b*) Input waveform (*c*) Output waveform.

Example 9.2 (a) If the triangular waveform shown in Fig. 9.8*a* is applied to the circuit of Fig. 9.6*a*, what will be the output waveform? (*b*) If noise is present as shown in Fig. 9.9*a*, find the effective input and output voltage waveforms.

Solution

(*a*) In this circuit the reference voltage $V_{ref} = 0$ and, therefore, the output makes a transition whenever the inputs passes through 0 V. The output waveform is shown in Fig. 9.8*b*.

(*b*) For simplicity, we assume the noise voltage to be sinusoidal. The waveform of effective input, v_i (combination of triangular waveform and noise) is shown in Fig. 9.9*b* and the corresponding output waveform is shown in Fig. 9.9*c*, which shows false transitions in the output waveform due to noise.

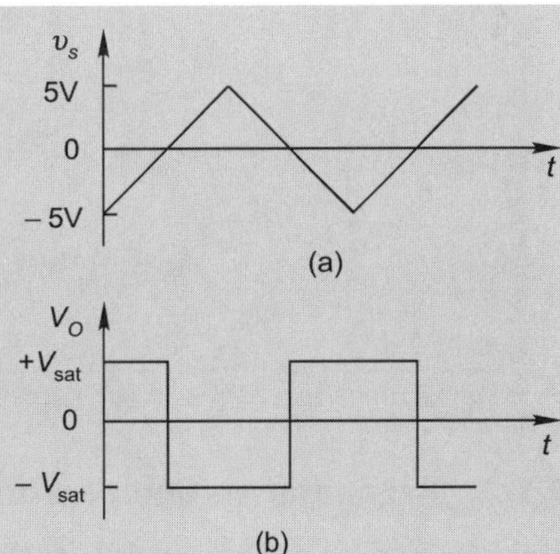

Fig. 9.8
(*a*) Triangular voltage waveform applied to the comparator of Fig. 9.6*a* (*b*) Its output waveform.

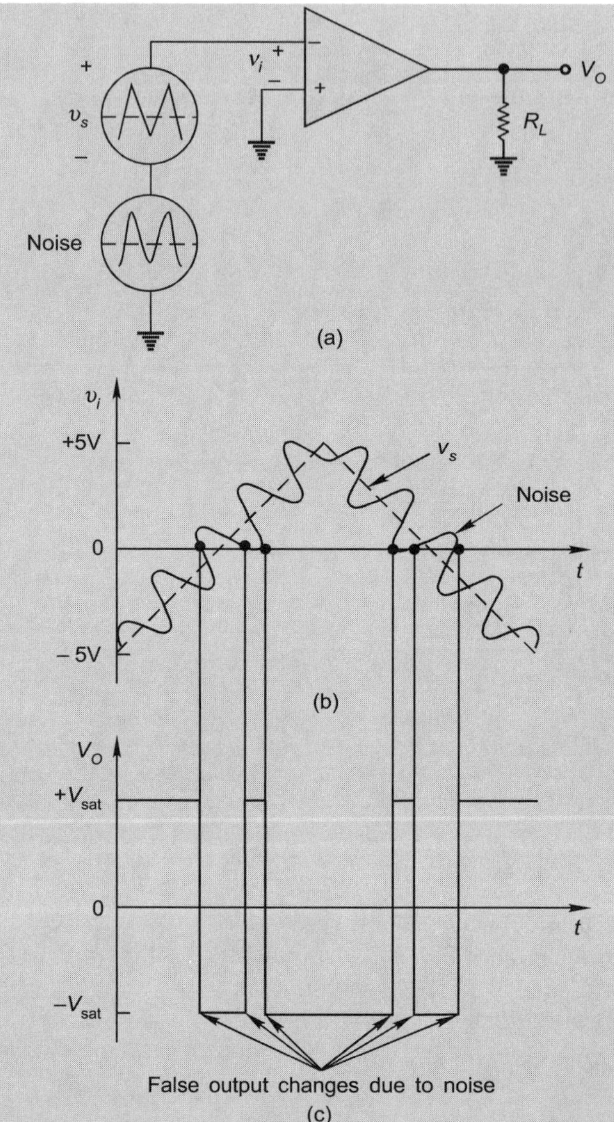

Fig. 9.9
(*a*) Circuit for Ex. 9.2*b* (*b*) Input waveform (*c*) Output waveform.

9.3.2 Regenerative Comparator (Schmitt Trigger)

Since it may not be possible to completely eliminate the noise voltage, therefore, we must prevent the circuit from sensing these false changes at the input. This is achieved by using positive feedback in the circuit and the resulting circuit is referred to as a *regenerative comparator* and is also known as *Schmitt trigger* (after the inventor of a vacuum-tube version of this circuit).

Figure 9.10*a* shows a Schmitt trigger circuit. Positive feedback is applied by taking a fraction of the output voltage, $v_f = R_2\, v_O/(R_1 + R_2)$ and applying it to the non-inverting input.

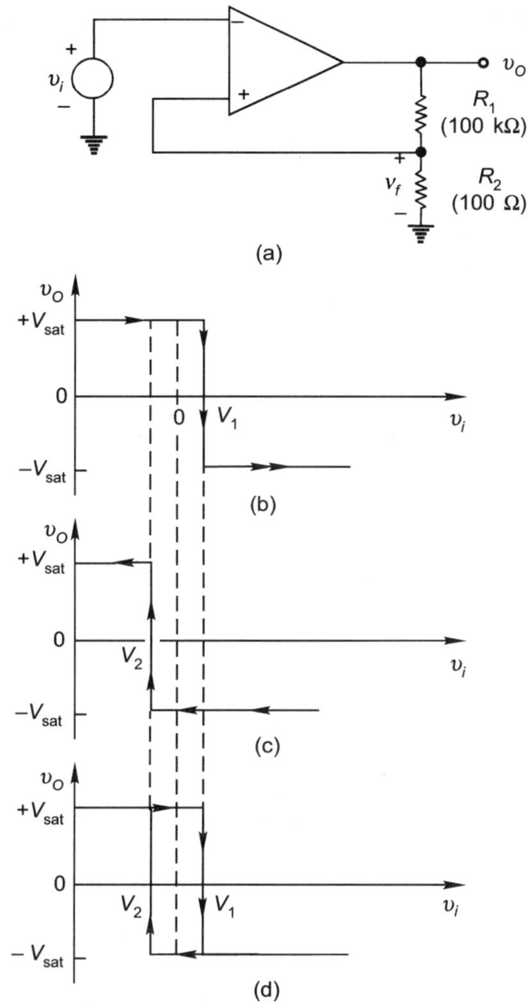

Fig. 9.10

(*a*) An inverting Schmitt trigger. The input–output characteristics for (*b*) Increasing v_i and (*c*) decreasing v_i. (*d*) The complete input–output characteristic.

Let $v_O = +V_{sat}$ and $v_i < v_f$. If v_i is now increased, then v_O remains constant at $+V_{sat}$, and $v_f = R_2/(R_1 + R_2) \cdot V_{sat}$ = constant until

$$v_i = V_1 = \frac{R_2}{R_1 + R_2} \cdot V_{sat} \tag{9.3}$$

At this voltage V_1, known as *upper-threshold voltage* or *upper-triggering voltage* (V_{UT}), the output switches to $-V_{sat}$ and remains at this value as long as $v_i > V_1$. This input–output characteristic is shown in Fig. 9.10*b*.

The voltage at the non-inverting input terminal for $v_i > V_1$ is

$$v_i = -\frac{R_2}{R_1 + R_2} V_{sat}$$

If we now decrease v_i, then the output remains at $-V_{sat}$ until v_i equals V_2, where

$$V_2 = -\frac{R_2}{R_1 + R_2} V_{sat} \tag{9.4}$$

and is known as *lower-threshold voltage* or *lower-triggering voltage* (V_{LT}). At this input voltage, the output switches to $+V_{sat}$ and remains at this value as long as $v_i < V_2$. This input–output characteristic is shown in Fig. 9.10*c*. The complete input–output characteristic is shown in Fig. 9.10*d*, where the portions without arrows may be traversed in either direction, but the segments with arrows can only be obtained if v_i varies as indicated by the arrows. This characteristic exhibits the hysteresis action in this circuit. The difference in voltage of V_{UT} and V_{LT} is called the *hysteresis voltage*, V_H.

Example 9.3

(*a*) In the Schmitt trigger circuit of Fig. 9.10*a*, if V_{sat} = 13 V, find V_{UT} and V_{LT}.

(*b*) If $v_i = 5 \sin wt$, find the waveform of the output voltage.

Solution

(*a*) $V_{UT} = \dfrac{0.1}{100.1} \times 13 \approx 13\,\text{mV}$

$V_{LT} = \dfrac{0.1}{100.1}(-13) \approx -13\ \text{mV}$

(*b*) The input and output voltage waveforms are shown in Fig. 9.11.

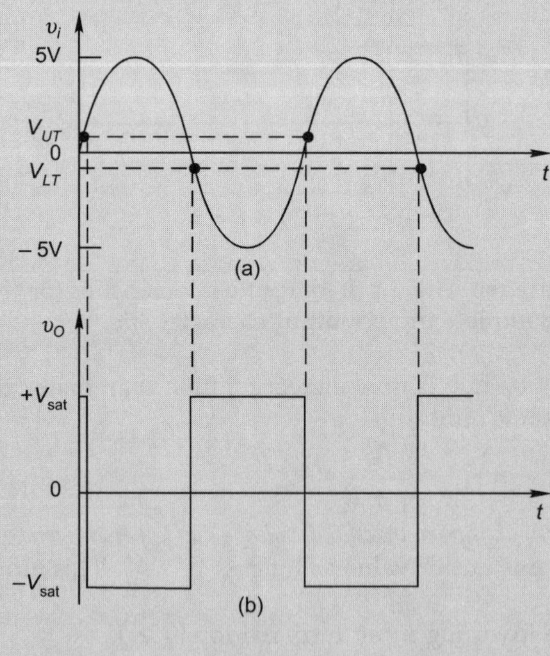

Fig. 9.11
Voltage waveforms for Ex. 9.3*b* (a) Input (b) Output.

Limiting output voltage

For an operational amplifier, the output voltage levels are limited to $+V_{sat}$ and $-V_{sat}$ and are dependent on supply voltages. These output voltages may not be compatible with voltage levels required by a particular load. For example, TTL requires input voltages that are approximately +5 V and 0 V.

To limit the output voltages, which are independent of the power supply voltages, two back-to-back Zener diodes are used at the output as shown in Fig. 9.12. The resistance R is used to limit the current in the Zener diodes.

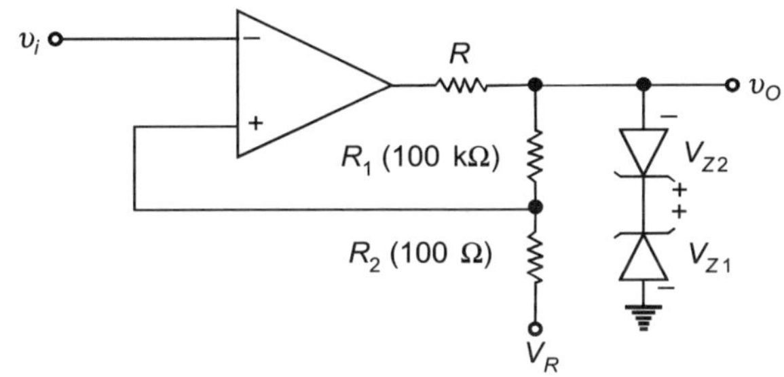

Fig. 9.12
A comparator circuit with output voltage limiting.

For this circuit, the upper and lower threshold voltages are given by (Problem 9.3)

$$V_{UT} = \frac{R_2}{R_1 + R_2}(V_{Z1} + V_D) + \frac{R_1}{R_1 + R_2} V_R \tag{9.5}$$

and

$$V_{LT} = -\frac{R_2}{R_1 + R_2}(V_{Z2} + V_D) + \frac{R_1}{R_1 + R_2} V_R \tag{9.6}$$

where V_D = voltage across a forward-biased diode (~0.7 V) and V_{Z1} and V_{Z2} are the Zener voltages.

The output voltages will be $(V_{Z1} + V_D)$ and $-(V_{Z2} + V_D)$. If the input signal v_i is applied in place of the reference voltage V_R, and V_R is applied at the inverting input terminal, then a non-inverting comparator is obtained.

9.3.3 Astable (or Free-Running) Multivibrator

An astable or free-running multivibrator is a square-wave generator. The circuit of Fig. 9.13 is a comparator circuit discussed above in which an RC low-pass circuit (combination of R_f and C) is used to integrate the output voltage v_O and the voltage across the capacitor, C, is applied to the inverting input terminal in place of the external signal.

The output voltage $v_O = V_Z + V_D = V_O$ if $v_i < 0$ and $v_O = -(V_Z + V_D) = -V_0$ if $v_i > 0$. Consider an instant of time when $v_i < 0$, therefore $v_O = V_O$. The voltage at the non-inverting input is $R_2 V_O/(R_1 + R_2) = \beta V_O$ where $\beta = R_2/(R_1 + R_2)$. The capacitor C charges exponentially towards V_O with the time constant $R_f . C$. The output voltage remains constant at V_O until $v_C = \beta V_0$ $(= . V_{UT})$ at which time the comparator output reverses to $-V_O$. Now v_C changes exponentially

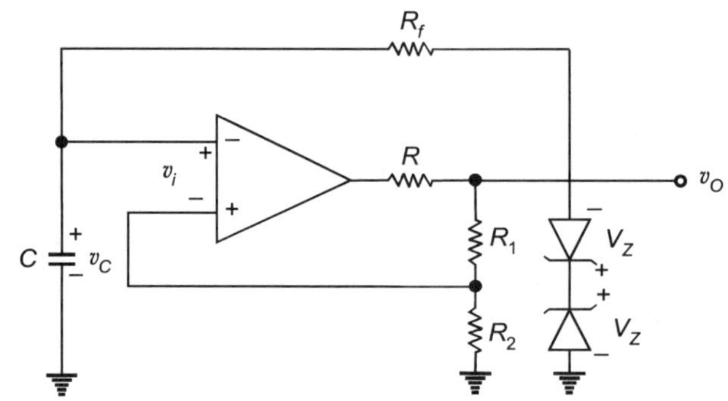

Fig. 9.13
Free-running multivibrator.

towards $-V_O$ with the same time constant and again the output makes a transition from $-V_O$ to $+V_O$ when $v_C = -\beta\ V_O\ (= V_{LT})$. The waveforms of the capacitor voltage, v_C, and the output voltage, v_O, are illustrated in Fig. 9.14.

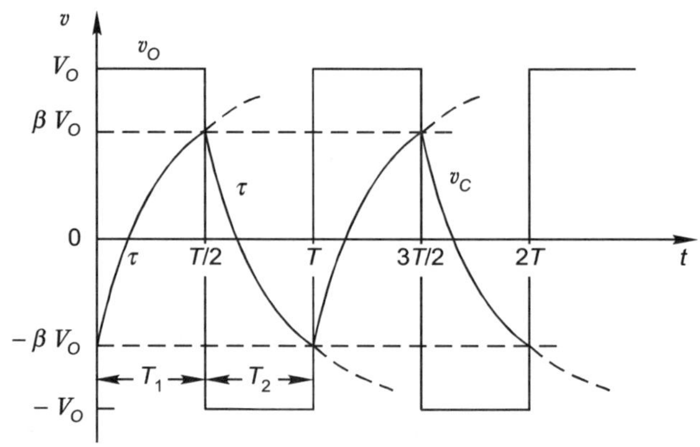

Fig. 9.14
Waveforms of output and capacitor voltage for free-running multivibrator.

The time period, T, of the output square waveform is determined using the charging and discharging of capacitor. The voltage across the capacitor, v_C, when it is charging from $-\beta\ V_O$ to $+V_O$ is given by

$$v_C = V_O\left[1-(1+\beta)e^{-t/\tau}\right] \tag{9.7}$$

where, $\tau = R_f . C$

At $t = T/2$, $v_C\left(\frac{T}{2}\right) = +\ \beta V_O$

Therefore, $T = 2\ \tau\ \ln\left(\frac{1+\beta}{1-\beta}\right) = 2\ R_f . C \ln\left(1+\frac{2R_2}{R_1}\right)$ (9.8)

The frequency, $f = 1/T$, of the square wave is independent of V_O. The circuit of Fig. 9.13 has two quasi-stable states. The output remains in one of these states for a time T_1 and then makes an abrupt transition to the second state and remains in that state for a time T_2. The cycle of period T $(= T_1 + T_2)$ repeats itself and hence this circuit is known as *free-running* or *astable* multivibrator.

This square-wave generator is useful in the frequency range of about 10 Hz to 10 kHz. At higher frequencies, the slew rate of the OP AMP limits the slope of the output square wave. The symmetry of the output waveform depends on the matching of the two Zener diodes (Problem 9.5). The unsymmetrical square wave $(T_1 \neq T_2)$ can be obtained by using different time constants for charging the capacitor to $+V_O$ and $-V_O$ (Problem 9.6).

Duty cycle

For an unsymmetrical square waveform, the percentage of time-period (T) for which the output is HIGH is referred to as *duty cycle* and is given by

$$\text{Percent duty cycle} = \frac{T_{\text{ON}}}{T_{\text{ON}} + T_{\text{OFF}}} \times 100 \qquad (9.9)$$

where T_{ON} and T_{OFF} are indicated in Fig. 9.15.

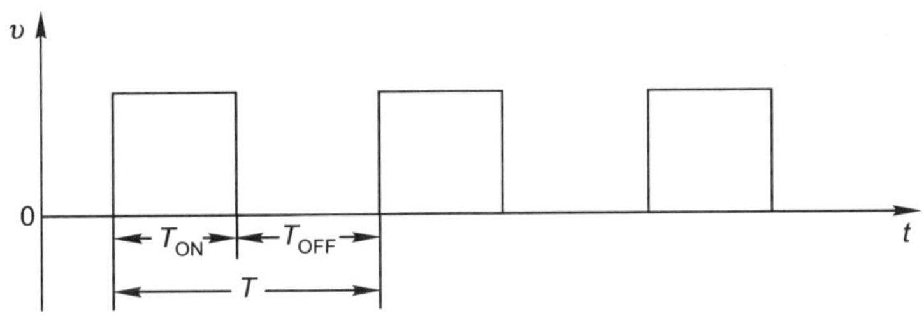

Fig. 9.15

Definition of T_{ON} and T_{OFF} of a repetitive square wave.

9.3.4 Monostable Multivibrator

A monostable multivibrator has one stable state and one quasi-stable state. The circuit remains in its stable state until an external triggering pulse causes a transition to the quasi-stable state. The circuit returns to its stable state after a time T. Hence it generates a single output pulse in response to an input pulse and is referred to as a *one-shot* or *single-shot.*

The astable multivibrator of Fig. 9.13 is modified to operate as a monostable multivibrator by connecting a diode (D_1) across C, so that v_C is clamped at V_D during positive excursion.

Under steady-state condition, this circuit will be in its stable state with the output v_O at $+V_O$. The capacitor C is clamped at the voltage V_D (ON voltage of diode $\approx$ 0.7 V). The voltage V_D must be less than $\beta\, V_O$ for $v_i < 0$. It can be verified that this circuit cannot remain in the other state ($v_O = -V_O$) under steady-state condition (Problem 9.7). This circuit can be switched to the other state by applying a negative trigger pulse with amplitude greater than $\beta\, V_O - V_D$ at the non-inverting input terminal. The triggering pulse is applied through a high-pass RC circuit (C_i R_i) and diode D_2. The complete circuit of the monostable multivibrator is shown in Fig. 9.16.

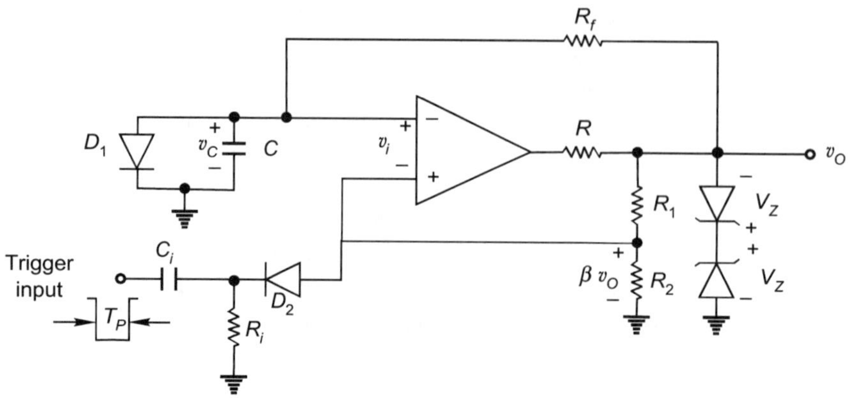

Fig. 9.16
Monostable multivibrator.

When a trigger pulse is applied v_i goes positive causing a transition in the state of the circuit ($v_O = -V_O$). The capacitor C now charges exponentially with a time constant $\tau = R_f \cdot C$ towards $-V_O$ (diode D_1 being reverse-biased). When v_C becomes more negative than $-\beta V_O$, v_i becomes negative and consequently the output swings back to $+V_O$ (steady-state output). From the above discussion, it is clear that the circuit has one stable state and the other state is quasi-stable. The width of the trigger pulse (T_P) must be much smaller than the duration T of the output pulse generated. The diode D_2 is used to avoid malfunctioning of the circuit due to any positive noise spikes present in the triggering line. The waveforms of v_C and v_O are shown in Fig. 9.17.

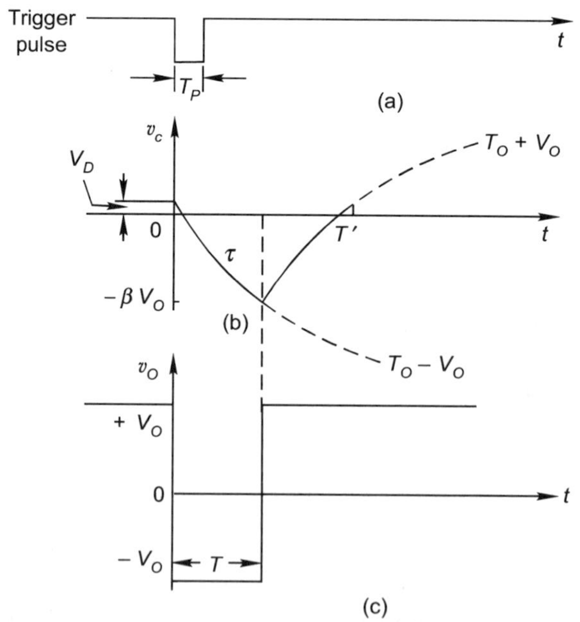

Fig. 9.17
(*a*) Negative trigger pulse (*b*) Waveforms of v_C (*c*) v_O.

During the quasi-stable state, the capacitor voltage v_C is given by

$$v_C = -V_O + (V_O + V_D)\, e^{-t/\tau} \tag{9.10}$$

$$\text{At } t = T,\ v_C = -\beta V_O$$

Therefore,

$$T = \tau \ln\left(\frac{1 + V_D/V_O}{1 - \beta}\right) \tag{9.11}$$

Usually $V_D \ll V_O$ and if $R_1 = R_2$, so that $\beta = \frac{1}{2}$, then

$$T \approx 0.69\, R_f \,.\, C \tag{9.12}$$

This circuit comes back to its steady-state condition at T' and the time interval $T' - T$ is referred to as *recovery time*. The circuit should not be triggered again before T' for reliable operation.

9.4 SCHMITT TRIGGER ICs

Some Schmitt trigger logic gate ICs are also available. These are given in Table 9.1. For the TTL gates, the threshold voltages V_{UT} and V_{LT} are 1.7 V and 0.9 V respectively, whereas for the CMOS gates these values are 3.6 V and 1.2 V respectively, when operated at $V_{CC} = 5$ V. These outputs are fast changing, similar to the outputs of other TTL circuits, but they can respond to slowly varying inputs.

Table 9.1 Available TTL Schmitt trigger ICs

IC No.	*Description*	*Logic symbol*
7413	Dual 4-input NAND Schmitt triggers	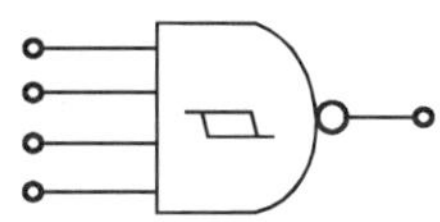
7414	Hex Schmitt trigger inverters	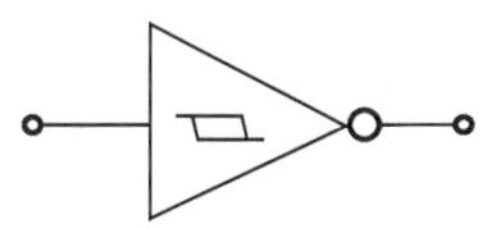
74132	Quad 2-input NAND Schmitt triggers	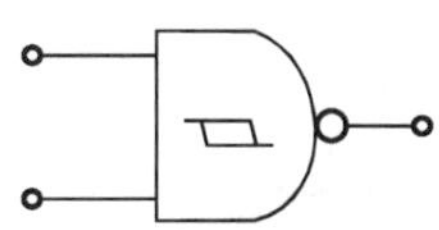

9.4.1 Schmitt Trigger Square-Wave Generator

A very simple square-wave generator can be made using a Schmitt trigger inverter as shown in Fig. 9.18. If its output is HIGH, the capacitor C charges with the time constant $\tau = RC$. When the capacitor voltage (v_C) reaches V_{UT}, the output voltage (v_O) goes LOW. Now the capacitor discharges through the output transistor of the gate which is in saturation. When v_C reaches V_{LT}

the output voltage goes HIGH. This process goes on and a square waveform shown in Fig. 9.19 is obtained at v_O. The time period of the square wave is given by

$$T = T_1 + T_2 = RC\left(\ln\frac{V(1)-V_{LT}}{V(1)-V_{UT}} + \ln\frac{V_{UT}}{V_{LT}}\right) \tag{9.13}$$

where $V(1)$ is the logic 1 output voltage.

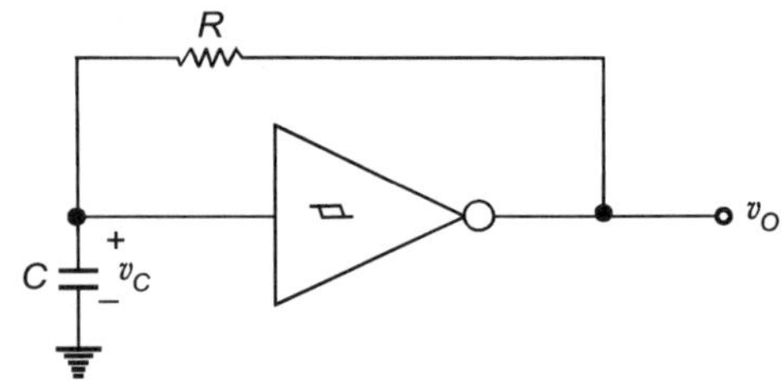

Fig. 9.18

A Schmitt trigger square wave generator.

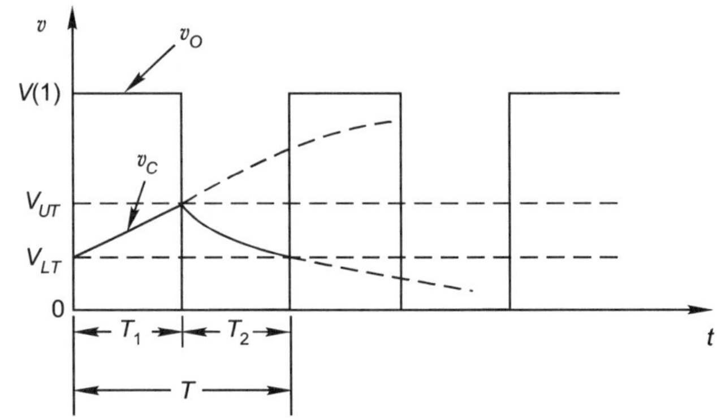

Fig. 9.19

Voltage waveforms of square wave generator.

9.5 MONOSTABLE MULTIVIBRATOR ICs

Some very useful monostable multivibrators are available in IC form. These are given in Table 9.2.

Table 9.2 Available monostable multivibrator ICs

IC No.	*Description*
74121	Monostable multivibrator (one-shot)
74122	Retriggerable monostable multivibrator with clear
74123	Dual retriggerable monostable multivibrator with clear
74221	Dual monostable multivibrator with clear

A brief description of these ICs is given below.

9.5.1 74121 Monostable Multivibrator

The functional diagram and the function table of the most popular and commonly used one-shot TTL IC 74121 are given in Figs 9.20 and 9.21 respectively.

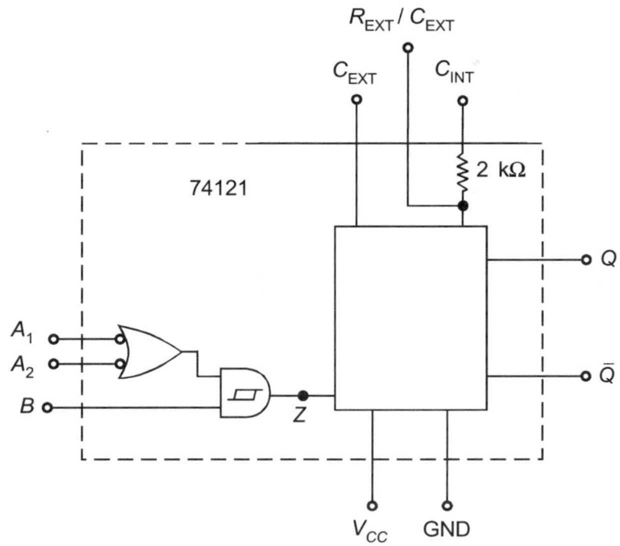

Fig. 9.20

Functional block diagram of monostable multivibrator IC 74121.

Inputs			*Outputs*	
A_1	A_2	B	Q	$\overline{Q}$
0	×	1	0	1
×	0	1	0	1
×	×	0	0	1
1	1	×	0	1
1	↓	1	⊓	⊔
↓	1	1	⊓	⊔
↓	↓	1	⊓	⊔
0	×	↑	⊓	⊔
×	0	↑	⊓	⊔

Fig. 9.21

Function tables of 74121.

In order to trigger the one-shot, there must be a rising pulse edge at point *Z*. This is possible in one of the following two ways:

1. One or both of the *A* inputs are at logic 0 and the *B* input makes a transition from logic 0 to 1 (↑).
2. The *B* input is at logic 1 and either one of the *A* inputs makes a transition from logic 1 to 0 (↓) while the other *A* input remains at logic 1 or both *A* inputs go from logic 1 to 0 (↓) simultaneously.

The duration of the output pulse is dependent upon the values of the resistor (R_{EXT} or R_{INT}) and capacitor (C_{EXT}) used. A timing capacitor (C_{EXT}) is to be externally connected between the terminals marked R_{EXT}/C_{EXT} and C_{EXT}. In case of electrolytic capacitor, + terminal is to be connected to R_{EXT}/C_{EXT}. The maximum allowable value of the external capacitor is 1,000 μF. If the external timing capacitor is not used, the stray capacitance between the pins of the IC will result in a very low pulse width output.

For timing resistor, there are two options: (*i*) an internal timing resistor (R_{INT}) of 2 kΩ becomes effective if the R_{INT} terminal is connected to V_{CC}, (*ii*) external timing resistor (R_{EXT}) is to be connected between R_{EXT}/C_{EXT} terminal and V_{CC}. The range of R_{EXT} is from 1.4 kΩ to 40 kΩ.

In any case, both R_{EXT} and R_{INT} must not be used simultaneously. The duration of the output pulse is given by

$$T_{ON} \approx 0.7\ RC \tag{9.14}$$

where R and C are the values of the timing resistor and capacitor used respectively.

The minimum possible pulse width is 30 to 35 ns (with R_{INT} and without C_{EXT}) and the maximum possible pulse width is about 28s (with R_{EXT} = 40 kΩ and C_{EXT} = 1,000 μF).

The maximum allowable duty cycle is 67% with R_{INT} and goes up to 90% with R_{EXT} of 40 kΩ.

The A inputs are applied to a normal TTL gate and therefore should make fast transitions, but the B input is a Schmitt trigger input that responds to very slowly changing inputs. Therefore, if slow waveform is required to trigger a one-shot it should be applied to the B input.

9.5.2 Retriggerable Monostable Multivibrators (74122 and 74123)

IC 74121 is a non-retriggerable one-shot, i.e. it responds to a trigger pulse only when it is in the quiescent (stable) state. A retriggerable one-shot responds to a trigger pulse even when it is in the quasi-stable (ON) state. If a trigger pulse accurs when it is in quasi-stable state, it resets the timing and does not go to stable state (OFF) until one pulse duration after the last trigger pulse. Its duty cycle is, therefore, unlimited. The functional block diagram and the function table of 74122 retriggerable monostable multivibrator are shown in Figs 9.22 and 9.23 respectively. It has an internal timing resistor (R_{INT}) of 10 kΩ and it is to be used in the same way as it is used in 74121. The limits of external resistor (R_{EXT}) is from 5 kΩ to 50 kΩ. The pulse width is to be determined in the following way:

1. If the timing capacitor, C_{EXT} < 1000 pF, the pulse width is to be determined from the graph of Fig. 9.26.
2. If the timing capacitor, $C_{EXT} \geq$ 1000 pF, then

$$T_{ON} \approx 0.3\ R \cdot C_{EXT}\ (1 + 0.7/R) \tag{9.15}$$

where T_{ON} is in nanoseconds, R in kilo ohms and C_{EXT} in picofarads.

74123 is a dual retriggerable one-shot IC. Its operation is similar to 74122 except that it does not have an internal timing resistor. Its block diagram is shown in Fig. 9.24 and the function table of each section is shown in Fig. 9.25.

These ICs have asynchronous clear inputs which are active-low and are used to clear ($Q = 0$) the one-shot.

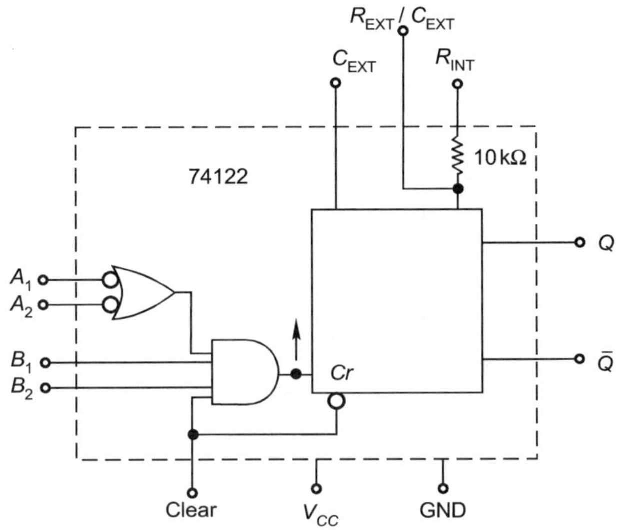

Fig. 9.22
Functional block diagram of 74122.

Inputs					*Outputs*	
Cr	A_1	A_2	B_1	B_2	Q	$\overline{Q}$
0	×	×	×	×	0	1
×	1	1	×	×	0	1
×	×	×	0	×	0	1
×	×	×	×	0	0	1
×	0	×	1	1	0	1
1	0	×	↑	1	⊓	⊔
1	0	×	1	↑	⊓	⊔
1	×	0	1	1	0	1
1	×	0	↑	1	⊓	⊔
1	×	0	1	↑	⊓	⊔
1	1	↓	1	1	⊓	⊔
1	↓	↓	1	1	⊓	⊔
1	↓	1	1	1	⊓	⊔
↑	0	×	1	1	⊓	⊔
↑	×	0	1	1	⊓	⊔

Fig. 9.23
Function table of 74122.

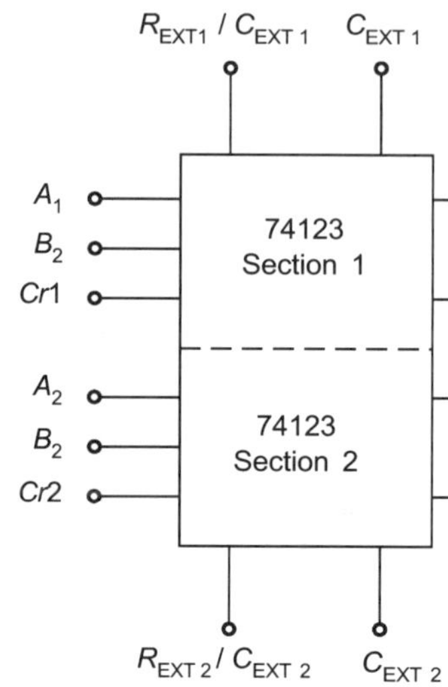

Fig. 9.24
Block diagram of 74123.

Inputs			*Outputs*	
Cr	*A*	*B*	*Q*	$\bar{Q}$
0	×	×	0	1
×	1	×	0	1
×	×	0	0	1
1	0	↑	⊓	⊔
1	↓	1	⊓	⊔
↑	0	1	⊓	⊔

Fig. 9.25
Function table of 74123.

9.5.3 Non-retriggerable Monostable Multivibrator with Clear (74221)

IC 74221 is a dual monostable multivibrator with performance characteristics identical to those of 74121 and its block diagram is same as that of IC 74123 (Fig. 9.24). Its function table is given in Fig. 9.27. There is an active-low asynchronous clear input for resetting the circuit.

Example 9.4 (*a*) A 74121 is used to generate 8 *m*s pulse with duty cycle of 80%. Its inputs are $A_1 = A_2 = 0$ and the trigger pulses shown in Fig. 9.28*a* are applied at the *B* input. Sketch the waveform of the *Q* output, (*b*) If the same trigger pulses are applied at the *B* input of one section of a 74123 (used to generate 8 *m*s pulse) with *A* = 0 and *Cr* = 1, what will be the *Q* output?

Solution

(*a*) The output waveform is shown in Fig. 9.28*b*.
(*b*) The output waveform is shown in Fig. 9.28*c*.

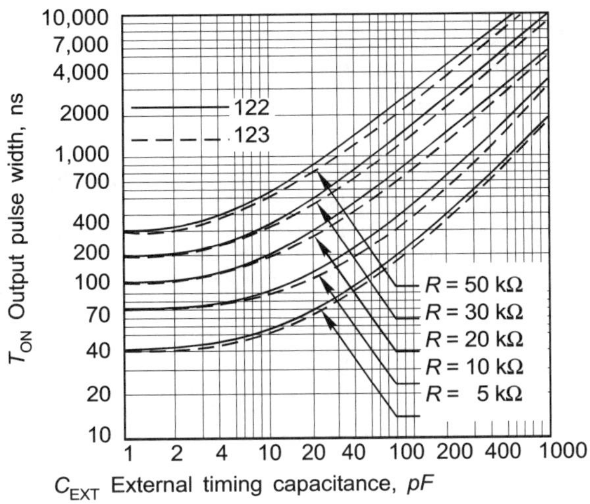

Fig. 9.26

Graph for determination of pulse width of retriggerable one-shots.

Inputs			Outputs	
Cr	A	B	Q	$\overline{Q}$
0	×	×	0	1
×	1	×	0	1
×	×	0	0	1
1	0	↑	⊓	⊔
1	↓	1	⊓	⊔

Fig. 9.27

Function table of 74221.

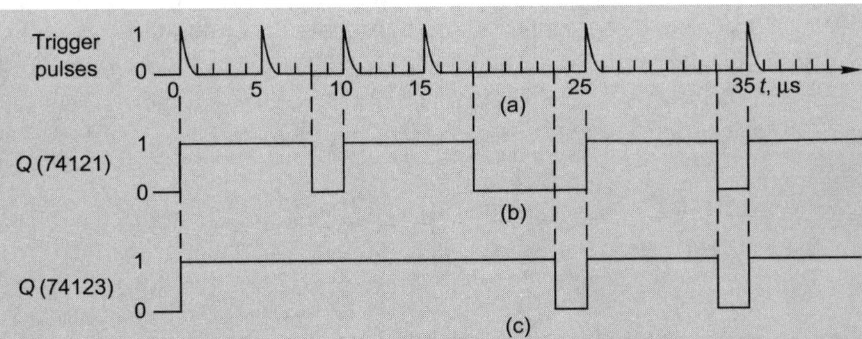

Fig. 9.28

(a) Tigger pulses (b) Q output of non-retriggerable one-shot 74121 (c) Q output of retriggerable one-shot 74123.

Example 9.5 (*a*) In the circuit of Ex. 9.4*b*, if clear input goes LOW every 12 μs (Figs. 9.29*a* and *b*), obtain the output waveform. (*b*) In the above circuit, if 74221 IC is used instead of 74123, obtain the output waveform.

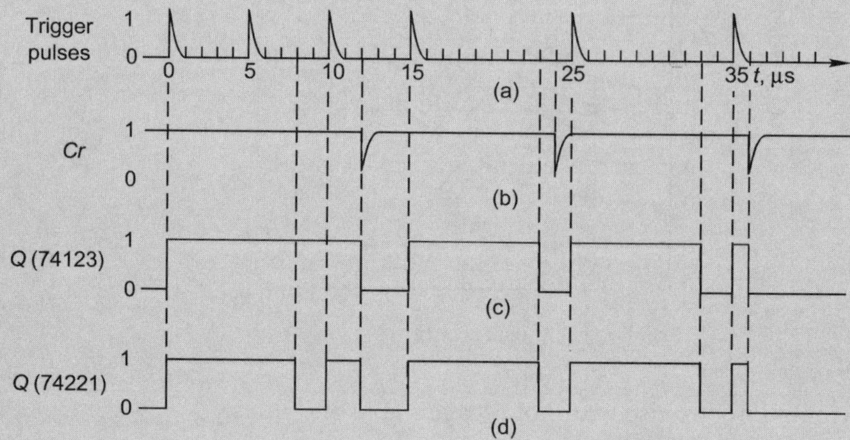

Fig. 9.29
Waveforms for Ex 9.5.

Solution

(a) The output waveform is shown in Fig. 9.29*c*.
(b) The output waveform is shown in Fig. 9.29*d*.

Example 9.6 Design a circuit to monitor a.c. mains voltage. If it ever misses a cycle, a buzzer should sound continuously until reset manually by depressing a push-button.

Solution

First the sinusoidal voltage is to be converted into a square wave which can be used as input for digital circuits. The circuit for this is shown in Fig. 9.30. The a.c. mains voltage is reduced to about 20 V peak-to-peak. The input to the gate is clamped to V_{CC} during positive cycle and 0 V during negative cycle of sinusoidal waveform with the help of the diodes D_1 and D_2 respectively.

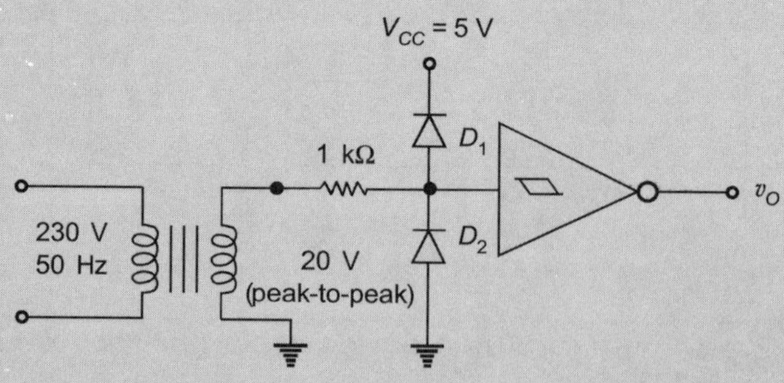

Fig. 9.30
Circuit for converting a.c. mains voltage to square waveform.

Since the output of the transformer is a slowly varying voltage, a Schmitt trigger circuit is used. The output of this circuit is a square wave of frequency 50 Hz.

Now, a retriggerable monostable multivibrator can be used to produce 100% duty cycle output as long as a.c. voltage is present. The circuit for this is shown in Fig. 9.31. The operation of this circuit is explained below.

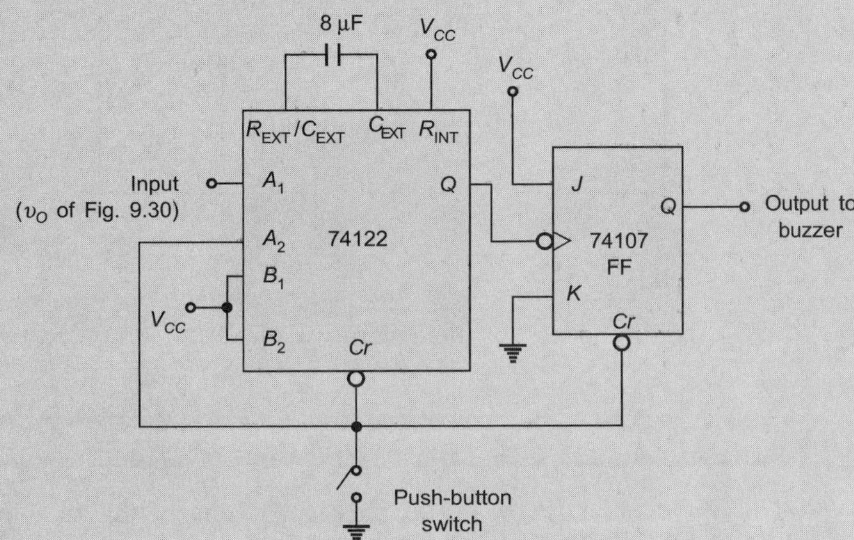

Fig. 9.31
Circuit for detecting missing pulse.

The output of the circuit of Fig. 9.30 produces a 50 Hz square wave, i.e. it produces pulse edges every 20 ms. These pulses are applied at the A_1 input of 74122 set for a pulse width of little more than 20 ms, say 25 ms, so that the one-shot is never reset as long as a.c. power is present.

With internal timing resistor of 10 kΩ, C_{EXT} required is ~ 8 μF (Eq. 9.15). When a.c. power goes OFF, the output of 74122 goes LOW which sets the FLIP-FLOP. This drives the buzzer. If the voltage and current requirements of the buzzer are high, then buffer/driver will be required.

When the push-button is depressed, it clears both the FLIP-FLOPs and the 74122. When the push-button is released, it triggers the 74122 and the circuit assumes normal operation if the a.c. power is resumed, otherwise it will sound the alarm again after 25 ms.

9.5.4 Astable Multivibrator Using One-Shots

Two one-shots can be coupled together as shown in Fig. 9.32 to make an astable multivibrator. Initially, Q_2 is LOW and when the switch SW is opened, the first multivibrator (IC1) goes to the quasi-stable state. When the output Q_1 goes LOW (↓) at a time T_1 (pulse width of IC1), IC2 is triggered and output Q_2 goes HIGH and remains HIGH for a time T_2 (pulse width of IC2) and then goes LOW (↓) which triggers IC1. This process is continuous and a square wave is

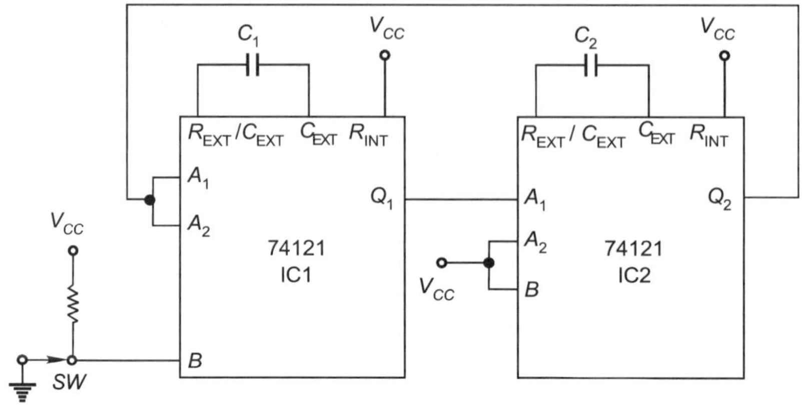

Fig. 9.32
An astable multivibrator using two 74121 ICs.

produced at Q_1 and Q_2. The frequency of the square wave can be controlled by the timing elements.

9.6 555 TIMER

IC 555 timer (available in 8 pin DIP or TO-99 package) is one of the most popular and versatile sequential logic devices which can be used in monostable and astable modes. Its inputs and output are directly compatible with both TTL and CMOS logic circuits. The functional block diagram of 555 timer is shown in Fig. 9.33.

It has two comparators which receive their reference voltages by a set of three resistances connected between the supply voltage V_{CC} and ground. The reference voltage for comparator 1 is $2V_{CC}/3$, and for comparator 2 is $V_{CC}/3$. These reference voltages have control over the timing which can be varied electronically, if required, by applying a voltage to the control-voltage input terminal. If this is not required, then a bypass capacitor (0.01 μF) should be connected between this terminal and ground to bypass noise and/or ripple voltages from the power supply.

On a negative-going excursion of the trigger input, when the trigger input passes through the reference voltage $V_{CC}/3$, the output of the comparator 2 goes HIGH and sets the FLIP-FLOP ($Q = 1$). On a positive-going excursion of the threshold input, the output of the comparator 1 goes HIGH when the threshold voltage passes through the reference voltage $2V_{CC}/3$. This resets the FLIP-FLOP ($\bar{Q} = 1$).

The FLIP-FLOP is cleared (irrespective of the S input) when the reset input is less than about 0.4 V. When this input is not required to be used, it is normally returned to V_{CC}.

An external timing capacitor, C is to be connected between the discharge terminal and ground. When the FLIP-FLOP is in the reset state, its $\bar{Q} = 1$, which drives T_1 to saturation thereby discharging the timing capacitor. The timing cycle starts when the FLIP-FLOP goes to set state and therefore T_1 is OFF. The timing capacitor charges with the time constant $\tau = R_A \cdot C$,

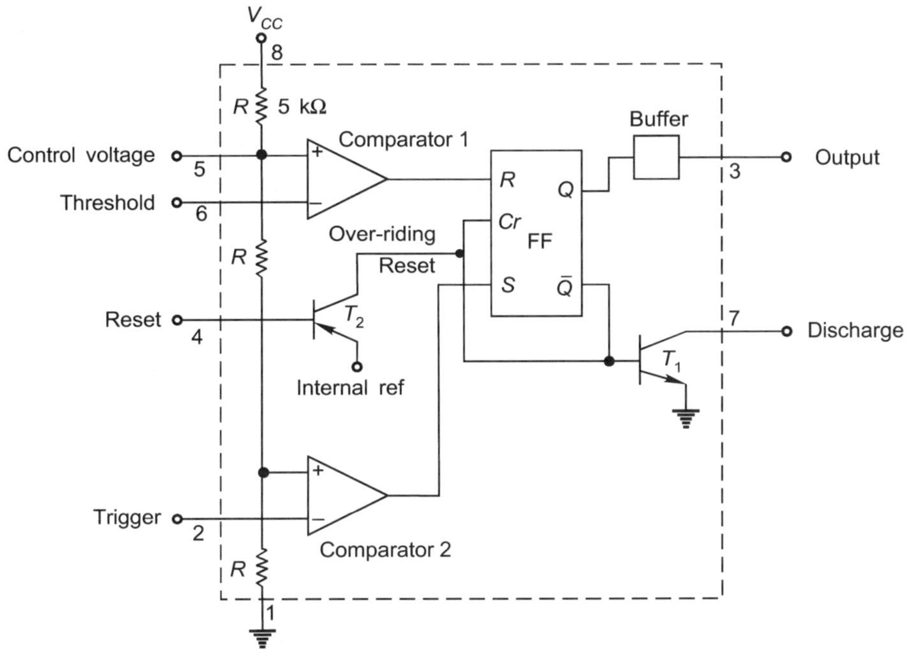

Fig. 9.33
Functional block diagram of 555 timer.

where C is the timing capacitor and R_A is an external resistor to be connected between the discharge terminal and V_{CC}.

The output is at logic 1 whenever the transistor T_1 is OFF and at logic 0 when T_1 is ON. The load can be connected either between the output terminal and V_{CC} or between the output and ground terminals. The four possible output connections are shown in Fig. 9.34. The maximum sink or source current is 200 mA. The voltage corresponding to HIGH output is approximately 0.5 V below V_{CC} and for LOW it is approximately 0.1 V. IC 556 timer contains two 555 timer circuits and is available in 14 pin DIP.

9.6.1 Monostable Multivibrator

A monostable multivibrator circuit using a 555 Timer is shown in Fig. 9.35. If the trigger input is held HIGH, then under steady-state condition, the transistor T_1 is ON, the discharge and output terminals are at LOW level. It can be verified that T_1 cannot be OFF under steady-state (Problem 9.15). When a negative pulse applied at the trigger input crosses the voltage $V_{CC}/3$, the output of the comparator 2 goes HIGH which sets the FLIP-FLOP and consequently T_1 turns OFF and the output goes HIGH. The capacitor C starts getting charged to V_{CC} with the time constant, $\tau = R_A\ C$. The circuit remains in this condition even after the trigger has returned to logic 1. When the increasing capacitor voltage reaches 2/3 V_{CC}, the output of the comparator 1 goes HIGH which resets the FLIP-FLOP. The transistor T_1 goes to saturation, thereby discharging the capacitor and the output goes LOW. The various waveforms are shown in Fig. 9.36. This is a non-retriggerable monostable multivibrator. This circuit can be converted

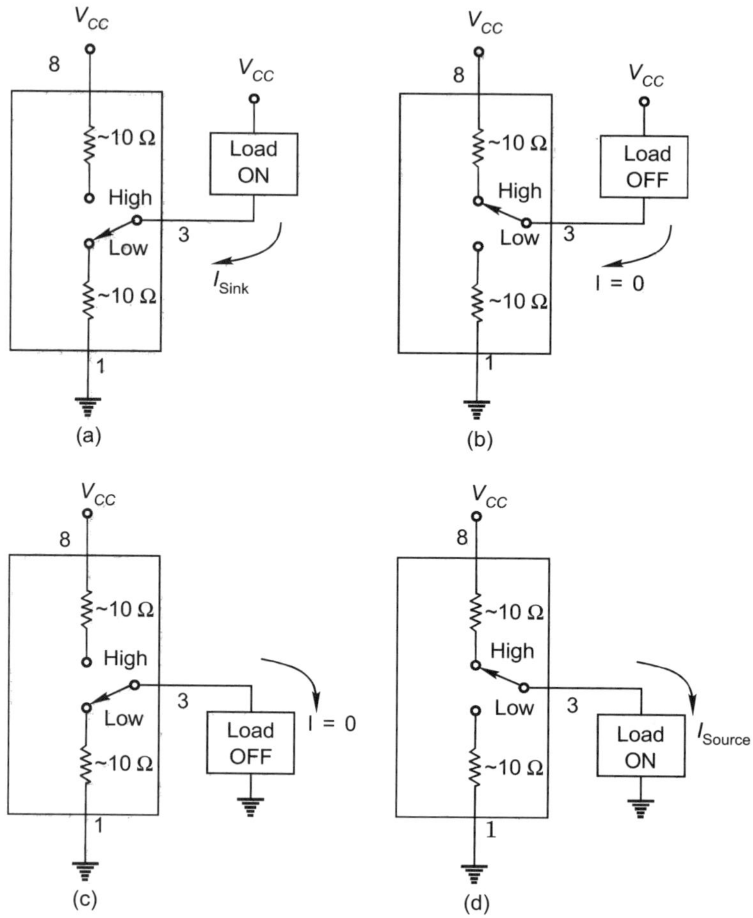

Fig. 9.34
555 Timer output terminal operation.

to a retriggerable type if reset is connected to trigger input instead of to V_{CC} (Problem 9.16) and it is triggered at the positive-edge of the trigger pulse.

The output timing interval, T is given by 1.1 $R_A \cdot C$.

9.6.2 Astable Multivibrator

An astable multivibrator using 555 timer is shown in Fig. 9.37. Let us assume that the output is in HIGH state and the capacitor C is charging through resistors R_A and R_B [time constant $\tau_1 = (R_A + R_B)\, C$]. When the voltage across $C(v_C)$ reaches 2/3 V_{CC}, the output goes LOW and C starts discharging through R_B with a time constant $\tau_2 \approx R_B \cdot C$. When v_C drops below $V_{CC}/3$, the timer is triggered and the output again goes HIGH. The capacitor C now again starts charging towards V_{CC} with the time constant τ_1. The various waveforms are shown in Fig. 9.38. The charging and discharging time intervals are given by

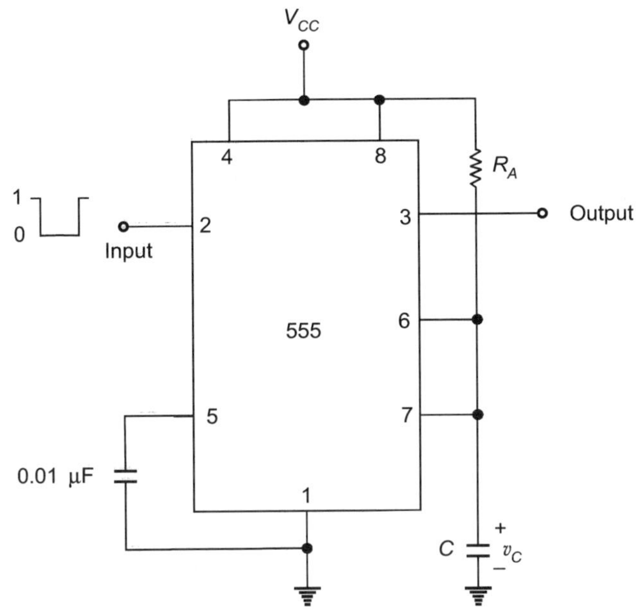

Fig. 9.35
A monostable multivibrator using 555.

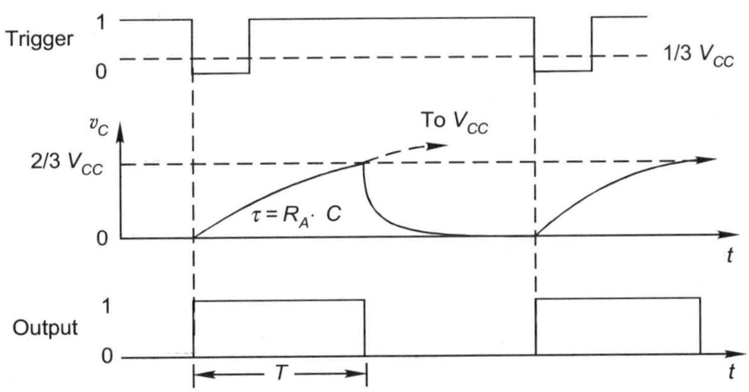

Fig. 9.36
Waveforms of monostable multivibrator.

$$T_1 \approx 0.7\ C\,(R_A + R_B) \tag{9.16}$$

and

$$T_2 \approx 0.7\ C\,R_B \tag{9.17}$$

Therefore,

$$f = \frac{1}{T} = \frac{1}{T_1 + T_2} = \frac{1.4}{C(R_A + 2R_B)} \tag{9.18}$$

and the duty cycle

$$D = \frac{R_A + R_B}{R_A + 2R_B} \times 100\% \tag{9.19}$$

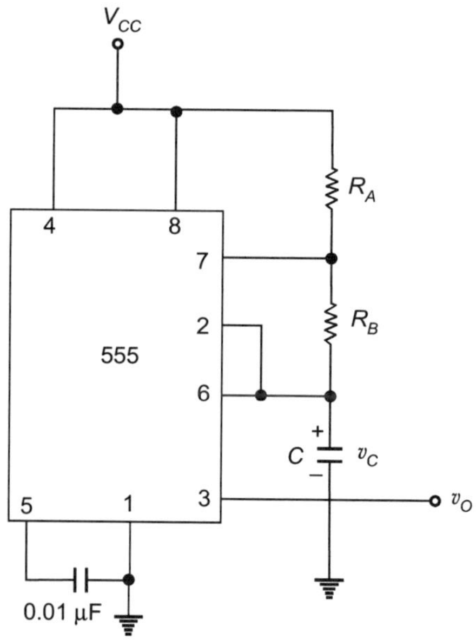

Fig. 9.37
An astable multivibrator using 555.

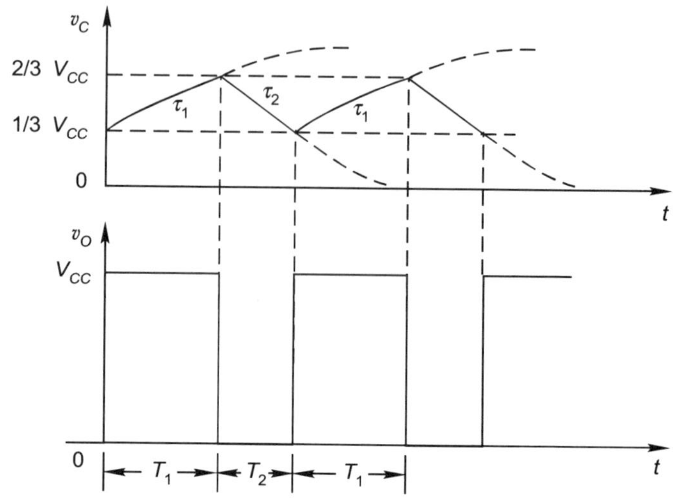

Fig. 9.38
Waveforms of 555 astable multivibrator.

From Eq. (9.19) we note that the duty cycle is always different from 50%. It can be made 50% (symmetrical square wave) by connecting a diode across R_B, which will clamp the voltage across R_B when the capacitor is charging and therefore if R_A and R_B are equal, τ_1 and τ_2 will be same. It is also possible to generate a symmetrical square wave by using the output of 555 as clock input of a *T*-type FLIP-FLOP with $T = 1$. The output of the FLIP-FLOP will be a symmetrical square wave.

9.7 SUMMARY

Some of the very useful circuits for generation of timing waveforms have been discussed in this chapter. These circuits can be designed using discrete devices but because of the availability of very useful ICs like gates, OP AMPS, monostable multivibrators, and 555 timer, circuits with discrete devices are no longer used. A variety of useful circuits can be designed using these ICs which are very convenient to use. A thorough understanding of the operations of these ICs is very essential for using them effectively in new applications.

Glossary

Analog comparator An analog circuit which compares two analog signals and produces an output whose polarity indicates the result of comparison.

Astable multivibrator A multivibrator circuit having both the states as quasi-stable states. It is a square wave oscillator. Also known as free-running multivibrator.

Bistable multivibrator A multivibrator circuit that has both the states as stable. It goes from one stable state to another when triggered. It is same as the FLIP-FLOP.

Free-running multivibrator Same as the astable multivibrator.

Hysteresis The diference between the upper and lower triggering voltages in a Schmitt trigger circuit.

Jitter When the successive pulse durations of a monostable multivibrator are not same it is said to jitter.

Monostable multivibrator An electronic circuit that produces an output pulse for a fixed time period in response to a trigger pulse and then returns to its quiscent state. The width of the output pulse produced depends upon the timing elements (R and C) used in the circuit. It is also known as one-shot circuit.

One-shot Same as the monostable multivibrator.

Operational amplifier (OP AMP) A high gain differential input amplifier.

Pulse stretcher Same as the monostable multivibrator.

Quasi-stable state The state which is not a stable state. When a circuit is forced to go to a quasi-stable state, it comes back to its stable state after a time depending upon the elements of the circuit.

Recovery time The time required for a timing circuit to come back to steady-state after having been triggered.

Regenerative comparator Same as the Schmitt trigger.

Retriggerable monostable multivibrator A monostable multivibrator that is capable of getting triggered even when it is in its non-quiscent (or quasi-stable) state.

Schmitt trigger An analog comparator circuit with different upper and lower triggering voltages. It exhibits hysteresis effect. It produces a single, sharp transition from a slowly changing input. It is same as regenerative comparator.

Review Questions

9.1 A free-running multivibrator has ________ quasi-stable states.

9.2 A ________ monostable multivibrator can be triggered even when it is not in stable state.

9.3 A pulse-stretcher circuit is same as a ________ multivibrator.

9.4 The symbol used for an inverting Schmitt trigger circuit is ________ .

9.5  represents a ________ Schmitt trigger circuit.

9.6 A sinusoidal waveform can be converted into square waveform by using a ________ .

9.7 The upper triggering voltage and the lower triggering voltage are not same for a ________ circuit.

9.8 A multivibrator circuit having one stable state and other quasi-stable state is known as a ________ multivibrator.

9.9 An astable multivibrator does not require ________ input.

9.10 ________ circuit is used for converting slowly varying signals suitable as inputs to logic circuits.

9.11 A multivibrator with two stable states is known as a __________ multivibrator.

9.12 False triggering in a comparator circuit due to the presence of noise can be eliminated by the use of __________ comparator.

9.13 The output timing interval T of a monostable multivibrator using 555 timer is __________.

9.14 The frequency of input sinusoidal waveform to a comparator circuit is 1 kHz. The frequency of the output square wave will be __________.

9.15 555 timer circuits can be used for making monostable and __________ multivibrators.

Problems

9.1 Verify that the circuit of Fig. 9.39 acts as an astable multivibrator.

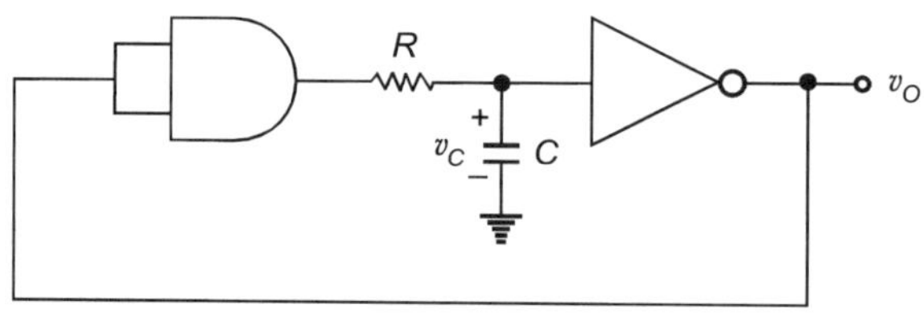

Fig. 9.39

Circuit for Problem 9.1.

9.2 Verify that v_O in logic 0 state is not a stable state for the circuit of Fig. 9.2.

9.3 Verify Eqs (9.5) and (9.6).

9.4 In the circuit of Fig. 9.12, the two Zener diodes are identical with $V_{Z1} = V_{Z2} = V_Z = 4.6$ V, voltage across conducting diode (V_D) = 0.6 V, $V_R = 1$ V. Find V_{UT} and V_{LT}.

Obtain the output voltage waveform if v_i is a sinusoidal voltage with 5 V peak value.

9.5 In the free-running multivibrator circuit of Fig. 9.13, the breakdown voltages of the Zener diodes are not same. Obtain the expression for the frequency of the square wave.

9.6 If the resistance R_f in Fig. 9.13 is replaced by the circuit of Fig. 9.40, obtain the output voltage waveform and its time period.

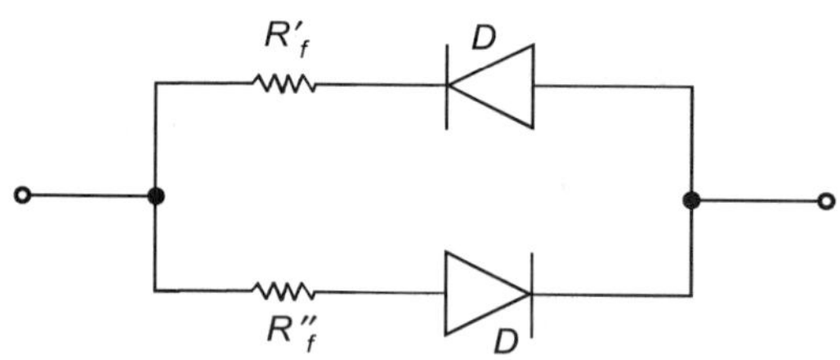

Fig. 9.40

9.7 Verify that the circuit of Fig. 9.16 cannot remain in the state $v_O = -V_O$ under steady-state condition.

9.8 Verify Eq. (9.13).

9.9 A traffic light is required to be red for 30s and green for 60s with automatic switching. Design a circuit for this purpose.

9.10 Design a one-shot to generate pulses of 0.2 μs, using
(*a*) 74121.
(*b*) 74122.

9.11 Design a one-shot to generate a 5 ms pulse, using
(*a*) The internal resistor,
(*b*) A 40 kΩ external resistor.
What is the maximum frequency of trigger pulses in each case. Use 74121.

9.12 Design a 100 kHz, 60% duty cycle square-wave generator using 555.

9.13 (*a*) An astable multivibrator using 555 timer is shown in Fig. 9.41. Explain its operation and sketch the relevant waveforms.
(*b*) Find the expression for the time period of the output waveform.
(*c*) Is it possible to get square waveform with 50% duty cycle? If yes, find out the condition under which it is possible.
(*d*) If $R_B = 20$ kΩ, find R_A for a 50% duty cycle.
(*e*) Find, if any, the restriction on the maximum value of R_B (in terms of R_A).

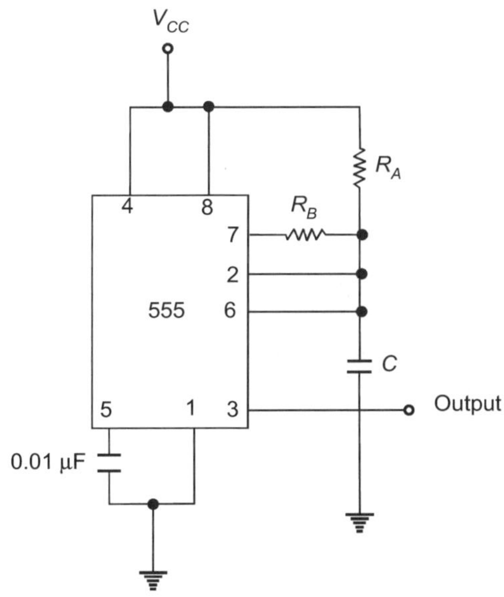

Fig. 9.41
Circuit for Problem 9.13.

9.14 (*a*) Explain how a 555 timer monostable multivibrator can be used as a frequency divider.
(*b*) Design a circuit using 555 timer to divide the frequency of a train of pulses by 3.

9.15 Verify that the circuit of Fig. 9.35 cannot have HIGH output in stable state.

9.16 Design a retriggerable monostable multivibrator using 555 timer and explain its operation.

10

A/D AND D/A CONVERTERS

10.1 INTRODUCTION

The advantages of processing signals using digital systems have been discussed in Chapter 1. Because of these advantages, digital systems are widely used for control, communication, computers, instrumentation, etc. In many such applications of digital systems, the signals are not available in the digital form.

An along voltage to be processed using digital techniques may be a fixed (d.c.) voltage or a time varying voltage $v(t)$. The process of converting an analog signal to a digital form involves a sequence of four processes. These are

- Sampling
- Holding
- Quantizing
- Encoding

These processes are not performed as separate operations. *Sampling* and *holding* operations are done simultaneously using a circuit known as sample-and–hold (S/H) circuit. These operations are required to be performed for the conversion of time varying analog signals and not for d.c signals. Quantizing and encoding processes are done simultaneously using a circuit referred to as an *analog-to-digital converter* (A/D *converter or ADC).* The process of conversion of an analog signal to digital signal is referred to as an *analog-to-digital conversion.*

The output of the system may be required to be in the analog form and, therefore, the digital output has to be converted back to the analog form. The process is referred to as a *digital-to-analog conversion* and the system used for this purpose is referred to as a *digital-to-analog converter* (*D/A converter or DAC*). Some of the examples where A/D and D/A converters are used are:

1. A digital system can be used to monitor the ambient temperature of an oven and if it exceeds a certain limit, it should reduce the fuel input. Here, an A/D converter is required to convert the output of the sensor (which converts temperature to an analog electrical signal) to digital form. If the temperature exceeds the specified limit, some digital output is produced which is to be converted to analog form in order to control the device which reduces the fuel input.
2. A digital voltmeter is used to measure an analog voltage and display the voltage in numerical form. In this, an A/D converter is required to convert the analog voltage into a digital signal. The required processing consists of determining its value. The output, in this case, is not required to be converted back to the analog form and hence a D/A converter is not needed.
3. A digital communication system is used to transmit messages which are in the form of analog electrical signals. This requires an A/D converter at the transmitting end and a D/A converter at the receiving end.

In microprocessor based process control systems, A/D and D/A converters are often used and are referred to as *peripherals* or I/O devices.

Since D/A converters are used as sub-systems in many A/D converters, D/A converters will be discussed first.

10.2 DIGITAL-TO-ANALOG CONVERTERS

The input to a D/A converter is an N-bit binary signal, available in parallel form. Normally, digital signals are available at the output of the latches or registers and the voltages, corresponding to logic 0 and logic 1, available to drive the converter are in general not precisely fixed voltages. Therefore, these voltages are not applied directly to the converter but are used to operate digitally controlled switches. The switch is thrown to one of the two positions depending upon the digital signal (1 or 0) which connects precisely fixed voltages $V(1)$ or $V(0)$ to the converter input, corresponding to 1 or 0, respectively.

The analog output voltage V_O of an N-bit straight binary D/A converter is related to the digital input by the equation

$$V_O = K(2^{N-1} b_{N-1} + 2^{N-2} b_{N-2} + \ldots + 2^2 b_2 + 2b_1 + b_0) \quad (10.1)$$

where K is a proportionality factor.

b_n = 1 if the nth bit of the digital input is 1
= 0 if the nth bit of the digital input is 0

Example 10.1 Find the analog output voltage of a 4-bit D/A converter for all possible inputs. Assume $K = 1$.

Solution

From Eq. (10.1), we obtain the output voltage for each input and these are given in Table 10.1.

Table 10.1

Digital input				Analog output
b_3	b_2	b_1	b_0	V
0	0	0	0	0
0	0	0	1	1
0	0	1	0	2
0	0	1	1	3
0	1	0	0	4
0	1	0	1	5
0	1	1	0	6
0	1	1	1	7
1	0	0	0	8
1	0	0	1	9
1	0	1	0	10
1	0	1	1	11
1	1	0	0	12
1	1	0	1	13
1	1	1	0	14
1	1	1	1	15

There are two types of commonly used D/A converters. These are:

1. Weighted-resistor D/A converter, and
2. R-$2R$ ladder D/A converter.

10.2.1 Weighted-Resistor D/A Converter

Let us assume an N-bit straight binary input to a resistor network (through digitally controlled electronic switches) which produces a current I corresponding to logic 1 at the most-significant bit, $I/2$ corresponding to logic 1 at the next lower bit, $I/2^2$ for logic 1 at the next lower bit and so on, and $I/2^{N-1}$ for logic 1 at the least-significant bit position. The total current thus produced will be proportional to the digital input. This current can be converted into a corresponding voltage, by using an OP AMP, which will be proportional to the digital input.

The circuit of Fig. 10.1 can be used for converting the digital input to analog output which operates according to the above principle. This circuit is referred to as a *weighted-resistor D/A converter* since the resistance values are weighted in accordance with the binary weights.

In the circuit Fig. 10.1, the digital inputs (1 or 0) operate the switches. A switch is thrown to position 1 or 0 for a digital input corresponding to that bit being 1 or 0, respectively. The voltage applied to a resistor is $V(1)$ if the switch connected to it is in position 1 and $V(0)$ if it is in position 0. The current, I_i is given by

$$I_i = I_{N-1} + I_{N-2} + I_{N-3} + \ldots + I_2 + I_1 + I_0 \tag{10.2}$$

where

$$\left.\begin{aligned} I_{N-1} &= V_{N-1}/R \\ I_{N-2} &= V_{N-2}/2R \\ I_{N-3} &= V_{N-3}/2^2 R \\ &\vdots \\ I_0 &= V_0/2^{N-1} R \end{aligned}\right\} \text{where } V_n = V(1) \text{ if } b_n = 1 \\ \qquad\qquad\qquad = V(0) \text{ if } b_n = 0 \tag{10.3}$$

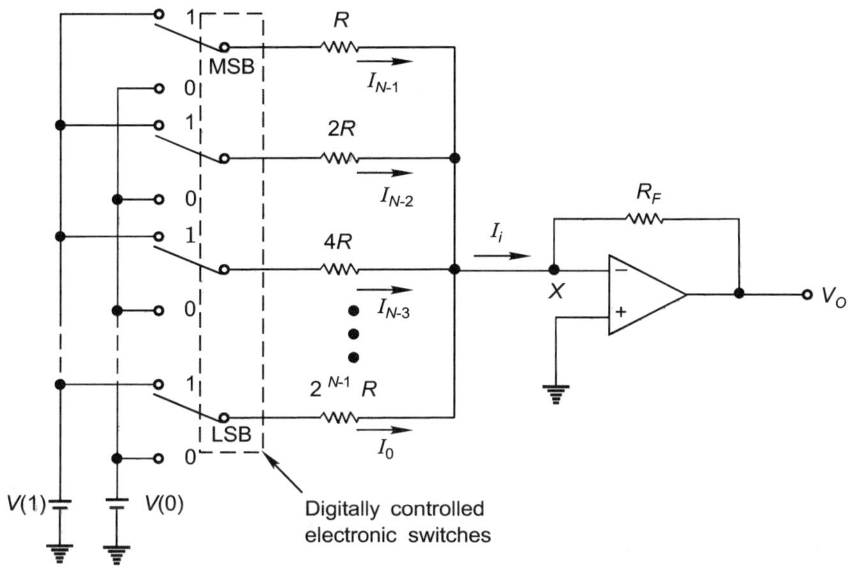

Fig. 10.1
Weighted-resistor D/A converter.

For straight binary input, $V(0) = 0$ and $V(1) = -V_R$, and the output voltage V_O is given by

$$V_O = -(-V_R)\left(\frac{R_F}{R}b_{N-1} + \frac{R_F}{2R}b_{N-2} + \frac{R_F}{2^2 R}b_{N-3} + \ldots + \frac{R_F}{2^{N-1} R}b_0\right) \tag{10.4}$$

which is of the same form as Eq. (10.1)

with
$$K = \frac{R_F}{2^{N-1} R} \cdot V_R \tag{10.5}$$

The output swings in only one direction and therefore is *unipolar.* If it is required to convert digital data in bipolar format such as in sign-magnitude, 1's complement or 2's complement format, then $V(0) \neq 0$. In such cases, $V(0)$ is used to *offset* the output swing.

With $V(1)$ and $V(0)$ as the voltages applied to the resistor network for 1 and 0, respectively, the output voltage V_O of Fig. 10.1 is given by

$$V_O = \frac{R_F}{2^{N-1} \cdot R}\left(2^{N-1} V_{N-1} + 2^{N-2} V_{N-2} + \ldots + 2^1 V_1 + 2^0 V_0\right) \tag{10.6}$$

An offset can also be produced in the output voltage V_O by using the circuit of Fig. 10.2. The offset voltage produced in this circuit is

$$-\frac{R_F}{R_{off}} \cdot V_{off}$$

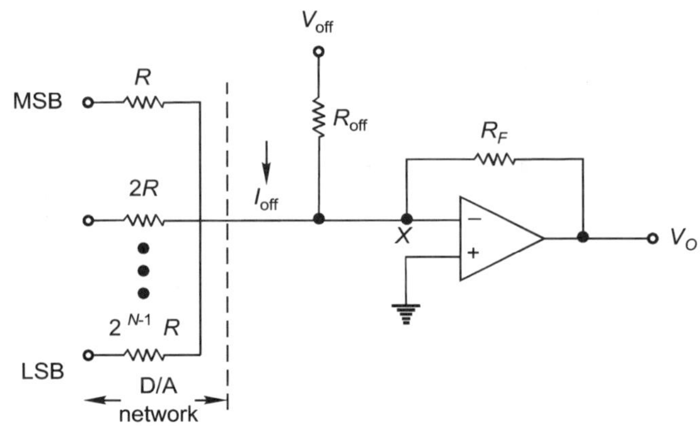

Fig. 10.2
Circuit used to offset the output voltage.

Example 10.2 (a) Consider a 4-bit unipolar D/A converter with $V(1) = -1$ V, $V(0) = 0$ V, and $R_F = 8R$. Obtain the analog output voltage for each of the digital inputs from 0000 to 1111. (*b*) Adjust the offset voltage, using the circuit of Fig. 10.2, such that $V_O = 0$ V for a digital input of 1000. With this offset, obtain the analog output voltage for each of the digital inputs. (*c*) With the offset used above, obtain the analog output voltages if the MSB is complemented before applying to the D/A converter.

Solution

(*a*) Using Eq. (10.6), we obtain the analog voltages which are same as those given in Table10.1.

(*b*) Without offset, the output voltage V_O is 8 V for the digital input 1000. Therefore, the offset must produce a voltage of – 8 V at the output, i.e.

$$-\frac{R_F}{R_{off}} V_{off} = -8 \text{ V}$$

We can use $R_{off} = R$ and $V_{off} = 1$ V. The analog voltages are given in Table 10.2.

(*c*) If the digital input to the D/A converter is $\bar{b}_3\ b_2\ b_1\ b_0$, then the analog output voltages can be obtained from Table 10.2, and these are given in Table 10.3.

From Table 10.3, we observe that this is a D/A converter which converts 2's complement format to analog signal.

Table 10.2

Digital input				Analog output
b_3	b_2	b_1	b_0	V
0	0	0	0	– 8
0	0	0	1	– 7
0	0	1	0	– 6
0	0	1	1	– 5
0	1	0	0	– 4
0	1	0	1	– 3

(Contd.)

Table 10.2 (*Contd.*)

0	1	1	0	− 2
0	1	1	1	− 1
1	0	0	0	0
1	0	0	1	+ 1
1	0	1	0	+ 2
1	0	1	1	+ 3
1	1	0	0	+ 4
1	1	0	1	+ 5
1	1	1	0	+ 6
1	1	1	1	+ 7

Table 10.3

Digital input				*Analog output*
b_3	b_2	b_1	b_0	V
1	0	0	0	− 8
1	0	0	1	− 7
1	0	1	0	− 6
1	0	1	1	− 5
1	1	0	0	− 4
1	1	0	1	− 3
1	1	1	0	− 2
1	1	1	1	− 1
0	0	0	0	0
0	0	0	1	+ 1
0	0	1	0	+ 2
0	0	1	1	+ 3
0	1	0	0	+ 4
0	1	0	1	+ 5
0	1	1	0	+ 6
0	1	1	1	+ 7

Example 10.3 Consider a 4-bit digital input in 1's complement format. Design a D/A converter for this.

Solution

Since the 1's complement representations of the positive numbers + 0 to + 7 are same as the representations of the unipolar binary numbers, no offset voltage is required for these inputs.

For the negative numbers 1111 to 1000, the output analog voltage is to be offset by –15 V. This can be achieved by operating a switch with MSB of input to introduce proper value of V_{off}.

An alternative circuit is given in Prob. 10.2.

The weighted-resistor D/A converter has the problem of having a wide range of resistor values (R to $2^{N-1} \cdot R$) with required precision and which track over a wide temperature range. It is difficult to fabricate such a wide range of resistance values in a monolithic IC. This difficulty is eliminated by R–$2R$ ladder D/A converter discussed below.

10.2.2 R–2R Ladder D/A Converter

An R–$2R$ ladder D/A converter is shown in Fig. 10.3. It uses resistors of only two values, R and $2R$. The inputs to the resistor network are applied through digitally controlled switches. A switch is in position 0 or 1 corresponding to the digital input for that bit position being 0 or 1, respectively. To analyse this circuit, for simplicity we consider a 3-bit R–$2R$ ladder D/A network shown in Fig. 10.4. In this circuit, we have assumed the digital input as 001.

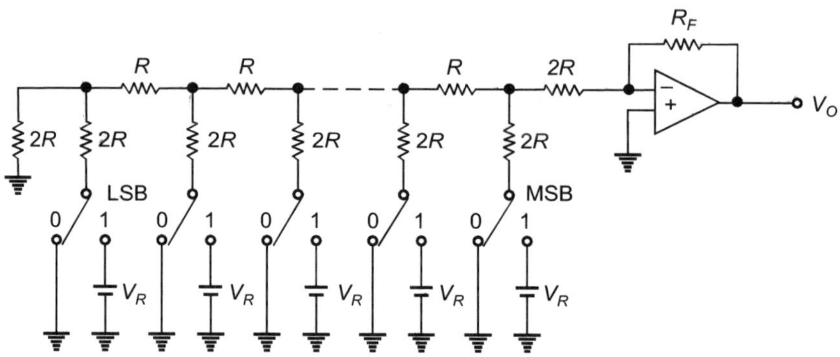

Fig. 10.3
R–2R ladder D/A converter.

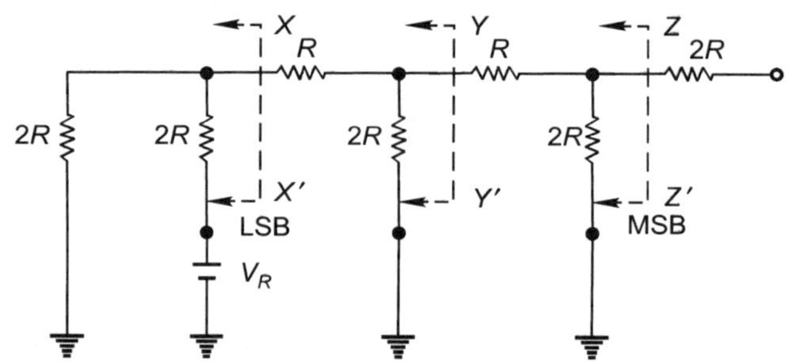

Fig. 10.4
3-bit R–2R ladder D/A network.

The circuit is simplified using Thevenin's theorem. Applying Thevenin's theorem at XX', we obtain the circuit of Fig. 10.5*a*. Similarly, applying Thevenin's theorem at YY' and ZZ', we obtain the circuits of Fig. 10.5*b* and *c*, respectively. Here, LSB has been assumed as 1 and the equivalent voltage obtained is $V_R/2^3$.

Similarly, for the digital input of 010 and 100, the equivalent voltages are $V_R/2^2$ and $V_R/2^1$, respectively. The value of the equivalent resistance is $3R$ in each case. Therefore, we obtain an equivalent circuit of Fig. 10.4 which is given in Fig. 10.5*d*. For the circuit of Fig. 10.5*d*, the output analog voltage V_O is given by

$$V_O = -\left(\frac{R_F}{3R}\cdot\frac{V_R}{2^3}b_0 + \frac{R_F}{3R}\cdot\frac{V_R}{2^2}b_1 + \frac{R_F}{3R}\cdot\frac{V_R}{2^1}b_2\right)$$

$$= -\left(\frac{R_F}{3R}\right)\cdot\left(\frac{V_R}{2^3}\right)\left[4b_2 + 2b_1 + 1b_0\right] \quad (10.7)$$

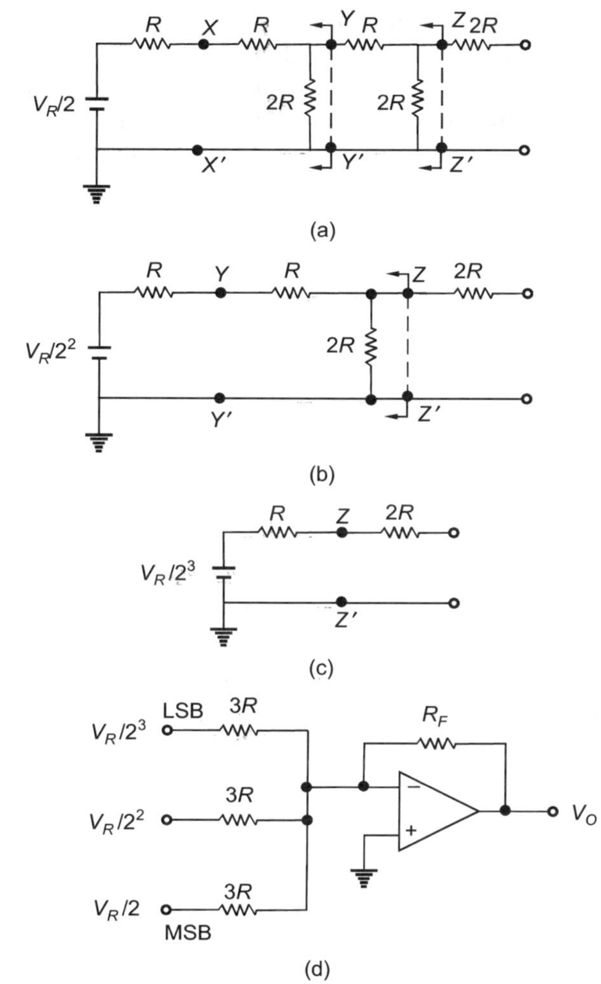

Fig. 10.5

Simplification of circuit of Fig. 10.4.

Equation (10.7) shows that the analog output voltage is proportional to the digital input. In general, for an N-bit D/A converter, the output voltage can be similarly determined and is given by

$$V_O = \left(2^{N-1}\, b_{N-1} + 2^{N-2}\, b_{N-2} + \dots + 2^2\, b_2 + 2^1\, b_1 + 2^0\, b_0\right) \tag{10.8}$$

where $R_F = 3\ R$ and $V_R = -\ 2^N$ V.

The number of resistors required for an N-bit D/A converter is $2N$ in the case of R–$2R$ ladder D/A converter, whereas it is only N in the case of a weighted-resistor D/A converter. But, because of the wide spread in the resistance values for large N, the weighted-resistor D/A converter is not suitable. However, the weighted-resistor network of Fig. 10.2 can be modified to accommodate a large number of bits without consequent spread in resistor values. One such circuit is shown in Fig. 10.6. Here, the bits are divided into groups of four. The most-significant four bits are applied in a manner similar to the one used in weighted-resistor D/A converter. The least-significant four bits are applied through an additional resistor *r*, in addition to weighted-resistors. This is done to produce input currents of OP AMP due to least-significant group of 4 -bits and the most-significant group of 4-bits in the ratio of 1: 16 $\left(\frac{b_3}{b_7}=\frac{b_2}{b_6}=\frac{b_1}{b_5}=\frac{b_0}{b_4}=1/16\right)$.

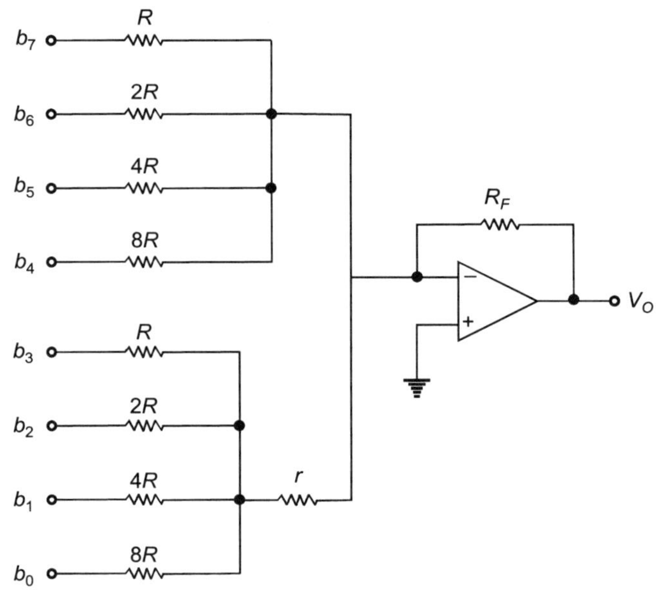

Fig. 10.6
Modified weighted-resistor D/A converter.

The resistor *r* is determined in the following way:

Let the b_3 bit be 1 and b_2, b_1, and b_0 bits be all 0. The portion of the circuit corresponding to this is shown in Fig. 10.7*a* and its simplified version is shown in Fig. 10.7*b*.

From Fig. 10.7*b*, we obtain I_{in} assuming virtual short at the input of the OP AMP.

$$I_{\text{in}} = \frac{V_R}{R + \dfrac{r\,(8/7\,R)}{(r + 8/7\,R)}} \times \frac{\left(\dfrac{8}{7}R\right)}{\left(r + \dfrac{8}{7}R\right)} \tag{10.9}$$

This current must be 1/16th of the current due to b_7, which is V_R/R. Therefore,

$$\frac{V_R\left(\dfrac{8}{7}R\right)}{R\left(r + \dfrac{8}{7}R\right) + r\left(\dfrac{8}{7}R\right)} = \frac{V_R}{16R}$$

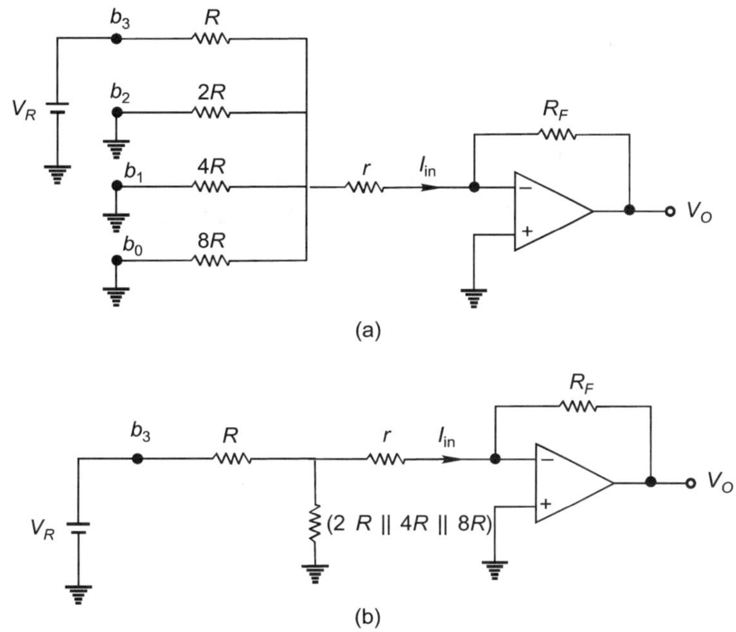

Fig. 10.7

A portion of modified weighted-resistor D/A converter.

or $$r = 8R \tag{10.10}$$

With this value of r, it can be verified that the currents due to b_2, b_1, and b_0 will be 1/16th of the currents due to b_6, b_5, and b_4, respectively (Problem 10.3).

The output analog voltage V_O of the modified D/A converter of Fig. 10.6 for $r = 8R$ is given by

$$V_O = -\left(\frac{V_R}{R} R_F\, b_7 + \frac{V_R}{2R} R_F\, b_6 + \frac{V_R}{4R} R_F\, b_5 + \frac{V_R}{8R} R_F\, b_4 + \frac{V_R}{16R} R_F\, b_3 + \frac{V_R}{32R} R_F\, b_2 + \frac{V_R}{64R} R_F\, b_1 + \frac{V_R}{128R} R_F\, b_0\right)$$

$$= -\left(\frac{V_R}{2^7} \cdot \frac{R_F}{R}\right) \cdot \left(2^7\, b_7 + 2^6\, b_6 + \ldots + 2^2\, b_2 + 2^1\, b_1 + b_0\right) \tag{10.11}$$

We observe from Eq. (10.11) that the analog output voltage is proportional to the digital input. The number of resistors in this circuit is less and also the spread in the resistor values is reduced. This configuration can be used for any number of bits.

Example 10.4 Design a 2-decade BCD D/A converter.

Solution

The circuit of Fig. 10.6 can be used for BCD D/A converter. The binary inputs corresponding to the least-significant digit are applied at b_3, b_2, b_1, and b_0 and those corresponding to the next digit, at b_7, b_6, b_5, and b_4. The value of r is chosen so as to make the input current of OP AMP corresponding to LSD as 1/10th of that of current due to MSD, and is given by

$$\frac{V_R\left(\frac{8}{7}R\right)}{R\left(r+\frac{8}{7}R\right)+r\left(\frac{8}{7}R\right)}=\frac{V_R}{10R}$$

or $$r = 4.8\,R \tag{10.12}$$

10.2.3 Specifications for D/A Converters

The characteristics of a D/A converter, which are generally specified by the manufacturers are:

1. Resolution,
2. Linearity,
3. Accuracy,
4. Settling time, and
5. Temperature sensitivity.

These are discussed below:

Resolution

This is the smallest possible change in output voltage as a fraction or percentage of the full-scale output range. For example, for an 8-bit converter, there are 2^8 or 256 possible values of analog output voltage, hence the smallest change in the output voltage is 1/255th of the full-scale output range. Its resolution is described as one part in 255, or 0.4 per cent. Alternatively, the number of bits accepted at the input can itself be used as the resolution. For example, an 8-bit D/A converter has an 8-bit resolution.

Linearity

In a D/A converter, equal increments in the numerical significance of the digital input should result in equal increments in the analog output voltage. In an actual circuit, the input–output relationship is not linear. This is due to the error in resistor values and voltage across the switches. The linearity of a converter is a measure of the precision with which the linear input–output relationship is satisfied.

The output voltages of a 3-bit unipolar D/A converter are shown in Fig. 10.8. The horizontal axis represents the input bit combinations with fixed-interval separations in order of numerical significance and the vertical axis represents the analog output voltage. The output voltage for each input is indicated by a dot. If the converter were ideal, the dots would fall on the straight line, as shown in Fig. 10.8.

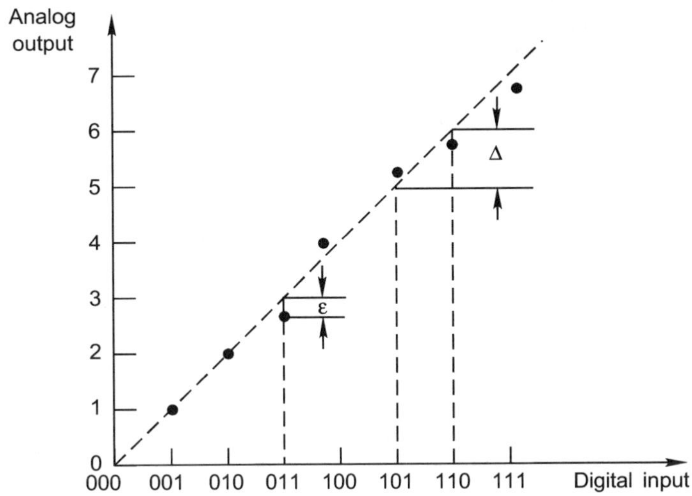

Fig. 10.8
Output vs input of a D/A converter.

The linearity error for a digital input is the difference between the voltage corresponding to the dot and the voltage obtained from the straight line (expected output). This is indicated by ε. The normal analog output voltage change corresponding to a digital input change equivalent to the least-significant bit is indicated by Δ.

The linearity of a D/A converter is generally specified by comparing ε with Δ. For example, the linearity of a commercial D/A converter unit is specified as "less than $\pm\frac{1}{2}$ LSB". This means that $.|\varepsilon| < \frac{1}{2}\Delta$.

Accuracy

The accuracy of a D/A converter is a measure of the difference between the actual output voltage and the expected output voltage. It is specified as a percentage of full-scale or maximum output voltage. For example, if a D/A converter has 10 V full-scale (maximum) output voltage and an accuracy of ± 0.2%, then the maximum error for any output voltage will be $0.002 \times 10 = 20$ mV.

Settling Time

When the digital input to a D/A converter changes, the analog output voltage does not change abruptly. Because of the presence of switches, active devices, stray capacitance, and inductance associated with the passive circuit components, the transients appear in the output voltage and oscillations may also occur. Typically, a plot of a change in the output voltage might be as shown in Fig. 10.9. The time required for the analog output to settle to within $\pm\frac{1}{2}$ LSB of the final value after a change in the digital input is usually specified by the manufacturers and is referred to as *settling time.* This imposes a limit on the frequency at which the digital input can

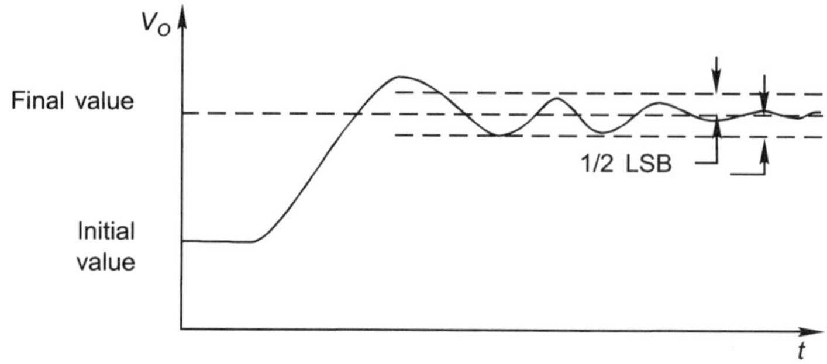

Fig. 10.9
Settling time of a D/A converter.

change. If it is operated at too high a frequency, it may not have time to settle to the correct output voltage before being switched on to the next digital input.

Temperature Sensitivity

The analog output voltage for any fixed digital input varies with temperature. This is due to the temperature sensitivities of the reference voltage source, resistors, OP AMP, etc. It is specified as ± ppm/°C.

10.3 AN EXAMPLE OF D/A CONVERTER IC

The AD DAC 80 is a 12-bit D/A converter, available in 24-pin DIP. It consists of matched bipolar switches, a precision resistor network, a low-drift, high-stability voltage reference with an optional output amplifier. The options available are 12-bit complementary binary (CBI) or three-digit BCD (complementary-coded-decimal, CCD) input codes, as well as current or voltage output modes. Its important performance characteristics are:

Resolution	:	12 bits (binary) or 3 digits (BCD)
Linearity error	:	± 0.12% (max)
Maximum gain drift	:	± 30 ppm/°C
Power dissipation	:	925 mW (max)
Maximum conversion time (Settling time to ± 0.01% of full scale range)	:	5 μs (voltage output) 1 μs (current output)
Digital input format	:	12-bit CBI or 3-digit CCD
Analog output voltage ranges	:	± 2.5 V, ± 5 V, ± 10 V, 0 to + 5 V, 0 to + 10 V (CBI), or 0 to + 10 V (CCD)
Output current	:	± 5 mA (min)
Output impedance (dc)	:	0.05 Ω
Analog output current ranges	:	± 1 mA, 0 to – 2 mA (CBI) or 0 to – 2 mA (CCD)
Output impedance	:	3.2 kΩ (bipolar) or 6.6 kΩ (unipolar)

The functional block diagram of DAC 80 is shown in Fig. 10.10.

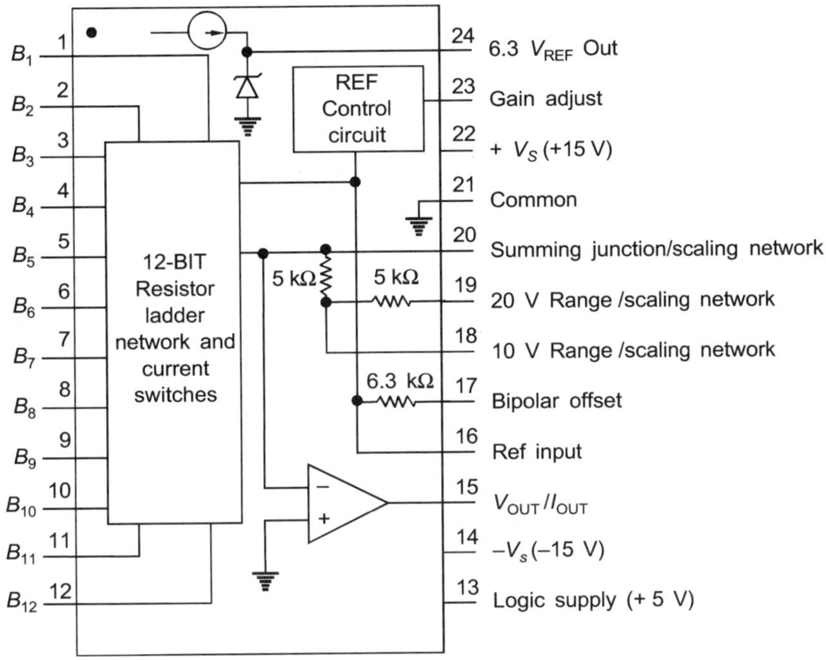

Fig. 10.10

Functional block diagram of DAC 80.

Most of the pins of the voltage and current models of ADC 80 have the same function. The pins marked with two functions have the first function for the voltage model and the second, for the current model.

10.3.1 Digital Input Codes

The DAC 80 accepts complementary digital input codes in either binary (CBI) or decimal (CCD) format. The CBI code may be any one of these codes: complementary straight binary (CSB), complementary offset binary (COB), or complementary 2's complement (CTC). These codes are given in Table 10.4. The 12-bit digital input is to be connected to pins marked B_1 (MSB) through B_{12} (LSB).

Table 10.4 Input codes for D/A converter

Digital input		*Analog output*			
Format	*Code*	*CSB*	*COB*	*CTC*	*CCD*
Binary	000000000000	+ Full scale	+ Full scale	– 1 LSB	—
	011111111111	+ 1/2 Full Scale	Zero	– Full scale	—
	100000000000	Mid-scale — 1 LSB	– 1 LSB	+ Full scale	—
	111111111111	Zero	– Full scale	Zero	—
BCD	011001100110	—	—	—	+ Full scale
	111111111111	—	—	—	Zero

10.3.2 Analog Output

The analog output voltage is available in different full-scale ranges (FSR). Table 10.5 gives the connections to be made for various output ranges.

Table 10.5 Output voltage range connections

Output range, V	*Digital input codes*	*Connect pin 15 to pin*	*Connect pin 17 to pin*	*Connect pin 19 to pin*
± 10 V	COB or CTC	19	20	15
± 5 V	COB or CTC	18	20	NC
± 2.5 V	COB or CTC	18	20	20
0 to +10 V	CSB	18	21	NC
0 to +5 V	CSB	18	21	20
0 to +10 V	CCD	19	NC	15

10.3.3 Calibration

The offset and gain potentiometers shown in Fig. 10.11 are used for calibration.

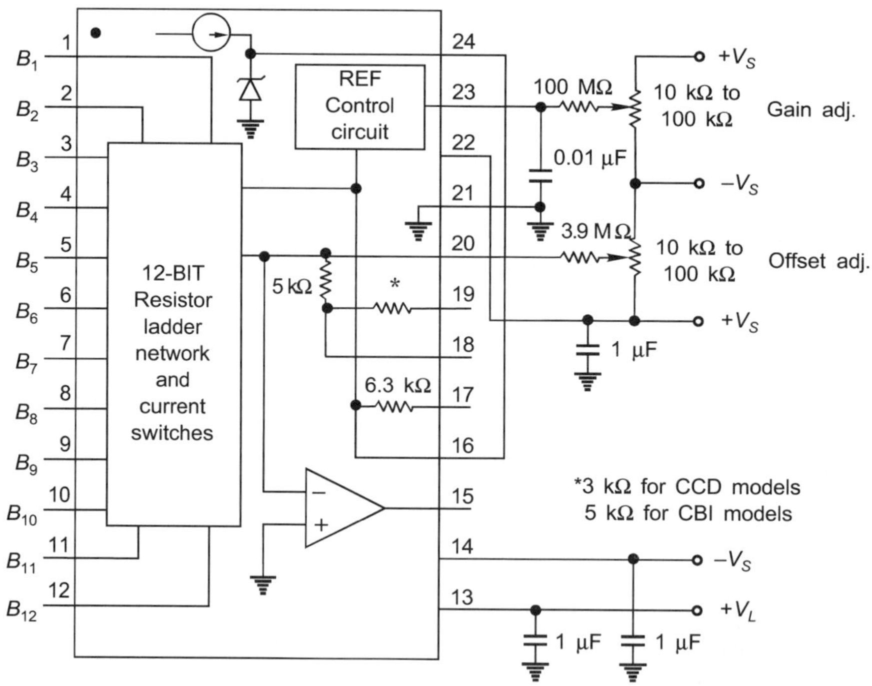

Fig. 10.11
Gain and offset adjustments.

Offset Adjustment

For unipolar configurations, apply the digital input code that should produce zero output potential and adjust the offset potentiometer for zero output. For bipolar configurations, apply the digital input code that would produce the maximum negative output voltage, and adjust the offset potentiometer for maximum negative output voltage.

Gain Adjustment

Apply the digital input that would give the maximum positive voltage output. Adjust the gain potentiometer for this positive full scale voltage.

The gain and offset adjustments for unipolar and bipolar D/A converters are illustrated in Fig. 10.12.

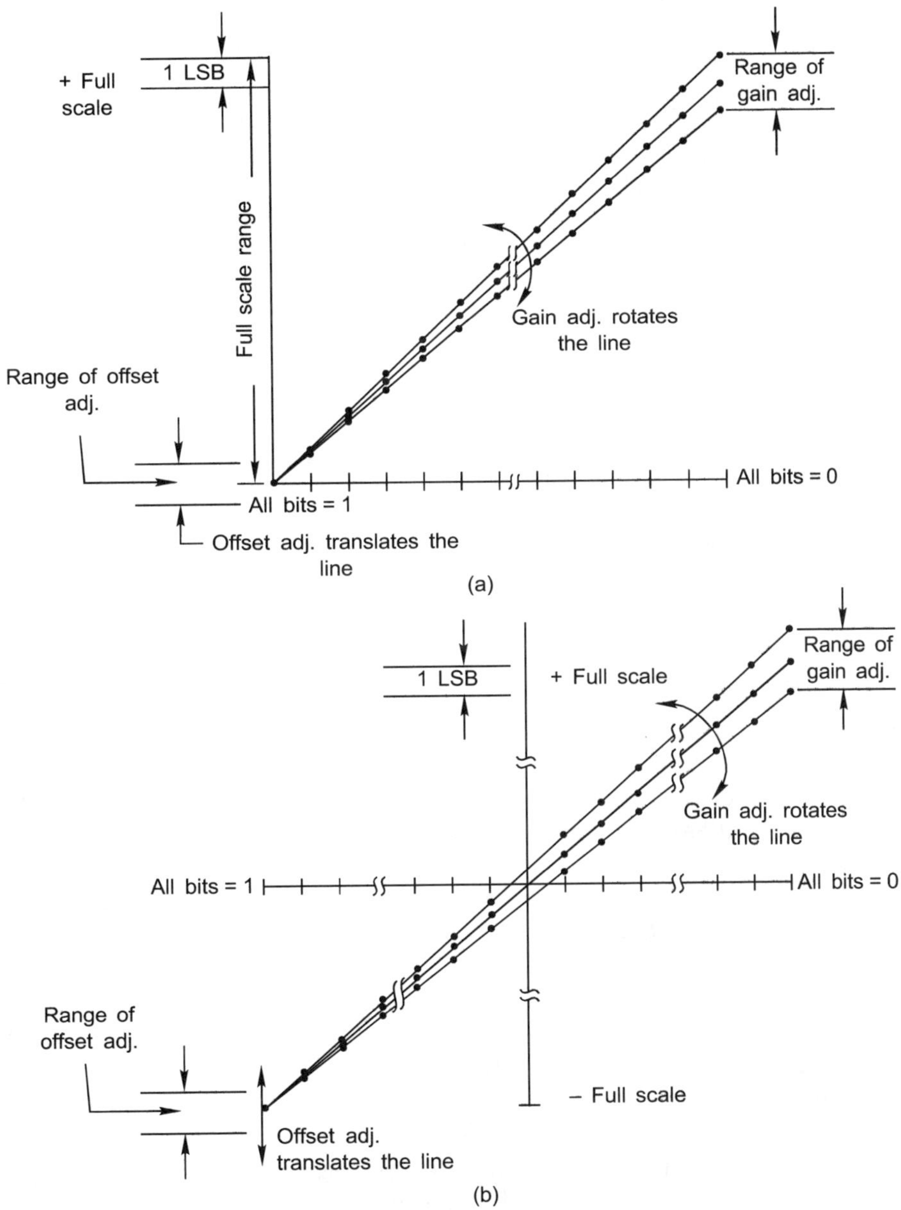

Fig. 10.12

Gain and offset adjustments for (a) Unipolar, (b) Bipolar, D/A converter.

10.4 SAMPLE-AND-HOLD

A time-varying analog signal can have any value in a given range. For example, $v(t) = V \sin \omega t$ shown in Fig. 10.13 will have different value of voltage between the range $+V$ and $-V$ at different instants of time.

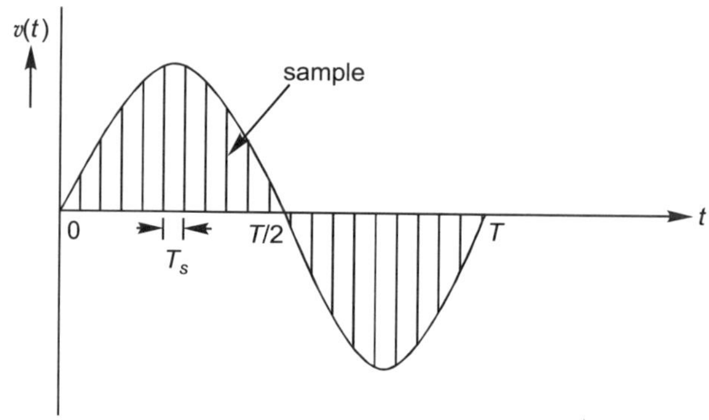

Fig. 10.13
A time varying analog signal $v(t) = V \sin \omega t$

If this signal is to be converted into digital form, we may think that the value of the signal at every instant of time is required to be converted into a group of n number of bits. This process will yield an infinite number of analog voltage values and thus is not at all practicable. On the other hand, if we take the samples of $v(t)$ at regular intervals, that is the signal is sampled regularly every T_s, then according to the sampling theorem, the signal can be uniquely determined and reconstructed from these samples with no error. The time T_s is called the sampling time and $f_s = \frac{1}{T_s}$ is sampling rate.

According to the sampling theorem,

$$f_s \geq 2f \qquad (10.13)$$

where f is the frequency of the analog signal. From Eq.(10.13), we conclude that the sampling rate be rapid enough for atleast two samples to be taken during the course of the time period of the analog signal.

The constant voltage corresponding to each sample is obtained using a sample-and-hold (S/H) circuit. The output of the S/H circuit is converted to digital signal by means of an analog-to-digital (A/D) converter circuit.

10.4.1 Sample-and-Hold Circuit

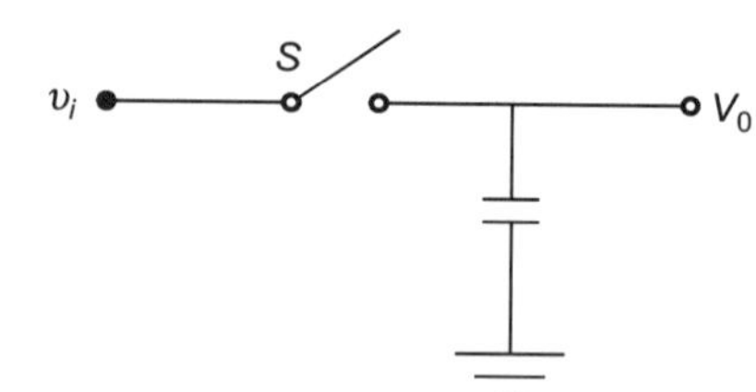

Fig. 10.14
Basic sample-and-hold circuit

A basic sample-and-hold circuit is shown in Fig. 10.14. In this circuit the voltage across the capacitor follows the input signal voltage v_i during the time switch S is closed. The capacitor holds the instantaneous value of the signal voltage attained just before the switch is opened. Thus for every T_s, the switch is closed for a short duration and then opened. The d.c voltage across the capacitor gives the value of the signal at the instant the switch is opened. This d.c voltage represents a sample of the signal and is converted to digital signal using A/D converter circuit during the *hold* period.

Figure 10.15 shows a practical S/H circuit. An enhancement mode MOSFET is used as a switch which is controlled by a control voltage V_s.

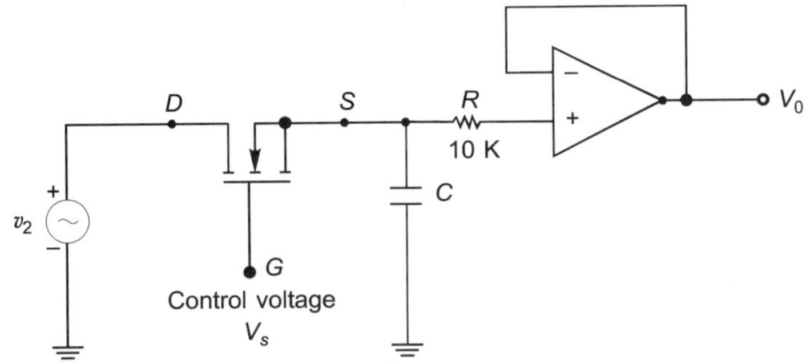

Fig. 10.15
A practical sample-and-hold circuit

An OP AMP is used to avoid discharge of capacitor due to the loading effect of A/D converter. During the positive values of V_s (higher than the threshold voltage), the MOSFET conducts, input voltage appears across *C* and at the output. When V_s is less than the threshold voltage, the MOSFET is cut-off (*S* open) and the voltage across *C* is retained because the input resistance of the OPAMP is very high.

The accuracy of the circuit depends upon the holding of the charge in the capacitor, therefore, a capacitor with a very low leakage must be used. A capacitor with polycarbonate, poly ethylene, or Teflon dielectric is prefered. Most of the other capacitors do not retain the stored charge for a sufficiently long duration due to polarization phenomenon.

10.5 ANALOG-TO-DIGITAL CONVERTERS

10.5.1 Quantization and Encoding

In a digital-to-analog converter, the possible number of digital inputs is fixed. For example, in a 3-bit D/A converter, there are 8 possible inputs. In contrast, in an analog-to-digital converter, the input analog voltage can have any value in a range, but the digital output can have only 2^N discrete values for an *N*-bit A/D converter. Therefore, the whole range of analog voltage is required to be represented suitably in 2^N intervals. This process is known as *quantization.* Each interval is then assigned a unique *N*-bit binary code, which is referred to as *encoding.*

Consider an analog voltage in the range of 0 to *V* and a 3-bit digital output for any voltage in this range. Let us divide the whole range of analog voltage in 8 intervals (3-bit output) of the size $S = V/8$. Each interval is assigned a 3-bit binary value. The intervals of the analog voltage and their corresponding digital values assigned are shown in Fig. 10.16. From this, we observe that the whole range of voltage in an interval is represented by only one digital value. Therefore, there is an error referred to as *quantization error,* involved in this process of quantization.

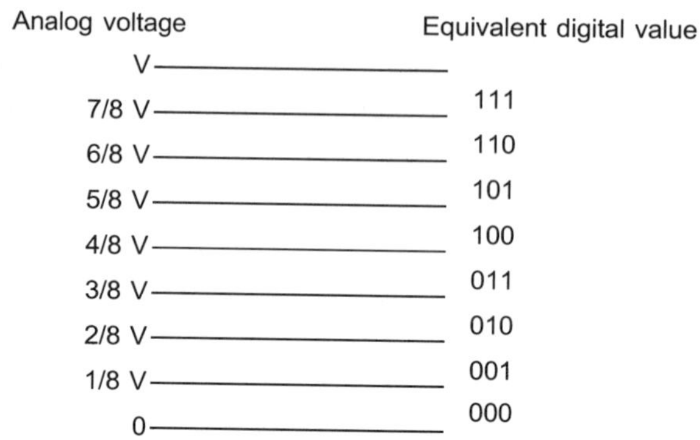

Fig. 10.16
Quantization and encoding

In this case, the maximum quantization error for any analog input voltage V_x in the given range is $V/8$.

The quantization error can be reduced if we choose the middle six intervals of size $S = V/7$ and the top and the bottom intervals of size $S/2 = V/14$. The intervals, along with their assigned digital values and equivalent analog output voltage if these digital signals are applied to a D/A converter, are shown in Fig. 10.17. This shows that the maximum quantization error will be $S/2 = V/14$ for any analog input voltage V_a in the range 0 to V. The quantization error can also be specified in terms of LSB. For example, in the above case, the maximum quantization error is $\pm \frac{1}{2}$ LSB.

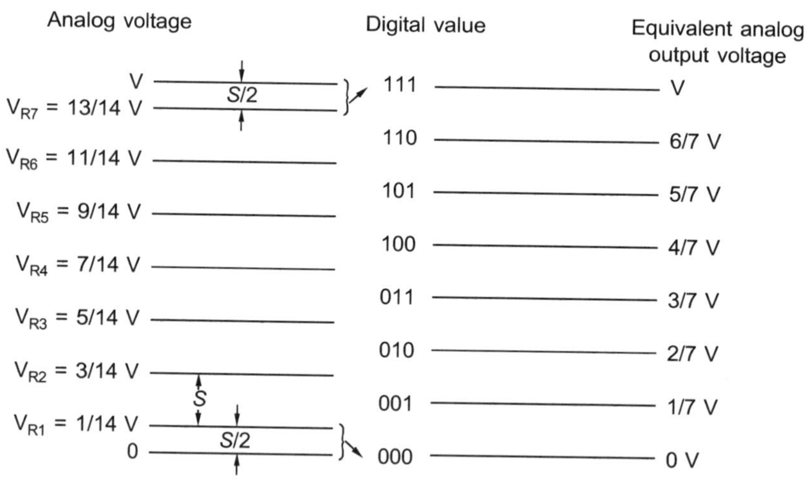

Fig. 10.17
Quantization with maximum error = $S/2$.

Example 10.5 An analog voltage in the range of $-V$ to $+V$ is required to be converted into a 3-bit 2's complement digital format. The digital value for 0 V should be 000 and the maximum quantization error should not exceed $\pm\frac{1}{2}$ LSB. Determine the quantization interval.

Solution

The digital value 000 should be assigned to the analog voltage interval 0 $V \pm S/2$, as shown in Fig. 10.18. Since in 2's complement representation, there is one more negative number than the number of positive numbers, the analog voltage from $-V$ to $+V$ should be divided in seven intervals, each of size $S = 2V/7$, and one digital value is to be assigned to each interval. The extra digital output 100 can be used to represent the interval $-V$ to $-9V/7$.

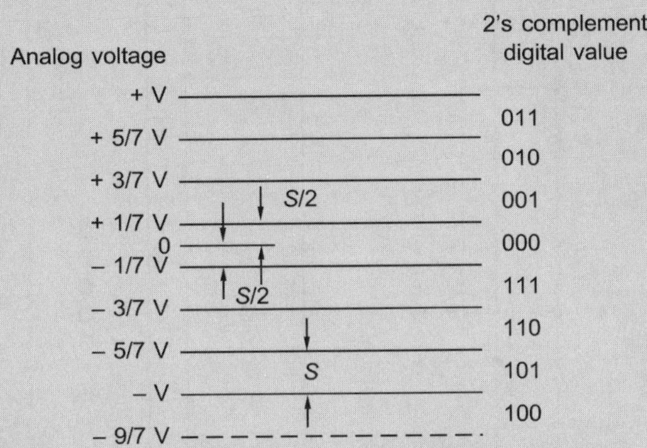

Fig. 10.18
Figure for Ex. 10.5.

Some of the commonly used A/D converters are discussed below:

10.5.2 Parallel-Comparator A/D Converter

A 3-bit parallel-comparator A/D converter is shown in Fig. 10.19. V_a is the analog voltage to be converted into digital form. The voltage corresponding to full scale is V from which the reference voltages V_{R1}, V_{R2}... (See Fig. 10.17) are generated using the resistor network. The voltage V_a is compared simultaneously with the reference voltages by using comparators. A 7-bit output is obtained from the comparators which is stored in latches. This 7-bit digital signal is converted to a 3-bit output by using a decoder circuit. The comparator outputs and the 3-bit digital output for each interval of the analog voltage are given in Table 10.6.

The principle of parallel-comparator A/D conversion is the simplest in concept and fastest. Its main disadvantages are rapid increase in the number of comparators with the number of bits [$(2^N - 1)$ comparators are required for an N-bit converter] and the corresponding complications of the decoder circuit.

10.5.3 Successive-Approximation A/D Converter

The principle of the successive-approximation A/D conversion can be explained by the following example. Consider an object of unknown weight in the range 0 to 1 kg. Suppose that a balance and a set of known weights of 1/2, 1/4, and 1/8 kg are available. These known weights are to be used in a succession of trials to determine the unknown weight.

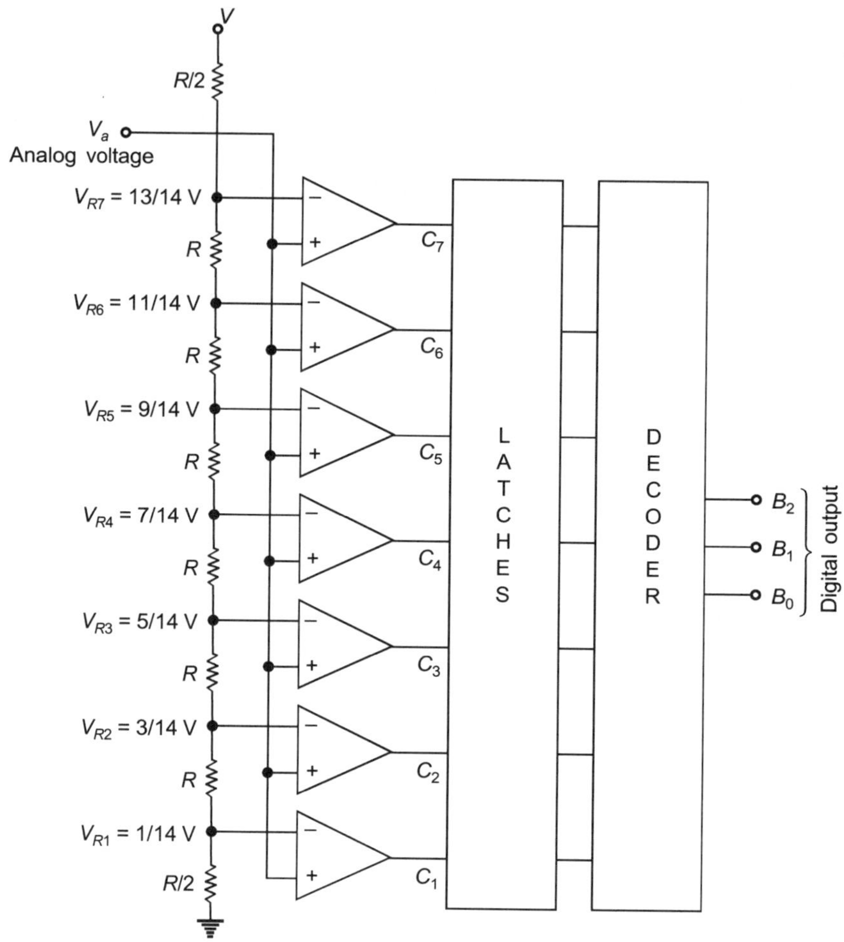

Fig. 10.19
A 3-bit parallel-comparator A/D converter.

Table 10.6 Comparator outputs and digital output of parallel- comparator A/D converter

Analog input	*Comparator outputs*							*Digital output*		
V_a	C_7	C_6	C_5	C_4	C_3	C_2	C_1	B_2	B_1	B_0
$0 \le V_a < V_{R1}$	0	0	0	0	0	0	0	0	0	0
$V_{R1} < V_a < V_{R2}$	0	0	0	0	0	0	1	0	0	1
$V_{R2} < V_a < V_{R3}$	0	0	0	0	0	1	1	0	1	0
$V_{R3} < V_a < V_{R4}$	0	0	0	0	1	1	1	0	1	1
$V_{R4} < V_a < V_{R5}$	0	0	0	1	1	1	1	1	0	0
$V_{R5} < V_a < V_{R6}$	0	0	1	1	1	1	1	1	0	1
$V_{R6} < V_a < V_{R7}$	0	1	1	1	1	1	1	1	1	0
$V_{R7} < V_a \le V$	1	1	1	1	1	1	1	1	1	1

We begin the process by placing the unknown weight W_a on one side of the balance and the 1/2 kg known weight on the other side and take a decision in the following way:

1. If $W_a \geq 1/2$ kg, retain the 1/2 kg weight on the scale and add 1/4 kg weight to its side. Also write a 1 as the most-significant bit.
2. If $W_a < 1/2$ kg, remove the 1/2 kg weight and place 1/4 kg weight in the scale. Corresponding to this, write a 0 as the most-significant bit.

We continue to try weights, successively smaller by a factor of 2. If a weight is finally retained, it is represented by a 1 and if removed then, by a 0. The complete process is illustrated in Fig. 10.20.

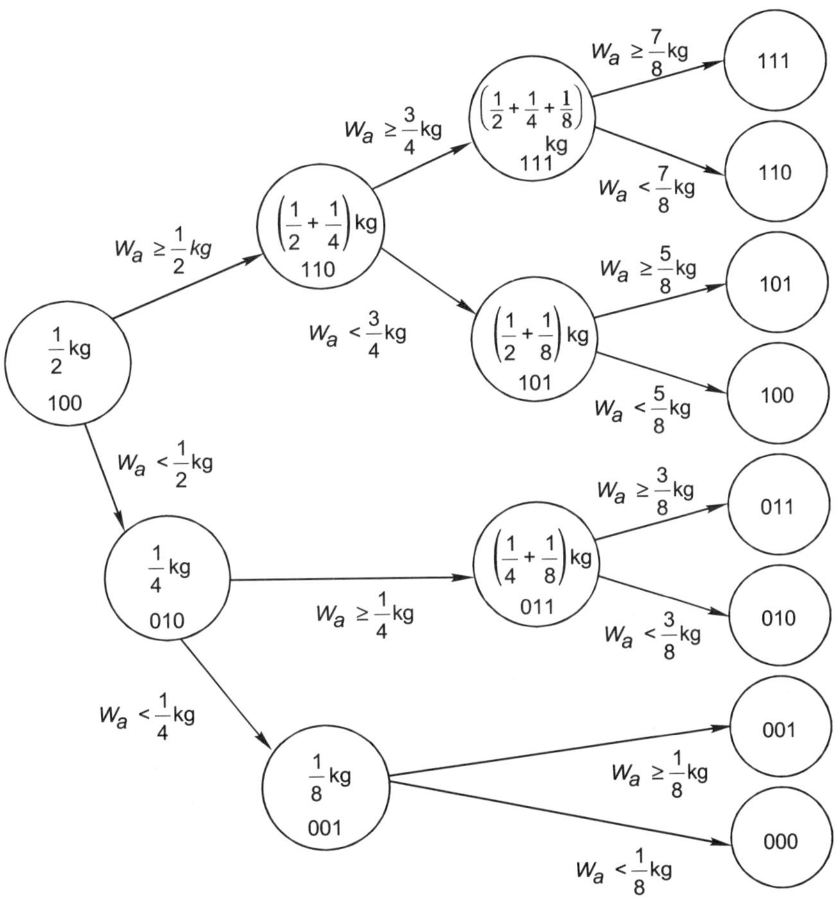

Fig. 10.20
Successive-approximation A/D conversion process.

The numerical significance assigned to various binary digits are: 1/2 kg to the most-significant bit, 1/4 kg to the next bit, etc. For example, the binary number 101 represents a weight of $\left(1\times\frac{1}{2}+0\times\frac{1}{4}+1\times\frac{1}{8}\right)$ kg $= 5/8$ kg.

Let us consider an unknown weight slightly less than $\frac{1}{2}$ kg. From Fig. 10.20, we find that the successive-approximation method gives a binary output of 011 for this, which represents a weight of $\frac{3}{8}$ kg. This shows a quantization error of $\frac{1}{8}$ kg. For reducing the quantization error, it is necessary to offset the scale, i.e. to tilt the scale in favour of the unknown weight. The magnitude of the offset must be equal to one-half of the smallest weight, i.e. $\frac{1}{16}$ kg in this case. With this offset, the interval corresponding to each binary output is shown in Fig. 10.21.

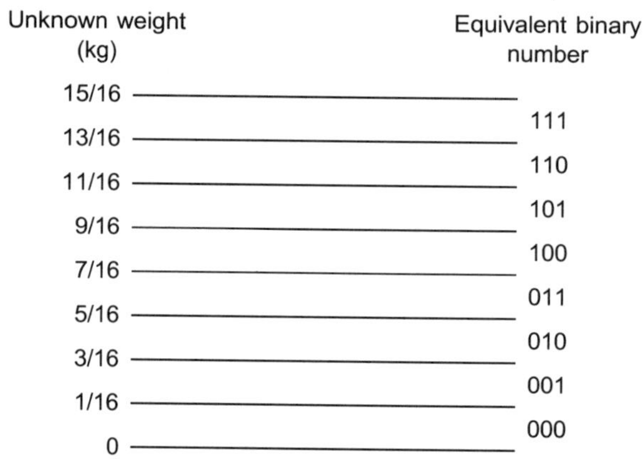

Fig. 10.21
Effect of offsetting the scale in successive-approximation weighing.

An A/D converter using the principle discussed above can be realized, as shown in Fig. 10.22. Here, the comparator serves the function of the scale, the output of which is used for

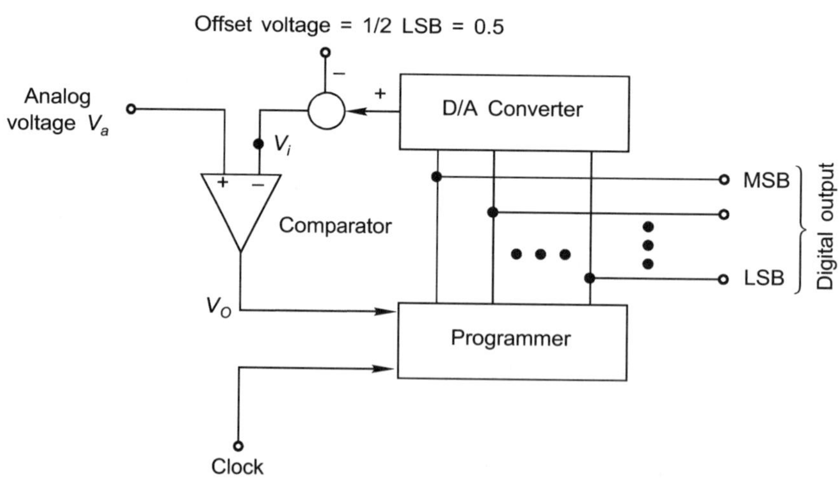

Fig. 10.22
Successive-approximation A/D converter.

setting/resetting the bits at the output of the programmer. This output is converted into equivalent analog voltage from which the offset voltage is subtracted and then applied to the inverting input terminal of the comparator. It should be noted that the offset weight was added on the side of the unknown weight, and therefore, it is to be subtracted from the known weight side for getting the equivalent effect. The outputs of the programmer will change only when the clock pulse is present. To start conversion, the programmer sets the MSB to 1 and all other bits to 0. This is converted into analog signal by the D/A converter and the comparator compares it with the analog input voltage. If the analog input voltage $V_a \geq V_{i,}$ the output voltage V_O of the comparator is HIGH which sets the next bit also. On the other hand, if $V_a < V_{i,}$ then V_O is LOW which resets the MSB and sets the next bit. Thus, a 1 is tried in each bit of the D/A converter until the binary equivalent of the analog input voltage is obtained. For an N-bit converter, the number of clock pulses required would be N and hence it is slower than the parallel-comparator A/D converter.

10.5.4 Counting A/D Converter

The successive-approximation A/D converter of Fig. 10.22 can be converted into a counting A/D converter if the programmer is replaced by an UP counter with a clear input, which must be clocked only as long as $V_O = 1$. To start the conversion process, the counter is reset to zero count using a clear pulse. The output of the counter is used as the input to the D/A converter, whose output (with offset voltage) is compared with the analog input voltage V_a. If $V_a > V_{i,}$ then $V_O = 1$. The output of the comparator is one input of a two-input AND gate, the other input being the clock pulses. The output of the AND gate is the clock input of the counter. Therefore, when $V_O = 1$, the clock pulses get applied to the clock input terminal of the counter and the counting proceeds. Since the number of pulses counted increases linearly with time, the voltage V_i increases, as shown in Fig. 10.23. As long as the analog input voltage V_a is greater than $V_{i,}$ the comparator output is HIGH and the clock pulses are transmitted to the counter. When V_i

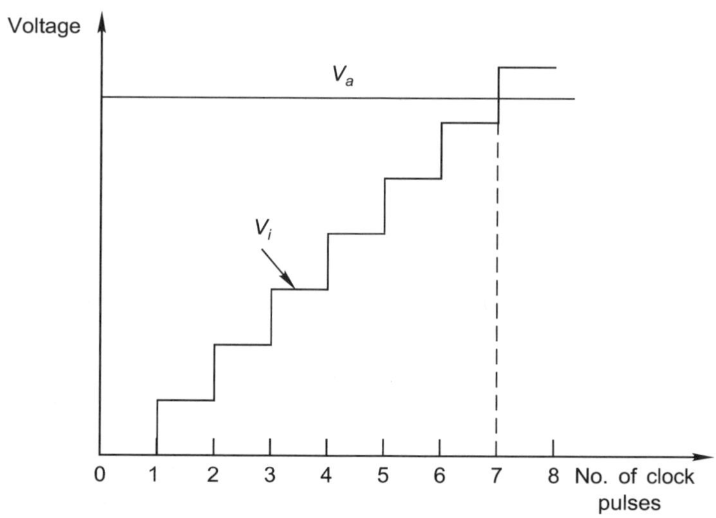

Fig. 10.23

Waveform of counting A/D converter.

exceeds V_a, the comparator output V_O changes to the LOW value disabling the AND gate and the counter stops counting. The output of the counter can be read out as the digital word, representing the analog input voltage. The maximum number of clock pulses required for conversion is 2^N for an N-bit A/D converter. Hence, this converter is slower than the other two converters discussed earlier.

10.5.5 Dual-Slope A/D Converter

The block diagram of a dual-slope A/D converter is shown in Fig. 10.24. It has four major blocks:

1. An integrator,
2. A comparator,
3. A binary counter, and
4. A switch driver.

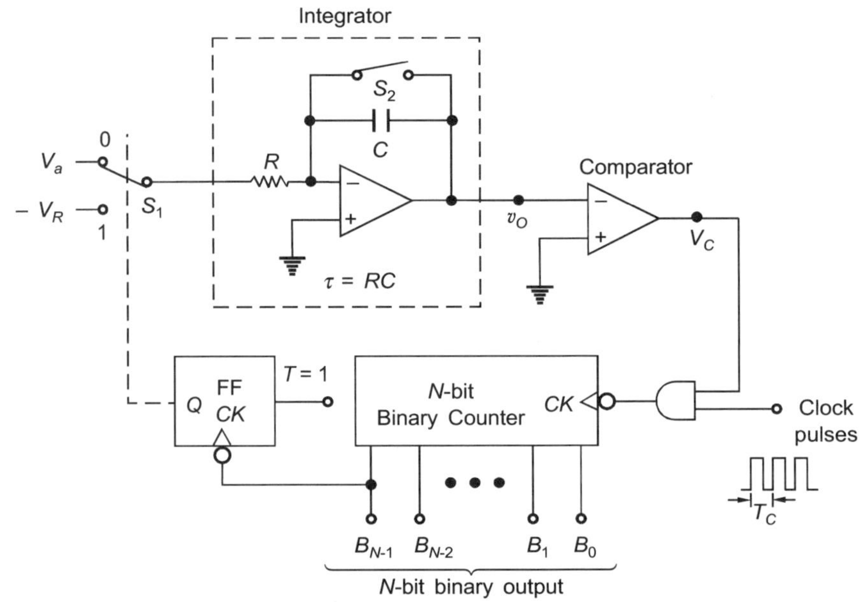

Fig. 10.24
Dual-slope A/D converter.

The conversion process begins at $t = 0$ with the switch S_1 in position 0 thereby connecting the analog voltage V_a to the input of the integrator. The integrator output

$$v_O = -1/\tau \int_0^t V_a \, dt = -(V_a/\tau)t.$$

This results in HIGH V_C, thus enabling the AND gate and the clock pulses reach the clock (CK) input terminal of the counter which was initially clear. The counter counts from 00 ... 00 to 111 ... 11 when $2^N - 1$ clock pulses are applied. At the next clock pulse 2^N, the counter is cleared and Q becomes 1. This controls the state of S_1 which now moves to position 1 at T_1, thereby connecting $-V_R$ to the input of the integrator. The output of the integrator now starts to move in the positive direction. The counter continues to count until $v_O < 0$. As soon as v_O goes positive at T_2, V_C goes LOW disabling the AND gate. The counter will stop counting in the absence of the clock pulses. The waveforms of voltages v_O and V_C are shown in Fig. 10.25.

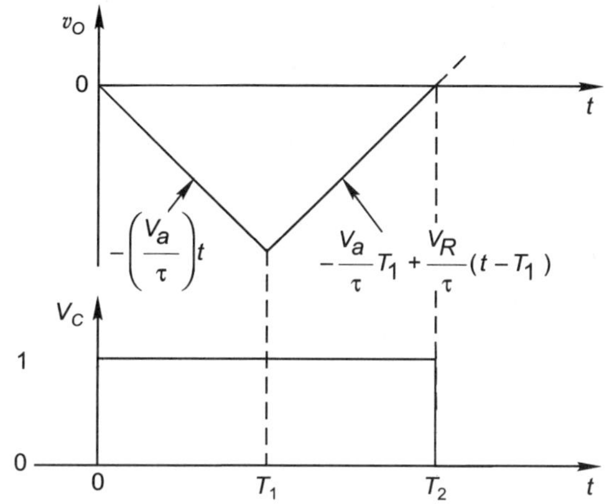

Fig. 10.25

Waveforms of dual-slope A/D converter.

The time T_1 is given by:

$$T_1 = 2^N T_C \tag{10.14}$$

where T_C is the time period of the clock pulses. When the switch S_1 is in position 1, the output voltage of the integrator is given by

$$v_O = -\frac{V_a}{\tau} T_1 + \frac{V_R}{\tau}(t - T_1) \tag{10.15}$$

$$v_O = 0 \text{ at } t = T_2$$

Therefore,

$$T_2 - T_1 = \frac{V_a}{V_R} T_1 = \frac{V_a}{V_R} 2^N T_C \tag{10.16}$$

Let the count recorded in the counter be n at T_2. Therefore,

$$T_2 - T_1 = n \cdot T_C = \frac{V_a}{V_R} 2^N T_C \tag{10.17}$$

which gives

$$n = \frac{V_a}{V_R} \cdot 2^N \tag{10.18}$$

This shows that the output of the counter is proportional to the analog voltage V_a. The count recorded in the counter is numerically equal to analog voltage V_a if $V_R = 2^N$.

This type of A/D converter is often used in digital voltmeters because of its good conversion accuracy and low cost. The disadvantage of the dual-slope A/D converter is its slow speed.

10.5.6 A/D Converter Using Voltage-to-Frequency Conversion

An analog voltage can be converted into digital form, by producing pulses whose frequency is proportional to the analog voltage. These pulses are counted by a counter for a fixed duration and the reading of the counter will be proportional to the frequency of the pulses, and hence, to the analog voltage.

A voltage-to-frequency converter is shown in Fig. 10.26*a*. The analog voltage V_a is applied to an integrator whose output is applied at the inverting input terminal of a comparator. The non-inverting input terminal of the comparator is connected to a reference voltage $-V_R$. Initially, the switch *S* is open and the voltage v_O decreases linearly with time ($v_O = -V_a t/\tau$), which is shown in Fig. 10.26*b*. When the decreasing v_O reaches $-V_R$ at $t = T$, the comparator output V_C goes HIGH. This is used to close the switch *S* through a monostable multivibrator. When the switch *S* is closed, the capacitor *C* discharges, thereby returning the integrator output v_O to 0. Since the pulse width of the waveform V_C is very small, a monostable multivibrator is used to keep the switch *S* closed for a sufficient time to discharge the capacitor completely. The rate at which the capacitor discharges depends upon the resistance of the switch.

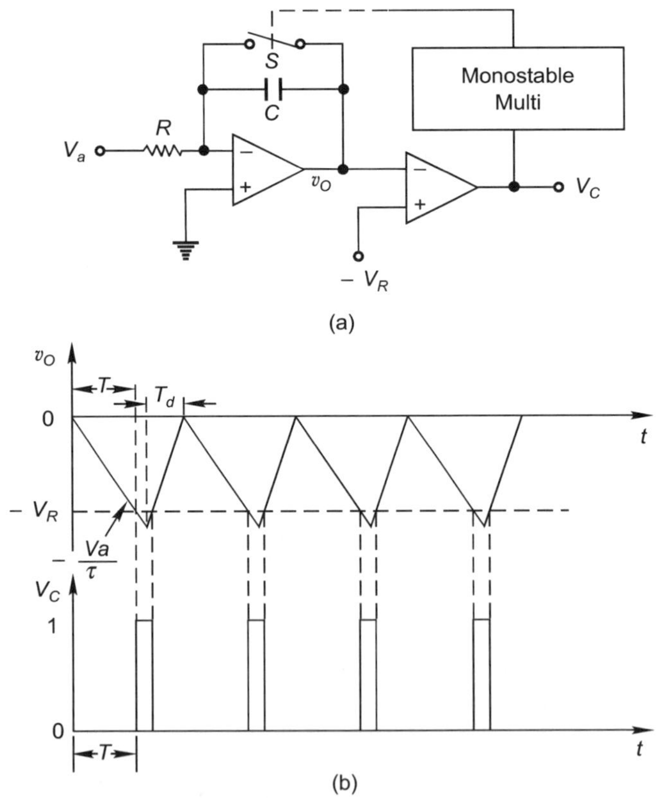

Fig. 10.26
Voltage-to-frequency converter (a) Schematic circuit, and (b) Waveforms.

Let the pulse width of the monostable multivibrator be T_d. Therefore, the switch S remains closed for T_d after which it opens and v_O starts decreasing again.

If the integration time $T \gg T_d$, the frequency of the waveforms v_O and V_C is given by

$$f = \frac{1}{T + T_d} \approx \frac{1}{T} = \frac{1}{\tau}\frac{V_a}{V_R} \tag{10.19}$$

Thus, we obtain an output waveform whose frequency is proportional to the analog input voltage.

An A/D converter using the voltage-to-frequency (V/F) converter is shown in Fig. 10.27. The output of the V/F converter is applied at the clock (CK) input of a counter through an AND gate. The AND gate is enabled for a fixed time interval T_1. The reading of the counter at $t = T_1$ is given by

$$n = fT_1 = \frac{1}{\tau}\frac{V_a}{V_R}T_1 \tag{10.20}$$

which is proportional to V_a.

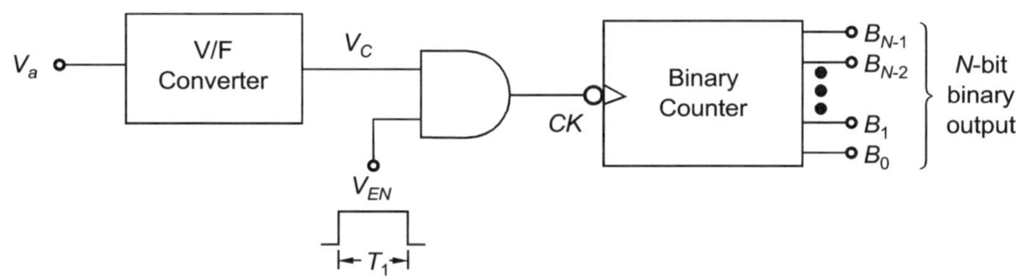

Fig. 10.27
An A/D converter using a V/F converter.

10.5.7 A/D Converter Using Voltage-to-Time Conversion

In an A/D converter using V/F converter, the cycles of a variable-frequency source are counted for a fixed period. Alternatively, it is possible to make an A/D converter by counting the cycles of a fixed-frequency source for a variable period. For this, the analog voltage is required to be converted to a proportional time period.

An A/D converter, which operates on this principle, is shown in Fig. 10.28*a*.

A negative reference voltage $-V_R$ is applied to an integrator, whose output is connected to the inverting input terminal of the comparator. The analog voltage V_a is applied at the non-inverting input terminal of the comparator. The output of the comparator V_C is at logical level 1 as long as the output of the integrator v_O is less than V_a. When v_O crosses V_a at $t = T$, V_C goes LOW. The AND gate is enabled when V_{EN} is LOW and switch S remains open. When V_{EN} goes HIGH, the switch S is closed, thereby discharging the capacitor. Also the AND gate is disabled. The various waveforms are shown in Fig. 10.28*b*. When the AND gate is enabled, the clock pulses will reach the clock (CK) input terminal of the counter. The output of the counter is the digital output corresponding to V_a.

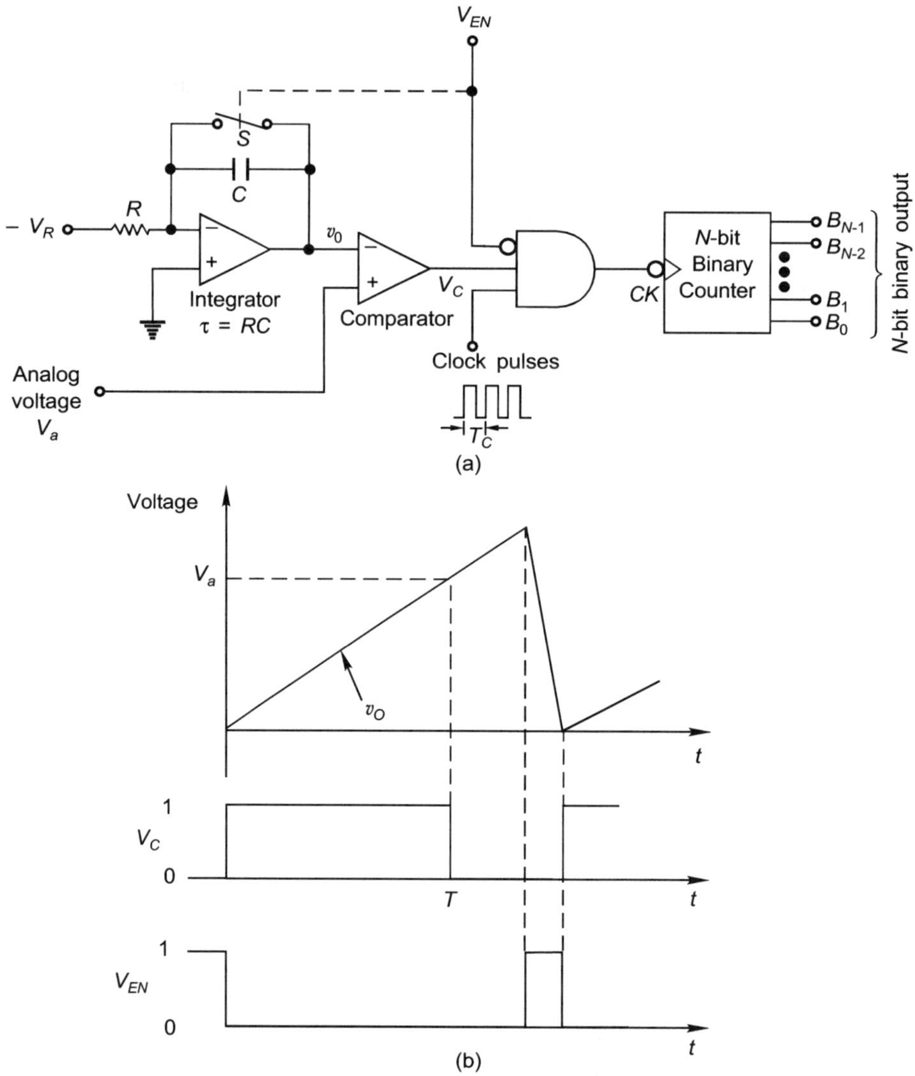

Fig. 10.28
An A/D converter using voltage-to-time converter (a) Schematic circuit, and (b) Waveforms.

The time T is given by

$$T = \frac{\tau}{V_R} V_a \tag{10.21}$$

which shows that T is proportional to V_a.

The counter reading is n at $t = T$, then

$$n = f_c \cdot T = \frac{f_c \, \tau}{V_R} V_a \tag{10.22}$$

where, f_c is the clock frequency. The count n is proportional to V_a.

10.5.8 Specifications of A/D Converters

The following specifications are usually specified by the manufacturers of A/D converters:

1. Range of input voltage,
2. Input impedance,
3. Accuracy,
4. Conversion time, and
5. Format of digital output.

10.6 AN EXAMPLE OF A/D CONVERTER IC

The AD ADC 80 is a 12-bit successive-approximation A/D converter available in a 32-pin DIP. Its important performance characteristics are:

Linearity error at + 25°C	:	± 0.012%
Maximum gain temperature		
Coefficient	:	30 ppm/°C
Power dissipation	:	800 mW
Maximum conversion time	:	25 μs
Digital output format	:	Unipolar and bipolar; parallel and serial
Output drive	:	2 TTL loads
Analog voltage ranges	:	± 2.5 V, ± 5 V, ± 10 V (bipolar) or 0 to 5 V, 0 to 10 V (unipolar)

The functional block diagram of ADC 80 is shown in Fig. 10.29.

10.6.1 Operation

When a convert start command is received, it converts the voltage at its analog input to an equivalent 12-bit binary number. The status flag is set during the time conversion is in progress. The status flag is reset after the conversion is over and the parallel output data becomes available. Serial data can be transferred by clocking it into a shift register bit by bit, starting from the decision taken about each bit.

10.6.2 Digital Output

The digital data is available in parallel as well as serial form. The parallel 12-bit data is available at pins marked B_1 through B_{12}, where B_1 is MSB and B_{12} is LSB. For unipolar input ranges, the output is in complementary straight binary code (CSB), whereas for bipolar input ranges the output is either in complementary offset binary (COB) or complementary 2's complement binary (CTC) code depending on whether B_1 (pin 6) or $\bar{B}_1$ (pin 8) is used as the MSB.

Serial data coding is in CSB for unipolar input ranges and COB for bipolar input ranges. The MSB is the first bit at the serial output pin and LSB is the last.

The circuit can be used for 12-, 10-, or 8-bit resolution by connecting the short-cycle terminal (pin 21) to pin 9, 28, or 30, respectively. The maximum conversion time is reduced to 21 and 17 μs for 10-bit and 8-bit resolutions, respectively.

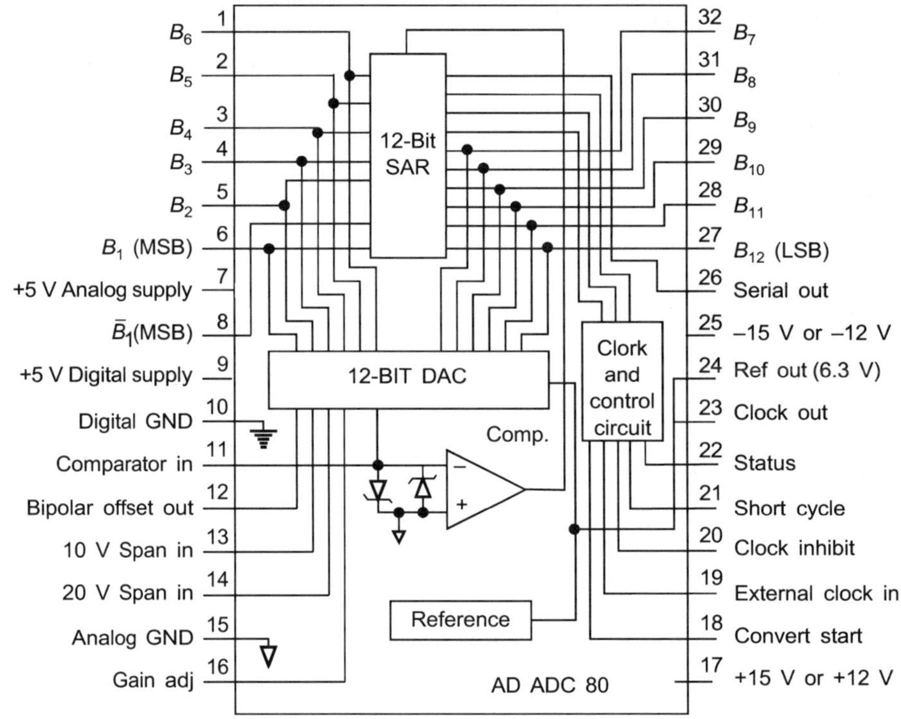

Fig. 10.29
Functional block diagram of AD ADC 80.

10.6.3 Analog Input

The analog input should be scaled as close to the maximum input signal range as possible in order to utilize the maximum signal resolution of the A/D converter. The input signal should be connected as given in Table 10.7.

Table 10.7 Input scaling connections

Input signal range	*Connect pin 12 to pin*	*Connect pin 14 to*	*Connect input signal to pin*
± 10 V	11	Input signal	14
± 5 V	11	Open	13
± 2.5 V	11	Pin 11	13
0 to + 5 V	15	Pin 11	13
0 to + 10 V	15	Open	13

The analog and digital grounds must be connected together at one point, usually the system power-supply ground.

10.6.4 Calibration

External zero adjustment and gain adjustment potentiometers are connected as shown in Fig. 10.30, for device calibration. To prevent interaction of these two adjustments, zero is

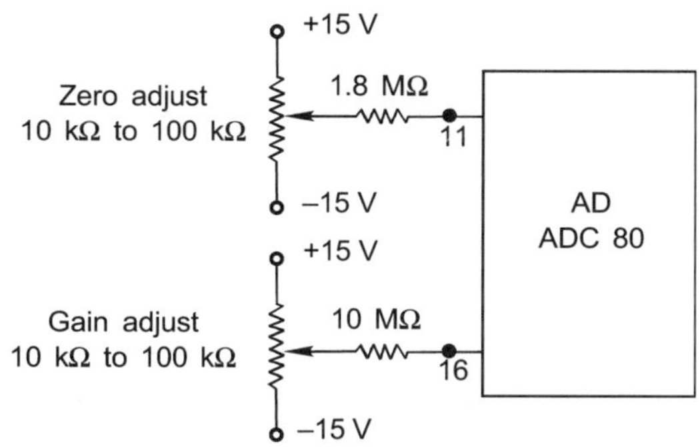

Fig. 10.30

Zero and gain adjustments for AD ADC 80.

always adjusted first and then gain. Zero is adjusted with the analog input voltage near the most negative end of the analog voltage range (0 for unipolar and – full scale for bipolar input ranges). Gain is adjusted with the analog input voltage near the most positive end of the analog range. The analog input voltages to be used for these adjustments and the corresponding digital outputs are given in Table 10.8, for two ranges. For other ranges, the corresponding analog input voltages can be similarly determined.

Table 10.8 Zero and gain adjustment settings

Input signal range	*Set analog input voltage to*	*Adjust zero for digital output*	*Adjust gain for digital output*
0 to + 10 V	+1 LSB = 0.0024 V	111111111110	–
	+FSR – 2 LSB = 9.9952 V	–	000000000001
	+5 V (Digital output = 011111111111)		
–10 V to +10 V	–9.9951 V	111111111110	–
	+9.9902 V	–	000000000001
	0.0 (Digital output = 011111111111)		

10.7 SUMMARY

Some of the commonly used techniques for digital-to-analog and analog-to-digital conversions have been discussed in this chapter. The cascade arrangement of Fig. 10.6 is usually preferred for accommodating a large number of bits in a D/A converter.

Of the A/D converters discussed, the fastest is the parallel comparator type. Therefore, this is the best choice where maximum speed is required. The speed of the successive-approximation converter is less than that of the parallel-comparator type. But this requires less hardware and is therefore quite popular.

One of the most popular A/D converters is the dual-slope converter, which is widely used in instruments such as digital voltmeters, where the conversion speed is not important.

The output voltage levels of A/D converters must often be adjusted to make them compatible with logic families such as TTL, CMOS, ECL, etc.

Chapter 10

Glossary

Accuracy The limits about the exact value within which the output value of a D/A converter is obtained.

Analog-to-digital converter (A/D converter) An electronic circuit that converts an analog signal to a corresponding digital signal.

CBI (Complementary binary input) Various binary codes such as complementary straight binary (CSB), complementary offset binary (COB), or complementary 2's complement (CTC).

CCD (Complementary coded decimal code) Complementary BCD code. For example, $(999)_{10} = 011001100110$. Here, each decimal digit is represented by complement of its BCD.

COB (Complementary offset binary code) A code obtained by finding out complementary straight binary (CSB) and then offsetting it by -2^{n-1}, where n is the number of bits used to represent the number. For example, $(2)_{10} = 1101 - 1000 = 0101$.

Conversion time of an A/D converter The time required by an A/D converter to convert one sample.

CSB (Complementary straight binary) Complement of straight binary. For example, $(2)_{10} = 1101$.

CTC (Complementary 2's complement code) Complement of 2's complement of a code. For example $(2)_{10} = 0001$.

Digital-to-analog converter (D/A converter) An electronic circuit that converts digital signals into proportional analog voltage or current.

Hold time The time between two samples in a sample-and-hold (S/H) circuit.

Linearity error It is a measure of the difference between the actual output voltage and the expected output voltage of a D/A converter. It is expressed as a percentage of full-scale or maximum output voltage.

Quantization The process of dividing the analog voltage into suitable intervals for converting analog signals into digital signals.

Quantization error The error involved in the quantization process.

Resolution The capacity to distinguish between two closely spaced values.

Sampling The process of taking samples of time varying signal for conversion to digital form.

Sampling rate The rate at which sampling takes place.

Sampling time The time between two consecutive samples.

Settling time The time required for the output of a D/A converter to come to and stay within $\pm \frac{1}{2}$ LSB of the final analog output voltage.

Review Questions

10.1 The resolution of a 12-bit D/A converter is __________ of the full-scale output.

10.2 The resolution of __________- bit D/A converter is approx. 0.4%.

10.3 The linearity of a D/A converter is specified as __________ LSB.

10.4 The time required for the analog output voltage to come to and settle to within $\pm\frac{1}{2}$ LSB of its final value after a digital input is applied is known as __________.

10.5 For an 8-bit D/A with 10V full scale output, the output voltage for a digital input 00000000 (CSB) is __________.

10.6 An 8-bit D/A converter's analog output voltages are full-scale voltage and 0 volt for digital inputs of 10000000 and 11111111 respectively. The digital input code for this converter is __________.

10.7 The minimum number of samples to be taken for conversion of a sinusoidal waveform of frequency 1 kHz to digital form in 2. ms is __________.

10.8 The maximum sampling time T_s of a 10 kHz sinusoidal voltage for conversion to digital form is __________.

10.9 The maximum quantization error for an A/D converter is __________ of maximum analog input voltage.

10.10 For __________-bit A/D converter the maximum quantization error is 1/2046 of maximum analog input voltage.

10.11 For maximum quantization error of an n-bit A/D converter to be $\pm\frac{1}{2}$ LSB, the step size S is __________ of the maximum analog voltage.

10.12 The most significant digit of a $4\frac{1}{2}$-digit digital voltmeter will have a maximum value of __________.

10.13 The number of comparators required for an 8-bit parallel-comparator A/D converter is __________.

10.14 The decimal equivalent of 0000 0000 in CSB code is __________.

10.15 The CCD code for decimal 325 is __________.

Problems

10.1 An 8-bit D/A converter provides an analog output which has a maximum value of 10 V. The output may have an error of ΔV due to drift in component values, temperature, etc. How large can ΔV be before the least-significant bit would no longer be significant?

10.2 For the circuit of Fig. 10.31, find the analog output voltage for each of the digital inputs and show that this circuit can be used for converting digital signal in 1's complement format to analog output.

10.3 Verify Eq. (10.10) for b_2, b_1, and b_0 bits.

10.4 In the circuit of Fig. 10.31, if the offset switch is removed and the switch S_2 is replaced by a switch $\bar{S}_2$ (with its resistor $R/4$ replaced by a resistor $R/3$), verify that the operation of the circuit will not change.

10.5 Design a D/A converter circuit similar to the circuit of Prob. 10.2, for 4-bit input.

10.6 Design a D/A converter circuit similar to the circuit of Prob. 10.4, for 4-bit input.

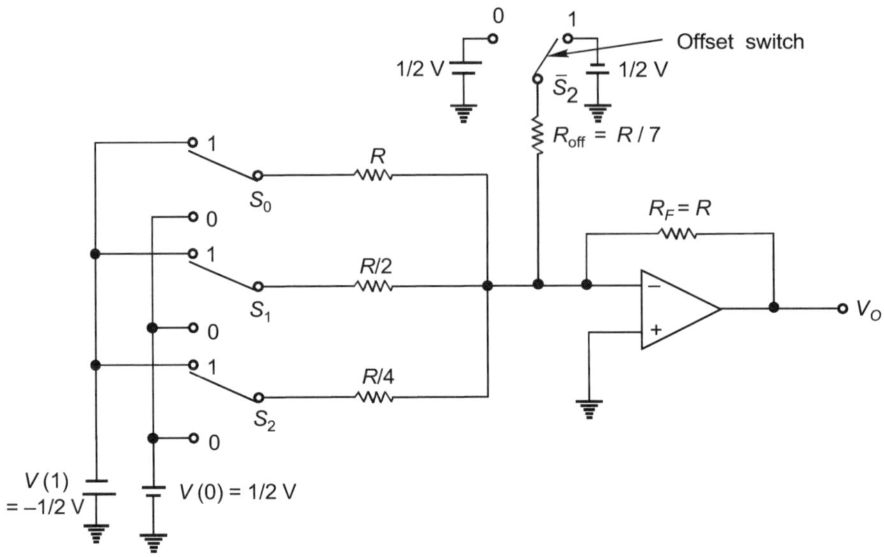

Fig. 10.31
Circuit for Prob. 10.2.

10.7 Design a 3-bit parallel-comparator A/D converter for 2's complement format.

10.8 A dual-slope A/D converter has a resolution of 12 bits. If the clock rate is 100 kHz, what is the maximum rate at which samples can be converted?

10.9 A D/A converter has a full scale analog output of 10 V and accepts six binary bits as inputs. Find the voltage corresponding to each analog step.

10.10 Find the analog voltage corresponding to LSB for each of the ranges of:
(*a*) DAC 80
(*b*) ADC 80

10.11 Explain the following codes and give examples using four bits:
(*a*) CSB
(*b*) COB
(*c*) CTC
(*d*) CCD

11

SEMICONDUCTOR MEMORIES

11.1 INTRODUCTION

A digital processing system invariably requires a facility for storing digital information. The information usually consists of instructions (processing steps) coded in binary form, data to be processed, intermediate and final results, etc. The sub-system of a digital processing system, which provides the storage facility, is referred to as the *memory*. Till recently, the memories used were mostly of magnetic type. With unprecedented developments in semiconductor technology, it has become possible to make semiconductor memories of various types and sizes. These memories have become very popular due to their small size, low cost, high speed, high reliability, and ease of expansion of the memory size. Therefore, it is necessary for a designer of digital processors to know thoroughly the principles of operation and limitations of various semiconductor memory devices.

11.2 MEMORY ORGANIZATION AND OPERATION

The basic element of a semiconductor memory is a `FLIP-FLOP` which has been discussed in Chapter 7. The information is stored in binary form. There are a number of locations in a memory chip, each location being meant for one word of digital information. The number of locations and the number of bits comprising the word vary from memory to memory. The size of a memory chip is specified by two numbers M and N as $M \times N$ bits. The number M specifies the number of locations available in the memory and N is the number of bits at each location. In other words, this means that M words of N bits each can be stored in the memory. The commonly used values of the number of words per chip are 64, 256, 512, 1024, 2048, 4096, etc.

whereas the common values for the word size are 1, 4, and 8, etc. Memories requiring higher number of words and/or larger word sizes can be formed by using these chips.

The block diagram of a memory device is shown in Fig. 11.1. Each of the M locations of the memory is defined by a unique *address* and, therefore, for accessing any one of the M locations, P inputs are required, where $2^P = M$. This set of lines is referred to as *address inputs* or *address bus*. The address is specified in the binary form. For convenience, octal and hexadecimal representations are commonly employed.

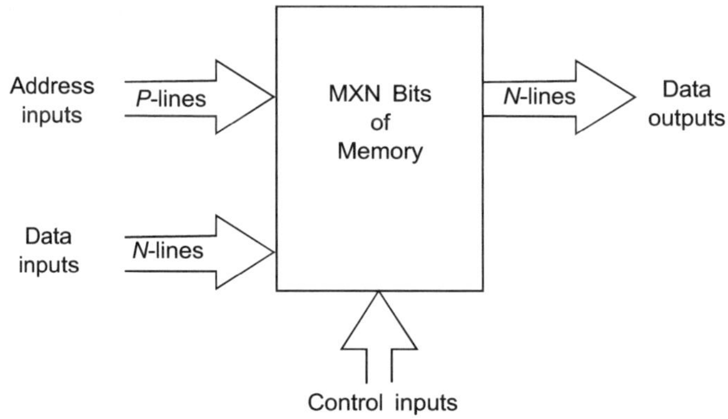

Fig. 11.1
Block diagram of a memory device.

In fact, the address input is applied to a P-to-M decoder circuit, which activates one of its M outputs depending on the address and, thus, the desired memory location is selected.

Example 11.1 Consider a memory of size 16 words. Find the binary address of each location.

Solution
Since $M = 16$, therefore, $2^P = M$ gives $P = 4$, i.e. for selecting one out of 16 words, a 4-bit address is required. The address is specified as $A_3\ A_2\ A_1\ A_0$, where A_3 represents the most-significant bit (MSB) and A_0 represents the least-significant bit (LSB) of the address. The address of each location is given in Table 11.1.

Table 11.1 Memory addresses for Ex. 11.1

Word number	*Binary address*			
	A_3	A_2	A_1	A_0
0	0	0	0	0
1	0	0	0	1
2	0	0	1	0
3	0	0	1	1
4	0	1	0	0
5	0	1	0	1

(*Contd.*)

(*Contd.*)

6	0	1	1	0
7	0	1	1	1
8	1	0	0	0
9	1	0	0	1
10	1	0	1	0
11	1	0	1	1
12	1	1	0	0
13	1	1	0	1
14	1	1	1	0
15	1	1	1	1

The number of inputs required to store the data into or read the data from any memory location is *N*. One set of *N* lines is required for storing the data into the memory, referred to as *data inputs* and another set of *N* lines is required for reading the data already stored in the memory, which is referred to as *data outputs*. The concept of *bus* is used to refer to a group of conductors carrying related set of signals. Therefore, the set of lines meant for data inputs is *input data bus* and for data outputs is *output data bus*. Input and output data buses are unidirectional, i.e. tha data can flow in one direction only. In most of the memory chips available, the same set of lines is used for data input as well as data output and is referred to as *bidirectional* bus. This means that the data bus is time multiplexed. It is used as input bus for some specific time and as output bus for some other time depending upon a Read/Write control input as shown in Fig. 11.2.

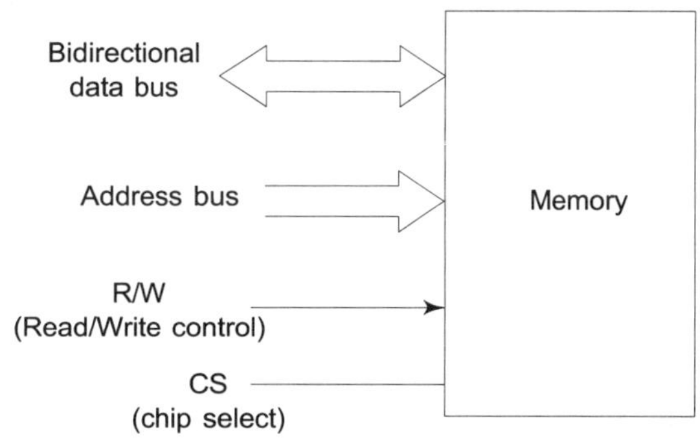

Fig. 11.2 Block diagram of memory with bidirectional data bus.

A number of control inputs are required to give commands to the device to perform the desired operation. For example, a command signal is required to tell the memory whether a read or a write ($R/\overline{W}$ is Fig. 11.2) operation is desired.

When $R/\overline{W}$ is HIGH, the data bus will be used for reading the memory (output bus) whereas when $R/\overline{W}$ is LOW, the bus will be acting in the input direction and the data on the bus will go into the memory. Other command include inputs chip enable (*CE*), chip select (*CS*), etc.

In addition to the above-mentioned functional pins, a minimum of two pins are required for power supply and ground. The internal organization of a 16 × 4 memory chip is illustrated in Fig. 11.3. The write and read operations are discussed below.

11.2.1 Write Operation

To write a word into the selected memory location requires logic 1 voltage to be applied to *CS* (Chip select), and Write inputs and logic 0 voltage to Read (*RD*) input. This combination of inputs gives outputs of AND gates A_1 and A_2 1 and 0 respectively. *A* 1 at the output of A_1 enables the input buffers so that the 4-bit word applied to the data inputs will be loaded (entered) into

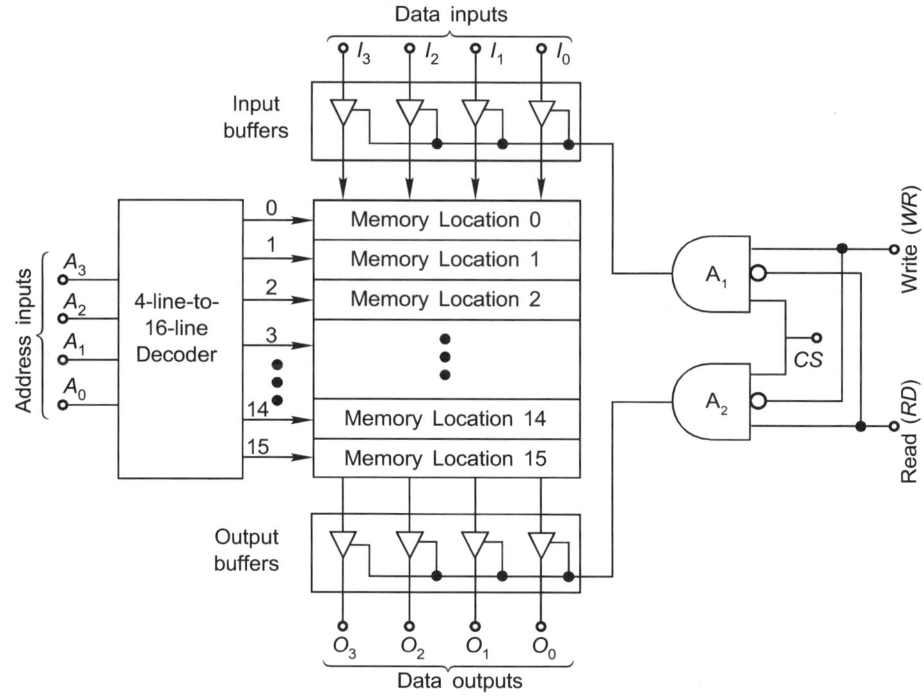

Fig. 11.3

Internal organization of a 16 × 4 memory chip.

the selected (addressed) memory location. *A* 0 at the output of A_2 disables (tristate) the output buffers so that the data outputs are not available.

For writing a word into a particular memory location, following sequence of operations is to be performed:

1. The chip select signal is applied to the *CS* terminal.
2. The word to be stored is applied to the data-input terminals.
3. The address of the desired memory location is applied to the address-input terminals.
4. A write command signal is applied to the write-control input terminal with $RD = 0$.

In response to the above operations, the addressed memory location is cleared of any word that might have been stored in it earlier, and the information presented at the data input terminal replaces it.

Figure 11.4 illustrates the various waveforms during the write operation. The important timing characteristics of the write cycle are:

Write Cycle Time (t_{WC})

This is the minimum amount of time for which the valid address must be present for writing a word in the memory. In other words, it is the minimum time required between successive write operations.

Write Pulse Time (t_W)

This is the minimum length of the write pulse.

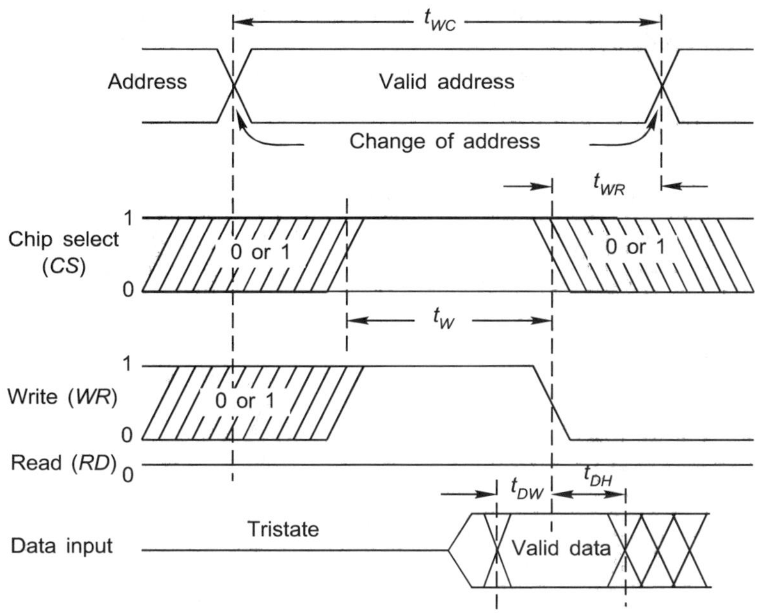

Fig. 11.4
Write-cycle waveforms.

Write Release Time (t_{WR})

This is the minimum amount of time for which the address must be valid after the write pulse ends.

Data Set Up Time (t_{DW})

This is the minimum amount of time for which the data must be valid before the write pulse ends.

Data Hold Time (t_{DH})

This is the minimum amount of time for which the data must be valid after the write pulse ends.

11.2.2 Read Operation

In order to read the contents of a selected memory location, the Read (*RD*), and the Chip select (*CS*) inputs must be at logic 1 level and *WR* at logic 0 level. This gives output of $A_2 = 1$ which enables the output buffers so that the contents of the selected (addressed) memory location will appear at the data outputs. $RD = 1$ tristates the input buffers so that the data inputs do not affect the memory during a read operation.

To read (or retrieve) a data word, known to be stored at a particular address, the following sequence of operations is required to be performed:

1. The chip-select signal is applied to the *CS* terminal.
2. The address of the desired memory location is applied to the address-input terminals.
3. A read-command signal is applied to the read control-input terminal.

In response to the above operations, the data word stored at the addressed location appears on the data-output terminals.

Figure 11.5 illustrates the various waveforms during the read operation. The important timing characteristics of the read cycle are:

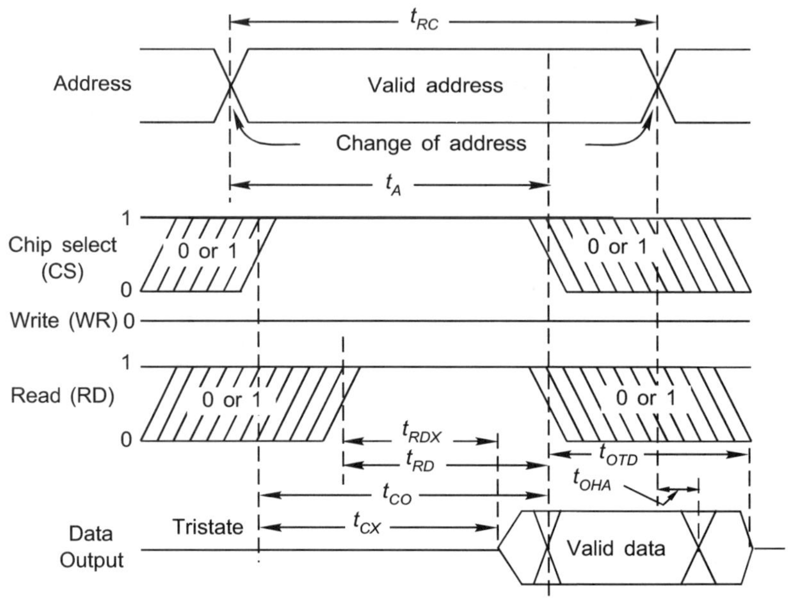

Fig. 11.5

Read-cycle waveforms.

Read Cycle Time (t_{RC})

This is the minimum amount of time for which the valid address must be present for reading a word from the memory. In other words, it is the minimum time required between successive read operations.

Access Time (t_A)

This is the maximum time from the start of the valid address of the read cycle to the time when the valid data is available at the data outputs. The access time is at most equal to the read-cycle time, i.e. $t_A \le t_{RC}$. In other words, the data outputs might be ready before the memory is actually ready for the next read operation.

Read To Output Valid time (t_{RD})

This is the maximum time delay between the beginning of the read pulse and the availability of valid data at the data outputs.

Read To Output Active Time (t_{RDX})

This is the minimum time delay between the beginning of the read pulse and the output buffers coming to active state (from the high-impedance state).

Chip-Select To Output Valid Time (t_{CO})

This is the maximum time delay between the beginning of the chip-select pulse and availability of valid data at the data outputs.

Chip-Select To Output Active Time (t_{CX})

This is the minimum time delay between the beginning of the chip-select pulse and the output buffers coming to active state.

Output Tristate From Read (t_{OTD})

This is the maximum time delay between the end of the read pulse and the output buffers going to high-impedance state.

Data Hold Time (t_{OHA})

This is the minimum time for which the valid data is available at the data outputs after the address ends.

The write- and read-cycle timings of a typical memory chip are given in Table 11.2.

Table 11.2 Timing parameters of a typical memory chip

Parameter	*Time (ns)*
t_{WC}	200
t_W	120
t_{WR}	0
t_{DW}	120
t_{DH}	0
t_{RC}	200
t_A	200
t_{RD}	70
t_{RDX}	20
t_{CO}	70
t_{CX}	20
t_{OTD}	60
t_{OHA}	50

Example 11.2 For the memory timing of Table 11.2, find the maximum rate (words/second) at which

(*a*) Data can be stored, and

(*b*) Data can be read.

Solution

(*a*) The maximum rate at which data can be stored is

$$\frac{1}{t_{WC}} = \frac{1}{200 \times 10^{-9}} = 5 \times 10^6 \text{ words/s}$$

(*b*) The maximum rate at which data can be read is

$$\frac{1}{t_{RC}} = \frac{1}{200 \times 10^{-9}} = 5 \times 10^6 \text{ words/s}$$

11.3 EXPANDING MEMORY SIZE

In many memory applications, the required memory capacity, i.e. the number of words and/or word size, cannot be satisfied by a single available memory IC chip. Therefore, several similar chips have to be combined suitably to provide the desired number of words and/or word size.

11.3.1 Expanding Word Size

If it is required to have a memory of word size n and the word size of the available memory ICs is N ($n > N$), then a number of similar ICs can be combined together to achieve the desired word size. The number of IC chips required is an integer, next higher to the value n/N. These chips are to be connected in the following way:

1. Connect the corresponding address lines of each chip individually, i.e. A_0 of each chip is connected together and it becomes A_0 of the overall memory. Similarly, connect other address lines together.
2. Connect the *RD* input of each IC together and it becomes the read input for the overall memory. Similarly, connect the *WR* and *CS* inputs.

Now, the number of data-input/output lines will be equal to the product of the number of chips used and the word size of each chip. The following example illustrates the above procedure clearly.

Example 11.3 Obtain a 16 × 8 memory using 16 × 4 memory ICs.

Solution

Since the word size required is $n = 8$ and the word size of the available IC is $N = 4$, therefore, $n/N = 2$ chips are required to obtain the desired memory.

Since each chip can store 16 4-bit words and we want to store 16 8-bit words, each chip is required to store half of each word. Figure 11.6 shows the relevant connections of the two chips. Here, we have assumed bidirectional input/output (*I/O*) lines which is common in many available memory chips. In this 16 × 8 memory, the higher order four bits (D_7, D_6, D_5, D_4) of each 8-bit word are located in memory M_1 and the lower order four bits (D_3, D_2, D_1, D_0) are located in memory M_0.

11.3.2 Expanding Word Capacity

Memory chips can be combined together to produce a memory, with the desired number of locations. To obtain a memory of capacity m words, using the memory chips with M words each, the number of chips required is an integer next higher to the value m/M. These chips are to be connected in the following way:

1. Connect the corresponding address lines of each chip individually.
2. Connect the RD input of each chip together. Similarly, connect the *WR* inputs.
3. Use a decoder of proper size and connect each of its outputs to one of the *CS* terminals of memory chips. For example, if eight chips are to be connected, a 3-line-to-8-line decoder is required to select one out of eight chips at any one time.

The following example clearly illustrates the above procedure.

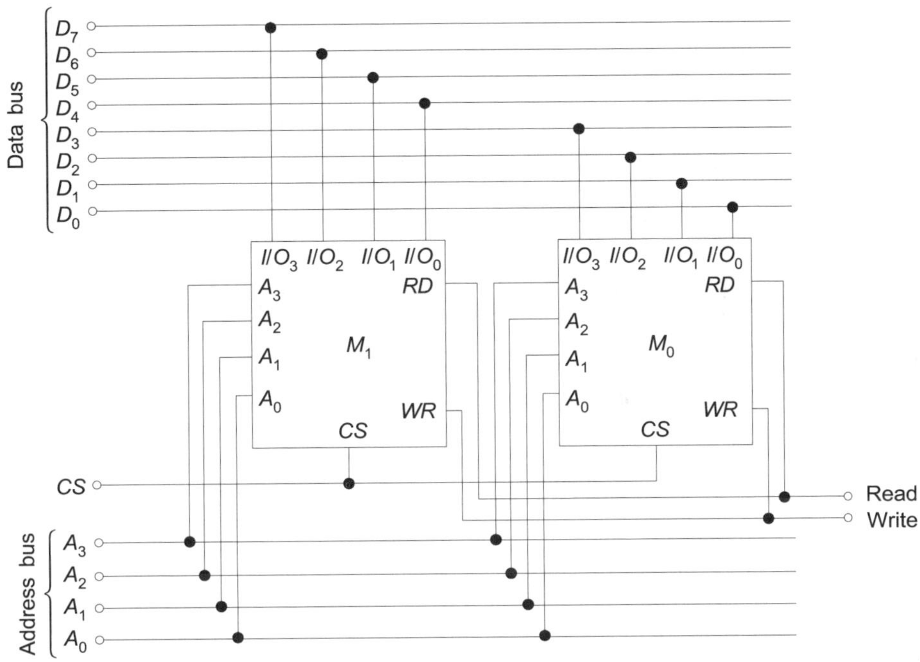

Fig. 11.6
A 16 × 8 memory obtained by combining two 16 × 4 memory chips.

Example 11.4 Obtain a 2048 × 8 memory using 256 × 8 memory chips.

Solution

The number of chips required is $\frac{2048}{256} = 8$. At any one time, only one of the 2048 locations is to be accessed, which will be in one of the eight chips. That means only one of the eight chips must get selected at a time.

For selecting one out of 2048 locations, the number of address lines required is 11 (2^{11} = 2048). The lower order eight bits of the address $A_7 - A_0$ will be same for each chip, and the higher order three bits of the address $A_{10} - A_8$ must select one out of the eight chips. For this purpose, a 3-line-to-8-line decoder is required. The memory connections are shown in Fig. 11.7. Here, we have assumed a common terminal ($R/\overline{W}$) for read and write. For the read operation, logic 1 is to be applied to $R/\overline{W}$, whereas logic 0 is to be applied for the write operation. The chip-select input is assumed to be active-low. The addresses of the chips are given in Table 11.3.

Table 11.3 Addresses of the memory chips

Memory chip	*Addresses (hex.)*
M_0	000 – 0FF
M_1	100 – 1FF
M_2	200 – 2FF

(*Contd.*)

(*Contd.*)

M_3	300 – 3FF
M_4	400 – 4FF
M_5	500 – 5FF
M_6	600 – 6FF
M_7	700 – 7FF

11.4 CLASSIFICATION AND CHARACTERISTICS OF MEMORIES

Various memory devices can be classified on the basis of their principle of operation, physical characteristics, mode of access, technology used for fabrication, etc.

11.4.1 Principle of Operation

Memories can be classified according to their principle of operation. The most commonly used memories are:

- Sequential memories
- Read-and Write memories (RWM or RAM)
- Read-only memories (ROM)
- Content addressable memories (CAM)

Sequential Memories

In the sequential memories, the memory locations are organised in sequence i.e. one after the other, such as a magnetic tape audio/video cassette. Writing into or reading from it is in sequential fashion. Therefore, the time required for accessing a memory location for writing or reading is different for different locations. There are two types of semiconductor sequential memories. These are:

- Shift registers, and
- Charge coupled devices (CCD)

Shift registers can be either *static* or *dynamic*. In a static memory, the contents of the memory location do not change with time as long as power is on. On the other hand, in dynamic memories, the information is stored in MOS capacitors which changes with time and, therefore, it has to be refreshed at regular intervals. The dynamic memories are simpler, less expensive, require less power, have high packing density in comparison to static memories, and are therefore widely used in digital systems. However, the cost of the additional circuitry required for refreshing may increase the system cost.

The charge coupled devices are implemented using MOS technology. These devices have high density and low cost.

The sequentially accessed memories using shift registers are discussed in detail in Section 11.5 and CCD memories in Section 11.9

Random Access Memories (RWM or RAM)

In this type of memories, the memory locations are organised in such a way so that any memory location requires equal time for writing or reading. This type of memory is also known as read-and-write memory (RWM) or RAM. RAMs can be static or dynamic and are fabricated

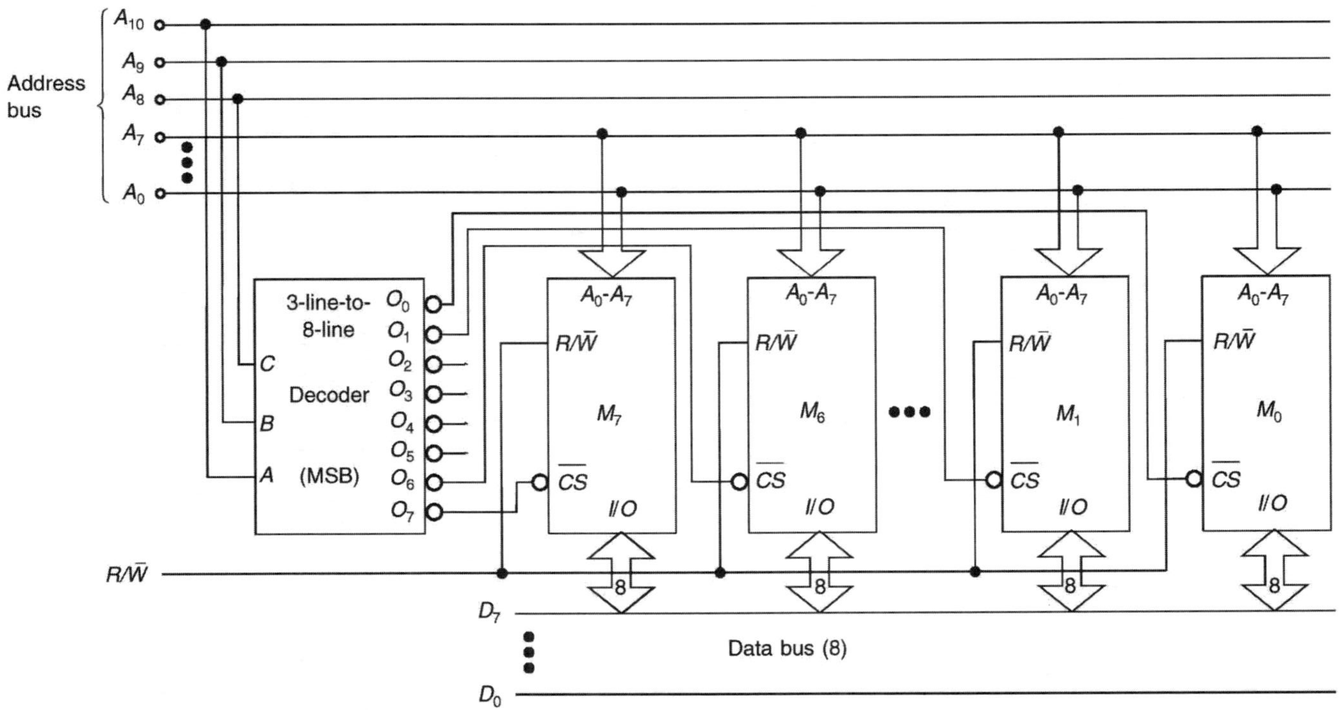

Fig. 11.7
A 2048 × 8 memory obtained by combining eight 256 × 8 memory chips.

using bipolar or unipolar technologies. The detailed operation of RAM is discussed in Section 11.7.

Read-Only Memories (ROM)

Read-only-memory (ROM), as the name suggests, is meant only for reading the information from it. This does not mean that information is not written into it, because unless some information is stored into it, there cannot be anything to read from. Actually, the process of entering information into this type of memory is much more complicated than for RAM and is done outside the system where it is used. Therefore, it is called as read-only memory. It is used to store information which is fixed, such as tables for various functions, fixed data and instructions. The ROMs are also organised so that every memory location requires equal time for reading the data already stored in that location.

The various types of read-only memories are further sub-classified on the basis of the technique employed for storing information into the memory (referred to as *programming*) or their erasability properties. These are:

Read- Only Memory It is programmed at the time of manufacturing, as the last process of fabrication, according to the information specified by the user. It is referred to as *custom programmed* or mask programmed. The data stored can not be changed after fabrication. Since, this process is quite costly, therefore, this type of ROM is suitable only for bulk requirements, of the order of millions of chips.

Programmable Read-Only Memory (PROM) This type of ROM is programmed by the user using a special circuit–a PROM programmer. A PROM can be programmed only once after which its contents are permanently fixed as in a ROM. This type of ROM is suitable for storage of data which is of permanent nature. The chip is available without any data stored from the vendor.

Erasable and Programmable ROM This type of ROMs are reprogrammable i.e. it can be programmed again and again. It is referred to as *erasable and programmable* ROM.

There are two techniques used for erasing; in one technique the chip is exposed to ultraviolet radiation, and in the other technique the contents are altered electrically. The erasable programmable ROM using ultraviolet radiation for erasing is known as EPROM, whereas the device using electrical voltage for erasing is known as *electrically alterable* ROM (EAROM).

The detailed operation of various types of ROMs is discussed in Section 11.6.
Content Addressable Memories

Content Addressable Memories (CAM)

Is a special purpose random access memory which performs association operation in addition to read/write operations. The detailed operation of the CAM is discussed in Section 11.8.

11.4.2 Physical Characteristics

Memories can be classified according to their physical characteristics, such as:

1. Erasable or non-erasable, and
2. Volatile or non-volatile.

Erasable or Non-erasable Memories

A memory in which the information stored can be erased and new information stored is called *erasable* memory. On the other hand, the information stored in the *non-erasable* memory cannot be erased, for example ROM is non-erasable.

The erasable memories can further be sub-classified as:

1. Location-by-location erasable, in which the desired memory locations can be erased one by one and new information can be entered. For example, in *electrically-alterable read-only-memory* (EAROM), the contents of a location have to be erased first before a new information can be entered. On the other hand, for *read and write memory* (RAM) and *content addressable memory* (CAM), the erasing operation is automatically performed whenever new information is entered in a memory location.
2. All memory locations are simultaneously erasable, in which the contents of all the locations of the memory chip get erased simultaneously by exposing the memory chip to ultraviolet (UV) radiation. Such a memory is referred to as *erasable–programmable read-only-memory* (EPROM).

Volatile or Non-volatile Memories

If the information stored in a memory is lost when electrical power is switched off, the memory is referred to as a *volatile memory*. For example, the RAM is a volatile memory. On the other hand, in a *non-volatile memory*, the information once stored remains intact until changed deliberately. All types of ROMs are non-volatile memories.

11.4.3 Mode of Access

Mode of access refers to the manner in which a memory location is accessed for reading or writing. In case of ROMs only reading is possible. There are two modes of access. These are:

- Sequential access, and
- Random access.

Sequential memories are referred to as sequentially accessed memories, whereas RAM, ROM, and CAM are random-access memories. In random- access memories any memory location requires equal time for accessing (*access time*) whereas the access time is different for different locations in the case of sequentially accessed memory.

11.4.4 Fabrication Technology

Memories can be classified on the basis of the fabrication technology used. The two broad categories of memories based on fabrication technology used are:

1. Bipolar, and
2. Unipolar (MOS).

These technologies have been discussed in detail in Chapter 4. Static RAM, ROM and PROM can be fabricated using either bipolar technology (TTL, ECL, etc.) or MOS technology, whereas dynamic RAM, EPROM, and EAROM can be fabricated only using unipolar devices (MOSFETs).

11.5 SEQUENTIAL MEMORY

In a sequential memory, the words are stored-in and read-out in sequence. For example, if the pth location is being accessed at a particular time, then the $(p + q)$th location is not accessible, unless all the intermediate locations have been accessed one-by-one, in sequence. In other words, the access to a particular location is obtained by waiting until the desired location is reached. This means that the access time is not same for all the locations. Shift registers discussed in Chapter 8 are examples of sequential memories. Another kind of shift registers gaining popularity in recent years are charge-coupled devices (CCDs) and bubble memories. The operation of these devices is entirely different from those using transistors.

A sequential memory of size $M \times N$ is shown in Fig. 11.8. It requires N shift registers, each of M stages. Each register holds one of the N bits of each of the M words. With each clock cycle, the bits in each register will advance towards the right by one bit position, and the stored words will appear sequentially at the outputs of the registers. This configuration is referred to as *first-in-first-out* (FIFO) sequential-memory system, since the word which is entered first will be the one read out first. In contrast, if the outputs are taken from the $(M - 1)$th stage instead of 0th stage, the resulting configuration is called as *last-in-first-out* (LIFO). The shift registers used in this configuration must have a provision of shifting the bits in either direction. In the FIFO sequential-memory system, the word being read may be transferred back to the left-most register position. This is known as a circulating shift register. The circulating feature is not employed in LIFO sequential-memory systems.

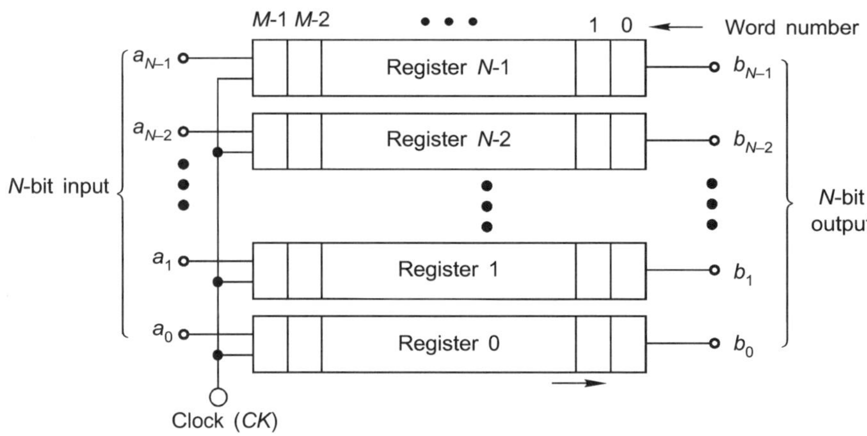

Fig. 11.8

$M \times N$ sequential memory.

11.5.1 Static Shift Register

As discussed in Section 8.2, a shift register is implemented with FLIP-FLOPs. The information is loaded into it or retrieved from it in a sequential manner in synchronism with clock pulses. The information will remain in the register as long as desired provided power is supplied to the circuit. Both bipolar and MOS devices can be used for implementing shift registers. This type of shift register is referred to as the static shift register.

It is impractical to fabricate static shift registers of large capacities because of excessively large requirements of power and silicon area. An alternative approach, to overcome these difficulties, is to fabricate an LSI dynamic shift register using MOS devices.

11.5.2 Dynamic Shift Register

In a dynamic memory system, the information is stored temporarily by charging the parasitic capacitances of the electronic devices. The MOS device is uniquely suited for this purpose, because at its input it looks like a capacitor (parasitic gate-substrate capacitance) which serves as temporary storage for the incoming data. These capacitors lose their charge due to leakage and, therefore, to maintain the required logic levels, they need to be refreshed continuously. Refreshing is achieved by simply recirculating the data held within the register. For this reason, dynamic registers must be operated at a minimum operating frequency of about 10 kHz, and these cannot be used for low-frequency applications.

The advantages of dynamic circuitry over static circuitry are:

1. The circuits are simpler and use MOS devices, resulting in higher density (number of bits in a given silicon area). Large capacity memories have become possible in this approach. For example, MM5027 is a 2048-bit IC shift register.
2. It requires less power than the static shift register. This is because, the input is sampled only during a fraction of the total time and the output is due to the charge stored on a capacitor rather than a constant voltage level. However, the power consumed increases with increasing operating frequency.

Dynamic MOS Inverter

A dynamic MOS inverter is shown in Fig. 11.9*a*. Here, *n*-channel enhancement mode MOSFETs have been used with voltage levels corresponding to logic 0 and 1 as 0 V and V_{DD}, respectively. The operation will be similar even if *p*-channel MOSFETs are used instead of *n*-channel.

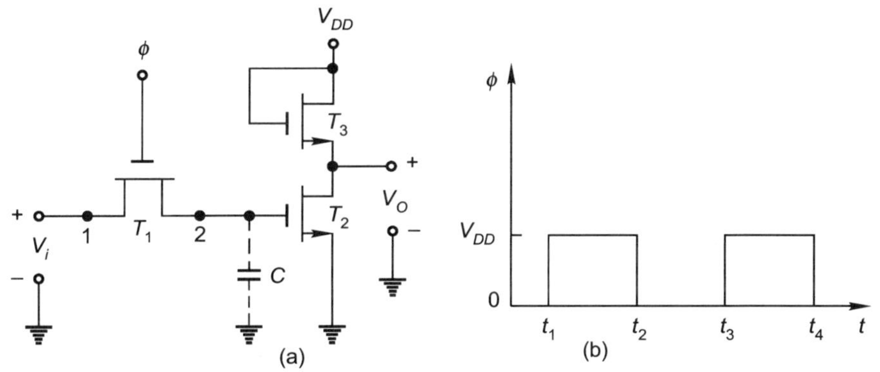

Fig. 11.9
(*a*) One-bit dynamic NMOS cell (*b*) Clock waveform ϕ.

A clock waveform such as shown in Fig. 11.9*b* is required for proper operation of the circuit. The capacitor C (~ 0.5 pF) represents the parasitic capacitance at the gate of the transistor.

When ϕ becomes 1 at $t = t_1$, the capacitor C gets charged to V_i through the conducting (ON) transistor T_1 which serves as a switch (also known as a transmission gate) and connects C to the input voltage V_i, As long as ϕ remains 1, the voltage across C is held at V_i. At $t = t_2$, ϕ becomes 0, consequently turning T_1 OFF. The inverter formed by T_2 and T_3 will remember the input data in the form of the charge stored on the capacitor, and serves the same function as a FLIP-FLOP.

The memory capability of this circuit is short lived. It is of the order of about 1 ms due to leakage of the charge through the insulation which supports the gate on the transistor T_2 and the reverse-biased junction formed between the substrate and the drain of T_1.

The transistor T_1 is a bidirectional switch. Terminal 2 acts as the source when C charges to V_i, whereas terminal 1 acts as the source when C discharges to ground through the driving transistor.

This inverter circuit is referred to as *ratio inverter* because its output voltage V_O depends upon the ratio of the ON resistance of T_2, and the effective load resistance of T_3. This ratio depends upon the relative physical sizes of T_2 and T_3 (typically, < 1: 5). Basically, there are two types of storage cell circuits used to construct dynamic shift registers. These circuits are referred to as the *ratio cell* and the *ratio-less cell*.

Two-Phase Ratio Memory Cell

The shift register stage shown in Fig. 11.10*a* is constructed with the inverter stages of Fig. 11.9*a*. For this circuit to behave in a manner similar to master–slave FLIP-FLOP, it is required that the transmission gates T_1 and T_4 must not be ON simultaneously. Therefore, we need two clock waveforms ϕ_1 and ϕ_2 as shown in Fig. 11.10*b*.

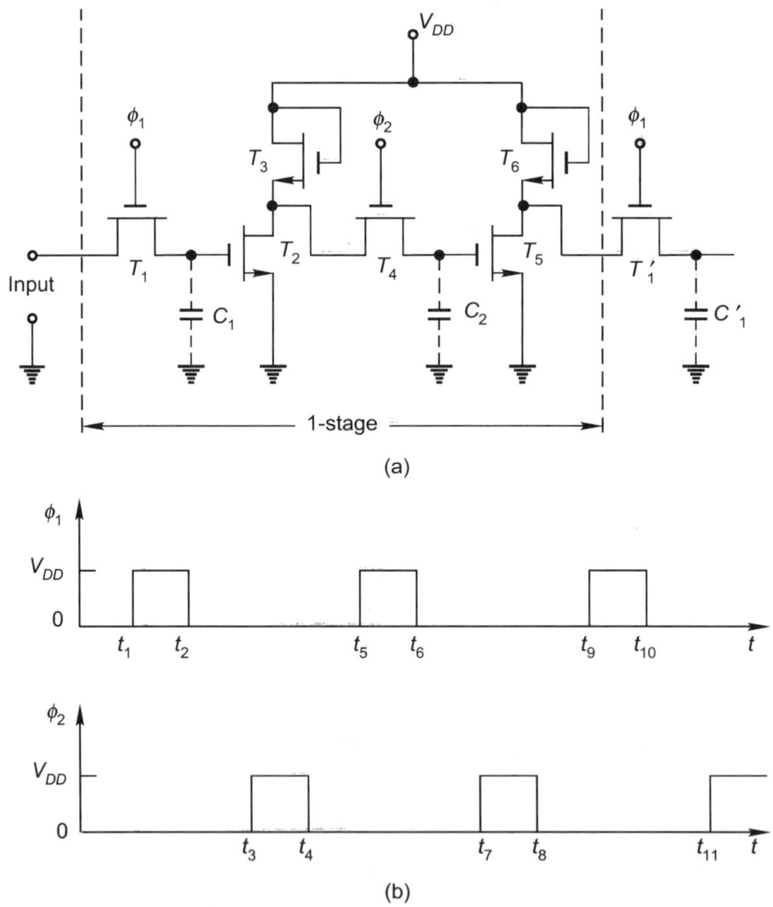

Fig. 11.10

(a) A two-phase ratio NMOS dynamic shift register stage.
(b) The two-phase clock waveforms ϕ_1 and ϕ_2.

At $t = t_1$, when ϕ_1 becomes 1, the transistor T_1 turns ON and the input-data bit to the stage is transferred to the capacitor C_1. At the same time, the transistor T'_1 of the next stage also becomes ON and, consequently, transfers the complement of the bit stored in C_2 to C'_1. At $t = t_3$, when ϕ_2 becomes 1, the complement of the bit on C_1 is transferred to C_2. From the above discussion, it is clear that the operation of this circuit is similar to that of the master–slave FLIP-FLOP. The first inverter stage (consisting of the transistors T_1, T_2, and T_3) acts in a manner similar to the master FLIP-FLOP, whereas the second inverter stage (consisting of the transistors T_4, T_5, and T_6) acts similar to the slave FLIP-FLOP.

The difference between a FLIP-FLOP and this circuit is that if the time between ϕ_1 attaining 0 and ϕ_2 attaining 1 is too great, the charge stored in the capacitor C_1 will leak off. To overcome this difficulty, the shift register is operated dynamically. Hence, it is referred to as the dynamic shift register.

The power dissipation of this circuit can be reduced considerably by using clocked loads. This requires the gates of the load devices to be connected to the gates of the succeeding transmission gates instead of connecting them to their respective drains (Problem 11.9).

The ratio shift registers discussed above have the following disadvantages:

1. The ratio of the channel resistance of the load device to the channel resistance of the driver device is required to be large. This means that the length-to-width ratio (L/W) of the load device must be much larger than that of the driver device. Correspondingly, the load device needs a larger area of the IC chip than the area required by the driver device. This affects adversely the density of components on the chip.
2. The resistance of the load device is involved in charging the gate capacitor of the succeeding inverter. A large value of this resistance adversely affects the speed of operation.

These shortcomings can be overcome by replacing each inverter with two low-resistance MOS devices of the same geometry. The resulting circuit is referred to as the ratio-less shift register.

Two-Phase Ratioless Shift Register

Figure 11.11 shows a two-phase ratio-less shift register. Let the input be at logic 1. At $t = t_1$ (Fig. 11.10*b*), when ϕ_1 becomes 1, the transistors T_1 and T_3 become ON. This will charge the capacitors C_1 and C_2 to V_{DD} (C_1 from input through T_1 and C_2 from ϕ_1 through T_3). During the charging of C_2, the transistor T_2 remains OFF, since at no time does the gate voltage of T_2 exceed its source voltage by the threshold voltage. This operation indicates that the combination of T_2 and T_3 does not operate as an inverter.

At $t = t_2$, when ϕ_1 returns to 0, T_1 and T_3 become OFF and T_2 becomes ON. This causes capacitor C_2 to discharge to 0 through T_2. Consequently, the bit stored in C_2 will be the complement of the input bit at the end of ϕ_1. This shows that the net effect of this type of operation is similar to the operation of an inverter, but since T_2 and T_3 do not conduct simultaneously, the output of the inverter is independent of the ratio of the resistances of the load and driver transistors. Hence, this circuit is a ratio-less circuit.

The inversion operation can be verified in a similar manner if the input data bit were 0.

The operation of the other half of the inverter consisting of transistors T_4, T_5, and T_6 is similar to the operation of the inverter consisting of T_1, T_2, and T_3 except that it is controlled by the clock ϕ_2. For proper operation of this inverter circuit, it is required to have $C_2/C_3 >> 1$ (Problem 11.10). In order to satisfy this condition, additional chip area is needed for C_2. There

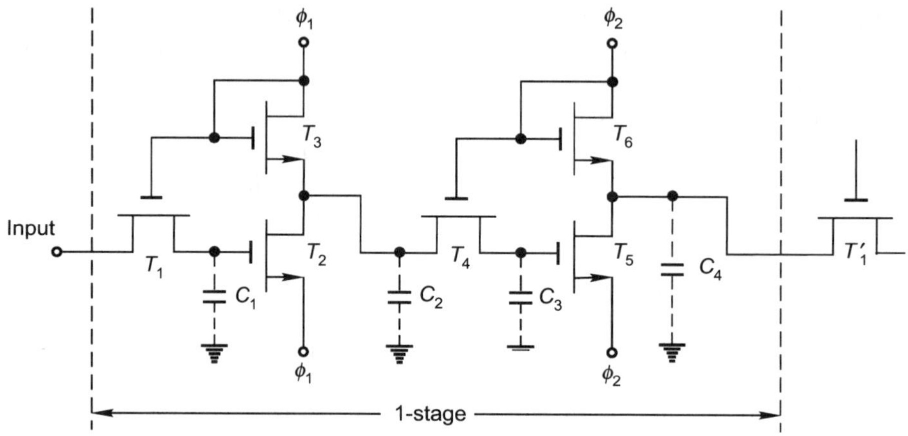

Fig. 11.11

A two-phase ratio-less shift register stage.

is another disadvantage of having a large C_2 value, viz. the speed of operation of this circuit is limited due to the large time constant of the charging circuit. The requirement of large C_2 can be overcome by using a four-phase register stage (Problem 11.11). This circuit does not require any capacitor to take charge from any other capacitor. However, this circuit is not very useful because of the complications of four-phase clock drivers.

The ratio-less dynamic shift register presents large capacitive loading on the clock drive circuitry because the clock must supply current to charge this capacitor. The excessive capacitive loading of the clock can be reduced by adding another transistor to the inverter stages, as shown in Fig. 11.12. In this circuit, the clock drivers for ϕ_1 and ϕ_2 are loaded only with the gate

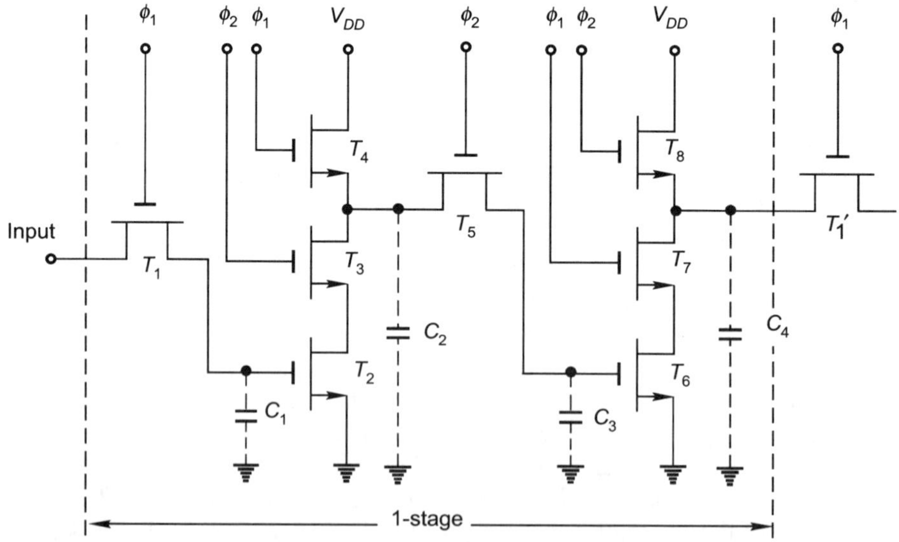

Fig. 11.12

Ratio-less shift register with reduced clock loading.

capacitance of individual transistors. The silicon area required for this cell is more than that required for the circuit of Fig. 11.11. This circuit can be analysed in a manner similar to the one used for that of Fig. 11.11.

Dynamic CMOS Shift Register

From the preceding discussion of dynamic shift registers, it is clear that, basically, a dynamic shift register needs inverters and transmission gates. The CMOS inverter and the transmission gate circuits of Figs 3.33 and 4.29, respectively, can be used to construct the dynamic shift registers. Figure 11.13 shows a dynamic shift register stage using CMOS. Here, *TG*1 and *TG*2 are the transmission gates (clock is represented by ϕ instead of *C*) and the inverters are I_1 and I_2.

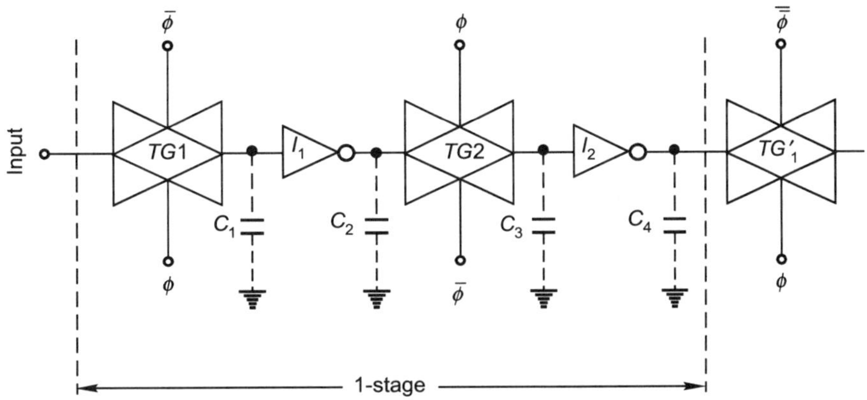

Fig. 11.13
Dynamic CMOS shift register stage.

The operation of the register stage closely parallels that of the ratio-less shift register.

The power dissipation in the CMOS shift registers is very low, since it does not draw a steady current from the V_{CC} supply. Power is drawn from the V_{CC} supply only for the transient charging of capacitors. This circuit behaves as a ratio-less shift register because the output voltage does not depend upon the ratio of the resistances of any of the devices.

As discussed in Section 4.14, this circuit requires large silicon chip area which reduces the memory capacity and increases the cost of the chip.

Dynamic Serial Memory ICs

Some commonly used dynamic serial memories with their characteristics are given in Table 11.4.

11.6 READ-ONLY MEMORY

A read-only memory (ROM) is a semiconductor memory device used to store information which is permanent in nature. It has become an important part of many digital systems because

Table 11.4 Dynamic serial memory ICs with their characteristics

IC No.	*No. of bits*	*Organization*	*No. of pins*	*Frequency*		*Max. Power dissipation mW*	*Input/ output levels*	*Clock*		*Power supplies V*
				Min kHz	*Max MHz*			*Phases*	*Levels*	
1402A	1024	Four 256-bit SISO shift-register	16	10	5	500	TTL	Two	MOS/TTL	5, – 5 or 5, – 9
1403A	1024	Two 512-bit SISO shift-registers	8	10	5	500	TTL	Two	MOS/TTL	5, – 5 or 5, – 9
1404A	1024	Single 1024-bit SISO shift-register	8	10	5	500	TTL	Two	MOS/TTL	5, – 5 or 5, – 9
1405A	512	Single 512-bit SISO recirculating shift-register	10	10	2	400	TTL	Two	MOS/TTL	5, – 5 or 5, – 9
2401	2048	Two 1024-bit SISO recirculating shift registers	16	25	1	350	TTL	One	TTL	+ 5
2405	1024	Single 1024-bit SISO recirculating shift-register	18	125	2	300	TTL	One	TTL	+ 5

of its low cost, high speed, system-design flexibility and data non-volatility. The read-only memory has a variety of applications in digital systems, such as implementation of combinational logic and sequential logic, character generation, look-up table, microprocessor programme storage, etc.

ROMs are well-suited for LSI manufacturing processes and are available in many forms. Two major semiconductor technologies are used for the manufacturing of ROM integrated circuits, viz. bipolar technology and MOS technology, which differ primarily in access time. In general, bipolar devices are faster and have higher drive capability, whereas MOS devices require less silicon area and consume less power. With improvements in MOS technology, it is now possible to make MOS memories with speeds comparable to those of bipolar memories.

The process of entering information into a ROM is referred to as *programming* the ROM. Depending on the programming process employed, the ROMs are categorized as:

1. *Mask programmable read-only memories*, which are referred to as ROMs. In these memories, the data pattern must be programmed as part of the fabrication process. Once programmed, the data pattern can never be changed. These are highly suited for very high volume usage due to their low cost.
2. Programmable read-only memories, which are referred to as PROMs. A PROM is electrically programmable, i.e. the data pattern is defined after final packaging rather than when the device is fabricated. The programming is done with an equipment referred to as *PROM programmer*. The programming techniques used will be discussed later.
3. *Erasable programmable read-only memories*, which are referred to as EPROMs. As the name suggests, in these memories, data can be written any number of times, i.e. they are reprogrammable. Reprogrammable ROMs are possible only in MOS technology. For erasing the contents of the memory, one of the following two methods are employed:
   (*a*) Exposing the chip to ultraviolet radiation for about 30 minutes.
   (*b*) Erasing electrically by applying voltage of proper polarity and amplitude. Electrically erasable PROM is also referred to as E^2PROM or EAROM.

11.6.1 ROM Organization

A read-only memory is an array of selectively open and closed unidirectional contacts. The address decoder of Fig. 11.3 is usually divided in two parts for simplifying the decoder design. One half of the address lines are decoded by one decoder used to energize one of the row lines, whereas the other half of the address lines are decoded by another decoder used to activate column lines. This method of addressing is referred to as two-dimensional, *X–Y*, or coincident-selection, addressing. A unidirectional switch is incorporated at the junction of every row and column.

A 16-bit ROM array is shown in Fig. 11.14. To select any one of the 16 bits, a 4-bit address (A_3 A_2 A_1 A_0) is required. The lower order two bits (A_1 A_0) are decoded by the decoder D_L which selects one of the four rows, whereas the higher order two bits (A_3 A_2) are decoded by the decoder D_H which activates one of the four-column sense amplifiers.

The diode matrix is formed by connecting one diode, along with a switch between each row and column. For example, the diode D_{21} is connected between row 2 and column 1.

The output is enabled by applying logic 1 at the chip select (*CS*) input.

Programming a ROM means to selectively open and close the switches in series with the diodes. For example, if the switch of the diode D_{21} is in closed position and if the address input

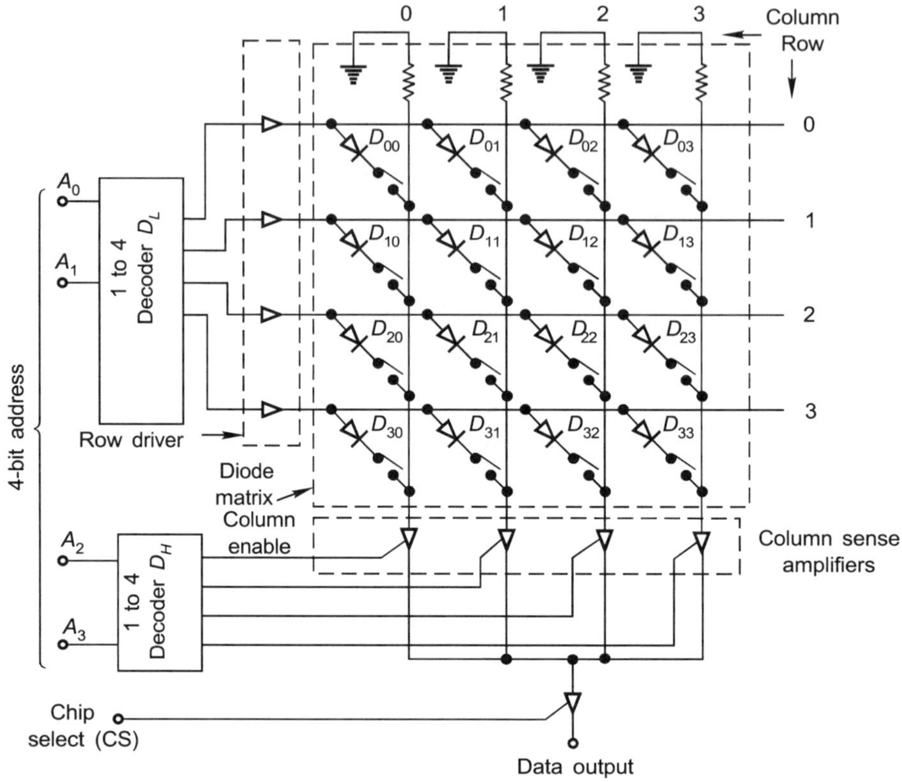

Fig. 11.14
16-bit ROM array.

is 0110, row 2 is activated connecting it to column 1. Also, the sense amplifier of column 1 is enabled which gives logic 1 output if the chip is selected ($CS = 1$). This shows that a logic 1 is stored at the address 0110. On the other hand, if the switch of diode D_{21} is open, logic 0 is stored at the address 0110.

Example 11.5 In the 16-bit ROM of Fig. 11.14, the switches of diodes D_{00}, D_{03}, D_{12}, D_{13}, D_{21} and D_{33} are programmed as closed. Find the bit stored in each location.

Solution

The bit stored in each location are given in Table 11.5.

The ROMs can also be implemented by using bipolar junction transistors or MOSFETs, instead of diodes. Each transistor is to be connected as shown in Fig. 11.15*a* for bipolar memories and as shown in Fig. 11.15*b*, for MOS memories. Each column line is to be connected to ground through a resistance for bipolar devices, whereas it is to be connected to the V_{DD} supply through a load MOSFET for MOS devices. The fuse used in the circuit of Fig. 11.15*a* is meant for programming, and is either retained intact or blown off depending upon whether 1 or 0 is to be stored in a selected location.

Table 11.5 Bit pattern of Ex. 11.5

Address				
A_3	A_2	A_1	A_0	*Bit stored*
0	0	0	0	1
0	0	0	1	0
0	0	1	0	0
0	0	1	1	0
0	1	0	0	0
0	1	0	1	0
0	1	1	0	1
0	1	1	1	0
1	0	0	0	0
1	0	0	1	1
1	0	1	0	0
1	0	1	1	0
1	1	0	0	1
1	1	0	1	1
1	1	1	0	0
1	1	1	1	1

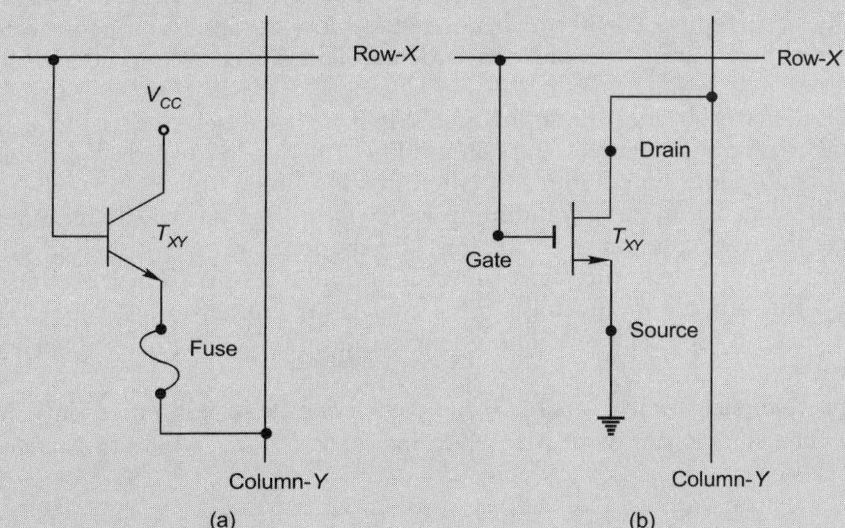

Fig. 11.15
Unidirectional switch (a) Bipolar (b) MOS.

11.6.2 Programming Mechanisms

Mask Programmable ROMs

Integrated circuits are fabricated through a number of processing steps, such as photomasking, etching, diffusion, etc. One of the final steps in the manufacturing process is to deposit a layer of aluminium on the entire surface of the silicon wafer. The desired interconnecting pattern is produced by selectively etching away portions of aluminium. The row-to-column contacts can be, retained or etched away, as desired, in the final aluminium etching process, in the manufacture of mask-programmed read-only memories.

Programmable ROMs

The most commonly used programming mechanisms for the bipolar memory devices are:

1. Fusible link process, and
2. Avalanche-induced migration (AIM) process.

In the fusible link process, a fuse is added in the emitter lead of every transistor (Fig. 11.15*a*). Usually, the fuse material used is either nichrome or polycrystalline silicon. At the time of the fabrication of these memory devices, all the fuse links are in place and can be selectively open circuited after packaging.

When nichrome is used as the fuse material, it is deposited as a very thin layer, which can be blown off (opening the connection between the row and column lines) by making a large current (20–50 mA) to flow through it. The fusing time varies between 5 and 200 μs. Typically, the programming rate is 5 ms per bit. It has power dissipation of about 170 μW per bit while operating from a 5 V supply. The nichrome fuse elements are to be left intact for storing 1s and blown off for storing 0s.

Another fuse material used is polycrystalline silicon. It is deposited as a thick layer (~ 3000Å) during the manufacturing process. The fuse is blown with a pulse train of successively wider pulses. Typically, a current of 20–30 mA is needed to blow the fuse. Temperatures of the order of 1400°C are reached during the blowing process. The silicon gets oxidized and forms an insulating material.

From Fig. 11.15*a*, we observe the following: When a row is selected (say X), the transistor T_{XY} is turned ON. If the fuse is intact, the column line Y is pulled towards V_{CC} (logic 1) whereas if the fuse is blown (open), the column bus is left floating (logic 0).

Another mechanism used for programming PROMs utilizes an avalanching approach. Figure 11.16 shows the arrangement used. In this, the diode D_1 is reverse-biased and the heavy flow of electrons in the reverse direction causes aluminium atoms from the emitter contact to migrate through the emitter to the base. This causes an emitter-to-base short. This process requires higher currents (200–300 mA) and voltages than the fusible link, but is faster, requiring 0.02–0.05 ms time.

The above mechanisms are irreversible, i.e. a device can be programmed only once in its life time. These mechanisms do not work with MOS memory devices, where resistance and current

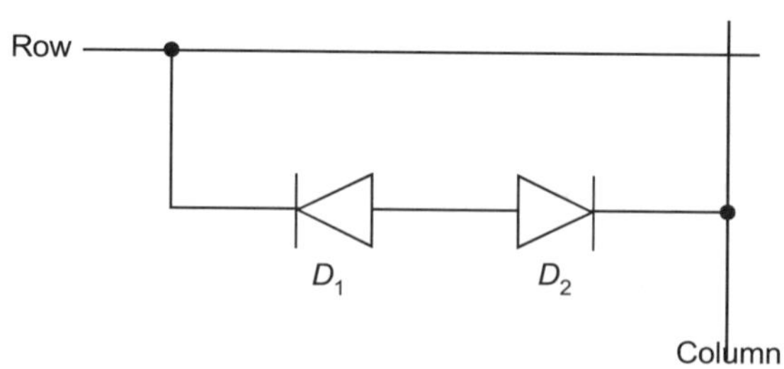

Fig. 11.16
A shorted junction cell.

levels required for the fusing process are incompatible with MOS impedance levels. Commonly used MOS technologies for the fabrication of programmable memories are:

1. Floating gate avalanche injection MOS (FAMOS), and
2. MAOS.

FAMOS PROM

In this, the storage device used is silicon gate MOSFET with no electrical connection to the gate, i.e. the gate is electrically floating in an insulating layer of silicon dioxide, as shown in Fig. 11.17*a*. The symbol of FAMOS is shown in Fig. 11.17*b*.

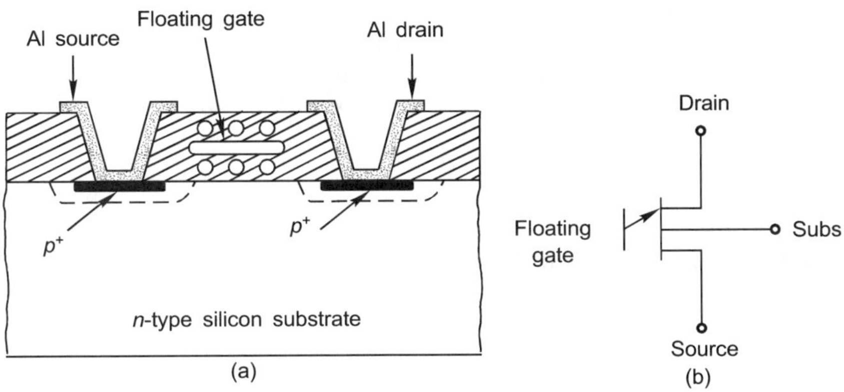

Fig. 11.17
p-channel FAMOS device: (*a*) Cross-sectional view, and (*b*) Symbol.

The operation of this device depends on the charge transport to the floating gate by avalanche injection of electrons from either the source or drain, caused by the application of high voltage (25–30 V) across the transistor. The amount of charge transferred to the floating gate is a function of the amplitude and duration of the applied voltage. The presence or absence of charge can be sensed by measuring the conductance between the source and the drain.

When the applied voltage is removed, the charge remains trapped in the gate, since no discharge path is available for the accumulated electrons because the gate is surrounded by a very low conductivity dielectric. The transistor now behaves as if an external voltage were permanently connected to the gate terminal.

The charge accumulated on the gate can be removed by illuminating the FAMOS device with ultraviolet light. This results in the flow of a photo current from the floating gate back to the silicon substrate, thereby discharging the gate to its initial condition.

These devices are packaged with a transparent quartz lid for exposing the device to ultraviolet radiation, for the purpose of erasing. Since erasing is possible in these devices, these are reprogrammable and are known as *erasable programmable read-only memory* (EPROM) devices. In case the device is packaged in inexpensive package without quartz lid, it works as a *one-time programmable* (OTP) ROM which is same as a PROM.

The use of second metal gate (*erase* gate) permits reprogramming without the use of ultraviolet radiation. Figure 11.18 shows a cross-sectional view of a *p*-channel device with an erase gate. The programming is done by applying a negative voltage to both the source (V_{SX}) and the drain

(V_{DX}), for inducing avalanche breakdown at both junctions. Simultaneously, a positive voltage applied to the second gate G_2 (V_{G2S}) increases the rate at which electrons accumulate on the floating gate. As electrons accumulate on the floating gate, the channel appears between the source and the drain and the device turns ON.

For the purpose of erasing, again a negative voltage is applied to both the source and drain junctions, for inducing avalanche breakdown similar to programming mode, but a negative voltage is applied to the gate G_2. This results in the accumulation of holes on the floating gate, which neutralizes the existing charge. Thus, the erasing is done electrically rather than with ultraviolet radiation.

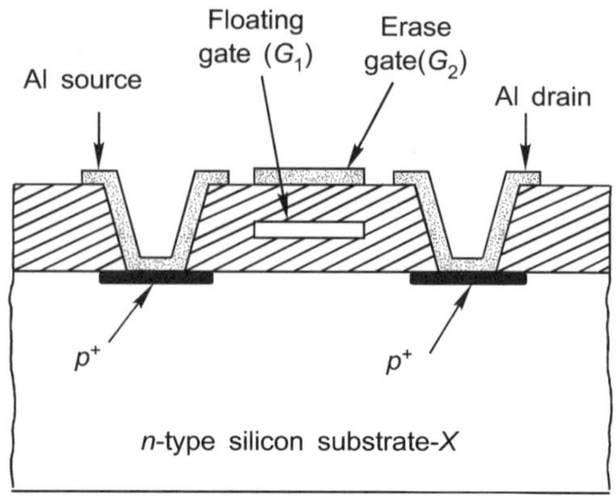

Fig. 11.18

Cross-sectional view of a FAMOS device with an erase gate.

These devices are referred to as *electrically alterable ROMs* (EAROMs) or *electrically erasable and programmable ROMs* (EEPROMs or E^2PROMs), and are reprogrammable. Sometimes, these memories are also referred to as *non-volatile RAM* (NOVRAM) or *flash-memory*.

MAOS PROM

Another PROM uses the gate dielectric, such as alumina (Al_2O_3) and silicon nitride itself, for charge storage and provide a reprogramming feature. A cross-sectional view of such a device using alumina dielectric is shown in Fig. 11.19. This device is referred to as MAOS memory element.

For a *p*-channel device, a positive gate voltage of about 50 V amplitude is required for 10–20 μs, for programming. Erasing requires a voltage of opposite polarity on the gate.

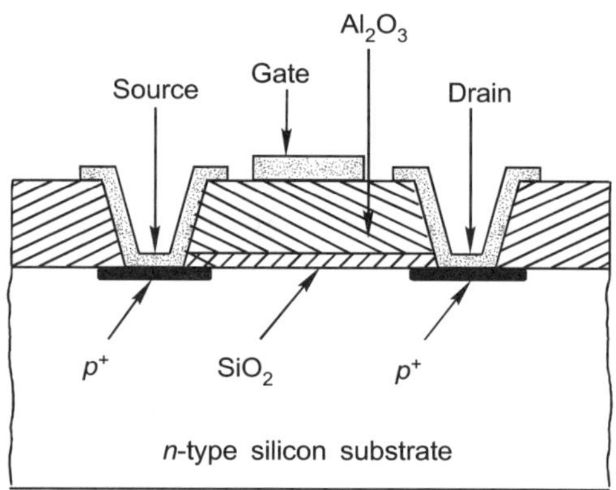

Fig. 11.19

Cross-sectional view of a MAOS device.

11.6.3 ROM ICs

Some of the commonly used ROM ICs with their characteristics are given in Table 11.6. The inputs and outputs are TTL compatible.

EPROMs 27128 (16K × 8), 27256 (32K × 8), 27512 (64K × 8) and 271024 (128K × 8) are also available which are commonly used in microprocessor-based systems.

11.7 READ AND WRITE MEMORY

Many digital systems require memories in which it should be possible to write into, or read from, any memory location with the same speed. In such memories, the data stored at any location can be changed during the operation of the system. This type of memory is known as a read/write memory and is usually referred to as RAM (random-access memory).

Table 11.6 ROM ICs with their characteristics

IC No.	*Organization*	*Output*	*Access time ns*	*Max. Power dissipation mW*	*Power supplies V*	*No. of pins*	*Technology*	*Type of ROM*
1702A	256 × 8	*TS*	1000	885	5, – 9	24	MOS	EPROM
2308	1024 × 8	*TS*	450	840	5, 12, – 5	24	MOS	ROM
2316E	2048 × 8	*TS*	450	630	5	24	MOS	ROM
2704	512 × 8	*TS*	450	800	5, 12, – 5	24	MOS	EPROM
2708	1024 × 8	*TS*	450	800	5, 12, – 5	24	MOS	EPROM
2716	2048 × 8	*TS*	450	525	5	24	MOS	EPROM
2732A	4096 × 8	*TS*	250	790	5	24	MOS	EPROM
2764	8192 × 8	*TS*	250	790	5	28	MOS	EPROM
2816	2048 × 8	*TS*	250	495	5	24	MOS	E^2PROM
3601	256 × 4	*OC*	70	685	5	16	Bipolar	PROM
3602A	512 × 4	*OC*	70	735	5	16	Bipolar	PROM
3604A	512 × 8	*OC*	70	998	5	24	Bipolar	PROM
3605	1024 × 4	*OC*	70	787	5	18	Bipolar	PROM
3608	1024 × 8	*OC*	80	998	5	24	Bipolar	PROM

The basic storage cell of a RAM is a FLIP-FLOP which simply consists of two cross-coupled inverters as shown in Fig. 7.3. A RAM is an array of these storage cells requiring as many FLIP-FLOPs as the bit storage capacity of the RAM, which is usually a large number. Therefore, simpler FLIP-FLOP circuits using bipolar or MOS transistors are used for RAM in order to save silicon chip area, reduce cost, increase speed of operation and reduce power dissipation.

Bipolar RAMs are static, whereas the MOS RAMs can be static or dynamic. In general, bipolar RAMs are faster (access time is of the order of a few tens of nanoseconds) and smaller in size (contains 1024 memory cells or less) than the MOS RAMs, which have access time of the order of a few hundreds of nano-seconds and storage capacity, upto 64K bits. With improvements in the MOS technology, it has become possible to make MOS RAMs with speeds comparable to those of bipolar RAMs.

11.7.1 Bipolar RAM Cell

The basic bipolar RAM storage cell is shown in Fig. 11.20. In this FLIP-FLOP, one transistor is ON and the other is OFF. When the OFF transistor is forced into the ON state by an external trigger pulse, the transistor that was initially ON turns OFF. Thus, it has two stable states which are used to store information in the form of logic 0 and 1. Special features are incorporated in this cell for addressing the cell, writing into it, and reading from it.

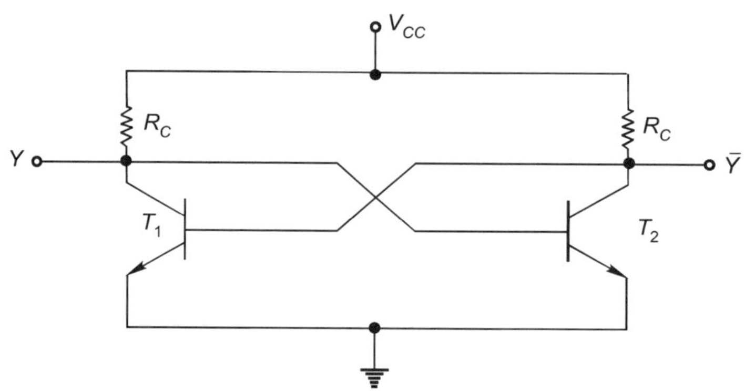

Fig. 11.20
Basic bipolar RAM storage cell.

Figure 11.21 illustrates a bipolar RAM cell with the required facilities incorporated. The FLIP-FLOP consists of the transistors T_1 and T_2. These transistors are provided with additional emitters to incorporate the facility of addressing them. The remaining circuitry provides a mechanism for writing and reading data. Signals A_X and A_Y are used to address the cell which are the outputs of the row-select and column-select address decoders, respectively. The cell is accessed for reading or writing when $A_X = A_Y = 1$.

Let A_X, A_Y, and $W/\bar{R}$ inputs be at logic 0. The outputs of gates G_1 and G_2 will be 1, which make transistors T_3 and T_4 ON. Therefore, the diodes D_1 and D_2 are non-conducting. If, say, the state of the FLIP-FLOP is such that T_1 is ON and T_2 is OFF, then the emitter current will flow in E_X and E_Y. A bias voltage of 0.5 V is applied through the resistor R_3 to emitter E_D. Therefore, E_D is more positive than E_X and E_Y and hence E_D does not conduct. The transistors T_5 and T_6 are also OFF, and the data output terminal is at logic 1 and will remain in this state, independent of the state of the FLIP-FLOP.

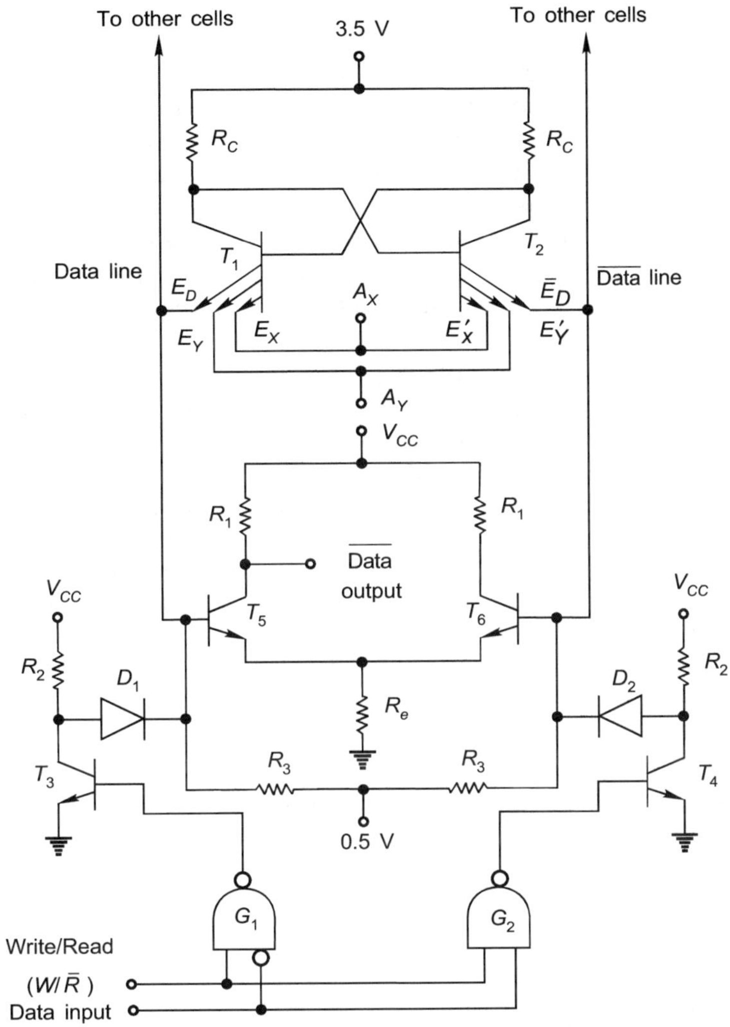

Fig. 11.21
Bipolar RAM cell.

Now, let the cell be addressed by raising both A_X and A_Y to logic 1. Then, the currents through E_X and E_Y will be diverted to E_D. A component of this current will flow into the base of T_5 and the $\overline{\text{data}}$ output terminal will assume the same logic level as present at the collector of T_1. Thus, with the $W/\bar{R}$ at logic 0, the operation of addressing the cell provides a reading of the cell.

Now, let $A_X = A_Y = 1$ and $W/\bar{R}$ input be at logic 1. If the data input is at logic 1, the output of G_1 will be 1 and the output of G_2 will be 0. Therefore, T_3 is ON and T_4 is OFF. The voltage at the collector of T_4 rises, D_2 conducts and raises the voltage at $\bar{E}_D$. Hence, irrespective of the original state of the FLIP-FLOP, T_2 cannot conduct. Hence, the logic level at the collector of T_2 will become the logic level of the data input. If the cell is not addressed, E_D and $\bar{E}_D$ will not be carrying current and the FLOP-FLOP will not respond to the writing operation.

11.7.2 MOS RAMs

MOS devices use two different techniques to store information. Depending upon the type of basic memory cell it can be categorized as either static or dynamic. Static MOS memories are slower but easier to drive than the dynamic memories, which generally require clock signals in addition to extra power supplies. Static memories use a FLIP-FLOP as a basic memory cell, whereas dynamic MOS circuits make use of temporary storage of data on the parasitic capacitances within the circuit. Due to the leakage associated with junctions within the circuit, the charges on these capacitances leak-off in a few milliseconds. To prevent loss of data, the charge must be restored (refreshed) by regenerating or recirculating the data. Refreshing of all bits is required about every 2 ms.

Dynamic circuits usually require externally generated clock voltages, which are usually too high to be generated using ordinary ICs. Thus, special ICs or discrete components are often used. Therefore, dynamic circuits are usually more difficult to interface to standard bipolar logic families than other types of circuits. Although more difficult to use, dynamic MOS circuits offer some very significant advantages. The chip area required per unit function is much smaller than for either bipolar or static MOS. Thus high functional density, combined with the higher yield of silicon gate MOS technology, makes the dynamic MOS circuit the most economical of the three approaches and permits the integration of large amounts of memory on a single chip. Speeds of MOS memories are slower than those of bipolar memories. The power dissipation per function is the lowest in dynamic MOS RAMs.

Static MOS RAM Cell

A RAM memory cell consisting of two cross-coupled MOS inverters is shows in Fig. 11.22. It is addressed by setting A_X and A_Y to 1. When $A_X = 1$, the cell is connected to the data and $\overline{\text{data}}$ line. When $A_Y = 1$, T_7 and T_8 are ON.

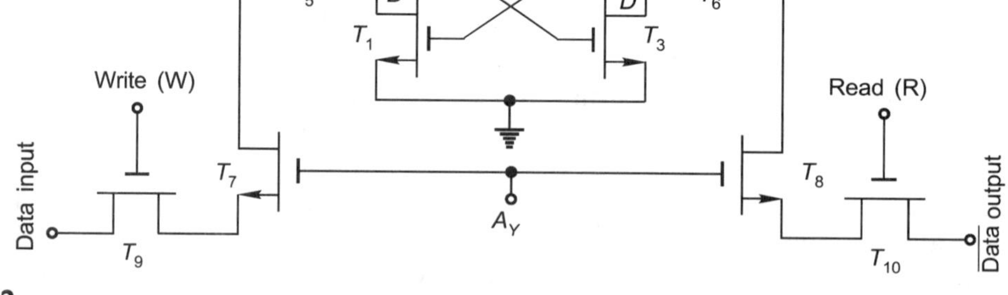

Fig. 11.22
Static MOS RAM cell.

To write into the cell, set $W = 1$, T_9 then becomes ON. If data input is 1, the voltage at node D will correspond to level 1 making T_3 ON and level at $\bar{D}$ will be 0. On the other hand, if the data input is at logic 0, then T_3 will be OFF and $\bar{D}$ would be at 1. To read the state of the FLIP-FLOP, we set $R = 1$. This connects the data output to $\bar{D}$. Thus, the complement of the data level written into the cell is read at data output.

Dynamic MOS RAM Cell

A dynamic cell uses four transistors in place of the six used in a static cell. This reduces the silicon chip area and results in saving of power. The circuit of a 4-transistor dynamic MOS RAM cell is shown in Fig. 11.23.

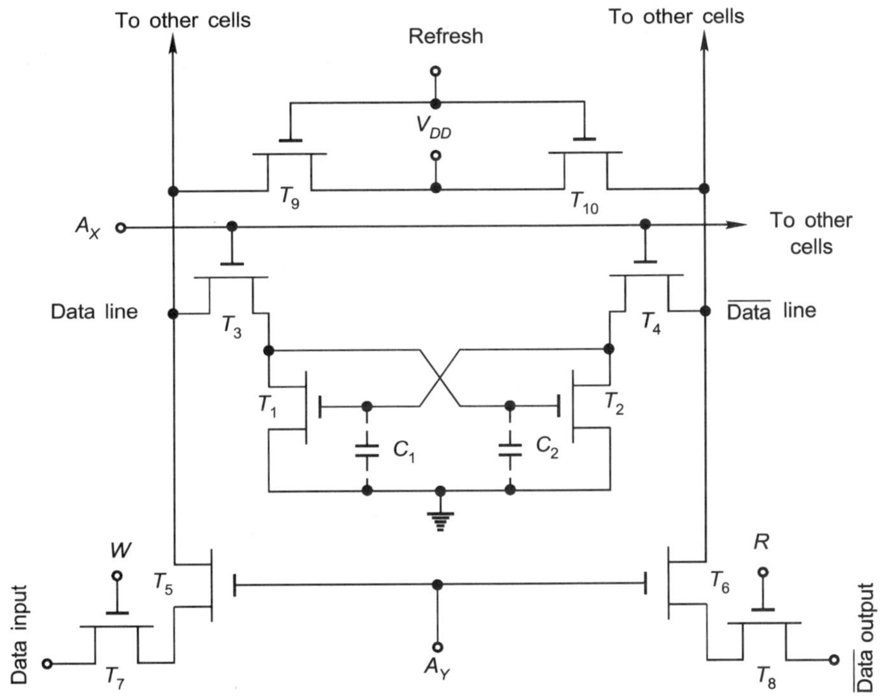

Fig. 11.23
Dynamic MOS RAM cell.

The state of the cell is stored on the stray capacitances C_1 and C_2, whose presence is essential. The cell is addressed by making $A_X = A_Y = 1$. In one state of the cell, the voltage across C_1 is large and T_1 is ON. Correspondingly, C_2 has zero voltage and T_2 is OFF. In the other state, the voltages on C_1 and C_2 and the conducting states of T_1 and T_2 are reversed. For writing into the cell, we set $W = 1$ and for reading from the cell we set $R = 1$.

It is necessary to refresh the cell periodically, otherwise the charge stored on the capacitors leak-off. The refreshing operation is accomplished by allowing brief access from the supply voltage V_{DD} to the cell. This is done by making $A_X = 1$ and the refresh terminal voltage corresponding to level 1. This makes T_3, T_4, T_9, and T_{10} ON. Suppose, initially T_1 is ON, T_2 is OFF. The voltage across C_1 is large and across C_2 it is zero. During the refresh interval, V_{DD} is applied through T_{10} and T_4 to C_1, since T_2 is OFF. Therefore, current from V_{DD} will flow through C_1, allowing C_1 to replenish any charge lost due to leakage. Since T_1 is ON, C_2 will not charge as rapidly as C_1.

Similarly, V_{DD} is applied to C_2, which is in parallel to T_1 when T_1 is OFF and T_2 is ON.

During the refresh interval, T_4 and T_{10} become the load for the driver T_2, and T_3 and T_9 becomes the load for T_1. Hence, during the refresh interval, the cell becomes a conventional FLIP-FLOP. In any event, whatever may be the initial state of the FLIP-FLOP, during the refresh interval this initial state is reinforced.

3-Transistor Dynamic Memory Cell

Figure 11.24 shows a dynamic memory cell using three transistors. In this circuit, the memory cell consists of the transistors T_1, T_2, and T_3. T_1 is used for writing, and T_2 and T_3 are used when the logic state of capacitor C is to be read. Refreshing is accomplished using a refresh amplifier consisting of transmission gate T_9 and the inverter formed by the capacitor C_r, and transistors T_{10} and T_{11}.

To write into the cell, disconnect the refresh circuit by setting $P = 0$. The cell is addressed by setting $A_X = A_Y = 1$. For writing into the cell, W is made 1. Through T_4 and T_1 the data input

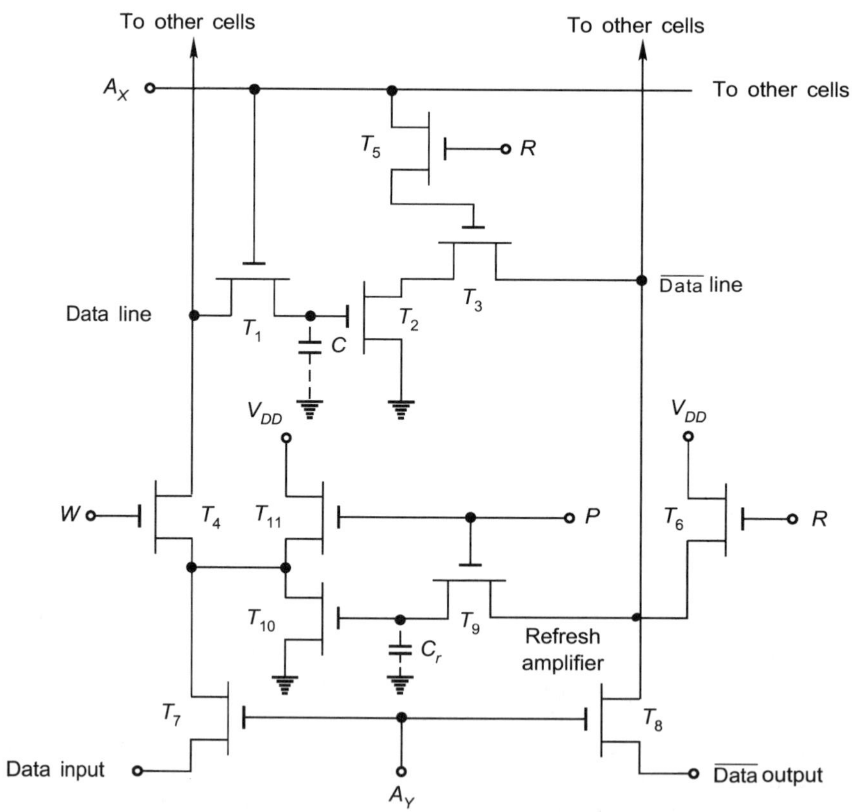

Fig. 11.24
Three-transistor dynamic MOS RAM cell.

terminal is connected to C and charges C to the state of the data input. To read from the cell, R is set to 1. Now, T_3, T_5, T_6, and T_8 are ON and connect the drain of T_2 to the $\overline{\text{data}}$ output terminal. Thus, the complement of the level stored on C is read at the $\overline{\text{data}}$ output terminal.

To refresh the cell, set $A_Y = 0$, $A_X = 1$, $P = 1$ and $R = 1$. The data input and the data output terminals are now disconnected from all cells in the memory. The complement of the logic level of capacitor C is now transferred through T_9 and is stored in capacitor C_r. The input terminal P is called the precharge input, because when $P = 1$, C_r precharges to the complement of the level on C. After this precharging is accomplished, R is set to 0 and W to 1. The output of the inverter T_{10} and T_{11} then refreshes the charge on capacitor C. Initially, the refresh-amplifier output is disconnected from T_1 and C to ensure that T_{11} does not initially load C until after C_r has been precharged to the correct level. If this precaution is not taken, C might be discharged erroneously.

For each column of the cell, there is a single refresh circuit. All cells having the same A_X address are refreshed simultaneously. Therefore, if there are m rows and n columns in the RAM, then n refresh circuits are needed and the entire memory is refreshed in time $m \times T_r$, where T_r is the time required to refresh one cell. The memory cell requires only three transistors, the other transistors are shared by other memory cells.

11.7.3 RAM ICs

Some of the commonly used RAM ICs with their characteristics are given in Table 11.7. The inputs and outputs are TTL compatible for all of them.

Static and dynamic RAMs of higher sizes are also available with considerable improvement in speed. SRAMs of 32K × 8, 64K × 8, 128K × 8, 32K × 32, etc. with access time of about 15 ns and DRAMs of 256K × 8 and 256K × 16, etc. with access time of about 45 ns are also available.

11.8 CONTENT ADDRESSABLE MEMORY

The content addressable memory (CAM) is a special purpose random access memory device that can be accessed by searching for data content. For this purpose, it is addressed by associating the input data, referred to as *key*, simultaneously with all the stored words and produces output signals to indicate the match conditions between the key and the stored words. This operation is referred to as *association* or *interrogation* and this type of memory is also known as *associative memory*.

After identifying the locations whose contents match the key, read or write operations can be performed to these locations. The key to be used may either consist of the entire data word or only some specific bits of the data word, i.e. the other bits can be masked.

A CAM differs from the conventional memory organization in that the addressing of a location in the latter has no relation to the memory content. A CAM has the ability to search out or interrogate stored data on the basis of its contents and, therefore, can be a powerful asset in many applications. For example, consider a list containing the names of persons, their ages, professions, and nationalities stored in a CAM. If one is interested in finding out engineers in the list, the CAM is able to check every memory location simultaneously by using the coded form for engineer as the key. On the other hand, if it is required to find the engineers of Indian nationality, the key will consist of the combination of the codes corresponding to engineer and

Table 11.7 RAM ICs with their characteristics

IC No.	*Organization*	*Access time (ns)*	*Power dissipation mW*	*Supplies V*	*No. of pins*	*Technology*	*Static/ Dynamic*	*Output*
1101A	256 × 1	1500	685	+ 5, – 9	16	MOS	Static	TS
2101A	256 × 4	350	300	+ 5	22	MOS	Static	TS
2104A	4096 × 1	150	420	+ 12, + 5, – 5	16	MOS	Dynamic	TS
2114-2	1024 × 4	200	300	+ 5	18	MOS	Static	TS
2142-2	1024 × 4	200	300	+ 5	20	MOS	Static	TS
2164	64K × 1	465	330	+ 5	16	MOS	Dynamic	TS
2167	16K × 1	100	625	+ 5	20	MOS	Static	TS
3101	16 × 4	60	525	+ 5	16	Schottky bipolar	Static	OC
5101	256 × 4	800	150	+ 5	22	CMOS	Static	TS
74S189A	16 × 4	50	550	+ 5	16	Schottky bipolar	Static	TS
74S289	16 × 4	35	250	+ 5	16	Schottky bipolar	Static	OC

Indian nationality. All the memory locations with engineers of Indian nationality will be identified and the remaining data (name and age) can then be retrieved by using the read operation. To do the same search process with a conventional memory, each memory word is to be read out and compared with the key. This search is a serial process and hence time consuming. Thus, CAMs are better suited for information retrieval than the conventional memories.

CAMs are manufactured using MOS, CMOS, or bipolar technologies. The most popular CAMs use ECL circuitry because of its high speed operation.

11.8.1 Operation of CAM

A CAM can perform three basic operations: read, write and associate. Figure 11.25 shows a block diagram of a CAM. Its storage capacity is $M \times N$ bits and is organized as M words of N bits each. It has N data input and N data output lines (one line for each bit of a word). The data input lines I_0 through I_{N-1} are used to input data to be written into the memory and for key word in case of associate operation. Data are read out of the CAM at the data output lines D_0 through D_{N-1}.

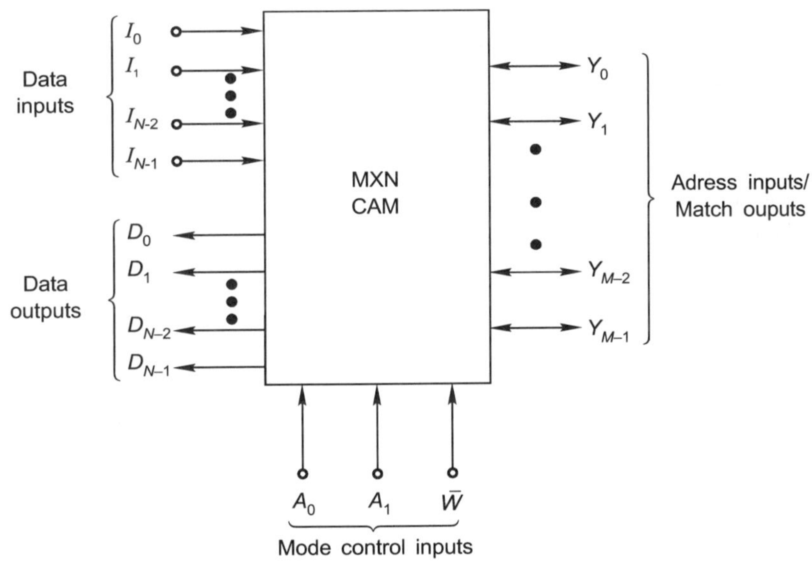

Fig. 11.25
Block diagram of a CAM.

The Y lines (Y_0 through Y_{M-1}) are bidirectional. During a read or write operation, these lines are used to select the storage location. There is one address input line for each word in the CAM. For example, Y_0 is the address line for memory location 0, Y_1, for memory location 1 and so on. Notice that linear selection addressing is used in CAMs rather than coincident selection addressing.

The Y lines serve as match output lines one for each memory location, when an association operation is performed. For example, if the keyword matches with the word stored in memory locations 5 and 8, lines Y_5 and Y_8 will become HIGH to indicate the match condition.

The mode control inputs are used to select the required operation. The read and write operations are performed in a manner similar to that used for RAM. However, during the write operation, the input data also appear at the data outputs. The reading of the data is non-destructive. The internal architecture of a typical 8×2 ECL CAM is shown in Fig. 11.26. It has eight 2-bit storage locations M_0 through M_7, each consisting of two parts, one for the higher order bit (M_{01} through M_{71}) and the other, for the lower order bit (M_{00} through M_{70}). The detailed circuitry of one of the locations M_0 is shown in the figure and all other locations have similar circuitry.

Each memory location consists of two *D*-type FLIP-FLOPs, one for each bit, and some logic gates to control the function of the CAM for read, write, and associate operations. The outputs Y_0 through Y_7 and the data outputs D_1 and D_0 are produced using wired-OR connections as shown in Fig. 11.26. The operation of this CAM is summarized in Table 11.8.

The word length and/or word size can be expanded by suitably connecting the available CAM chips in a manner similar to the one used for RAMs and ROMs expansion.

Table 11.8 Operation of the 8×2 CAM

Operation	*Control inputs*			*Data inputs*		*Data outputs*		Y_n	
	A_1	A_0	$\overline{W}$	I_1	I_0	D_1	D_0	*Input/output*	*Value*
Associate	1	1	×	1/0	1/0	0	0	Output	1 = Mismatch 0 = Match
Associate (higher bit masked)	0	1	1	×	1/0	D_1	0	Output	1 = Lower bit mismatch 0 = Lower bit match
Associate (lower bit masked)	1	0	1	1/0	×	0	D_0	Output	1 = Higher bit mismatch 0 = Higher bit match
Read	0	0	1	×	×	D_1	D_0	Input	0 = Selected address
Write	0	0	0	1/0	1/0	I_1	I_0	Input	0 = Selected address
Associate and write at match addresses	0	1	0	1/0	1/0	I_1	0	Output	1 = Lower bit mismatch 0 = Lower bit match
	1	0	0	1/0	1/0	0	I_0	Output	1 = Higher bit mismatch 0 = Higher bit match

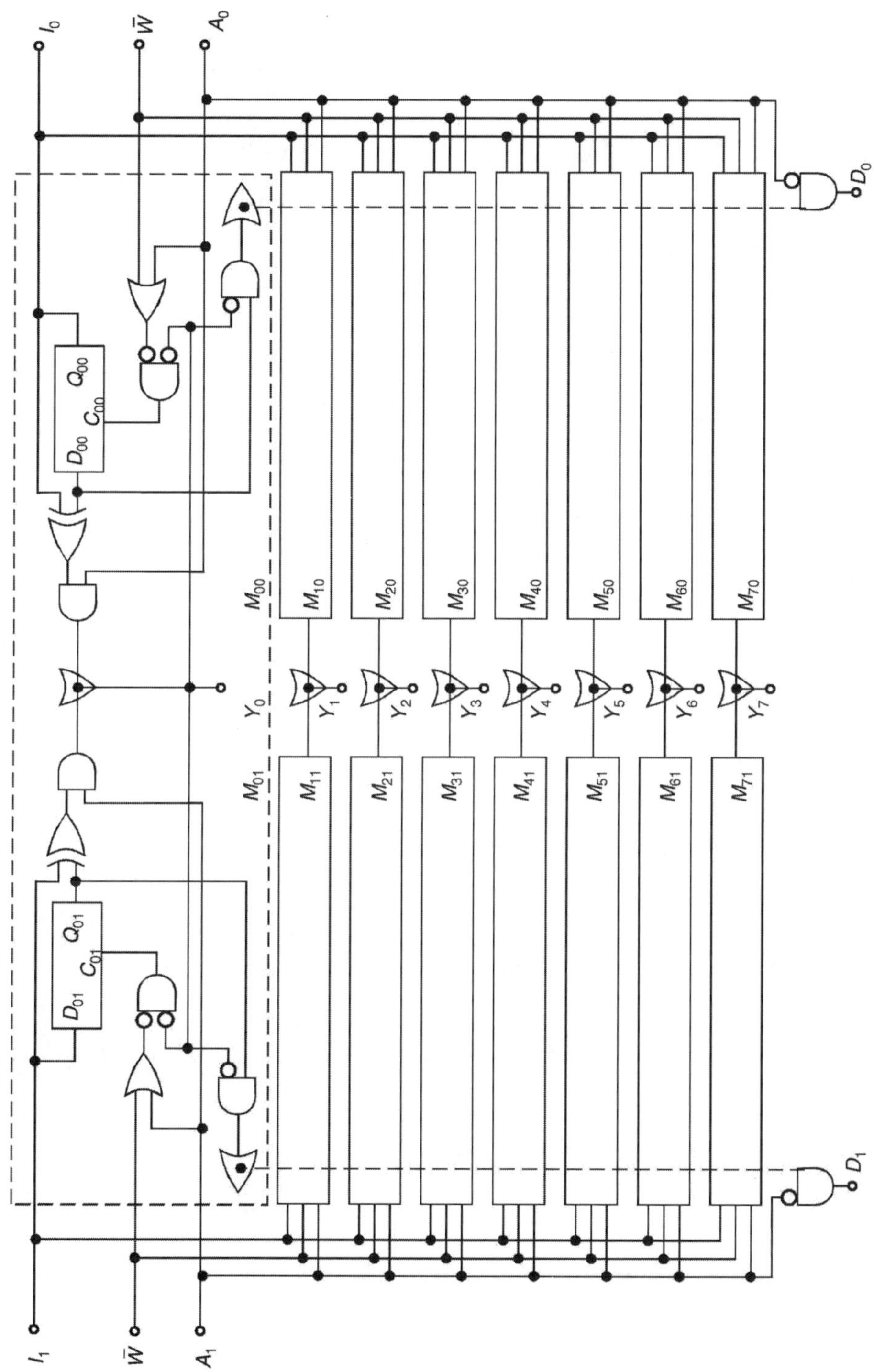

Fig. 11.26
Internal architecture of an ECL 8 × 2 CAM.

Example 11.6 Consider the CAM of Fig. 11.26. Assume that it contains data given in Table 11.9

Table 11.9 Contents of CAM

Memory location	*Content*
0	00
1	01
2	00
3	11
4	10
5	01
6	11
7	01

What is the word at data outputs D_1 and D_0 for each of the following conditions? Also find if there is any change in the memory contents.

(*a*) $A_1\,A_0 = 00$, $\overline{W} = 1$
$Y_7\,Y_6\,Y_5\,Y_4\,Y_3\,Y_2\,Y_1\,Y_0 = 11011111$

(*b*) $A_1\,A_0 = 00$, $\overline{W} = 1$
$Y_7\,Y_6\,Y_5\,Y_4\,Y_3\,Y_2\,Y_1\,Y_0 = 11001111$

(*c*) $A_1\,A_0 = 00$, $\overline{W} = 0$
$Y_7\,Y_6\,Y_5\,Y_4\,Y_3\,Y_2\,Y_1\,Y_0 = 11110111$
$I_1\,I_0 = 00$

(*d*) $A_1\,A_0 = 00$, $\overline{W} = 0$
$Y_7\,Y_6\,Y_5\,Y_4\,Y_3\,Y_2\,Y_1\,Y_0 = 11010111$
$I_1\,I_0 = 10$

Solution

(*a*) Since $Y_5 = 0$, the memory location 5 is selected for read out, i.e.
$D_1\,D_0 = 01$
The memory contents do not change.

(*b*) Since $Y_5 = Y_4 = 0$, the memory locations 4 and 5 are selected for read out. The output is obtained by ORing the contents of these locations, i.e.
$D_1\,D_0 = 11$
The memory contents do not change.

(*c*) Since $Y_3 = 0$, write operation is performed in memory location 3. The input data is stored in this location and also appears at the output.
$D_1\,D_0 = 00$
Contents of memory location 3 = 00.

(*d*) In this case, the memory locations 3 and 5 are selected for writing since $Y_3 = Y_5 = 0$. The contents of these locations will become 10 and also $D_1\,D_0 = 10$.

Example 11.7 Assume that the CAM of Fig. 11.26 contains the data given in Table 11.9.
For each of the following conditions, find the Y outputs. Also find if there is any change in the memory contents.

(*a*) $A_1\ A_0 = 11$, $I_1\ I_0 = 01$
(*b*) $A_1\ A_0 = 01$, $\overline{W} = 1$, $I_1\ I_0 = 01$
(*c*) $A_1\ A_0 = 10$, $\overline{W} = 1$, $I_1\ I_0 = 01$
(*d*) $A_1\ A_0 = 01$, $\overline{W} = 0$, $I_1\ I_0 = 01$
(*e*) $A_1\ A_0 = 10$, $\overline{W} = 0$, $I_1\ I_0 = 01$

Solution

(*a*) The association operation is performed with keyword 01. The memory locations 1, 5, and 7 match the keyword giving out logic 0 at the corresponding Y outputs. Therefore,

$$Y_7\ Y_6\ Y_5\ Y_4\ Y_3\ Y_2\ Y_1\ Y_0 = 01011101$$

(*b*) In this case, the association operation is performed between the lower bit of the key with the lower bits of the stored words. Therefore,

$$Y_7\ Y_6\ Y_5\ Y_4\ Y_3\ Y_2\ Y_1\ Y_0 = 00010101$$

(*c*) In this case, the association operation is performed between the higher bit of the key with the higher bits of the stored words. Therefore,

$$Y_7\ Y_6\ Y_5\ Y_4\ Y_3\ Y_2 Y_1\ Y_0 = 01011000$$

(*d*) The association operation is performed between the lower bit of the key and the lower bits of the stored words. The Y outputs are

$$Y_7\ Y_6\ Y_5\ Y_4\ Y_3\ Y_2\ Y_1\ Y_0 = 00010101$$

The higher bit I_1 is stored in the higher bit positions of the memory locations 7, 6, 5, 3, and 1. Therefore, the contents of the memory locations will be as given in Table 11.10.

Table 11 .10

Memory location	*Contents*
0	00
1	01
2	00
3	01
4	10
5	01
6	01
7	01

(*e*) In this case, the association operation is performed between the higher bit of the key and the higher bits of the stored words. The Y outputs are

$$Y_7\ Y_6\ Y_5\ Y_4\ Y_3\ Y_2\ Y_1\ Y_0 = 01011000$$

The lower bit I_0 is stored in the lower bit positions of the memory locations 7, 5, 2, 1, and 0. The new memory contents are given in Table 11.11.

Table 11.11

Memory location	*Contents*
0	01
1	01
2	01
3	11
4	10
5	01
6	11
7	01

11.9 CHARGE COUPLED DEVICE MEMORY

The charge coupled device (CCD), a new concept for storage of digital information, was announced in early 1970 by Bell Telephone Laboratories of U.S.A. It is an array of MOS capacitors operating as a dynamic shift register. CCDs are simple, versatile, and low cost devices and can be used wherever a serially accessed memory is required.

The operation of CCDs involve the following steps:

1. Conversion of digital input signal into charge,
2. Transfer of charge through various stages in sequential manner, and
3. Conversion of charge at the output into digital signal.

During each charge transfer step, a small amount of charge is lost. Also, due to thermal effects, undesirable charge may be generated which is known as *dark current*. To overcome these defects, the charge is recirculated around the shift register for refreshing.

11.9.1 Basic Concept of CCD

Consider a *p*-type silicon substrate covered with a thin oxide layer and closely spaced metallic electrodes, as shown in Fig. 11.27. Each metallic electrode (gate) and the substrate form a MOS capacitor which can store charge. If a positive voltage is applied at a gate electrode, a depletion

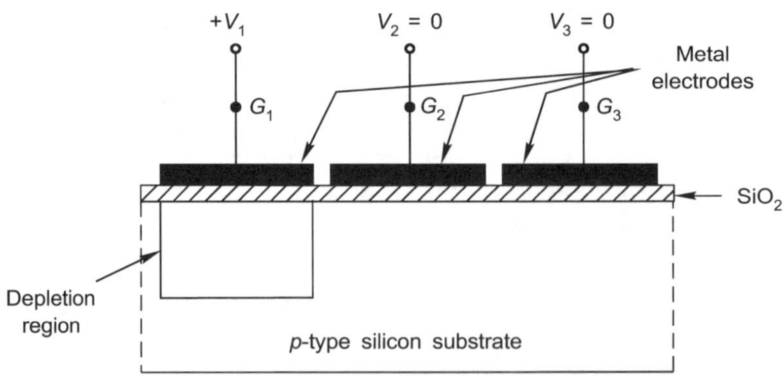

Fig. 11.27
Basic CCD structure.

region is formed in the substrate immediately under the metallic electrode. This happens due to the repulsion of free holes in the substrate because of the positive voltage at the gate electrode. These holes are driven downward away from the oxide layer and consequently immobile negative ions are exposed and a depletion region comes into existence. In Fig. 11.26, a positive voltage V_1 is applied at the G_1 gate and the other two gates are held at the same potential as the substrate. The depletion region is indicated below the gate G_1. This plot also represents the potential-energy barrier (well) for electrons, which are the minority charge carriers. Now, if a packet of negative charge is injected into the depletion region, these charges can move freely within the well, but cannot penetrate the potential-energy walls of the well. This means that as long as the voltage V_1 is present, the negative charge will be held (trapped) there.

In a conventional capacitor, the charges are held on conducting plates, whereas in a MOS device the charges are held on a conductor and in the depletion region under the conductor. The stored charge can be moved from left to right down the channel by applying voltages at the gates in a proper sequence. We assume that a logic 1 is stored when a negative charge is held in the depletion region and a logic 0 is stored when the depletion region is empty. This type of operation makes it possible to make long shift registers using these devices.

11.9.2 Operation of CCD

A portion of the structure of a 4-phase charge-coupled device (CCD) shift register, along with the charge transfer plots, is shown in Fig. 11.28*a*. The four clock waveforms ϕ_1, ϕ_2, ϕ_3, and ϕ_4 used to drive the circuit are shown in Fig. 11.28*b*. An array of four adjacent electrodes driven by the 4-phase clock constitutes a single dynamic `FLIP-FLOP`. The 4-phase arrangement is required to give this dynamic `FLIP-FLOP` the operating features of a master–slave `FLIP-FLOP`, which allows shift register operation and assures that the data move in only one direction. The whole device is a long array of MOS devices in which all the ϕ_1 electrodes are connected together. A similar arrangement exists for the ϕ_2, ϕ_3, and ϕ_4 electrodes.

During the time interval t_1 only ϕ_1 is at a positive voltage, so that depletion regions are formed only under ϕ_1. The charge indicated in the depletion region under ϕ_1 is injected either from an outside source or from the preceding ϕ_4 gate.

During the interval t_2, the depletion regions under ϕ_1 gates persist while new depletion regions are formed under ϕ_3 gates because the clock ϕ_3 becomes positive. In the interval t_3, the clock ϕ_2 also becomes positive while the clocks ϕ_1 and ϕ_3 are held positive. Therefore, depletion regions are generated extending from the ϕ_1 gates to the ϕ_3 gates. As a result, the charge is now able to spread throughout the extended region. During the interval t_4, the clock ϕ_1 becomes 0 thereby eliminating the depletion regions under the ϕ_1 gates. Similarly, during the interval t_5, the depletion regions under ϕ_2 gates vanish and the charge (or no charge) originally under the ϕ_1 gates is pushed laterally to regions under the ϕ_3 gates.

Continuing the above logic, we observe that in the succeeding intervals t_6 through t_8, and finally back to t_1, the charge under the gate ϕ_3 will be moved to the region under the next ϕ_1 gate. Hence, altogether, after eight intervals, charge (or no charge) under a ϕ_1 gate will shift to the next ϕ_1 gate.

Special arrangement must be made to inject charge into the first depletion region as required, and to detect the presence (logic 1) or absence (logic 0) of charge at the last depletion region. The injection and detection of charge must be done in synchronism with the clock waveform.

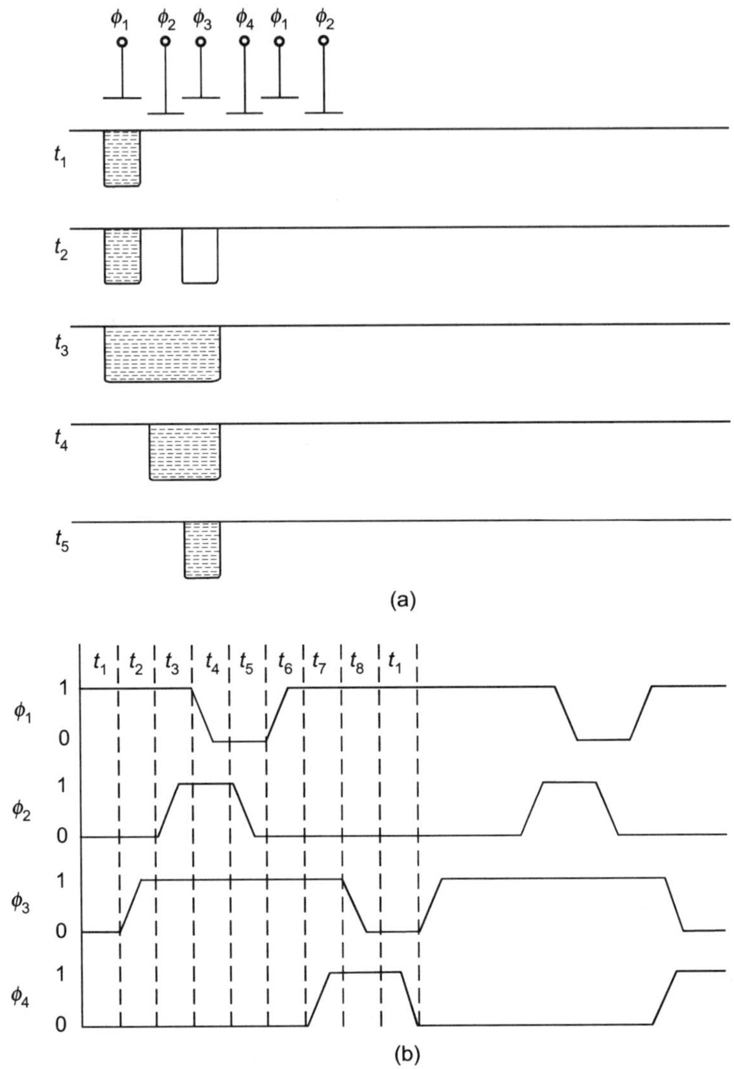

Fig. 11.28

(*a*) A portion of four-phase CCD structure along with the charge transfer plots and (*b*) Four-phase clock waveform.

When the charge transfer takes place down the shift register, there is some loss of charge. Therefore, it is necessary to incorporate provision for refreshing the charge at periodic intervals along the length of the CCD structure. In fact, the charge transfer efficiency is so high (~99.999%) that 100 or more FLIP-FLOPs can be cascaded before refreshing becomes necessary.

11.9.3 A Practical CCD Memory Device

The basic organization of the intel 2416 CCD memory is shown in Fig. 11.29. It is a 16 384 × 1 bit serial memory, organized as 64 independent recirculating shift registers of 256 bits each.

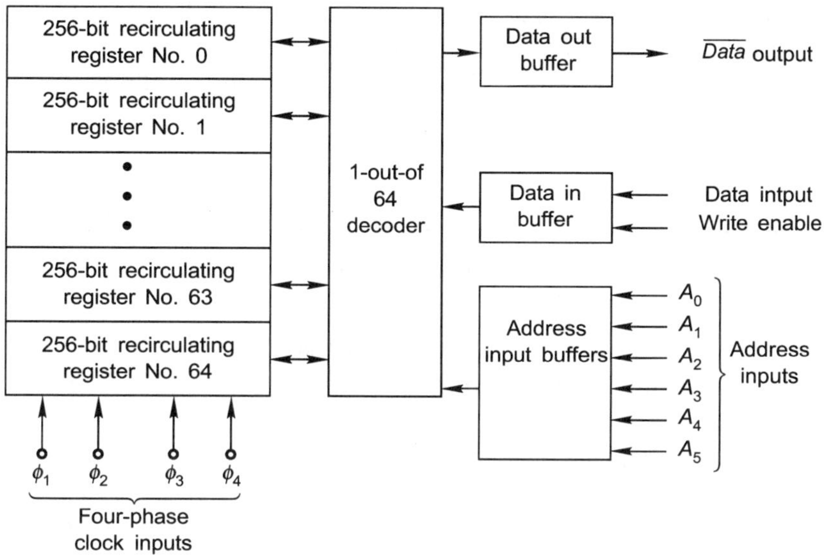

Fig. 11.29

Basic organization of Intel 2416 CCD memory.

Any one of the 64 registers can be accessed by applying the appropriate 6-bit code at the address inputs. The data in the shift registers is simultaneously shifted by using the 4-phase clock signals ϕ_1 through ϕ_4. After a shift cycle, each of the 64 registers can be selected for an input/output operation by applying the appropriate 6-bit address code.

When addressed, one bit is written into or read from the memory. If the address input is fixed, then as shifting progresses, the bit positions in the addressed register will be presented serially for reading or writing. The output is open-drain which allows wired-OR connection. During the interval between shifts, we can have access to a bit from each of the registers by changing the address. The 64 bits, which can be accessed by changing address, are available on a random access basis. The 256 bits in a single register are available only in the serial mode.

In serial mode operation, access to a desired bit may require no shift or it may require upto a maximum of 256 shifts. On an average, 256/2 = 128 shifts are required for accessing a bit. This access time in serial operation is known as *latency* or *latency time.*

The Intel 2464 is a 65536-bit CCD memory organized in a manner similar to the 2416 chip, but with 256 independent circulating registers of 256 bits each. It has an active-low chip enable signal $\overline{CE}$.

The bit capacity and word lengths can be expanded by using appropriate circuitry (Probs. 11.19 and 11.20).

11.10 SUMMARY

Semiconductor memories have become very popular in recent years. Various types of semiconductor memories, such as ROMs, RAMs, CAM, CCD, shift register, etc. have been introduced. Their organization and operation have been discussed, which will be very useful in the optimum utilization of these devices for system design. The methods used for the expansion of memory size will be of great help to the system designer.

The various methods for the fabrication and programming technologies of PROMs have been discussed in detail.

Glossary

Access time Time required for reading or writing a memory location.

Address The binary code of a memory location.

Address bus A parallel array of conductors used for accessing a memory location.

Address decoder A n-line-to-2^n lines decoder used to select a specific memory chip or memory location.

Bidirectional bus A bus (group of electrical lines) capable of transmitting data in both the directions.

CAM (Content addressable memory) A special purpose RAM device which can be accessed by its contents. It is also known as 'associative memory'.

CCD (Charge coupled device) It is serial-access semiconductor memory device with very high bit packing density.

Chip A piece of silicon or other semiconductor material on which an IC is fabricated.

Chip enable An input control signal which when activated enables the chip.

Chip select An input control signal that allows the chip to be selected.

Control bus A bus used for handling control signals.

Cycle time The minimum time between successive read or write cycles in a memory.

Data bus A bus used for carrying data.

Dynamic memory A memory in which data needs to be refreshed periodically. Data is stored on MOS capacitors.

EAROM (Electrically alterable read-only memory) A read-only memory in which the stored words can be erased electrically. Also known as E^2PROM (Electrically erasable and programmable read-only memory).

EPROM (Erasable programmable read-only memory) A read-only memory which can be erased by exposure to ultraviolet light, and then reloaded with new information.

Erasable memory A memory contents of which can be erased.

Fusible link A link of nichrome, or some other materials, used in PROMs and other programmable devices which can be either burnt or kept intact while storing 0s and 1s in these devices.

Non-erasable memory A memory contents of which can not be erased, such as a ROM.

Non-volatile memory A memory that does not loose its contents when power is turned off, such as ROMs.

Programming (of Memory) To store the desired data in a programmable memory which is of read-only type.

PROM (Programmable read-only memory) A read-only memory that can be programmed only once by the user by selectively opening the fusible links.

PROM programmer An equipment that is used to program a programmable ROM.

RAM (Random-access memory) A read-and-write semiconductor memory in which any memory location can be accessed for reading or writing at random.

Refresh The act of restoring the data in a dynamic memory.

ROM (Read-only memory) A semiconductor memory in which data can only be read.

Semiconductor memory A memory fabricated using semiconductor material.

Serial-access memory A memory in which the access time of a stored bit or word depends on its location in the memory.

Static memory A memory in which the data does not change with time.

Volatile memory A memory that looses its contents when power is turned off.

Review Questions

11.1 Information in a memory chip is stored in ________ form.

11.2 The maximum number of bytes which can be stored in a memory of size 1024×8 is ________ .

11.3 The number of address lines required in a memory of 128 KB is ________ .

11.4 While specifying the memory size, the letter *K* stands for ________ .

11.5 An EPROM is a ________ access memory.

11.6 A shift register is a ________ memory.

11.7 The contents of location *0A00H H* of RAM is 01001100, its contents after a read operation will be ________ .

11.8 The contents of location *BAC0H* of EPROM is 11100101, its contents after a read operation will be ________ .

11.9 The number of IC chips of memory size 1024×4 required to have 16 KB memory will be ________ .

11.10 The contents of location *FEE0H* of a RAM is *BEH*, its contents after a write operation is performed with *ECH* data at the data bus will be ________ .

11.11 The access-time of a RAM is 10 ns, the minimum time which must elapse between two read operations will be ________ .

11.12 An EAROM is ________ erasable.

11.13 An EPROM is erased by ________ .

11.14 A dynamic RAM is fabricated using ________ technology.

11.15 CAM stands for ________ .

Problems

11.1 For a memory with M words storage, find the number of pins required for addressing and the address range in binary format for each of the following cases:

(*a*) $M = 4$
(*b*) $M = 16$
(*c*) $M = 64$
(*d*) $M = 256$
(*e*) $M = 1024 = 1$ K
(*f*) $M = 2048 = 2$ K
(*g*) $M = 64$ K
(*h*) $M = 1$ M

11.2 Express the address range for each of the cases of Prob. 11.1 in

(*a*) Hexadecimal format.
(*b*) Octal format.

11.3 The access time and cycle time for a set of memories are given in Table 11.12. Determine the maximum rate at which data can be accessed in each case.

Table 11.12

Memory	*Access time* *ns*	*Cycle time* *ns*
A	1500	1500
B	300	580
C	450	450
D	200	200
E	60	60
F	800	800

11.4 The block diagram of Intel 2142, a 1024 × 4 bit static RAM, is shown in Fig. 11.30. Find the number of 2142 and other ICs required to obtain

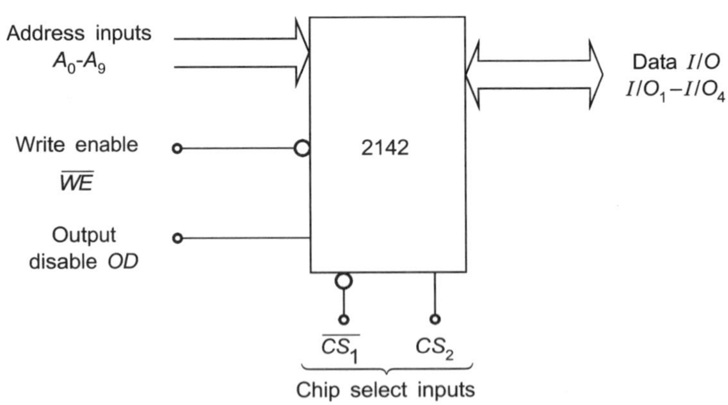

Fig. 11.30
Block diagram of 2142 RAM.

(*a*) 4096 × 4 bit RAM
(*b*) 1024 × 8 bit RAM
(*c*) 16 K bytes of RAM.

11.5 Implement the RAMs of Prob. 11.4.

11.6 The block diagram of Intel 2716 2K × 8 EPROM is shown in Fig. 11.31. Find the number of 2716 and other ICs required to obtain.
(*a*) 4 K bytes of ROM
(*b*) 2 K × 16 ROM
(*c*) 4 K × 16 ROM

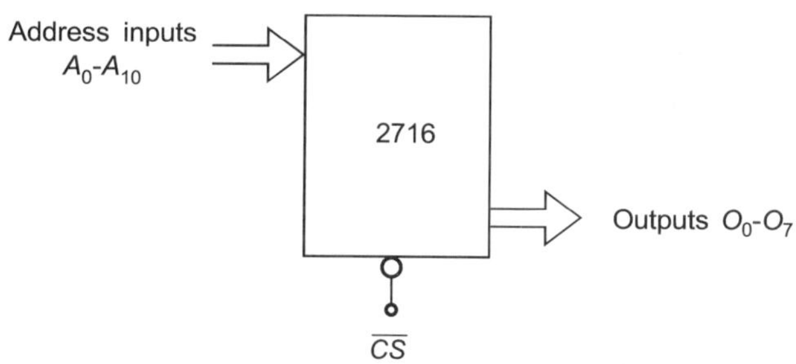

Fig. 11.31
Block diagram of 2716 EPROM.

11.7 Implement the ROMs of Prob. 11.6.

11.8 Explain the following:
(*a*) Linear selection addressing.
(*b*) Coincident selection addressing.

11.9 A MOS dynamic shift register stage with load clocking is shown in Fig. 11.32. Explain its operation and show that the power dissipation of this circuit is less than that of the circuit of Fig. 11.10*a*.

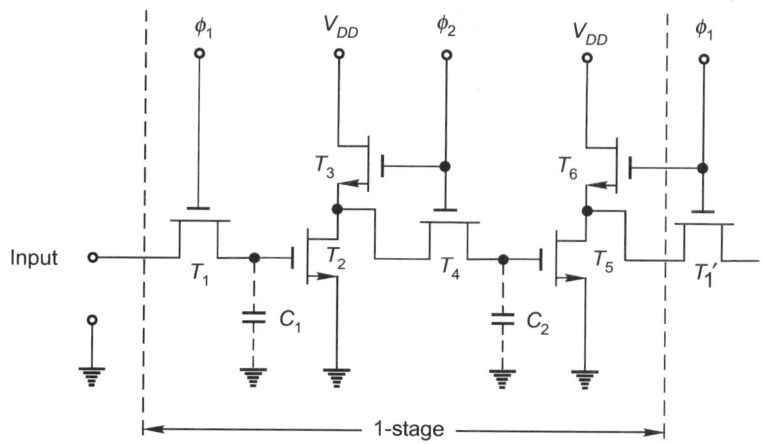

Fig. 11.32
Circuit for Prob. 11.9.

11.10 Show that $C_2 \gg C_3$ for the proper operation of the two-phase ratio less shift register stage of Fig. 11.11.

11.11 Explain the operation of the 4-phase ratioless shift register stage shown in Fig. 11.33*a*. The 4-phase clock waveform is shown in Fig. 11.33*b*.

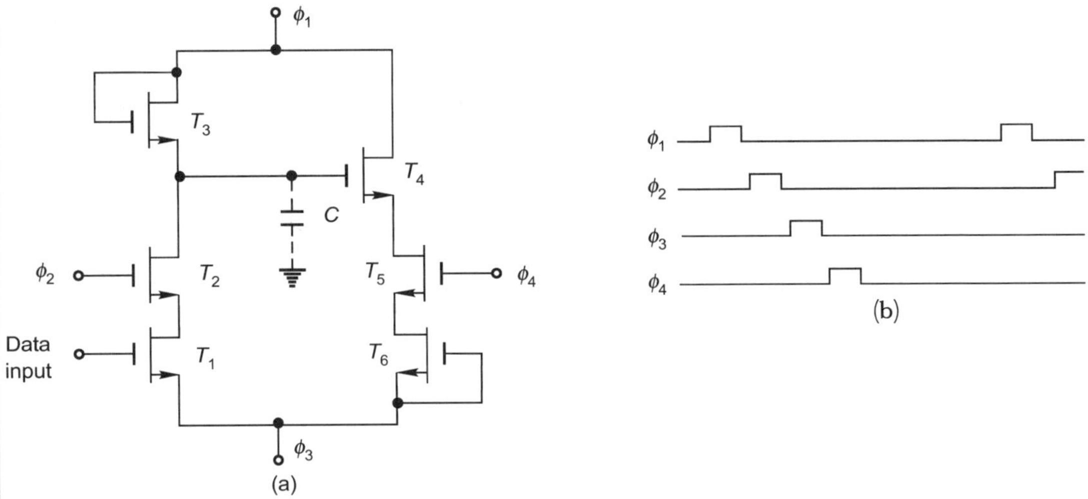

Fig. 11.33

(a) A 4-phase ratio less shift register stage and (b) Four-phase clock waveform.

11.12 Verify the operations given in Table 11.8.

11.13 Design a 16 × 2 CAM using two 8 × 2 CAM chips.

11.14 Design an 8 × 8 CAM using four 8 × 2 CAM chips.

11.15 Design a 16 × 8 CAM using 8 × 2 CAM chips.

11.16 It is desired to find the maximum valued number stored in a CAM of size 16 × 8. Suggest a suitable method. Compare this method with the method used if a RAM is used instead of a CAM.

11.17 Repeat Prob. 11.16 to Find the minimum valued number.

11.18 It is desired to design a memory system for storing information which is not already stored in it. This type of memory is known as *learning memory*. Will you prefer to use RAM or CAM for this purpose, why?

11.19 Suggest a suitable arrangement for expanding the bit capacity of CCDs.

11.20 Suggest a suitable arrangement for expanding the word length of CCDs.

12

PROGRAMMABLE LOGIC DEVICES

12.1 INTRODUCTION

The combinational and sequential digital circuits have been discussed in earlier chapters. Various ICs for performing basic digital operations and other functions, such as multiplexers, demultiplexers, adders, comparators, code converters, shift registers and counters, etc. have also been discussed. These ICs are referred to as *fixed-function* ICs, i.e. each one of them performs a specific , fixed function. These devices are designed by their manufacturers and are manufactured in large quantities to meet the needs of a wide variety of applications and are readily available.

To design a circuit, a designer can select from the available ICs most appropriate for the circuit, usually working from a block diagram design concept. The design may have to be modified to meet the special requirements of these devices. The advantages of this method are:

1. Low development cost,
2. Fast turn around of designs, and
3. Relatively easy to test the circuits

Some of the disadvantages of this method are:

1. Large board space requirements,
2. Large power requirements,
3. Lack of security, i.e. the circuits can be copied by others, and
4. Additional cost, space, power requirements, etc. required to modify the design or to introduce more features.

To overcome the disadvantages of designs using fixed-function ICs, *application specific integrated circuits* (ASICs) have been developed. The ASICs are designed by the users to meet the specific requirements of a circuit and are produced by an IC manufacturer (*foundry*) as per the specifications supplied by the user. Usually, the designs are too complex to be implemented using fixed-function ICs.

The advantages of this method are:

1. Reduced space requirement,
2. Reduced power requirement,
3. If produced in large volumes, the cost is considerably reduced,
4. Large reduction in size through the use of high level of integration, and
5. Designs implemented in this form are almost impossible to copy.

The disadvantages of this method are:

1. initial development cost may be enormous, and
2. testing methods may have to be developed which may also increase the cost and effort.

Another approach which has the advantages of both the above methods is the use of *programmable logic devices* (PLDs). A programmable logic device is an IC that is user configurable and is capable of implementing logic functions. It is an LSI chip that contains a 'regular' structure and allows the designer to customize it for any specific application, i.e. it is programmed by the user to perform a function required for his application.

PLDs have the following advantages of fixed function ICs:

(i) Short design cycle
(ii) Low development cost

The advantages over fixed-function ICs are:

(i) Reduction in board space requirements
(ii) Reduction in power requirements
(iii) Design security
(iv) Compact circuitry
(v) Higher switching speed

PLDs have many of the advantages of ASICs as given below:

(i) Higher densities
(ii) Lower quantity production costs
(iii) Design security
(iv) Reduced power requirement
(v) Reduced space requirement

The PLDs allow designers more flexibilities to experiment with designs because these can be reprogrammed in seconds. The design deficiencies and modifications etc. can be carried out in short time thereby reducing the possibility of huge cost over-runs.

PLDs are also useful for prototyping ASIC designs since foundry produced ASICs may require months of costly development.

Because of various advantages of PLDs mentioned above, a large number of PLDs have been produced by IC manufacturers with variety of flexibilities and options available for a circuit designer and have become very popular. The architecture and various other features of PLDs such as ROMs, programmable logic arrays (PLAs), programmable array logic (PAL), simple programmable logic devices (SPLDs), complex programmable logic devices (CPLDs), and field-programmable gate arrays (FPGA) have been discussed in this chapter. The use of these devices require changes in the traditional design methods, although the basic concepts remain the same.

12.2 ROM AS A PLD

Read-only memories (ROMs) have been discussed in Section 11.6. A read-only memory is basically a combinational circuit and can be used to implement a logic function. A ROM of size $M \times N$ has M number of locations and N number of bits can be stored at each location. The number of address inputs is P, where $2^P = M$ and the number of data output lines is N. It can also be considered as a logic device with P inputs and N outputs as shown in Fig. 12.1.

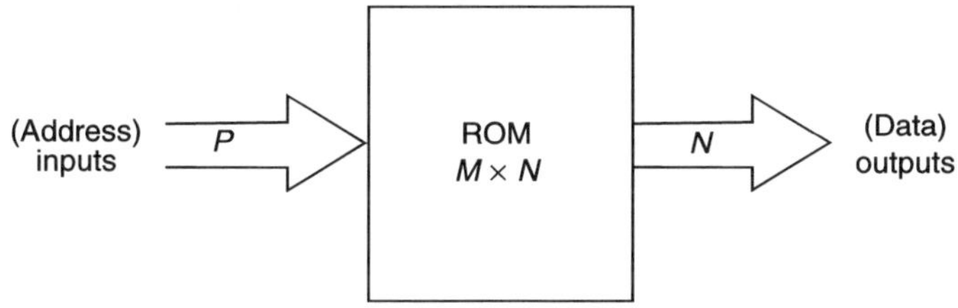

Fig. 12.1

ROM as a combinational circuit.

The 16-bit ROM array shown in Fig. 11.14 has four inputs and one output, i.e. $M = 16$, $N = 1$ and $P = 4$. The bit pattern stored, as given in Table 11.5, can be considered as truth table with $A_3\ A_2\ A_1\ A_0$ as 4-bit input and Y as the output which is same as the bit stored. The logic function corresponding to this is

$$Y = \sum m(0, 6, 9, 12, 13, 15)$$

In general, a P variable, N outputs logic function can be implemented using a ROM of size $2^P \times N$ since all the possible 2^P minterms are effectively generated as is clear from the above discussion.

If a mask programmable ROM is used, the user can specify the bit pattern to be stored according to the requirements of the logic function, whereas the user can himself program it in case of PROM, EPROM and E^2PROM. Since programmable ROMs can be used for logic design, therefore, it is also referred to as a programmable logic device (PLD).

The advantages of using ROM as a programmable logic device are:

1. Ease of design since no simplification or minimization of logic function is required.
2. Designs can be changed, modified rapidly.
3. It is usually faster than discrete SSI/MSI circuit.
4. Cost is reduced.

There are few disadvantages also of ROM-based circuits, such as non-utilization of complete circuit, increased power requirement, and enormous increase in size with increase in number of input variables making it impractical.

12.3 PROGRAMMABLE LOGIC ARRAY

A programmable logic device (PLD) usually consists of programmable array of logic gates and interconnections with array inputs and outputs connected to the device pins through fixed logic elements, such as inverting/non-inverting buffers and FLIP-FLOPs. The logic gates used may be two-level AND-OR, NAND-NAND or NOR-NOR configuration. In some cases, AND-OR-EX-OR configuration is also used. Basically, there are two types of PLDs, *programmable logic array* (PLA) and *programmable array logic* (PAL). These are suitable for implementing logic functions in SOP form.

A PLA consists of two-level AND-OR circuits on a single chip. The number of AND and OR gates and their inputs are fixed for a given PLA chip. The AND gates provide the product terms, and the OR gates logically sum these product terms and thereby generate a SOP expression. It has M inputs, n product terms, and N outputs with $n < 2^M$, and can be used to implement a logic function of M variables with N outputs. Since all of the possible 2^M minterms are not available, therefore, logic minimization is required to accommodate a given logic function.

The internal architecture of a PLA is shown in block diagram form in Fig. 12.2.

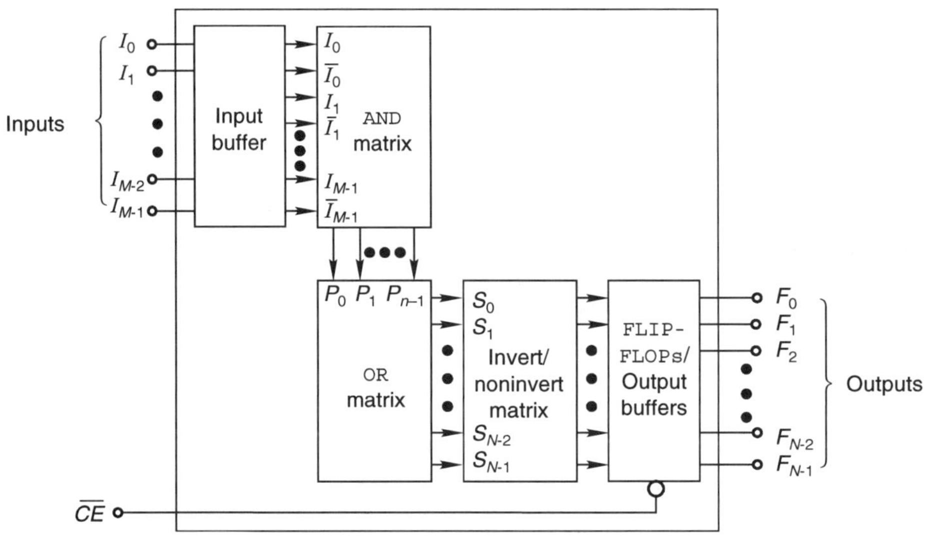

Fig. 12.2
Block diagram of a PLA device.

12.3.1 Input Buffer

The buffer circuits at the input are required to limit loading of the sources that drive the inputs. It produces inverted as well as non-inverted inputs at the output as shown in Fig. 12.3 for one input. Similar buffers are there for each of the M inputs.

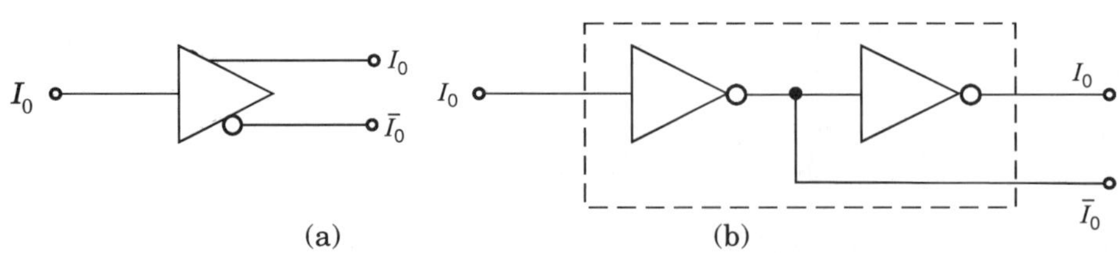

Fig. 12.3
An input buffer.

12.3.2 AND Matrix

An AND matrix is used to form product terms. A typical AND matrix is shown in Fig. 12.4. It has n AND gates with outputs P_0 through P_{n-1} and $2M$ inputs (I_0 through I_{M-1} and $\bar{I}_0$ through $\bar{I}_{M-1}$) for each AND gate. This shows that each AND gate has all the input variables in complemented and uncomplemented form. There is a nichrome fuse link in series with each diode. All the links are intact in an unprogrammed PLA device. Each AND gate generates one product term which is given by

$$P = I_0 \cdot \bar{I}_0 \cdot I_1 \cdot \bar{I}_1 \dots I_{M-1} \cdot \bar{I}_{M-1}$$

and it is logic 0 in an unprogrammed device. For generating a required product term the unwanted links are opened through the method of programming by a programmer device in case of a field-programmable logic device (FPLA). A section of the AND matrix for P_0 output is

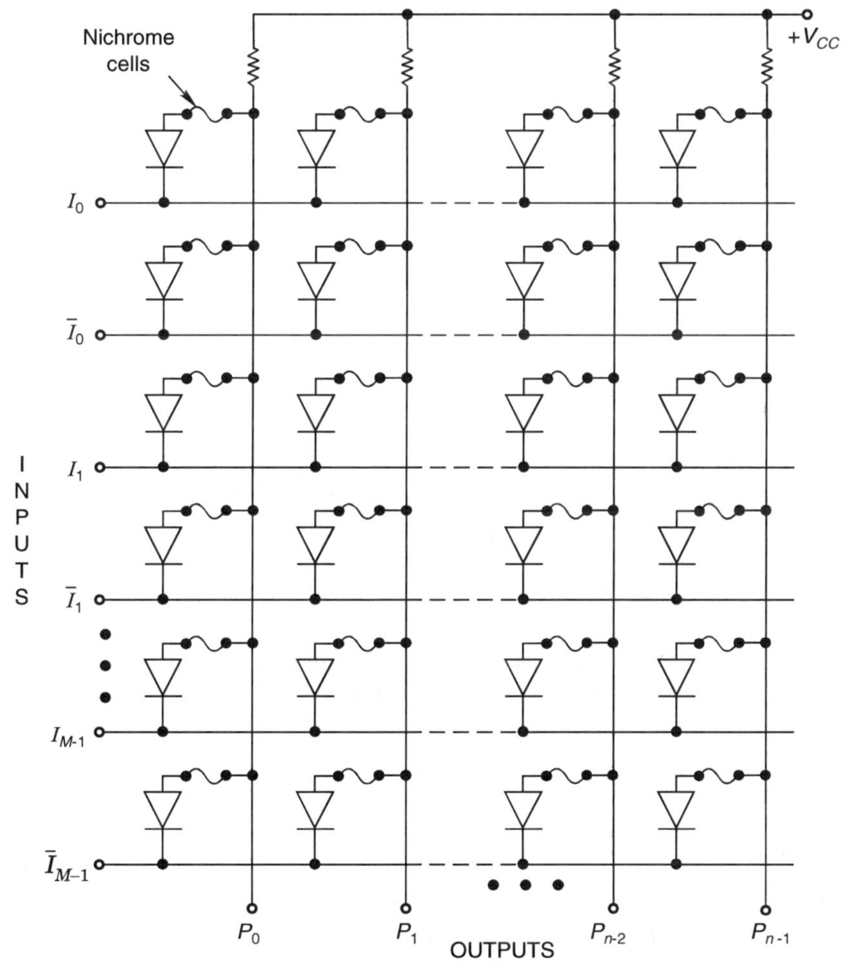

Fig. 12.4

An AND matrix.

shown in Fig. 12.5*a* and its equivalent logic gate representation is shown in Fig. 12.5*b*. A similar arrangement exists for each of the other outputs.

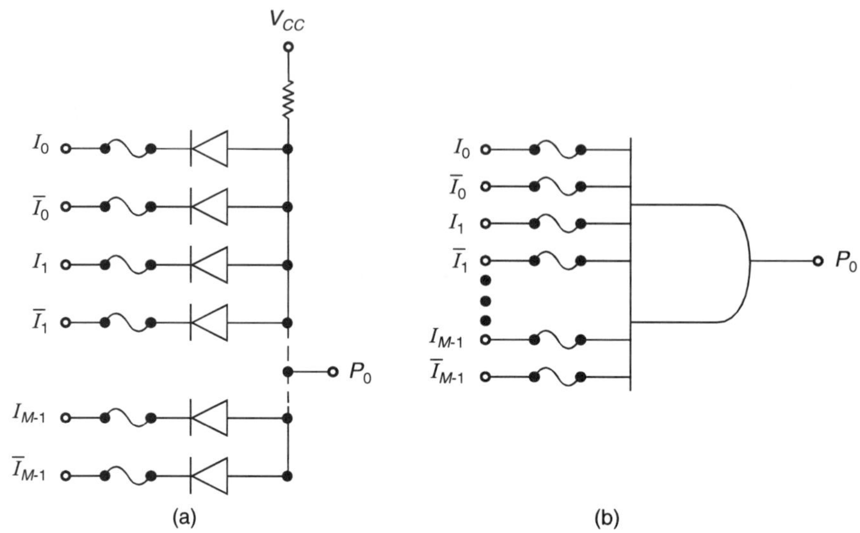

Fig. 12.5
A section of the AND matrix.

> **Example 12.1** In Fig. 12.5, if all the links except the ones present for inputs I_0 , $\overline{I}_2$, $\overline{I}_3$,and I_6 are opened, find the product term P_0.
>
> *Solution*
> For an open link, the input to the AND gate is logic 1, whereas for a closed link the corresponding input to the AND gate is same as the voltage applied at that input, therefore,
>
> $$P_0 = I_0 \cdot \overline{I}_2 \cdot \overline{I}_3 \cdot I_6$$

Figure 12.6 illustrates a convenient method of showing the input buffers and the AND matrix with the interconnections marked as *X*s. Each AND gate has 2*M* inputs which is indicated by a single line in the figure. When an array is programmed to implement a particular logic function, the desired interconnections are left with *X* marks and the unwanted interconnections without *X* marks.

12.3.3 OR Matrix

The OR matrix is used to produce the logical sum of the product term outputs of the AND matrix. Figure 12.7 shows an OR matrix using RTL circuitry. An OR gate consists of parallel connected transistors with a common emitter load. The outputs of the OR matrix are obtained at S_0 through S_{N-1}. Consider the output S_0 when all the fuse links are intact, which is given by

$$S_0 = P_0 + P_1 + \ldots + P_{n-1}$$

The required sum terms can be generated by opening the unwanted fuse links. For example,

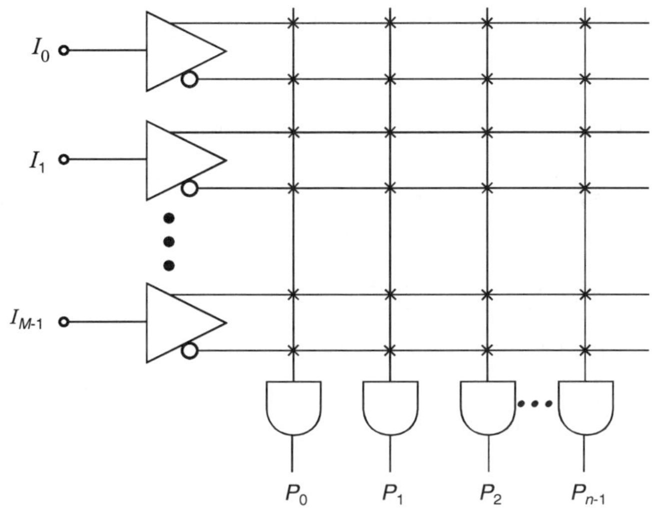

Fig. 12.6

Representation of input buffers and AND matrix.

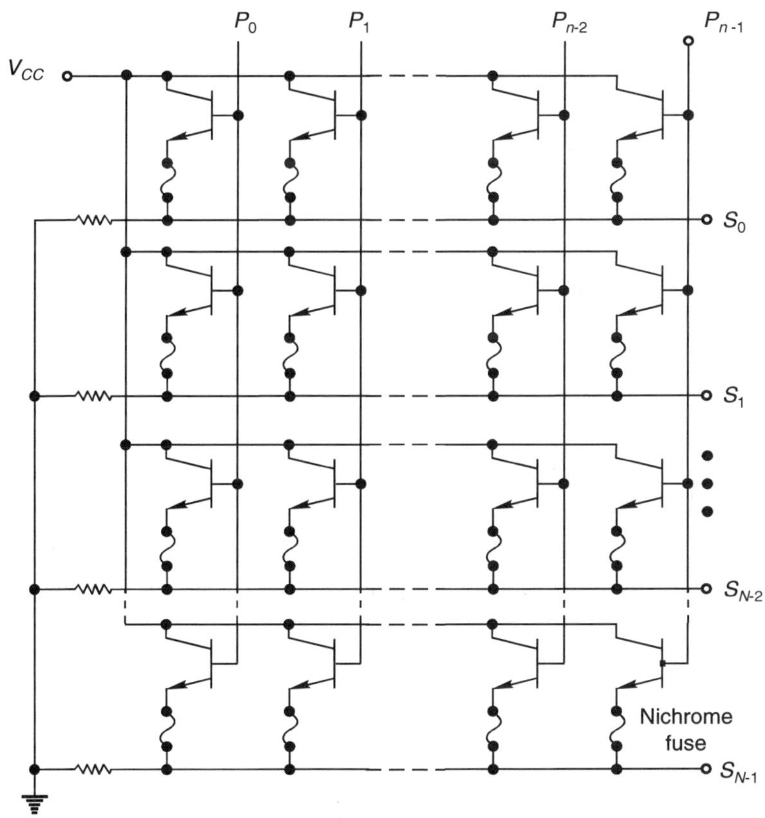

Fig. 12.7

An OR matrix.

if all the fuse links, except the ones for the product terms P_0 and P_1, are blown off for the output S_0, then

$$S_0 = P_0 + P_1$$

Thus an OR matrix can be programmed by opening the unwanted fuse links, which effectively makes logic level 0 at the corresponding OR gate inputs.

Figure 12.8 gives logic symbol for one of the OR gates, and Fig.12.9 illustrates the relevant portion of the PLA containing OR matrix in a way similar to Fig. 12.6.

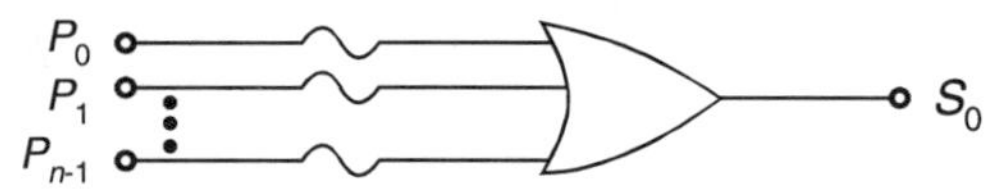

Fig. 12.8

Logic symbol of a section of OR matrix.

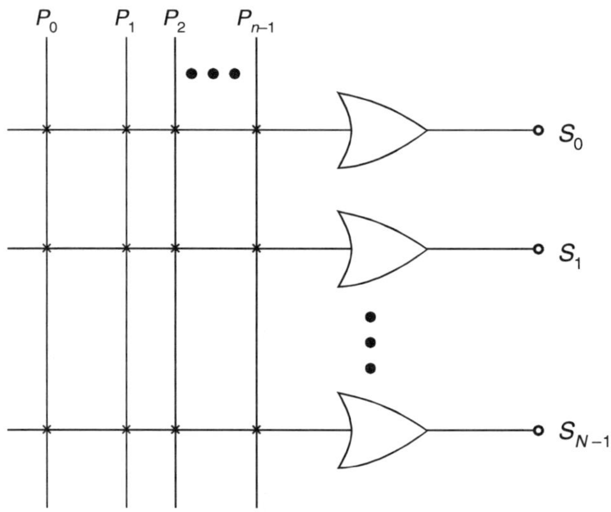

Fig. 12.9

Representation of OR matrix.

12.3.4 INVERT/NON-INVERT Matrix

This is a programmable buffer that can be set for INVERTING or NON-INVERTING operation corresponding to active-low or active-high output, respectively. Typical circuits for this operation are shown in Fig. 12.10.

In the case of an EX-OR gate, if the fuse is intact, the output is S, whereas the output is $\bar{S}$ if the fuse is blown off. Similarly, the output in case of Fig. 12.10*b* is S or $\bar{S}$ depending upon whether the fuse is intact or open.

12.3.5 Output Buffer

Output buffers are required to increase the driving capability of the PLA. Usually, the outputs are TTL compatible. The outputs may be totem-pole, open-collector, or three-state. Figure 12.11 shows the three-state output buffers.

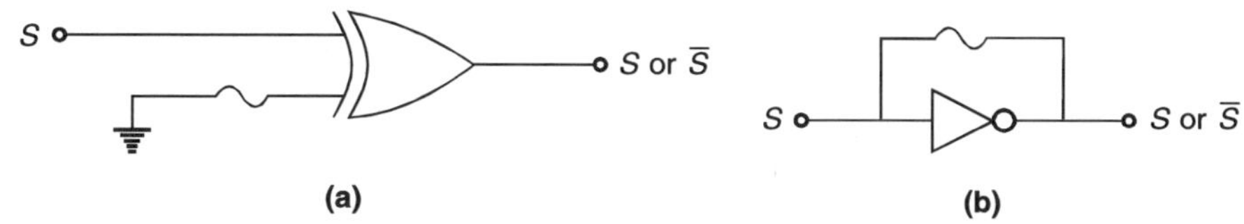

Fig. 12.10
Inverting/non-inverting circuits.

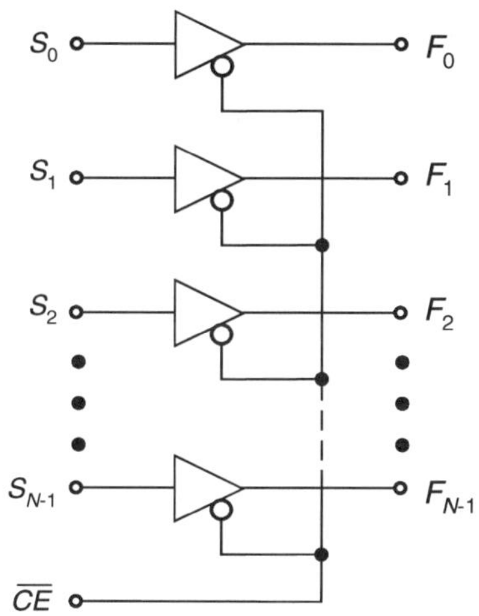

Fig. 12.11
Output buffers.

12.3.6 Output through FLIP-FLOPs and Buffers

The PLA devices suitable for state machine applications have FLIP-FLOPs and buffers in the output circuitry as shown in Fig. 12.12.

In this positive-edge triggered *S-R* FFs are connected at the outputs of the OR gates, and the device outputs are available through tristate buffers.

12.3.7 Programming the PLA

A PLA device is to be programmed for the desired input–output relationship similar to the programming of ROMs. For a mask-programmable PLA device, the data pattern is to be specified by the customer. The appropriate masks are designed by the manufacturers and the data pattern are built in during the manufacturing process.

An FPLA has all its nichrome fuse links intact at the time of manufacturing. The unwanted links are electrically open circuited during programming. The links to be opened are accessed by applying voltages at the inputs and outputs of the device, The FPLAs are not reprogrammable.

12.3.8 Expanding PLA Capacity

Some applications require a capacity that exceeds the capacity of a single PLA. The required capacity can be achieved by suitably connecting several identical devices together. For increas-

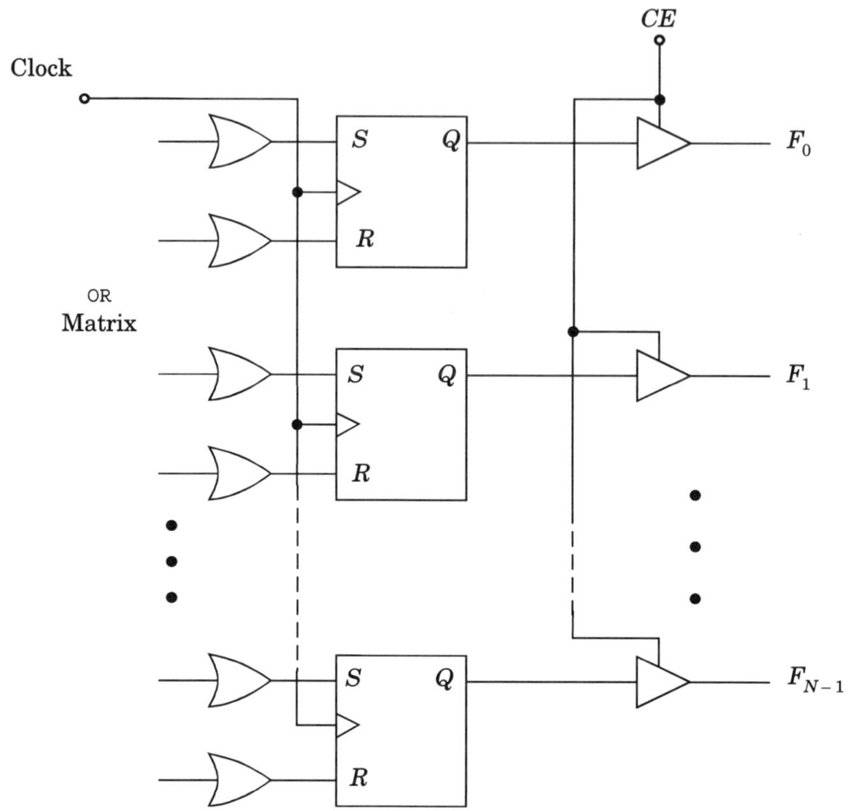

Fig. 12.12
A section of PLA with FLIP-FLOPs in the output.

ing the number of outputs, the inputs of two or more devices are to be connected individually in parallel. This connection does not change the number of inputs and the product terms.

For increasing the number of product terms keeping the number of inputs and outputs unchanged, the inputs and outputs of two or more devices are to be individually connected in parallel.

The number of inputs can be increased by making the connections as shown in Fig. 12.13. This circuit has $(M + Q)$ inputs and N outputs. This connection also increases the number of product terms. The number of product terms is $P \times (2^Q - 1)$. The outputs are allowed to be connected together for the devices with passive pull-up only.

12.3.9 Applications of PLAs

The PLAs can be used to implement combinational and sequential logic circuits. For the design of combinational circuits, PLAs with output circuit as shown in Fig. 12.10 or Fig. 12.11 are used, whereas the devices having FLIP-FLOPs in the output circuit as shown in Fig. 12.12 are required for the design of sequential circuits.

The following steps can be used for implementing combinational logic functions:

1. Prepare the truth table.
2. Write the Boolean equations in SOP form.

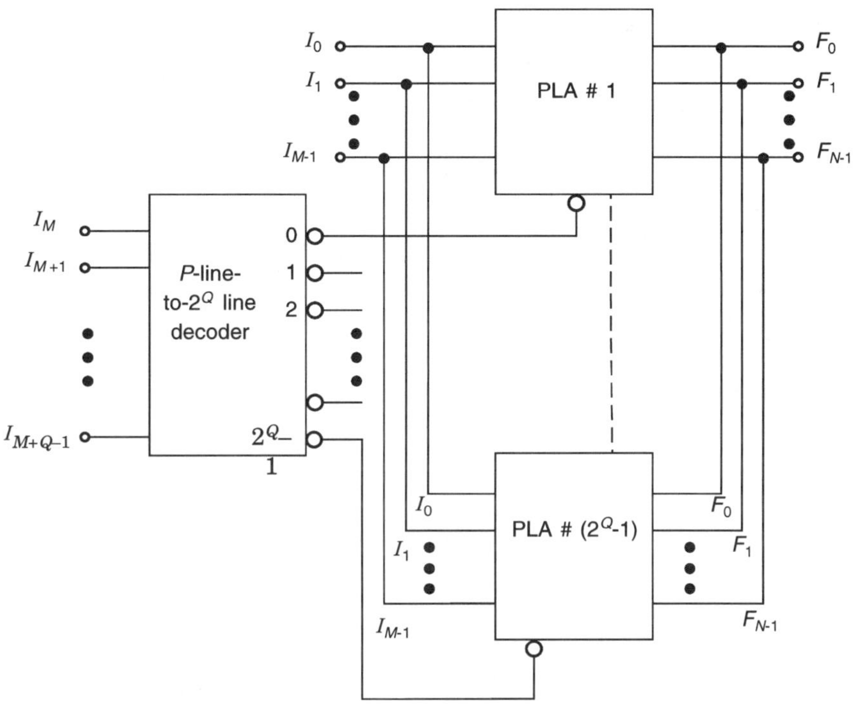

Fig. 12.13
Expanding the input word length of PLA.

3. Simplify the equations to obtain minimum SOP form. The main criterion is to minimize the number of product terms.
4. Determine the input connections of AND matrix to generate the required product terms.
5. Determine the input connections of OR matrix to generate the required sum terms.
6. Determine the connections required for INVERT/NON-INVERT matrix to set the active logic levels of the outputs.
7. Program the PLA.

Example 12.2 Design a 4-input, 5-output combinational circuit using PLS100 PLA. The input variables are *A,B,C,* and *D.*

$$Y_1 = \Sigma m(0, 3, 5, 6, 9, 10, 12, 15)$$
$$Y_2 = \Sigma m(0, 1, 2, 3, 11, 12, 14, 15)$$
$$Y_3 = \Sigma m(0, 4, 8, 12)$$
$$Y_4 = \Sigma m(0, 2, 3, 5, 7, 8, 12, 13)$$
$$Y_5 = \Sigma m(0, 1, 3, 4, 5, 6, 11, 13, 14, 15)$$

Solution

The PLS 100 PLA device (Fig. 12.14) has 16 inputs, 8 outputs and 48 product terms. The output is through EX-OR gate, for controlling the polarity of the output, and a 3-state buffer enabled through pin 19 of the chip. A low signal is required to be applied to this pin for

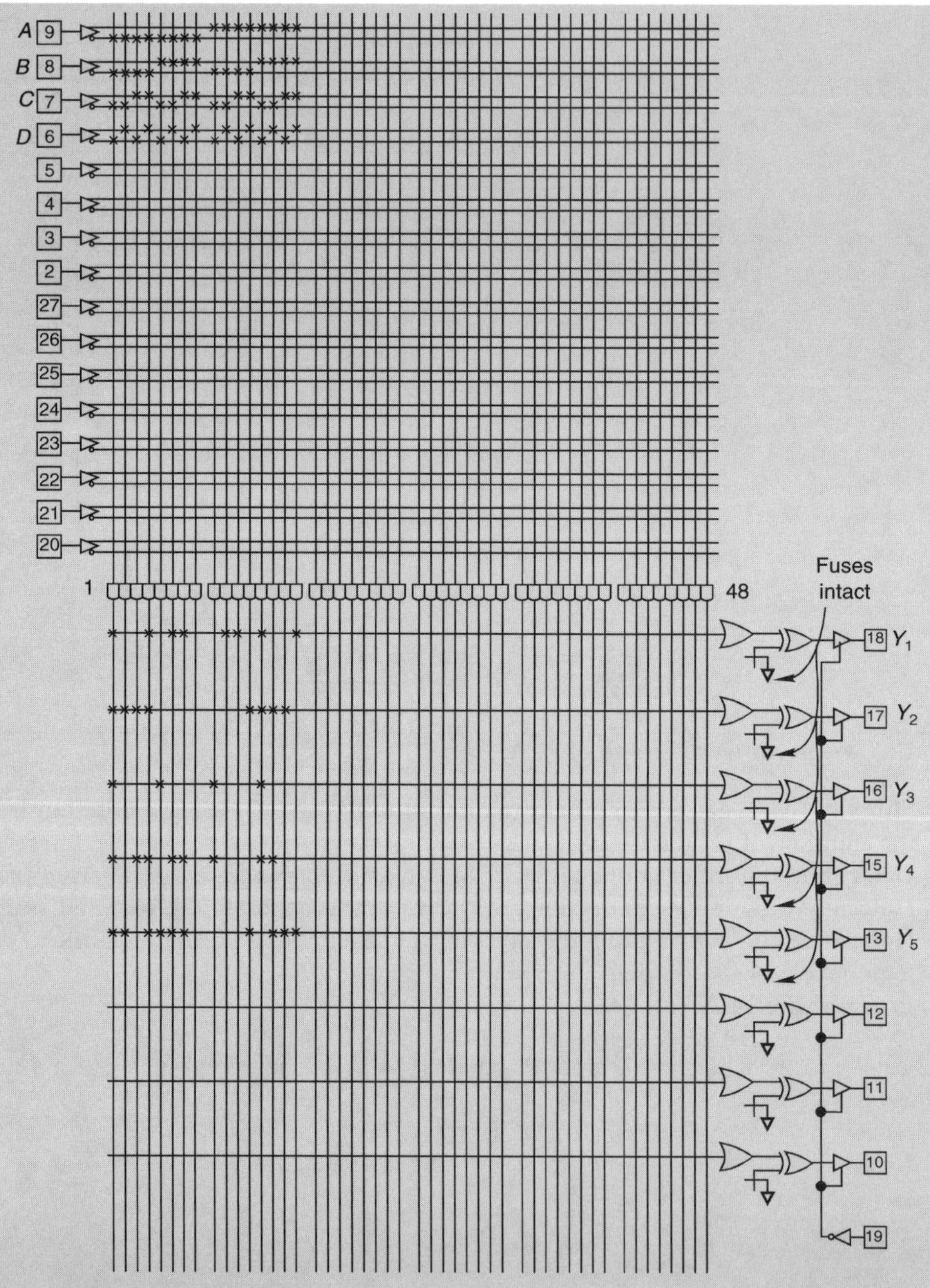

Fig. 12.14
PLS device programmed for Ex. 12.2.

enabling the output buffers. The fuses of EX-OR gates are to be kept intact to maintain these inputs at ground (0) level for active-high outputs.

The functions $Y_1 - Y_5$ are given in minterm form which are 16 in number. This requires only 16 product terms (or 16 AND gates) which are available in the device and, therefore, no simplification or minimisation is required.

The inputs *A, B, C,* and *D* are applied at pins 9, 8, 7 and 6, respectively, and the AND gates are numbered from 1 to 48 starting from the left most gate.

Gates 1 to 16 are used for generating 0 to 15 minterms. The relevant *X* are indicated in Fig. 12.14 in the rows corresponding to the inputs. The output pins used for the outputs Y_1 to Y_5 are 18, 17, 16, 15 and 13, respectively. The relevant *X* are indicated in the figure. One input of each of the output EX-OR gate is connected to ground through a fuse.

The design of sequential circuits follow the same method as discussed in chapter 8 and then the PLA is configured accordingly.

Example 12.3 Design a 4-bit UP/DOWN counter with $U/\overline{D}$ control.

Solution

The sequential PLS 105 PLA device can be used for this. It has 14 FLIP-FLOPs, 6 of which are buried within the device and are not available for outputs. The outputs of these 6 FFs are fedback to the AND array. The remaining 8 FFs are associated with the device outputs and are not used for feedback to the AND array. Pin 1 is used for clock input and 19 for providing preset input (*P*) to the FLIP-FLOPs and enable input to the output buffers.

Table 12.1 gives the count sequence and the inputs required for the FFs. The K-maps for S_3, R_3, S_2, R_2, S_1, R_1, S_0, and R_0 are given in Fig. 12.15, and the expressions for these inputs in minimized form are also given.

The device is programmed using these equations and is shown in Fig. 12.16.

Table 12.1

UP/DOWN	Counter state				S-R FFs inputs							
Control $U/\overline{D}$	Q_3	Q_2	Q_1	Q_0	S_3	R_3	S_2	R_2	S_1	R_1	S_0	R_0
0	0	0	0	0	0	×	0	×	0	×	1	0
0	0	0	0	1	0	×	0	×	1	0	0	1
0	0	0	1	0	0	×	0	×	×	0	1	0
0	0	0	1	1	0	×	1	0	0	1	0	1
0	0	1	0	0	0	×	×	0	0	×	1	0
0	0	1	0	1	0	×	×	0	1	0	0	1
0	0	1	1	0	0	×	×	0	×	0	1	0
0	0	1	1	1	1	0	0	1	0	1	0	1
0	1	0	0	0	×	0	0	×	0	×	1	0
0	1	0	0	1	×	0	0	×	1	0	0	1
0	1	0	1	0	×	0	0	×	×	0	1	0
0	1	0	1	1	×	0	1	0	0	1	0	1
0	1	1	0	0	×	0	×	0	0	×	1	0
0	1	1	0	1	×	0	×	0	1	0	0	1
0	1	1	1	0	×	0	×	0	×	0	1	0

(*Contd.*)

$U/\bar{D} = 0$

Q_1Q_0 \ Q_3Q_2	00	01	11	10
00	0	0	×	×
01	0	0	×	×
11	0	1	0	×
10	0	0	×	×

$U/\bar{D} = 1$

Q_1Q_0 \ Q_3Q_2	00	01	11	10
00	1	0	×	0
01	0	0	×	×
11	0	0	×	×
10	0	0	×	×

$$S_3 = \bar{Q}_3 \cdot Q_2 \cdot Q_1 \cdot Q_0 \cdot (\overline{U/\bar{D}}) + \bar{Q}_3 \cdot \bar{Q}_2 \cdot \bar{Q}_1 \cdot \bar{Q}_0 \cdot (U/\bar{D})$$

(*a*) K-map for S_3

Q_1Q_0 \ Q_3Q_2	00	01	11	10
00	×	×	0	0
01	×	×	0	0
11	×	0	1	0
10	×	×	0	0

Q_1Q_0 \ Q_3Q_2	00	01	11	10
00	0	×	0	1
01	×	×	0	0
11	×	×	0	0
10	×	×	0	0

$$R_3 = Q_3 \cdot Q_2 \cdot Q_1 \cdot Q_0 \cdot (\overline{U/\bar{D}}) + Q_3 \cdot \bar{Q}_2 \cdot \bar{Q}_1 \cdot \bar{Q}_0 \cdot (U/\bar{D})$$

(*b*) K-map for R_3

Q_1Q_0 \ Q_3Q_2	00	01	11	10
00	0	×	×	0
01	0	×	×	0
11	1	0	0	1
10	0	×	×	0

Q_1Q_0 \ Q_3Q_2	00	01	11	10
00	1	0	0	1
01	0	×	×	0
11	0	×	×	0
10	0	×	×	0

$$S_2 = \bar{Q}_2 \cdot Q_1 \cdot Q_0 \cdot (\overline{U/\bar{D}}) + \bar{Q}_2 \cdot \bar{Q}_1 \cdot \bar{Q}_0 \cdot (U/\bar{D})$$

(*c*) K-map for S_2

Q_1Q_0 \ Q_3Q_2	00	01	11	10
00	×	0	0	×
01	×	0	0	×
11	0	1	1	0
10	×	0	0	×

Q_1Q_0 \ Q_3Q_2	00	01	11	10
00	0	1	1	0
01	×	0	0	×
11	×	0	0	×
10	×	0	0	×

$$R_2 = \bar{Q}_2 \cdot Q_1 \cdot Q_0 \cdot (\overline{U/\bar{D}}) + Q_2 \cdot \bar{Q}_1 \cdot \bar{Q}_0 \cdot (U/\bar{D})$$

(*d*) K-map for R_2

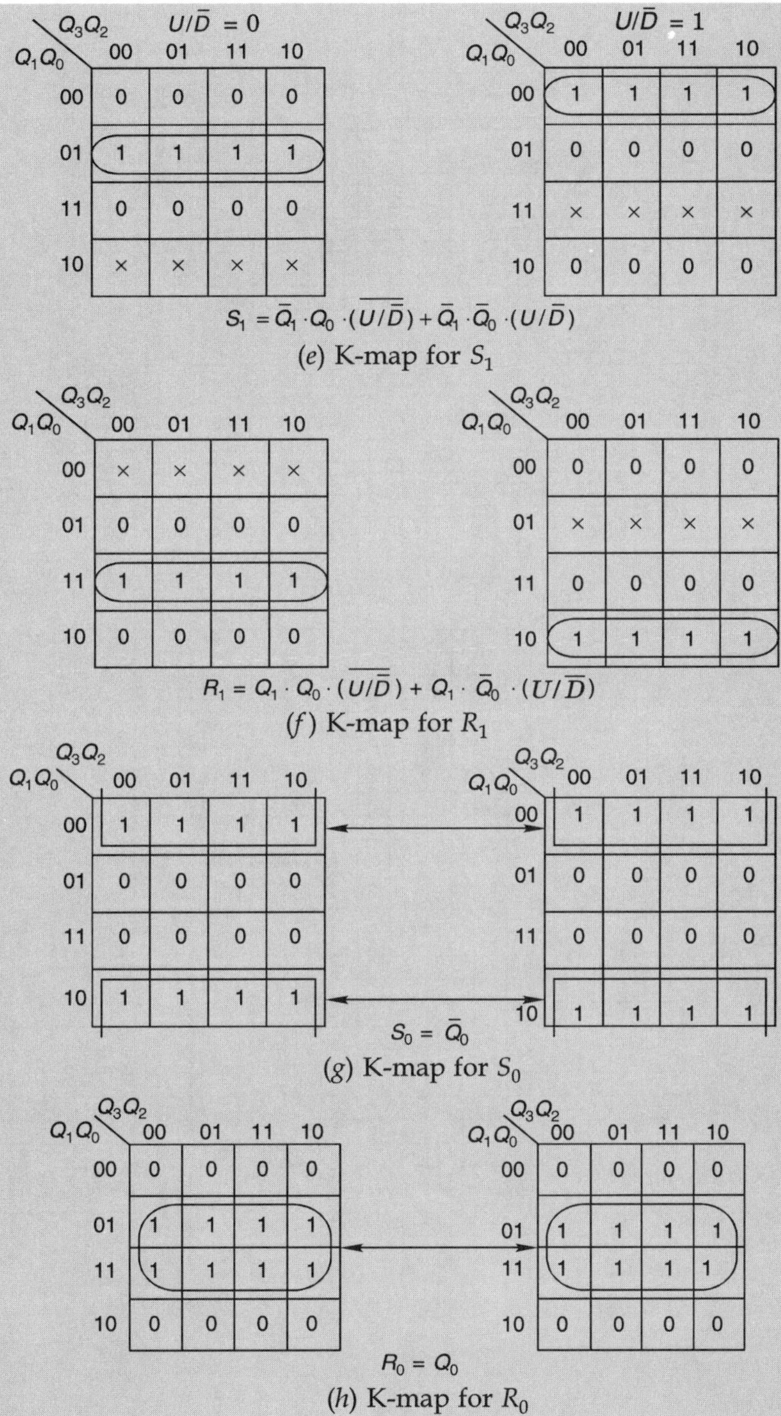

$S_1 = \bar{Q}_1 \cdot Q_0 \cdot (\overline{U/\bar{D}}) + \bar{Q}_1 \cdot \bar{Q}_0 \cdot (U/\bar{D})$

(*e*) K-map for S_1

$R_1 = Q_1 \cdot Q_0 \cdot (\overline{U/\bar{D}}) + Q_1 \cdot \bar{Q}_0 \cdot (U/\bar{D})$

(*f*) K-map for R_1

$S_0 = \bar{Q}_0$

(*g*) K-map for S_0

$R_0 = Q_0$

(*h*) K-map for R_0

Fig. 12.15

K-maps for Ex. 12.3.

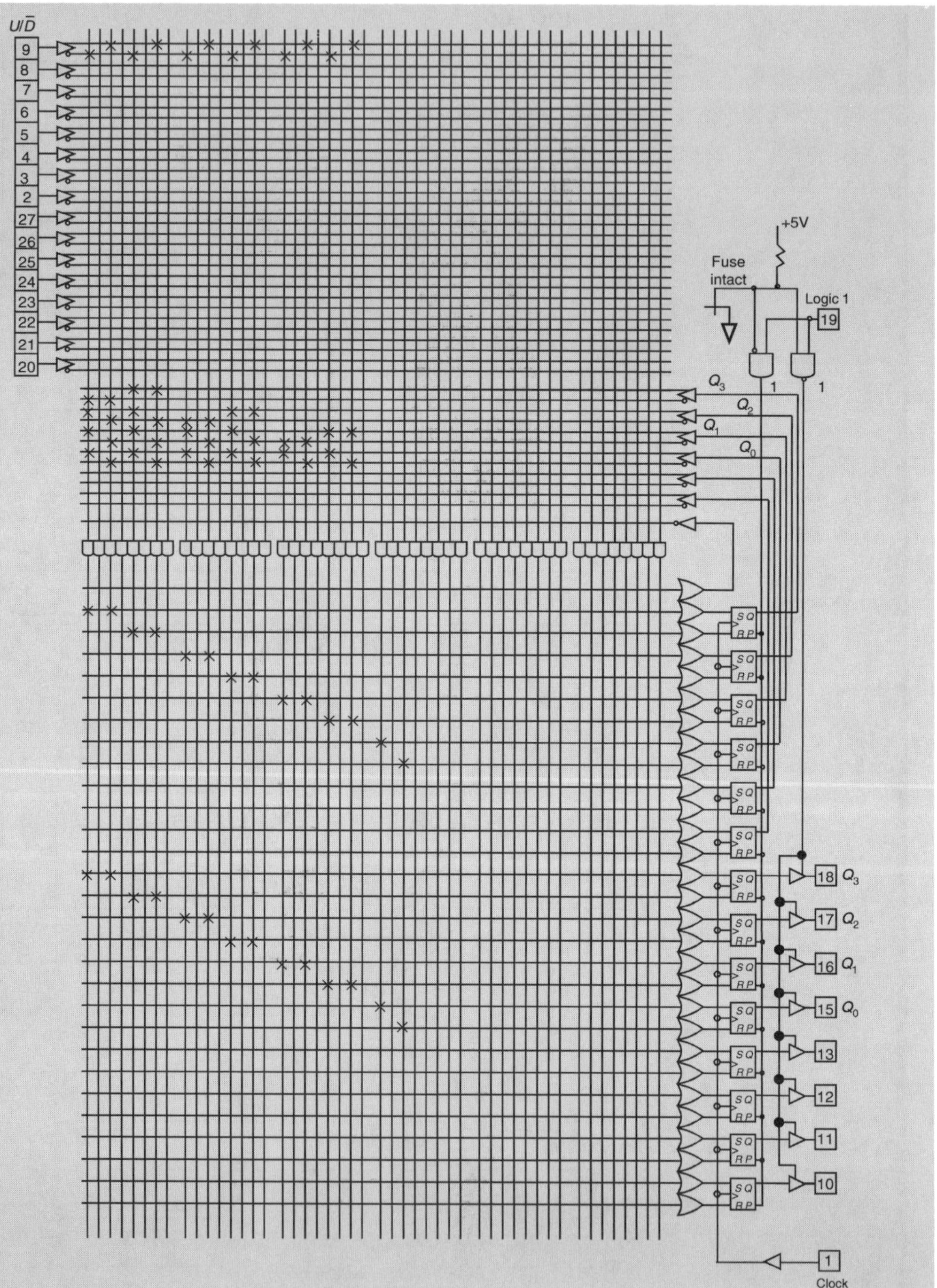

Fig. 12.16

PLS 105 PLA Device Programmed for Ex. 12.3.

Table 12.1 *(Contd.)*

UP/DOWN	Counter state				S-R FFs inputs							
Control $U/\overline{D}$	Q_3	Q_2	Q_1	Q_0	S_3	R_3	S_2	R_2	S_1	R_1	S_0	R_0
0	1	1	1	1	0	1	0	1	0	1	0	1
1	0	0	0	0	1	0	1	0	1	0	1	0
1	1	1	1	1	×	0	×	0	×	0	0	1
1	1	1	1	0	×	0	×	0	0	1	1	0
1	1	1	0	1	×	0	×	0	0	×	0	1
1	1	1	0	0	×	0	0	1	1	0	1	0
1	1	0	1	1	×	0	0	×	×	0	0	1
1	1	0	1	0	×	0	0	×	0	1	1	0
1	1	0	0	1	×	0	0	×	0	×	0	1
1	1	0	0	0	0	1	1	0	1	0	1	0
1	0	1	1	1	0	×	×	0	×	0	0	1
1	0	1	1	0	0	×	×	0	0	1	1	0
1	0	1	0	1	0	×	×	0	0	×	0	1
1	0	1	0	0	0	×	0	1	1	0	1	0
1	0	0	1	1	0	×	0	×	×	0	0	1
1	0	0	1	0	0	×	0	×	0	1	1	0
1	0	0	0	1	0	×	0	×	0	×	0	1
	0	0	0	0								

12.3.10 Available PLAs

Some of the commercially available PLA ICs with their features are given in Table 12.2. The 82S200 and 82S201 are pin-for-pin mask programmable replacements for the 82S100 and 82S101, respectively. The PLS 100 and PLS 105 PLAs are equivalent to 82S100 and 82S105, respectively. The DM 7575 PLA has totem-pole output, whereas DM7576 and IM5200 have passive pull-up. The devices with passive pull-up are useful for expanding functions by wire-ANDing the outputs of similar other devices.

12.4 PROGRAMMABLE ARRAY LOGIC

A most commonly used type of PLD is *programmable array logic* (PAL). It is programmable array of logic gates on a single chip in AND-OR configuration. In contrast to PLA, it has programmable AND array and a fixed OR array in which each OR gate gets inputs from some of the AND gates, i.e. all the AND gate outputs are not connected to any OR gate. Figure 12.17 illustrates the configuration of AND and OR arrays for a PAL with 5 inputs, 8 programmable AND gates and 4 fixed OR gates. Each AND gate has all the 10 inputs (in complemented and uncomplemented form) with fusible links intact which can be programmed to generate 8 product terms. Each OR gate gets inputs from the outputs of only two AND gates shown by •. The input and output circuits of PALs are similar to those of PLAs. The number of fusible links in a PAL is the product of $2M$ and n, where M is the number of input variables and n is the number of product terms.

Table 12.2 Some available PLAs with their features

IC No.	PLA/ FPLA	Number of			Output	Access time ns	Supply voltage V	Power dissipation mW	Packaging	Input/output logic levels
		Inputs	Product-terms	Outputs						
82S200	PLA	16	48	8	*TS*	80	+5	600	28-pin DIP	TTL compatible
82S201	PLA	16	48	8	*OC*	80	+5	600	28-pin DIP	TTL compatible
82S100	FPLA	16	48	8	*TS*	80	+5	600	28-pin DIP	TTL compatible
82S101	FPLA	16	48	8	*OC*	80	+5	600	28-pin DIP	TTL compatible
DM7575	PLA	14	96	8	*TS*	150	+5	550	24-pin DIP	TTL compatible
DM7576	PLA	14	96	8	*OC*	150	+5	550	24-pin DIP	TTL compatible

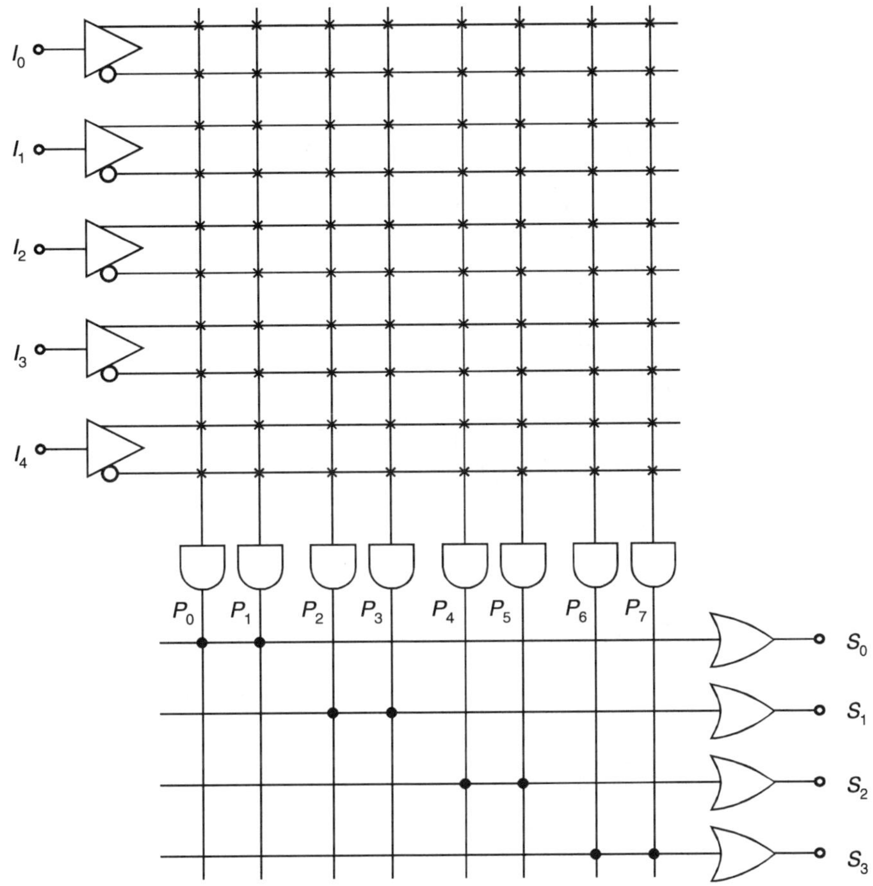

Fig. 12.17
Programmable array logic.

One of the most commonly used PAL 16L8 is shown in Fig. 12.18. Its programmable AND array consists of 64 gates. It has 16 input variables and 8 outputs, therefore, each AND gate has $2 \times 16 = 32$ inputs. Eight AND gates are associated with each output pin, seven of them provide inputs to a fixed 7-input OR gate and the eighth is connected to the enable input terminal of the output buffer. Thus, any output can perform only logic functions that can be expressed as sums of seven or fewer product terms. Each product term can be a function of up to all 16 inputs but only 7 such product terms are available. Since, in a PAL the inputs to the OR gates are fixed, therefore, no two OR gates can share a product term. Where a product term is needed by two OR gates, it must be generated twice.

There is a three-state inverter between the output of each OR gate and the output pin of the device, therefore, the output may be programmed as always enabled, always disabled, or enabled by a product term involving the inputs (Problem 12.11).

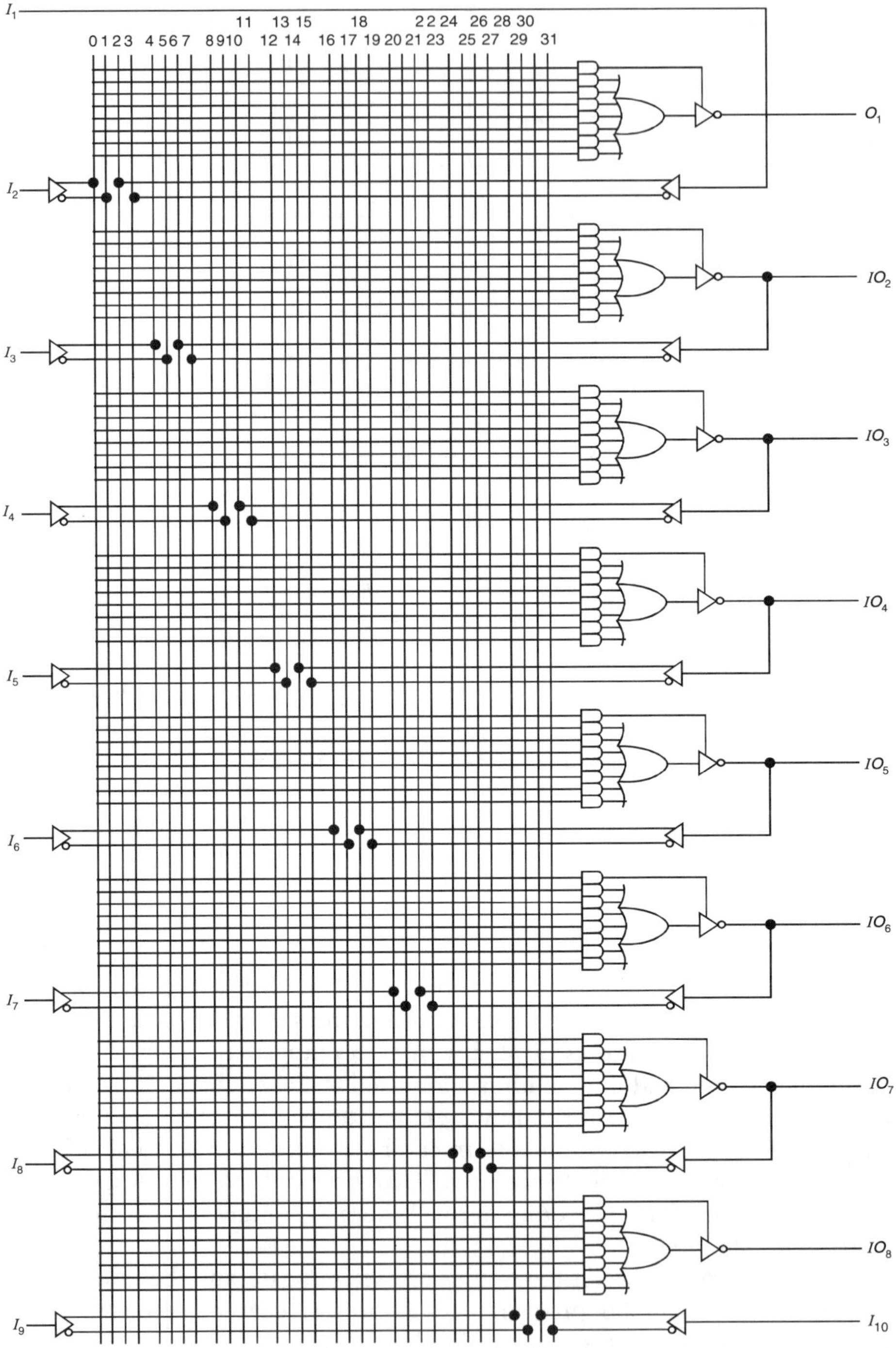

Fig. 12.18

Logic diagram of the PAL 16L8.

This PAL has 10 pins (I_1 to I_{10}) for 10 inputs, the other six pins are used for inputs as well as for six of the outputs. These are IO_2 to IO_7. O_1 and O_8 are dedicated output pins. The device can be programmed for the following options:

1. If an *IO* pin's output inverter is disabled, then the pin can be used only as an input.
2. If an *IO* pin is not required to be input, then it can be used as an output. The output inverter can be enabled either continuously or for certain input conditions by using a product term. In case it is enabled continuously, its output can also be used as an input.
3. This can be used either for the implementation of logic functions that can not be accommodated in 7 product terms by using two pass AND-OR configuration or for implementing sequential circuits by providing feedback.

Since limited number of product terms are available for each output, logic minimization techniques are required when logic circuits are implemented in PAL devices.

PAL 16H8 is same as 16L8 except that it does not have inverters in the output. Another PAL 16P8 has programmable output polarity using EX-OR gates as shown in Fig. 12.10*a*. The alphabets *L,H,* and *P* used in device numbers stand for active-low, active-high, and programmable outputs, respectively, and these are all *combinational* programmable array logic devices, i.e. these do not have memory elements.

12.4.1 Registered PALs

Sequential digital circuits use FLIP-FLOPs in addition to combinational circuits and, therefore, for the design of sequential circuits, PALs have been developed with FLIP-FLOPs in the outputs and these devices are known as *registered* PALs. Some of the available registered PALs are 16R4, 16R6, 16R8, etc.

Figure 12.19 illustrates the structure of a 16R6 registered PAL. This device has the following features:

- Eight primary (external) inputs, I_1 to I_8.
- Two pins common for input/output similar to the device 16L8, IO_1 and IO_8.
- Six outputs (O_2 to O_7) through positive-edge-triggered D-FFs and three state inverters. $\bar{Q}$ outputs of these FFs are routed back to the AND array, making total number of inputs to the AND array as 16 and 8 outputs.
- The FFs are all controlled by a common clock through a pin and another dedicated input pin is available for output enable control of inverters.

12.4.2 Configurable PALs

Developments in the design of programmable array logic devices have led to the introduction of configurable outputs enhancing the output capabilities of such devices. The *configurable* (also known as *generic*) device architecture is provided by equipping the device with enhanced special circuitry, known as *output macrocells*. A macrocell has circuitry with fuses which can be configured for a variety of output options, giving flexibility to the device. A configurable PAL can replace a large number of simpler PAL type devices with a 'one size fits all' device and allows designs to be implemented that are challenging or simply impossible for the simpler PAL devices to handle.

Figure 12.20 gives logic diagram of 22V10 configurable PAL. It has 22 inputs, 10 outputs, and 120 product terms. Out of 22 inputs, 12 are dedicated inputs (I_1 to I_{12}) and ten may be used

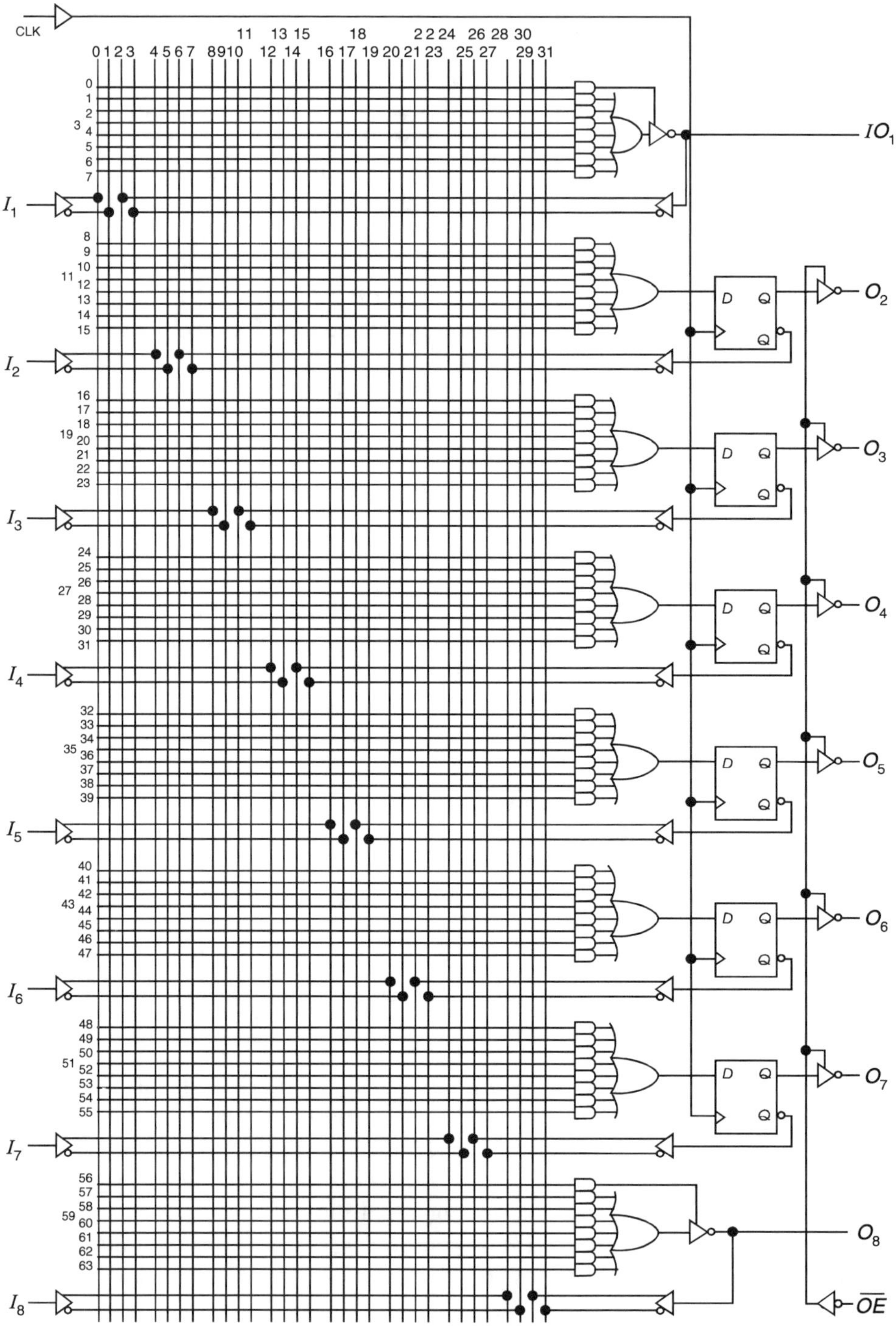

Fig. 12.19

Logic diagram of the Registered PAL 16R6.

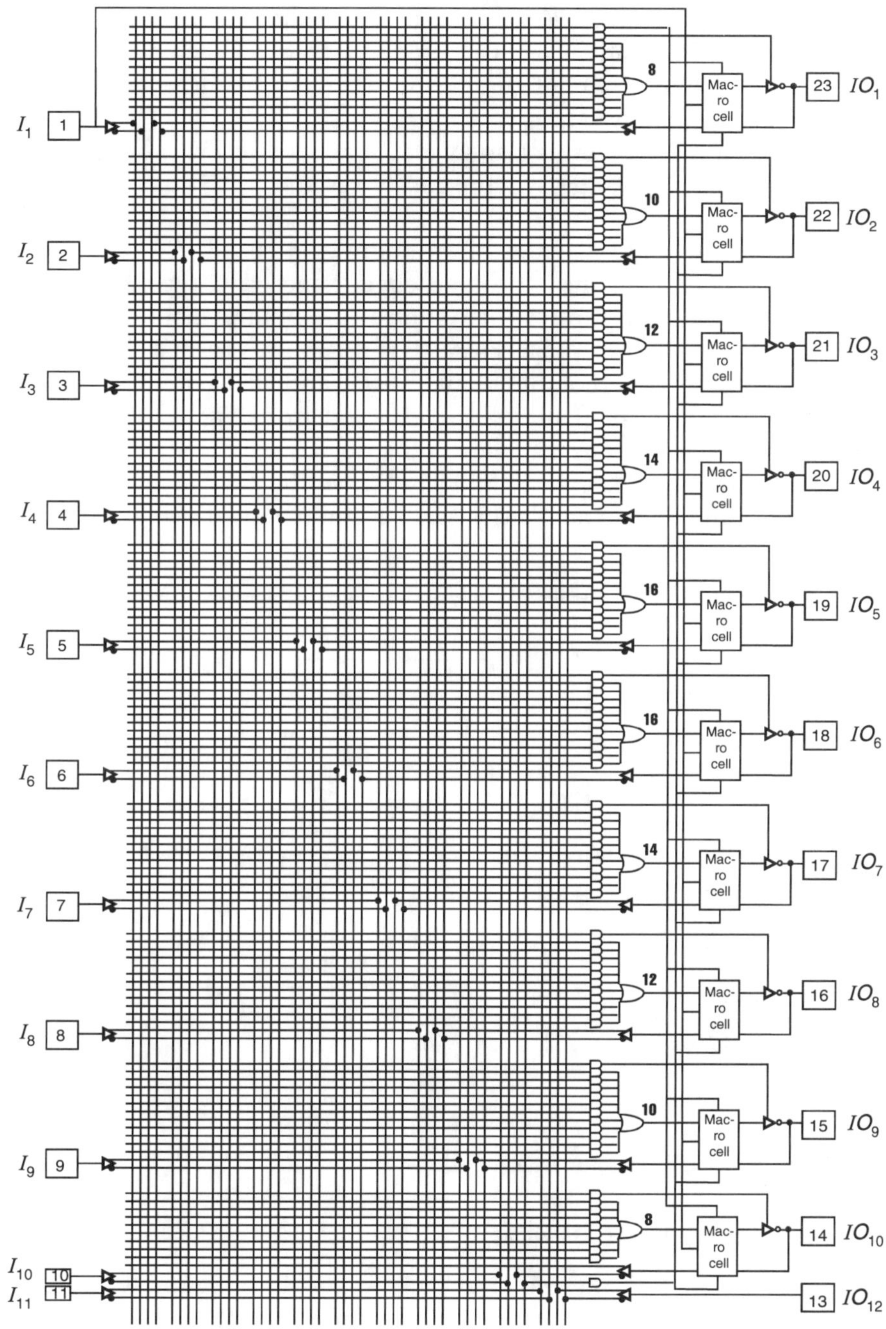

Fig. 12.20

22V10 Configurable PAL.

for inputs as well as outputs (IO_1–IO_{10}). One of the dedicated inputs I_1 also functions as the common clock input to the positive-edge-triggered D-type FF of each of the ten output macrocells. All the ten outputs are through configurable macrocells and three-state inverters. The output pins can also be used as inputs.

There are two extra product terms also available as seen in Fig. 12.20 which can be used for synchronously presetting (SP) and asynchronously resetting (AR) the D-type FFs in the output macrocells.

Figure 12.21 gives the logic diagram of an output macrocell. All the output macrocells are identical. Each macrocell contains a positive edge-triggered D-type FF and fuse-configurable multiplexers. The two fuses, S_1 and S_0 that control the multiplexers can be configured in four different ways, as shown in Fig. 12.22. From this, it is observed that this configurable PAL can function as registered PAL or combinational PAL, and in each case the output may be in inverted or non-inverted form. It may also be noted from Fig. 12.20 that all the number of product terms available to various OR gates in the device are not same. The number of product terms associated with each OR gate is indicated near the OR gate. Because of this, the logic functions of significantly more complexity can be implemented using 22V10.

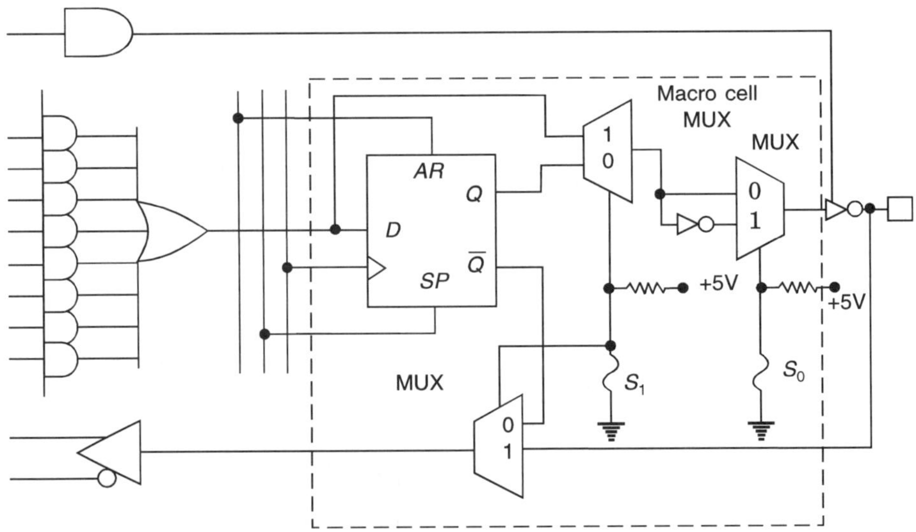

Fig. 12.21
Logic diagram of a 22V10's output macrocell.

12.4.3 Generic Array Logic Devices

The *Generic array logic* (GAL) device is another type of configurable PAL. GAL devices are intended as pin-for-pin replacements for a wide variety of PAL devices. It is designed to be compatible, all the way to the fuse level, for any simpler PAL which can be directly implemented in the GAL device. The GAL 16V8 replaces most 20-pin PAL devices, while the 20V8

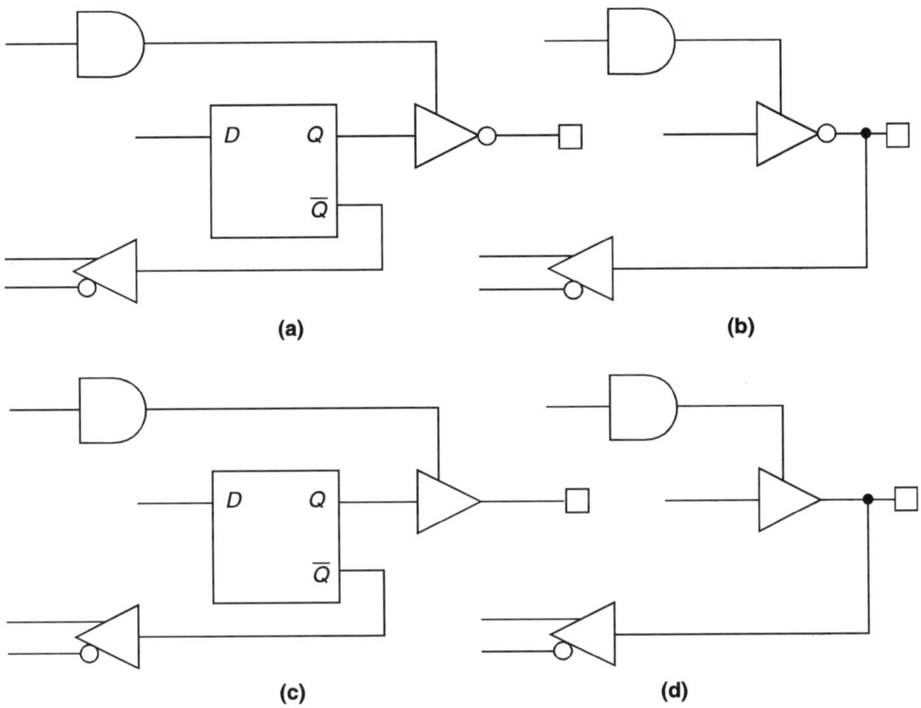

Fig. 12.22

Output configurations for fuses (*a*) S_1 and S_0 intact (*b*) S_0 intact and S_1 open (*c*) S_0 open and S_1 intact (*d*) S_1 and S_0 open.

replaces most 24-pin PAL devices. Figure 12.23 shows 16V8 GAL device. Its macrocell can be configured in one of the three basic configurations. In this device, the OR gate is considered to be a part of the macrocell to obtain the various types of I/O configurations found in the PAL devices that the GAL is designed to replace. The three modes of 16V8 GAL's output macrocell are shown in Fig. 12.24.

Another family of devices that are intended as PAL replacements are programmable electrically erasable logic (PEEL) devices. Figure 12.25 shows 18CV8 PEEL device. Its output macrocell shown in Fig. 12.26 and can be configured in twelve different configurations (Problem 12.10).

12.4.4 EX-OR PALs

The use of EX-OR gates in PALs to implement fuse configurable output polarity has already been discussed. EX-OR gates are also used in AND-OR-EX-OR configuration as shown in Fig. 12.27 in the EX-OR PAL device. Here, the EX-OR outputs are fed to the inputs of the FLIP-FLOPs. Each of these EX-OR gates is fed in turn by two sum-of-products arrays of two product terms each. This configuration is used to reduce the amount of logic required for many

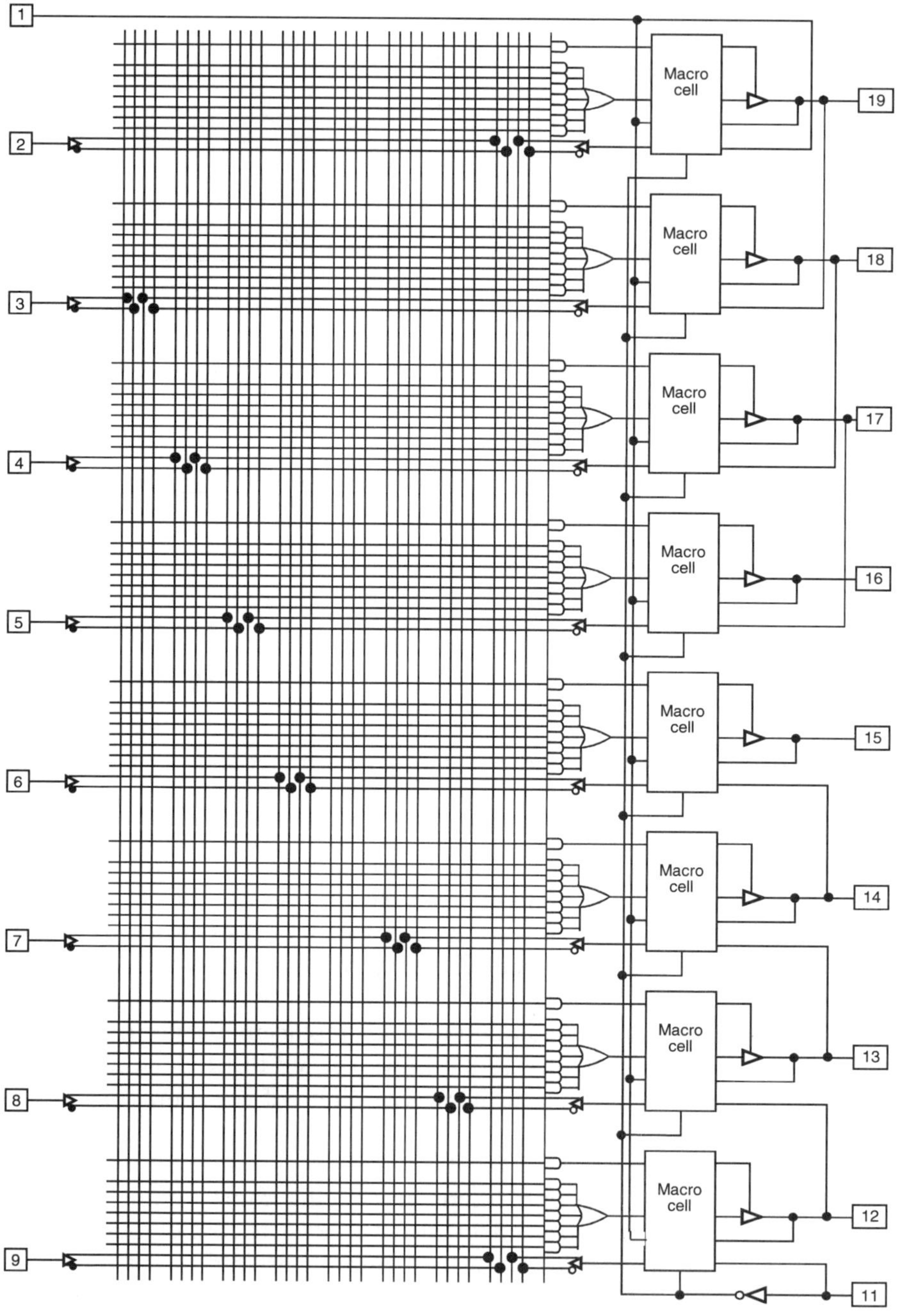

Fig. 12.23

16V8 GAL device.

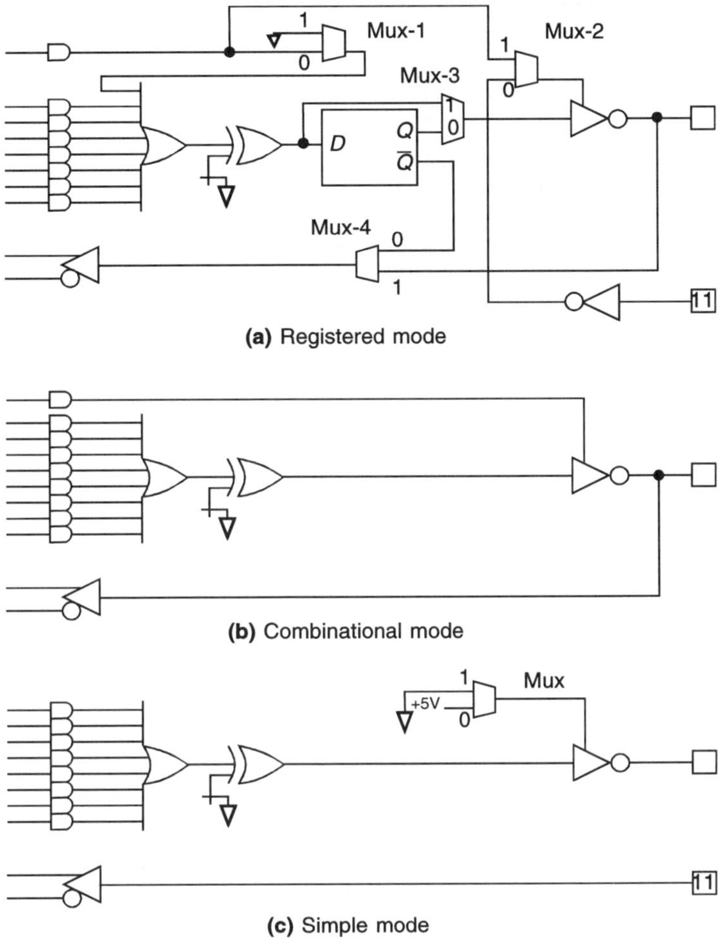

Fig. 12.24

Output Macrocell details of 16V8 GAL.

applications, particularly counters which allow complex designs to be implemented using very few product terms.

PALs are also available which have independently clocked FFs useful for many applications, such as asynchronous state machine applications. In these devices, D-type of FFs are controlled by product terms including their asynchronous preset and clear operations.

More and more powerful output macrocells are being introduced in configurable PALs to extend their capabilities.

12.4.5 Available PALs

Some of the commercially available PAL devices with their features are given in Table 12.3.

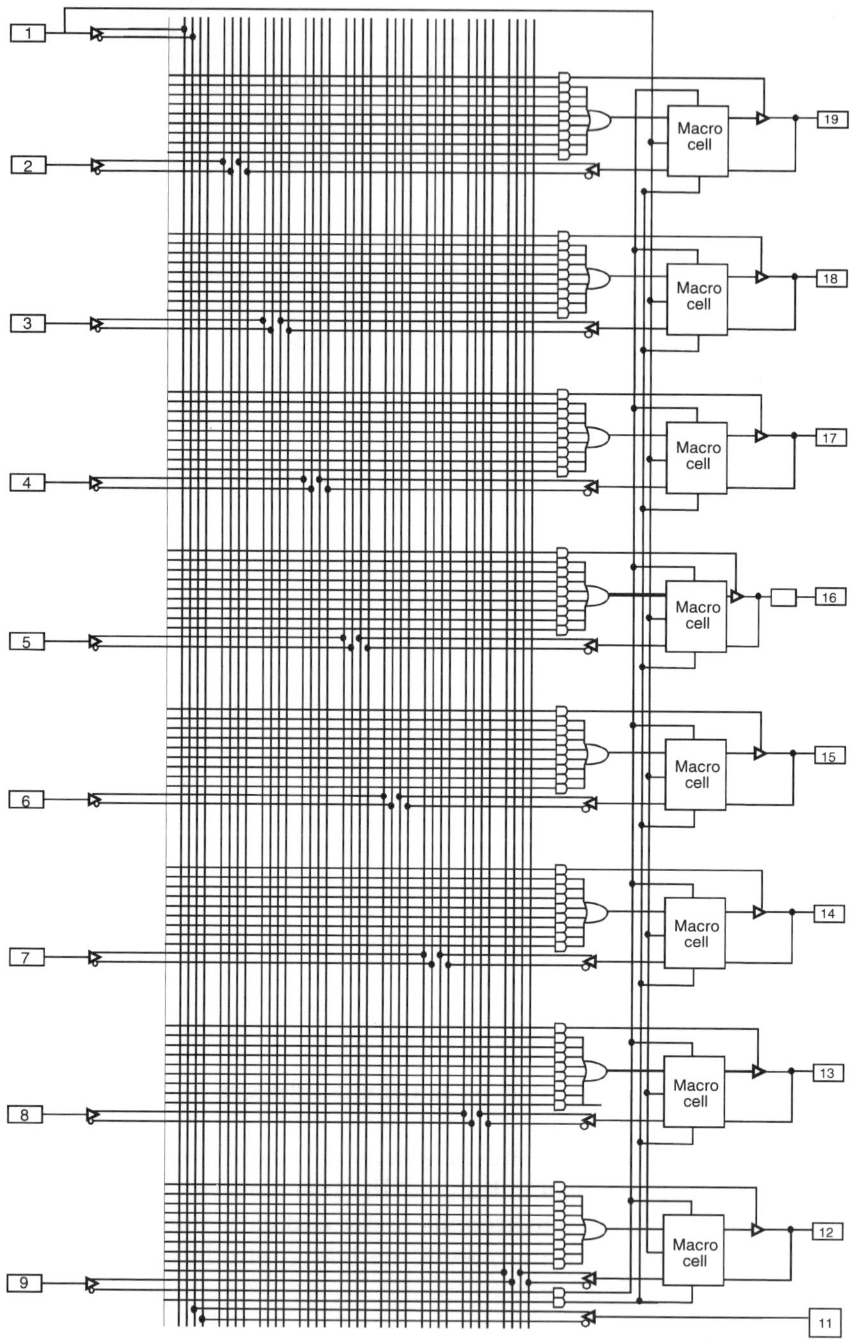

Fig. 12.25
18CV8 PEEL device.

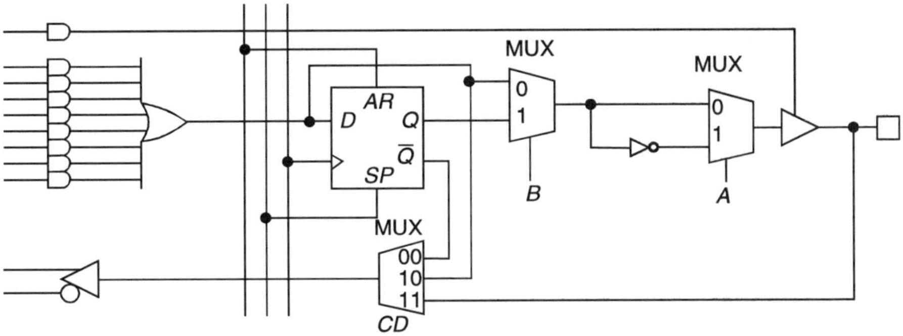

Fig. 12.26
Macrocell details of 18CV8.

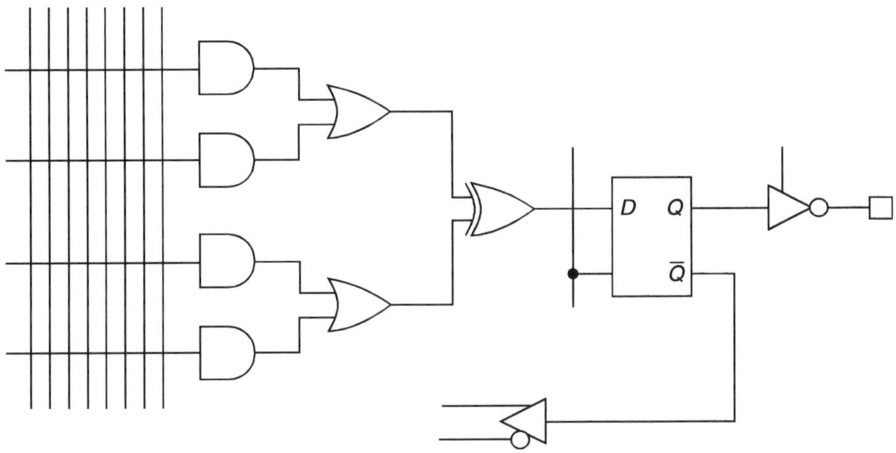

Fig. 12.27
EX-OR PAL.

Table 12.3 Some available PAL devices

Device	*Inputs*			*Outputs*				
	External	*Feedback*	*Total*	*Bidirectional combinational*	*Registered*	*Combinational*	*Total*	*Package pins*
PAL 10L8	10	0	10	0	0	8	8	20
PAL 16L8	10	6	16	6	0	2	8	20
PAL 20L8	14	6	20	6	0	2	8	24
PAL 20L10	12	8	20	8	0	2	10	24
PAL 16R4	8	8	16	4	4	0	8	20
PAL 16R6	8	8	16	2	6	0	8	20
PAL 16R8	8	8	16	0	8	0	8	20
PAL 20R4	12	8	20	4	4	0	8	24
PAL 20R6	12	8	20	2	6	0	8	24

(Contd.)

Table 12.3 (*Contd.*)

Device		*Inputs*			*Outputs*				
		External	*Feedback*	*Total*	*Bidirectional combinational*	*Regi-stered*	*Combi-national*	*Total*	*Package pins*
PAL	20R8	12	8	20	0	8	0	8	24
PAL	20X4	10	10	20	6	4	0	10	24
PAL	20X8	10	10	20	2	8	0	10	24
PAL	20X10	10	10	20	0	10	0	10	24
PAL	22V10	12	10	22	10	10	0	10	24
GAL	16V8	8	8	16	8	8	8	8	20
GAL	20V8	12	8	20	8	8	8	8	24
PEEL	18CV8	10	8	18	8	8	8	8	24

12.4.6 Simple PLDs (SPLDs)

The PLDs such as PLAs, PALs and other similar type of devices discussed above are simple PLDs (SPLDs). Some of the major manufacturers of SPLDs, alongwith their some of the SPLD products and their WWW locators are given in Table 12.4

Table 12.4 SPLD manufacturers

Manufacturer	*SPLD Products*	*WWW Locator*
Altera	Classic	http://www.altera.com
Atmel	PAL	http://www.atmel.com
Cypress	PAL	http://www.cypress.com
Lattice	GAL	http://www.latticesemi.com
Philips	PLA,PAL	http://www.philips.com
Vantis	PAL	http://www.vantis.com

12.5 COMPLEX PROGRAMMABLE LOGIC DEVICES (CPLDs)

The programmable logic devices (PLDs) such as PLAs and PALs have limited number of inputs, product terms, and outputs. These devices, therefore, can support up to about 32 total number of inputs and outputs only (see Table 12.3).

For implementation of circuits that require more inputs and outputs than that are available in a single SPLD chip, either multiple SPLD chips can be employed as discussed in Sec. 12.3.8 or more sophisticated type of chip, referred to as *complex programmable logic device* (CPLD) can be used.

The complexity of any digital IC chip can be specified in terms of number of equivalent 2-input NAND gates. A typical PAL has 8 macrocells, if each macrocell represents about 20 equivalent gates, then the PAL can accommodate a circuit that needs up to about 160 gates. For circuits requiring a very large number of gates, CPLDs having large number of macrocells (say 512 macrocells) can implement circuits of up to about 10,000 equivalent gates. There are a number of manufacturers of CPLDs manufacturing a wide range of products with different features. Some of the major manufacturers of CPLDs are given in Table 12.5.

Table 12.5 CPLD manufacturers

Manufacturer	*CPLD Products*	*WWW Locator*
Altera	MAX 3000, 7000 and 9000	http://www.altera.com
Atmel	ATF, ATV	http://www.atmel.com
Cypress	FLASH 370, Ultra 37000	'http://www.cypress.com
Lattice	isp LSI 1000 to 8000	http://www.lattice.com
Philips	XPLA	http://www.philips.com
Vantis	MACH 1 to 5	http://www.vantis.com
Xiliux	XC9500	http://www.xilinx.com

12.5.1 Block Diagram

Figure 12.28 gives block diagram of a complex programmable logic device (CPLD). It consists of a number of *PAL-like blocks, I/O blocks*, and a *set of interconnection* wires. The PAL-like blocks are connected to a set of interconnection wires and each block is also connected to an I/O block to which a number of chip's input and output pins are attached.

A PAL-like block usually consists of about 16 macrocells. Each macrocell consists of an AND-OR configuration, an EX-OR gate, a FLIP-FLOP, a multiplexer, and a tri-state buffer. A typical macrocell is shown in Fig. 12.29. Each AND-OR configuration usually consists of 5-20 AND gates and an OR gate with 5-20 inputs. An EX-OR gate is used to obtain the output of OR gate in inverted or non-inverted form depending upon its other input being 1 or 0 respectively. A D-FF stores the output of the EX-OR gate, a multiplexer selects either the output of the D-FF or the output of the EX-OR gate depending upon its select input (1 or 0). The tri-state buffer acts as a switch which enables the chip's pin to be used either as an output (tri-state enabled) or as an input (tri-state disabled). In case the chip's pin is used as an input pin, an external source

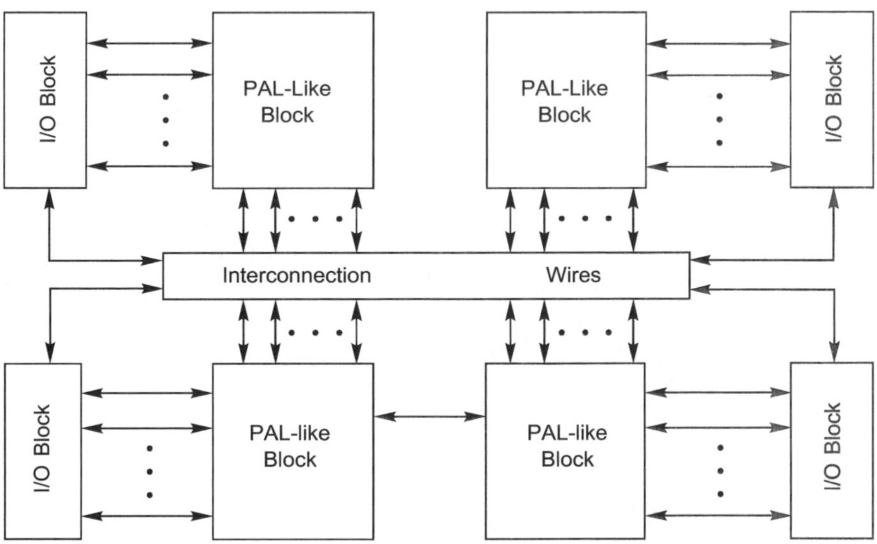

Fig. 12.28
Block diagram of a CPLD

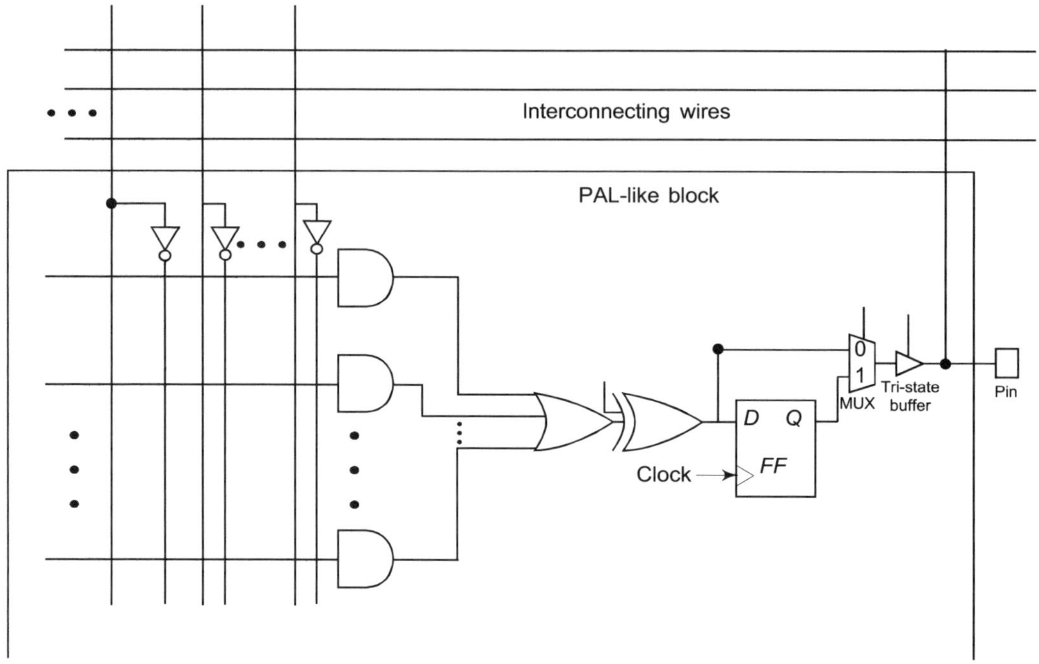

Fig. 12.29
A typical macrocell of a CPLD

can drive a signal on to the pin which can be connected to other macrocells using the interconnection wiring. When used as an input pin, the macrocell becomes redundant and it is wasted.

12.5.2 Programming

Programmable logic devices, SPLDs and CPLDs, are implemented using *electrically erasable programmable read-only memory (EEPROM)* technology. These are programmed in the same way as EEPROMs. The SPLD chips have a small number of pins and can therefore be taken out of the circuit board, without much of inconvenience, and put in a programming unit. In the case of CPLDs, instead of relying on a programming unit to configure a chip, it would be very convenient and advantageous if it is possible to perform the programming with the chip remaining attached to the circuit board itself. This method of programming is known as *in-system programming* (ISP). There are two main reasons for employing ISP.

- CPLDs have large number of pins (may even exceed 200) on the chip package, and these pins are fragile and easily bent.
- A socket is required to hold the chip in a programming unit. For large CPLDs the packages used are very expensive, sometimes more expensive than the CPLD device itself.

For the reasons mentioned above, CPLD devices usually support the ISP technique. For programming SPLDs and CPLDs a large number of programmable switches are required to be configured, hence it is not practically feasible for a user of these chips to specify manually the desired state of each switch. For this purpose computer-aided design (CAD) systems are employed. Once the user has completed the design of a circuit using CAD tools, a *programming file* or *fuse map* is generated, that specifies the state of each switch in the target PLD required to realize the designed circuit. A computer system that runs the CAD tools is connected by a cable to the programming unit. In the case of the ISP technique a small connector is included on the printed circuit board (PCB) that houses the CPLD and the computer system is connected by a cable to this connector. The programming involves transferring the programming file generated by the CAD system from the computer into the CPLD through this cable. The circuitry on the CPLD that allows in-system programming has been standardized by the IEEE and is usually called a *JTAG port.* The abbreviation JTAG stands for Joint Test Action Group. It uses four wires to transfer information between the computer and the device being programmed. Figure 12.30 illustrates the use of a JTAG port for programming two CPLDs on a circuit board.

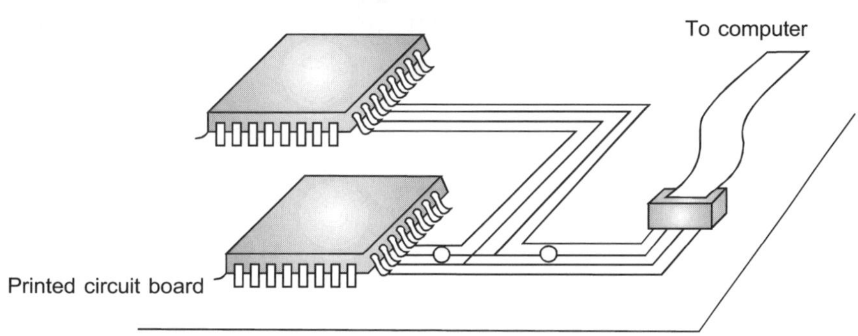

Fig. 12.30
JTAG ISP programming

12.5.3 Packaging

CPLDs have a large number of pins, making it impractical to use dual-in-line packaging (DIP). Some of the commonly used packages for CPLDs are:

Plastic-Leaded Chip Carrier (PLCC) A PLCC package has pins on all the four sides that 'wrap around' the edges of the chip, rather than extending straight down as in the case of a DIP. The IC socket of PLCC is soldered to the PCB, and the chip is held in the socket by friction. Figure 12.31 illustrates a PLCC package with socket.

Quad flat pack (QFP) A QFP package also has pins on all four sides like a PLCC package, but with pins extending outward from the package with a downward-curving shape as shown in Fig. 12.32. The QFP's pins are much thinner than those on a PLCC, making it suitable for supporting a larger number of pins. QFPs are available with more than 200 pins, whereas PLCCs are limited to fewer than 100 pins. Some of the varieties of QFPs available are: Plastic quad flat pack (PQFP), power quad flat pack (RQFP), and 1.0 mm thin quad flat pack (TQFP).

Ceramic pin grid array (PGA) It has pins extending straight outwards from the bottom of the package in a grid pattern. It can accommodate a few hundred pins in total. Figure 12.33 illustrates bottom view of a PGA package.

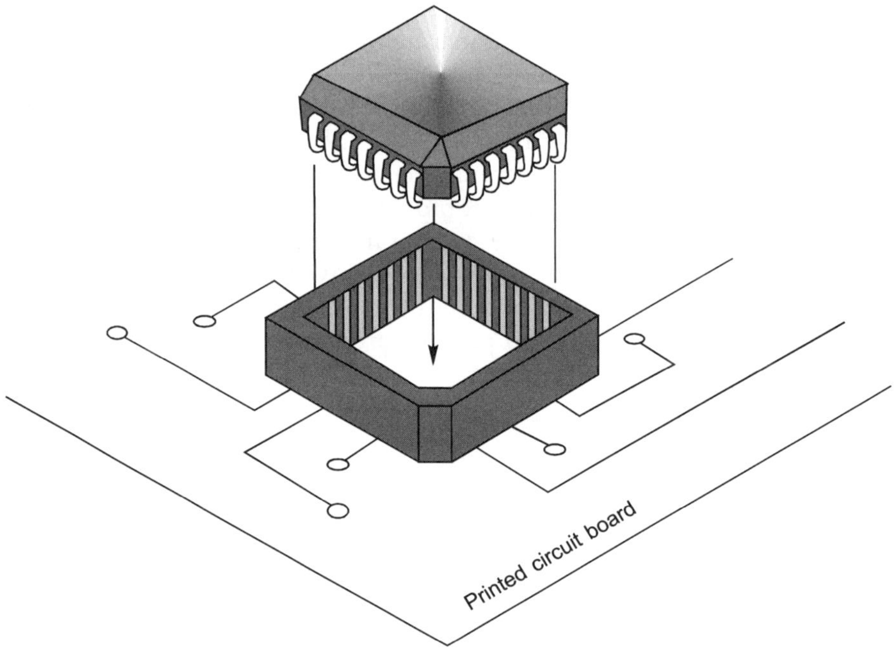

Fig. 12.31
A PLCC package with socket

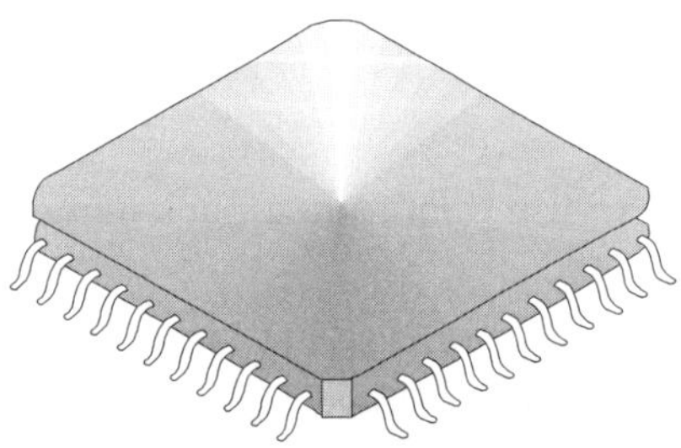

Fig. 12.32
A QFP package

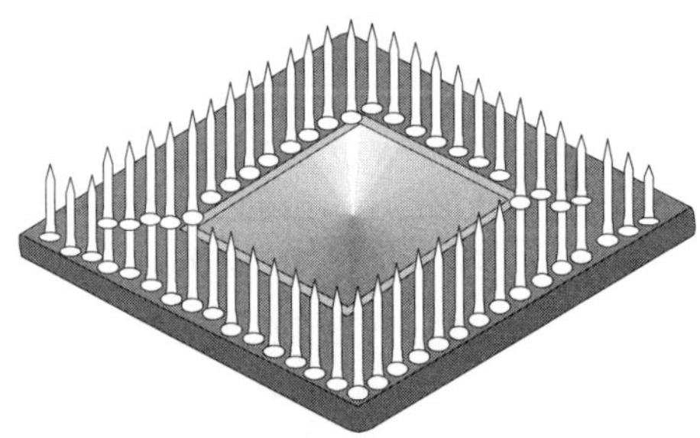

Fig. 12.33
Bottom view of a PGA package

Ball grid array (BGA) The ball grid array (BGA) packaging is similar to the pin grid array (PGA) packaging except that the pins are small round balls, instead of posts. The pins in a BGA package are very small, hence more pins can be provided on the package.

12.5.4 Available CPLDs

A number of CPLDs are available from various manufacturers. These devices are fabricated using CMOS EEPROM fabrication technology. Salient features of ALTERA's popular families, MAX 3000 A, MAX 7000, and MAX 9000 are given in Table 12.6. Data sheets for these devices and other devices manufactured by other manufacturers can be consulted for details.

Table 12.6 Features of ALTERA CPLDs

Features	*MAX 3000 A Device Family*	*MAX 7000 Device Family*	*MAX 9000 Device Family*
Usable gates	600-10,000	600-5,000	6,000-12, 000
Macrocells	32-512	32-256	320-560
Logic array blocks	2-32	2-16	
Maximum user I/O pins	34-208	36-164	168-216
Pin-to-pin logic delay (tpd) ns	4.5-7.5	6-12	10-15
Maximum counter frequency (f_{CNT}) MHz	116.3-227.3	90.9-175.4	118-144
Pin/package options	44–Pin PLCC 44-Pin TQFP 100-Pin TQFP 144-Pin TQFP 208-Pin PQFP 256-Pin BGA	44-Pin PLCC 44-Pin PQFP 44-Pin TQFP 68-Pin PLCC 84-Pin PLCC 100-Pin PQFP 100-Pin TQFP 160-Pin PQFP 160-Pin PGA 192-Pin PGA 208-Pin PQFP 208-Pin RQFP	84-Pin PLCC 208-Pin RQFP 240-Pin RQFP 280-Pin PGA 304-Pin RQFP 356-Pin BGA
ISP via JTAG Interface	3.3V operation available	5.0 operation available in S-devices only	available
No. of product terms/macrocell	32	32	

12.6 FIELD-PROGRAMMABLE GATE ARRAY (FPGA)

The programmable logic devices (SPLDs and CPLDs) are based on similar basic architecture–the programmable array logic. Over the years, programmable arrays have increased in size and complexity, and highly configurable output macrocells have been added to enhance their flexibility and expandability. To increase the effective size and to add more functionality in a single programmable device, alternative architectures have been developed which are known as *field-programmable gate arrays* (FPGAs). The logic densities of FPGAs are much higher than those of CPLDs. They range in size from 10,000 to a few hundreds of thousands equivalent gates. From modern standards digital circuits with hundreds of thousands of gates is not too large. FPGA devices support implementation of relatively large logic circuits.

The FPGAs do not contain AND, OR planes, instead they provide logic blocks for implementation of the required digital functions.

An FPGA is composed of a number of relatively independent configurable logic modules, configurable I/O cells, and programmable interconnection paths (known as *routing channels*). All the resources of the device are uncommitted and that these must be selected, configured and interconnected by a user to form a logic circuit for his application.

FPGAs differ primarily in the size and configuration of their logic modules and interconnection requirements. FPGAs with larger logic modules may not be sufficiently utilized, thereby wasting the logic modules for performing simple logic functions. Use of smaller logic modules leads to much larger number of interconnect paths in the device causing significant propagation delay as well as consuming large percentage of the FPGA's area. The optimal logic module size and interconnect requirements are highly dependent on the application that is being implemented in FPGA. For a given FPGA device, there are many possible ways to interconnect the resources required by an application.

There are a number of different types of FPGAs available which differ in their architectures, technologies, programming techniques, and packaging. These devices are available in PLCC, QFP, PGA, and BGA packages discussed in Sec. 12.5.3 and are usually programmed using in-system programming (ISP) technique discussed in Sec. 12.5.2. There are a number of manufacturers of FPGAs manufacturing a wide range of products with different features. Some of the major manufacturers of FPGAs are given in Table 12.7.

Table 12.7 FPGA manufacturers

Manufacturer	*FPGA Products*	*www locator*
Actel	Act 1,2,3, MX, SX	http://www.actel.com
Altera	Flex 6000, 8000, 10K, APEX 20K	http://www.altera.com
Atmel	At 6000, AT 40K	http://www.atmel.com
Lucent	ORCA 1,2, and 3	http://www.lucent.com
Quick Logic	pASIC 1,2, and 3	http://www.quicklogic.com
Vantis	VF1	http://www.vantis.com
Xiliux	XC 2000, XC 3000, XC 4000, XC 5200, Virtex	http://www.xilinx.com

12.6.1 Logic Cell Array

The first family of FPGAs, developed by Xilinx Corporation of America, announced in 1984, is known as *logic cell array* (LCA) device. It is composed of a number of small programmable logic elements called configurable logic blocks (CLBs). Each CLB contains logic elements and a memory element in the form of a configurable FLIP-FLOP that can be used as a transparent latch or a positive edge-triggered D-type FLIP-FLOP. Special purpose logic cells, called I/O blocks (IOBs), are arranged around the device, and each can be configured as an input, three-state output or bidirectional I/O pin. The IOBs can also be configured with FLIP-FLOPs and can also be accessed by the CLBs for non-I/O functions. These configurable cells can be interconnected using the programming routing channel. Figure 12.34 illustrates a portion of the Xilinx XC2064 FPGA device.

The XC2064 LCA device contains 64 CLBs arranged in an 8×8 matrix. Each CLB contains a 4-input combinational logic section, similar in many respects to a PROM, that is capable of implementing any logic function of up to four variables. In addition, it contains a configurable FLIP-FLOP, internal routing and control circuitry. Figure 12.35 shows a CLB. There are four general purpose inputs, A, B, C and D, a special clock input (K), and two outputs, X and Y. The delay through the logic cell is constant regardless of the logic function being implemented. The CLB can also be configured for use as two three-input logic functions or for multilevel applications including some 5-input functions.

When it is used in four-variable form, a single-output function is available, i.e. F and G outputs of the combinational logic block are same as shown in Fig. 12.36. Two independent output functions F and G are obtained if used as shown in Fig. 12.37. In this case, both the

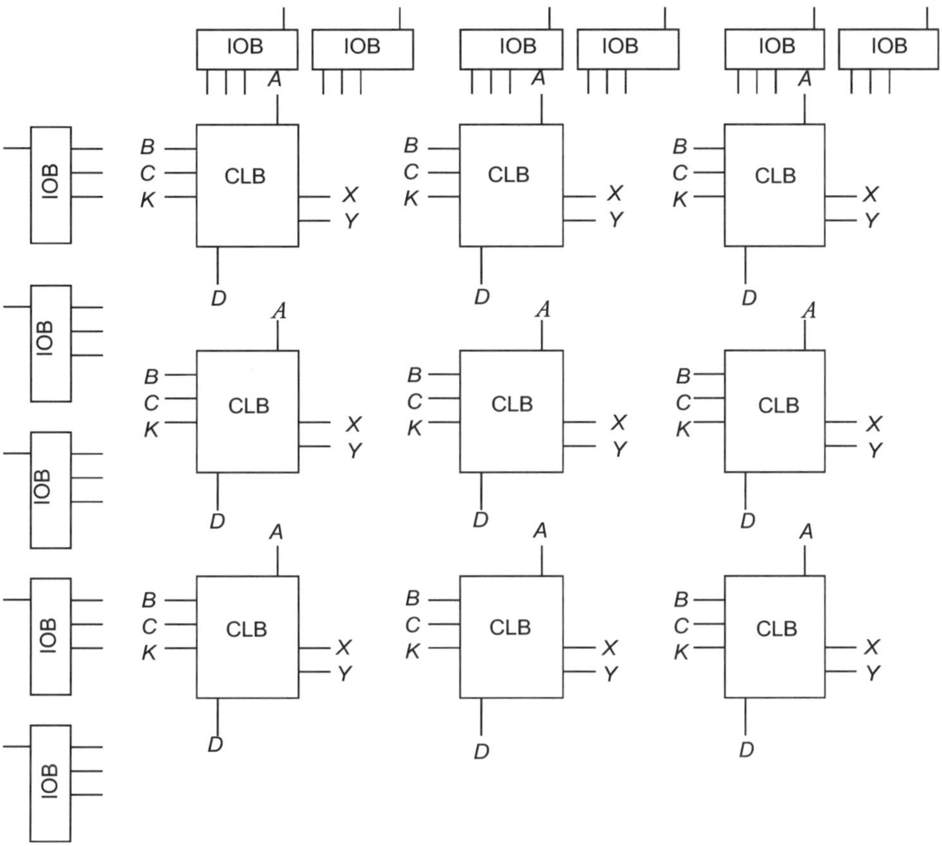

Fig. 12.34
Xilinx XC2064 FPGA.

outputs are of three input variables. The input variables may be selected from the four inputs to the CLB and the Q-output of the CLBs FLIP-FLOP.

When used in the third form as shown in Fig. 12.38, the output of the combinational logic block $F = G$ switches between two functions dynamically with B input selecting the input to the multiplexer which is available at F and G. This operation allows many five-input logic functions to be implemented in a single CLB.

The FLIP-FLOP of each CLB can be configured to be an edge-triggered D-type FLIP-FLOP, a level sensitive latch or it can be disabled. S and R are asynchronous set and reset terminals, respectively. Clock may be provided from the special clock input to the CLB, from input C or from the output G.

The XC2064 contains 58 IOBs, arranged around the periphery of the device. The I/O blocks provide the interface between the external pins of the device and internal logic. Each I/O block can be configured as an input, an output (3-state) or a bidirectional buffer as shown in Fig. 12.39. The I/O blocks can also be configured for operation with either TTL or CMOS logic level signals.

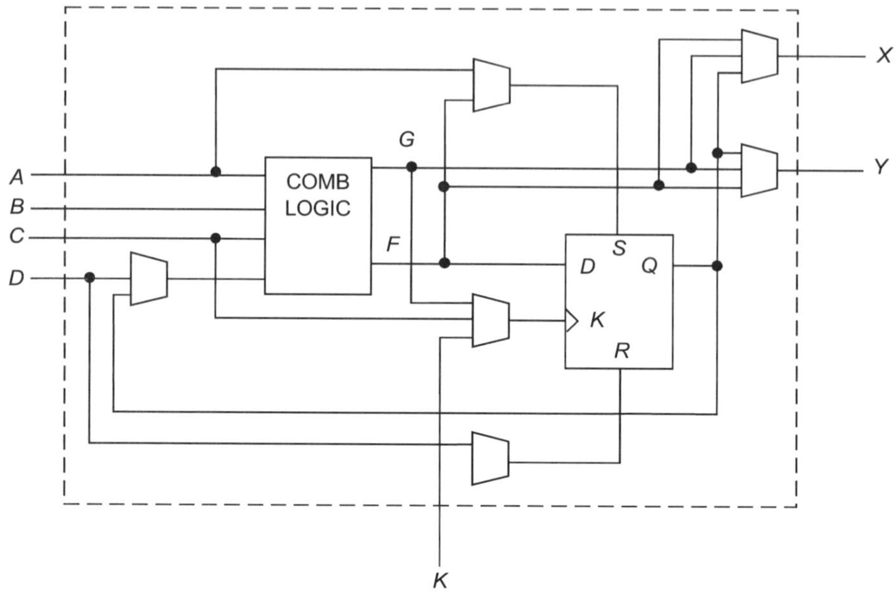

Fig. 12.35
CLB of XC 2064 LCA.

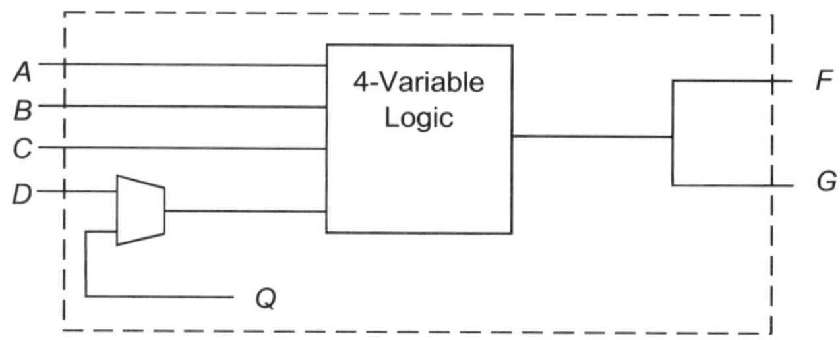

Fig. 12.36
Use of CLB as 4-variable single-output function.

The programmable interconnection channels provide routing paths to connect inputs and outputs of I/O blocks and logic cells in different configurations. Three types of such channels are provided:

(1) *The general purpose interconnections* run horizontally and vertically between the rows and columns of logic cells and I/O blocks. At the intersection of a row and column, a switching matrix is provided to allow interconnection of adjacent row or column oriented segments.
(2) *Long lines* run vertically and horizontally completely across the device. These lines bypass the switch matrices and are used for signals that must travel a large distance across the device.
(3) *Direct interconnect lines* are the most efficient way to interconnect adjacent CLBs or I/O blocks. Propagation delay time is minimum for the signals routed through direct lines.

The interconnection channels are shown in Fig. 12.40.

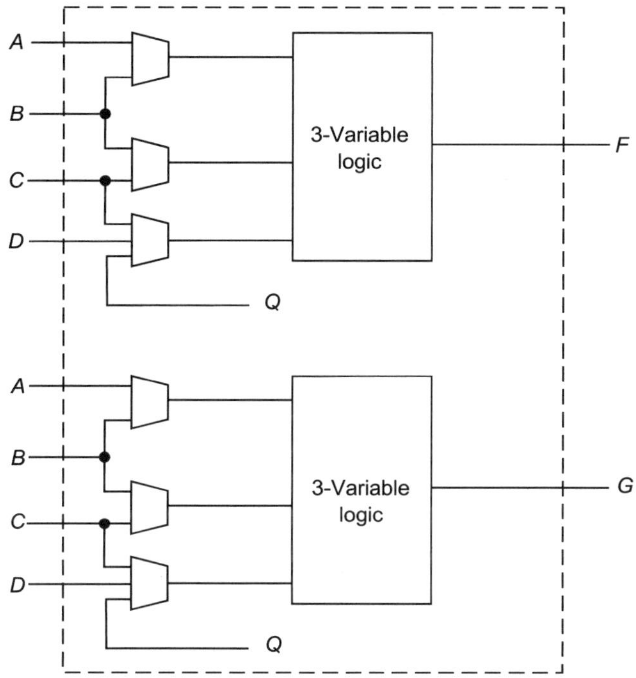

Fig. 12.37
Use of CLB for two-output function.

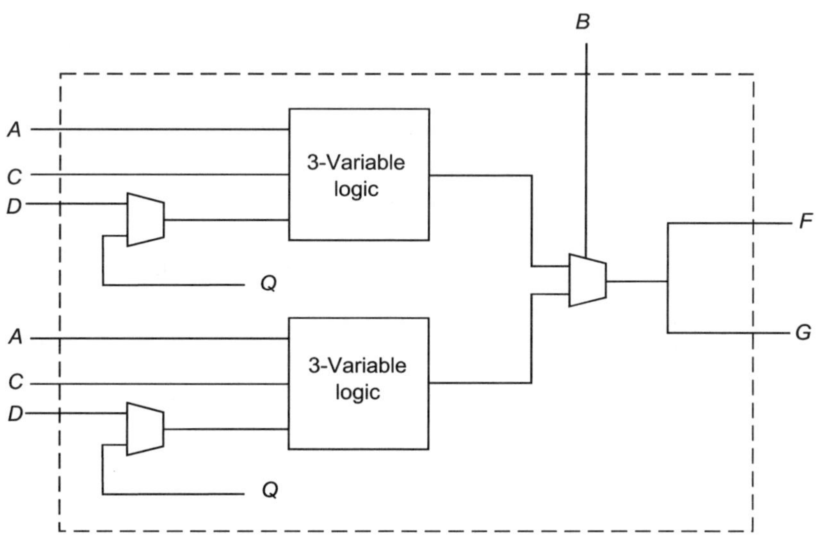

Fig. 12.38
Use of CLB for 5-input function.

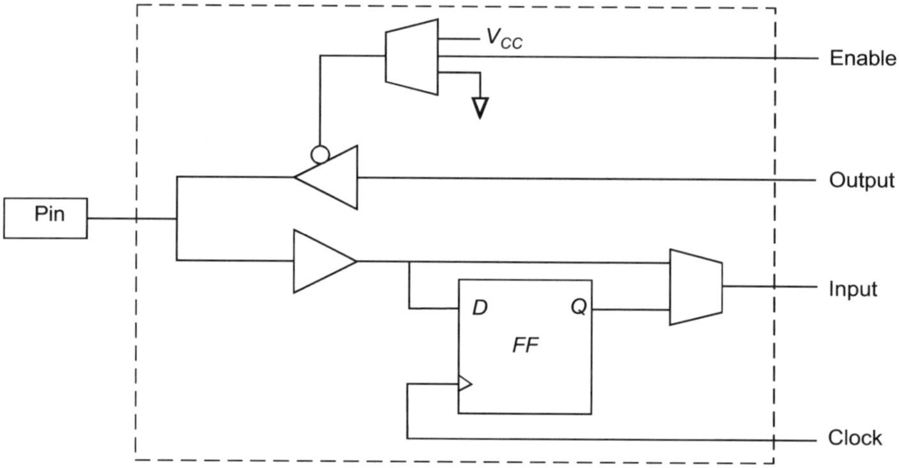

Fig. 12.39
I/O block.

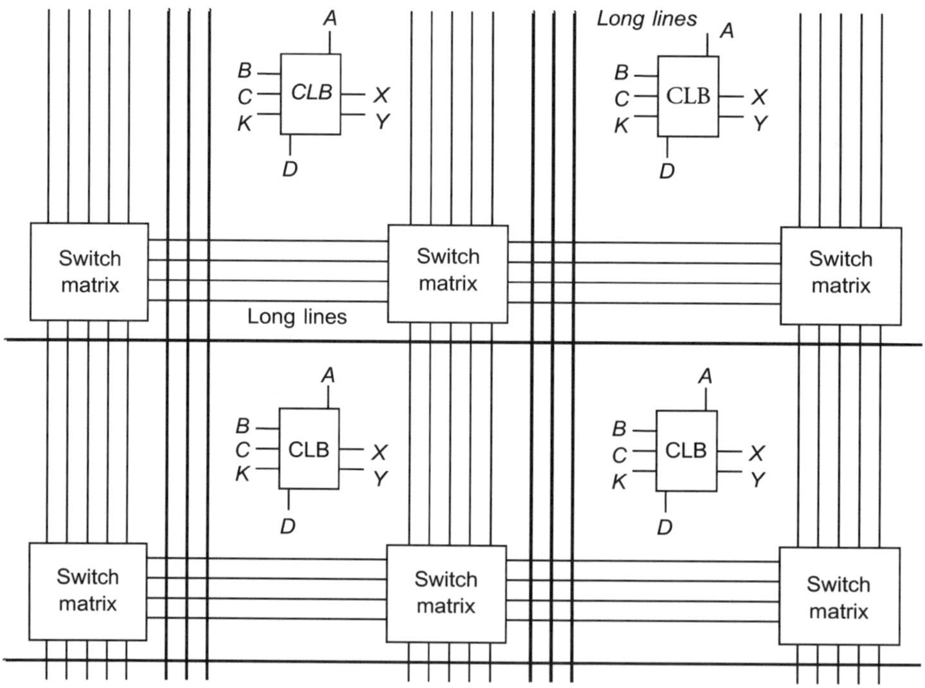

Fig. 12.40
LCA interconnection channels.

The programming methods used in LCA devices are quite different from those used in PLDs. The PLDs are programmed with fuses or erasable PROM technology, whereas the LCA's programming scheme is based on *static RAM technology*. In this technique, the device is programmed with a configuration pattern while in circuit, and new configurations can easily be loaded into the chip even while it is operating in a system. In most cases, the LCA configuration pattern is stored in a configuration PROM that is installed on the board with the LCA device. The configuration pattern is read from this PROM automatically by the LCA upon power-up.

12.6.2 Actel ACT

The FPGAs of Actel Corporation of America, known as ACT devices consist of uncommitted logic blocks separated by configurable routing channels as shown in Fig. 12.41. Its logic module consists of three 2-input multiplexers and an *OR* gate. There are no built-in registers, these are constructed using logic modules wherever necessary and, therefore, there is no wastage of FLIP-FLOPs. The logic modules in the ACT devices are of small sized internally non-configurable type as shown in Fig. 12.42. These are configured with other modules through programmable interconnections.

The programming technique used in these devices is different from those of PROMs, PLDs and LCAs. The programmable element in these devices is *programmable low impedance circuit element* (PLICE), which is an open connection in its unprogrammed state and the connection is established when programmed. The PLICE programming element is known as *antifuse* which is of extremely small size, 50 times smaller than the static RAM element of LCAs and 10 times

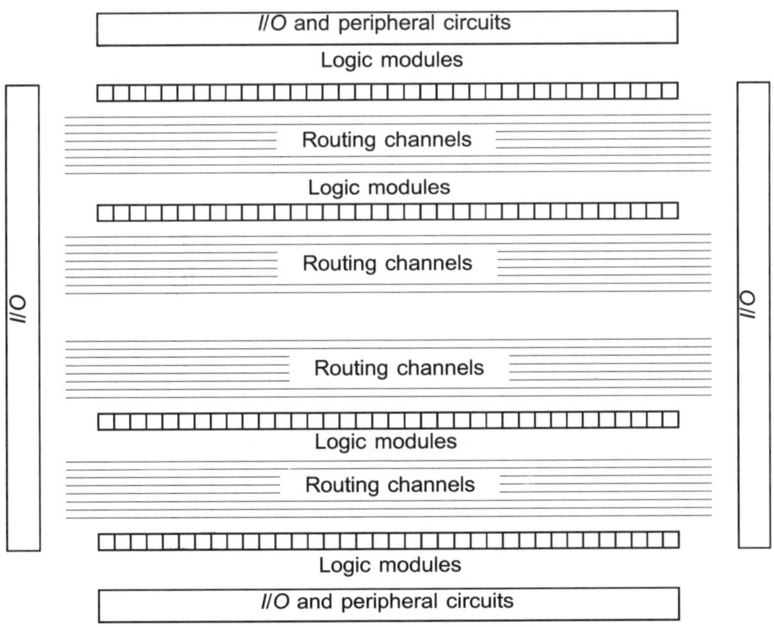

Fig. 12.41
Actel ACT FPGA.

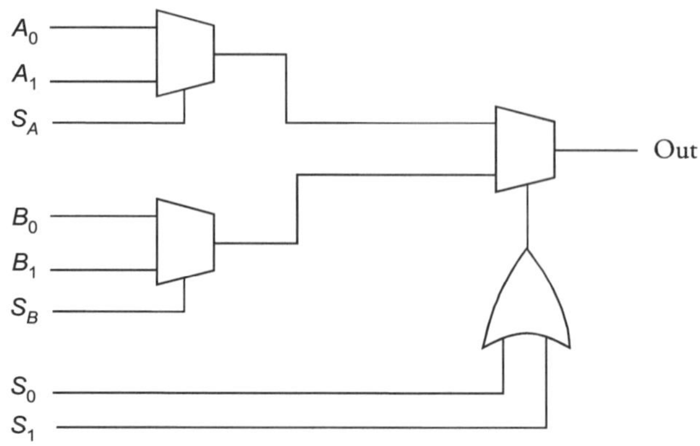

Fig. 12.42
ACT logic module.

smaller than a CMOS EPROM cell. The antifuses are not erasable and therefore, ACT devices are non-reprogrammable. The PLICE antifuse technology has made possible higher device densities and speed because of their small size and metallic interconnect paths.

12.6.3 Antifuse

An antifuse is a programmable element invented by Actel Corporation and is named as programmable low-impedance circuit element (PLICE). It is normally an open device in which electrical connection is established by the application of a programmable voltage. The antifuse is a vertical, two-terminal structure. It consists of polysilicon layer on top, N+ doped silicon on the bottom, and an oxide–nitride–oxide dielectric layer in-between. The size of the antifuse is about 1.8 mm^2. It has a low resistance, typically 500 Ω when programmed and over 100 MΩ in the unprogrammed state. The antifuse is a very attractive alternative to EPROM, EEPROM, or RAM for use as a programming element in a large programmable gate array. It is a onetime programmable device.

12.6.4 Plessey ERA

The *electrically reconfigurable array* (ERA) device of Plessey Semiconductors is similar in some respects to the LCA. Unlike the LCA, ERA is composed of a large number of simple 2-input NAND gates. Other functions and FLIP-FLOPs are constructed using NAND gates. In addition to the central array of NAND gates, a series of 84 configurable I/O cells and a 10-line configurable bus surrounding the main array for interconnections, are provided in ERA devices. Each of the NAND gate logic modules can be used either for logic functions or for interconnection routing. The 10-line bus can be used to route signals from one section of the array to another or to and from the configurable I/O cells or fixed I/O pads. The programming technique used in the ERA devices is the same as that used in the LCA devices, i.e. the static RAM technology. The ERA devices can be partially reprogrammed while they are operating. For this, one segment of logic within the ERA can control the dynamic reconfiguration of a different segment in the device.

12.7 SUMMARY

The basic concepts of programmable logic devices and programmable gate arrays have been introduced. With the development of these devices, it has become possible to design complex digital systems. However, high-level design techniques and computer-aided tools are required to produce efficient PLD and FPGA implementations. Testing of PLD and FPGA implementations also require computer-assisted test tools. The design and test tools for these programmable devices are beyond the scope of this book.

The emergence of these devices has revolutionized the design of digital systems similar to the emergence of microprocessor. The programmable logic concept has emerged as a technology that has given the power to design one's own custom ICs which cannot be copied by others.

Glossary

Antifuse A programmable element invented by Actel Corporation named as PLICE (programmable low-impedance circuit element).

ASIC (Application specific integrated circuit) An IC configured by the manufacturer as per the specifications supplied by the user for a specific application.

BGA (Ball grid array) An IC package used for ICs requiring large number of pins. The pins are of small round ball shapes.

CPLD (Complex programmable logic device) A programmable logic device containing a large number of equivalent gates.

Fixed-function ICs ICs meant for general purpose specific functions.

Foundary IC manufacturer.

FPGA (Field-programmable gate array) A programmable logic device containing very large number of equivalent gates.

FPLA (Field-programmable logic array) A PLA programmable by the user.

GAL (Generic array logic) A type of configurable PAL.

ISP(in-system programming) A technique for programming large programmable logic devices such as CPLDs and FPGAs in which the device is programmed on the circuit board itself rather than in a programming unit.

JTAG (Joint test action group) An IEEE standard for ISP.

Macrocell The basic logic function provided in large programmable logic devices which are used to design logic circuits by configuring them.

PAL (Programmable array logic) A programmable AND array that provides product terms that feed fixed OR gates.

PGA (Pin-grid array) An IC package used for ICs requiring large number of pins.

PLA (Programmable logic array) A programmable AND array whose product term outputs feed a programmable OR array.

PLCC (Plastic-leaded chip carrier) An IC package meant for ICs requiring large number of pins.

PLD (Programmable logic device) An IC which is programmed (configured) for designing digital circuits.

PLICE (Programmable low-impedance circuit element) A programmable element invented by Actel Corporation.

QFP (Quad flat package) An IC package meant for ICs requiring large number of pins.

Programming (PLDs) Configuring the programmable logic devices to obtain the desired logic function.

SPLD (Simple programmable logic device) A programmable logic device which can be used for relatively smaller logic circuits, such as PLA, PAL, etc.

Review Questions

12.1 A PLA is comprised of programmable ______ arrays.

12.2 For designing a digital circuit of 32 variables ______ PLAs with 16 inputs and 8 outputs are required.

12.3 A PAL is comprised of programmable ______ gates.

12.4 An ASIC is configured by the ______.

12.5 For maximum possible security of a digital circuit ______ design is preferred.

12.6 For designing digital circuits of the complexity of a few hundreds of gates ______ is preferred.

12.7 A PAL has 10 input pins, the number of inputs to each of the AND gates is ______.

12.8 For CPLDs and FPGAs ______ programming technique is the most suitable.

12.9 The in-system programming uses ______ IEEE standard.

12.10 The IC package used for ICs with more than 200 pins is ______.

12.11 QFP stands for ______.

12.12 PLICE is used for programming ______ FPGAs.

12.13 A ______ device can be used for digital circuits requiring more than 200,000 equivalent gates.

12.14 CPLDs and FPGAs are programmed using ______ tools.

12.15 For the design of a digital circuit ASIC is ______ expensive than designing using FPGA.

Problems

12.1 Design a BCD-to-Excess-3 code converter using a (a) PROM, (b) PLA, (c) PAL.

12.2 Design an Excess-3-to-BCD code converter using a (a) PROM, (b) PLA, (c) PAL.

12.3 Design a BCD-to-seven segment decoder using a (a) PROM, (b) PLA, (c) PAL.

12.4 How will you obtain 16-bit output word using 82 S100 FPLAs?

12.5 Explain the function of the circuit of Fig. 12.13.

12.6 What is meant by the term 'architecture of a PLD'? Give some suitable examples.

12.7 For 16L8 PAL device shown in Fig. 12.18, find out the input lines (columns) corresponding to the inputs I_1 through I_{10} and IO_2 through IO_7.

12.8 For 16R6 Registered PAL shown in Fig. 12.19, find out the input lines (columns) corresponding to the inputs I_1 through I_8, IO_1, IO_8, and feedback connections corresponding to D-FFs in O_2 through O_7 lines.

12.9 Consider the output macrocell of 16V8 GAL shown in Fig. 12.24*a*. Determine the output circuit configuration for various select inputs configurations of the four multiplexers which are controlled by fusible links. Comment on each case.

12.10 For the output macrocell of 18CV8 PEEL device shown in Fig. 12.26, determine the twelve output configurations for the select inputs $A = 0, 1$; $B = 0, 1$ and $CD = 00, 10, 11$ of the multiplexers.

12.11 In the PAL 16L8 device, program the output as
(a) Always enabled
(b) Always disabled
(c) Enabled by a product term $ABCD\bar{E}\bar{F}GH$ of eight variables.

12.12 A 2-input look-up table (LUT) of an FPGA's logic block is shown in Fig. 12.43. Determine its truth table.

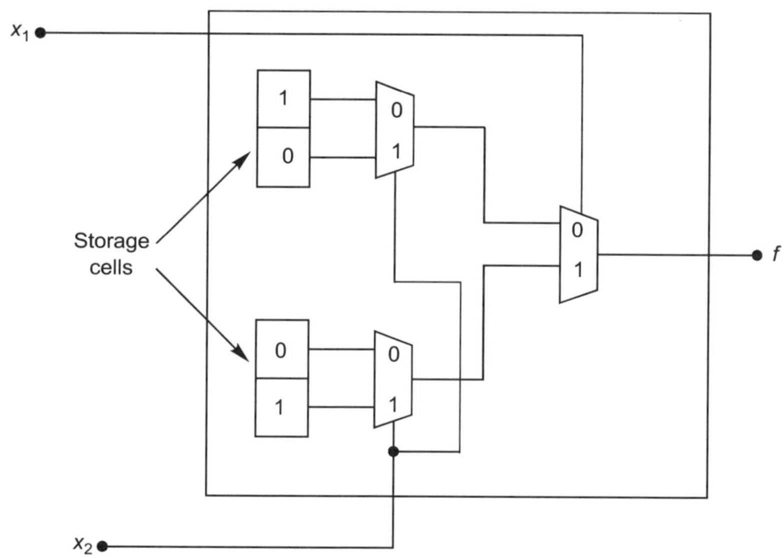

Fig. 12.43

12.13 A 3-input LUT is shown in Fig. 12.44. Determine the bits to be stored in the storage cells to realize the logic function.

$$f = \bar{x}_1\bar{x}_2x_3 + \bar{x}_1x_2\bar{x}_3 + x_1\bar{x}_2\bar{x}_3 + x_1x_2x_3$$

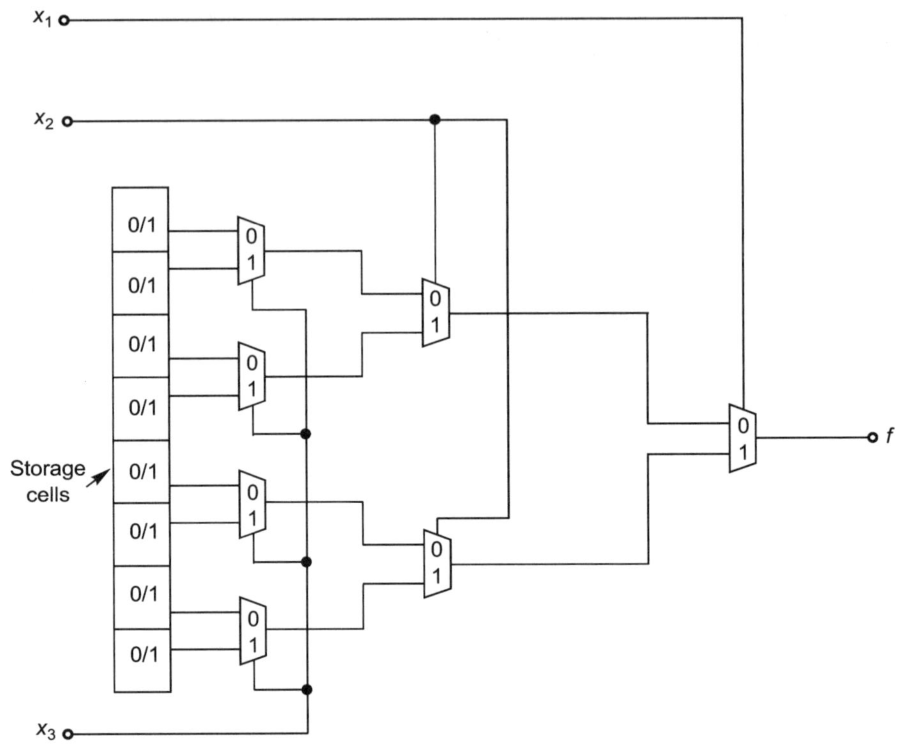

Fig. 12.44

13

FUNDAMENTALS OF MICROPROCESSORS

13.1 INTRODUCTION

The most important technological invention of recent times is the *microprocessor.* The developments in the technology of the integrated circuits made it possible for the engineers of the Intel Corporation of America to develop a microprogrammable computer on a chip in 1971. This device (Intel 4004) consists of about 2300 transistors on a chip, which was fabricated using silicon-gate *p*-channel MOS technology. It was later named as microprocessor.

Since the introduction of Intel 4004, a 4-bit microprocessor, in November 1971, a large number of other microprocessors have been developed by various companies which have found applications in a large variety of products, such as laboratory instruments, consumer products, pocket calculators, process-control systems, aircraft flight control systems, toys, games, computers, etc. In fact, it has brought in a technological revolution.

Table 13.1 gives some of the important landmarks in the evolution of microprocessors. Some of the popular 8-bit and 16-bit microprocessors, with their features, are given in Table 13.2.

The developments in the field of microprocessors have resulted in tremendous growth in the power of microprocessors. The data-bus width and the memory size, which can be used with some of the popular Intel family of microprocessors, is given in Table 13.3. The clock-frequency has increased from about 3 MHz for 8085 A μP to 3.06 GHz for the latest Pentium 4 Intel microprocessor announced in November, 2002. More and more powerful and faster microprocessors are expected to be available in the near future.

Table 13.1 Evolution of microprocessors

Microprocessor name	*Manufacturer*	*Distinction*
4004	INTEL	The first microprocessor (1971)
8008	INTEL	First 8-bit microprocessor (1972)
8080A	INTEL	First *n*-channel, second generation microprocessor (1974)
6800	MOTOROLA	First + 5 V-only microprocessor (1974)
PACE	NATIONAL SEMI-CONDUCTOR	First 16-bit microprocessor (1974)
1802	RCA	First CMOS microprocessor (1974)
8048	INTEL	First 8-bit single-chip microcomputer (1976)
8088	INTEL	First 8-bit processor with 16-bit internal architecture (1979)
2920	INTEL	First analog-signal processor (1979)
80386	INTEL	First 32-bit microprocessor (1982)
PENTIUM	INTEL	First 64-bit microprocessor (1993)

13.2 AN IDEAL MICROPROCESSOR

Although there is nothing like an ideal microprocessor (μP), we shall still introduce this hypothetical device to explain the function of a microprocessor. Figure 13.1 shows an ideal μP with n inputs and m outputs. It is assumed that a ROM is included in this μP which stores the sequence of operations, referred to as the *program*, in the form of binary codes.

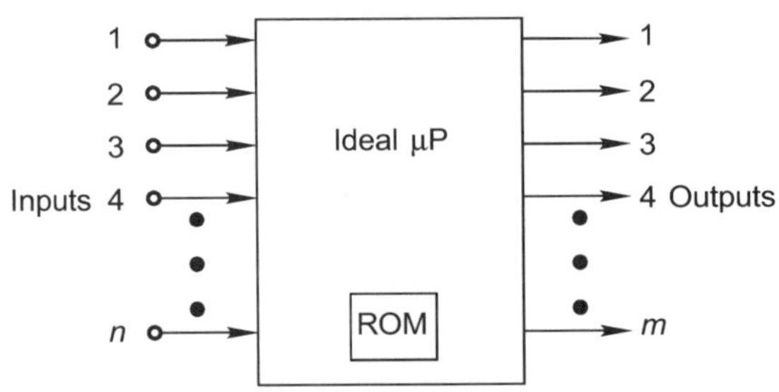

Fig. 13.1
An ideal microprocessor.

Input signals in the binary form, applied at the input terminals, are processed according to the dictates of the program and the outputs in binary form are available at the output terminals. The inputs are applied from an input device, which may be:

1. Switches,
2. Sensors,
3. A/D converters,
4. Key-boards,
5. Tape readers, etc.

Table 13.2 Popular 8- and 16-bit microprocessors with their features

Features/μP	*8080A*	*8085*	*MC6800*	*Z80*	*8748*	*8086*	*MC68000*	*Z8000*
Year	1974	1976	1974	1976	1977	1978	1979	1979
Process technology	NMOS	NMOS	NMOS	NMOS	NMOS	HMOS (*n*-channel)	NMOS	NMOS
Data bus width (bits)	8	8	8	8	8	16	16	16
Address bus width (bits)	16	16	16	16	12	20	24	16/23
Memory space (bytes)	64K	64K	64K	64K	4K	1M	16M	64K/8M
Power-supply voltages	+12 V, +5 V, –5 V	+5 V	+5 V	+5 V	+5 V	+5 V	+5 V	+5 V
Packaging	40-pin DIP	40-pin DIP	40-pin DIP	40-pin DIP	40-pin DIP covered with transparent quartz lid	40-pin DIP	64-pin DIP	40/48-pin DIP
Number of basic instructions	78	80	72	158	96	97	61	110 +
Instruction time (μs)	2	1	1	1	2.5	0.5	0.5	0.75
Resident program memory (bytes)	None	None	None	None	1K ROM 64 RAM	None	None	None

Table 13.3 Popular Intel microprocessors

Microprocessor	*Data bus width*	*Memory size*
8085A	8	64K
8086	16	IM
8088	8	IM
80186	16	IM
80188	8	IM
80286	16	16M
80386 SX	16	16M
80386 SL	16	32M
80386 DX	32	4G
80486 SX	32	4G
80486 DX	32	4G
Pentium	64	4G

The outputs are fed to an output device, which may be:

1. Actuators,
2. Lamps,
3. Alarms,
4. Indicators,
5. D/A converters,
6. CRT displays, etc.

The basic operation of a μP, explained above, can be realized by using an ALU (arithmetic logic unit) such as 74181, with some control circuitry and ROM. As discussed earlier, the 74181 can perform various arithmetic and logical operations on two 4-bit numbers. The operation to be performed depends on the signals applied at the mode control and the 4-bit function select inputs. By using proper control circuitry and program stored in ROM, it is possible to perform a sequence of operations according to the program.

13.3 THE DATA BUS

An ideal μP, as defined in Sec. 13.2, needs a large number of pins for input and output, whereas a real μP chip cannot afford to have a large number of input and output pins because of the limitations on the number of pins available on any practical IC package. Therefore, the values of m and n are limited and usually $m = n$. Also, the number of pins get reduced if it is possible to use the same pins for inputs as well as outputs. This is made possible by using bidirectional pins. The number of such pins is referred to as the *data path width*.

Collectively, these pins are referred to as *data bus*. The word *bus* is derived from the Latin *omnibus*, meaning *for all*, and is commonly used in digital systems for a group of lines. The data bus in μP is bidirectional and is represented as shown in Fig. 13.2. The number of lines forming the bus may be written by its side.

The number of lines in the data bus shown in Fig. 13.2 is 8. This number is also known as *bus width*. The organization of bidirectional bus is shown in Fig. 13.3, which is self-explanatory. Here, the direction (input/output) is specified with respect to the microprocessor.

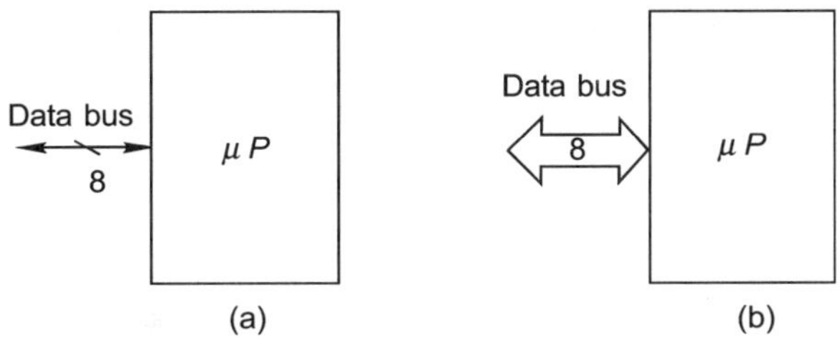

Fig. 13.2
Representation of bidirectional bus.

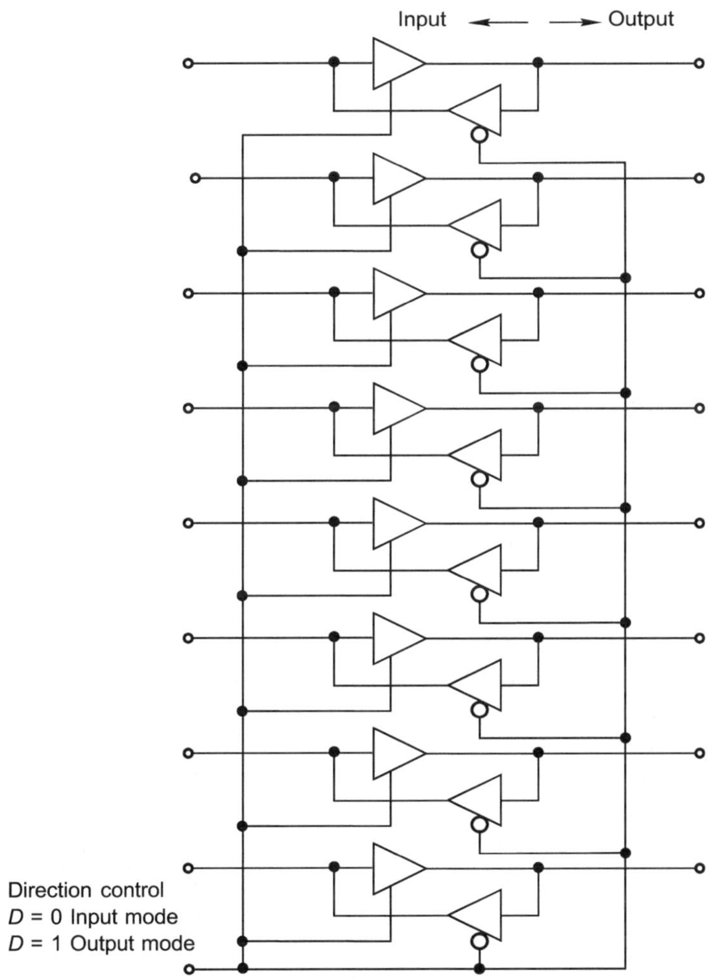

Fig. 13.3
Organization of an 8-bit bidirectional bus.

A microprocessor processes data in the form of binary digits (bits). The number of bits that it can process at a time is referred to as the *word length* or the *word size* of the microprocessor which is same as the data-bus width. Word size is an important factor in measuring the power of a microprocessor, and a microprocessor is referred to as 8-bit, 16-bit, or 32-bit microprocessor according to its word size.

13.4 THE ADDRESS BUS

In the ideal μP, we have assumed an internal memory (ROM) in which the program is stored. In real μPs, either the internal memory is not present or it is very small, that is, only a small program can be stored in the internal memory. Therefore, we have to use an external memory, which can be RAM and/or ROM. Instructions and data are stored in memory at a set of memory locations. Each memory location has a unique address. The μP must be able to store information in this memory and fetch information from it. For this purpose, both the μP and the memory must have a set of lines known as the *address bus.* The size of the memory which can be addressed with P lines in the address bus is 2^P, referred to as *address space* or *memory space.* Therefore, for connecting a large memory to the μP, the address bus width has to be large. In many real μPs, the address bus width is 16, which can address up to 2^{16} = 65 536 (= 64K, where 2^{10} = K) memory locations. Usually, the memories are arranged to store information in the form of bytes, that is, at every memory location one byte is stored. The address can be specified in the normal binary form or in a more compact form using the hexadecimal notation. Figure 13.4 depicts the interfacing of a memory with μP.

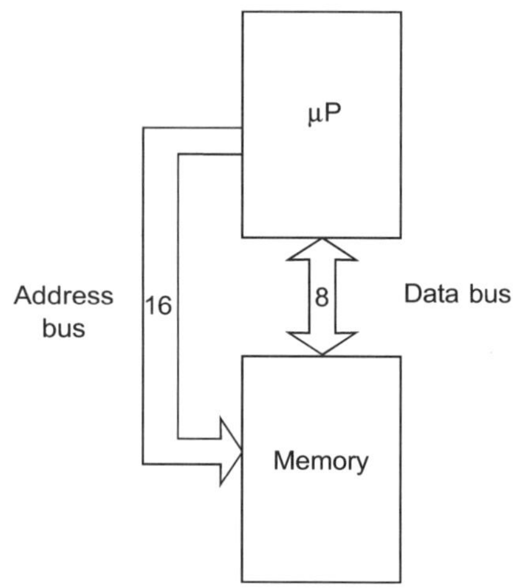

Fig. 13.4
Interfacing of memory with μP.

Before reading from or writing into a memory location, the μP must first select the desired memory location by sending signals on the address bus. The address must be present as long as the reading or writing operation is not complete. In many real situations, latches are present at the address input lines of the memory, which are used to latch the address as soon as it arrives. Once the address is latched, the address bus is no longer required for the rest of the operation of the *memory read* or *memory write* cycle and can be used for any other purpose. Earlier microprocessors have the *dedicated* address bus, which means the address bus is used exclusively for transmitting addresses only. Some of the latest microprocessors, such as Intel 8085A and 8086, use the same bus for transmitting addresses as well as for transferring data. This means that a bus is used for dual purpose: it is used as an address bus when an address is to be sent and as a data bus when data transfer is to take place. For example, in the Intel 8085A μP, the lower byte of the address is sent over the address/data (AD) bus with pins marked as AD_0 – AD_7, and the higher byte of the address is sent over a dedicated 8-bit address bus with pins marked as A_8 – A_{15}. This type of operation, where a single bus is used for two different functions, is known as *multiplexing*, which results in a saving of pins on the IC chip.

A microprocessor is accessed by the outside world through the input and the output devices. These devices are referred to as *I/O devices* and are interfaced with the μP through *I/O ports*. A number of such devices may be connected to a μP, the number depending on the address bus width, and are addressed by the same address bus to have access to a particular *I/O* device. Figure 13.5 illustrates the interfacing of *I/O* devices with a μP. The address bus width may not be same for memory and *I/O* devices. For example, in the 8085A μP, the address bus width for memory addressing is 16 bits, whereas for *I/O* addressing it is only 8 bits. This means that the number of *I/O* devices (*I/O* address space) which can be addressed is $2^8 = 256$. In fact, the same 8-bit address is available on $AD_7 - AD_0$ and $A_{15} - A_8$ pins. A μP with separate address space for memory and *I/O* devices must indicate whether the address at a particular instant is meant for a memory location or an *I/O* device. In contrast, the 6800 μP has an address space of 64K bytes which is shared by both the memory and the *I/O* devices. A particular address will be either for a memory location or an *I/O* device.

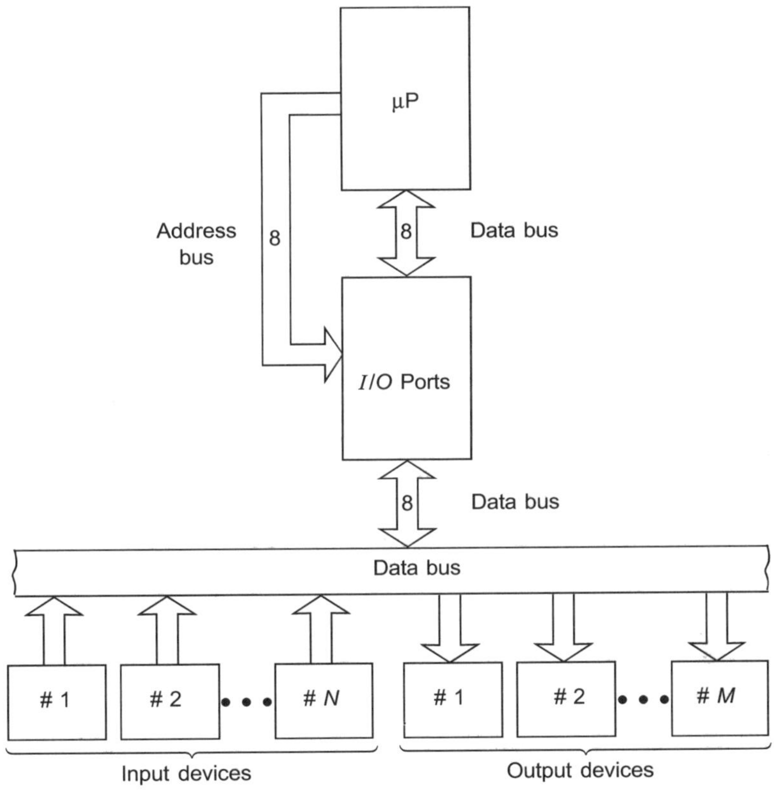

Fig. 13.5
Interfacing of I/O devices with μP.

13.5 THE CONTROL BUS

For proper operation, a microprocessor must have a set of control lines, both input and output control lines. This set of lines is referred to as the *control bus*, which is used to synchronize the operation of the μP with the operation of the external circuitry. For example, a control line is required to inform whether the address on the address bus at a particular time is meant for a memory location or on *I/O* device, since the address bus is common for both. In 8085A μP, there is a line . If the signal on this line is 1, the address is meant for *I/O*, whereas a 0 on the line means that the address is for the memory. Another example of control signals is the facility of *interrupting* the execution of the normal sequence of the program. For this purpose, there is a control input line, say *INTR*. When an *I/O* device wants to interrupt the μP, it sends out 1 over this line. The device could be an analog-to-digital (A/D) converter, which informs the μP of the fact that the conversion process is over and the data is available, or could be an indication of exceeding the safe limit of a process parameter such as temperature or pressure. The μP, after completing the current instruction, sends out a 0 on the (interrupt acknowledgement) line, acknowledging the request of the interrupting device and then services the interrupt. Here, *INTR* is an input control line whereas is an output control line. This communication protocol of making a request and then waiting for a response from the μP before proceeding is very common in μP-based systems, and is referred to as *handshaking*.

Another very useful operation is *direct memory access* (DMA). Normally, data transfer between the *I/O* devices and memory is through the μP, according to the instructions stored in the memory. This process of data transfer is very slow and is highly inconvenient if a large number of bytes are to be transferred at a time. For this purpose a μP is provided with some special arrangement by which a direct link is established between the *I/O* device and the memory. A *HOLD* input line informs the μP of the intention of an *I/O* device for having direct access to the memory. This is acknowledged by the μP by giving a logical 1 signal on *HLDA* (hold acknowledgement) line. The μP releases the address and data buses and the DMA operation starts.

13.6 MICROPROCESSOR BASED SYSTEM—BASIC OPERATION

Figure 13.6 shows the block diagram of a simple μP-based system. Before getting into a detailed discussion of a μP-based system, the steps involved in performing a simple task will be described. This will help the reader understand clearly how the μP, memory, and *I/O* ports function as a system.

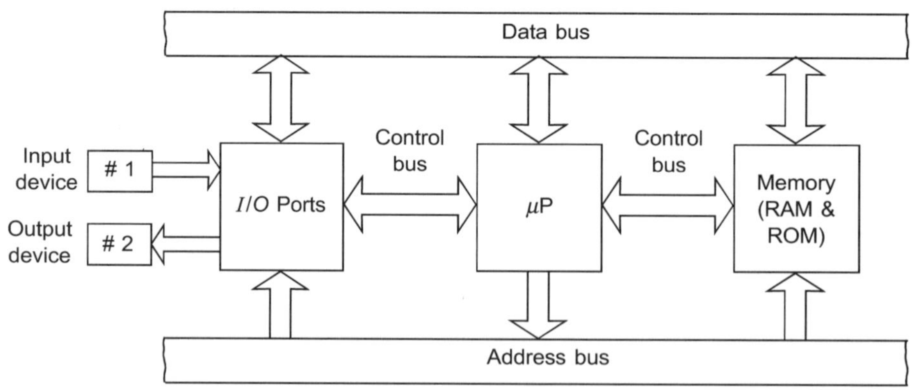

Fig. 13.6

Block diagram of a simple μP-based system.

Consider the following simple task:

1. Input a number from input port # 01.
2. Add to this the contents of memory location 0*A*2*F* (hex).
3. Output the result to output port # 02.

Since a microprocessor is a digital device consisting of various digital circuits, it performs various operations on digital signals. The operations, which a microprocessor is capable of performing, are also required to be given in the form of digital signals. Therefore, every operation which a μP can perform is to be given in the form of a binary code which is specified by the manufacturer of the μP. For example, if one byte is used for specifying codes of operations in a μP, there are $2^8 = 256$ different codes possible for the μP. These are known as ***operation codes*** (Op codes). Similarly, addresses of various memory locations and *I*/*O* devices are also specified using binary numbers.

The task to be performed is required to be specified in terms of Op codes, addresses of memory and *I*/*O* devices, and data, in a proper sequence, known as ***program*** and this process is known as ***programming.***

Let us now examine the above task step by step.

1. "Input a number from input port # 01." It has two pieces of information. The first one indicates an 'input' operation from an input port and the second one specifies the address of the port (01H). This information is specified using two bytes: the first byte for the Op code and the second byte for the port number. This two-byte sequence is known as *instruction.*
2. "Add the contents of the memory location 0*A*2*F*H." This again has two pieces of information: 'addition' operation and location of the data to be added, the other data is available in one of the registers (accumulator) in the μP. Here, it is assumed that the address of the memory location is stored in another special register in the μP, for which no separate information is required to be provided. This means that only the Op code for the operation 'addition from the memory location' is required. Therefore, this instruction has just one byte.
3. "Output the result to output port # 02." It has a two-byte instruction: the first one is for the operation 'output' and the second, for the port address.

Using the above procedure, the program is written and then stored in contiguous memory locations from the initial memory location 0000 (hex.). Table 13.4 gives the contents of the memory locations for this sample program.

Table 13.4 Contents of memory for sample program

Memory address (hex)	*Memory contents (hex)*	*Remarks*
0000	*DB*	Input Op code
0001	01	Input port number
0002	86	Add memory
0003	*D*3	Output Op code
0004	02	Output port number
0005		
0*A*2*F*	03	Data

The sequence of operations performed for the execution of this program is illustrated in Fig. 13.7. The execution of the program starts with the μP sending out the address of the first memory location 0000H (H stands for hexadecimal) to fetch the first instruction (step 1*A*). In response to this, the memory returns the contents of the memory location 0000H to the μP (step 1*B*). The μP decodes this word (11011011) as an *INPUT* instruction and generates control signals for the next memory-fetch operation, and sends out the address of the next memory location 0001H to get the number of the desired input port (step 2*A*). In response to this, the memory returns the contents of the memory location 0001H to the μP (step 2*B*). This is recognized by the μP as the port number 1. Now, the first instruction is completely available in the μP. The μP will now execute this instruction by sending out the address of the input port number 1 (step 2*C*). In response to this, the binary number available at the input port 1 is received by the μP (step 2*D*) and stored in an internal register (accumulator). With this, the first instruction is complete and the μP sends out the address of the next memory location 0002*H* to fetch the next instruction (step 3*A*). In response to this, the contents of the memory location 0002*H* are received by the μP (step 3*B*). This is recognized by the μP as: *ADD* the contents of the memory location pointed to by the register in the μP meant for this purpose. This instruc-

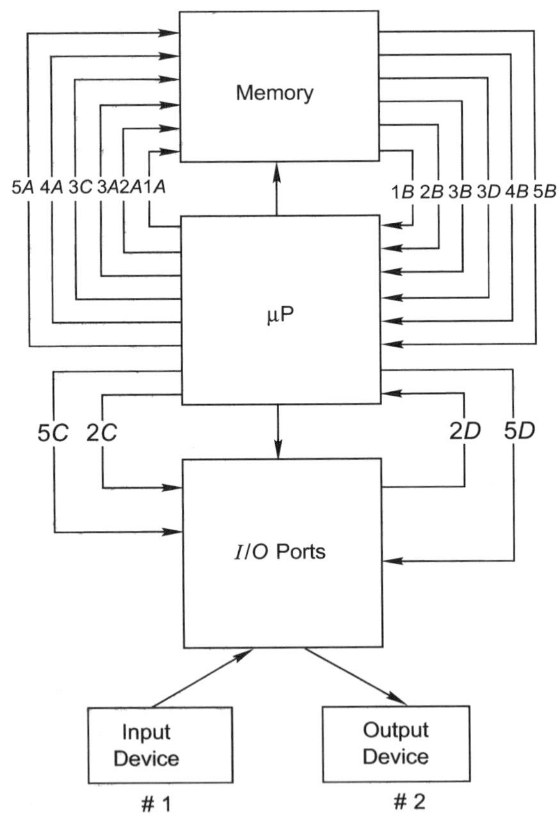

Fig. 13.7
Execution sequence of the sample program.

tion is executed by the μP by sending out the address of the memory location (0*A2FH*), the contents of which are to be added to the contents of the accumulator (step 3*C*). The memory returns the contents of this location to the μP (step 3*D*) which are then added to the contents of the accumulator. The sum is stored in the accumulator. This completes the execution of the second instruction.

Next, the address of the next memory location 0003H is sent out by the μP (step 4*A*), and in response to this, the memory returns the contents of the memory location 0003H to the μP (step 4*B*). This is recognized by the μP as an *OUTPUT* instruction. It will then send out the address of the next memory location to fetch the number of the output port (step 5*A*) and receives this number (step 5*B*). Now, the third instruction is completely available in the μP and its execution starts with the μP sending out the address of the port (step 5*C*), and after selecting the port, the contents of the accumulator are sent out to port number 2 (step 5*D*). This completes the execution of the program.

The above steps can be summarized as follows:

Step	*Operation*
1*A*	The μP sends out the address (0000H) of the first instruction.
1*B*	The contents of 0000H memory location are received by the μP.
2*A*	The μP sends out the next address (0001H).
2*B*	The contents of 0001H memory location are received by the μP.
2*C*	The μP sends out the address (01) of the input port.
2*D*	The μP receives the data from the input port (01).
3*A*	The μP sends out the next memory address (0002H).
3*B*	The contents of 0002H memory location are received by the μP.
3*C*	The address of the required memory location is sent out by the μP.
3*D*	The data from memory location 0*A2FH* is received by the μP.
4*A*	The μP sends out the next memory address (0003H).
4*B*	The contents of 0003H memory location are received by the μP.
5*A*	The μP sends out the next memory address (0004H).
5*B*	The contents of 0004H memory location are received by the μP.
5*C*	The μP sends out the address (02) of the output port.
5*D*	The μP sends out the data to the output port (02).

The above procedure, for the execution of such a simple program, may seem to be tedious and time consuming. However, a μP can run through all these steps in a few microseconds. While a μP is very fast, it is not capable of thinking and it does whatever it is asked to do by a program. Therefore, writing a correct program is most essential.

13.7 MICROPROCESSOR OPERATION

From the above discussion, it is amply clear that basically a microprocessor performs the following two operations:

1. FETCH an instruction from memory, and
2. EXECUTE the instruction.

The various steps involved in performing these operations are as follows:

FETCH

1. The μP places the address of the first byte of instruction on the address bus, along with a control signal, to read from the addressed memory location.
2. The μP gets this byte on the data bus. This byte is referred to as the *operation code* (*Op code*) and the operation of getting this byte from the memory as the *Op-code fetch.*
3. The Op code is decoded and the necessary control signals are generated.
4. If the instruction is a multi-byte instruction, the second and the subsequent bytes are read from the memory one by one by following steps similar to 1 and 2.

The operation of getting instruction bytes from the memory is known as *instruction fetch.*

EXECUTE

After the μP gets the complete instruction (all the bytes of the instruction), it performs the operation specified by the instruction. This is referred to as *execution.*

Thus, a microprocessor can be defined as a digital device on a chip which can fetch instructions from a memory, decode and execute them, i.e. perform certain arithmetic and logical operations, accept data from input devices, and send results to output devices.

It should be clearly understood that a μP alone cannot do anything. It can be used meaningfully only when it is interfaced with memory and *I/O* devices.

13.8 MICROPROCESSOR ARCHITECTURE

For performing the various operations discussed above, a microprocessor must consist of the following components:

1. Data bus,
2. Address bus,
3. Control bus,
4. Arithmetic logic unit (ALU),
5. Registers,
6. Program counter,
7. Flags, and
8. Timing and control section.

13.8.1 System Bus

The various sub-systems of the microprocessor system are interconnected by system bus, which includes the data bus, address bus, and control bus. Their functions have been discussed earlier. The data bus is used to transfer data from one part of a μP to its another part, between μP and memory, and between μP and *I/O* devices. The address bus is used to carry the address of memory locations or *I/O* devices. The address bus is unidirectional. A number of lines are required for the control operation, such as reading from memory, writing into memory, interrupt, DMA, etc.

13.8.2 Arithmetic Logic Unit (ALU)

The arithmetic logic unit is the heart of a microprocessor. It is used to perform certain arithmetic operations, such as addition, subtraction, etc. and logical operations such as AND, OR, EX-OR, etc.

13.8.3 Registers

Registers are used for storage of small data in the microprocessor. Some of the registers are accessible to the user through instructions, whereas others are not. Out of the accessible registers, are general purpose registers, referred to as *scratch-pad registers*; accumulator, used for arithmetic, logical and many other operations; and a special register known as the *stack pointer*, used to keep track of a portion of the RAM, which is used as a stack. The operation of stack will be discussed later in detail.

The registers which are not accessible to the user include the instruction register, which gets the instruction from the memory, and some temporary registers.

13.8.4 Program Counter (PC)

A program counter is required to keep track of the address of the next instruction to be fetched from the memory for execution. It always holds the address of either the first byte of the next instruction to be fetched for execution or the address of the next byte of a multi-byte instruction which has not yet been completely fetched. In either case, it gets automatically incremented one by one as the instruction bytes get fetched.

One of the control inputs to the microprocessor is the reset input. When the microprocessor is reset, the program counter sets to zero. This is the memory address from which the first instruction is to be fetched.

13.8.5 Flags

Flags are single-bit registers used to store certain conditions which arise as a result of execution of certain instructions. Some of the commonly used flags are:

Carry

If a carry is generated from MSB as a result of certain operation, the carry flag is set (1), otherwise it is reset (0).

Zero

If the result of an operation is zero, the zero flag is set, otherwise it is reset.

Sign

If the result of an operation produces 1 as MSB in the accumulator, the sign flag is set, otherwise it is reset.

Parity

If the result of an operation makes the parity of the bits in the accumulator even, the parity flag is set, otherwise it is reset.

Auxiliary Carry (half-carry)

If an operation produces carry out from the lower order four bits, the auxiliary carry flag is set, otherwise it is reset. It is used for BCD arithmetic.

Overflow

Subtraction is performed using 2's complement representation of numbers. If the result of an operation produces overflow then the overflow flag is set, otherwise it is reset.

13.8.6 Timing and Control Unit

It is used to generate proper timing and control signals which control and synchronize all the operations performed by the various sections of the microprocessor and other devices in the system.

The way in which all the above units are organized to construct a microprocessor is referred to as its architecture.

13.9 INSTRUCTION SET

Basically, the microprocessor must receive data in binary form, from the outside world (from an input device); process the data in accordance with the program stored in the memory (in binary form); and send the results in binary form, back to the outside world (to an output device). The microprocessor itself is capable of performing certain specified actions only in response to instructions in binary form. The sequence of instructions that causes the microprocessor to perform a complete task is referred to as a *program.* The collection of instructions that the microprocessor recognizes is known as its *instruction set* which is specified by the manufacturers. A user must fully understand the instruction set of the microprocessor, he/she wishes to use.

Each instruction must contain sufficient amount of information to perform the desired operation. It must contain the following information implicitly or explicitly:

1. The operation to be performed. It is known as Op code. For convenience of the user, it is specified by suitable names, known as *mnemonics.* For example, ADD for addition, ANA for logical AND operation, etc.
2. The source(s) of the operands to be used. Some instructions require only one operand, which may be in the memory, or the register, or in the instruction itself. For example, the instruction MOV *A*, *B* transfers the contents of the register *B* to the accumulator. Here, the operand is in the register *B*, which is the source of data.

   Some instructions require two operands, one of which is normally in the accumulator and the source of the other operand is to be specified in the instruction. For example, ADD *C*, adds the contents of the register *C* to the contents of the accumulator. Here, one of the source operands is in the register *C* and the other operand is implicitly contained in the accumulator.
3. The destination of the result. After the operation has been performed, the result goes to the destination, specified in the instruction. For example, MOV *A*, *M* instruction uses the accumulator as the destination. The destination may be a register, memory location or an output device.

In some instructions, the destination is not required to be specified and it is implicitly taken as the accumulator. For example, in the instruction ADD *C*, the result of addition goes to the accumulator, that is, the result replaces one of the operands.

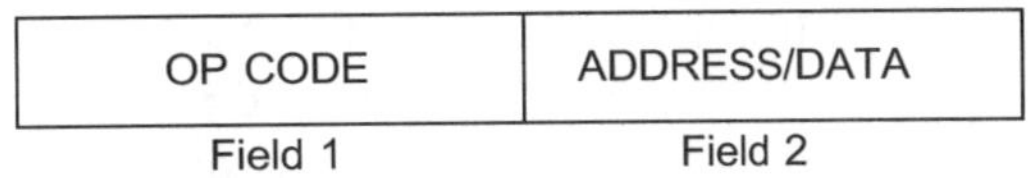

Fig. 13.8

General instruction format.

In general, an instruction has one or two fields, as shown in Fig. 13.8.

Usually, the fields are in the form of bytes. The Op code field is a single byte field, whereas the address/data field may be absent in some cases, for example, ADD *B*, else it may have an integral number of bytes. For example, in the 8085A microprocessor, an instruction may be one, two, or three bytes long. The three possible arrangements of the bytes are shown in Fig. 13.9. The instruction bytes are stored in contiguous memory locations starting from the first byte.

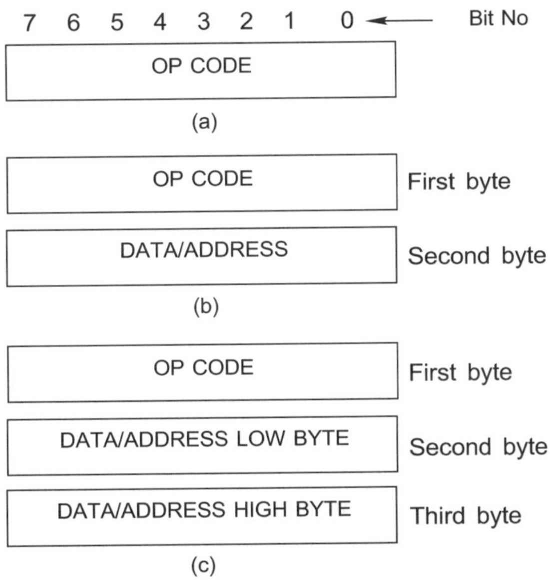

Fig. 13.9

Arrangement of instruction bytes in 8085A microprocessor: (a) Single-byte instruction (*b*) Two-byte instruction (*c*) Three-byte instruction.

Example 13.1 It is required to add the contents of memory locations 2*C*4*A* (*hex*) and 3*DA*6 (hex) and store the sum in memory location 7*AFF* (hex).

Solution

The required program will have the following sequence of instructions for 8085A microprocessor.

```
LDA    2C4AH
LXI    H,3DA6H
ADD    M
STA    7AFFH
```

The number of bytes in these instructions are 3, 3, 1, and 3, respectively. The program is to be stored in the memory before it can be executed by the microprocessor. Let us assume that the program is stored as shown in Table 13.5.

Table 13.5 Sequence of bytes for the program of Ex. 13.1

Memory address (hex)	*Contents (binary)*	*Remarks*
0000	00111010	Op code of LDA
0001	01001010	Lower eight bits of address (4*A*)
0002	00101100	Higher eight bits of address (2*C*)
0003	00100001	Op code of LXI H
0004	10100110	Lower byte of address (*A*6)
0005	00111101	Higher byte of address (3*D*)
0006	10000110	Op code of ADD *M*
0007	00110010	Op code of STA
0008	11111111	Lower byte of address (*FF*)
0009	01111010	Higher byte of address (7*A*)

It should be clearly understood that the design of microprocessor-based systems require a knowledge of both hardware (circuitry) and software (program).

13.10 THE 8085A MICROPROCESSOR

The basic concepts involved in the architecture and operation of a microprocessor have resulted in a large number of different microprocessor chips with varying flexibilities. Although the basic concepts are more or less same, they differ in architecture, operation, and instruction set. If one tries to learn about all the microprocessors simultaneously, he/she will simply end up in confusion. On the other hand, if one can master one type of microprocessor he/she can easily understand the other microprocessors. Therefore, we have chosen the most popular microprocessor, Intel 8085A, for discussion.

13.10.1 Organization and Operation

The Data and Address Buses

The Intel 8085A is an 8-bit microprocessor. Its block diagram is shown in Fig. 13.10. It has an 8-bit bidirectional data bus which is used for data input as well as data output in the form of 8-bit ($D_0 - D_7$) words.

The address bus is 16-bit wide, which means that memory address space is 64K bytes. The address bus is divided into two sections. The lower-order eight bits of the address (A_0–A_7) are transferred on the data bus itself, and hence, the bus is referred to as AD (AD_0–AD_7) bus. The higher-order eight bits of the address are transferred on A_8–A_{15} lines. The time-multiplexed bus system is used to save the number of pins in the IC package. The address and data information is present on the bus in different time slots.

Arithmetic Logic Unit

It has an 8-bit ALU. The arithmetic and logical operations are performed in the ALU on 8-bit words.

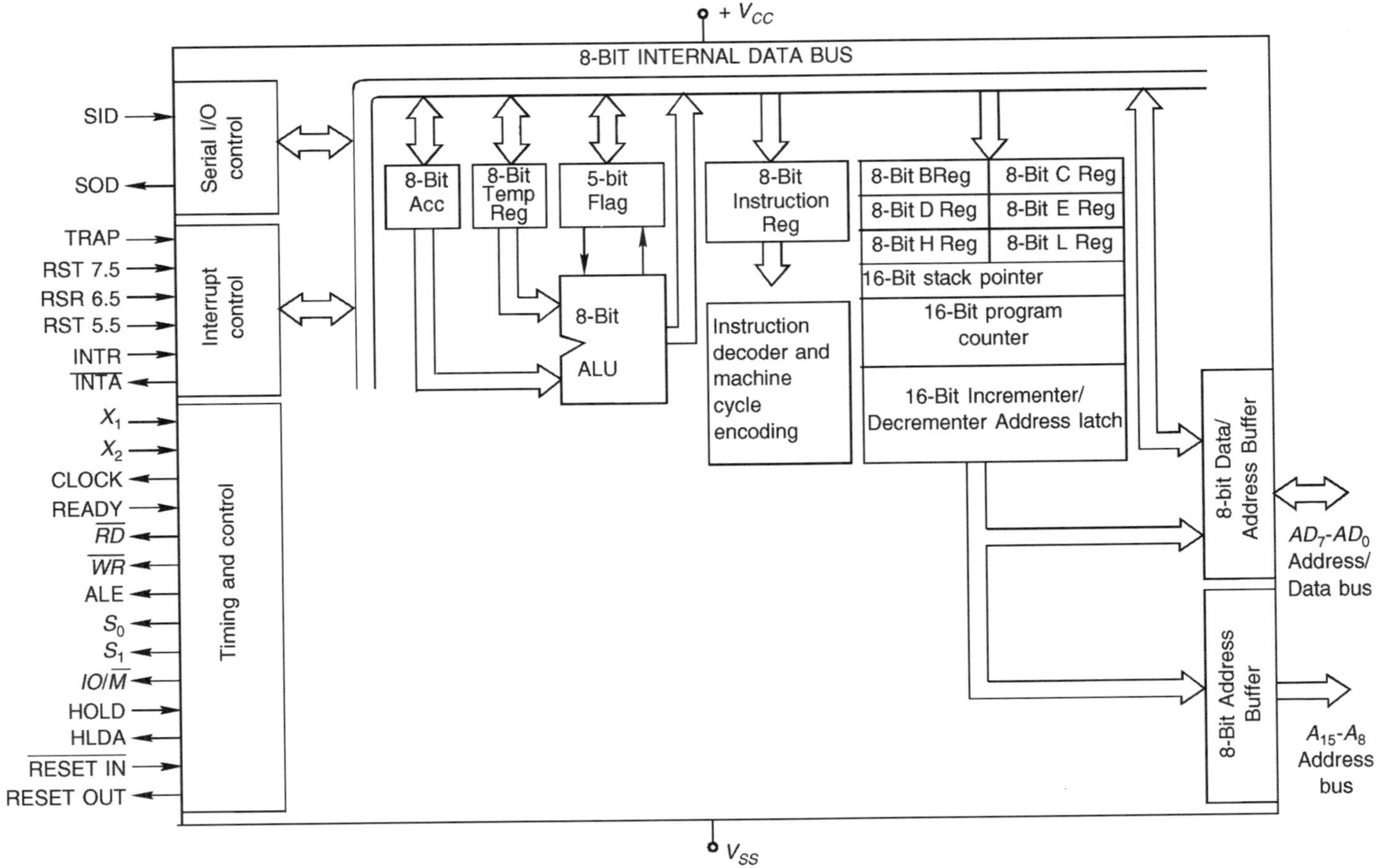

Fig. 13.10
Functional block diagram of 8085A microprocessor

Accumulator Register

There is an 8-bit register, known as the accumulator. For arithmetic operations, such as addition, subtraction, etc. and logical operations, such as AND, OR, EX-OR, etc. one of the operands is contained in the accumulator. The result of these operations is also stored in the accumulator. This register is labelled as A.

General-purpose Registers

There are six 8-bit general-purpose registers, labelled as B, C, D, E, H, and L. These registers can be used as single 8-bit registers or in pairs, as 16-bits registers. When used in pairs, the allowed pairs are B and C; D and E; and H and L. The higher-order eight bits (or higher byte) are stored in the first register of the pair (B, D, or H) and the lower-order eight bits (or lower byte) are stored in the second register of the pair (C, E, or L). For example, if $1AC2$ (hex) is to be stored in H – L pair, $1AH$ will be stored in the H register and $C2H$ in the L register. A multiplexer is used to route the data on the internal data bus to either the higher-order registers column or the lower-order registers column.

Stack Pointer

The stack pointer (SP) is a 16-bit register used to store the address of the stack top. The use of stack and stack pointer are discussed later in detail.

All these registers are accessible to the user through instructions. They are not accessible physically. The names and binary codes of the 8-bit registers are given in Table 13.6.

Table 13.6 Names and binary codes of 8-bit registers of 8085 A

Register name	*Binary code*
A	111
B	000
C	001
D	010
E	011
H	100
L	101

The names and binary codes for register pairs are given in Table 13.7.

Table 13.7 Names and binary codes of register pairs of 8085A

Register pair	*Name*	*Binary code*
B–C	B	00
D–E	D	01
H–L	H	10
Stack pointer (16-bit register)	SP	11

The 16-bit H–L register pair is used to provide the implied memory address for most memory-reference instructions, for example,

MOV *A*, *M*
ADD *M*
ANA *M*

etc, A limited number of memory-reference instructions use the B–C and D–E register pairs, for example,

LDAX *B*
LDAX *D*
STAX *B*
STAX *D*

Flags

There are five FLIP-FLOPs (one-bit registers), which serve to indicate certain conditions that arise during arithmetic and logical operations and are referred to as *flags.* These are:

1. Zero (*Z*)
2. Sign (*S*)
3. Parity (*P*)
4. Carry (*CY*)
5. Auxiliary carry (*AC*)

A flag is *set* by forcing the bit to 1 and *reset* by forcing the bit to 0. When an instruction affects a flag, it affects it in the manner discussed in Sec. 13.8. These flags may be accessed by some instructions along with the accumulator, as a single processor status word (PSW). PSW bits are assigned as shown in Fig. 13.11.

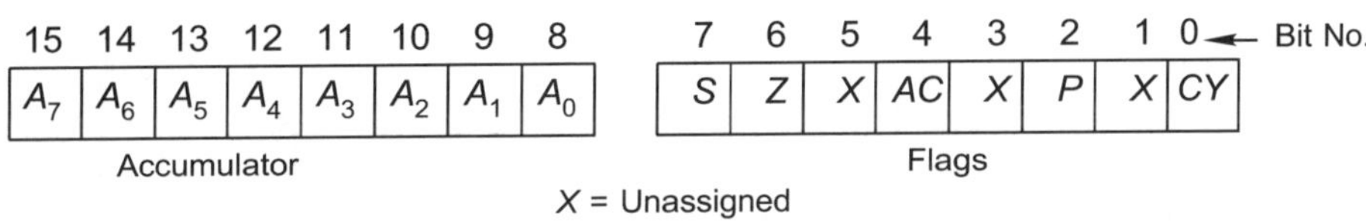

Fig. 13.11
Organization of PSW.

There are two instructions STC (set carry) and CMC (complement carry) which directly address the carry flag.

The flags are examined in conditional instructions to determine if the specified condition is true or false. The names and the binary codes of the conditions are given in Table 13.8.

Table 13.8 Names and binary codes for conditions

Condition name	*Flag status*	*Binary code*
NZ – Not zero	$Z=0$	000
Z – Zero	$Z=1$	001
NC – No carry	$CY=0$	010
C – Carry	$CY=1$	011
PO – Parity odd	$P=0$	100
PE – Parity even	$P=1$	101
P – Plus	$S=0$	110
M – Minus	$S=1$	111

Program Counter

The instructions to be executed are stored in contiguous memory locations. The address of the next instruction to be fetched from memory is contained in the 16-bit program counter (PC). For example, consider an instruction LDA 02*C*4H. This instruction means load the accumulator with the contents of the memory location 02*C*4H. Let us assume that the instruction code (LDA) of this instruction is stored in the memory location 0100H. The next two memory locations 0101H and 0102H will contain *C*4H and 02H, respectively, and the next instruction will be stored starting from the memory location 0103H.

When the microprocessor has completed the execution of the instruction prior to LDA 02*C*4H, the PC will contain 0100H. It goes on incrementing automatically as it fetches the instruction bytes from the memory. At the end of the execution of this instruction, the contents of the PC will be 0103H.

In addition to the above registers, which can be accessed through instructions, there are some registers which cannot be accessed. The instruction register gets the operation code of the instruction in Op code fetch machine cycle and passes it on to the decoder. The function of the decoder is to interpret the instruction and enable the control section to produce proper signals to carry out the tasks required to be performed in the instruction.

Several other registers, such as temporary register, are involved only in internal operations and are not accessible to the user.

Clock

A crystal or *RC* network connected between pins X_1 and X_2 determines the frequency of the internal clock generator which synchronizes the operation of the 8085A. The clock frequency generated is one half of the crystal frequency. This is also available at the clock output pin which can be used for synchronizing external devices.

Power Supply

The 8085A operates on a single +5 V power supply, connected between V_{CC} and V_{SS} pins.

Interrupt System

In many applications, it may be necessary to postpone the routine job that the microprocessor is doing to attend to some urgent task. For example, a microprocessor-based system may be used for monitoring the temperature and pressure of a process. If the temperature and/or pressure goes beyond a prespecified safe limit, the microprocessor should be able to initiate proper actions to avoid any damage to life and equipment.

For this purpose, a signal indicating such an emergency must be able to interrupt the microprocessor. In fact, the interrupting system of the microprocessor allows signals of higher priority to usurp and steer the processor into a different program.

The interrupt system of the 8085A microprocessor consists of five different input signals, which have a fixed priority unless changed by SIM instruction (discussed later). These are given below according to their priorities.

TRAP Highest priority
RST 7.5
RST 6.5

RST 5.5
INTR Lowest priority

In the case of the first four interrupts, the control is transferred to the memory location as given below, whereas for INTR, a CALL address instruction is supplied to the microprocessor by an external device, known as *interrupt controller.*

Interrupt	*Branching address*
TRAP	$4.5 \times 8 = 24H$
RST 5.5	$5.5 \times 8 = 2CH$
RST 6.5	$6.5 \times 8 = 34H$
RST 7.5	$7.5 \times 8 = 3CH$

13.10.2 Programming

The instruction set of the 8085A microprocessor have been grouped into the following five groups:

1. Data transfer group,
2. Arithmetic group,
3. Logical group,
4. Branch group, and
5. Stack, *I/O* and machine control group.

Data Transfer Group Instructions

This includes instructions to move data between registers, between memory and registers, and the data contained in the instruction itself to memory or registers.

Move Instructions These are one-byte instructions for the transfer of a byte of data from one 8-bit register to another 8-bit register, or from a memory location to an 8-bit register, or from an 8-bit register to a memory location. The mnemonic for these instructions is MOV. The destination and the source follow the mnemonic and are separated by comma. The destination is always listed before the comma and the source, after. For example, the instruction MOV *A*, *B* moves the contents of the register *B* to the register *A*. In fact, the contents of register *B* are copied in register *A*. These instructions do not change the contents of the source (register or memory location).

Whenever memory location is involved as the source or destination of data, the address of the memory location is taken as the contents of the H–L pair. Therefore, the desired address must first be loaded into the H–L register pair. For example, the MOV *A*, *M* instruction, copies the contents of the memory location, whose address is stored in the H–L register pair into the accumulator. The H–L register pair is used as a pointer to a memory location, as shown in Fig. 13.12.

Similarly, the instruction MOV *M*, *B* copies the contents of register *B* into the memory location pointed to by the H–L register pair.

Move Immediate Instructions The move immediate instructions (MVI) transfer the byte of data specified within the instruction itself to the register or the memory location (whose address is stored in the H–L register pair). For example, MVI *C*, 02H, transfers the data byte (02H) to

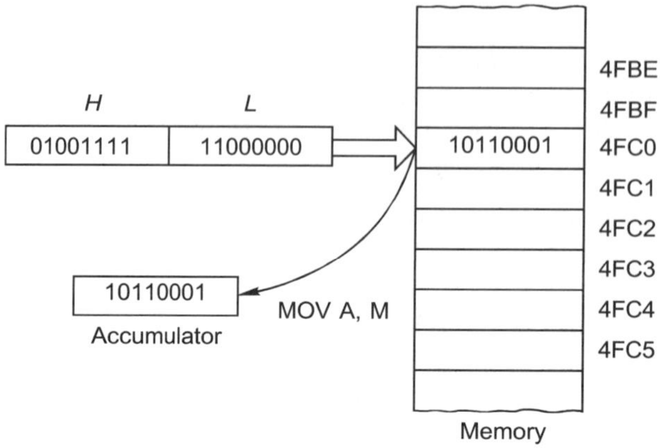

Fig. 13.12

Use of H–L register pair as memory pointer.

the register *C*, whereas, MVI *M*, 3*A*H, transfers the data byte (3*A*H) to the memory location pointed to by the H–L register pair. These instructions are two bytes long, requiring two memory locations to store them. The first byte is for the Op code which is stored in the first memory location, followed by the byte corresponding to the data contained in the instruction.

Load Extended Immediate This group of instructions is used to transfer a 16-bit number, specified in the instruction in the form of two bytes, to the specified register pair. For example, LXI H, 23*A*7H loads 00100011 into the H register and 10100111, into the L register. These instructions are three bytes long, one for the Op code, and two for the data. The instruction bytes are stored in three consecutive memory locations, as shown in Fig. 13.13.

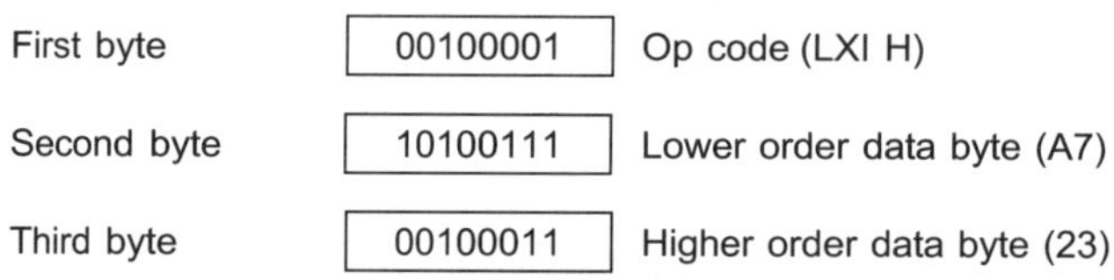

Fig. 13.13

Instruction bytes for LXI H, 23*A*7H.

Example 13.2 Write a program to transfer one byte of data from the memory location 0010H to 1000H.

Solution

Instruction	*Comments*
LXI *H*, 0010H	Load H–L pair with 0010H.
MOV *A*, *M*	Transfer the contents of memory location pointed to by the H–L pair to the *A* register (accumulator).
LXI *H*, 1000H	Load the H–L pair with 1000H.
MOV *M*, *A*	Transfer the contents of the register *A* to the memory location addressed to by the H–L pair.

We note that there is no instruction to directly transfer the contents of one memory location to another. The transfer is possible only through one of the registers of the microprocessor.

Load/Store Accumulator Direct Load accumulator direct (LDA) loads the accumulator with the byte stored at the memory location whose address is specified in the instruction. Similarly, store accumulator direct (STA) stores the contents of the accumulator in the memory location whose address is specified in the instruction. These are three-byte instructions. The first byte is for the Op code, second, for the lower-order 8 bits of the address and the third, for the higher-order 8 bits of the address.

Considering Ex. 13.2, the same program can be written as:

LDA	0010H	Load the accumulator with the contents of memory location 0010H.
STA	1000H	Store the contents of the accumulator in the memory location 1000H.

Load/Store H and L Direct The instruction LHLD or load H and L direct, loads the contents of the memory location, whose address is specified in the second and third bytes of the instruction, into the L register and the byte in the next memory location is loaded into the H register. The instruction SHLD, or store H and L direct, is the reverse operation. It transfers the contents of the L register in the memory location specified by the second and third bytes of the instruction, and the contents of the H register is transferred into the next memory location. For example, the LHLD 0A22H instruction transfers the contents of the memory location 0A22H to the L register and 0A23H, to the H register. This is illustrated in Fig. 13.14.

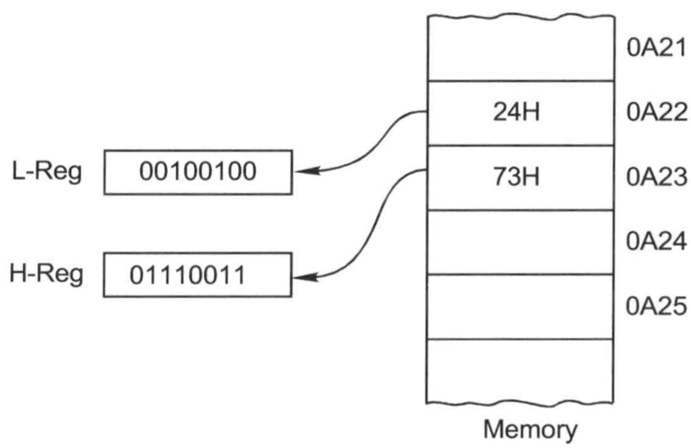

Fig. 13.14
Operation performed by LHLD instruction.

Similarly, the operation performed by the SHLD instruction is shown in Fig. 13.15.

Load/Store Accumulator Indirect The instruction LDAX B loads the accumulator with the contents of the memory location whose address is in the B–C register pair. Similarly, LDAX D uses the contents of the register pair D–E as the address of the memory location.

The instructions STAX B and STAX D store the contents of the accumulator at the address pointed to by register pairs B–C or D–E, respectively. All four instructions are single-byte instructions.

Example 13.3 Explain the program given below:

```
LXI    B, 2475H
LXI    D, 3794H
LDAX   B
```

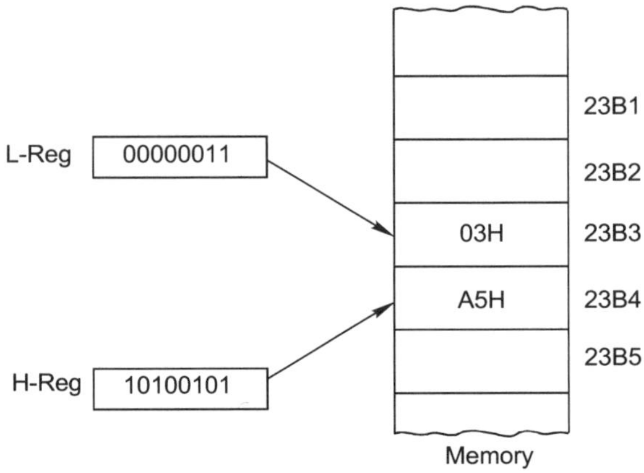

Fig. 13.15
Operation performed by the instruction SHLD 23B3 H.

```
MOV     L, A
LDAX    D
STAX    B
MOV     A, L
STAX    D
```

Solution

The operation performed by each instruction is given below.

Instruction	*Operation*
LXI B, 2475H	Loads the B–C pair with 2475H.
LXI D, 3794H	Loads the D–E pair with 3794H.
LDAX B	Loads A with the contents of memory location pointed to by the B–C pair (2475H).
MOV L, A	Moves the contents of A to L register
LDAX D	Loads A with the contents of memory location pointed to by the D–E pair (3794H).
STAX B	Stores A at the memory location whose address is in the B–C pair (2475H).
MOV A, L	Moves the contents of L to A.
STAX D	Store the contents of A at the memory location whose address is in the D–E pair.

In short, this program stores the addresses 2475H and 3794H of two memory locations in register pairs B–C and D–E, respectively, and exchanges the contents of these memory locations.

The same job can be performed by writing a program in a different way. Try and make sure that there is no unique way of writing a program.

Exchange Instruction The instruction XCHG, a single byte instruction, exchanges the contents of H–L and D–E register pairs.

None of the five flags are affected by the instructions of the data-transfer group.

All the instructions of the data-transfer group, in terms of their mnemonics, and operands along with their Op code (hex), are given in Table 13.9.

Arithmetic Group Instructions

Add Instructions Addition instructions are used to add the contents of the specified register, memory location, or the data-byte in the instruction itself, to the contents of the accumulator. The result is stored in the accumulator. If the addition produces any carry, the carry flag is set ($CY = 1$), otherwise it is reset. All the other flags are also affected, depending upon the results. In the case of an operand being in the memory, the memory address is the contents of the H–L register pair. The addition instructions are:

ADD r	Add the contents of the register r.
ADD M	Add the contents of the memory location (address in the H–L pair).
ADI data	Add the data byte.

There is another group of ADD instructions in which the carry bit (CY) is also added in addition to the specified register, memory location or the data byte. These are:

ADC r	Add the register with the carry.
ADC M	Add the memory with the carry.
ACI data	Add the data-byte with the carry.

Example 13.4 Assume the contents of the accumulator and register C as 2*E*H and 6*C*H, respectively. The instruction ADD *C* performs the addition as follows:

2*E* H = 00101110
6*C* H = 01101100
---
9 *A* H = 10011010

The contents of registers A, C and flags, following execution of the ADD *C* instruction, are given in the Table 13.10.

Table 13.10 *Table for Ex. 13.4*

Register or Flag	*Contents*		*Remarks*
	Before ADD C	*After ADD C*	
A	00101110	10011010	Result of addition
C	01101100	01101100	No change
CY	Don't Care	0	No carry
S	"	1	MSB of result
Z	"	0	Result not zero
P	"	1	Even parity
AC	"	1	Carry from bit 3

Table 13.9 Data transfer group instructions

Instruction Mne-monic	Instruction Oper-ands	Op Code (hex)	Instruction Mne-monic	Instruction Oper-ands	Op Code (hex)	Instruction Mne-monic	Instruction Oper-ands	Op Code (hex)	Instruction Mne-monic	Instruction Operands	Op Code (hex)
	A,A	7F		D,A	57		L,A	6F		B, 16-bit	01
	A,B	78		D,B	50		L,B	68		D, 16-bit	11
	A,C	79		D,C	51		L,C	69	LXI	H,16-bit	21
MOV	A,D	7A	MOV	D,D	52	MOV	L,D	6A		SP,16-bit	31
	A,E	7B		D,E	53		L,E	6B			
	A,H	7C		D,H	54		L,H	6C	LDA	16-bit addr	3A
	A,L	7D		D,L	55		L,L	6D			
	A,M	7E		D,M	56		L,M	6E	STA	16-bit addr	32
	B,A	47		E,A	5F		M,A	77	LHLD	16-bit addr	2A
	B,B	40		E,B	58		M,B	70			
	B,C	41		E,C	59		M,C	71			
	B,D	42		E,D	5A		M,D	72	SHLD	16-bit addr	22
MOV	B,E	43	MOV	E,E	5B	MOV	M,E	73			
	B,H	44		E,H	5C		M,H	74		B	0A
	B,L	45		E,L	5D		M,L	75	LDAX	D	1A
	B,M	46		E,M	5E						
	C,A	4F		H,A	67		A, byte	3E		B	02
	C,B	48		H,B	60		B, byte	06	STAX	D	12
	C,C	49		H,C	61		C, byte	0E			
MOV	C,D	4A	MOV	H,D	62		D, byte	16			
	C,E	4B		H,E	63	MVI	E, byte	1E			
	C,H	4C		H,H	64		H, byte	26	XCHG		EB
	C,L	4D		H,L	65		L, byte	2E			
	C,M	4E		H,M	66		M, byte	36			

Example 13.5 Assume that the accumulator contains 14 H and that the carry bit is set (CY = 1). The instruction ACI 42H will have the following effect:

Register A	=	14H	=	00010100
Immediate data	=	42H	=	01000010
Carry (CY)	=	1	=	1
				01010111 = 57H

The contents of the accumulator following ACI 42H instruction will be 57H and the carry flag will be reset (CY = 0). All other flags will also be affected.

Subtract Instructions Subtraction instructions are used to subtract the contents of the specified register, memory location or data-byte in the instruction itself, from the contents of the accumulator. The result of a subtraction is stored in the accumulator. If the result is negative, the carry flag will be set (CY= 1). The setting of the carry flag indicates that the result in the accumulator is in 2's complement form. The subtraction is actually performed by adding 2's complement of the subtrahend to the minuend. All the flags are affected by these instructions. The subtraction instructions are given below:

SUB r Subtract the contents of the register r from the accumulator.

SUB M Subtract the contents of the memory location (pointed to by H–L register pair) from the accumulator.

SUI data Subtract the data-byte from the accumulator.

Example 13.6 Consider the instruction SUB M. Let the contents of the relevant memory location, registers and flags be as given below:

Accumulator = 3EH
Register H = 00H
Register L = 7CH
Memory location (007CH) = 9FH

CY = 0
S = 0
P = 1
Z = 0
AC = 1

The subtraction operation performed, following the execution of SUB M instruction, is as follows:

Accumulator = 3EH = 00111110
Memory location contents = 9FH = 01100001 (2's complement)

9FH = 10011111

The contents of the memory location, registers H, and L are not affected by this instruction. The contents of the accumulator will be 10011111 and the condition flags are set as follows:

CY = 1 Indicates negative result.
S = 1 Indicates MSB of result is 1.
Z = 0 Indicates result is non-zero.

$P = 1$ Indicates even parity.
$AC = 0$ Indicates absence of auxiliary carry.

Since the carry flag is set indicating negative result, the contents of the accumulator is in 2's complement form and, therefore, the actual result is – 61H.

Note carefully the setting of the carry flag in the subtraction operation. In fact, it is named as *borrow flag* during the subtraction operation. If the addition of 2's complement of the subtrahend does not produce a final carry output then the flag is set ($CY = 1$), whereas if a carry is produced then the carry flag is reset ($CY = 0$).

Another group of subtraction instructions subtracts the contents of memory location, register or immediate data-byte and the borrow flag (CY) from the contents of the accumulator, and the result is stored in the accumulator. These are:

SBB r — Subtract the register r with the borrow.
SBB M — Subtract the memory with the borrow.
SBI data — Subtract the data-byte with the borrow.

All the flags are affected in the same manner as discussed above, for normal subtraction operation.

Example 13.7 In Ex. 13.6, if $CY = 1$ and SBB M instruction is executed, the subtraction operation will be performed as shown below:

Accumulator = 3*E*H =	0011 1110	
Subtract *CY*	1	
	0011 1101	
Memory location = 9*F*H contents	0110 0001	(2's complement)
	1001 1110	

Since the addition process does not produce a carry, the borrow flag is set ($CY = 1$), following the execution of this instruction, indicating a negative result. You can find the settings of other condition flags.

Increment/Decrement Instructions The increment instructions increment the contents of the specified register or memory location by one and the decrement instructions decrement the contents by one. The results of increment/decrement are left in the same register or memory location.

These instructions affect all the flags except the carry flag. Since the carry flag is preserved by these instructions, these can be used within multi-byte routines for incrementing/decrementing counts and similar purposes. The increment/decrement instructions are given below:

INR r — Increment the contents of the register r by one.
INR M — Increment the contents of the memory location (pointed to by H–L register pair) by one.
DCR r — Decrement the contents of the register r by one.
DCR M — Decrement the contents of the memory location by one.

Increment/Decrement Register Pair The instruction INX rp adds one to the contents of specified register pair (rp) and DCX rp decrements the contents of the specified register pair by one. Any one of the four (B, D, H) register pairs or SP may be incremented or decremented by these instructions.

Since these instructions do not affect the condition flags, they can be used to modify the address in any instruction sequence.

Double Precision Addition The instruction DAD rp adds the contents of the specified register pair (rp) to the H–L register pair. The 16-bit result is stored in the H–L register pair and the carry flag is set if there is a carry-out of the 16-bit addition. No other flags are affected.

Example 13.8 Assume the following contents of various registers before the DAD B instruction is executed. Find the contents after the execution of this instruction.

(PC) = 0F78H (H) = 02H
(B) = 3FH (L) = 46H
(C) = BAH $(PSW)_L$ = 17H

Solution

The contents after the execution of the instruction DAD B are given below:

(PC) = 0F79H
(B) = 3FH
(C) = BAH
(H) = 42H
(L) = 00H
$(PSW)_L$ = 16H

Decimal Adjust Accumulator Instruction The instruction DAA is used when adding decimal numbers. When decimal numbers are added, the 8-bit value in the accumulator may not be a valid BCD. This instruction operates as follows:

1. If the least-significant four bits of the accumulator is not a valid BCD digit (greater than 9), or if the auxiliary carry flag is set ($AC = 1$), then 0110 is added to the accumulator.
2. If the most-significant four bits of the accumulator is not a valid BCD digit, or if the carry flag is set ($CY = 1$), then 0110 0000 is added to the accumulator.

This is the only instruction which requires the use of the auxiliary carry (AC) flag. All the flags are affected by this instruction.

All the instructions of arithmetic group, in terms of their mnemonics and operands along with their Op code (hex), are given in Table 13.11.

Logic Group Instructions

Logic Instructions There are instructions for performing AND, OR, EX-OR and comparison operations. The bits of the specified register, memory location or data-byte in the instruction and the corresponding bits of the accumulator are involved in the operation on a bit-by-bit basis. The result of AND, OR, and EX-OR operation is placed in the accumulator. All the flags are affected, but the carry (CY) and the auxiliary carry (AC) flags setting depend upon the operation being performed rather than on the results of operation. These are given in Table 13.12.

In the case of comparison instruction, the contents of the specified register, the memory location, or the immediate byte are subtracted from the contents of the accumulator. The accumulator contents do not change, but the comparison is indicated by the zero and carry flags. The zero flag is set ($Z = 1$) if both the numbers are equal and the carry flag is set ($CY = 1$)

Table 13.11 Arithmetic group instructions

Instruction Mnemonic	Operands	Op code (hex)	Instruction Mnemonic	Operands	Op code (hex)	Instruction Mnemonic	Operands	Op code (hex)
ADD	A	87	SBB	A	9F	INX	B	03
	B	80		B	98		D	13
	C	81		C	99		H	23
	D	82		D	9A		SP	33
	E	83		E	9B	DCX	B	0B
	H	84		H	9C		D	1B
	L	85		L	9D		H	2B
	M	86		M	9E		SP	3B
ADC	A	8F	INR	A	3C	DAD	B	09
	B	88		B	04		D	19
	C	89		C	0C		H	29
	D	8A		D	14		SP	39
	E	8B		E	1C	ADI	byte	C6
	H	8C		H	24	ACI	byte	CE
	L	8D		L	2C	SUI	byte	D6
	M	8E		M	34	SBI	byte	DE
SUB	A	97	DCR	A	3D	DAA		27
	B	90		B	05			
	C	91		C	0D			
	D	92		D	15			
	E	93		E	1D			
	H	94		H	25			
	L	95		L	2D			
	M	96		M	35			

Table 13.12 Flag setting for logic instructions

Operation	Instructions	Flags settings
AND	ANA r ANA M ANI byte	$CY = 0$, $AC = 1$
EX-OR	XRA r XRA M XRI byte	$CY = 0$, $AC = 0$
OR	ORA r ORA M ORI byte	$CY = 0$, $AC = 0$
Compare	CMP r CMP M CPI byte	$Z = 1$ if (A) = (byte) $CY = 1$ if (A) < (byte)

if the contents of the accumulator is less than the byte being compared. The other flags are affected as in the usual subtraction process.

Example 13.9 Suppose the accumulator contains *AAH* and the register B contains 0*F*H. Assume the following flag settings:

$$Z = 1, \quad CY = 1, \quad S = 1, \quad P = 1, \quad AC = 0$$

Find the contents of the accumulator, register *B*, and flag settings, following execution of each of the following instructions:

```
ANA   B
XRA   B
ORA   B
CMP   B
```

Solution These are given in Table 13.13.

Table 13.13 Table for Ex. 13.9

Instruction	*Contents of registers*		*Flags*				
	A	*B*	*CY*	*AC*	*Z*	*S*	*P*
ANA B	00001010	00001111	0	1	0	0	1
XRA B	10100101	00001111	0	0	0	1	1
ORA B	10101111	00001111	0	0	0	1	1
CMP B	10101010	00001111	0	0	0	1	0

Note that the contents of the accumulator do not change, following the execution of the CMP instruction. Also, $CY = 0$ and $Z = 0$ and the contents of A are greater than those of B.

Rotate Instructions The rotate or shift instructions have two variations which differ in the way the carry bit is used. Operands are not permitted with these instructions, and only the carry flag is affected following their execution. The instructions are explained below.

Rotate Accumulator Left (RLC) RLC sets the carry flag equal to the most-significant bit (MSB) of the accumulator and then rotates the contents of the accumulator one bit position to the left with the MSB transferring to the LSB. This is shown in Fig. 13.16*a*.

Rotate Accumulator Right (RRC) RRC sets the carry flag equal to the LSB of the accumulator and then rotates the contents of the accumulator one bit position to the right with the LSB transferring to the MSB. This operation is shown in Fig. 13.16*b*.

Rotate Left through Carry (RAL) RAL rotates the contents of the accumulator and the carry flag one bit position to the left. The carry flag is transferred to LSB and is set by the MSB. This operation is shown in Fig. 13.16*c*.

Rotate Right through Carry (RAR) RAR rotates the contents of the accumulator and the carry flag one bit position to the right. The carry flag is transferred to the MSB and is set by the LSB. This operation is shown in Fig. 13.16*d*.

Special Instructions The instruction CMA complements or inverts each bit in the accumulator and the flags are not affected. STC sets the carry flag to one ($CY = 1$) and CMC complements the carry flag. These instructions do not affect the other flags.

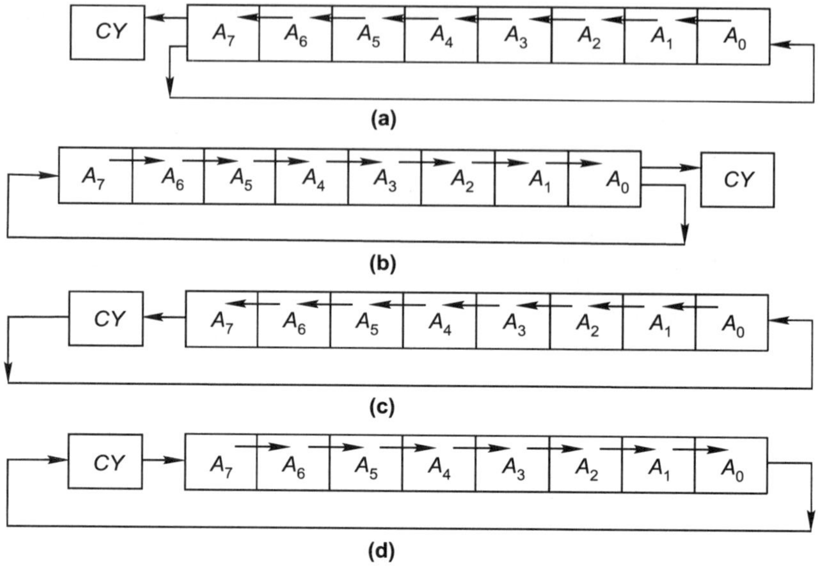

Fig. 13.16
Rotate instructions.

All the instructions of logic group in terms of their mnemonics and operands along with their Op code (hex) are given in Table 13.14.

Table 13.14 Logical group instructions

Instruction		*Op code*	*Instruction*		*Op code*
Mnemonic	*Operand*	*(hex)*	*Mnemonic*	*Operand*	*(hex)*
	A	A7		A	BF
	B	A0		B	B8
	C	A1		C	B9
ANA	D	A2		D	BA
	E	A3		E	BB
	H	A4	CMP	H	BC
	L	A5		L	BD
	M	A6		M	BE
	A	AF	ANI	byte	E6
	B	A8	XRI	byte	EE
	C	A9	ORI	byte	F6
XRA	D	AA	CPI	byte	FE
	E	AB	RLC		07
	H	AC	RRC		0F
	L	AD	RAL		17
	M	AE	RAR		1F
			CMA		2F
	A	B7	STC		37

(*Contd.*)

Table 13.14 *(Contd.)*

Instruction		*Op code*	*Instruction*		*Op code*
Mnemonic	*Operand*	*(hex)*	*Mnemonic*	*Operand*	*(hex)*
	B	*B*0	CMC		3*F*
	C	*B*1			
	D	*B*2			
ORA	E	*B*3			
	H	*B*4			
	L	*B*5			
	M	*B*6			

Branch-Group Instructions

The instructions in this group change the normal sequence of program execution. As discussed earlier, the microprocessor fetches instructions from contiguous memory locations. The address of the next byte to be fetched from the memory is in the program counter (PC), which automatically gets incremented by one when the byte is fetched from the memory. The sequence of fetching bytes from the memory can be altered by changing the contents of the PC. For this purpose, the various instructions are: jump, call, return, restart, and move H–L to the PC.

The jump, call, and return operations may be unconditional or conditional. For conditional instructions, the operation is carried out if the specified condition is true. The conditions that may be specified are as follows:

NZ – Not zero $(Z = 0)$
Z – Zero $(Z = 1)$
NC – No carry $(CY = 0)$
C – Carry $(CY = 1)$
PO – Parity odd $(P = 0)$
PE – Parity even $(P = 1)$
P – Plus $(S = 0)$
M – Minus $(S = 1)$

The condition flags are not affected by the branch-group instructions.

Jump Instructions The jump instructions are 3-byte instructions with the first byte containing the Op code and the second and third bytes containing a 16-bit address. The lower byte of the address is the second byte and the higher byte is the third byte.

When an unconditional jump instruction (JMP addr) is executed, the address in the instruction is loaded into the program counter. The CPU will therefore fetch its next instruction from the memory at this new address. The microprocessor then continues to execute the instructions sequentially from this address. This instruction may be used to keep a program running over and over again in a continuous loop. For this, the last instruction in the program is a JMP. The address bytes of the instruction are loaded with the starting address of the program. Each time the microprocessor reaches this instruction, it sets the PC back to the starting address. The program then runs again from the starting point.

Conditional jump instructions, such as JC (jump if carry), will cause the PC to be loaded with the address bytes, specified in the instruction only if the specified condition is true ($CY = 1$), otherwise, the execution of instructions will continue in its normal sequence. The conditional jump instructions can be used to keep running around a program loop, until the specified condition is satisfied.

The various conditional jump instructions are:

JNZ	Jump if zero flag is not set	$(Z = 0)$
JZ	Jump if zero flag is set	$(Z = 1)$
JNC	Jump if carry flag is not set	$(CY = 0)$
JC	Jump if carry flag is set	$(CY = 1)$
JPO	Jump if the parity of the byte is odd	$(P = 0)$
JPE	Jump if the parity of the byte is even	$(P = 1)$
JP	Jump if sign flag is not set	$(S = 0)$
JM	Jump if sign flag is set	$(S = 1)$

Example 13.10 Write a program to transfer ten bytes of data from one area of the memory to another. Assume that the bytes are in contiguous memory locations starting from 0100 (hex) address, and are to be tran sferred to starting from A200 (hex) address.

Solution

The program is given below:

```
      LXI D,  0100H     Load the register pair D–E with 0100H.
      LXI H,  A200H     Load the register pair H–L with A200H.
      MVI C,  0AH       Load the register C with decimal 10.
LOOP: LDAX  D           Load the accumulator.
      STAX  H           Store the accumulator.
      INX   D           Increment the D–E register pair.
      INX   H           Increment the H–L register pair.
      DCR   C           Decrement the register C.
      JNZ   LOOP        Jump not zero to LOOP.
```

In the above program, register C has been used as a counter. Let LOOP be the label of the address at which the instruction LDAX D has been stored. The conditional jump instruction JNZ will test the zero flag and, as long as it is not set, the control will be transferred to the address LOOP.

Call Instructions A call instruction is used to jump to the starting address of a subprogram or subroutine in the memory. A subroutine is a short program section which does a specific task, for example, multiplying two numbers. If, in a program, it is required to multiply two numbers a number of times, the sequence of instructions for performing the multiplication can be stored once in memory and then just called each time it is needed. This approach uses the memory space only once rather than storing this sequence of instructions in the main program again and again whenever it is needed.

A call instruction contains the starting address of the subroutine in the second byte (lower-order address) and third byte (higher-order address) of the instruction. Upon execution of the call instruction, the program counter is loaded with the address specified in the instruction.

The execution returns to the main program at the end of the subroutine by using a return (RET) instruction. This is made possible by saving the contents of the PC in a special section of RAM, the stack, during the execution of the call instruction.

Example 13.11 Let the instruction CALL 0*A*2*F*7H be stored starting at the memory location 147*A*H and the contents of the stack pointer (SP) be *FFFF*H. Find the contents of the PC and relevant memory locations following the execution of the call instruction.

Solution These are given in Table 13.15.

Table 13.15 Table for Ex. 13.11

Register/Memory location	*Contents*	
	Before CALL	*After CALL*
PC	147*A* H	*A*2*F*7 H
SP	*FFFF* H	*FFFD* H
147*A*H	CALL (op code)	CALL
147*B*H	*F*7H (lower byte)	*F*7H
147*C*H	*A*2H (higher byte)	*A*2H
*FFFF*H	Any byte	no change
*FFFE*H		14 (higher byte of next instruction)
*FFFD*H		7D (lower byte of next instruction)

Note that the stack grows downward in the memory.

A call can be conditional. In this case, the execution jumps to the address specified in the instruction only if the condition specified is true. The various call instructions with their meanings are given below:

CNZ	Call if	$Z = 0$
CZ	Call if	$Z = 1$
CNC	Call if	$CY = 0$
CC	Call if	$CY = 1$
CPO	Call if	$P = 0$
CPE	Call if	$P = 1$
CP	Call if	$S = 0$
CM	Call if	$S = 1$

Return Instructions The return instruction (RET) at the end of a subroutine transfers the program execution back to the next instruction after the call instruction in the main program. The return instruction does not contain the address and the program counter will be loaded with the top two bytes of the stack following the execution of this instruction. Now, the execution of the main program sequence will continue.

Return also can be unconditional or conditional. The various conditional return instructions with their meanings are given below:

RNZ	Return if	$Z = 0$
RZ	Return if	$Z = 1$
RNC	Return if	$CY = 0$

RC	Return if	$CY = 1$
RPO	Return if	$P = 0$
RPE	Return if	$P = 1$
RP	Return if	$S = 0$
RM	Return if	$S = 1$

The subroutine can be simple or nested. The program flow in a simple subroutine is shown in Fig. 13.17*a* and that in a nested subroutine, in Fig. 13.17*b*. It is clear from Fig. 13.17*b* that a call instruction in a subroutine can call another sub-routine.

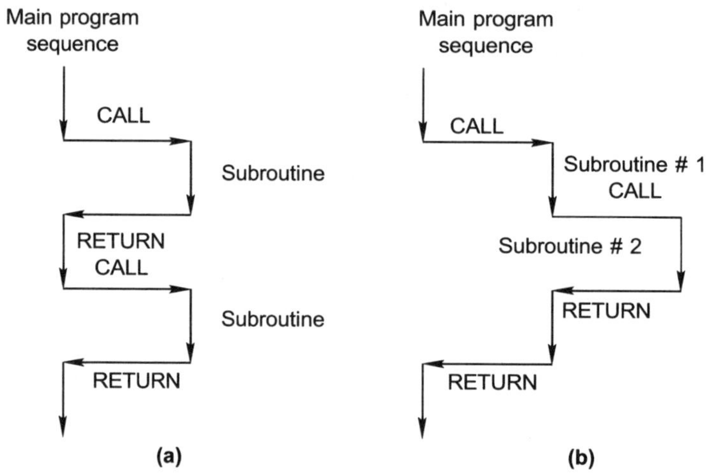

Fig. 13.17

Program flow with call instructions: (*a*) simple subroutines (*b*) nested subroutines.

The microprocessor executes the program until it encounters a call instruction, then goes to subroutine # 1. It executes subroutine # 1 until it encounters another call instruction, then it goes to subroutine # 2. When subroutine # 2 is completed, the return instruction at the end of this subroutine transfers the control to subroutine # 1 to the instruction, next to the call instruction. When subroutine # 1 is completed, the return instruction at the end of this subroutine transfers the control to the instruction next to the call instruction in the main program.

Restart Instruction Restart instruction (RST *n*) is a special purpose CALL instruction, designed primarily for use with interrupts. It pushes the contents of the program counter onto the stack to provide a return address and then jumps to one of the eight addresses, determined by the value of *n* (a three-bit binary code). It is a single-byte instruction and seldom appears as an instruction in the program. More often, the peripheral devices seeking interrupt service pass this one-byte instruction to the processor. The eight restart instructions, and the corresponding addresses to which the control is transferred following their execution are given below:

Instruction	*Address to which control is transferred*
RST 0	0000H
RST 1	0008H
RST 2	0010H

RST 3	0018H
RST 4	0020H
RST 5	0028H
RST 6	0030H
RST 7	0038H

PCHL Instruction This instruction moves the contents of the H–L pair to the program counter. Following the execution of this instruction, the program execution jumps to the address which was stored in the H–L register pair.

Condition flags are not affected by any instruction in this group. All the instructions of branch group, in terms of their mnemonics and operands along with their Op code (hex), are given in Table 13.16.

Table 13.16 Branch group instructions

Instruction		*Op code*	*Instruction*		*Op code*
Mnemonic	*Operand*	*(hex)*	*Mnemonic*	*Operand*	*(hex)*
JMP	addr	C3	RET		C9
JNZ	addr	C2	RNZ		C0
JZ	addr	CA	RZ		C8
JNC	addr	D2	RNC		D0
JC	addr	DA	RC		D8
JPO	addr	E2	RPO		E0
JPE	addr	EA	RPE		E8
JP	addr	F2	RP		F0
JM	addr	FA	RM		F8
PCHL		E9			
CALL	addr	CD		0	C7
CNZ	addr	C4		1	CF
CZ	addr	CC		2	D7
CNC	addr	D4	RST	3	DF
CC	addr	DC		4	E7
CPO	addr	E4		5	EF
CPE	addr	EC		6	F7
CP	addr	F4		7	FF
CM	addr	FC			

Stack, I/O and Machine Control Group Instructions

This group of instructions manipulates the stack, performs I/O operations, and alters internal control flags.

Stack Operations A small portion of the RAM is set aside to store return addresses for subroutine calls and restart instructions. It can also be used to store the contents of the register pairs (B–C, D–E, or H–L) or the program status word (PSW), so that they will not be changed or destroyed by a subroutine. For example, data stored in the register B in the main program will be altered if the register *B* is also used in the subroutine. The PUSH instruction is used to push the contents of B–C, D–E, H–L, or PSW to the stack.

Data has to be stored in the stack on a last-in-first-out (LIFO) basis. The stack-pointer register (SP) holds the address of the stack top, which is the last occupied location of the stack. The stack grows downward in memory from the starting address, which means that the SP decrements after each byte is pushed onto the stack. It increments after each byte is popped off the stack, following the execution of POP instruction. The popped-off bytes can be stored in the B–C, D–E, H–L register pairs or in the accumulator and flags (PSW). The organisation of a PSW is shown in Fig. 13.11.

Example 13.12 Assume that the stack begins at the memory location *FFFFH* and the bottom-most two locations are full, as shown in Fig. 13.18*a*.

Solution

Now, if the instruction PUSH H is executed, the contents of the register pair H–L will be pushed in the stack as shown in Fig. 13.18*b*.

Now, if POP H instruction is executed, the stack will be as shown in Fig. 13.18*a*. The byte from the memory location *FFFC* will be transferred to the lower-order register (L) and the byte from the next location *FFFD* will be transferred to the higher-order register (H). The SP gets incremented by two and it now contains *FFFE*.

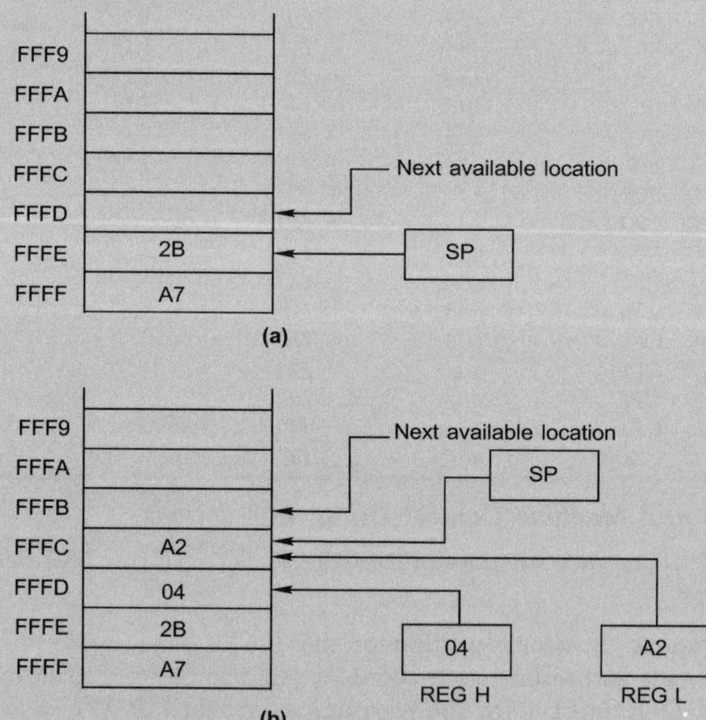

Fig. 13.18

Stack (*a*) before, and (*b*) after the PUSH instruction.

There are two more instructions involving the stack. These are XTHL and SPHL. XTHL exchanges two bytes of data from the top of the stack with the two bytes in the H and L registers. The contents of the L register is exchanged with the contents of the memory location whose address is stored in SP, and the contents of the H register is exchanged with the contents of the memory location whose address is one more than the contents of the SP.

The instruction SPHL moves the contents of the registers H and L into the stack-pointer register.

All instructions, except POP PSW, involving stack operation do not affect the condition flags.

I/O Instructions The IN instruction inputs a byte of data from an input port, specified in the instruction, to the accumulator. The OUT instruction outputs a byte of data from the accumulator to the port specified in the instruction. These are two-byte instructions, the second byte being the port address. Since the port address is 8-bit, the microprocessor can address upto 256 input and output devices. The condition flags are not affected by these instructions.

Control Instructions

These are single-byte instructions without any operand. The various control instructions are explained below.

Interrupt Enable (EI) The interrupt system is enabled following the execution of the next instruction after EI.

Disable Interrupt (DI) The interrupt system is disabled immediately, following the execution of the DI instruction.

Halt (HLT) The halt instruction stops the processor. The registers and flags are unaffected.

There are two ways in which the microprocessor comes out of the halt state. These are: hold and interrupt. If the signal on the hold line is a logical 1 then the processor leaves the halt state as long as the hold is 1. As soon as the hold line goes LOW, it again enters the halt state.

The processor permanently comes out of the halt state, if interrupted, provided the interrupt system is enabled before the HLT instruction.

No Operation (NOP) During the execution of this instruction, the processor does nothing. The registers and flags are unaffected. The NOP instructions are used in a program for the processor to wait for a certain time. Also, they are put in to be substituted by other instructions when some instructions are missed due to some mistake in programming, to avoid shifting of the entire program in the memory.

Set Interrupt Mask (SIM) The SIM instruction is used to perform the following functions:

1. Set the interrupt mask (disable) for the hardware interrupts RST 5.5, RST 6.5, and RST 7.5.
2. Reset RST 7.5 interrupt.
3. Output bit 7 of the accumulator to the serial-output data latch.

The desired bit-configuration is to be loaded into the accumulator before executing the SIM instruction. Figure 13.19 shows the interpretation of each bit of the accumulator by the SIM instruction. Bit 3 is used for enabling the interrupt mask and resetting the operation of RST 7.5 FLIP-FLOP. Unless it is a 1, bits 0, 1, 2, and 4 will have no effect.

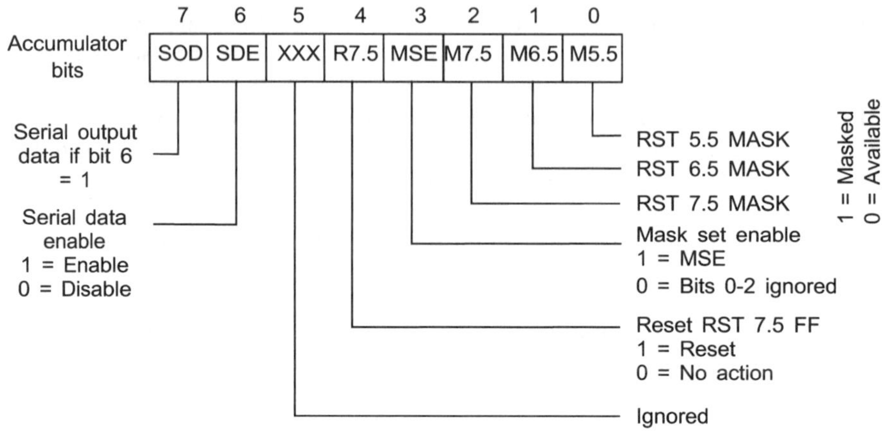

Fig. 13.19
Bit format of SIM instruction.

If it is desired to send a serial output bit, without affecting the interrupt mask, the bit 3 is made 0. If bit 3 is a 1 and, for example, bit 1 is a 1, then the RST 6.5 interrupt input will not respond to an interrupt signal. This means that RST 6.5 interrupt is masked.

The DI instruction overrides the SIM instruction. This means, whether masked or not, RST 5.5, RST 6.5, and RST 7.5 interrupts are all disabled when the DI instruction is in effect.

The RST 7.5 is sensed via a processor `FLIP-FLOP`, that is, set when a peripheral device issues a pulse with a rising edge. This `FLIP-FLOP` is not affected by the setting of the interrupt mask or the DI instruction and, therefore, can be set at any time. However, the interrupt cannot be serviced when RST 7.5 is masked or the DI instruction is in effect.

If bit 6 is set to 1, the serial-output data function is enabled. The processor latches the accumulator bit 7 into the SOD output where it can be accessed by a peripheral device. If bit 6 is 0, then bit 7 is ignored. A hardware reset will set all RST masks, reset/disable all interrupts, and sets the SOD latch to zero.

> **Example 13.13** Assume that the accumulator contains 00011100 when SIM instruction is executed. The SIM instruction resets the RST 7.5 interrupt. If an RST 7.5 interrupt is pending when this SIM instruction is executed, it is overridden without being serviced. Also, any subsequent RST 7.5 interrupt is masked and cannot be serviced until the interrupt mask is serviced.

> **Example 13.14** Assume that the accumulator contains 11001111. The SIM instruction masks out the RST 5.5, RST 6.5, and RST 7.5 level interrupts and loads the accumulator bit 7 into the SOD latch.

Read Interrupt Mask (RIM) The RIM instruction loads the accumulator indicating the following:

1. current setting of the interrupt mask,
2. setting of the interrupt flag,
3. pending interrupts, and
4. one bit of serial input data, if any.

Figure 13.20 shows the interpretation of each bit of the accumulator following the execution of the RIM instruction.

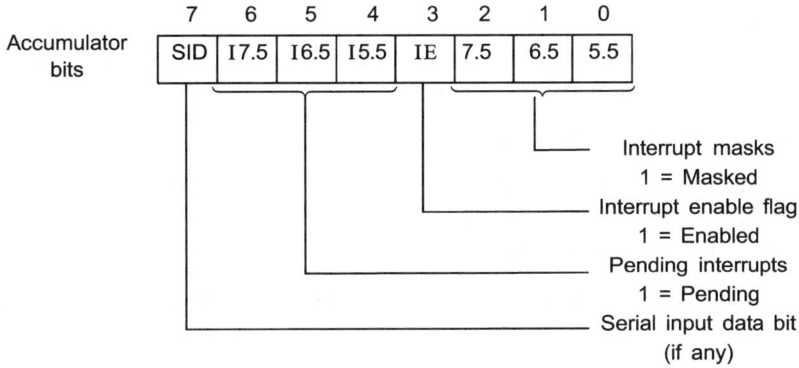

Fig. 13.20
Bit format of RIM instruction.

The mask and pending flags refer to the RST 5.5, RST 6.5, and RST 7.5 interrupts. The IE flags refer to the entire interrupt system. Bit 7 contains data brought into the accumulator from the SID input.

All the instructions of the stack, *I/O*, and machine control group in terms of their mnemonics and operands along with their Op codes (hex) are given in Table 13.17.

Table 13.17 Stack, I/O, and machine control group instructions

Instruction		*Op code*	*Instruction*		*Op code*
Mnemonic	*Operand*	*(hex)*	*Mnemonic*	*Operand*	*(hex)*
	B	*C5*	OUT	byte	*D3*
PUSH	D	*D5*	IN	byte	*DB*
	H	*E5*	DI		*F3*
	PSW	*F5*	EI		*FB*
			NOP		00
	B	*C1*	HLT		76
POP	D	*D1*	RIM		20
	H	*E1*	SIM		30
	PSW	*F1*			
XTHL		*E3*			
SPHL		*F9*			

13.11 THE 8086 MICROPROCESSOR

The 16-bit 8086 microprocessor was introduced by Intel in 1978. It was fabricated using silicon-gate H-MOS process, containing 29,000 transistors on a chip of about 225 mils square. The high performance MOS technology resulted in a 5 MHz clock rate, making it faster than any μP chip available at that time.

The 8086 μP has the attributes of both 8- and 16-bit μPs. It has capabilities not available in 8-bit microprocessors, such as, 16-bit arithmetic, signed 8- and 16-bit arithmetic, including multiply and divide, efficient interruptible byte–string operations, and improved bit manipulation. It also includes operations such as reentrant code, position-independent code, and dynamically relocatable programs. The 8086 μP can directly address upto 1 megabyte (MB) of memory and can support multiple-processor configurations.

13.11.1 The 8086 Architecture

Figure 13.21 shows internal block diagram of the 8086 microprocessor. It is divided into two units, the *bus interface unit* (BIU), and the *execution unit* (EU). Since the operation in this microprocessor is divided into two units, its speed of operation increases.

The Bus Interface Unit The bus interface unit handles all transfers of data and addresses on the buses for fetching instructions from the memory, read and write operations for the memory

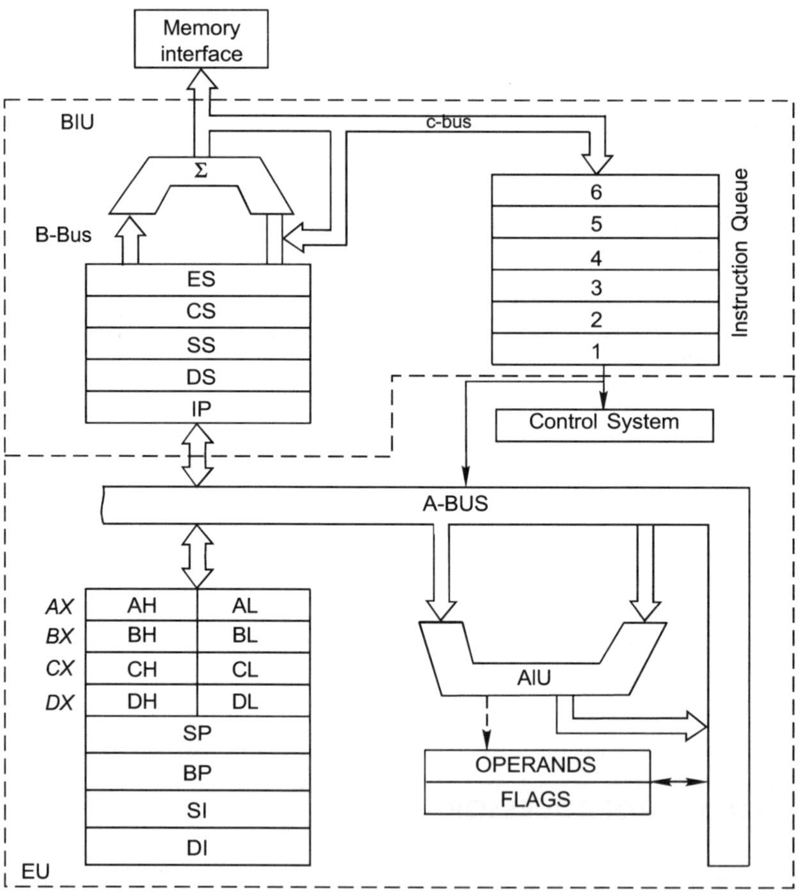

Fig. 13.21
Internal block diagram of Intel 8086 microprocessor.

and the I/O devices. It has a queue of six bytes for holding prefetched instruction bytes operating in the FIFO configuration. Due to this facility, the BIU can fetch instruction bytes while the EU is decoding or executing an instruction, which does not require use of the buses.

The 16-bit data bus is time multiplexed with the sixteen bits of address for both memory and *I/O* (AD_{15}–AD_0). The width of the address bus for memory referencing is 20-bit, consisting of A_{19}–A_{16} and AD_{15}–AD_0, which can address 2^{20} = 1,048,576 = 1 Megabyte (MB) of memory. The memory is organised in the form of 64K byte (KB) segments. The method of addressing memory is given below.

The BIU contains four 16-bit segment registers: The code segment (CS) register, the stack segment (SS) register, the extra segment (ES) register, and the data segment (DS) register. These segment registers are loaded with the most-significant 16 bits of the starting addresses of the corresponding memory segments which the 8086 is working with at a particular time. The BIU inserts zeros for the lowest four bits of the 20-bit starting address for a segment. For example, if the code segment register contains 73*AEH*, the code segment will start at 73*AE*0 H.

If the first byte of the word is at an even address, 8086 reads the entire word in one operation. On the other hand, if the first byte of the word is at an odd address, 8086 will read the first byte in one operation and the second byte, in another operation. For an even starting address of the word, the data byte with even address is transferred on the D_7–D_0 bus (lower byte), while odd-addressed data byte is transferred on the D_{15}–D_8 bus lines (higher byte). When the 8086 μP is reset, it always begins execution at the location *FFFF*0 H, where the jump instruction must be stored in memory.

Instruction-pointer (IP) register is a 16-bit register which holds the address of the next byte of the instruction code within the code segment of the memory. The value contained in the IP register is added to the 20-bit base address formed using CS register and it generates 20-bit physical address of the code byte. For example, if the CS contains 124*A*H and IP contains 0728H, the 20-bit physical address of the code byte generated will be:

$$\begin{array}{r} 1\;2\;4\;A\;0\text{H} \\ +\;0\;7\;2\;8\text{H} \\ \hline 1\;2\;BC\;8\text{H} \end{array}$$

The Execution Unit The execution unit (EU) decodes the instructions fetched from the memory and executes the instructions. It tells the BIU about the locations from where to fetch instructions or data at any time. It contains control circuitry, 16-bit ALU, general purpose and index registers, flags, etc.

It contains eight 8-bit general purpose registers, which can also be used in pairs to form four 16-bit registers. The 16-bit registers are labelled as AX (AH– AL), BX (BH–BL), CX (CH–CL), and DX (DH–DL) registers. The AX register is called the accumulator.

The stack-segment register (SS) contains the 16-bit (upper) part of the starting address of the stack and the stack pointer register (SP) contains the 16-bit offset value. The physical address in the stack segment is obtained by shifting the contents of the SS register towards left by four bit positions, adding four zeros in the least-significant four bit positions and adding the 16-bit contents of the SP in this 20-bit number.

The base pointer (BP), source index (SI) and destination index (DI) registers are used to hold the 16-bit offset of a data word in data-segment (DS) register or extra-segment (ES) register. Various addressing modes use these registers in different ways.

The 8086 μP contains nine flags. Six of them are conditional flags which indicate some condition produced by the execution of an instruction and the remaining three are used to control certain operations of the processor. The conditional flags are:

1. Carry flag: *CF*
2. Parity flag: *PF*
3. Auxiliary carry flag: *AF*
4. Zero flag: *ZF*
5. Sign flag: *SF*
6. Overflow flag: *OF*

The control flags are:

1. Trap flag (TF): Used for single stepping through a program.
2. Interrupt flag (IF): Used to allow/prohibit the interruption of a program.
3. Direction flag (DF): Used with string instructions.

13.12 PROGRAMMING LANGUAGES

As is evident from the way microprocessors operate and information is stored and transferred between various subsystems, it understands only the binary signals (0 and 1). Therefore, the program is to be written in binary form which is referred to as the *machine language.* Machine language alone is directly executable by a microprocessor. The microprocessor decides whether the binary number corresponds to an instruction, part of an address, or simply a piece of data, according to what it expects at a particular point in the instruction cycle. Therefore, the binary codes corresponding to the sequence of instructions in a program must be stored in proper sequence, otherwise the information will be misinterpreted by the microprocessor, resulting in serious errors.

For programming in machine language, the programmer needs a table of binary instruction codes. Also, he/she has to keep track of the contents of memory locations, registers, and flags. Of course, he/she should be thoroughly familiar with the details of the microprocessor.

Consider the program of Ex. 13.3 written in machine language:

```
00000001
01110101
00100100
00010001
10010100
00110111
00001010
01101111
00011010
00000010
01111101
00010010
```

In this, it is difficult to determine whether a particular byte corresponds to an instruction, a data, or an address. Also, working with binary numbers is not convenient and we are liable to commit mistakes.

The machine-language program listing can be simplified somewhat by converting the binary codes to octal or hexadecimal format. Although such a shorthand notation alleviates some of the drudgery of writing the codes, it does not improve the understandability of the program. Also, the program has to be converted back to the machine language by some means before it is loaded into memory.

A better alternative to machine language is *assembly language.* In this, names are assigned to instructions, registers, data, and memory locations. These names are usually the abbreviations of the names or descriptions of the instructions (mnemonic), addresses, or data and are used to aid the designer's memory.

The assembly-language program is much easier to write, modify, understand and interpret than the corresponding machine-language program.

The program written in assembly language has to be translated into machine language for loading into the memory. This translation process is referred to as *assembly.* While short programs can be hand-assembled (manual translation) without much difficulty, for long programs, another program, referred to as *assembler*, is used to perform automatic translation.

Example 13.15 Consider the following program written in assembly language:

```
       LXI H, BUFER
       MVI C, 0AH
LOOP:  IN DATA
       MOV M,A
       INX H
       DCR C
       JNZ LOOP
```

Here, BUFER has been used as the symbolic name for 16-bit data. LOOP is the symbolic address of the memory location at which the Op code of IN is stored and DATA is the symbolic address of the input port.

The ease of programming is further enhanced by using high-level languages such as, PASCAL, FORTRAN, BASIC, etc. A single instruction of high-level language is equivalent to several machine-language instructions. High-level language programs must be translated to machine language for microprocessor execution. This is done by using a *compiler* program.

High-level languages have some drawbacks which make them unsuitable for some applications. The main drawbacks are:

1. They require more memory than machine- or assembly-language program.
2. The execution time required for high-level language programs is usually more than that required for machine- or assembly-language program.
3. They require a compiler program in the memory which is fairly expensive and need large memory space.

13.13 SUMMARY

The vast topic of microprocessors has been introduced in such a way so as to enable a novice to understand clearly the basic concepts involved. Since the introduction of the first microprocessor in 1971, a variety of microprocessors have been evolved which are commercially available. Although their principle of operation is more or less similar, they have wide differences as far as their architecture and instruction sets are concerned. It is not advisable for a novice to try to understand the details of different microprocessors simultaneously. The best course is to understand one microprocessor completely and then, without much difficulty, the other microprocessors can be learnt.

With this in view, the most popular 8-bit microprocessor, Intel's 8085A, has been selected for detailed discussion. Brief discussion of Intel's 16-bit microprocessor has also been included. The reader should write a number of programs to fully understand the options and the flexibilities available. It is not possible to cover all aspects of hardware and software of microprocessor-based systems in just one chapter, nevertheless, the basic concepts have been clearly discussed which will enable a beginner to initiate the study of microprocessor without any difficulty.

Glossary

Absolute address A memory address whose exact value is specified, and is not required to be computed or inferred from any other information.

Accumulator A register in the CPU of a computer or in a microprocessor that is used for most of the arithmetic and logic operations.

Addressing modes The methods used for specifying the address of hardware registers and/or memory to be used in an instruction.

Address space The number of memory locations and or I/O ports that a μP is capable of addressing.

Assembler A computer program that converts assembly language program into executable binary codes, i.e. machine language.

Assembler directive An instruction to the assembler. It is not translated into machine code.

Assembling The process of translating an assembly language program into machine language.

Assembly language It is a low-level programming language in which mnemonic instruction codes, labels, and names for data and addresses are used to refer directly to their binary equivalents.

Borrow flag A status bit indicating whether a borrow bit is produced or not as a result of a subtraction operation.

Branching instruction A μP instruction which causes transfer to a nonsuccessive memory address for the next instruction.

Bus, time-multiplexed A bus used for different tasks at different timings in a system. For example, a bus may be used to transmit address in one time slot and data in another time slot.

Call instruction An instruction that transfers control from one program segment to another program segment.

Carry flag A status bit indicating whether a carry bit is produced or not as a result of certain operation.

Compiler A computer program that translates commands in high-level language into machine code.

Computer architecture The specific interconnections of registers and logic blocks that define the design of a CPU or μP.

Conditional branching instruction A μP (or computer) instruction which causes transfer to a nonsuccessive memory address for the next instruction only when a specific condition is satisfied.

CPU (Central processing unit) The subsystem of a computer which is responsible for performing various arithmetic and logic operations and controls the overall operation of the computer. The microprocessor performs all these operations and hence it acts as CPU of a computer.

Data transfer instruction An instruction that causes data to be moved between internal registers of μP, or between μP and memory or μP and I/O devices.

Flag A single bit register in the μP that indicates certain conditions which may be used for taking decisions.

General purpose register A register that can be used for temporary data storage. It is also known as scratch pad register.

Hand assembly The process of translating an assembly language program into a machine language program mannually rather than by an assembler.

Hardware interrupt An interrupt caused by an external device connected to one of the interrupt pins of a μp.

High-level language A programming language in which the statements represent procedures rather than single instructions, for example, PASCAL, BASIC, C++ etc.

Input port A circuit that is used to interface an input device with the μP.

Instruction A group of bits that tells the μP what to do. It is a part of the instruction set of the μP that is specified by its manufacturers.

Instruction code Same as the op code.

Instruction cycle The process of fetching, decoding, and executing an instruction.

Instruction cycle time The time required to fetch, decode, and execute an instruction.

Instruction decoder A logic circuit used to interpret the instruction in μP.

Instruction execution Carrying out the tasks specified in the instruction after it has been fetched from the memory.

Instruction fetch Getting the instruction from the memory to the μP.

Instruction mnemonic A short and convenient name used to represent the operation in an instruction.

Instruction set The instructions which are allowed in a μP. It is specified by its manufacturer.

Interrupt A μP input that causes the temporary suspension of the normal sequence of operations and transfers control to a special routine meant to service the interrupt.

I/O address space The number of I/O devices which a μP is capable of addressing as standard I/O.

I/O device The device used to establish link between the μP and the outside world.

I/O mapped I/O I/O devices addressed as I/O rather than as memory locations. It is same as the standard I/O.

I/O port The circuit used to interface I/O devices with the μP. The data from μP or I/O device comes into and goes out to I/O device or μP respectively through the I/O port.

Jump instruction An instruction that causes the control to be transferred to a new location, whose value gets loaded in the program counter.

Logical shift A shift operation in which zeros are placed in empty bit locations.

Machine language The programming language that a computer or μP can directly understand. It is in the form of 1s and 0s.

Memory address space The number of memory locations which a μP is capable of addressing.

Memory mapped I/O I/O devices connected as memory locations.

Microcomputer A computer which uses a μP as its CPU.

Microprocessor architecture Same as the computer architecture.

Mnemonic Symbolic names or abbreviations for instructions, data, registers, addresses etc.

Op code (Operation code) A binary number that serves as a code for a particular operation in a μP.

Operand The quantity that is operated on.

Output port A circuit that is used to interface an output device with the μP.

Peripheral Same as the I/O device.

Program A sequence of instructions that direct a computer or a μP based system to perform a specific task.

Program counter (PC) A register in the μP that contains the address of the next instruction byte to be fetched from the memory.

Programming (of computer or μP) To program a computer or a μP based system.

Scratch-pad memory Memory locations or registers that are used to store temporary or intermediate results.

Stack A set of contiguous memory locations used in LIFO configuration.

Stack pointer *(SP)* A register in μP whose contents point to the top of the stack.

Standard I/O Same as I/O mapped I/O.

Unidirectional bus A bus that can transmit signals only in one direction.

Word size (or length) The number of bits that a μP or a computer is capable of processing at a time. It is usually same as the data bus width.

Review Questions

13.1 The word size of 8085 μP is ________ .

13.2 The 8085 μP has 8-bit address bus and 8-bit AD bus, it can address ________ memory.

13.3 A *μP* with an address bus width of ________ can address 1M bytes of memory.

13.4 I/O devices are interfaced with μP using ________ .

13.5 Stack resides in ________ .

13.6 The stack pointer gets ________ after PUSH instruction.

13.7 There are ________ number of flags in 8085 μP.

13.8 The I/O address space of 8085 μP is ________ .

13.9 NOP instruction of 8085 μP performs ________ operation.

13.10 The number of bytes of memory locations available for handling RST interrupts in 8085 μP based system is ________ .

13.11 For an 8085 μP operating at 3.125 MHz will have a crystal of frequency ________ .

13.12 The contents of 8085 μP's PC after resetting the μP will be ________ .

13.13 The 8086 μP can address ________ of memory.

13.14 In an 8086 μP based system if SS = 1234 H and SP = 0100H, the current address of the stack is ________ .

13.15 In an 8086 μP based system if IP = 1000 H and CS = 3000 H, the physical address of the next instruction will be ________ .

Problems

13.1 What is the size of the memory address space for the following microprocessors?

Microprocessor	*Data-bus width*	*Address-bus width*
8080A	8-bit	16-bit
6800	8-bit	16-bit
8086	16-bit	20-bit
9900	16-bit	16-bit
Z8000	16-bit	23-bit

13.2 How many different instructions are possible in an 8-bit microprocessor? Compare this number with the number of instructions of an 8085A.

13.3 For a particular 8085A microprocessor-based system, it is required to use 2K bytes of EPROM and 4K bytes of RAM. Assume 2142 RAMs and 2716 EPROMs are available.
(a) Find the number of RAM and EPROM chips required.
(b) Connect the memory chips to the microprocessor and show all the connections clearly. You can make suitable assumptions wherever required.
(c) Find the addresses of EPROMs and RAMs and express them in hex notation.

13.4 It is desired to clear the accumulator of 8085A. Explain the possible instructions for this purpose.

13.5 Write a program for an 8085A to complement the bytes stored in locations 0*F*00H through 0*FFF*H and store them in locations starting from 1*F*00H.

13.6 Write a program for an 8085A to find the larger of the two numbers stored in locations 0*A*00H and 0*A*01H and store it in memory location 0*A*02H.

13.7 Write a program to clear the memory location 01*A*0H in an 8085A microprocessor system.

13.8 Write a program to find the smallest element in a block of 8-bit unsigned binary data, whose number is stored in memory location *A*001H, and the data are stored in memory locations begining from *A*002*H*. Store the smallest number in the memory location *FF*00H. The μP used is 8085A.

13.9 It is desired to mask (set to 0) the most-significant four bits of a byte in the accumulator. Suggest an instruction to accomplish this in an 8085A μP system.

13.10 An 8085A program is given below. Find the task performed by this program.

```
START:   LXI H, BUFR
         MVI C, 0BH
LOOP:    DCR C
         JZ FINIS
         IN DATA
         MOV M,A
         INX H
         JMP LOOP
FINIS:   MOV B,A
```

13.11 Find the memory requirement of the program of Ex. 13.3.

13.12 Assume that the accumulator contains 08 and a register B contains 93 (both in BCD format). Find the contents of the accumulator after the ADD B instruction is executed.

Is the result a valid BCD number? If not, what should be done to obtain the result in BCD format? Explain clearly.

13.13 Suggest a suitable system to display a BCD number entered through a keyboard to the µP system on a 7-segment LED. Using an 8085A microprocessor, write the program. Make the necessary assumptions.

13.14 The first four instructions of a subroutine are:

```
PUSH  B
PUSH  D
PUSH  H
PUSH  PSW
```

What will be the last six instructions of the subroutine? Explain clearly.

13.15 Write an 8085A program to convert a decimal digit in memory location 00*F*0 to an ASCII character and store it in memory location 00*F*1H. If memory location 00*F*0H does not contain a decimal digit, an ASCII question mark should be placed in memory location 00*F*1H. See Table 2.10 for decimal to ASCII conversion.

13.16 What is the address-bus width of the following microprocessors?

8086
80186
80286
80386 SL
80386 DX
80486 DX
Pentium

13.17 How many 16-bit general-purpose registers are available in the 8086 μP?

13.18 16-bit Segment registers are used to address a 64K byte block of memory in an 8086 μP, whereas the address-bus width is 20 bits. Is it possible? Explain.

13.19 If IP = 1*A*00H and CS = 2000H, find the address of the next instruction to be fetched.

13.20 If SS = 2400 H and SP = *A*000 H, find the current address of the stack.

14

COMPUTER AIDED DESIGN OF DIGITAL SYSTEMS

14.1 INTRODUCTION

The importance of digital techniques, circuits, and systems have been well recognized in all walks of human life. In fact the term 'digital' has now entered every sphere of our activities. This has created a need for designing and developing more and more powerful and complex digital circuits and systems. As we have seen through all the earlier chapters, any digital circuit consists of only a few basic circuits; AND, OR, and NOT gates and a memory element FLIP-FLOP, irrespective of the size and complexity of the circuit. We have also discussed the design of digital circuits using manual methods, such as; simplification of Boolean expressions using Boolean algebraic theorems, graphical method, tabular method, using available SSI and MSI devices, etc. These design (synthesis) methods or tools have served well for understanding the concepts of digital circuits and their design for systems which are relatively small in size and are not complex enough in today's context.

However, the ever increasing size and the complexity of digital systems require design methods which make use of computers, which themselves are complex digital systems. These methods are known as computer aided design (CAD) methods and a number of CAD tools have been developed for this purpose.

One may wonder how the earlier computers were designed having complex electronic circuitry without the help of CAD tools. It was a result of human ingenuity which made it possible to develop a system which has made tremendous effect on the way we think, work, and live. Since, there were no computers at that time, therefore, the question of having CAD tools did not arise. Now with the tremendous progress in semiconductor technology, design of computers and other digital systems have become a very complex process requiring CAD tools without which we can not think of designing powerful and efficient complex digital systems including computers.

CAD tools have not only made it possible to design modern complex logic circuits but have also made the design work much easier. Many tasks in the design process are performed automatically by the CAD tools resulting in faster and efficient design. However, the importance of learning the theory of digital circuits can not be minimized even when CAD tools are employed for design.

A number of hardware description languages (HDLs) have been developed for describing the structure and behaviour of complex digital circuits and a number of HDL based CAD tools have been developed for the design of digital systems. Proliferation of hardware description languages for the design of *very high speed integrated circuits* (VHSIC) led to the efforts to evolve a common standard language to be used by all designers. As an outcome of these efforts VHDL (VHSIC Hardware Description Language), an industry standard hardware description language was developed, which was originally adopted in 1987 as IEEE Standard 1076. A revised standard IEEE 1164 was adopted in 1993 which is the most popular HDL in use.

Computer aided design concepts, CAD tools, and VHDL have been discussed here which will help the readers to enter into the most fascinating and useful field of digital systems design.

14.2 COMPUTER AIDED DESIGN (CAD) CONCEPTS

Figure 14.1 illustrates a basic digital design process to design a digital system that performs certain tasks and meets certain specifications. The design process begins with the specifications such as functionality and input-output behaviour of the system to be designed. From a complete set of specifications, a designer with his/her considerable design experience and intuition makes an initial design by defining the general structure of the intended hardware. The initial design idea has to go through several design stages before its hardware implementation is obtained. At each stage of design process the designer evaluates and verifies the results of that design stage. For the evaluation of the results of a design stage, the design is to be tested by applying test data (stimuli) at the input and finding the output and checking this output and verifying it against the original system specifications. If there is any error then the design must be modified to eliminate the error. The modified design is again put to evaluation and verification. This process is repeated until there is no error in the designed system.

The above procedure indicates that the designed system needs to be implemented in hardware at every stage of design.

This is tedious, time consuming, and costly and it may not always be possible to implement it. The development of various CAD tools have made it very convenient and easy to go through this process. CAD tools enable the designers to simulate the behaviour of a design without its hardware implementation. Various *simulation* CAD tools are available. These are used to simulate the design and are also used to determine whether the obtained design meets the required specifications or not. If not, then appropriate modifications are made in the design and the verification of the modified design is repeated through simulation till the correct design is obtained. Once the functionally correct design is obtained, logic synthesis and optimization CAD tools are used to obtain optimized logic expression to suit the target hardware technology to be used for implementing the design. Using these optimal logic expressions, physical design tool is used for obtaining gate level or transistor level design.

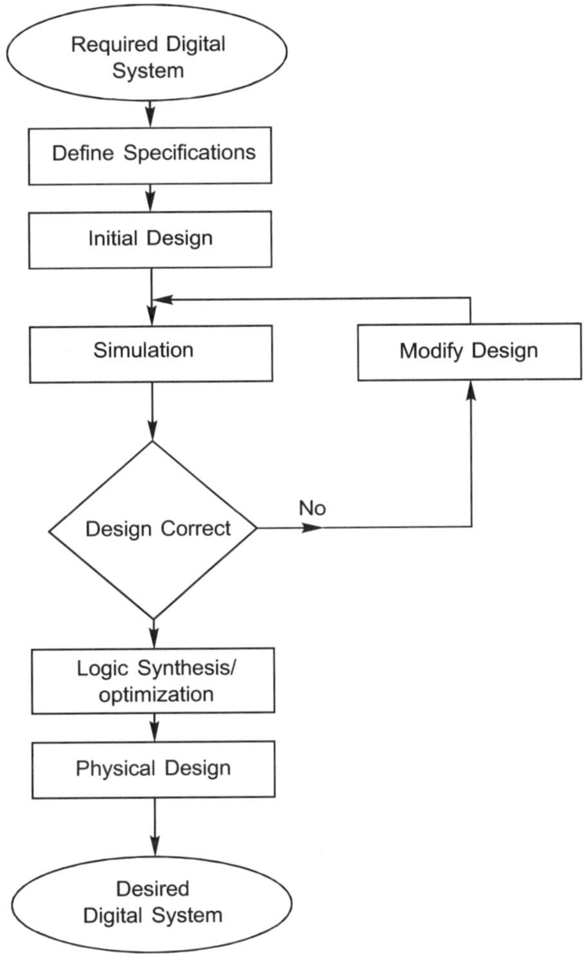

Fig. 14.1
Basic digital design process

As the size and complexity of a digital system to be designed increases, more and more computer-aided design tools are introduced in the process.

14.3 CAD TOOLS

Various computer aided tools have been developed for aiding the designer of digital systems in performing different design tasks. These tools make the design process easy, convenient, accurate and fast. A variety of such tools are available. In some cases a compatible set of such tools are packaged together forming a CAD system. The CAD tools for performing various steps in the design process are discussed as follows:

14.3.1 Design Entry

The first stage in the design of any digital system is the conception of what the intended system is required to do and the formulation of its general description. This stage requires designer's intuition and experience and is essentially a manual process. The description of the system conceived is to be entered into the CAD system and this process is known as *design entry*. The system to be designed may be described in one or more of the three methods, accordingly design entry tools are required to handle each one of these descriptions. The three design entry methods are:

- Truth table
- Schematic diagram
- Hardware description language (HDL)

Design Entry Using Truth Table

The truth table of a logic function or a timing waveforms diagram may be specified for designing the necessary hardware, using the components available, in an optimum way. A CAD system capable of transforming the truth table or the timing waveforms into a network of logic gates is required for this purpose. The CAD tool used for drawing timing diagram and transforming it automatically into a network of logic gates is known as the *waveform editor*.

This method of design entry can be used only for a small number of variables, since truth table method of description is possible only for logic functions with a few variables. For a larger circuit involving relatively large number of variables, this method can be employed for smaller logic functions that are parts of the larger circuits. These smaller circuits can be interconnected by using another CAD tool known as *schematic capture tool*.

Design Entry using Schematic Capture

The term *schematic* refers to a circuit diagram in which the circuit elements are represented by their graphical symbols and the connections between the circuit elements are drawn as lines. A schematic capture tool known as a *graphic editor* available in a CAD system can be used to draw a schematic diagram. It uses the graphics capabilities of a computer and allows a designer to draw a schematic diagram using a computer mouse.

The schematic tool provides a library of graphical symbols that represent gates of various types with different number of inputs. Some of the commonly required circuits using these gates can be created by a designer and these can also be represented by graphical symbols. These special symbols of the circuits can also be made available in the library and can be imported into the user's schematic. These small circuits and other sub-circuits or sub-systems can be created by the user, using either different entry methods or the schematic tool itself.

The schematic capture design entry method uses the gates and the other circuits created by the user and connects them to create the desired logic circuit. This is a hierarchical design method and is, therefore, a convenient way of dealing with the complexities of larger circuits.

Design Entry Using Hardware Description Languages

Similar to various programming languages, a number of languages have been developed to describe hardware rather than a program to be executed on a computer. The two most commonly used *hardware description languages (HDLs)* in industry are: VHDL (Very High-Speed

Integrated Circuit Hardware Description Language) and *Verilog HDL*. Both these HDLs are IEEE (Institution of Electrical and Electronics Engineers, USA) standards. Using a HDL a logic circuit is represented in its code which is used for design entry. This method of design entry is the most commonly used method for the design of digital systems. A number of CAD tools are commercially available for the design that uses HDLs. This method can be used for efficiently designing small as well as large systems.

Both the IEEE standard HDL languages, VHDL and Verilog HDL, widely used in the industry differ in many ways but they offer similar features. Since VHDL is more popular than Verilog HDL, therefore, we have chosen VHDL here, although the concepts discussed for VHDL can easily be applied when using Verilog HDL.

The three design entry methods can be mixed in a system design for different subcircuits, for example, a schematic capture tool can be used in which a subcircuit in the schematic is described using VHDL.

14.3.2 Initial Synthesis

Synthesis is the process of transforming design entry information of the circuit into a set of logic equations. By using synthesis CAD tools logic equations that describe the logic functions needed to realize the circuit are automatically generated. In general, these logic equations produced by the initial synthesis tools are not in the optimal form, because these equations are generated on the basis of designer's input to the CAD tools. It is difficult for a designer to manually produce optimal results, especially for large circuits. *Optimization* techniques are used to obtain logic equations which are not only functionally correct but are optimized for the design goal. The process of optimization is performed after functionally correct design is obtained.

14.3.3 Functional Simulation

After the design entry and the initial synthesis are complete, it is necessary to verify the circuit function of the designed circuit with the expected function. For this purpose, a *functional simulator* CAD tool is used. The functional simulator simulates the circuit function from the logic equations obtained from the synthesis tool and the inputs which are applied. The output of the simulator, which is obtained either in the truth table form or as the timing diagram are examined to verify whether the designed circuit operates as required or not. If the output from the simulation process is not the desired output then the design is to be modified suitably. This process is repeated unless the functionally correct design is obtained as shown in Fig. 14.2.

14.3.4 Logic Synthesis and Optimization

After the functionally correct design is obtained logic synthesis and optimization CAD tools are used to obtain optimized design to achieve design goal such as cost, speed, or the technology of implementation. Use of logic synthesis and optimization tools before the functionally correct design is obtained does not serve any purpose. The CAD synthesis tools automatically implement the synthesis techniques. In case the logic synthesis tool is *technology independent*, the optimization is independent of the resources available in the target chip, whereas the *technology mapping* synthesis tool ensures that the circuit designed by it uses the logic resources available

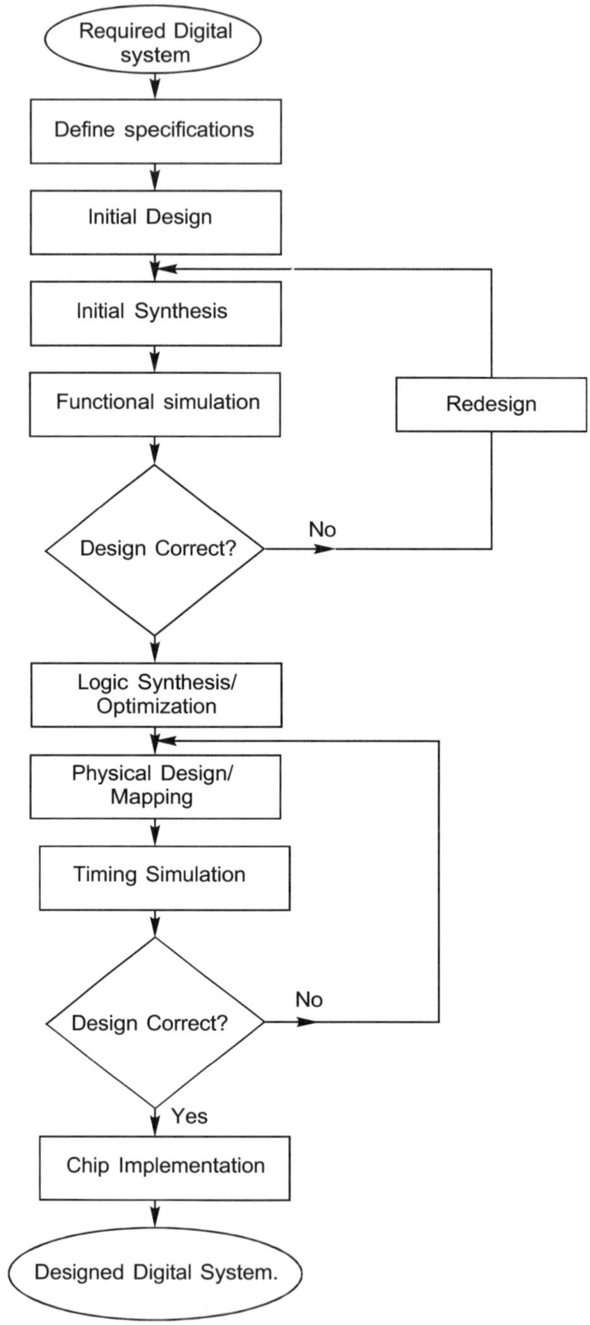

Fig. 14.2
Digital design process

in the target chip. For example, if the target chip is a CPLD, then each logic function in the circuit is expressed in terms of the gates available in its macrocells. Similarly, for an FPGA target chip the number of inputs to each logic function is constrained by the size of the look-up tables (LUTs) or type of logic cells avaiable as the case may be.

14.3.5 Physical Design

After the logic expressions are optimized, the next step is to design the circuit using the available logic resources in the target chip. This step is known as *physical design*, or *layout synthesis.* The physical design consists of two operations: *placement* of logic functions in the optimized circuits in the target chip (CPLD or FPGA) and interconnecting the components in the chip, referred to as *routing.* The *placement* and *routing* CAD tools are used for this purpose.

14.3.6 Timing Simulation

The functional simulation process assumes negligible propagation delay time of the logic gates, which is not true for practical circuits. There is always some time needed for signals to propagate through logic gates, therefore, it is necessary to take care of propagation delays while designing digital circuits. For this purpose timing simulation is used, which simulates the actual propagation delays in the technology chosen for implementation of the design. The model to be used for timing simulation must take care of delays associated with the chip, macrocells if the target device is a CPLD and logic cells for an FPGA target device, and the delays through the interconnection wires. *Timing simulator* CAD tools are used for this.

14.3.7 Summary

A typical CAD system for digital design comprises tools for performing the following tasks:

- *Design entry* It allows a designer to enter a description of the desired circuit/system in the form of truth tables, schematic diagrams, or HDL codes such as VHDL.
- *Initial Synthesis* It generates logic expressions based on data entered during the design entry stage.
- *Functional Simulation* It is used to simulate and verify the functionality of the circuit, using the inputs (stimuli) provided by the designer.
- *Logic Synthesis and optimization* It is used to design optimized circuit.
- *Physical design* It is used to implement the optimized circuit in a given technology, for example, in a CPLD chip or in a FPGA chip.
- *Timing Simulation* It is used to include the effect of propagation delays that are expected in the target technology and ensures that the designed circuit meets the required performance.
- *Chip configuration* It configures the actual chip to realize the designed circuit.

The design process is illustrated in Fig. 14.2.

14.4 INTRODUCTION TO VHDL

VHDL is a Hardware Description Language used for modelling digital systems made of interconnection of components. The complexity of the digital system being modeled may vary from that of a simple gate to a complete electronic circuit/system. VHDL is an acronym for VHSIC

Hardware Description Language and VHSIC is an acronym for Very High Speed Integrated Circuits.

VHDL is an industry standard language used to describe hardware from the abstract to the concrete level and has become the universal communication medium of design and for specifiying input and output from various design tools, such as simulation tools, synthesis tools, placement and routing tools, etc. from vendors of CAD work stations, ASICs, CPLDs, and FPGAs etc.

VHDL contains elements which can be used to describe the *behaviour* (*concurrent* or *sequential*) or structure, with or without timing, of any digital system irrespective of the complexity of the sytem. Since it is the most widely used hardware description language, therefore, it is a commonly used method of documenting circuits by the designers, thus making it possible to understand circuits designed by other designers.

For computer aided design of digital systems, VHDL is a very convenient and commonly used HDL for design entry. It supports hierarchical modeling of the system and top-down and bottom-up methodologies of design. Models written can be verified using a VHDL simulator.

When a digital circuit is to be designed, it is required to be described in VHDL. For this we must specify an *entity* and an *architecture* at the top level, and also specify an entity and architecture for each of the component modules that are part of the circuit. Each entity declaration includes a list of interface signals and their types that can be used to connect it to other modules or to the outside world. The behaviour of the circuit and each of its components is described in the architecture declaration. The behaviour may be described in structural form, i.e. interconnection of components, or as a set of concurrent or sequential statements.

A VHDL source code file has two main sections: an entity and an architecture. Similar to any other programming language, VHDL has some rules and characteristics. Some of these are given below:

- There are a number of reserved words (or keywords) which have specific meanings and these can not be used for identifiers or as any other name. These are given in Appendix-A1.
- VHDL is a free-form language. Carriage returns, blank lines, and additional blank spaces may be included between words for clarity without any ill effects.
- Long statements can continue over more than one line.
- A line starting with two adjacent hyphens (– –) is treated as comment.
- Concluding semicolon (;) is syntactically required. Its absence will cause error message during the compilation of the code.
- TIME is a predefined type, any of the following time units are available in VHDL:

  fs – femtosecond
  ps – picosecond
  µs – microsecond
  ms – millisecond
  s – second
  min – minute
  hr – hour

  ns is built in VHDL. It need not be specified.

- A basic identifier in VHDL is composed of a sequence of one or more alphanumeric characters and underscore (_) character. The first character must be an alphabet, and the last character must not be an underscore. Two consecutive underscores are not allowed.
- VHDL is case insensitive, i.e. upper and lower case letters are considered identical.
- Boolean operators AND, OR, NOT, NAND, NOR, XOR, and XNOR are built in VHDL.
- A *system library* ieee library is provided which stores packages for standard logic gates and other commonly used functions. A *user library* can be created by the user.
- Special notation symbols and syntax provided in VHDL are given in Appendix-A2.

14.4.1 Entity

The term *design entity* or just *entity* in VHDL refers to any digital device that possesses some form of intercommunication characteristic. It is a hardware abstraction of an actual digital hardware device. The device to be modeled (entity) may be a single gate, a central processsing unit (CPU), board containing discrete components, or even a complete digital system.

An entity may be decomposed into its constituent lower level entity, or subcomponents or alternatively it may be treated as a building block to construct a high level entity. This means an entity X can be used as a component of another entity Z, or an entity W can be decomposed into its lower level entities P,Q,R, . . . etc. For example, an EX-OR gate shown in Fig. 14.3a is an entity which can be decomposed into its lower level enitites AND, OR, and NOT gates as shown in Fig. 14.3b, or conversely an EX-OR entity can be considered as composed of AND, OR, and NOT entities.

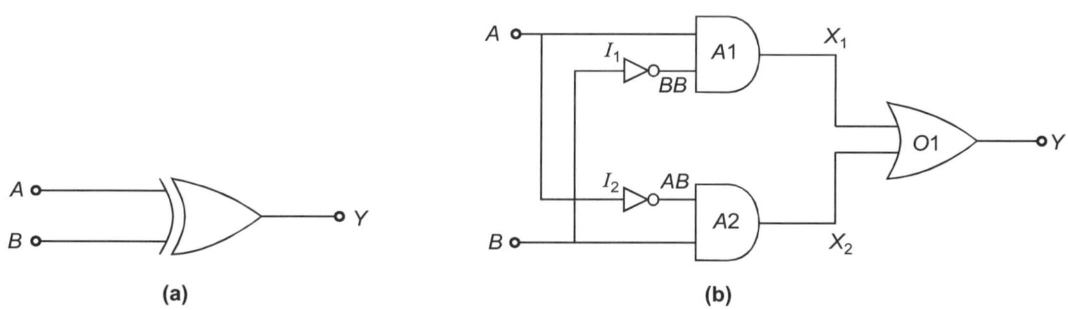

Fig. 14.3
(a) An EX-OR gate (b) An EX-OR function using AND, OR, and NOT gates.

In VHDL all designs are created using entities, which is the most basic building block in a digital design and since an entity can be decomposed into its lower level entities or can be treated as a building block for higher level entity, therefore, VHDL can support both top-down and bottom-up design methodologies. This methodology is referred to as *hierarchical design* and it is very useful in dealing with the complex circuits.

The term ENTITY is a key-word in VHDL and is a construct used in VHDL code. The keyword ENTITY signifies the start of an entity statement. Each entity is uniquely declared describing its external communication links to the outside world. It specifies the number of ports, the directions of the ports, and the type of ports. Some more information such as timing can be put into the entity.

An entity is assigned a name and the corresponding construct declares its input and output signals, known as *ports*. A port is indentified by the keyword PORT. Each port has an associated *mode* that specifies whether it is an input port (keyword IN) or an output port (keyword OUT) for the entity. Since each port represents a signal hence it has an associated *type*.

The syntax of an entity declaration is:

ENTITY *entity-name* IS
PORT (List of input, output port names and their types);
END *entity-name;*

The words: ENTITY, IS, PORT, IN, OUT, and END used above are the *reserved* words or *keywords* in VHDL. These words have specific meanings for the VHDL compiler, and they can not be used for any other purpose.

An *entity-name* is composed of a sequence of one or more alphanumeric characters and the underscore (_) character. The first character of a name must be an alphabet (upper-or-lower case) and the last character must not be an underscore. Also a name can not have two successive underscores and it must not be a VHDL reserved word. VHDL is not case sensitive, which means upper and lower case letters can be freely used, for example, SET_CK_HIGH, Select_Signal, ROM_address are all valid names. Entry name should preferrably be assigned as a word meaningfully related to the function performed by the entity so as to make it convenient for the designer for indentification.

The list of input, output port names and their types describes the input and output signals of the entity and their types. For example, for the circuit of Fig. 14.3a, it will appear as

```
PORT  (A,B: IN BIT;
            Y: OUT BIT);
```

which means ports *A* and *B* are input ports and the signals are of BIT type (0 or 1), and *Y* is an output port and the signal is of type BIT.

The direction of the port can be input (IN), output (OUT), or bidirectional (INOUT) and is referred to as *mode.*

The reserved word END signifies the end of the ENTITY declaration. The symbols colon(:) and semicolon(;) are used as separator and terminator respectively.

From the entry declaration illustrated above, it is clear that an entity declaration describes only the input and output signals of the entity. It does not give any details about the behaviour of the entity or the way the circuit components are connected, i.e. the structure of the circuit. In fact the entity declaration does not and cannot model a device's behaviour. It is merely an outer shell without any functional core.

Example 14.1 Write entity declaration constructs in VHDL for the AND, OR, NOT, and EX-OR gates shown in Fig. 14.4.

Solution

(a) For AND gate

```
ENTITY and2 IS
PORT (A, B: IN BIT;
Y: OUT BIT);
END and2;
```

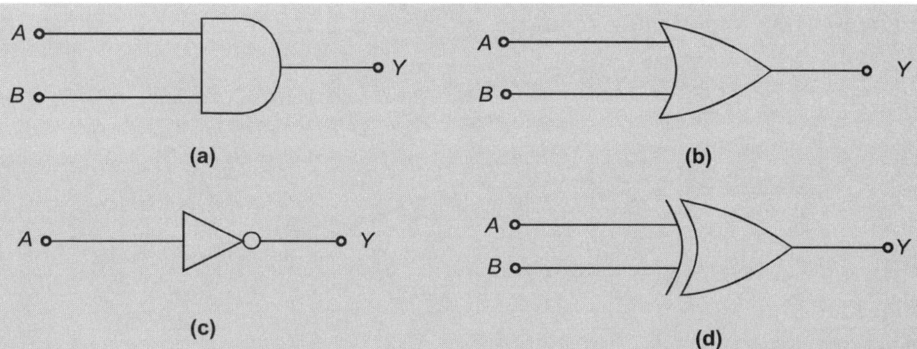

Fig. 14.4
(a) AND (b) OR (c) NOT (d) EX-OR gates.

Here, the name of the entity is chosen as *and2* (a 2-input AND gate), *A* and *B* are the input ports, and *Y* is the output port. The type of signal is BIT.

(b) For OR gate

```
ENTITY or2 IS
PORT (A, B: IN BIT;
Y: OUT BIT);
END or2;
```

(c) For NOT gate

```
ENTITY not IS
PORT (A: IN BIT; Y: OUT BIT);
END not;
```

(d) For EX-OR gate

```
ENTITY xor2 IS
PORT (A, B: IN BIT; Y: OUT BIT);
END xor2;
```

Example 14.2 Write entity construct for the EX-OR circuit of Fig. 14.3b.

Solution

Let the name of the entity be Circuit_Fig. It has two input ports *A* and *B* and one output port *Y*. The entity declaration for this circuit will be

```
ENTITY Circuit_Fig IS
PORT (A, B: IN BIT; Y: OUT BIT);
END Circuit_Fig;
```

From this entity declaration, we observe that although this circuit consists of AND, OR, and NOT gates, the circuit itself is an entity and the entity declaration gives no information about the structure or behaviour of the circuit.

Example 14.3 Write entity construct for the R-S FLIP-FLOP circuit shown in Fig. 14.5.

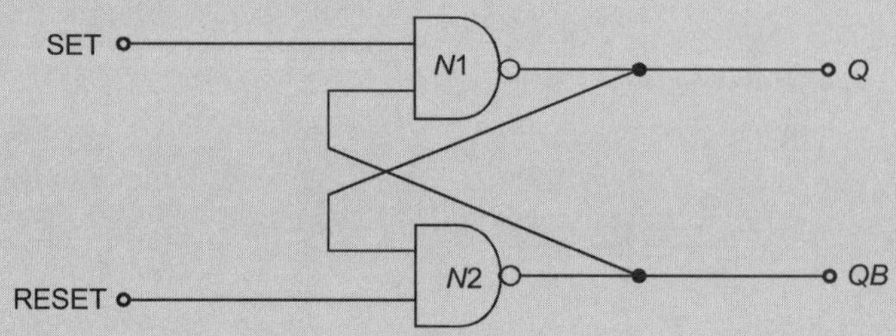

Fig. 14.5

R-S FLIP-FLOP

Solution

Let the name of the entity be rsff. It has two input ports SET and RESET and two bidirectional ports *Q* an *QB*. The entity construct will be

```
ENTITY rsff IS
PORT (SET, RESET: IN BIT;
Q, QB: INOUT BIT);
END rsff;
```

14.4.2 Architecture

A VHDL design entity is modeled using an entity declaration and its atleast one architecture body. The entity declaration describes the external view of the entity, whereas the architecture body contains the internal description of the entity. The internal description can be specified in the following ways:

- A set of interconnected components that represent the structure of the entity.
- A set of concurrent statements that represent the behaviour of the entity.
- A set of sequential statements that represent the behaviour of the entity.

Each of the above methods of representation can be specified in a different architecture body or mixed within a single architecture body. An architecture body is composed of two parts: *declarative part* and *statement part.* The syntax of architecture body is:

```
ARCHITECTURE architecture-name OF entity-name IS
[declarative part]
BEGIN
[statement part]
END architecture-name
```

An architecture must be given a name consisting of a text string which should be assigned by a designer in a way meaningful to the design.

The words ARCHITECTURE, OF, BEGIN are keywords in VHDL.

The declarative part appears before the keyword BEGIN. It can be used to declare signals, user-defined types, constants, and components. The statement part of the architecture is contained between the keywords BEGIN and END. All the statements are *concurrent* statements, which means, these are executed concurrently (simultaneously) and not sequentially as in the case of any programming language.

Structural Modeling

In structural modeling, an entity is described as a set of interconnected components in the architecture body. The architecture body for the EX-OR gate shown in Fig. 14.3b is given below:

```
ARCHITECTURE XOR_STRUCTURE OF xor2 IS
COMPONENT and2
PORT (X, Y: IN BIT; Z: OUT BIT);
END COMPONENT;
COMPONENT or2
PORT (P, Q: IN BIT; R: OUT BIT);
END COMPONENT;
COMPONENT not
PORT (I: IN BIT; J OUT BIT);
END COMPONENT
SIGNAL AB, BB, X1, X2 : BIT;
BEGIN
I1: not PORT MAP (B, BB);
I2: not PORT MAP (A, AB);
A1: and PORT MAP (A, BB, X1);
A2: and PORT MAP (AB, B, X2);
O1: or PORT MAP (X1, X2, Y);
END XOR_STRUCTURE;
```

In this the architecture body has been named as *XOR_STRUCTURE* and it is associated with the entity declaration with the name *xor2* and therefore it inherits the list of interface ports from that entity declaration. This circuit has three types of components, two 2-input AND gates, one 2-input OR gate and two NOT gates. For each type of component, component declaration is required. The component declarations are shown in the declarative part of the architecture body. The architecture body also contains a signal declaration that declares four signals *AB*, *BB* X_1, and X_2, of type BIT. These signals which represent wires, are used to connect the various components that form the EX-OR circuit. The scope of these signals is restricted to the architecture body and are not visible outside the architecture body.

For each component, component instantiation statement is required in the statement part. This requires port map, i.e. input and output ports to be specified. A component instantiation statement is a concurrent statement, therefore, these statements can appear in any order. The structural modeling describes only interconnection of components, without specifying behaviour of the components or the entity they collectviely represent.

The component instantiation statements include a label, such as *I*1, *I*2, *A*1, *A*2 . . . etc.; component name, such as and2, or2, not . . . etc.; and association between the actual signals that are visible in the architecture body and the ports of the component being instantiated. The mapping of ports specifies the association between the ports of a component with the ports in the architecture body, which may consist of external input, output ports and internal ports declared through signal statement. For example, *I*1 is a NOT gate with port *B* as input and *BB* signal as output, similarly *I*2, *A*1, *A*2, and *O*1 are instantiated.

Example 14.4 Write architecture body of R-S FLIP-FLOP shown in Fig. 14.5 using structural modeling.

Solution
It consists of two 2-input NAND gates. Let us use the name of the architecture also rsff which is same as the entity name. It is allowed for the architecture body to have same name as entity. The architecture body will be

```
ARCHITECTURE rsff OF rsff IS
COMPONENT nand2
PORT (A, B : IN BIT; Y: OUT BIT);
END COMPONENT;
BEGIN
N1: nand2
PORT MAP (SET, QB, Q);
N2: nand 2
PORT MAP (RESET, Q, QB);
END rsff;
```

In this example, there are no signals to be declared.

Data Flow Modeling

In this modeling, the flow of data through the entity is expressed using concurrent signal assignment statements. As the name implies, the statements contained in the model assign values to signals. These statements execute concurrently, i.e. in parallel, not serially, as in the case of a programming language. The structure of the entity is not explicitly specified in this modeling, but it can be implicitly deduced.

A signal assignment statement is of the form:

$$A \Leftarrow B; \tag{14.1}$$

It means *A* gets the value of *B*, i.e. the current value of signal *B* is assigned to signal *A*. This statement is executed whenever signal *B* changes value. Signal *B* is in the sensitivity list of this statement. A signal assignment statement is executed whenever a signal in its sensitivity list changes value. A transaction is generated when a signal assignment statement is executed. The target signal may or may not change following the execution of the signal assignment statement. If a new value of target signal is generated, then an event is scheduled for the target signal.

Time delay can also be introduced in the signal assignment statements as given below:

$$A \Leftarrow B \text{ AFTER } 10 \text{ ns}; \tag{14.2}$$

Here, signal *B* is in the sensitivity list of this signal assignment and according to this statement signal *A* gets the value of signal *B* after a lapse of 10 ns.

Example 14.5 Write the architecture body of each of the entities of Example 14.1 using concurrent signal assignments, i.e. the data flow model. Assume 10 ns delay.

Solution

(a) For AND gate

```
ARCHITECTURE df_ and 2 OF and2 IS
BEGIN
Y ⇐ A AND B AFTER 10ns;
END df_and2;
```

(b) For OR gate

```
ARCHITECTURE df_or2 OF or2 IS
BEGIN
Y ⇐ A OR B  AFTER  10ns;
END df_or2;
```

(c) For NOT gate

```
ARCHITECTURE INV OF not IS
BEGIN
Y ⇐ NOT A AFTER 10ns;
END INV;
```

(d) For EX-OR gate

```
ARCHITECTURE  df_xor OF xor2 IS
BEGIN
 ⇐ Y (A AND (NOT B)) OR (B AND (NOT A));
END df_xor;
```

Since VHDL does not assume any precedence of logic operators, therefore, parentheses must be used in the expression to avoid compile-time error for the expression.

Example 14.6 Write architecture body of *R-S* FLIP-FLOP shown in Fig. 14.5 using data flow model. Assume 5ns delay time.

Solution

```
ARCHITECTURE df_rsff OF rsff IS
BEGIN
Q ⇐ NOT (QB AND SET) AFTER 5 ns;
QB ⇐ NOT (Q AND RESET) AFTER 5 ns;
END df_rsff;
```

Behavioural Modeling

In this type of modeling, the behaviour of an entity is expressed using statements which are executed sequentially similar to that of a high-level programming language. A process statement is the main mechanism used to model the behaviour of an entity. The functionality of an entity is described in an algorithmic representation in the process statement.

The process statement starts with keyword PROCESS and ends with the keyword END PROCESS. All the statements between the keywords PROCESS and END PROCESS are considered part of the process statement.

The process statement consists of three parts: sensitivity list, declarative part, and statement part.

Sensitivity List

A process statement is always active and executes at all times if not suspended. Following PROCESS keyword, the sensitivity list (signals) in parentheses is specified. The process is activated when an event occurs on any one of these signals. When the program flow reaches the last sequential statement, the process becomes suspended until another event occurs on a signal that it is sensitive to. Regardless of the events on the sensitivity list signals, processes are executed once at the beginning of the simulation run.

Declarative Part

The declarative part is used to declare local variables or constants that can be used only inside the process. i.e. such objects are visible only to the process within which they are declared. Signals and constants declared in the declarative part of an architecture that encloses a process statement can be used inside a process. Such signals are the only means of communication between different processes.

Initialization of objects declared in a process is done only once at the beginning of a simulation run. The initializations in a subprogram are performed each time the subprogram is called.

The process declarative part consists of the area between the sensitivity list and the keyword BEGIN.

Statement Part

The statement part of the process consists of the area between the keywords BEGIN and END PROCESS. All the statements are sequential and are executed one after the other in a sequential order.

The statement part of a process is always active. When the program flow reaches the last sequential statement of this part, the execution returns to the first statement in the statement part and continues.

Example 14.7 Write architecture body of *R-S* FLIP-FLOP shown in Fig. 14.5 using behavioural model.

Solution

```
ARCHITECTURE behave_rsff OF rsff IS
BEGIN
PROCESS (SET, RESET)
BEGIN
IF SET = '1' AND RESET = '0' THEN
Q ⇐ '0' AFTER 10 ns;
QB ⇐ '1' AFTER 20 ns;
ELSEIF SET = '0' AND RESET = '1' THEN
Q ⇐ '1' AFTER 10 ns;
QB ⇐ '0' AFTER 20 ns;
```

```
ELSEIF SET = '0' AND RESET = '0' THEN
Q <= '1' AFTER 10 ns;
QB <= '1' AFTER 10 ns;
END IF;
END PROCESS;
END behave_rsff;
```

Let us examine the execution of this architecture when SET changes to a '0' and RESET remains at a '1'. Since, SET is in the sensitivity list of the process statement, the process is invoked and each statement in the process is executed sequentially. The first statement is an IF statement. This statement yields a negative result because SET = '0' and RESET = '1', therefore, the following two signal assignments are not executed. Then the next check is performed which succeeds and therefore, the next two signal assignments are executed.

Architecture Selection

An entity can have more than one architecture. Three different architectures have been described above. One or more than one (mixed) architecture styles can be used in a single architecture body. It depends on the designer as to which architecture should be used to model the entity. If the model is going to be used to drive a layout tool, then the structural architecture will probably be most appropriate. If a structural model is not wanted, then a more efficient model, could be data flow or sequential, from the point of view of memory space required and speed of execution. Depending upon the preference of the designer for concurrent or sequential code, one of these architectures can be selected.

14.4.3 Configuration Declaration

An entity can have multiple architectures. Which of these architecture bodies is to be used in a simulation? The VHDL language has two options available for this purpose. In one option, a simulation run will, by default, use an architecture body that was most recently compiled. Use of default option in not a preferred way because of the requirement of continual recompilation of architecture bodies in order to ensure their inclusion in a simulation and for difficulty in future references. In the other option we can explicitly pronounce as to which one of the architectures is to be simulated. This is done via a configuration statement. The reserved word `CONFIGURATION` announces a user's intention to create a configuration declaration. The binding of component instantiations to design entities is performed by configuration specifications. For example, we consider a configuration statement using structural architecture of the *rsff* entity (Ex. 14.4). The configuration statement is given below.

```
CONFIGURATION rsff_str OF rsff IS
FOR rsff
FOR N1, N2: nand2 USE ENTITY WORK. mynand
(Version 1);
END FOR;
END rsff_str;
```

Here, *rsff_str* is name given to configuration. Following the reserved word `OF` name of the design entity (*rsff*) that we are configuring is written. For the top most entity, architecture *rsff* is used. The two components *N*1 and *N*2 are instrantiated in the *rsff* architecture. The architecture of 2-input NAND gate is used from library called WORK with entity mynand and architecture version 1. All of the entities now have architectures specified for them.

This binds the architecture body *rsff* to the design entity *rsff*. The first END terminates the pronounced binding and the last END terminates the configuration declaration. The configuration name is optional with the last END reserved word.

A simulation model can be created by compiling the entities, architectures, and the configuration. In case we want to use another architecture for simulation, we need not recompile the complete design, only the new configuration needs to be recompiled.

14.4.4 Generic

It is often necessary to pass certain information to an entity in VHDL. For instance, if an entity is a gate level model with a rise and a fall delay, values for the rise and fall delays could be passed into the entity with generics.

Consider a positive-edge triggered D-type FLIP-FLOP with asynchronous set and reset inputs shown in Fig. 14.6.

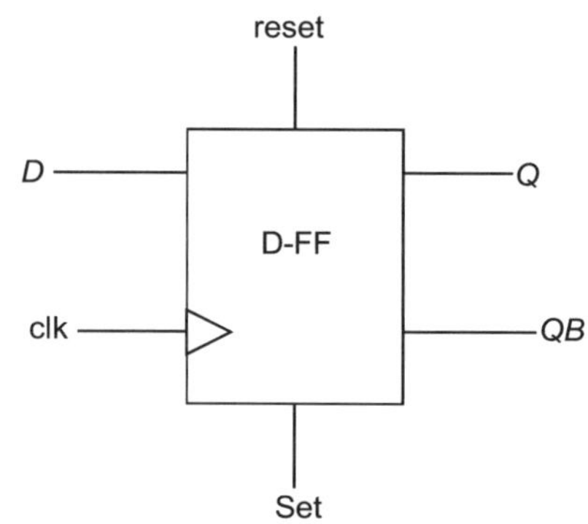

Fig 14.6

Positive-edge triggered D-FF

Its entity declaration using generics will be

```
ENTITY D_FLIP FLOP IS
GENERIC (sq_delay, rq_delay, cq_delay: TIME: = 5 ns);
PORT (D, SET, RESET, clk: IN BIT; Q, QB: OUT BIT);
END D_FLIP FLOP;
```

Here, sq_delay is delay from set to *Q*, rq_delay is delay from reset to *Q*, and cq_delay is delay from clock to *Q*. The delay time is 5 ns. The delay parameters are passed on to the entity through the generic.

14.4.5 Data Objects

In VHDL code information is represented as signals, constants, and variables. These are referred to as *data objects.* The data objects are given names. The rules for the names are same as the rules for identifiers.

SIGNAL data objects are used to represent individual logic signals in a circuit (specified as '0' or '1'), multibit logic symbols (specified as "1101" or "11100110"), and binary numbers (integers). The value of CONSTANT or VARIABLE data objects are specified in the same way as for SIGNAL data objects.

Signal data objects represent the logic signals, or wires, in a circuit. A signal has to be declared with an associated type, as

```
SIGNAL signal_name: type_name;
```

Some of the commonly used signal types are:

BIT and BIT_VECTOR: These are predefined in VHDL standard IEEE 1164

STD_LOGIC and STD_LOGIC-VECTOR: These are used by including the two statements LIBRARY IEEE; USE IEEE. STD_LOGIC_1164. ALL

INTEGER

ENUMERATION

The data object type BIT can assume only values 0 and 1, whereas the type STD_LOGIC can have several different values. This is a standard data type for representation of logic signals. For logic functions, STD_LOGIC can assume 0, 1, *Z* and –, where, *Z* stands for high-impedance state (tristate) and – stands for don't care condition.

To use this type, following two lines are included at the beginning of the VHDL code which serve as directive to the compiler.

```
LIBRARY ieee;
USE ieee. std_logic_1164.all;
```

STD_LOGIC is provided in a set of files known as *library*. It can be used with VHDL code when compiled. In the second line above std_logic_1164 refers to the package which contains STD_LOGIC file, and ALL is used to specify that all legal uses are to be used.

14.4.6 Examples of VHDL Codes

A VHDL source code file has two main sections: an entity and an architecture. Basic concepts of entity and architecture have been discussed above. Some examples of VHDL codes are given here.

Example 14.8 Write VHDL code for the circuit of Fig. 5.19 using

(a) Structural architecture
(b) Data flow architecture. Assume gate delay of 5 ns for AND and 10 ns for XOR.

Solution

(a) The VHDL code is given below:

```
– – We make use of IEEE. STD_LOGIC_1164. ALL
LIBRARY IEEE;
USE IEEE.STD_LOGIC_1164.ALL;
ENTITY H_A IS
PORT (A, B: IN STD_LOGIC; S, C: OUT  STD_LOGIC);
END H_A;
- - H_A is name of the half-adder circuit.
ARCHITECTURE  HA_STR OF H_A IS
- - HA_STR is name of the architecture for entity H_A.
COMPONENT XOR2
PORT (X, Y: IN  STD_LOGIC; Z: OUT STD_LOGIC);
END COMPONENT;
COMPONENT AND2
PORT (P, Q: IN STD_LOGIC; R: OUT STD_LOGIC);
END COMPONENT;
BEGIN
X1: XOR2 PORT MAP (A, B, S);
A1: AND2 PORT MAP (A, B, C);
END HA_STR;
```

(b) The entity construct will be same as that of (a). The VHDL code is given below:

```
LIBRARY IEEE;
USE IEEE.STD_LOGIC_1164. ALL;
ENTITY H_A IS
PORT (A, B: IN STD_LOGIC; S, C : OUT STD_LOGIC);
END H_A;
ARCHITECTURE df_HA OF H_A IS
BEGIN
S <= A XOR B AFTER 10 ns;
C <= A AND B AFTER 5 ns;
END df_HA;
```

Example 14.9 Write VHDL code for a 2-to-4 decoder circuit shown in Fig. 14.7. Use sequential architecture model.

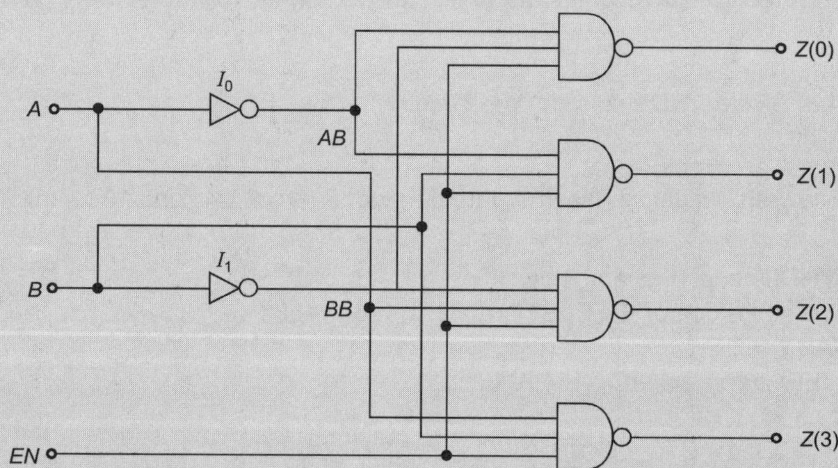

Fig. 14.7
2-to-4 decoder circuit

Solution

```
-- Let the entity name be DECODER 2 × 4
-- Three input ports A, B, and EN
-- Four output ports Z(0), Z(1), Z(2), and Z(3)
-- Signal type is STD_logic for input ports and
-- STD_LOGIC_VECTOR for output ports
LIBRARY IEEE;
USE IEEE.STD_LOGIC_1164.ALL;
ENTITY DECODER 2 × 4 IS
PORT A, B, EN : IN STD_LOGIC; Z : OUT STD_LOGIC_VECTOR (0 TO 3);
END DECODER 2 × 4;
```

```
ARCHITECTURE behave_DEC OF DECODER 2 × 4 IS
BEGIN
PROCESS (A, B, EN)
VARIABLE AB, BB : STD_LOGIC;
BEGIN
AB: = NOT A;
BB: = NOT B;
IF EN = '1' THEN
Z(3) ⇐ NOT (A AND  B);
Z(2) ⇐ NOT (A AND BB);
Z(1) ⇐ NOT (AB AND B);
Z(0) ⇐ NOT (AB AND BB);
ELSE
Z ⇐ "1111";
END IF;
END PROCESS;
END behave_DEC;
```

The signals *A, B*, and *EN* shown in parenthesis after the keyword process is the sensitivity list. *AB* and *BB* are declared as variables, starting with the key word variable. A variable is different from a signal in that it is always assigned (assignment operator: =) a value instantaneously. Whereas a signal is assigned (assignment operator ⇐) a value always after a certain delay.

14.5 SUMMARY

The computer aided design of digital systems is widely used in industry for designing complex digital systems. Therefore, it has become necessary to learn the various concepts involved in computer aided design techniques and the CAD tools. The basic principles of digital theory and practice are essentially required even when CAD tools are used. The most popular hardware description language VHDL has been briefly introduced in this chapter. Since it is a vast subject and therefore it is not possible to cover all the details of VHDL in this book. However, the basic concepts introduced here will be helpful in learning more details of the language from texts devoted to VHDL.

Glossary

Architecture body A body associated with an entity declaration to describe the internal organization or operation of a design entity. It is used to describe the behaviour, data flow, or structure of a design entity.

Binding The process of associating a design entity and, optionally, an architecture with an instance of a component. It can be specified in an explicit or default binding indication.

CAD (Computer-aided design) Design methodologies using computers.

CAD tools Various programs used to design systems using computers. Each tool performs a specific task in the design process.

Character literal A character of the literal type formed by enclosing one of the graphic characters (including a space and non breaking space characters) between two apostrophe (') characters.

Character type An enumeration type with atleast one character literal among its enumeration literals.

Concurrent statement A statement that executes asynchronously, with no defined relative order. Concurrent statements are used for data flow and structural descriptions.

Configuration A construct that defines how component instances in a given block are bound to design entities in order to describe how design entities are put together to form a complete design.

Constant An object whose value may not be changed.

Declaration A construct that defines a declared entity and associates an identifer with it.

Declarative part A syntactic component of certain declarations or statements such as entity declarations and architecture bodies.

Declarative region A semantic component of certain declarations or statements.

Design entity An entity declaration together with an associated architecture body. Different design entities may share the same entity declaration.

Design library A host-dependent storage facility for intermediate form representations of analyzed design units.

Design unit A construct that can be independently analyzed and stored in a design library. A design unit can be an entity declaration, an architecture body, a configuration declaration, a package declaration, or a package body declaration.

Entity declaration A definition of the interface between a given design entity and the environment in which it is used. It may also specify declarations and statements that are part of the design entity. A given entity declaration may be shared by many design entities, each of which has a different architecture.

Enumeration literal A literal of an enumeration type. It may be either an identifier or a character literal.

Enumeration type A type whose values are defined by listing (enumerating) them. The values of the types are represented by enumeration literals.

Event A change in the current value of a signal, which occurs when the signal is updated with its effective value.

Generic An interface constant declared in the block header of a block statement, a component declaration, or an entity declaration.

HDL (Hardware description language) A coding language used to describe hardware rather than a programming language.

Identifier It defines an alias for some other name.

Instance A subcomponent of a design entity whose prototype is a component declaration, design entity, or configuration declaration. A component instantiation statement whose instantiated unit denotes a component creates an instance of the corresponding component.

Integer literal A literal of the type integer.

Integer type A discrete scalar type whose values represent integer numbers with in a specified range.

Layout synthesis Same as physical design.

Library See design library.

Mode The direction of information flow through the port or parameter.

Numeric literal A literal of a physical type.

Numeric type An integer type, a floating point type, or a physical type.

Optimization Process of designing a circuit making optimum use of resources available.

Routing Process of interconnecting logic blocks.

Sensitivity set The set of signals to which a statement is sensitive.

Sequential statements Statements that execute in sequence in the order in which they appear. These are used for algorithmic descriptions.

Synthesis Same as design.

Target chip A semiconductor chip such as a CPLD or FPGA which is to be used for physical design.

Technology mapping Transforming design on the technology of the device being used.

Verilog HDL A hardware description language.

VHDL Very high speed integrated circuit Hardware Description Language.

VHSIC Acronym for Very High Speed Integrated Circuit.

Review Questions

14.1	In a CAD system, the most convenient method of design entry for a large system use ______.
14.2	_____ tool is used after functionally correct simulation of the circuit is obtained.
14.3	The CPLD or FPGA device to be used for design implementation is known as ______ chip.
14.4	`IN`, `OUT`, and `INOUT` are three possible ______ of a port.
14.5	A line starting with – – signifies ______ in VHDL.
14.6	An entity can have more than ______ architecture body.
14.7	Data type BIT can have values ______.
14.8	______ declaration specifies only the outer description of a circuit/component.
14.9	VHDL stands for ______.
14.10	VHSIC stands for ______.
14.11	______ specifies the behaviour of a circuit to be designed.
14.12	In data flow modeling statements are executed ______.
14.13	The statements with in PROCESS are executed ______.
14.14	The signals written in parentheses after the keyword PROCESS is ______.
14.15	______ is used to bind a component instance to an entry-architecture pair.

Problems

14.1 Are the following names valid for entities? Justify your answer.

(a) rp_jain (b) R.P. JAIN (c) 3_input

(d) Third_Edition (e) Edition-3 (f) Circuit __4

14.2 Write entity declarations for the gates shown in Fig. 14.8.

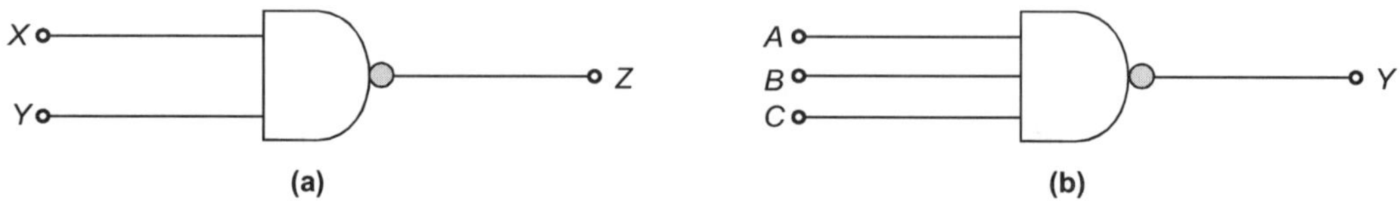

Fig. 14.8

14.3 Write entity declaration for a 4:1 multiplexer circuit

14.4 Write architecture declaration for the gates shown in Fig. 14.8.

14.5 Write VHDL code for a full adder shown in Fig. 14.9. Refer to Fig. 5.21 for its NAND gate realization.

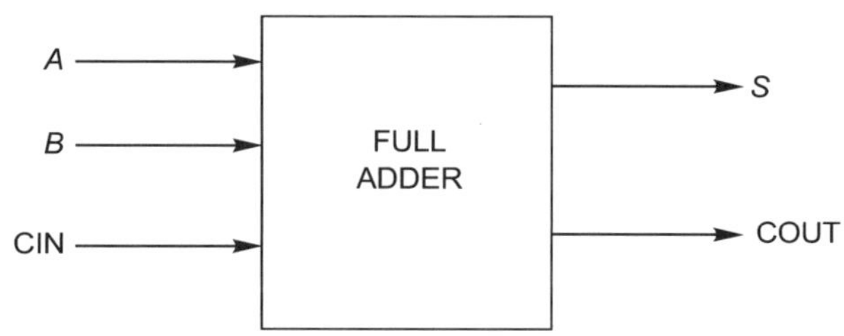

Fig. 14.9

Block diagram of a full adder

14.6 Write VHDL code for the full adder shown in Fig. 14.9. Refer to logic expressions 5.35 for data flow architecture.

APPENDIX A1

RESERVED WORDS IN VHDL

ABS
ACCESS
AFTER
ALIAS
ALL
AND
ARCHITECTURE
ARRAY
ASSERT
ATTRIBUTE

BEGIN
BLOCK
BODY
BUFFER
BUS

CASE
COMPONENT
CONFIGURATION
CONSTANT

DISCONNECT
DOWNTO

ELSE
ELSEIF
END
ENTITY
EXIT

FILE
FOR
FUNCTION

GENERATE
GENERIC
GROUP
GUARDED

IF
IMPURE
IN
INERTIAL
INOUT
IS

LABEL
LIBRARY
LINKAGE
LITERAL
LOOP

MAP
MOD

NAND
NEW
NEXT
NOR
NOT
NULL

OF
ON
OPEN
OR
OTHERS
OUT

PACKAGE
PORT
POSTPONED
PROCEDURE
PROCESS
PURE

RANGE
RECORD
REGISTER
REJECT
REM
REPORT
RETURN
ROL
ROR

SELECT
SEVERITY
SHARED
SIGNAL
SLA
SLL
SRA
SRL
SUBTYPE

THEN
TO
TRANSPORT
TYPE

UNAFFECTED
UNITS
UNTIL
USE

VARIABLE

WAIT
WHEN
WHILE
WITH

XOR
XNOR

APPENDIX A2

SYMBOLS DEFINED IN VHDL

<table>
<tr><th>Symbol</th><th>Meaning</th></tr>
<tr><td>+</td><td>Addition, or positive number</td></tr>
<tr><td>–</td><td>Subtraction, or negative number</td></tr>
<tr><td>/</td><td>Division</td></tr>
<tr><td>=</td><td>Equality</td></tr>
<tr><td><</td><td>Less than</td></tr>
<tr><td>></td><td>Greater than</td></tr>
<tr><td>&</td><td>Concatenator</td></tr>
<tr><td>|</td><td>Vertical bar</td></tr>
<tr><td>;</td><td>Terminator</td></tr>
<tr><td>#</td><td>Enclosing based literals</td></tr>
<tr><td>(</td><td>Left parenthesis</td></tr>
<tr><td>)</td><td>Right parenthesis</td></tr>
<tr><td>.</td><td>Dot notation</td></tr>
<tr><td>:</td><td>Separates data object from type</td></tr>
<tr><td>”</td><td>Double quote</td></tr>
<tr><td>’</td><td>Single quote or tick mark</td></tr>
<tr><td>**</td><td>Exponentiation</td></tr>
<tr><td>⇒</td><td>Arrow meaning “then”</td></tr>
<tr><td>⇒</td><td>Arrow meaning “ gets”</td></tr>
<tr><td>:=</td><td>Variable assignment</td></tr>
<tr><td>/=</td><td>Inequality</td></tr>
<tr><td>>=</td><td>Greater than or equal to</td></tr>
<tr><td><=</td><td>Less than or equal to</td></tr>
<tr><td><=</td><td>Signal assignment</td></tr>
<tr><td><></td><td>Box</td></tr>
<tr><td>--</td><td>Comment</td></tr>
</table>

APPENDIX B

BIBLIOGRAPHY

Actel – ACT Family Field Programmable Gate Array Data Book, April 1990.

Analog Devices, Inc., *Data Acquisition Data Book*, 1982.

Bhasker, J., *VHDL Primer*, Addison Wesley Longman, Singapore, 1999.

Blakeslee, T.R., *Digital Design with Standard MSI and LSI*, John Wiley, New York, 1979.

Brey, Barry B., *The Intel Microprocessors* 8086/8088, 80186, 80286, 80386, and 80486, 3rd Edition, Prentice-Hall, New Delhi, 1995.

Brown, Stephen and Zvonko Vranesic, *Fundamentals of Digital Logic with VHDL Design*, McGraw-Hill International, 2000.

Coughlin, R.F. and F.F. Driscoll, *Operational Amplifiers and Linear Integrated Circuits*, Prentice-Hall, Englewood Cliffs, N.J., 1977.

Fletcher, W.L., *An Engineering Approach to Digital Design*, Prentice-Hall, Engle-wood Cliffs, New Jersey, 1980.

Garland, H., *Introduction to Microprocessor System Design,* McGraw-Hill Book Company, New York, 1979.

Garrett, L.S., "Integrated-Circuit Digital Logic Families, Part I, RTL, DTL, and HTL Devices", *IEEE Spectrum*, vol. 7, pp. 46-56, October, 1970.

________ "Integrated-Circuit Digital Logic Families, Part II, TTL Devices", *IEEE Spectrum*, vol. 7, pp. 63-72, November, 1970.

________ "Integrated-Circuit Digital Logic Families, Part III, ECL and MOS Devices", *IEEE Spectrum*, vol. 7, p. 41, December, 1970.

Hall, D.V., *Microprocessors and Digital S*ystems, McGraw-Hill, New York, 1980.

________ *Microprocessors and Interfacing – Programming and Hardware*, McGraw-Hill, New York, 1986.

Hill, F.J. and G.R. Peterson, *Introduction to Switching Theory and Logical Design,* John Wiley, New York, 1974.

Hnatek, E.R., *A User's Handbook of Semiconductor Memories*, John Wiley, New York, 1977.

Hnatek, E.R., *A Users Handbook of Integrated Circuits*, John Wiley, New York, 1973.

Intel Corporation, *Memory Design Handbook*, 1977.

________ *Data Catalog*, 1977, 1981.

________ *MCS-85 Users Manual,* 1978.

________ *MCS-86 Product Description*, 1979

________ 8080/8085 *Assembly Language Programming Manual*, 1977.

IEEE Standard VHDL Language Reference Manual, New York, June 1994.

Jain, R.P., *Digital Electronics and Microprocessors*, Tata McGraw-Hill, New Delhi, 1987.

Jain, R.P. and M.M.S. Anand, *Digital Electronics Practice Using Integrated Circuits*, Tata McGraw-Hill, New Delhi, 1983.

Kohavi, Zvi, *Switching and Finite Automata Theory*, Tata McGraw-Hill, New Delhi, 1978.

Leventhal, L.A. *Introduction to Microprocessors: Software, Hardware, Programming*, Prentice-Hall, Englewood Cliffs, N.J., 1980.

Mano, M.M., *Computer Logic Design*, Prentice-Hall, Englewood Cliffs, New Jersey, 1972.

Marcovitz, Alan B., *Introduction to Logic Design*, Tata McGraw-Hill, New Delhi, 2002.

Millman, J. and C.C. Halkias, *Integrated Electronics,* McGraw-Hill, New York, 1972.

Millman, J. and H. Taub, *Pulse, Digital, and Switching Waveforms*, McGraw-Hill, New York, 1965.

Millman, J. and Grabel, A., *Microelectronics*, 2nd Edition, McGraw-Hill, New York, 1987.

Motorola, *The Integrated Circuit Data Book*, 1968.

______ *Semiconductor Data Library*, vol. 4, 1974.

National Semiconductor, *TTL DATA BOOK*, February, 1976.

______ *CMOS Integrated Circuits*, March, 1975.

Navabi, Zainalabedin, *VHDL Analysis and Modeling of Digital Systems,*, McGraw-Hill International, 1998.

Noyce, Robert N. and Marcian E. Hoff, Jr., "A History of Microprocessor Development at Intel" *IEEE MICRO*, vol. 1, No. 1, pp. 8-21, February, 1981.

Pellerin, David and Holley, Michael, *Practical Design Using Programmable Logic*, Prentice-Hall, Inc., New Jersey, 1991.

Perry, Douglas, *VHDL*, 3rd Edition, Tata McGraw-Hill, New Delhi, 2001.

Roth, Charles H., *Fundamentals of Logic Design*, Jaico, New Delhi.

Semiconductor Industry Association, *National Technology Roadmap for Semiconductors*, http://www.semichips.org/

Short, K.L., *Microprocessors and Programmed Logic*, Prentice-Hall, Inc., Englewood Cliffs, N.J., 1981.

Signetics, *Linear Data Book*, vol. 2, 1972.

Signetics, *Logic –TTL Data Manual*, 1978.

Taub, H. and D. Schilling, *Digital Integrated Electronics*, McGraw-Hill, New York, 1977.

Texas Instruments, Inc., *The TTL Data Book for Design Engineers*, 1976.

Texas Instruments Staff, *Designing with TTL Integrated Circuits,* McGraw-Hill, New York, 1971.

Triebel, W.A. and A.E. Chu, *Handbook of Semiconductor and Bubble Memories,* Prentice-Hall, Englewood Cliffs, N.J. 1982.

Unger, Stephen H., *The Essence of Logic Circuits,* Prentice-Hall, New Jersey, 1989.

Wakerly, John F., *Digital Design Principles and Practices,* Prentice-Hall, New Jersey, 1990.

Yarbrough, John M., *Digital Logic Applications and Design,* Vikas Publishing House, New Delhi, 2001.

APPENDIX C

ANSWERS TO REVIEW QUESTIONS

CHAPTER 1

1.1 digital
1.2 digital
1.3 high
1.4 high
1.5 Inverse of other input
1.6 0
1.7 1
1.8 1
1.9 1
1.10 active-high
1.11 active-low
1.12 16
1.13 active-high
1.14 3
1.15 8

CHAPTER 2

2.1 2; 0 and 1
2.2 Hexadecimal
2.3 8
2.4 1; 128
2.5 sign
2.6 the number itself.
2.7 8
2.8 1011
2.9 7
2.10 128
2.11 Even
2.12 2
2.13 3
2.14 2's complement method
2.15 non-weighted

CHAPTER 3

3.1 Impurity
3.2 Excess minority
3.3 Higher
3.4 Common-emitter
3.5 Rise time
3.6 Excess minority
3.7 Saturation
3.8 Lower
3.9 CMOS
3.10 MOS

CHAPTER 4

4.1 Bipolar
4.2 MOS
4.3 Propagation delay time × power dissipation
4.4 fan-out
4.5 number of inputs
4.6 deeper or harder
4.7 higher
4.8 not be
4.9 Logic 1
4.10 not be left open
4.11 lower
4.12 0° ot 70°C
4.13 0, 1, and high-impedance
4.14 Open collector or passive pull-up
4.15 ECL

CHAPTER 5

5.1 Literal
5.2 memory
5.3 four
5.4 SOP
5.5 NAND-NAND
5.6 IC package
5.7 minterm
5.8 maxterm
5.9 2^n
5.10 Gray
5.11 0 or 1
5.12 3
5.13 two
5.14 essential prime implicant
5.15 essential prime implicant

CHAPTER 6

6.1 multiplexer
6.2 demultiplexer
6.3 5
6.4 *BCD*-to-7-segment decoder
6.5 0110
6.6 digital comparator
6.7 power
6.8 16 : 1
6.9 6
6.10 arithmetic and logic unit
6.11 active-low
6.12 active-low input
6.13 47
6.14 9's
6.15 adder

CHAPTER 7

7.1 memory
7.2 n
7.3 is not
7.4 1, 0
7.5 $\bar{Q}$
7.6 race around condition
7.7 does not change
7.8 2
7.9 bit
7.10 falling-edge or negative-edge or trailing-edge
7.11 low
7.12 0
7.13 latch
7.14 excitation table
7.15 FLIP-FLOPs

CHAPTER 8

8.1 4
8.2 asynchronous
8.3 16
8.4 1111
8.5 6, 2
8.6 0010
8.7 8
8.8 asynchronous
8.9 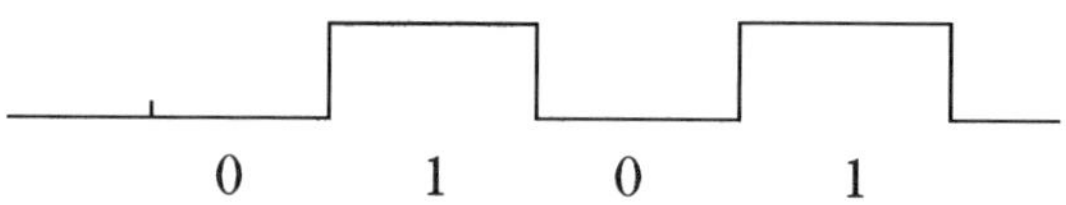

8.10 lower
8.11 1
8.12 1
8.13 universal
8.14 4
8.15 32

CHAPTER 9

9.1 2
9.2 Retriggerable
9.3 Monostable
9.4 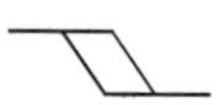
9.5 non-inverting
9.6 comparator circuit
9.7 regenerative comparator (Schmitt-trigger)
9.8 monostable
9.9 external (triggering)
9.10 Schmitt-trigger
9.11 bistable
9.12 regenerative
9.13 0.7 RC
9.14 1 kHz
9.15 astable

CHAPTER 10

10.1 1/4095
10.2 8
10.3 less than $\pm \frac{1}{2}$
10.4 settling time

10.5 10 V
10.6 CTC
10.7 4
10.8 0.05 ms
10.9 1/510
10.10 10
10.11 $1/2^n - 1$
10.12 1
10.13 255
10.14 255
10.15 1100 1101 1010

CHAPTER 11

11.1 binary
11.2 1024
11.3 17
11.4 1024
11.5 random
11.6 serial
11.7 01001100
11.8 11100101
11.9 32
11.10 ECH
11.11 10 ns
11.12 electrically
11.13 ultraviolet radiation
11.14 MOS
11.15 content addressable memory

CHAPTER 12

12.1 AND and OR
12.2 two
12.3 AND
12.4 manufacturer
12.5 ASIC
12.6 PAL (SPLD)
12.7 20
12.8 ISP
12.9 JTAG
12.10 BGA
12.11 quad flat pack
12.12 ACT
12.13 FPGA
12.14 CAD
12.15 more

CHAPTER 13

13.1 8
13.2 64 K Bytes
13.3 20
13.4 I/O ports
13.5 RAM
13.6 decremented by 2
13.7 5
13.8 256
13.9 no
13.10 4
13.11 6.25 MHz
13.12 0
13.13 1 M byte
13.14 12440 H
13.15 31000 H

CHAPTER 14

14.1 HDL
14.2 logic synthesis and optimization
14.3 target
14.4 modes

14.5 comment
14.6 one
14.7 0 and 1
14.8 Entity
14.9 VHSIC Hardware Description Language
14.10 very high speed integrated circuit
14.11 Architecture body
14.12 concurrently
14.13 sequentially
14.14 sensitivity list
14.15 configuration declaration

APPENDIX D

ANSWERS TO SELECTED PROBLEMS

CHAPTER 1

1.1 (a) Analog (b) Digital (c) Analog (d) Digital
(e) Analog (f) Digital (g) Digital

1.2 (c) (i) AND (ii) OR (iii) NOT (iv) EX-OR

1.4 (a)

Inputs		*Output*
A	*B*	*Y*
0	0	1
0	1	0
1	0	0
1	1	0
NOR		

(b)

Inputs		*Output*
A	*B*	*Y*
0	0	1
0	1	1
1	0	1
1	1	0
NAND		

(c)

Inputs		*Output*
A	*B*	*Y*
0	0	0
0	1	0
1	0	0
1	1	1
AND		

(d)

Inputs		*Output*
A	*B*	*Y*
0	0	0
0	1	1
1	0	1
1	1	1
OR		

1.6 (a) NAND, NOR (b) AND (c) NAND (d) OR

1.13

(a)

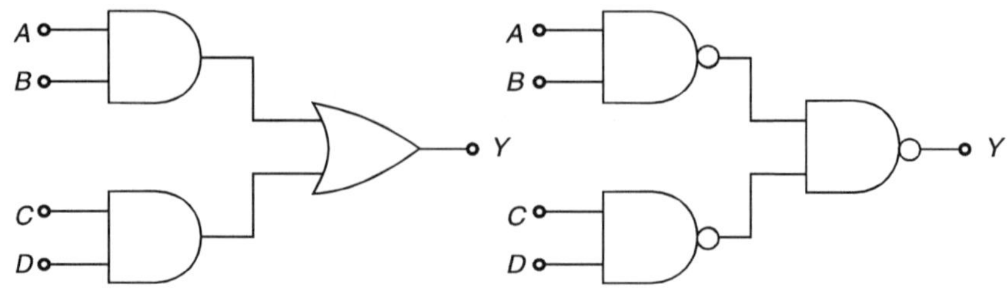

(i) **(ii)**

(b)

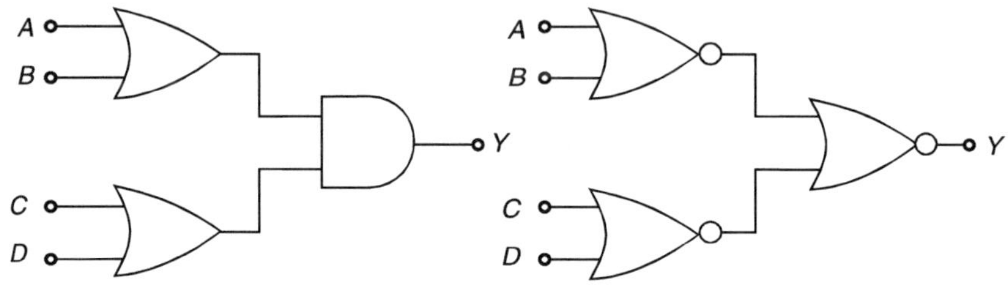

(iii) **(iv)**

1.14 (a) $A \cdot B = B \cdot A$ Commutative

$(A \cdot B) \cdot C = A \cdot (B \cdot C)$ Associative

(b) $A + B = B + A$ Commutative

$(A + B) + C = A + (B + C)$ Associative

(c) $A \oplus B = B \oplus A$ Commutative

$(A \oplus B) \oplus C = A \oplus (B \oplus C)$ Associative

1.16

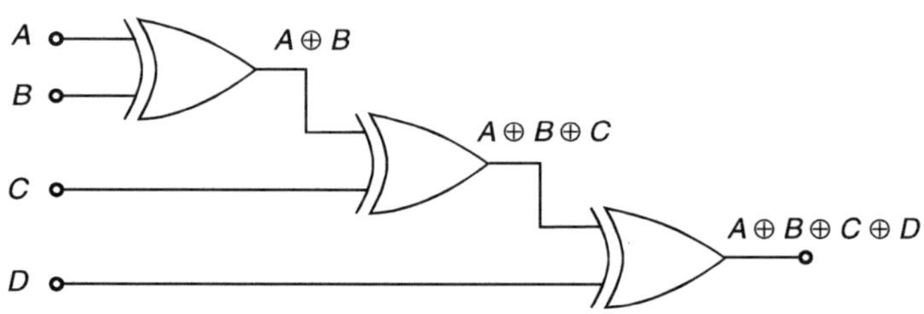

1.18 2^N

1.19 (a) – (i) 14

1.20 (a) (i) 7408, 7432 (ii) 7400

(b) (i) 7432, 7408 (ii) 7402

1.21 *Logic Circuit A*

0.4 V = 0

2 V = 1

Logic Circuit B

−0.75 V = 1

−1.55 V = 0

1.22

Inputs			*Output*			
A	*B*	*C*	*AND* (*a*)	*OR* (*b*)	*NAND* (*c*)	*NOR* (*d*)
0	0	0	0	0	1	1
0	0	1	0	1	1	0
0	1	0	0	1	1	0
0	1	1	0	1	1	0
1	0	0	0	1	1	0
1	0	1	0	1	1	0
1	1	0	0	1	1	0
1	1	1	1	1	0	0

1.23 (a)

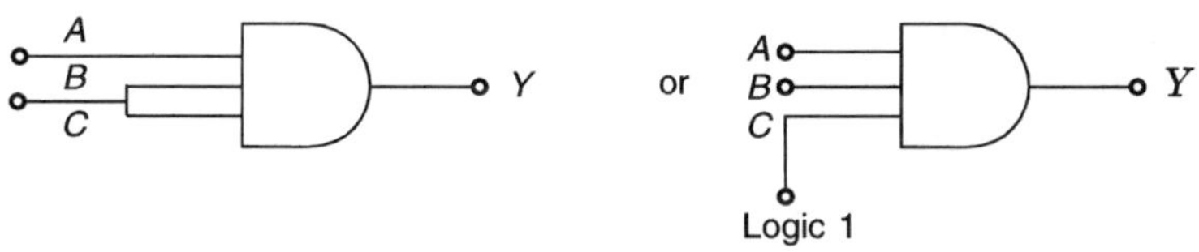

(b)

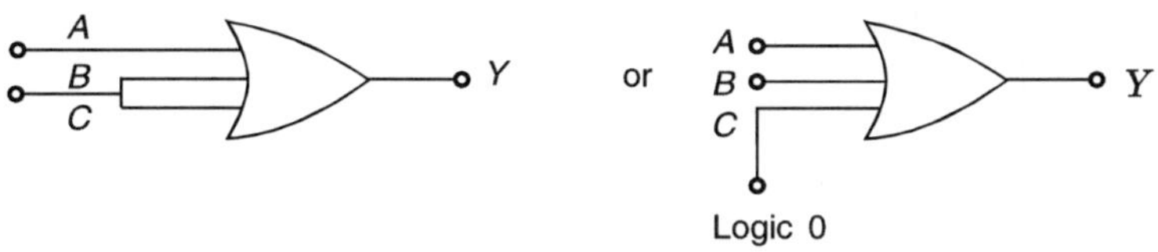

(c)

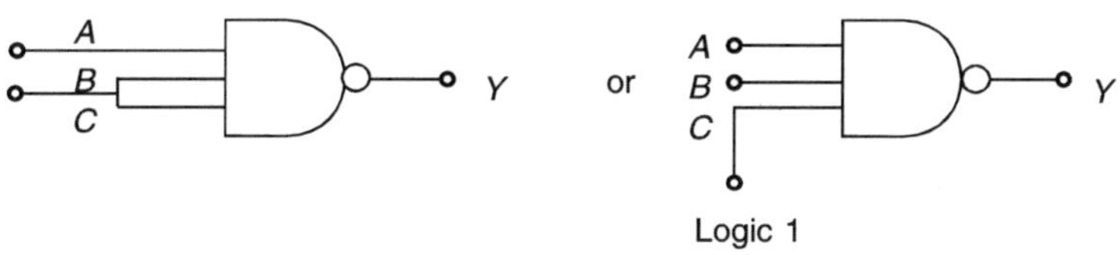

(d)

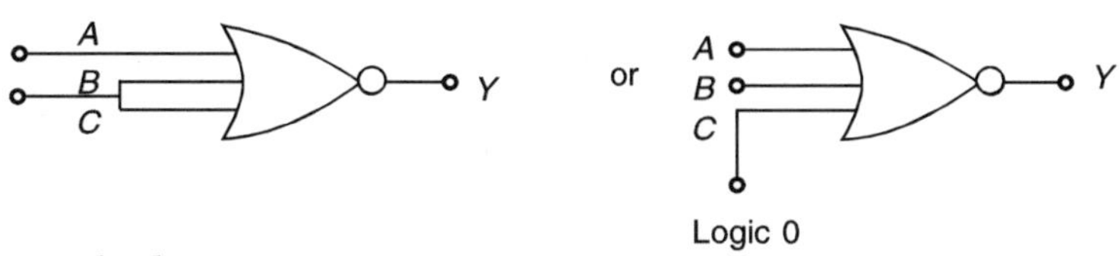

1.25 (a) Active-high
(b) Active-low
(c) Active-high
(d) Active-low

1.26 (a) Active-low
(b) Active-high
(c) Active-low
(d) Active-high

CHAPTER 2

2.1 (a) 57 (b) 41 (c) 254 (d) 100
(e) 13.1875 (f) 10.625 (g) 0.875
2.2 (a) 100101 (b) 11111111 (c) 1111
(d) 11010.01 (e) 1011.11 (f) 0.00011001
2.3 (a) 11000 (b) 10000.0001
2.5 (a) $(567)_8 = (101110111)_2$
(b) $(371)_8 = (011111001)_2$
(c) $(33.1)_8 = (011011.001)_2$
2.6 (a) $(334.52)_8 = (220.65625)_{10}$
(b) $(123.25)_8 = (83.328125)_{10}$
(c) $(263)_8 = (179)_{10}$
2.7 (a) $(177)_{16} = (000101110111)_2$
(b) $(F9)_{16} = (11111001)_2$
(c) $(1B.2)_{16} = (00011011.0010)_2$
2.9 (a) 01000110 (b) 001100100111.10001001
(c) 00100000.001100000101
2.10 (a) 01111001 (b) 011001011010.10111100
(c) 01010011.011000111000
2.11 111001
2.12 10011100101100001110110010100
2.16 (a) 6 (b) 8
2.19 100 × 20 × 8 bits
2.20 924 bits
2.21

Decimal digit	*Position →*	*1*	*2*	*3*	*4*	*5*	*6*	*7*
		p_1	p_2	n_1	p_3	n_2	n_3	n_4
0		1	1	0	1	0	0	0
1		0	0	0	0	0	0	1
2		1	0	0	0	0	1	0
3		0	1	0	1	0	1	1
4		0	1	0	0	1	0	0
5		1	0	0	1	1	0	1
6		0	0	0	1	1	1	0
7		1	1	0	0	1	1	1
8		0	0	1	1	0	0	0
9		1	1	1	0	0	0	1

CHAPTER 3

3.2 (a) 36 mV (b) 5.14%
3.3 (a) 5.232 mA (b) 161.6%
3.4 119.73 mV

3.5 (a) 10^{-9} C (b) 1.099 μs
3.6 (a) Active region; Saturation region
(b) 4.7 kΩ (c) 127.27 kΩ
3.7 (a) $V_{BB} \leq 0.5$ V (b) 0.5 V $< V_{BB} <$ 3.25 V
(c) $V_{BB} \geq 3.25$ V
3.8 233.3
3.9 Active region
$I_C = 8.43 \times 10^{-2}$ mA, $I_E \approx -8.43 \times 10^{-2}$ mA
$I_B = 8.43 \times 10^{-4}$ mA
3.11 (a) 1.05 V (b) 3.8 V (c) 0.3 mA
3.12 (a) 5 V (b) 0 V (c) 0 V (d) NOR
3.13 (a) Saturation region (b) Saturation region
(c) Active region
3.14 NO

3.16 $\dfrac{R_C}{2} \cdot C_O$

CHAPTER 4

4.1 16.48 mW

4.2 (a)
(b)

	$h_{FE} = 10$		$h_{FE} = 20$	
N	V_O	Noise margin $\lvert\Delta 1\rvert$	V_O	Noise margin $\lvert\Delta 1\rvert$
5	1.14	0.1	1.14	0.22
6	1.09	0.05	1.09	0.17
7	1.055	0.015	1.055	0.135
8	< 1.04	Load gate	1.026	0.106
9	< 1.04	transistors	0.997	0.077
10	< 1.04	not in saturation	0.984	0.064

(c) fan-out and noise margin increases with h_{FE}.
(d) Noise margin decreases with N.
4.3 (a) 10
4.4 (c) 32 mA
4.6 26
4.9 (a) 35 (b) $\Delta 1 = 0.7$ V, $\Delta 0 = -3.4$ V
(c) 12.045 mW
4.10 (a) $\Delta 1 = 7$ V, $\Delta 0 = -7.2$ V
(b) 76 (c) 29.48 mW
4.13 (a) 3.532 mA (b) 1.025 mA (c) 42.385 mA

4.18 0.72 kΩ $\le R_C \le$ 1.74 kΩ
4.33 Yes
4.34 Yes

CHAPTER 5

5.1 (a)

Switch		Light
S_1	S_2	L
0	0	0
0	1	1
1	0	1
1	1	0

5.2 (b) $f = BC + BD$

5.3 (b) $f_1 = (\bar{B}+C+D)(\bar{A}+\bar{B}+C)(\bar{A}+B+D)$
$(A+B+\bar{D})(A+\bar{B}+\bar{C})$

$f_2 = (A+\bar{C})(A+B)(\bar{A}+C+\bar{D})(B+\bar{D})$

5.4 (a)

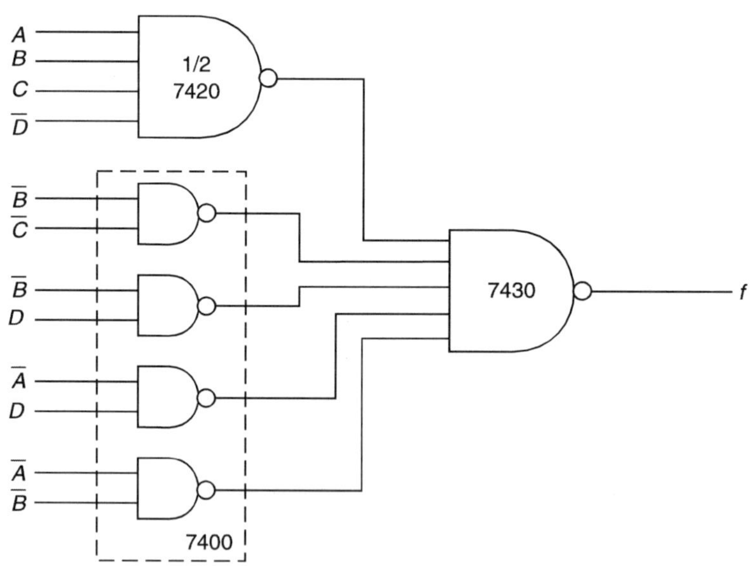

5.4(a)

(b)

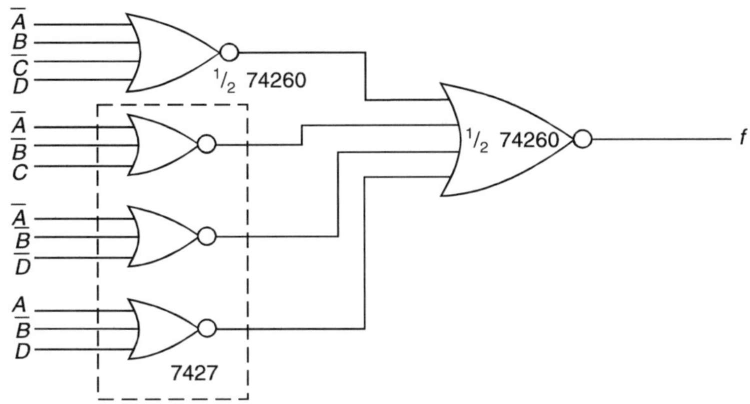

5.4(b)

5.5 (a) one – 7410

(b) $1\frac{1}{3} - 7427$

5.6 One – 7427

5.7 (c) $f = A + C$

5.8 (a) $f_1 = (A \oplus B) \odot (C \oplus D)$

(b) $f_2 = \overline{A}\,\overline{B} + AB\overline{D} + ACD$

5.9 $E_0 = \overline{A}$

$E_1 = BA + \overline{B}\,\overline{A}$

$E_2 = C\overline{B}\,\overline{A} + \overline{C}A + \overline{C}B$

$E_3 = D + CA + CB$

5.10 $A = \overline{E}_0$

$B = \overline{E}_1 E_0 + E_1 \overline{E}_0$

$C = \overline{E}_2\,\overline{E}_1 + E_2 E_1 E_0 + E_3 E_1 \overline{E}_0$

$D = E_3 E_2 + E_3 E_1 E_0$

5.11 (a) $f_1 = \overline{C + D}$

(b) $f_2 = (A + \overline{B} + D)(B + C + \overline{D})(\overline{A} + \overline{C})$

(c) $f_3 = (A + \overline{B} + \overline{C} + \overline{D})(B + \overline{C} + D)(\overline{A} + B + C)(\overline{A} + C + D)$

5.12 $f_1 = \overline{A}BE + A\overline{C}\,\overline{E} + ABD + B\overline{C} + A\overline{B}C\overline{D}E$

$f_2 = \overline{C}\,\overline{E} + A\overline{B}D + \overline{A}\,\overline{D}\,\overline{E} + ADE + \overline{B}CE + CDE + \overline{A}\,\overline{B}E$

5.14 (a) $f_1 = A\overline{B}\,\overline{C} + \overline{C}D + \overline{B}D + AD$

(b) $f_2 = \overline{A}\,\overline{C}\,\overline{D} + B\overline{C} + \overline{A}B$

5.15

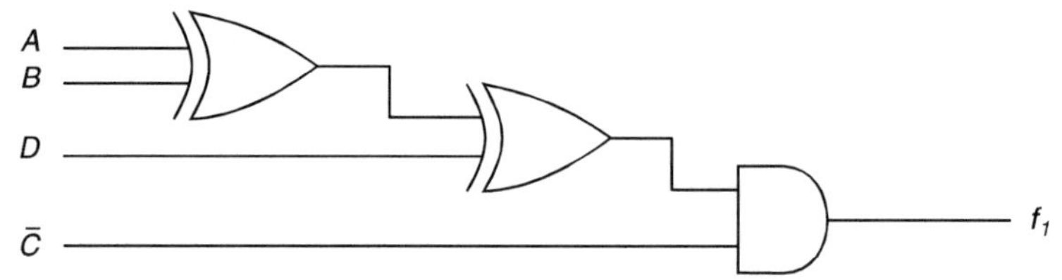

(a)

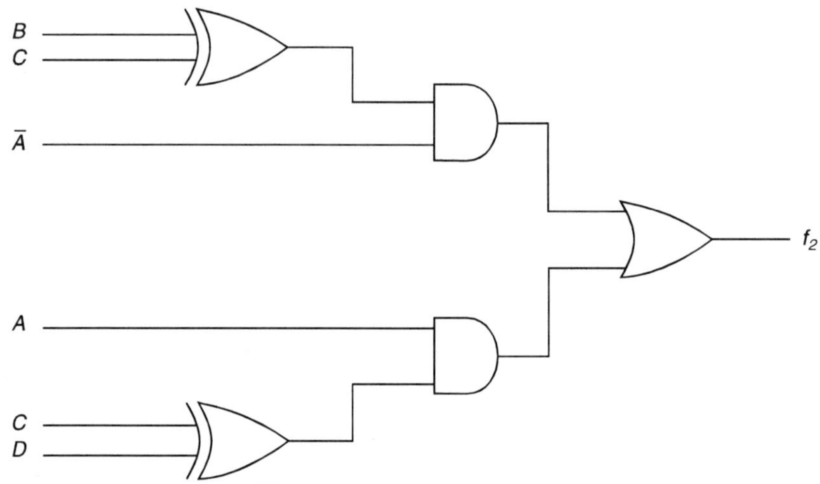

(b)

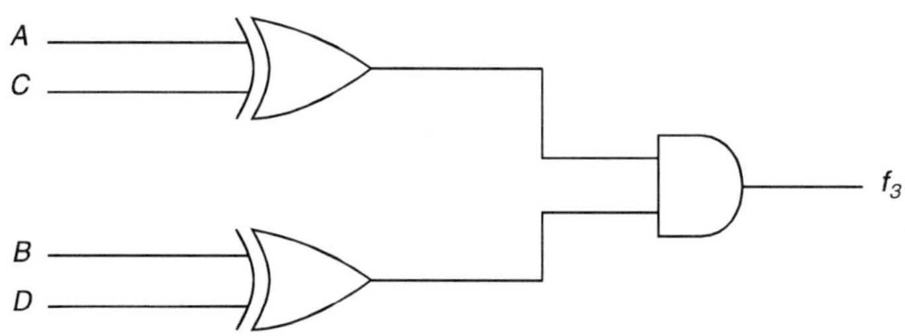

(c)

5.16

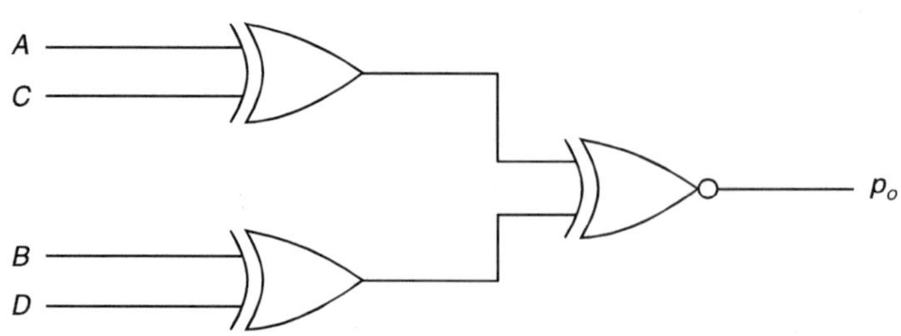

5.17

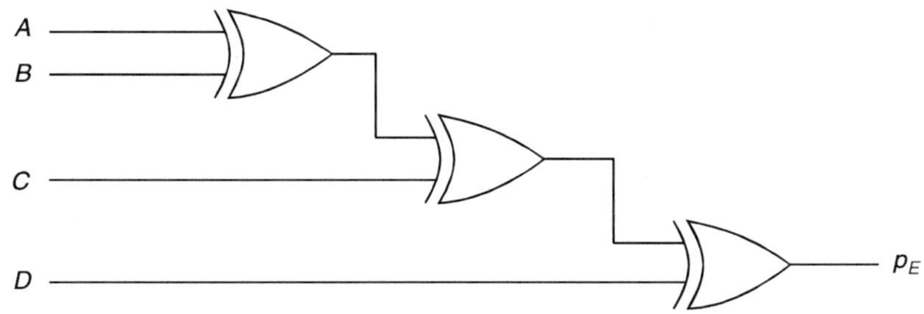

5.18 (a) $f_1 = \overline{A}B\overline{C}\overline{D}E + \overline{A}BC\overline{D}F + \overline{A}\overline{B}\overline{C}DE\overline{F}$

(b) $f_2 = (A + \overline{B} + \overline{C} + D + E + \overline{F})\,(\overline{A} + B + D + E + F)$
$(\overline{A} + B + C + \overline{E} + \overline{F})\,(A + \overline{C} + \overline{D} + E + F)\,(A + B + \overline{C} + E + F)$
$(A + \overline{B} + C + E + F)\,(A + \overline{B} + C + \overline{E} + F)\,(A + B + C + \overline{D})$
$(A + B + \overline{D} + E)\,(B + C + \overline{D} + E)\,(B + C + \overline{D} + F)$
$(A + \overline{B} + C + D)$

5.19 (a)

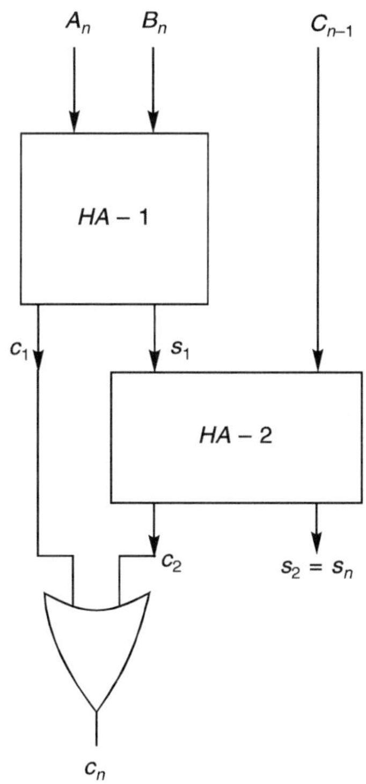

(b)

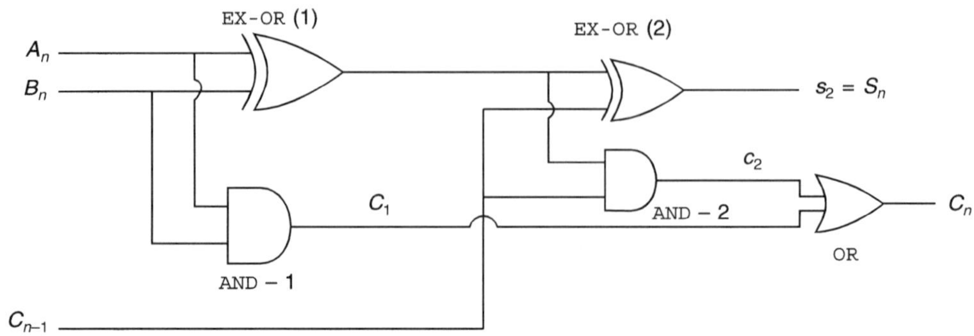

5.20 $t_{pd}(S_n) = 40\ ns$

$t_{pd}(C_n) = 40\ ns$

5.21 $f(A, B, C, D) = \bar{A}\bar{C} + \bar{A}\bar{B}D + \bar{A}B\bar{D} + \bar{B}CD + B\bar{C}D + BC\bar{D} + ACD$

5.22 $f(A, B, C, D) = \bar{B}D + \bar{C}D + AD + A\bar{B}\bar{C}$

5.23 $f(A, B, C, D, E) = A\bar{B}C\bar{D}E + A\bar{C}\bar{E} + \bar{A}BE + ABD + B\bar{C}$

CHAPTER 6

6.1

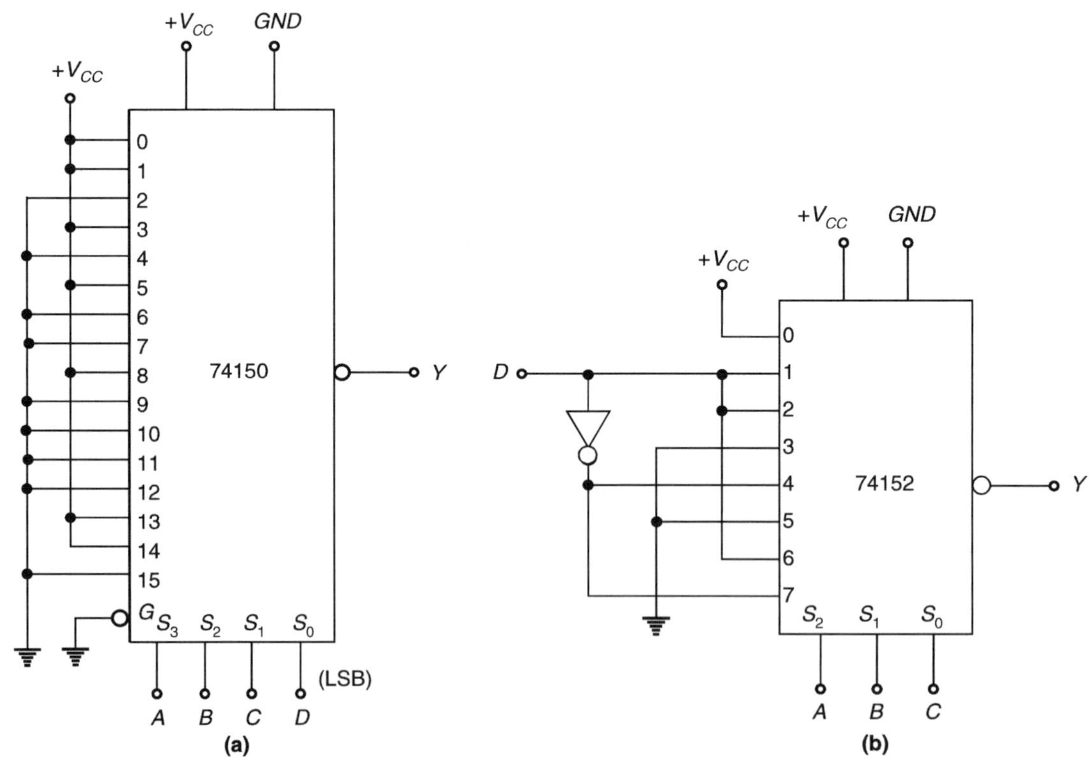

6.5 (b)

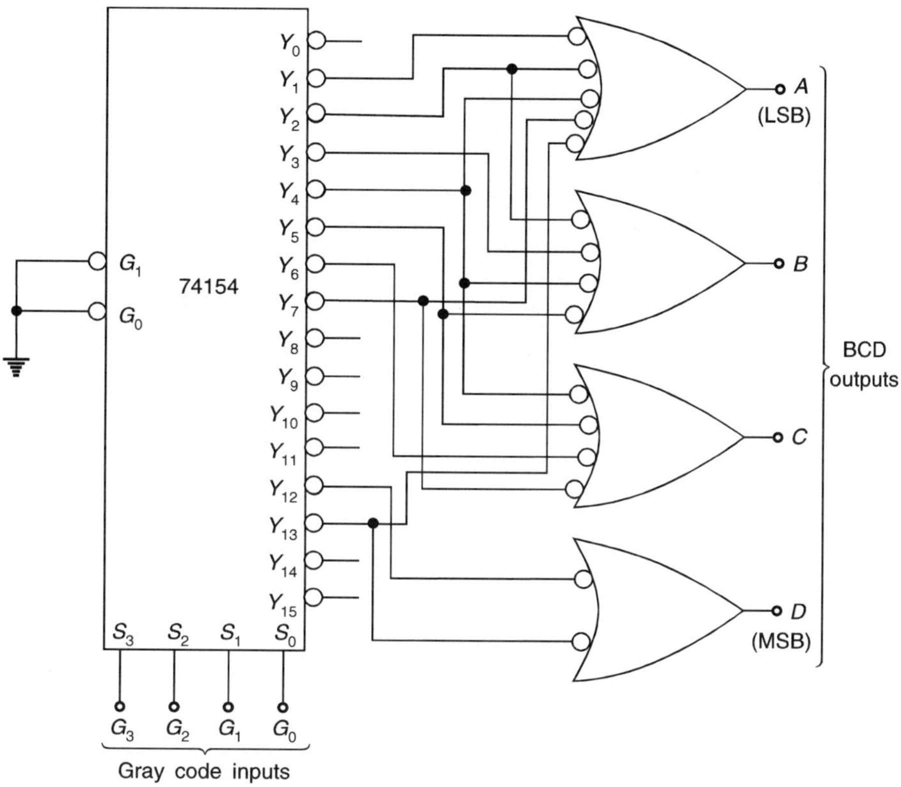

6.11 $t_{pd} = 40\ n$ ns

6.16

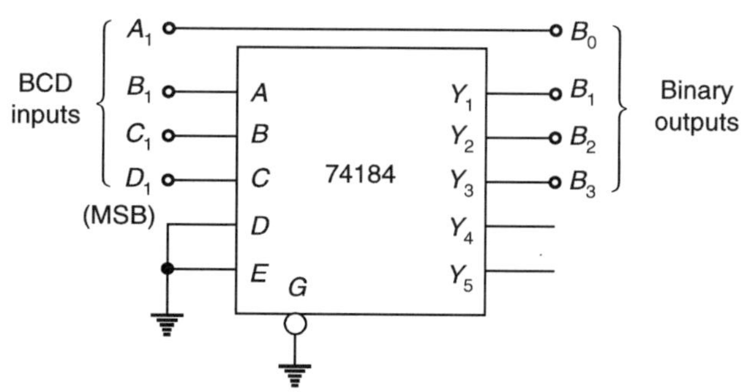

6.21

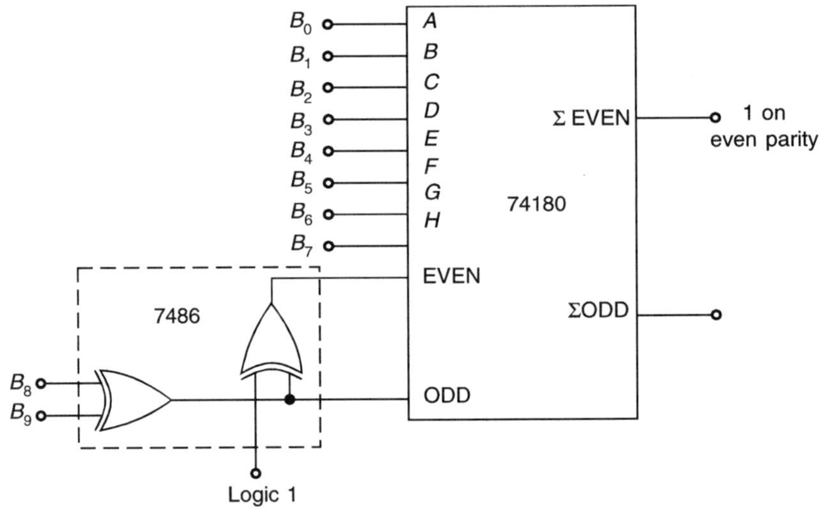

CHAPTER 7

7.14 (a) $Y_1 = \overline{\bar{Q} \cdot J \cdot CK}$

$Y_2 = \overline{Q \cdot K \cdot CK}$

(b) $Y_1 = \overline{CK \cdot D}$

$Y_2 = \overline{CK \cdot \bar{D}}$

(c) $Y_1 = \overline{CK \cdot T \cdot \bar{Q}}$

$Y_2 = \overline{CK \cdot T \cdot Q}$

7.15 (a) $S = D, R = \bar{D}$

(b) $J = D, K = \bar{D}$

(c) $D = J \cdot \bar{Q} + \bar{K} \cdot Q$

(d) $S = T \cdot \bar{Q}, R = T \cdot Q$

(e) $J = K = T$

(f) $T = J \cdot \bar{Q} + K \cdot Q$

(g) $T = D \oplus Q$

(h) $D = S + \bar{R} \cdot Q$

(i) $D = T \oplus Q$

(j) $T = S\bar{Q} + RQ$

(k) $J = S, K = R$

7.17

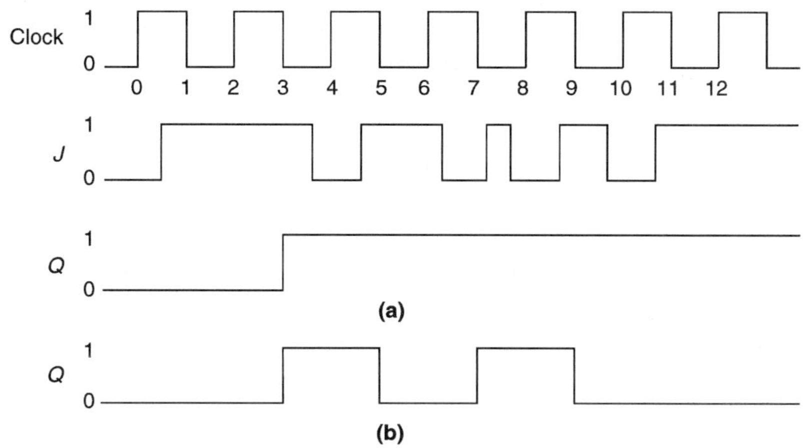

7.20

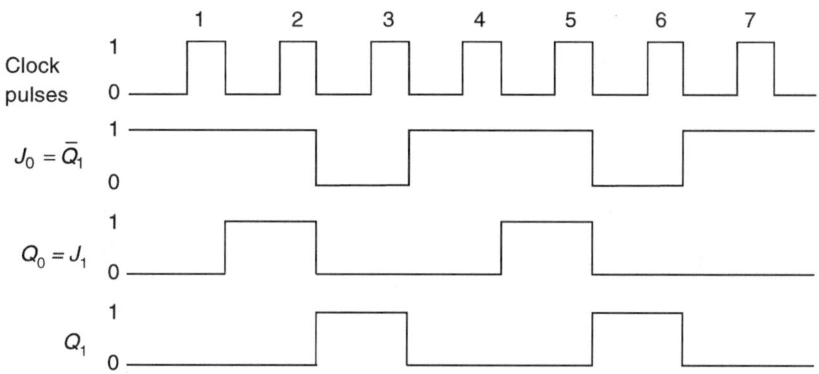

CHAPTER 8

8.5 Count sequence:

Q_2	Q_1	Q_0
0	0	0
1	1	1
1	1	0
1	0	1
1	0	0
0	1	1
0	1	0
0	0	1

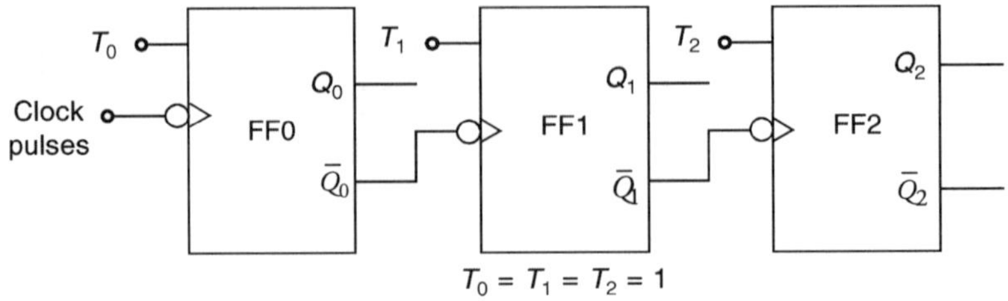
T_0
Clock pulses
FF0
Q_0
$\bar{Q}_0$
T_1
FF1
Q_1
$\bar{Q}_1$
T_2
FF2
Q_2
$\bar{Q}_2$
$T_0 = T_1 = T_2 = 1$

8.9 (a)

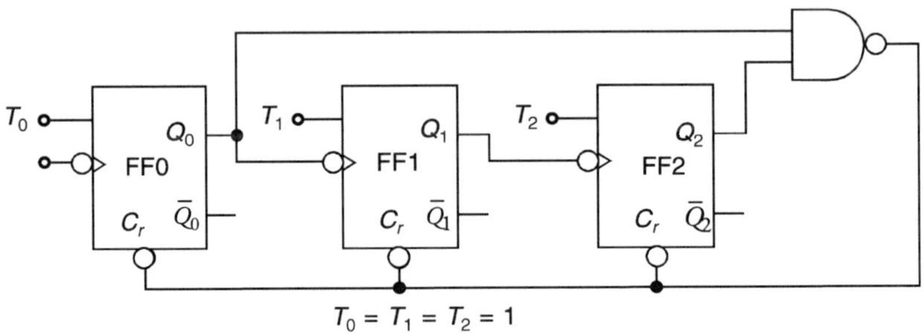
T_0
FF0
Q_0
C_r
$\bar{Q}_0$
T_1
FF1
Q_1
C_r
$\bar{Q}_1$
T_2
FF2
Q_2
C_r
$\bar{Q}_2$
$T_0 = T_1 = T_2 = 1$

8.10

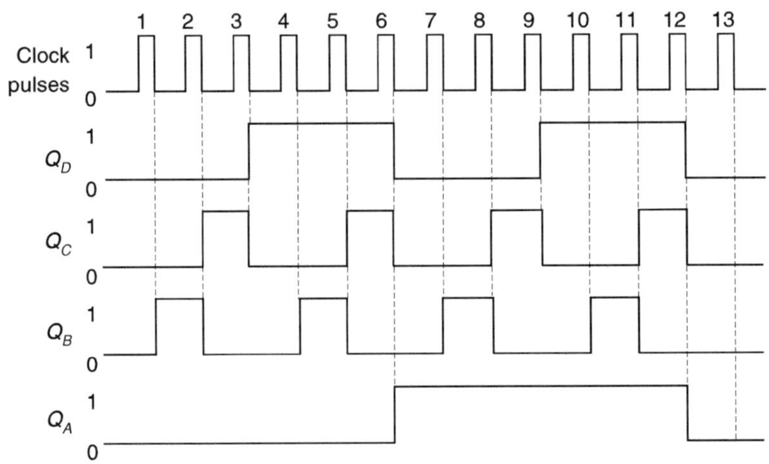
1 2 3 4 5 6 7 8 9 10 11 12 13
Clock pulses
Q_D
Q_C
Q_B
Q_A
1
0

8.21

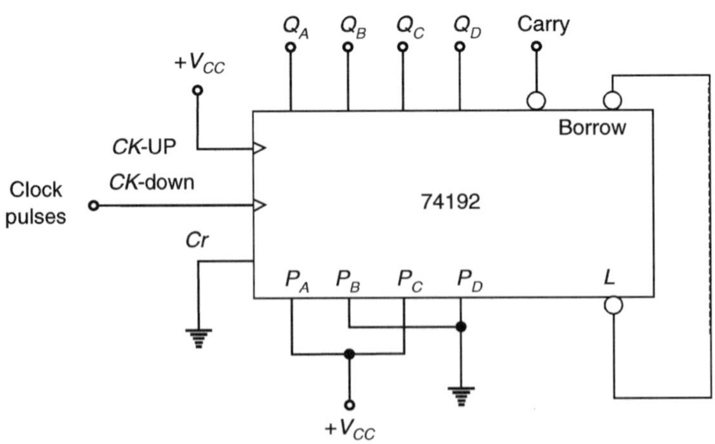
Q_A Q_B Q_C Q_D
Carry
$+V_{CC}$
Borrow
CK-UP
CK-down
Clock pulses
74192
Cr
P_A P_B P_C P_D
L
$+V_{CC}$

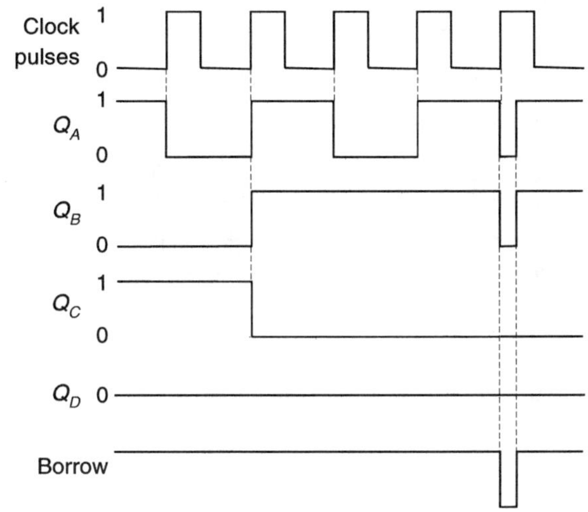

Waveforms

8.25

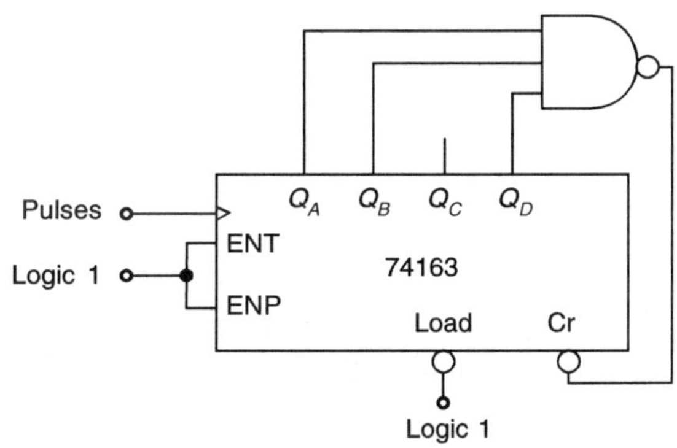

8.26 (b)

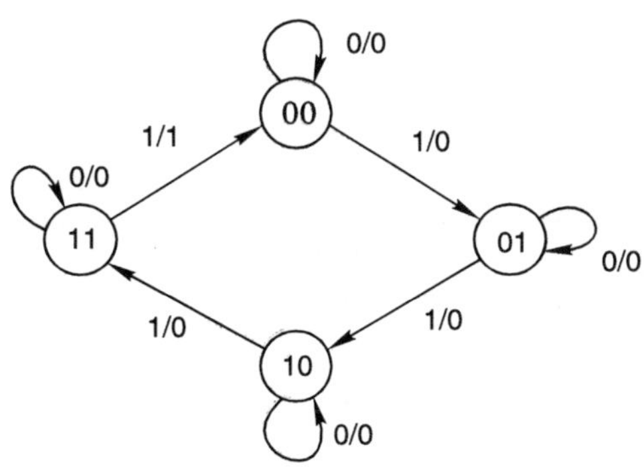

8.27 (a) $D_1 = Q_1 \oplus X$

$D_0 = Q_0 \oplus Q_1$

$Z = \bar{Q}_1 \cdot \bar{X} + Q_0$

(b) 001110

8.28

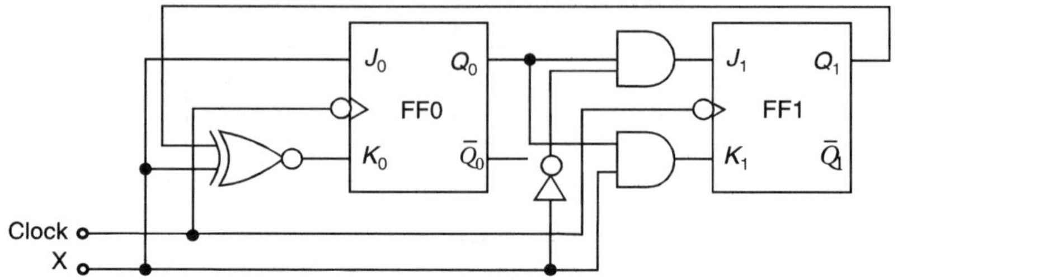

8.33 01000111010

8.34 (b) (a)-(i) yes, critical
(a)-(ii) yes, non-critical

CHAPTER 9

9.4 $V_{UT} \approx 1.0042$ V
$V_{LT} \approx 0.9938$ V

9.10 (a) R = 1.5 kΩ, $C = 200$ pF
(b) R = 10 kΩ, $C = 35$ pF

9.11 (a) 134 Hz
(b) 180 Hz

9.12 $R_A = 1$ kΩ
$R_B = 2$ kΩ
$C = 4.67$ nF

9.13 (b) $T = 0.7\, R_A C + \dfrac{R_A\, R_B}{R_A + R_B} C \ln\left(\dfrac{2R_A - R_B}{R_A - 2R_B}\right)$

(d) 48 kΩ

(e) $R_B < R_A/2$

CHAPTER 10

10.1 5/255 V

10.2

Digital input			*Analog output*
S_2	S_1	S_0	V
0	0	0	0
0	0	1	1
0	1	0	2
0	1	1	3
1	0	0	– 3
1	0	1	– 2
1	1	0	– 1
1	1	1	0

10.8 Less than 12/s
10.9 10/63 V

CHAPTER 11

11.1 (a) 2 (b) 4 (c) 6 (d) 8
(e) 10 (f) 11 (g) 16 (h) 20

11.2 (a) (i) 0 – 3 (ii) 0 – *F* (iii) 00 – 3*F* (iv) 00 – *FF*
(v) 000 – 3*FF* (vi) 000 – 7*FF* (vii) 0000 – *FFFF*
(viii) 00000 – *FFFFF*
(b) (i) 0 – 3 (ii) 00 – 17 (iii) 00 – 77 (iv) 000 – 377
(v) 0000 – 1777 (vi) 0000 – 3777 (vii) 000000 – 177777
(viii) 0000000 – 3777777

11.4 (a) 4 RAM chips and one 2-line-to-4-line decoder IC
(b) 2 RAM chips
(c) 32 RAM chips and one 4-line-to-16-line decoder IC

11.6 (a) 2 (b) 2 (c) 4

11.13

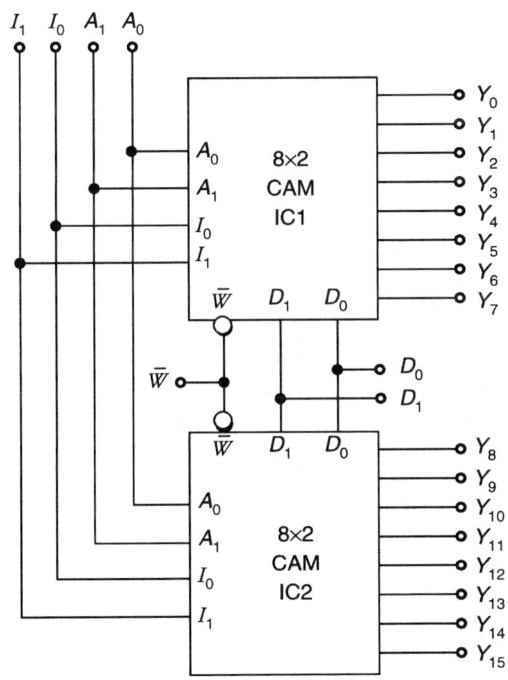

CHAPTER 12

12.1 (b)

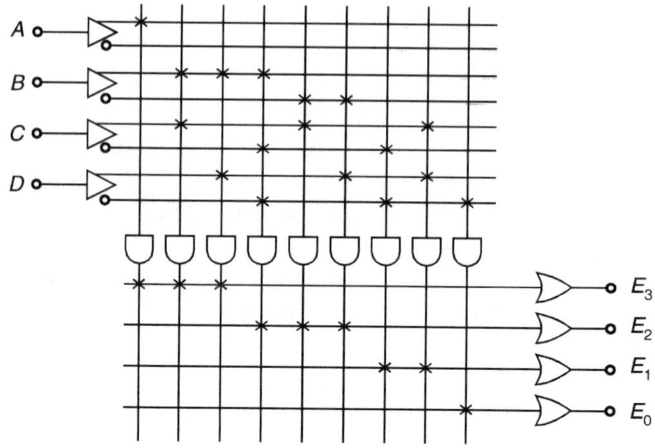

12.1(b)

(c)

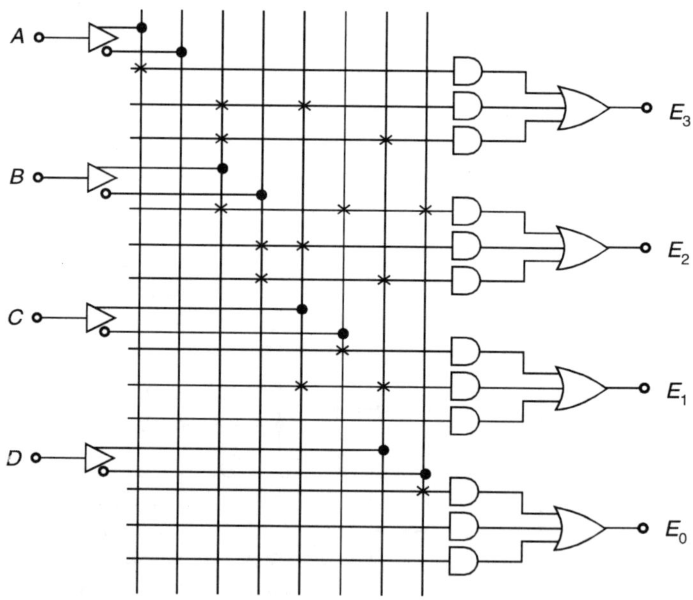

12.1(c)

12.4 The inputs of two 82S100 devices are to be connected in parallel to get 16 outputs.

12.12

x_1	x_2	f
0	0	1
0	1	0
1	0	0
1	1	1

12.13

0
1
1
0
1
0
0
1

CHAPTER 13

13.1

Microprocessor	*Address space*
8080A	64 K bytes
6800	64 K bytes
8086	1 M bytes
9900	64 K bytes
Z8000	8 M bytes

13.2 256

13.3 (a) 8 RAMs and 1 EPROM

13.4 MVI A, 00H; SUB A; ANI 00H; XRA A

13.6

```
        LXI H,  0A02H
        LDA     0A00H
        MOV     B, A
        LDA     0A01H
        CMP     B
        JZ      FINIS
        JC      GREAT
        MOV     M, A
        JMP     FINIS
GREAT:  M0V     M, B
FINIS:
```

13.7

```
LXI     H,01A0H
MVI     M,00H
```

13.9

```
ANI     0FH
```

13.11 12 bytes

```
13.14 POP    PSW
      POP    H
      POP    D
      POP    B
      EI
      RET
```

CHAPTER 14

14.1 (a) Yes. It contains all the allowed characters.
(b) No. Character '•' is not permitted.
(c) No. Starting character can not be a numeral.
(d) Yes. Upper and lower case characters can be mixed.
(e) No. Hyphen (–) is not allowed.
(f) No. Two consecutive underscores are not allowed.

14.2 (a)
```
ENTITY NAND2 IS
PORT (X, Y: IN BIT; Z: OUT BIT);
END NAND2;
```
(b)
```
ENTITY NAND3 IS
PORT (A, B, C: IN BIT; Y: OUT BIT);
END NAND3;
```

14.3 A 4:1 multiplexer is shown in Fig. below. It has four data inputs I_0, I_1, I_2, and I_3 and two select inputs A and B. There is one output Y.

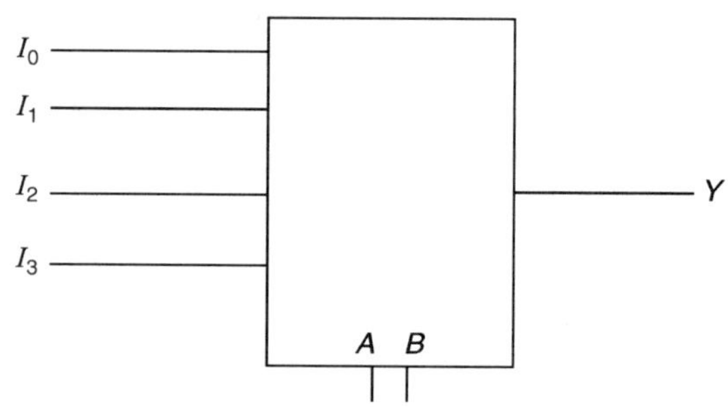

Fig. Prob. 14.3

The entity declaration is
```
LIBRARY IEEE;
USE IEEE STD_LOGIC_1164.ALL;
ENTITY MULTI_4 IS
PORT (I0, I1, I2, I3, A, B: IN STD_LOGIC; Y: OUT STD_LOGIC);
END MULTI_4;
```

14.4 (a) For 2-input NAND gate
```
ARCHITECTURE df_nand2 OF NAND2 IS
```

```
BEGIN
Z <= NOT (X AND Y) AFTER 10 ns;
END df_nand2;
```

(b) For 3-input NAND gate

```
ARCHITECTURE df_nand3 OF NAND3 IS
BEGIN
Y <= NOT (A AND B AND C) AFTER 10 ns;
END df_nand3;
```

14.5

```
LIBRARY IEEE;
USE IEEE. STD_LOGIC_1164.ALL;
- - Name of entity chosen is F-A
ENTITY F_A IS
PORT (A, B, CIN: IN STD_LOGIC; S, COUT : OUT STD_LOGIC);
END F_A;
ARCHITECTURE FA_STR OF F_A IS
COMPONENT NAND3
PORT (X1, X2, X3 : IN STD_LOGIC; Y: OUT STD_LOGIC);
END COMPONENT;
COMPONENT NAND4
PORT (X4, X5, X6, X7 : IN STD_LOGIC; Z OUT STD_LOGIC);
END COMPONENT;
COMPONENT INV
PORT (P: IN STD_LOGIC; Q: OUT STD_LOGIC);
END COMPONENT;
COMPONENT NAND2
PORT (X8, X9 : IN STD_LOGIC; R : IN STD_LOGIC);
END COMPONENT;
SIGNAL AB, BB, CINB, S1, S2, S3, S4, S5, S6, S7: STD_LOGIC;
BEGIN
I1 : INV PORT MAP (A, AB);
I2 : INV PORT MAP (B, BB);
I3 : INV PORT MAP (CIN, CINB);
N1 : NAND3 PORT MAP (AB, B, CINB, S1);
N2 : NAND3 PORT MAP (AB, BB, CIN, S2);
N3 : NAND3 PORT MAP (A, BB, CINB, S3);
N4 : NAND3 PORT MAP (A, B, CIN, S4);
N5 : NAND4 PORT MAP (S1, S2, S3, S4, S5);
N6 : NAND2 PORT MAP (A, B, S5);
N7 : NAND2 PORT MAP (B, CIN, S6);
N8 : NAND2 PORT MAP (A, CIN, S7);
N9 : NAND3 PORT MAP (S5, S6, S7, COUT);
END FA_STR;
```

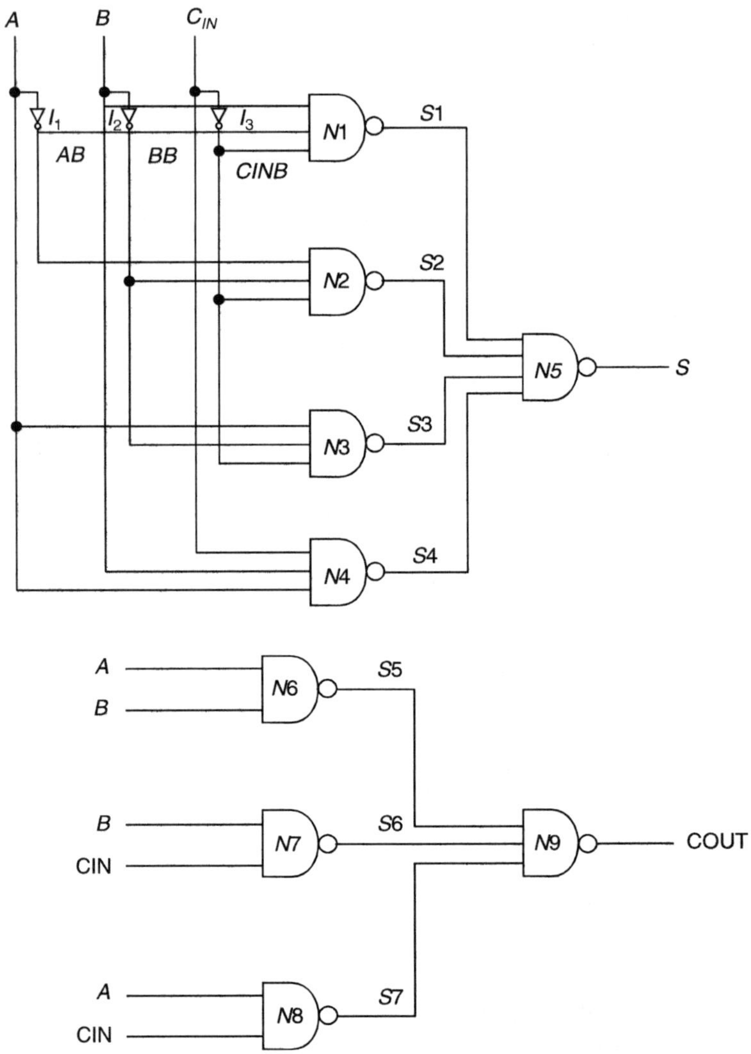

Fig. Prob. 14.5

14.6

```
LIBRARY IEEE;
USE IEEE.STD_LOGIC_1164.ALL;
ENTITY F_A IS
PORT (A, B, CIN: IN  STD_LOGIC; S, COUT : OUT STD_LOGIC);
END F_A;
ARCHITECTURE  FULL_ADDER OF F_A IS
BEGIN
S <=    ((NOT A) AND B AND (NOT CIN)) OR
        ((NOT A) AND (NOT B) AND CIN) OR
```

```
        (A AND (NOT B) AND (NOT CIN)) OR
        (A AND B AND CIN) AFTER 15 ns;
COUT <= (A AND B) OR (B AND C IN) OR
        (A AND C IN) AFTER 10 ns;
END FULL_ADDER;
```

INDEX